W9-DFR-319

Foundations of Probability and Statistics

William C. Rinaman

Le Moyne College
Syracuse, New York

Saunders College Publishing
Harcourt Brace Jovanovich College Publishers
Forth Worth   Philadelphia   San Diego   New York   Orlando   Austin   San Antonio
Toronto   Montreal   London   Sydney   Tokyo

Text Typeface: Times Roman
Compositor: Technique Typesetting
Acquisitons Editor: Robert Stern
Managing Editor: Carol Field
Project Editor: Nancy Lubars
Copy Editor: Zanae Rodrigo
Manager of Art and Design: Carol Bleistine
Art Director: Doris Bruey
Art & Design Coordinator: Caroline McGowan
Text Designer: Alan Wendt
Cover Designer: Lawrence R. Didona
Text Artwork: Tech Graphics
Director of EDP: Tim Frelick
Production Manager: Robert Butler
Marketing Manager: Monica Wilson

Cover Credit: Illustration produced by James T. Hoffman of the Center for Geometry Analysis Numerics and Graphics, University of Massachusetts at Amherst. For more information, contact him via the INTERNet at jim@gang.umass.edu

Printed in the United States of America

Foundations of Probability and Statistics

0-03-071806-6

Library of Congress Catalog Card Number: 92-050763

3456 016 987654321

Preface

Introduction   A sound introduction to probability and statistics theory is required for many college majors. This text, written for junior and senior level undergraduates and beginning graduate students, provides that solid instruction. Designed for students with no previous experience in probability and statistics, this text uses many real-life examples to show how mathematical topics are related. Theories are introduced once solid relationships have been established.

This text encourages students to draw on knowledge gained in previous math classes. Many students entering their junior year have been exposed to a number of mathematical ideas, but they may have little experience applying these ideas to new subjects. Therefore, I have endeavored to show connections between material learned in earlier courses, such as calculus, and how they relate to the topics discussed here. Many junior level undergraduates need some help applying axioms and theorems to more complex mathematical ideas. For this reason, I have tried, in the earlier chapters, to provide some insight into why one would choose to prove a result in a particular manner. Later chapters offer less support so that students can develop their own skills.

Juniors and seniors in the mathematical and engineering sciences have the requisite background needed for this text. This means that for all chapters except Chapters 13 and 14 the only background assumed is three semesters of calculus. In Chapters 13 and 14 some rudimentary knowledge of matrix algebra is required. The first seven chapters provide more than enough material for a one-semester course in probability theory. Chapters 8 through 12 form the core of a one-semester course in statistical theory. Additional topics may be added, as desired, from Chapters 13 and 14.

Chapter Synopses

A brief synopsis of each chapter follows.

Chapter 1: The idea of random experiments is introduced. This is followed by the axioms of probability. From there some elementary consequences are discussed. Set theory and combinatorics are reviewed for later application. Probabilities for finite sample spaces of equally likely events are covered.

Chapter 2: The basic concepts of conditional probability and independent events are covered. In addition, Bayes' Rule is discussed.

Chapter 3: Random variables are defined along with the distribution function. The differences between discrete and continuous random variables are presented. Random vectors and the concept of independent random variables are introduced. Conditional distributions are also covered.

Chapter 4: Expected value is discussed. The expected values of functions of random variable, especially moments, are covered. Moment generating functions and their properties are considered. Conditional expectation is also briefly addressed.

Chapter 5: Coverage of the most commonly encountered discrete and continuous random variables is provided. The bivariate normal distribution is also considered.

Chapter 6: Standard methods for deriving the distributions of random variables, which are functions of other random variables, are introduced. The uses of distribution functions, moment generating functions, and Jacobians are discussed. These provide a basis for deriving the distribution of sample statistics.

Chapter 7: The general concepts of convergence in probability and distribution are discussed. Chebyshev's Inequality is introduced as a method for showing convergence in distribution. The Central Limit Theorem for independent and identically distributed random variables is given along with additional results for sum, products, and quotients of sequences of random variables.

Chapter 8: The role of descriptive statistics in data analysis is introduced. The quantile–quantile plot is emphasized as a method of determining the adequacy of a statistical model. Simple random sampling is covered.

Chapter 9: The chi-square, t and F distributions are defined. The independence of $\bar{X}$ and s^2 is proven. Order statistics are discussed in a very complete fashion for a text at this level.

Chapter 10: The point estimation problem is covered. Maximum likelihood and method of moments estimation are discussed as alternative approaches to estimation. The ideas of unbiasedness, efficiency, sufficiency and completeness are covered to show how it is possible to derive unbiased estimators having minimum variance. The large sample properties of estimators, such as consistency and being best asymptotically normal, are discussed. Robustness is introduced to address the possibility that the assumed model may differ from the actual one. Bayes estimators are introduced.

Chapter 11: The most standard methods for deriving confidence intervals are covered. One procedure for finding a nonparametric confidence interval for the median is discussed.

Chapter 12: Hypothesis testing is introduced. Types I and II errors are given as a method to discern between competing critical regions. Neyman-Pearson tests, uniformly most powerful tests, and likelihood ratio tests are discussed as methods to derive "good" tests. Chi-square testing is introduced. Nonparametric tests are covered to illustrate the case when the underlying distribution is not known.

Chapter 13: Least squares estimators are introduced and their properties analyzed for a simple regression model with normally distributed error terms. The calculus of matrices is covered to facilitate the discussion of multiple regression. Here a basic knowledge of matrix arithmetic is required. Multiple regression is presented in a matrix algebra setting. Correlation analysis is discussed. One emphasis here is to highlight the differences between regression and correlation. Nonparametric methods for simple regression and correlation are also considered.

Chapter 14: The basic concepts of experimental design are introduced. Single-factor, randomized complete block, and two-factor factorial designs are considered. The theoretical discussion here introduces the idea of idempotent matrices. A basic understanding of the rank of a matrix is assumed for this material. Nonparametric alternatives for the single-factor and randomized complete block cases are discussed.

Supplementary Material

An Instructors Manual is available with this text. It provides detailed solutions for all of the exercises in the text.

Acknowledgments

I wish to express my thanks to many people who have either directly and indirectly contributed to the completion of this text. The following reviewers are to be thanked for their helpful and insightful comments.

Dennis L. Young, Arizona State University

George R. Terrell, Virginia Polytechnic Institute

Basil P. Korin, The American University

William J. Studden, Purdue University

Ennis D. McCune, Stephen F. Austin State University

At Le Moyne College I wish to thank Lifang Hsu and Cathleen Zucco for their help and suggestions. I would also like to thank my students for their help in making the presentation more accessible. Another student, Jill Seamon, is to be thanked for her help during preparation of the manuscript. Finally, there is my wife Debra, whose patience and understanding made the whole thing possible.

Table of Contents

1 Random Experiments and Probability

Introduction

Most people know intuitively what the word "probability" means. Many of us have encountered probabilities in our daily activities. For example, if we flip a coin it is our understanding that heads should result about half the time. When a local weather forecaster states that there is a 40 percent chance for rain tomorrow we have a general understanding that what is meant is that, given the current conditions, rain may occur 4 times out of 10. If we play bridge, we may draw on past experience to assess the likelihood that the cards held by our opponents have been distributed in a certain way in order to plan how we play a current hand. In fact, anyone who is consistently successful at games involving chance such as poker and backgammon is well aware of the relative frequency of the various possible outcomes in the play of the game.

The situations mentioned above have some things in common. In each we face a situation that is, at least in principle, well defined and repeatable. The possible outcomes are known, but the particular outcome that will take place this time cannot be predicted. What probability attempts to do is to determine a numerical value that tells what proportion of the time each possibility will occur.

One could correctly argue that the weather example is not like the others. It does not involve chance occurrences but rather represents a less than complete understanding of the physics of the atmosphere. This is the case in a number of areas where probability may be applied. Many physical situations are so complicated that it is not feasible to develop a mathematical model that accounts for all of the variables. In such cases it is common to develop models that use the main variables and then lump together the unused variables in a term that represents the unpredicted part of the model called the "noise." The noise in the model is what injects randomness into predictions.

Our goal is to investigate the mathematical structures that can be used to describe these types of situations, to define precisely what a probability is, and to study its properties. These concepts can be studied on a number of levels ranging from an intuitive approach, such as would be appropriate for a person mainly interested in applications, to a highly abstract approach, where probability is a corner of an area of real analysis known as measure theory. We aim for a middle-of-the-road approach—rigorous enough to give some appreciation for the mathematics involved but grounded enough in real-life applications to show how probability applies to real-life situations.

The origins of the study of probability theory can be traced to the 17th century where it was used to predict games of chance. A professional gambler named Chevalier de Mere posed the following problem to the French mathematician Pascal, now referred to as the *problem of the points*. Two players put up stakes and play a game of chance with the understanding that the winner takes all the money. How should the stakes be divided if the game is stopped before either player has won? In attacking this problem Pascal began and carried on a lengthy correspondence with another famous French mathematician, Fermat. At about the same time other famous mathematicians such as Christian Huygens and James Bernoulli worked on another problem that has become a classic, the *gambler's ruin problem*. In this problem two players, A and B, begin with some coins and play a game of chance. On each play of the game A wins a coin from B p percent of the time, and B wins a coin from A $100 - p$ percent of the time. The game runs until one player has won all of the coins. What proportion of the games will be won by A? We shall return to these problems in Chapter 2.

For a long time, probability theory concentrated in the area of games of chance, mainly because the areas of applications in the natural sciences such as errors of measurements, genetics, and statistics were not very well developed. Beginning in the 18th century the natural sciences began to use probabilistic ideas. The more advanced analytical methods were developed by mathematicians of the time such as DeMoivre, Laplace, Gauss, and Poisson. In the middle of the 19th century, the Russian mathematicians Chebyshev, Markov, and Liapunov introduced the random variable. This concept had an enormous impact on the study of probability.

In the 20th century, probability theory became a topic of study for its own sake. As we shall see in Section 1.2, the Soviet mathematician Andrei Kolmogorov developed an axiomatic basis for the subject. In addition, Emil Borel demonstrated connections between probability and measure theory. This connection has served as the basis for many recent developments in probability theory.

Today, probability theory is an important component in many areas of study. For example, the subjects of quantum mechanics and thermodynamics in physics are based heavily on the idea of random events. Statistics uses probability as its primary tool for assessing the quality of procedures. The topics of reliability, quality control, inventory theory, and queuing theory are applications of probability theory in operations research and industrial engineering. And in medicine, the subject of epidemiology models the spread of disease using probabilistic ideas. The application of probability theory to real-life problems is widespread, as will be demonstrated in this text.

1.2 ═══════
Set Theory

Probability theory is based on the occurrence of events in a random experimen
Random experiments and events are most conveniently described mathematicall
by using sets and set operations. Random experiments will be formally definec
and discussed in detail in Section 1.3. For now all we need to know is that an
experiment is taken to be any activity that is conducted according to a well-defined
set of rules where an observation of the outcome is made.

This section reviews set theories and the ideas that are useful in describing
these sets. You may be familiar with the basic ideas in set theory from other mathematics courses. For more complete coverage, consult any discrete mathematics
text. A more rigorous approach to the topic is given in *Naive Set Theory* by P. R.
Halmos.

Definition 1.1

> An *event* is a well-defined collection of outcomes of an experiment. The
> outcomes that belong to the set are called its *elements*. The collection of all
> possible outcomes of an experiment is called the *sample space*.

The term "well defined" simply means that if you are given a description of
an event and an outcome it is possible to determine definitely if that outcome is
an element of the event. Events are typically denoted by using capital letters such
as A, B, and so on. The sample space is commonly denoted by S. Outcomes are
denoted by using lowercase letters. If an object x is an outcome in the event A
we would write this fact as

$$x \in A.$$

Similarly, if x is not an outcome in A we would write

$$x \notin A.$$

The exact composition of a given event is usually described in one of two
ways. If the event has relatively few outcomes it is often convenient simply to
describe the event by listing them in braces ($\{,\}$) and separating them by commas.
For example, in an experiment where a number is selected from the digits the
sample space S of digits would be described by

$$S = \{0, 1, 2, 3, 4, 5, 6, 7, 8, 9\}.$$

The event O of observing an odd digit would be described by

$$O = \{1, 3, 5, 7, 9\}.$$

The order in which elements appear in the list is meaningless. Thus, we could
just as correctly define S by

$$S = \{7, 4, 2, 3, 5, 8, 1, 0, 9, 6\}.$$

In addition, if an outcome is in an event it appears only once in the list of elements.
This means that $\{1, 2, 2\}$ is the same as $\{1, 2\}$ for the purpose of defining an event.

The other common way to define an event is to list those properties that characterize the outcomes in the event. Using this method we would describe S as follows.

$$S = \{x : x \text{ is an integer}, 0 \leq x \leq 9\}$$

This would read as "S is the set of outcomes x such that x is an integer and x is greater than or equal to 0 and less than or equal to 9." The colon is translated to read "such that," and the comma is read as "and." It is important to note that

$$E = \{x : x \text{ is an integer}\}$$

does not define the same sample space. The integer -1 is an element E but not of S. Thus, when defining a sample space or an event in this manner it is critical to give a list of properties that are satisfied by and only by each of the outcomes of the set.

We next look at the concept of 2 events being equal. The idea is quite intuitive. We would certainly be inclined to state that the event $A = \{1, 2, 3\}$ is the same as the event $\{3, 1, 2\}$. This notion is made formal by the following definition.

Definition 1.2

> Two events, A and B, are said to be *equal* if they contain exactly the same elements.

In formal set theory this is known as the *axiom of extension*. If the event A contains the same elements as the event B we should state that "A equals B" and write

$$A = B.$$

Similarly, if one event contains an outcome that is not in the other event then we would say "A is not equal to B" and write

$$A \neq B.$$

Example 1.1

Suppose we select a letter from the English alphabet. Let $V = \{a, e, i, o, u\}$. Here the sample space would be the letters of the English alphabet. We could define

$$A = \{x : x \text{ is a letter in the English alphabet}, x \text{ is a vowel}\}.$$

In this case $A = V$. ∎

The opposite of the sample space, which contains all of the outcomes of an experiment, is the *empty set*. This event contains no outcomes and is denoted by $\varnothing$. An example of an event specification that results in the empty set is

$$\varnothing = \{x : x < 1, x > 1\}.$$

There are no real numbers that are simultaneously greater than and less than unity. Thus, the set is empty. It is logical to ask if there can be more than one empty set. The answer is no due to the axiom of extension. If A is empty and B is empty, then there are no elements in A and none in B. Hence there is no element that is in A that is not in B, which means that they are the same set. This kind of reasoning may seem a little strange, but arguments involving the empty set are typically carried out in this manner.

Example 1.2

Suppose we select a positive integer. The sample space S would be the positive integers. Then

$$A = \{x : x^2 \leq 5\} = \{1, 2\} \quad \text{while} \quad B = \{x : x^2 = 5\} = \varnothing.$$

∎

If two events A and B are not equal, 3 things are possible. One event may be completely contained in the other such as $A = \{1, 2, 3, 4\}$ and $B = \{2, 3\}$. The 2 events have no outcomes in common such as $A = \{1, 2\}$ and $B = \{3, 4\}$. The last possibility is that some but not all outcomes are common to both sets such as $A = \{1, 3\}$ and $B = \{2, 3, 4\}$. When an event is completely contained in another, then the event that is contained is said to be a *subset* of the other. In the first example given above we would say that B is a subset of A and write it as

$$B \subset A.$$

It is common practice when A has outcomes that are not in B to refer to B as being a *proper subset* of A and to A as being a *superset* of B. A formal definition of subset is given below. If B is a subset of A, but not necessarily a proper subset, then we would write

$$B \subseteq A.$$

Definition 1.3

If, for any $x \in A$, we have $x \in B$ then $A \subset B$. If A is not a subset of B we would write

$$A \not\subset B.$$

All determinations of whether one event is a subset of another proceed in the same manner. Suppose we wish to determine if A is a subset of B. The method is to pick a general element from A and show that its belonging to A necessarily implies that it also belongs to B. Some basic properties of subsets are given by the following theorem.

Theorem 1.1

Let A, B, and C be any three events from the sample space S. Then the following are true.

1. $\varnothing \subseteq A \subseteq S$
2. $A \subseteq A$
3. If $A \subseteq B$ and $B \subseteq C$, then $A \subseteq C$.
4. $A = B$ if and only if $A \subseteq B$ and $B \subseteq A$.

Proof

1. Since there is no element in $\emptyset$ that is not contained in A, we see that $\emptyset \subseteq A$. Also, since A consists entirely of elements from S, we find that if $x \in A$ then $x \in S$. Thus, $A \subseteq S$.

2. If $x \in A$ then $x \in A$. Therefore, by definition $A \subseteq A$.

3. Let x be any element of A. Since $A \subseteq B$, $x \in A$ implies $x \in B$. Since $B \subseteq C$, this outcome x that we have shown belongs to B is, by definition, also in C. Therefore, for any $x \in A$ we see that $x \in C$. Therefore, $A \subseteq C$.

4. First assume that $A = B$. Since events that are equal contain exactly the same outcomes, if $x \in A$ then $x \in B$. Therefore, $A \subseteq B$. Also, if $x \in B$ then $x \in A$, and $B \subseteq A$.

   Now assume that $A \subseteq B$ and $B \subseteq A$. This implies that every outcome in A is an outcome in B and every outcome in B is also an outcome in A. Thus, there is no outcome in either event that is not also an outcome in the other. This implies that they have the same outcomes. Thus, $A = B$. ∎

In simple terms, Part 1 of the theorem states that the empty set is a subset of every event, and every event is a subset of the sample space. Part 2 says that any event is an improper subset of itself. Part 3 says that the subset relationship is transitive. Part 4 gives an alternative characterization for event equality. This characterization is particularly useful in proving the equality of two events. That is, the statement $A = B$ is almost always proven by first showing $A \subseteq B$ followed by showing $B \subseteq A$.

Example 1.3

Suppose that a real number is selected, and let intervals of real numbers be events. Let $a \leq b$ and define

$$A = \{x : x = a \quad \text{or} \quad x = b\} = \{a, b\},$$
$$B = \{x : a < x < b\} = (a, b), \quad \text{and}$$
$$C = \{x : a \leq x \leq b\} = [a, b].$$

Here we see that $A \subset C$ and $B \subset C$ but $A \not\subseteq B$. In addition, if $a = b$ we note that $A = C = \{a\}$ and $B = \emptyset$. ∎

A convenient way of showing the relationship between events and the result of operations performed on events is a pictorial representation called a *Venn diagram*. Usually the sample space is drawn as a rectangle and each event is drawn as a circle within that rectangle. If the circle of one event overlaps that of another, then they have elements in common. If one event is a subset of another, then its circle is completely inside that of the other. Some sample Venn diagrams are illustrated in Figure 1.1.

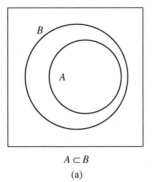 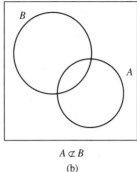

$A \subset B$ $A \not\subset B$

(a)   (b)

Figure 1.1

In dealing with real numbers we are familiar with certain *operations* that allow us to perform arithmetic. These operations include addition, subtraction, multiplication, division, and negation. They allow us to combine numbers so as to obtain new ones. There is a corresponding collection of *set operations*. This definition of the set operations will be followed by an example.

Definition 1.4	Given any two events A and B from a sample space, the *union of A and B*, denoted by $A \cup B$, is the event consisting of all outcomes that belong to either A or B. That is, $$A \cup B = \{x : x \in A \quad \text{or} \quad x \in B\}.$$ In some sense the union of two events is analogous to addition. Therefore, the alternative notation of $A + B$ is sometimes seen.

Definition 1.5	Given any two events A and B from a sample space, the *intersection of A and B*, denoted by $A \cap B$, is the set of all elements that belong both to A and to B. That is, $$A \cap B = \{x : x \in A, x \in B\}.$$ Another commonly used notation is AB. If two sets A and B are such that $A \cap B = \varnothing$, then A and B are said to be *mutually exclusive*.

Definition 1.6	Given any event A from a sample space S, the *complement of A*, denoted by A', is the set of all outcomes in S that are not members of A. That is, $$A' = \{x : x \in S, x \notin A\}.$$ An alternative notation is A^c.

A Venn diagram for each of these operations is illustrated in Figure 1.2. In each case, the set resulting from the indicated operation is shaded.

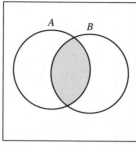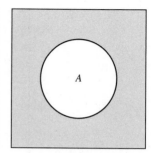

$A \cup B$ $A \cap B$ A'

Figure 1.2

Example 1.4

Suppose we select a digit from the numbers 0–9. The sample space will be $S = \{0,1,2,3,4,5,6,7,8,9\}$. Define the two events $A = \{1,3,5,7,9\}$ and $B = \{0,1,2,3,4\}$. Then

$$A \cup B = \{0,1,2,3,4,5,7,9\},$$

$$A \cap B = \{1,3\}, \quad \text{and}$$

$$A' = \{0,2,4,6,8\}. \qquad \blacksquare$$

The following theorem shows the relationship between the subset relation and the operations of union and intersection. The proof is left as an exercise.

Theorem 1.2

Given two events A and B from a sample space, the following are equivalent.

1. $A \subset B$,
2. $A \cup B = B$, and
3. $A \cap B = A$.

There are a number of identities involving events and set operations that are commonly referred to as the *laws of the algebra of sets*. We state them as a theorem. The proofs of these laws are omitted. Some of them are covered in the exercises.

Theorem 1.3

Let A, B, and C be any events from a sample space S. Then these events obey the following laws.

1. Commutative laws.
    a. $A \cup B = B \cup A$
    b. $A \cap B = B \cap A$
2. Associative laws.
    a. $(A \cup B) \cup C = A \cup (B \cup C)$
    b. $(A \cap B) \cap C = A \cap (B \cap C)$

3. Distributive laws.
   a. $A \cup (B \cap C) = (A \cup B) \cap (A \cup C)$
   b. $A \cap (B \cup C) = (A \cap B) \cup (A \cap C)$
4. Identity laws.
   a. $A \cup \varnothing = A$
   b. $A \cup S = S$
   c. $A \cap \varnothing = \varnothing$
   d. $A \cap S = A$
5. Idempotent laws.
   a. $A \cup A = A$
   b. $A \cap A = A$
6. Complementation laws.
   a. $(A')' = A$
   b. $A \cup A' = S$
   c. $A \cap A' = \varnothing$
   d. $S' = \varnothing$
   e. $\varnothing' = S$
7. DeMorgan's laws.
   a. $(A \cup B)' = A' \cap B'$
   b. $(A \cap B)' = A' \cup B'$

The reader probably has noticed that many of the laws can be obtained from other laws by simply exchanging $\cup$ for $\cap$. This is known as the *principle of duality* for set algebra.

The Associative laws state that unions and intersections can be taken in any order. Thus, using parentheses to specify a particular order is not necessary. Therefore, we use

$$(A \cup B) \cup C = A \cup B \cup C \quad \text{and}$$

$$(A \cap B) \cap C = A \cap B \cap C.$$

There will be times when we will need to refer to the union and intersection of a large number of events. The following notations serve as a convenient shorthand.

$$E_1 \cup E_2 \cup \cdots \cup E_n = \bigcap_{i=1}^{n} E_i \tag{1.1}$$

$$E_1 \cap E_2 \cap \cdots \cap E_n = \bigcap_{i=1}^{n} E_i \tag{1.2}$$

In plain language, Equation 1.1 refers to the event that consists of all the outcomes contained in at least one of the events $E_1, \ldots, E_n$, and Equation 1.2 is the event that consists of those outcomes that are contained in all of the events $E_1, \ldots, E_n$.

There will also be occasions when we will need to refer to the union and intersection of an infinite number of events. For this we shall use the following notation.

$$\lim_{n \to \infty} \bigcup_{i=1}^{n} E_i = \bigcup_{i=1}^{\infty} E_i \tag{1.3}$$

$$\lim_{n \to \infty} \bigcap_{i=1}^{n} E_i = \bigcap_{i=1}^{\infty} E_i \tag{1.4}$$

In using the Distributive laws we cannot neglect to write the proper parentheses. This is because, in general,

$$(A \cup B) \cap C \neq A \cup (B \cap C).$$

We illustrate this with Venn diagrams in Figure 1.3.

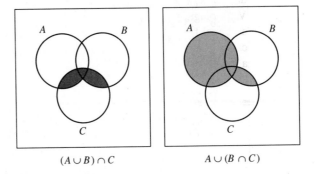

$(A \cup B) \cap C$ $A \cup (B \cap C)$

Figure 1.3

Example 1.5

We wish to prove DeMorgan's laws. The approach will be to prove $(A \cup B)' = A' \cap B'$ directly and then use it to prove $(A \cap B)' = A' \cup B'$.

1. $(A \cup B)' = A' \cap B'$: A Venn diagram for this identity is shown below.

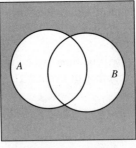

$(A \cup B)' = A' \cap B'$

2. Let $x \in (A \cup B)'$. Then $x \notin A \cup B$. Therefore, $x \notin A$ because if $x \in A$ then $x \in A \cup B$. Hence, $x \in A'$. Similarly, $x \notin B$ implies $x \in B'$. Since $x \in A'$ and $x \in B'$, we have that $x \in A' \cap B'$.

$A' \cap B' \subset (A \cup B)'$: Let $x \in A' \cap B'$. Then $x \in A'$ and $x \in B'$. This implies that $x \notin A \cup B$. Therefore, $x \in (A \cup B)'$.

3. $(A \cap B)' = A' \cup B'$: With Part 1 proven all we need to do is to note that

$$(A' \cup B')' = (A')' \cap (B')' = A \cap B.$$

Then by using law 6a of Theorem 1.3 (page 8) we have

$$((A' \cup B')')' = A' \cup B' = (A \cap B)'.$$

∎

Many introductory texts indicate that an appropriate method for proving set theorems is by using Venn diagrams. It must be pointed out that there is some danger in this approach. The problem is that it is quite possible that a particular Venn diagram might inadvertently impute some properties to the sets that are not assumed by the statement of the theorem. For this reason, proof by Venn diagram cannot be recommended. They are, however, useful in gaining insight when doing a rigorous proof.

Exercises 1.2

1. Let the sample space be the letters of the English alphabet.

$$U\{a, b, c, d, e, \dots, x, y, z\}$$

List the elements in the following events. Which events, if any, are equal?

$$A = \{x : x \text{ is a vowel}\}$$

$$B = \{x : x \text{ is a letter in the word "later"}\}$$

$$C = \{x : x \text{ is lexicographically larger than } r\}$$

$$D = \{x : x \text{ is a letter in the word "rattle"}\}$$

2. Let $S = \{x : x \text{ is an integer}\}$. List the outcomes in the following events.
   a. $A = \{x : x \in S, |x| < 2\}$
   b. $B = \{x : x \in S, x \text{ odd}, 0 < x < 15\}$
   c. $C = \{x : x \in S, 4x + 6 = 3\}$
   d. $D = \{x : x \in S, x^2 + x = 12\}$

3. Consider the events $A = \{0, 1, 2\}$, $B = (0, 2)$, and $C = [0, 2]$. Here $U = \{x : x \text{ is a real number}\}$. Use the correct symbol, $\in$, $\notin$, $\subset$, or $\not\subset$ for each of the following pairs of events.
   a. A, B
   b. A, C
   c. A, U
   d. $\emptyset, B$
   e. $(0, 1], B$

4. Let $S = \{0, 1, 2, 3, 4, 5, 6, 7, 8, 9\}$ and A–F equal each of the following.

$$A = \{0, 2, 4, 6, 8\} \qquad B = \{1, 2, 3\} \qquad C = \{5, 6, 7, 8\}$$
$$D = \{1, 3, 5, 7, 9\} \qquad E = \{0, 1, 8, 9\} \qquad F = \{2, 3, 4, 5\}$$

List the outcomes in the following events.

a. $A \cup B$ b. $A \cap B$ c. $A \cap D$

d. $A \cup (B \cap E)$ e. $A' \cap B'$ f. F'

g. $D' \cup A'$ h. $A \cap E$ i. $(A \cap E) \cup (B \cap E)$

5. Using the Venn diagrams provided below, shade the following sets.
   a. $A \cap (B \cup C)$
   b. $(A \cap B) \cup (B \cup C)$
   c. $A \cap (B \cup C)'$

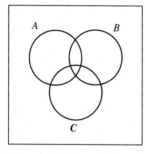 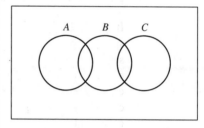

6. Show that $A \cup (B \cap C) = (A \cup B) \cap (A \cup C)$.

7. Show that $A \subset B$ if and only if $B' \subset A'$.

8. Prove or find a counterexample for each of the following.
   a. If $A \cup B = A \cup C$ then $B = C$.
   b. If $A \cap B = A \cap C$ then $B = C$.
   c. If $A \cup B = A \cap B$ then $A = B$.

9. Show that $((A \cup B) \cup C)' = (A' \cap B') \cap C'$. (Hint: Use DeMorgan's laws.)

10. Prove each of the following. Make as much use of the theorems and definitions given in this section as possible.
    a. $A \cap (A \cup B) = A$
    b. $(A \cap B) \cup (A \cap B') = A$
    c. $((A \cap B') \cap (A \cup B)) \cap A' = \varnothing$
    d. $(A \cup B) \cap (A \cap B)' = (A \cap B') \cup (A' \cap B)$

1.3

Random Experiments

Probability theory is based on the concept of a *random experiment*. In this section we wish to make it clear what a random experiment is and illustrate the idea with a number of examples. In addition, we shall introduce terminology that is in common use in probability. To begin, let us ask the question, what is it that makes an experiment a random one and another not? An example of a random experiment is the tossing of a coin. An example of a nonrandom experiment is the swinging of a pendulum. In both cases we are doing something according to a precise set of directions. Also, in both cases we know precisely what will happen when the experiment is conducted. The difference lies in our ability to predict what the outcome of the experiment will be. In the case of the coin toss it is impossible to know beforehand whether a toss will result in a head or a tail showing. With the pendulum, however, we can use the laws of physics to describe the motion of the pendulum and predict its position at any given time. This example of a random experiment describes the necessary conditions, which are formalized in the following definition.

Definition 1.7

> A *random experiment* is an experiment that has a known sample space and the property that the particular outcome of any given run of the experiment cannot be known with certainty beforehand.

It should be pointed out that a random experiment can have more than one sample space depending on what is being observed. Events, which are subsets of the sample space, are usually denoted by capital letters like A, B, and so forth. Those events that are sets containing a single outcome for the experiment are commonly called *elementary events*. Elementary events are to probability what atoms are to chemistry. Since they contain single outcomes, it must be true that the intersection of two elementary events is the empty set. In addition, any event in the sample space can be obtained by taking the union of the appropriate elementary events. Since events are sets we must use set operations such as union, intersection, and complementation to manipulate them. Suppose that A is an event in the sample space S. If the outcome from conducting the random experiment is an element of the set A, then the event A is said to have *occurred*. We illustrate these ideas with some examples.

Example 1.6

Consider an experiment where a coin is tossed 3 times with the side facing up being noted each time. It is easy to verify that this experiment meets the requirements of Definition 1.7 and, hence, qualifies as a random experiment. An outcome of the experiment is an ordered triple of the form (result of toss 1, result of toss 2, result of toss 3). There are 8 possible triples, and each of these is an elementary event. They are listed below.

$$E_1 = \{(HHH)\} \quad E_2 = \{(HHT)\} \quad E_3 = \{(HTH)\} \quad E_4 = \{(THH)\}$$
$$E_5 = \{(TTH)\} \quad E_6 = \{(THT)\} \quad E_7 = \{(HTT)\} \quad E_8 = \{(TTT)\}$$

The sample space S is the union of these 8 events. So

$$S = E_1 \cup E_2 \cup E_3 \cup E_4 \cup E_5 \cup E_6 \cup E_7 \cup E_8.$$

Now E_1 through E_8 are not the only events that can be defined for this sample space. Some other events are

$$F = \{\text{exactly 2 heads}\} = E_2 \cup E_3 \cup E_4,$$
$$G = \{\text{all 3 the same}\} = E_1 \cup E_8, \quad \text{and}$$
$$H = \{\text{exactly 2 the same}\} = E_2 \cup E_3 \cup E_4 \cup E_5 \cup E_6 \cup E_7 = (E_1 \cup E_8)'.$$

Note that these events are all formed by taking the union of elementary events. Now, if a coin is tossed 3 times and the triple (HHH) occurs then we would say that the elementary event E_1 has occurred. We could also say that the event G, or any other event that contains E_1, occurred.   ■

Example 1.7   In a classic experiment on genetics, Mendel crossed peas from plants having smooth yellow seeds with peas from plants having wrinkled green seeds. Now seeds from such a crossing must be either green or yellow and smooth or wrinkled. Growing pea plants is a well-defined, repeatable experiment. The types of peas that can result are known; yet, which type of pea will result on a given plant is not known beforehand. Therefore, this is a random experiment.

There are four elementary events that reflect the possible pairing of color and seed type. They are given below.

$$E_1 = \{\text{smooth green seed}\}$$
$$E_2 = \{\text{wrinkled green seed}\}$$
$$E_3 = \{\text{smooth yellow seed}\}$$
$$E_4 = \{\text{wrinkled yellow seed}\}$$

The sample space for this experiment is $S = E_1 \cup E_2 \cup E_3 \cup E_4$. Some other events that can be defined on this experiment are shown below.

$$F_1 = \{\text{green seed}\} = E_1 \cup E_2,$$
$$F_2 = \{\text{yellow seed}\} = E_3 \cup E_4,$$
$$F_3 = \{\text{smooth seed}\} = E_1 \cup E_3, \quad \text{and}$$
$$F_4 = \{\text{wrinkled seed}\} = E_2 \cup E_4.$$

We note that $S = F_1 \cup F_2 \cup F_3 \cup F_4$. Even though it is true that the union of the second set of events is the sample space, it would not be correct to say that they are a set of elementary events. Remember that elementary events are sets containing a single outcome for the experiment. An outcome for this experiment is described by pea color and skin type. Also, $F_1 \cap F_2 = \varnothing$, but $F_1 \cap F_3 = E_1 \neq \varnothing$.   ■

Example 1.8   In many board games a spinner is used to determine a player's next move. If we define spinner position in terms of the angle measured counterclockwise from a reference point, the sample space can be viewed as consisting of the half-open interval $[0, 2\pi)$. What constitutes an event in this sample space is a little trickier than in the previous examples. There are some highly technical ways to describe a set of events, but we wish to avoid them. It is sufficient to say at this point that events are intervals that are subsets of $[0, 2\pi)$. In addition, the method for assigning probabilities to these subintervals must be consistent with the axioms of probability that we shall discuss in Section 1.4.   ■

The thing that makes Example 1.8 different from the others is that the sample spaces for Examples 1.6 and 1.7 contained only a finite number of elements. The intervals $[0, 2\pi)$ contain an *uncountable* number of points. Roughly speaking a set is uncountable if it is not possible to assign each point to an integer. That is, there are more points than there are integers. We conclude with some sample spaces that contain a countable number of outcomes.

Example 1.9

In physics, the radioactive decay of elements such as radium is a random process. It is impossible to predict when a particular atom will decay. However, it is possible to determine quite accurately what proportion of a large number of radium atoms will decay in a given period of time. Suppose that we select an interval of time, say one hour, and count the number of radium atoms in the original quantity that undergo radioactive decay. From this description we see that it is possible that no atoms will decay in that hour, one atom will, two atoms will, and so on. Therefore, if we assign the number i to the result of observing i atoms of radium decay in the one-hour time interval, then the sample space becomes all of the nonnegative integers. The integers $0, 1, 2, \ldots$, constitute the elementary events in this sample space. Using these elementary events it is then possible to formulate other events such as $A = \{$an odd number of atoms in an hour decay$\}$ and $B = \{$no more than five atoms decay in an hour$\}$.   ■

Example 1.10

We again toss a coin. This time, however, we keep tossing the coin until a head appears. Thus, it is possible to toss the coin once, observe a head and then stop. Another outcome would be to toss the coin, observe a tail, toss it a second time and get a head and then stop. Therefore, the elementary events in this experiment consist of tuples whose length is equal to the number of times the coin is tossed. We can describe them as follows.

$$E_1 = \{H\},$$
$$E_2 = \{(TH)\},$$
$$E_3 = \{(TTH)\},$$
$$E_4 = \{(TTTH)\},$$
$$\vdots$$

Clearly, it is possible that we may never stop tossing this coin. If each toss is a tail, then there will be an infinite number of possible tuples in our sample space. The event $A = \{$at most three tosses$\}$ would look like

$$A = \{\{H\}, \{TH\}, \{TTH\}\} = E_1 \cup E_2 \cup E_3.$$ ■

Exercises 1.3

1. Let A, B, and C be events in the sample space S. Use the set operations of union, intersection, and complementation to construct the following events.

   a. A, B, and C occur.

   b. At least one of the events A, B, or C occur.

   c. None of the events A, B, or C occur.

   d. At least two of the events A, B, or C occur.

   e. A and at least one of the events B or C occur.

   f. Either A occurs or B occurs but A and B do not occur together.

2. A box contains 4 balls, numbered 1, 2, 3, and 4. Two balls are selected at random from the box.

   a. Specify the sample space for this random experiment when the first ball is returned to the box before choosing the second.

   b. Specify the sample space when the first ball is not returned to the box before choosing the second.

   c. For each sample space specify the event $A = \{$sum of the 2 balls is 4$\}$. That is, list the elementary events that are in A.

3. Two dice are tossed. Let $A = \{$sum is $\geq 4\}$, $B = \{$sum is even$\}$, and $C = \{$second die is 3$\}$. Give the outcomes that are in each event along with a verbal description of the event.

   a. $B \cap C$

   b. $C \cap A'$

   c. $C \cup (A \cap B')$

   d. $A' \cap B' \cap C'$

4. A box contains 6 white balls and 4 red balls. Consider the following experiment.

   i. Choose a ball at random from the box.

   ii. If the ball is red, stop. Otherwise, discard the ball and go back to Step i.

   a. List the elementary events in the sample space.

   b. List the elementary events in the event that an odd number of balls are removed from the box.

5. Bowl I contains 2 pennies, 1 nickel, and 1 dime. Bowl II contains 2 nickels and 2 dimes. A coin is selected at random from each bowl.

   a. List the elementary events in the sample space.

   b. List the elementary events in the event that at least 1 nickel is obtained.

   c. List the elementary events in the event that the 2 coins total less than 10 cents.

6. In a certain game, Sam and Joe alternate selecting a ball from a box containing one white ball and one black ball and replacing it until the white ball is drawn. The person getting the white ball wins. Assume that Sam selects first.

   a. Give the sample space for this game.

   b. In terms of your sample space give the event $S = \{$Sam wins$\}$.

   c. In terms of your sample space give the event $J = \{$Joe wins$\}$.

   d. Use De Morgan's laws to find $(S \cup J)'$ in terms of your sample space.

7. A point is selected at random from a line segment of length L. Let X denote the distance from one end of the line segment to the point.

a. Give the sample space in terms of values for X.

b. Describe the event that the point is in the middle third of the line segment in terms of values for X.

1.4
The Axioms
of Probability

We have stated that a probability is a numerical measure of the proportion of the time that a particular event would be expected to occur in a random experiment. The initial question that must be addressed concerns the manner in which these numerical values are assigned to each of the events in the sample space. Since we wish to measure a proportion, it is certainly reasonable that any method of assigning probabilities to events produces a result that looks and acts like a proportion. As an example consider the experiment of Example 1.6. Recall that a coin was tossed 3 times, and the sample space consisted of the following 8 ordered triples.

$$E_1 = \{(HHH)\} \quad E_2 = \{(HHT)\} \quad E_3 = \{(HTH)\} \quad E_4 = \{(THH)\}$$
$$E_5 = \{(TTH)\} \quad E_6 = \{(THT)\} \quad E_7 = \{(HTT)\} \quad E_8 = \{(TTT)\}$$

Intuitively, there is no reason to expect that any of these elementary outcomes should occur more often than the others. Therefore, it is reasonable to anticipate that, on average, each elementary event will occur $\frac{1}{8}$ of the time. Suppose we let this proportion be the probability of occurrence for each elementary event. The notation for this would be

$$P(E_i) = \frac{1}{8}, \quad \text{for } i = 1, 2, \ldots, 8.$$

This is read as "the probability of occurrence of E_i is one-eighth." We note that the elementary events constitute a set of mutually exclusive events whose union is the sample space. Each time this experiment is conducted one of the elementary events must occur. This means that, if we consider the sample space to be an event, $P(S) = 1$. We further note that

$$P(S) = P(E_1) + P(E_2) + P(E_3) + P(E_4) + P(E_5) + P(E_6) + P(E_7) + P(E_8).$$

Now consider the event that exactly one head occurs. For 3 outcomes out of the 8 possible occurrences it is reasonable to assign a probability of three-eighths to this event. Again, note that the event

$$F = \{\text{exactly one head}\} = E_5 \cup E_6 \cup E_7, \quad \text{and}$$
$$P(F) = \frac{3}{8} = P(E_5) + P(E_6) + P(E_7).$$

Note the following.

1. Since probability is a proportion, probabilities are between 0 and 1.

2. Since something must take place each time the experiment is run, the probability that some elementary event in the sample space will take place is 1. That is, $P(S) = 1$.

3. If events are mutually exclusive the probability of their union is the sum of the individual probabilities for each event.

Mendel's pea growing experiment in Example 1.7 is different. The sample space would have to consist of the results of an infinite number of plants grown from crossing the two types of peas. All we have are the results of the plants that Mendel grew. They constitute a *sample* from the sample space of all possible pea crossings. This means that it is not always possible to define probability in terms of the proportion of outcomes in the sample space. To do that you must be able to list what is in the sample space. In such a case, the best we can hope for is to conduct the experiment a large number of times and observe the result of each pea cross. In this way we can obtain a proportion of crosses that result in, say, smooth yellow seeds. Having done this it is then convenient to view the true probability of observing smooth yellow seeds as being the limit of this proportion as the number of crosses becomes infinitely large. Thus, if we let A be the event of getting a smooth yellow seed and $n(A)$ be the number of times we observe a smooth yellow seed in crossing n seeds, then we can consider

$$P(A) = \lim_{n \to \infty} \frac{n(A)}{n}.$$

As in the coin tossing example we note the following things about $P(A)$.

1. $P(A)$ is the limit of a sequence of numbers that is always between 0 and 1. Therefore, if the limit exists it must also be in that range.

2. Each cross must result in some kind of seed. Therefore,

$$P(S) = \lim_{n \to \infty} \frac{n(S)}{n} = \lim_{n \to \infty} \frac{n}{n} = 1.$$

3. Let B be the event of getting a wrinkled yellow seed. Then its probability would be

$$P(B) = \lim_{n \to \infty} \frac{n(B)}{n}.$$

These are mutually exclusive events whose union is the event of getting a yellow seed. If we let C be this event then

$$P(A \cup B) = P(C) = \lim_{n \to \infty} \frac{n(C)}{n}$$
$$= \lim_{n \to \infty} \frac{n(A) + n(B)}{n} = \lim_{n \to \infty} \frac{n(A)}{n} + \lim_{n \to \infty} \frac{n(B)}{n}$$
$$= P(A) + P(B).$$

Again the probability of the union of mutually exclusive events is the sum of their individual probabilities.

The view of probability illustrated by Mendel's experiment is called the *long run proportion* interpretation of probability. It does have some practical problems if we choose to use it to define probability. First, as we hinted in the previous example, this interpretation is the limit of a sequence. As the reader knows from calculus not all sequences converge to a limiting value. So how do we know that $P(A)$ when computed in this manner even exists? Second, even if the limit does exist, how does one go about evaluating it? Despite these questions we shall see that long run proportions can be a useful concept for interpreting probabilities when the sample space is not finite. Also experience shows that these sequences of proportions do seem to converge.

To illustrate the last statement regarding evaluation, suppose we toss a fair coin a large number of times. We would expect approximately one-half of the tosses to be heads. This is popularly known as the *law of averages*. To see how this works we conducted a computer simulation of a large number of such coin tosses and observed the following.

Number of Tosses	Proportion of Heads
10	0.4000
100	0.4700
1000	0.5180
10,000	0.5020
100,000	0.5003

This gives some empirical evidence that these long run proportions do converge to the true probability of an event. Unfortunately, it doesn't prove that this must always happen.

So far we have discussed two useful views—proportion and long run proportion—of what a probability represents. In addition, we have discovered some properties that appear to be common to both interpretations. We would now like to define probability in precise mathematical terms in such a manner that the two interpretations are still meaningful and consistent with the properties that underlie all interpretations of probability. Such a definition, based on a set of axioms, was put forward in 1933 by the Soviet mathematician Andrei Kolmogorov. Historically, this was nearing the end of a period when mathematicians were quite interested in trying to find axiomatic bases for all areas of mathematics. As the reader may already know, areas such as geometry, set theory, and the real number system all have axioms that define the basic properties of each. The axiomatic definition for probability is given below.

Definition 1.8

Axioms of Probability   Let S be the sample space of a random experiment and let E be any event in S. Then the *probability of E*, denoted by $P(E)$, is the value of a function which assigns a unique real number to E in accordance with the following axioms.

Axiom 1: $P(E) \geq 0$.

Axiom 2: $P(S) = 1$.

Axiom 3: If A and B are mutually exclusive events, then $P(A \cup B) = P(A) + P(B)$.

Axiom 4: For a sequence of mutually exclusive events, $A_1, A_2, \ldots$, (i.e., $A_i \cap A_j = \emptyset$, if $i \neq j$),

$$P\left(\bigcup_{i=1}^{\infty} A_i\right) = \sum_{i=1}^{\infty} P(A_i).$$

Axiom 1 states that the probability assigns a nonnegative real number to any event E. This is in line with the behavior of proportions. Axiom 2 gives the value of the probability function for one specific event—the sample space. It also gives it a value that is consistent with the proportion interpretation for probabilities. Axiom 3, coupled with induction, gives a way to compute the probability of the union of a finite number of mutually exclusive events. These are properties deduced from our examples. Axiom 4 states that, given a countable collection of mutually exclusive events, the probability that at least one of them will occur is the sum of their individual probabilities. This extension to an infinite collection is necessary to derive all of the other properties of the probability function.

The probability function $P(E)$ is unlike functions that we are familiar with in calculus courses. There, a point in the domain and range of a function was always a real number or a vector from some Euclidean space of real numbers. Here, a point in the domain of the probability function is a *set*—namely, an event in the sample space. So, rather than pairing one real number with another, the probability function pairs a set with a real number. This is of practical importance because this means that we have to be careful to perform real number arithmetic on the output of the probability function and set arithmetic—union, intersection, and complementation—on the input of the probability function. It is improper to use addition, for example, when dealing with events, but it is perfectly proper to use addition for the probability of events. The statement of Axiom 3 is an excellent illustration of this point.

If the reader recalls Euclidean geometry, the general development was to list some undefined terms such as "point," then to state some definitions and follow these with a set of axioms. At this stage there still was not much to go on. Therefore, the next thing to do was to use these terms, definitions, and axioms to prove theorems that developed all of the properties of geometric objects. We are currently at the point where we have some undefined objects such as "outcome." We have some definitions like "random experiment" and "event." Definition 1.8 has given us a set of axioms that describe the properties that any probability must have. We still, however, only know the exact probability for a single event—the sample space. Our task now is to use these things to derive some of the other properties of probability functions. As was done in geometry, these properties will be stated as theorems. We shall prove some of them and leave the other proofs as exercises.

Since we can compute the probability of the union of a finite collection of mutually exclusive events it is desirable to discuss a particular set of mutually exclusive events—an event and its complement. The relationship between the probability of an event and the probability of its complement is very useful in a practical sense. Many times it is less effort to compute the probability of the complement of an event than to find the probability of the original event. Suppose we toss a coin three times and wish to find the probability of getting at least one head. As we shall see, to compute the probability for the event $A = \{$at least one head$\}$, it is necessary to break it up into a set of three mutually exclusive events as follows.

$$\{\text{at least one head}\} = \{\{\text{exactly one head}\} \cup \{\text{exactly two heads}\}$$

$$\cup \{\text{exactly three heads}\}\}$$

The complement, however, is much simpler in that $A' = \{\text{no heads}\}$. Thus, we would prefer to compute $P(A')$ and use it to get $P(A)$. The following theorem gives us the tool we need.

Theorem 1.4

> Let A be an event. Then
>
> $$P(A') = 1 - P(A). \qquad (1.5)$$

Proof   We note that A and A' are mutually exclusive. In addition $A \cup A' = S$. Therefore, we can use Axiom 2 and Axiom 3 to show that

$$P(A \cup A') = P(S) = 1 = P(A) + P(A').$$

This is Equation 1.5.   ∎

This theorem is intuitively obvious. After all, in any run of an experiment, either the event A will occur or it will not occur. Therefore, the probability that A or A' occurs must be 1.

We can now use Theorem 1.4 to arrive at the probability of the empty set.

Theorem 1.5

> $$P(\varnothing) = 0 \qquad (1.6)$$

Proof   This seems like an obvious result. If the event contains no outcomes it can never happen. The problem is that all we know about actual values of probabilities is given in the axioms. So the issue becomes how to use the fact that the sample space has a probability of 1. This is made simple by noting that the complement of the sample space is the empty set, or $S' = \varnothing$. Therefore, Theorem 1.4 and Axiom 2 give the following.

$$P(\varnothing) = 1 - P(S)$$
$$= 1 - 1$$
$$= 0 \qquad ∎$$

When we introduced the axioms of probability we mentioned that Axiom 3 coupled with mathematical induction let us conclude that the probability of any finite union of mutually exclusive events could be computed by summing their individual probabilities. The next theorem makes this formal. The proof is left as an exercise.

Theorem 1.6

> Let $A_1, A_2, \ldots, A_n$ be mutually exclusive events. Then
>
> $$P\left(\bigcup_{i=1}^{n} A_i\right) = \sum_{i=1}^{n} P(A_i). \qquad (1.7)$$

The next theorem confirms another fact that should be clear from an intuitive standpoint. If one event contains another, the larger event should have a probability that is at least as big as that of the smaller event. For example, in the three coin toss experiment the event {at least one head} contains the event {exactly one head}. It seems clear that it should be true that $P(\{\text{at least one head}\}) \geq P(\{\text{exactly one head}\})$. We state the result and leave the proof as an exercise.

Theorem 1.7

> Let A and B be two events. If $A \subset B$ then $P(A) \leq P(B)$.

We now know the probability for two special events and how to compute the probability of the union of mutually exclusive events. The discussion now turns to how we handle events when they are not mutually exclusive. In the example where 3 coins are tossed, consider the events

$$A = \{\text{at least 2 heads}\} = \{\{HHT\}, \{HTH\}, \{THH\}, \{HHH\}\}, \quad \text{and}$$

$$B = \{\text{first coin is heads}\} = \{\{HTT\}, \{HTH\}, \{HHT\}, \{HHH\}\}.$$

Since $A \cap B = \{\{HHT\}, \{HTH\}, \{HHH\}\}$, we could not use Theorem 1.2 to compute $P(A \cup B)$. Assuming that each elementary event has a probability of one-eighth we see that $P(A) + P(B) = \frac{1}{2} + \frac{1}{2} = 1$. However,

$$A \cup B = \{\{HHT\}, \{HTH\}, \{THH\}, \{HHH\}, \{HTT\}\},$$

which should have a probability of $\frac{5}{8}$. The problem is that summing the two probabilities counts the elementary events in the intersection twice instead of just once. Therefore, one way to fix this problem would be to compute the probability of the union by adding $P(A)$ and $P(B)$ and then subtracting $P(A \cap B)$. The fact that this is appropriate for the union of any 2 events is established in the following theorem.

Theorem 1.8

> Let A and B be any two events. Then
>
> $$P(A \cup B) = P(A) + P(B) - P(A \cap B). \qquad (1.8)$$

Proof The proof uses an idea that occurs fairly often. The only unions we can handle at this point require that the events be mutually exclusive. Therefore, we must find a collection of mutually exclusive events whose union is equal to $A \cup B$ and then apply Theorem 1.6. A glance at the Venn diagram in Figure 1.4 shows that such a collection is

$$E = A \cap B', \quad F = A' \cap B, \quad \text{and} \quad G = A \cap B.$$

Now
$$A \cup B = E \cup F \cup G,$$

$$A = E \cup G, \quad \text{and}$$

$$B = F \cup G.$$

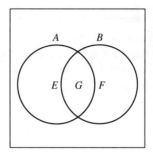

Figure 1.4

Thus, application of Theorem 1.6 gives

$$P(A \cup B) = P(E) + P(F) + P(G),$$

$$P(A) = P(E) + P(G), \quad \text{and}$$

$$P(B) = P(F) + P(G).$$

From the last two equations we get

$$P(E) = P(A) - P(G), \quad P(F) = P(B) - P(G).$$

If we substitute these in the equation for $P(A \cup B)$ we get

$$P(A \cup B) = P(A) - P(G) + P(B) - P(G) + P(G)$$
$$= P(A) + P(B) - P(G)$$
$$= P(A) + P(B) - P(A \cap B). \qquad \blacksquare$$

Now that we know how to obtain the probability of the union of any two events, it is possible to use induction to find the union of any number of events. For example, the probability of the union of three events can be found as follows.

$$P(A \cup B \cup C) = P[(A \cup B) \cup C]$$
$$= P(A \cup B) + P(C) - P[(A \cup B) \cap C]$$
$$= P(A) + P(B) + P(C) - P(A \cap B)$$
$$- P[(A \cap C) \cup (B \cap C)]$$
$$= P(A) + P(B) + P(C) - P(A \cap B) - P(A \cap C) - P(B \cap C)$$
$$+ P[(A \cap C) \cap (B \cap C)]$$
$$= P(A) + P(B) + P(C) - P(A \cap B) - P(A \cap C) - P(B \cap C)$$
$$+ P(A \cap B \cap C)$$

In practical terms, this formula indicates that, if the probabilities of the three events are added, the two way intersections are overcounted. After subtracting the probabilities of these, the three way intersection is then undercounted and must be added back in. In combinatorics, this procedure is known as the *inclusion–exclusion principle*.

Example 1.11   A box contains 4 balls, 2 of which are white and 2 of which are black. Two balls are removed from the box. If we label the balls W_1, W_2, B_1, and B_2, the sample space will be unordered pairs indicating which balls were chosen. The elementary events are as follows.

$$E_1 = \{W_1, W_2\} \quad E_2 = \{W_1, B_1\} \quad E_3 = \{W_1, B_2\}$$
$$E_4 = \{W_2, B_1\} \quad E_5 = \{W_2, B_2\} \quad E_6 = \{B_1, B_2\}$$

Assume that $P(E_i) = \frac{1}{6}, i = 1, 2, \ldots, 6$. Let A be the event that a white ball is chosen, B be the event that a black ball is chosen, C be the event that both balls are the same color, and D be the event that the 2 balls are of different color. We can use Theorem 1.6 to obtain the following probabilities.

$$P(A) = P(E_1 \cup E_2 \cup E_3 \cup E_4 \cup E_5) = \frac{1}{6} + \frac{1}{6} + \frac{1}{6} + \frac{1}{6} + \frac{1}{6} = \frac{5}{6},$$

$$P(B) = P(E_2 \cup E_3 \cup E_4 \cup E_5 \cup E_6) = \frac{1}{6} + \frac{1}{6} + \frac{1}{6} + \frac{1}{6} + \frac{1}{6} = \frac{5}{6}, \quad \text{and}$$

$$P(C) = P(E_1 \cup E_6) = \frac{1}{6} + \frac{1}{6} = \frac{1}{3}.$$

We can use Theorem 1.4 to compute

$$P(D) = 1 - P(C) = 1 - \frac{1}{3} = \frac{2}{3}.$$

Finally, Theorem 1.8 can be used along with the fact that $A \cap D = \{E_2 \cup E_3 \cup E_4 \cup E_5\}$ to obtain

$$P(A \cup D) = P(A) + P(D) - P(A \cap D)$$
$$= \frac{5}{6} + \frac{2}{3} - \frac{2}{3} = \frac{5}{6}. \qquad \blacksquare$$

Example 1.12   A fair coin is tossed twice. The sample space for this experiment is $S = \{(H,H), (H,T), (T,H), (T,T)\}$. Suppose we define 2 events A and B such that $A = \{\text{first toss heads}\}$, and $B = \{\text{second toss heads}\}$. Further assume that each elementary event S is assigned a probability of $\frac{1}{4}$. Therefore, by Axiom 3

$$P(A) = P(B) = \frac{1}{4} + \frac{1}{4} = \frac{1}{2}.$$

Furthermore,

$$P(A \cap B) = \frac{1}{4}.$$

We now use Theorem 1.8 to compute $P(A \cup B)$ as follows.

$$P(A \cup B) = P(A) + P(B) - P(A \cap B) = \frac{1}{2} + \frac{1}{2} - \frac{1}{4} = \frac{3}{4} \qquad \blacksquare$$

As we indicated, Theorem 1.8 can be generalized by using induction to compute the probability of unions of any number of events. We state the result as a theorem and leave the proof as an exercise. The notation is somewhat complicated, but careful reading shows that it is a straightforward continuation of the process we used to proceed from a union of two events to a union of three events.

Theorem 1.9

Let $A_1, A_2, \ldots, A_n$ be arbitrary events. Then

$$P(A_1 \cup A_2 \cup \cdots \cup A_n) = \sum_{i=1}^{n} P(A_i) - \sum_{i<j} P(A_i \cap A_j) + \sum_{i<j<k} P(A_i \cap A_j \cap A_k)$$

$$- \sum_{i<j<k<l} P(A_i \cap A_j \cap A_k \cap A_l) + \cdots$$

$$+ (-1)^{n+1} P(A_1 \cap A_2 \cap \cdots \cap A_n). \tag{1.9}$$

Notes:

1. The sum for the intersection of r events will have $\binom{n}{r}$ components.
2. The indices of summation are shorthand for multiple summations. For example,

$$\sum_{i<j<k} P(A_i \cap A_j \cap A_k) = \sum_{i=1}^{n-2} \sum_{j=i+1}^{n-1} \sum_{k=j+1}^{n} P(A_i \cap A_j \cap A_k).$$

3. The indices are selected in this way to ensure that each r-way intersection is included only once. Recall from set theory that intersection is commutative. That is, $A \cap B = B \cap A$.
4. Theorem 1.4 and Theorem 1.8 are special cases of Theorem 1.9.

Example 1.13

We wish to use Theorem 1.9 directly to obtain the formula for computing the probability of the union of 4 events. Let the events be $A_1, A_2, A_3,$ and A_4. Using Equation 1.9 we obtain

$$P\left(\bigcup_{i=1}^{4} A_i\right) = \sum_{i=1}^{4} P(A_i) - \sum_{i=1}^{3} \sum_{j=i+1}^{4} P(A_i \cap A_j) + \sum_{i=1}^{2} \sum_{j=i+1}^{3} \sum_{k=j+1}^{4} P(A_i \cap A_j \cap A_k)$$

$$- P\left(\bigcap_{i=1}^{4} A_i\right)$$

$$= P(A_1) + P(A_2) + P(A_3)P(A_4) - P(A_1 \cap A_2) - P(A_1 \cap A_3)$$

$$- P(A_1 \cap A_4) - P(A_2 \cap A_3) - P(A_2 \cap A_4) - P(A_3 \cap A_4)$$

$$+ P(A_1 \cap A_2 \cap A_3) + P(A_1 \cap A_2 \cap A_4)$$

$$+ P(A_1 \cap A_3 \cap A_4) + P(A_2 \cap A_3 \cap A_4)$$

$$- P(A_1 \cap A_2 \cap A_3 \cap A_4).$$

∎

Exercises 1.4

1. Let A and B be two events. Prove that if $A \subset B$ then $P(B) = P(A) + P(A' \cap B)$.

2. Let A and B be two events. Prove that if $A \subset B$ then $P(A' \cap B) = P(B) - P(A)$.

3. Let A and B be two events. Show that the probability that exactly one of them occurs is found by $P(A) + P(B) - 2P(A \cap B)$.

4. Express the following probabilities in terms of $P(A), P(B)$, and $P(A \cap B)$.
   a. $P(A' \cap B')$
   b. $P[A \cup (A' \cap B)]$
   c. $P[(A' \cap B)']$

5. Let A and B be two events. Prove that

$$P(A \cup B) \leq P(A) + P(B).$$

6. Use induction and Axiom 3 to prove Theorem 1.6.

7. Prove Theorem 1.7.

8. Use induction to prove Boole's Inequality.

$$P\left(\bigcup_{i=1}^{n} A_i\right) \leq \sum_{i=1}^{n} P(A_i)$$

9. Prove Bonferroni's Inequality.

$$P(A \cap B) \geq P(A) + P(B) - 1$$

10. Use induction to generalize Bonferroni's Inequality of Exercise 1.8 to

$$P(A_1 \cap A_2 \cap \cdots \cap A_n) \geq P(A_1) + P(A_2) + \cdots + P(A_n) - (n - 1).$$

11. Use induction and Theorem 1.8 to prove Theorem 1.9.

12. Let A and B be 2 events in a given sample space with $P(A) = \frac{1}{2}$, $P(B) = \frac{2}{3}$, and $P(A' \cap B) = \frac{1}{4}$. Find each of the following.
    a. $P(A')$
    b. $P(A \cap B)$
    c. $P(A \cup B')$
    d. $P[(A \cup B)']$

13. A university has 2 campuses. Let A denote the event that there is still space in a probability course at the first campus and B denote that there is still space in the probability course at the second campus. In a randomly chosen semester, $P(A) = 0.3$, $P(B) = 0.4$, and $P(A \cap B) = 0.2$. A student wishes to enroll in the course. Find the probability that there is room in

    a. the course on at least 1 campus.

    b. the course on neither campus.

    c. the course on exactly 1 of the campuses.

14. A company has 2 copiers. The probability that the first copier is busy is 0.5, the second copier is busy with probability 0.4, and the probability that at least 1 of the copiers is busy is 0.7.

a. What is the probability that both copiers are busy?

b. What is the probability that the second copier is busy but not the first one?

c. What is the probability that exactly 1 of the copiers is busy?

15. At a small engine repair shop 60% of all repairs are done on lawn mower engines and 10% are done on chain saw engines. Let A be the event that the next engine to be repaired belongs to a lawn mower and B be the event that the next engine to be repaired is a chain saw engine. Find each of the following probabilities.

   a. $P(A \cup B)$

   b. $P(B')$

   c. $P(A' \cap B')$

   d. $P(A' \cap B)$

1.5

Counting Elementary Events

In computing probabilities, we are often required to count the number of elementary events in a given event. For example, in tossing a coin 8 times, we may need to know how many sequences of tosses have 5 heads and 3 tails. We could always list all possible sequences of 8 coin tosses ($\{HHHHHHHH\}$), $\{HHHHHHHT\}$, $\{HHHHHHTH\}$, etc.) and search the list to count those with 5 H's and 3 T's. This can be an arduous task at best. After all, there are 256 different sequences of tosses in our example. With 256 items to be correctly enumerated the potential for error is great. More importantly, we are only interested in how many outcomes have 5 heads and 3 tails. Knowing what they look like is useless information as far as we are concerned. The subject of combinatorics is directed at counting the number of objects in a set without having to enumerate the elements in that set. It is our purpose to give a brief introduction to the subject in order to facilitate the computation of probabilities for events in finite sample spaces. The interested reader can find a more extensive treatment of the topics discussed here in texts such as *Introduction to Combinatorics,* by G. Berman and K. D. Fryer.

Most of the ideas we shall discuss derive from what is known as the *basic accounting principle.* Briefly stated, if an experiment is conducted in 2 stages with m possible outcomes for the first stage and n possible outcomes for the second, then this principle states that there are mn possible outcomes for the entire experiment. For example, if we roll a pair of dice, we can view the number of dots appearing on one die as a stage having 6 possible outcomes. The outcome on the other die would be another stage also with 6 possible outcomes. Then the basic counting principle gives that there are $6 \cdot 6 = 36$ possible outcomes for both dice. A formal statement of this idea is given in the following theorem.

Theorem 1.10

> **Basic Counting Principle**   Suppose that an experiment consists of 2 subexperiments. Further suppose that subexperiment 1 can result in any of m possible outcomes and subexperiment 2 can result in any of n possible outcomes. If an outcome for the experiment is represented by an ordered pair (x, y), where x is an outcome for subexperiment 1 and y is an outcome for subexperiment 2, then there are mn possible ordered pairs or outcomes for the experiment.

Proof   The theorem is proved by enumeration. We let (i, j) be an outcome where subexperiment 1 had its ith outcome and subexperiment 2 had its jth outcome. Now arrange the possible ordered pairs in a matrix with row i being the ordered pairs of the form (i, y) and column j being those of the form (x, j) to obtain

$$
\begin{matrix}
(1,1) & (1,2) & (1,3) & \ldots & (1,n) \\
(2,1) & (2,2) & (2,3) & \ldots & (2,n) \\
\vdots & \vdots & \vdots & \ddots & \vdots \\
(m,1) & (m,2) & (m,3) & \ldots & (m,n).
\end{matrix}
$$

Since this is a matrix with m rows and n columns, it contains mn entries. Thus, there are mn possible outcomes for the experiment.   ■

Example 1.14   Suppose we toss a coin 2 times. Subexperiment 1 is the first toss and subexperiment 2 is the second toss. Each toss has 2 possible outcomes—heads or tails. Thus, in the notation of the basic counting principle, we have $m = n = 2$. Therefore, there are $2 \cdot 2 = 4$ possible outcomes. We could have enumerated these as (H, H), (H, T), (T, H), (T, T), but hopefully it is clear that this would have become unmanageable as the number of tosses got larger.   ■

Example 1.15   There are 12 juniors and 9 seniors majoring in mathematics at a particular college. The mathematics faculty is interested in involving the students in departmental planning and decides to select 1 junior and 1 senior to participate in department meetings. If we consider selecting a junior to be subexperiment 1 and selecting a senior to be subexperiment 2, then $m = 12$ and $n = 9$. Therefore, by the basic counting principle, there are $12 \cdot 9 = 108$ possible ways to select 1 junior and 1 senior.   ■

The basic counting principle can be generalized to cover experiments with more than two subexperiments and with differing numbers of outcomes for each subexperiment. It is the logical extension of Theorem 1.10.

Theorem 1.11

> **Generalized Counting Principle**   Suppose an experiment consists of k subexperiments, and for each subexperiment there are n_i possible outcomes for subexperiment i, where $i = 1, 2, \ldots, k$. If an outcome for the experiment is represented as an ordered k-tuple, $(x_1, x_2, \ldots, x_k)$, where x_i is an outcome for subexperiment i, then there are $n_1 \cdot n_2 \cdots n_k$ possible outcomes for the experiment.

Proof   We shall use induction on k, the number of subexperiments. The base case, $i = 1$, is obvious. If there is one subexperiment with n_1 outcomes, then the experiment has n_1 possible outcomes. We now assume that there are $n_1 \cdot n_2 \cdots n_k$ possible outcomes for experiments consisting of k subexperiments and consider an experiment consisting of $k + 1$ subexperiments. Our decision to view the experiment as consisting of $k + 1$ subexperiments is arbitrary. For example, we can view an experiment consisting of 3 coin tosses as being composed of 3 subexperiments of 1 coin toss each or as being 2 subexperiments, where subexperiment 1 is the first 2 tosses and subexperiment 2 is the third. Therefore, we can consider our $k + 1$ stage experiment to have 2 subexperiments. Subexperiment 1 would be the first k stages, and subexperiment 2 would be the $(k + 1)$st stage. By the induction hypothesis subexperiment 1 has $n_1 \cdot n_2 \cdots n_k$ possible outcomes and by assumption subexperiment 2 has n_{k+1} possible outcomes. Since this way of viewing the experiment is like a 2-stage experiment it is covered by Theorem 1.10 with $m = n_1 \cdot n_2 \cdots n_k$ and $n = n_{k+1}$. Therefore, this 2-stage experiment has $(n_1 \cdot n_2 \cdots n_k) \cdot n_{k+1} = n_1 \cdot n_2 \cdots n_{k+1}$ possible outcomes. ∎

Example 1.16

We wish to determine how many possible outcomes there are from 8 coin tosses. In this case we have 8 subexperiments, each of which has 2 possible outcomes. Therefore, in the notation of Theorem 1.11, $k = 8$ and $n_1 = n_2 = \cdots = n_8 = 2$. Therefore, Theorem 1.11 gives that there are

$$\underbrace{2 \cdot 2 \ldots 2}_{\text{8 times}} = 256$$

possible outcomes for 8 coin tosses. ∎

Example 1.17

Automobile license plates for a certain state consist of 3 letters followed by 3 digits. We wish to determine how many different license plates are possible if it is assumed that all letter and number combinations are permitted. This can be viewed as a 6-stage experiment with the first 3 stages being letters and the last 3 being numbers. We allow a number like 000. Thus, $k = 6$ with $n_1 = n_2 = n_3 = 26$ and $n_4 = n_5 = n_6 = 10$. Therefore, by Theorem 1.11 we see that there are $26 \cdot 26 \cdot 26 \cdot 10 \cdot 10 \cdot 10 = 17{,}576{,}000$ different possible license plates. ∎

Example 1.18

In a race, the first 5 finishers from a field of 49 runners receive a medal. We wish to determine how many possible ways there are to award the medals. We can view this as being an experiment consisting of 5 subexperiments. Let subexperiment i be awarding a medal to the ith finisher, $i = 1, 2, 3, 4, 5$. Since there are 48 possibilities for the first place runner, we get that $n_1 = 48$. Once first place is chosen there are 47 runners left to come in second. Thus, $n_2 = 47$. If we proceed in the same manner we get that $n_3 = 46, n_4 = 45$, and $n_5 = 44$. Therefore, Theorem 1.11 gives that there are $48 \cdot 47 \cdot 46 \cdot 45 \cdot 44 = 205{,}476{,}480$ possible ways to award medals for the race. ∎

The generalized counting principle can be used to count what are known as *permutations*. We begin with a formal definition.

Definition 1.9

A *permutation* of a set of objects is a mapping of the set onto' itself.

This is a rigorous way of saying that a permutation is a way to generate re-arrangements of a set. Given a set in a particular order, the permutation mapping shows exactly how to get the next ordering of the set. Suppose the set in question is $\{a, b, c\}$. One possible permutation begins with the set in the order (a, b, c) and rearranges it into the order (b, c, a).

In what follows we shall use *factorial notation*. This is an operation that is denoted by the symbol $r!$, where r is a nonnegative integer. It represents the product $r! = r(r-1)(r-2)\cdots 3\cdot 2\cdot 1$. Thus $4! = 4\cdot 3\cdot 2\cdot 1 = 24$. It is convention to define $0! = 1$.

We wish to determine the number of different permutations that are possible for a set consisting of n elements. That is, how many different ways can the members of a set be ordered? We can use the generalized counting principle. For a set of size n we can view a mapping as consisting of n subexperiments with subexperiment i being the mapping of the ith elements in the original ordering to its new position. Let x_i be the element that is initially in position $i, i = 1, 2, \ldots, n$. If we begin with x_1, there are n possibilities for its next position. Once x_1 is moved there are $n-1$ possibilities for relocating x_2. After this x_3 can be assigned to any of the remaining $n-2$ places, and so forth. Thus, Theorem 1.11 gives that there are $n(n-1)(n-2)\cdots 2\cdot 1 = n!$ different orderings possible. This paragraph constitutes a proof of the following.

Theorem 1.12

> The number of different permutations of a set of n objects is $n!$

The foregoing really says nothing more than there are $n!$ possible orderings of n distinct objects. That is, given the set $\{a, b, c\}$ we can arrange them in $3! = 6$ different ways. These are the following.

$$(a,b,c), \quad (a,c,b), \quad (b,a,c), \quad (b,c,a), \quad (c,a,b), \quad \text{and} \quad (c,b,a)$$

Example 1.19

The mathematics department at a certain college has 8 full-time faculty members. It is allocated a block of 8 offices in a particular building. Theorem 1.12 tells us that there are $8! = 40{,}320$ possible ways to assign faculty members to offices. ∎

Example 1.20

A course in combinatorics consists of 4 seniors, 3 juniors, and 8 sophomores. The classroom contains 15 desks.

1. Suppose students can sit in any of the desks. Theorem 1.12 tells us that there are $15!$ possible ways to assign students to desks.

2. Now assume that seniors sit in desks 1 through 4, juniors in desks 5 through 7, and sophomores in desks 8 through 15. We now have an experiment consisting of 3 subexperiments. Therefore, by using Theorem 1.11 and Theorem 1.12 we see that there are $(4!)(3!)(8!)$ different ways to assign students to desks under this scheme.

3. Suppose the students are asked to stand along one wall of the classroom grouped by class. We now have an experiment containing 4 subexperiments. Subexperiment 1 is to choose the order in which classes are to stand. Subexperiments 2 through 4 determine how students stand within their class group. There are 3! ways to arrange the classes, 4! ways to arrange the seniors, 3! ways to arrange the juniors, and 8! ways to arrange the sophomores. Therefore, there are (3!)(4!)(3!)(8!) possible ways for the class to stand.

■

Suppose we have a collection of n objects from which we wish to select a subset of size r and place them in order. That is, the first object selected goes in position 1, the second object selected goes in position 2, and so on. Such an arrangement is referred to as an *r-arrangement* or a *permutation of n objects taken r at a time*. We are interested in determining how many such permutations are possible. The number of permutations of n objects taken r at a time is denoted by $P(n, r)$. A straightforward application of the generalized counting principle gives the following. We leave the proof as an exercise.

Theorem 1.13

> The number of permutations of n objects taken r at a time, where $r \leq n$ is
>
> $$P(n, r) = n(n - 1)(n - 2) \cdots (n - r + 1) = \frac{n!}{(n - r)!}. \qquad (1.10)$$

Example 1.21

A game in a state lottery selects 3 numbers from the set of digits, $\{0, 1, 2, 3, 4, 5, 6, 7, 8, 9\}$, with no number being repeated, to obtain a 3-digit number. For example, if the digits are selected in the order 0 followed by 7 followed by 3, then the 3-digit number is 073. A 3-digit number selected in this manner is a permutation of 10 objects taken 3 at a time. Therefore, by Theorem 1.13, there are

$$P(10, 3) = 10 \cdot 9 \cdot 8 = \frac{10!}{7!} = 720$$

possible numbers.

■

If it is permitted for objects to be selected any number of times, the generalized counting principle shows how to determine the *number of permutations of n objects taken r at a time with repetition*. We have an r stage experiment with n possible outcomes in each subexperiment. Thus we obtain the following.

Theorem 1.14

> There are n^r distinct permutations of n objects taken r at a time with repetition.

Example 1.22   If we allow the numbers to repeat in the lottery game of Example 1.21, then Theorem 1.14 shows that there are

$$(10)^3 = 1000$$

possible 3-digit numbers.   ∎

In some situations an experiment can result in outcomes that are indistinguishable from other outcomes. Suppose we wish to determine how many distinct arrangements exist using the letters from the word "onion." Since there are 5 letters there are $5! = 120$ permutations of the letters. However, some will look the same. For example, suppose we are interested in how many result in *oonni*. By subscripting the letters as $o_1 n_1 i o_2 n_2$ to make them distinct we see that

$$o_1 o_2 n_1 n_2 i, \quad o_2 o_1 n_1 n_2 i, \quad o_1 o_2 n_2 n_1 i, \quad \text{and} \quad o_2 o_1 n_2 n_1 i$$

are permutations of the original 5 letters that are the same sequence of letters. In fact, given any placement of the two o's there are $2!$ possible orderings for them that are identical in appearance. Similarly there are $2!$ identical ways to place the n's. Thus, out of the $5!$ possible arrangements of the letters each distinct arrangement occurs $2!2! = 4$ times. This means that only $5!/4$ arrangements are distinct. This idea is generalized as follows.

Theorem 1.15

Suppose that a collection of n objects is composed such that there are k distinct objects. Let n_i be the number of objects of type $i, i = 1, 2, \ldots, k$ in the collection, where $n_1 + n_2 + \cdots + n_k = n$. Then there are

$$\frac{n!}{n_1! n_2! \cdots n_k!} \tag{1.11}$$

different permutations of the n objects.

Example 1.23   We wish to determine the number of distinct permutations that can be obtained from *MISSISSIPPI*. Here there are 11 letters of which 1 is M, 2 are P, 4 are I, and 4 are S. Thus by Theorem 1.15 we find that there are

$$\frac{11!}{(1!)(2!)(4!)(4!)} = 35,650$$

distinct permutations of the letters.   ∎

Example 1.24   A group of n objects are arranged in a circle. We are interested in the relative position of the objects. For example, given four objects the following two arrangements would be considered identical.

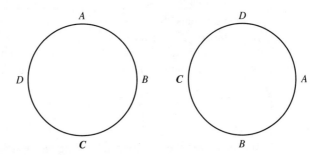

The number of different arrangements is obtained first by arbitrarily placing the first object. Now we have $n-1$ objects left to arrange. Theorem 1.12 gives that there are $(n-1)!$ ways to do this. Thus, there are $(n-1)!$ different arrangements of n objects in a circle. Another way to view this problem is that there are $n!$ ways to place the objects in the n positions around the circle, but in terms of relative placement there are n copies of each different ordering. Thus, there are really only $\frac{n!}{n} = (n-1)!$ distinct arrangements.   ∎

Many times we are interested in knowing the number of ways a given subset of objects can be selected. In such cases, the order in which they are chosen is of no importance. For example, suppose 2 letters are selected from the set $\{a, b, c, d, e\}$ with no repetitions. How many of these arrangements are distinct subsets of $\{a, b, c, d, e\}$? The $P(5,2) = 20$ possible selections are given below.

$$(a,b) \quad (a,c) \quad (a,d) \quad (a,e) \quad (b,a)$$
$$(b,c) \quad (b,d) \quad (b,e) \quad (c,a) \quad (c,b)$$
$$(c,d) \quad (c,e) \quad (d,a) \quad (d,b) \quad (d,c)$$
$$(d,e) \quad (e,a) \quad (e,b) \quad (e,c) \quad (e,d)$$

Of these 20 permutations of the 5 letters taken 2 at a time, there are $10 = P(5,2)/$ $(2!)$ different subsets. They are shown below.

$$\{a,b\} \quad \{a,c\} \quad \{a,d\} \quad \{a,e\} \quad \{b,c\}$$
$$\{b,d\} \quad \{b,e\} \quad \{c,d\} \quad \{c,e\} \quad \{d,e\}$$

A *subset* of r objects from a set containing n objects is known as a *combination*. The number of combinations that are possible when selecting r objects from a set of n objects is denoted by $C(n,r)$ and computed according to the following theorem.

Theorem 1.16

The number of combinations of r objects taken from n objects with no repetitions, where $r \leq n$, is

$$C(n,r) = \frac{P(n,r)}{r!}. \qquad (1.12)$$

Proof   We can view choosing a permutation as an experiment consisting of 2 subexperiments. Subexperiment 1 chooses a subset of r objects which by our notation has $C(n, r)$ possible outcomes. Subexperiment 2 then permutes these r objects for which Theorem 1.15 gives that there are $r!$ possible outcomes. The generalized counting principle then shows that $P(n, r) = (r!)C(n, r)$. Therefore,

$$C(n, r) = \frac{P(n, r)}{r!}.$$

■

If we rewrite Equation 1.12 using factorial notation, a commonly used formula for computing the number of combinations is found.

$$C(n, r) = \frac{n!}{r!\,(n - r)!} \tag{1.13}$$

Furthermore, we can use this notation to show that

$$C(n, n - r) = \frac{n!}{(n - r)!\,(n - [n - r])!}$$

$$= \frac{n!}{r!\,(n - r)!}$$

$$= C(n, r). \tag{1.14}$$

This identity, which shows a symmetry in combinations, can be useful from a computational standpoint. For example, if we know the value for $C(10, 2)$ we also know $C(10, 8)$. It is also sometimes useful in algebraic manipulations involving combinations.

An alternative notation that will be used extensively in this book is $C(n, r) \equiv \binom{n}{r}$. When written in this form, $\binom{n}{r}$ is referred to as a *binomial coefficient* because it is the coefficient of the x^r term in the expansion of $(x + y)^n$. We shall return to this later.

Example 1.25

A final examination is designed where the student is to answer any 7 questions from a group of 10 questions. We wish to determine how many possibilities there are for questions to answer. Clearly the order in which questions are answered is not relevant. Thus, we are interested in the number of combinations of 7 objects chosen from a collection of 10. By Equation 1.13 this is

$$\binom{10}{7} = \frac{10!}{7!\,3!} = 120.$$

■

Example 1.26

An instructor's manual lists sample examination questions of the subject of combinations. Out of a total of 15 questions 6 are theoretical and 9 are computational. We wish to know how many examinations can be divised that consist of 2 theoretical and 4 computational questions. This can be viewed as an experiment consisting of 2 subexperiments. Subexperiment 1 chooses the theoretical questions, and subexperiment 2 picks the computational ones. Since the order in which questions are chosen is unimportant, we are again interested in combinations. Theorem 1.16 gives that there are

$$\binom{6}{2} = \frac{6!}{2!\,4!} = 15$$

possible sets of theoretical questions and

$$\binom{9}{4} = \frac{9!}{4!\,5!} = 126$$

sets of computational questions. Therefore, the generalized counting principle gives that there are

$$\binom{6}{2} \cdot \binom{9}{4} = (15) \cdot (126) = 1890$$

different examinations.   ∎

We mentioned the relationship between combinations and the coefficients in the expansion of $(x + y)^n$. We shall state and prove this formally after we prove an identity that will be useful.

Theorem 1.17

$$\binom{n}{r} = \binom{n-1}{r-1} + \binom{n-1}{r} \tag{1.15}$$

Proof   We shall prove the result both algebraically and combinatorially. Algebraically:

$$\binom{n-1}{r-1} + \binom{n-1}{r} = \frac{(n-1)!}{(r-1)!(n-r)!} + \frac{(n-1)!}{r!(n-1-r)!}$$

$$= \frac{(n-1)!}{(r-1)!(n-r-1)!}\left(\frac{1}{n-r} + \frac{1}{r}\right)$$

$$= \frac{(n-1)!}{(r-1)!(n-r-1)!} \cdot \frac{r+(n-r)}{r(n-r)}$$

$$= \frac{n!}{r!(n-r)!} = \binom{n}{r}.$$

Combinatorially: Select an element from the set of n objects. Call it x. In every combination it is the case that x is either in that combination or it is not. For every combination containing x we select the remaining $r - 1$ objects from the other $n - 1$. Thus, there are $\binom{n-1}{r-1} \cdot 1$ combinations of the n objects taken r at a time that contain x. Now for every combination that does not contain x we select r objects from the other $n - 1$ objects. Thus, there are $\binom{n-1}{r}$ combinations that do not contain x. We get Equation 1.14 by adding these 2 together.   ∎

Although many students are initially uncomfortable with the combinatorial method of proof, it has the advantage of being easier to follow on an intuitive level than the algebraic one. Theorem 1.17 is often referred to as *Pascal's formula* because it is the basis for Pascal's Triangle. It gives an inductive way for generating the binomial coefficients. A small segment of the triangle is given in Figure 1.5.

$$
\begin{array}{ccccccccccccc}
 & & & & & & 1 & & & & & & \\
 & & & & & 1 & & 1 & & & & & \\
 & & & & 1 & & 2 & & 1 & & & & \\
 & & & 1 & & 3 & & 3 & & 1 & & & \\
 & & 1 & & 4 & & 6 & & 4 & & 1 & & \\
 & 1 & & 5 & & 10 & & 10 & & 5 & & 1 &
\end{array}
$$

Figure 1.5

The triangle gives binomial coefficients in the following manner. Each row represents a particular value of n. The top row is $n = 0$, the next is $n = 1$, the third is $n = 2$, and so on. The leftmost entry in any row corresponds $r = 0$, the next entry is for $r = 1$, and so forth. Therefore, to find $\binom{3}{2}$ in Pascal's Triangle we simply go to the fourth row from the top and look at the third entry starting from the left to see that $\binom{3}{2} = 3$. To see how Pascal's formula works select any interior entry in the triangle. That entry is $\binom{n}{r}$. By using our scheme we see that the entry in the triangle above and to the left of $\binom{n}{r}$ is the entry for $\binom{n-1}{r-1}$ and the entry above and to the right is $\binom{n-1}{r}$. Therefore, for our example we see that $\binom{2}{1} + \binom{2}{2} = \binom{3}{2} = 3$. This is how Pascal's formula generates Pascal's Triangle.

Example 1.27

We wish to use Pascal's formula to produce the next row in Pascal's Triangle after that given in Figure 1.5. If we let each entry be the sum of the elements that are above and to the left and that which is above and to the right with the understanding that an empty entry is equal to zero we obtain

$$1 \quad 6 \quad 15 \quad 20 \quad 15 \quad 6 \quad 1.$$

We leave it to the reader to verify that these are, in fact, the correct values for the binomial coefficients for $n = 6$. ■

We now give a combinatorial proof of the Binomial theorem and leave the inductive proof as an exercise. The inductive proof uses Theorem 1.17.

Theorem 1.18

> **Binomial Theorem**   If n is a positive integer then
>
> $$(x + y)^n = \sum_{i=0}^{n} \binom{n}{i} x^i y^{n-i}. \tag{1.16}$$

Proof   Consider the product $\underbrace{(x + y)(x + y) \cdots (x + y)}_{n \text{ times}}$. Its expansion consists of
2^n terms. For example, if $n = 2$ we get $xx + xy + yx + yy$. Then for $n = 3$ this
becomes $xxx + xxy + xyx + yxx + yyx + yxy + xyy + yyy$. In this expansion we see
that $x^i y^{n-i}, i = 0, 1, \ldots, n$, occurs whenever x is chosen from i of the n $(x + y)$
factors. Therefore, there are $C(n, i)$ terms of the form $x^i y^{n-i}$, and the proof is
completed.   ∎

Compare this proof with the tortuous calculations involved in carrying out the
proof by induction to appreciate the efficiency of combinatorial arguments.

Example 1.28

The expansion of $(x - y)^4$ is by Theorem 1.18

$$(x + y)^4 = \binom{4}{0} x^0 (-y)^{4-0} + \binom{4}{1} x^1 (-y)^{4-1} + \binom{4}{2} x^2 (-y)^{4-2}$$

$$+ \binom{4}{3} x^3 (-y)^{4-3} + \binom{4}{4} x^4 (-y)^{4-4}$$

$$= y^4 - 4y^3 x + 6y^2 x^2 - 4yx^3 + x^4.$$

■

Example 1.29

The power set of a set A is the set of all subsets of A. For example, the power
set of $\{a, b, c\}$ is

$$\{\varnothing, \{a\}, \{b\}, \{c\}, \{a, b\}, \{a, c\}, \{b, c\}, \{a, b, c\}\}.$$

It contains 8 elements, each of which is a subset of A. We can use the Binomial
theorem to determine the number of elements in the power set of a set A when A
has n elements. Recall that $\binom{n}{i}$ counts the number of subsets of i elements that
can be selected from a set of size n. Subsets may contain $0, 1, 2, \ldots, n$ elements.
Therefore, the number of possible subsets is

$$\sum_{i=1}^{n} \binom{n}{i} = \sum_{i=0}^{n} \binom{n}{i} 1^i 1^{n-i} = (1 + 1)^n = 2^n.$$

■

In some cases we begin with n objects, and it is desired to divide them into r
different groups of size $n_1, n_2, \ldots, n_r$. where $\sum_{i=1}^{r} n_i = n$. We wish to determine
how many such divisions are possible. For example, suppose we wish to investigate
the difference between the performance of students who commute from home, live
in school dormitories, and live elsewhere. We wish to assign 12 students to living
arrangements such that 3 of them are commuters, 4 are dormitory residents, and
let the remaining 5 live elsewhere. How many different such assignments are
possible? In the binomial case we considered only 2 possible assignments of n

objects—those in a particular set and those not in that set. Here there are more possibilities. We can treat this as an r stage experiment. In subexperiment 1 we choose n_1 objects to be assigned to group 1. This can be done $\binom{n}{n_1}$ ways. Of the remaining $n - n_1$ objects, we assign n_2 in $\binom{n-n_1}{n_2}$ ways, and so forth. This gives that the number of possible assignments is

$$\binom{n}{n_1}\binom{n-n_1}{n_2}\cdots\binom{n-n_1-n_2-\cdots-n_{r-1}}{n_r}$$

$$= \frac{n!}{n_1!(n-n_1)!} \cdot \frac{(n-n_1)!}{n_2!(n-n_1-n_2)!} \cdots \frac{n-n_1-n_2-\cdots-n_{r-1}}{n_r!0!}$$

$$= \frac{n!}{n_1!n_2!\cdots n_r!}.$$

$$(1.17)$$

The number of ways of dividing n objects in this manner is known as a *multinomial coefficient* because of the Multinomial theorem that we state without proof.

Theorem 1.19

> **Multinomial Theorem**   If n is a positive integer, then
>
> $$(x_1 + x_2 + \cdots + x_r)^n = \sum \binom{n}{n_1\ n_2\ \cdots\ n_4} x_1^{n_1} x_2^{n_2} \cdots x_r^{n_r},$$
>
> where the summation is taken over all possible nonnegative integer values such that $n_1 + n_2 + \cdots + n_r = n$ and
>
> $$\binom{n}{n_1\ n_2\ \cdots\ n_r} = \frac{n!}{n_1!\,n_2!\cdots n_r!}.$$

Example 1.30   In our student assignment example cited at the beginning of this section, Theorem 1.19 shows that there are

$$\frac{12!}{3!\,4!\,5!} = 27720$$

different ways to assign students to living arrangements.   ■

Example 1.31   In an art class of 10 students, it is possible for 2 to work on a sculpture project, 3 to do an oil painting, and 5 to do a pen and ink drawing. Theorem 1.19 gives that there are

$$\frac{10!}{2!\,3!\,5!} = 2520$$

different ways to assign students to projects.   ■

A number of probability problems can be viewed as placing balls in urns. Consider a problem where an elevator stops at r floors. If the elevator begins with n passengers, we may need to know

1. how many ways the n people can get off on the r floors, or

2. how many ways the n people can get off on the r floors under the condition that at least one person departs on each floor.

Here we can equate the floors with urns and the people with balls and view this as a ball and urn problem. In a more numerical setting we can interpret Problem 1 as being that of determining how many nonnegative integer solutions exist to the problem

$$x_1 + x_2 + \cdots + x_r = n.$$

The value of x_i is the number of balls assigned to urn i. Problem 2 is that of determining how many positive integer solutions there are to the equation

$$x_1 + x_2 + \cdots + x_r = n,$$

where $r \leq n$. We give the answer to Problem 1 in the following theorem.

Theorem 1.20

> Given n identical balls and r different urns there are $\binom{n+r-1}{n}$ different ways to distribute the balls among the urns.

Proof   Let $|$ represent an urn boundary and o represent a ball. Therefore, there are $r + 1$ $|$'s and n o's for a total of $n + r + 1$ symbols. Now an arrangement such as $||o|oo\cdots|o|$ represents an assignment of the balls to the urns. Any assignment must begin and end with a $|$, which means we are free to assign symbols to the remaining $n + r - 1$ locations. Once we have selected the n locations for placement of the o's the other assignments are fixed. Therefore, there are $\binom{n+r-1}{n}$ different assignments of balls to urns. ∎

Example 1.32

Consider the problem of placing 15 identical balls in 3 urns. Theorem 1.20 gives that there are

$$\binom{15 + 3 - 1}{15} = 136$$

different ways to do this. ∎

Example 1.33

How many solutions are there to $x_1 + x_2 + x_3 = 15$ when x_1, x_2, and x_3 must be positive integers? Since x_1, x_2, and x_3 must be at least equal to one, let $y_1 = x_1 - 1, y_2 = x_2 - 1$, and $y_3 = x_3 - 1$. Therefore,

$$x_1 + x_2 + x_3 = y_1 - 1 + y_2 - 1 + y_3 - 1 = 15 \quad \text{or}$$

$$y_1 + y_2 + y_3 = 12,$$

where y_1, y_2, and y_3 are nonnegative integers. Therefore, Theorem 1.20 gives that there are

$$\binom{12 + 3 - 1}{12} = 91$$

possible solutions. This solution technique is equivalent to determining the number of ways we can assign 15 balls to 3 urns under the condition that each urn contains at least 1 ball. The counting argument begins by first assigning 1 ball to each urn which leaves us free to assign the remaining 12 balls in any manner. Now we can appeal to Theorem 1.20 to obtain the result with 12 balls and 3 urns.   ■

Exercises 1.5

1. An experiment consists of 2 subexperiments. Subexperiment 1 has i possible outcomes. If subexperiment 1 results in outcome j, for $j = 1, 2, \ldots, i$, then there are n_j possible outcomes for subexperiment 2. How many possible outcomes are there for the overall experiment?

2. How many integers between 1000 and 10,000 have no digits other than 2, 4, or 6?

3. A neighborhood pet show consists of four dogs and three cats. The pets are placed in a row.

   a. How many different arrangements of pets by type are possible?

   b. How many arrangements by type of pet are possible if no two pets of the same species may be placed next to each other?

4. How many consecutive trailing zeros occur in (20!)? As an example $(10!) = 3,628,800$ and, thus, has 2 trailing zeros.

5. Prove Theorem 1.13.

6. How many seating arrangements are possible for eight people at a round table if

   a. there are no restrictions, and

   b. two of the eight must be seated next to each other?

7. How many distinct permutations are there of the letters in the word structure? Statistics?

8. How many 13-card bridge hands are possible from a standard deck of 52 cards?

9. Prove that $\sum_{i=0}^{n} \binom{n}{i} (-1)^i = 0$.

10. Suppose there are r urns and n balls with $r > n$. How many ways can balls be assigned to urns

    a. if there are no restrictions, and

    b. if no more than one ball can be placed in an urn?

11. What is the coefficient of the x^7 term in $(x + 2x^2)^5$?

12. A ship has 3 lifeboats of capacity 5, 10, and 15. How many different assignments are possible for a crew of 30 people?

13. Use induction on n to prove that

$$(x + y)^n = \sum_{i=0}^{n} \binom{n}{i} x^i y^{n-i}.$$

14. Show that

$$\binom{n}{1} + \binom{n}{3} + \cdots = \binom{n}{0} + \binom{n}{2} + \cdots = 2^{n-1}.$$

15. Show that

$$\binom{n}{0} + 2 \cdot \binom{n}{1} + \binom{n}{2} + 2 \cdot \binom{n}{3} + \cdots = 3 \cdot 2^{n-1}.$$

16. Twelve points lie in a plane with no three lying on a straight line.
    a. How many different line segments are determined by these points?
    b. How many different triangles are determined by these points?

17. Show that

$$\binom{n+m}{r} = \sum_{i=0}^{r} \binom{n}{i}\binom{m}{r-i}$$

by considering the number of possible ways of choosing r objects from a group consisting of n objects of one type and m of another type.

18. Use the result of Problem 17 to show that

$$\binom{2n}{n} = \sum_{i=0}^{n} \binom{n}{i}^2.$$

19. What is the sum of all nonnegative integers a, b, c of the form $\frac{8!}{a!\,b!\,c!}$?

20. A class of 12 students is to be divided into project teams of size 3 each. Each team is assigned to a different project. How many different team assignments are possible?

21. A class of 12 students is to be divided into project teams of size 3 each. Each team does the same project. How many different team assignments are possible?

22. In how many ways can 10 students be assigned to 3 advisors? What if each advisor must get at least 2 students?

23. What is the coefficient of x^4 in $(a + bx + cx^2)^5$?

24. Consider all positive integers consisting of 3 digits that are greater than or equal to 600.
    a. How many even numbers are there?
    b. How many have distinct digits?
    c. How many are divisible by 4?

25. A group consists of four married couples.
    a. How many ways can the group be seated in a row?
    b. How many ways can the group be seated in a circle?
    c. How many ways can the group be seated in a row if no man sits next to his wife?
    d. How many ways can the group be seated in a circle if each man sits next to his wife?

1.6
Probabilities for Finite Sample Spaces with Equally Likely Outcomes

A surprisingly large number of random experiments are such that there are only a finite number of possible outcomes and each outcome has the same probability of occurrence. For example, if we toss a fair coin there are two possible outcomes and the fairness of the coin insures that a head is just as likely to occur as a tail. In such a case the two elementary events would be said to be *equally likely*. Contrast this with the experiment of tossing a thumb tack. Again there are just two possible outcomes, but a little experimentation will show that the probability of the tack landing with its point up is not the same as that of landing with its point down. Therefore, these two elementary events are not equally likely. Random experiments that have a sample space consisting of a finite number of elementary events each of which has the same probability of occurrence are particularly nice to deal with. In such cases, probabilities can be found by computing the proportion of elementary events in the sample space that are in the event of interest. To be a little more precise, the term "equally likely" means that if there are N elementary events, $E_1, E_2, \ldots, E_N$, in a sample space S, then the fact that elementary events are mutually exclusive allows us to apply Axiom 2 and Theorem 1.6 to find

$$\sum_{i=1}^{N} P(E_i) = P\left(\bigcup_{i=1}^{N} E_i\right) = P(S) = 1. \tag{1.18}$$

Now, if the elementary events are such that $P(E_1) = P(E_2) = \ldots = P(E_N)$, then Equation 1.18 implies that

$$P(E_i) = \frac{1}{N}, i = 1, 2, \ldots, N.$$

This is what we mean by equally likely outcomes.

Given a finite sample space consisting of equally likely elementary events, we compute the probability of any event, say A, in the following manner. Let $E_1, E_2, \ldots, E_N$ be the elementary events as before. Then we can write A as the union of some of the elementary events, say $E_1, E_2, \ldots, E_{n(A)}$. Then

$$P(A) = P\left(\bigcup_{i=1}^{n(A)} E_i\right) = \sum_{i=1}^{n(A)} P(E_i) = \frac{n(A)}{N}. \tag{1.19}$$

More informally, this means that

$$P(A) = \frac{\text{number of elementary events in } A}{\text{number of elementary events in } S}. \tag{1.20}$$

Note that this is the proportion of outcomes in S that are in A that we discussed in Section 1.4 as an interpretation of probability. From a practical standpoint this means that the computation of probabilities in this type of experiment is reduced to a problem of counting elementary events. Since we are only interested in how many elementary events are in a given event it will, in most cases, be to our advantage if we can avoid having to enumerate the sample space and count events directly. For this reason we will want to make use of the techniques of counting theory that were discussed in Section 1.5.

Some experiments that result in finite sample spaces consisting of equally likely outcomes are coin tossing, dice rolling, drawing cards from a well-shuffled deck, placing balls in urns, and drawing balls from urns. In addition, any experiment that can be reworded to become one of these will produce a sample space of this type. As an example of how this is done, consider an experiment where people board an elevator in the basement of a building and depart the elevator on randomly selected floors. Here we can view the people as being balls and the floors of the building as being urns. Therefore, this experiment is logically equivalent to that of placing balls in urns, and we can use the equally likely model for the sample space.

The rest of the section will be devoted to examples of the techniques used to deal with equally likely sample spaces. The basic idea is to break complicated events into unions of simpler events. Ideally, the events in the unions will be mutually exclusive, but this is not always the case. This process is continued until the events are simple enough for us to count the number of elementary events that are in them. Once this is done we use Theorem 1.6 or Theorem 1.9, whichever is appropriate, to combine these probabilities.

Before turning to examples, however, we need to establish some terminology. There are two different ways to conduct random experiments where objects are chosen from a given set. When the experiment is such that each object selected is returned to the set before choosing the next one the experiment is said to be *sampling with replacement*. Tossing coins and rolling dice fall into this category. When the selection of an object means that it is ineligible for future selection then it is called *sampling without replacement*. An example of this case is when we are dealing out poker hands. Ball and urn experiments can be of either type. Balls selected from urns can either be put back in the urn after being chosen or not. When one is placing balls in urns the same urn can receive more than one ball or not depending on the experiment's rules. Whether sampling is done with or without replacement will have a direct effect on how we go about counting elementary events.

Example 1.34

Consider an experiment where 2 fair dice are rolled. The sample space consists of 36 elementary events that can be represented as ordered pairs where the first entry is the number of dots on the first die and the second entry is the number of dots on the second die. In our terminology, this experiment can be viewed as sampling 2 numbers from the set $\{1, 2, 3, 4, 5, 6\}$ with replacement.

$$
\begin{array}{cccccc}
(1,1) & (1,2) & (1,3) & (1,4) & (1,5) & (1,6) \\
(2,1) & (2,2) & (2,3) & (2,4) & (2,5) & (2,6) \\
(3,1) & (3,2) & (3,3) & (3,4) & (3,5) & (3,6) \\
(4,1) & (4,2) & (4,3) & (4,4) & (4,5) & (4,6) \\
(5,1) & (5,2) & (5,3) & (5,4) & (5,5) & (5,6) \\
(6,1) & (6,2) & (6,3) & (6,4) & (6,5) & (6,6)
\end{array}
$$

Let $A = \{\text{sum of both dice is 7}\}$. Since there are 6 elementary events whose sum is 7, we find that $P(A) = \frac{6}{36}$. Let $B = \{\text{first die has 2 dots}\}$. There are 6 elementary events where the first die shows a 2. Therefore, $P(B) = \frac{6}{36}$. Similarly, the event $A \cap B$, which is the event where the first die is 2 and the sum of the 2 dice is 7, has a single elementary event in it. Therefore, $P(A \cap B) = \frac{1}{36}$. We can now use Theorem 1.8 to determine the probability that either the first die shows 2 or the sum is 7 as follows.

$$
P(A \cup B) = P(A) + P(B) - P(A \cap B) = \frac{6}{36} + \frac{6}{36} - \frac{1}{36} = \frac{11}{36}
$$

We can also use Theorem 1.4 to compute the probability that the sum of the 2 dice will not be 7 as follows.

$$P(A') = 1 - P(A) = 1 - \frac{6}{36} = \frac{5}{6}$$

■

Example 1.35   We have 2 boxes. Box I contains 5 red balls and 5 white balls. Box II contains 7 red balls and 3 white balls. One ball is selected from each box. Since this can be viewed as a 2-stage experiment, we can use the basic counting principle to determine that there are $10 \cdot 10 = 100$ elementary events in the sample space. The sample space will consist of ordered pairs with the first entry being the ball drawn from Box I and the second entry being the ball drawn from Box II. Let A be the event that at least 1 white ball is drawn and B be the event that both balls are the same color. We shall compute $P(A)$ first. Here it is useful to view A as

$$A = \{\{\text{exactly 1 white ball}\} \cup \{\text{exactly 2 white balls}\}\}.$$

We can further break the event of drawing exactly 1 white ball as follows.

$$\{\text{exactly 1 white ball}\} = \{\{\text{white ball from Box I and red ball from Box II}\}$$
$$\cup \{\text{red ball from Box I and white ball from Box II}\}\}$$

These 3 events are mutually exclusive. We also have broken A into sufficiently small pieces where it is possible to apply the basic counting principle to determine the number of elementary events in each one. These are as follows.

$$P(\text{white ball from Box I and red ball from Box II}) = \frac{5 \cdot 7}{100} = \frac{35}{100},$$

$$P(\text{red ball from Box I and white ball from Box II}) = \frac{5 \cdot 3}{100} = \frac{15}{100}, \quad \text{and}$$

$$P(\text{exactly 2 white balls}) = \frac{5 \cdot 3}{100} = \frac{15}{100}.$$

We now use Theorem 1.6 to find that

$$P(A) = \frac{35}{100} + \frac{15}{100} + \frac{15}{100} = \frac{65}{100}.$$

We must point out that this was the hard way to work this problem. Less work is involved in computing the probability of A', which is the event of getting exactly 2 red balls and applying Theorem 1.4. $P(A')$ can be found immediately by the basic counting principle as follows.

$$P(A') = \frac{5 \cdot 7}{100} = \frac{35}{100}$$

We then see that

$$P(A) = 1 - P(A') = 1 - \frac{35}{100} = \frac{65}{100},$$

as before.

To compute $P(B)$ we note that

$$\{\text{both balls same color}\} = \{\{\text{both balls white}\} \cup \{\text{both balls red}\}\}.$$

These 2 events are mutually exclusive, and we can use the basic counting principle to determine their probabilities.

$$P(\text{both balls white}) = \frac{5 \cdot 3}{100} = \frac{15}{100}, \quad \text{and}$$

$$P(\text{both balls red}) = \frac{5 \cdot 7}{100} = \frac{35}{100}.$$

Therefore, Axiom 3 gives that

$$P(B) = \frac{15}{100} + \frac{35}{100} = \frac{1}{2}.$$

Example 1.36   A shipment of 100 bags of sugar is rumored to have some bags that contain illegal drugs. Local customs officials decide to screen shipments by randomly testing 4 bags. If any are found to contain drugs, the whole shipment will be seized. We wish to compute the probability that a shipment consisting of 95 bags of sugar and 5 bags of illegal drugs will be detected in this manner. Let A be the event that drugs are found in some of the 4 bags tested. This means that at least 1 of the 4 bags was found to contain drugs. Therefore, we want to compute the probability that at least 1 of the bags will be chosen from the group of 5 bags which originally contained the drugs. Here, it seems to be less work to deal with A', the event where all 4 bags selected contain sugar. We can use either combinations or permutations to count the number of elementary events in the sample space and the number of elementary events in A'. Since we are selecting 4 bags from a set of 100 bags we see that

$$N = \binom{100}{4}.$$

In A' we are selecting 4 bags from the set of 95 bags that contain sugar. Therefore,

$$n(A') = \binom{95}{4}.$$

Therefore, we use Theorem 1.3 to find

$$P(A) = 1 - P(A') = 1 - \frac{\binom{95}{4}}{\binom{100}{4}} = 1 - 0.812 = 0.188.$$

One would hope to use a screening procedure with a higher chance of success. ∎

Example 1.37   Poker hands consist of 5 cards taken from a standard 52-card deck. In the history of the western United States there is a story that Wild Bill Hickock was shot to death while holding a hand consisting of a pair of aces and a pair of eights. For this reason "aces and eights" has come to be known as the "dead man's hand." What is the probability of getting a dead man's hand by randomly selecting 5 cards? What is the probability of getting any 2 pairs by randomly selecting 5 cards? To get an idea of how calculations like this proceed we shall look at a slightly easier poker hand—a flush. A flush is where all 5 cards are from the same suit. We shall further simplify the case by ignoring the distinction between flushes and straight flushes where the cards, in addition to being from the same suit, are also in order. We begin by letting $F = \{$flush$\}$. Since we do not care about the order of the cards, combinations appear to be a logical way to count elementary events. Since we are selecting 5 cards from a set of 52 without replacement, there will be $\binom{52}{5}$ different poker hands. Now getting 5 cards all from the same suit can be broken up as follows.

$$\{\text{flush}\} = \{\{5 \text{ spades}\} \cup \{5 \text{ hearts}\} \cup \{5 \text{ diamonds}\} \cup \{5 \text{ clubs}\}\}$$

These 4 events are mutually exclusive. In addition, since there are 13 cards in each suit we can select 5 cards from any one of the suits in $\binom{13}{5}$ ways. Therefore,

$$P(\text{flush}) = P(5 \text{ spades}) + P(5 \text{ hearts}) + P(5 \text{ diamonds}) + P(5 \text{ clubs})$$

$$= \frac{\binom{13}{5}}{\binom{52}{5}} + \frac{\binom{13}{5}}{\binom{52}{5}} + \frac{\binom{13}{5}}{\binom{52}{5}} + \frac{\binom{13}{5}}{\binom{52}{5}}$$

$$= 4\frac{\binom{13}{5}}{\binom{52}{5}}.$$

This probability could have been computed in a different manner. This technique will be useful for the more complicated poker hands that follow. It is perfectly valid to view choosing 5 cards from the same suit as being a 2-stage experiment. In the first stage, we select the suit from the set $\{$spades, hearts, diamonds, clubs$\}$. This can be done in $\binom{4}{1}$ ways. In the second stage, we select the 5 cards from that suit. This can be done in $\binom{13}{5}$ ways. Therefore, by the basic counting principle, the product of these is the number of ways to select 5 cards from the same suit. Therefore we obtain

$$P(\text{flush}) = \frac{\binom{4}{1} \cdot \binom{13}{5}}{\binom{52}{5}} = 4\frac{\binom{13}{5}}{\binom{52}{5}}$$

as before.

We now turn to the probability of drawing "aces and eights." We can view this as being a 3 stage experiment. In stage 1, we select 2 aces from the available 4. This can be done in $\binom{4}{2}$ ways. In stage 2, we select 2 of the 4 eights. This can also be done in $\binom{4}{2}$ ways. In stage 3, we select the fifth card from the remaining 44, which are neither aces nor eights. This can be done in $\binom{44}{1} = 44$ ways. Then, by the generalized counting principle the number of ways to draw "aces and eights" is $\binom{4}{2} \cdot \binom{4}{2} \cdot 44$, and

$$P(\text{aces and eights}) = \frac{\binom{4}{2} \cdot \binom{4}{2} \cdot 44}{\binom{52}{5}}.$$

The problem of computing the probability of drawing 5 cards such that there are 2 from each of 2 separate denominations and a fifth card that has a different denomination from the pairs is slightly more complicated. Again the plan is to break the experiment into stages. One way to do this is to view it as being a 4-stage experiment as follows.

1. Select 2 denominations in $\binom{13}{2}$ ways.

2. Select 2 cards from 1 denomination in $\binom{4}{2}$ ways.

3. Select 2 cards from the other denomination in $\binom{4}{2}$ ways.

4. Select 1 card from the remaining denominations in $\binom{44}{1}$ ways.

Then the generalized counting principle shows that the number of ways to draw 2 pairs is the product of these. Therefore,

$$P(\text{two pair}) = \frac{\binom{13}{2} \cdot \binom{4}{2}^2 \cdot \binom{44}{1}}{\binom{52}{5}}.$$

■

Example 1.38   *The Birthday Problem*   What is the probability that, in a group of n randomly chosen people, at least 2 of them will have the same birthday? This is a classic problem, mainly because the answer is somewhat surprising. Most people would think that, with 365 possible birthdates, n would have to be rather large before the probability would exceed $\frac{1}{2}$. We shall assume that the probability that any given person would have a particular birthday is $\frac{1}{365}$ and make the additional simplification of ignoring leap years. Let $A = \{$at least 2 have the same birthday$\}$. The first question is whether it is more convenient to compute $P(A)$ or $P(A')$. Since $A' = \{$all n people have different birthdays$\}$ this event can be viewed as selecting a set of n values from the set $\{1, 2, \ldots, 365\}$ without replacement. In dealing with A we would have to account for the possibility that 2 people have the same birthday, or 3 people have the same birthday, or 4 people have the same birthday, and so on. To compute $P(A')$ it is desirable to view this as an n stage experiment. The number of elementary events in the sample space is obtained by noting that at each stage it is possible to select a birthday from any of the 365 possibilities. Therefore, the generalized counting principle gives that $N = (365)^n$. In selecting n differing birthdays, we see that the first birthday can be chosen from any of the 365 possibilities, the second birthday must be chosen from one of the 364 other dates, the third birthday must come from the remaining 363, and so forth. This means that the generalized counting principle will give that there are $365 \cdot 364 \ldots (365 - n + 1)$ ways to select n different birthdates from the 365. Therefore,

$$P(A) = 1 - P(A') = 1 - \frac{365 \cdot 364 \cdot (365 - n + 1)}{(365)^n}.$$

Note that the probability of A increases as n increases. It is equal to 1 when $n \geq 365$ and is equal to 0 when $n = 1$. The interesting thing to note is that when $n = 22$, $P(A) = 0.475$, and, when $n = 23$, $P(A) = 0.507$. Thus, it does not take very many people before it becomes more likely for 2 or more people to have the same birthday than it does for all to have a different birthday.   ∎

Example 1.39   Four married couples are randomly assigned to chairs that are arranged in a row. What is the probability that each man will be seated next to his wife? What is the probability that no 2 women sit next to each other? Let $A = \{$each man sits next to his wife$\}$ and $B = \{$no women occupy adjacent seats$\}$. Since it matters who sits in which seat, the most convenient way to count elementary events appears to be by using either the generalized counting principle or permutations. Recall that they are essentially the same thing. Since we are assigning 8 people to 8 seats, it is plain that there are 8! different ways to seat the people. Therefore, $N = 8!$. To compute the number of elementary events in A it is helpful to view the experiment as consisting of 2 stages as follows. Group the chairs in 4 pairs and assign the 4 couples to the pairs. Next, in each pair of chairs, assign each member of the couple to a seat. Since there are only 4 different positions to place the 4 couples there are 4! different arrangements for the 4 couples. Now for each couple they can be seated in 2! ways—(husband, wife) or (wife, husband). Therefore, there will be $4! \cdot (2!)^4$ different ways for A to occur. Thus,

$$P(A) = \frac{4! \cdot (2!)^4}{8!} = \frac{1}{150}.$$

To compute the probability of B occurring we note that men and women will alternate seats in 2 distinct patterns.

$$M\,W\,M\,W\,M\,W\,M\,W \qquad \text{or} \qquad W\,M\,W\,M\,W\,M\,W\,M.$$

These 2 patterns are mutually exclusive outcomes. Therefore,

$$P(B) = P(\{M\,W\,M\,W\,M\,W\,M\,W\}) + P(\{W\,M\,W\,M\,W\,M\,W\,M\}).$$

Consider $\{M\,W\,M\,W\,M\,W\,M\,W\}$. There are 4 different ways to seat the leftmost man. Then there will be 4 different ways to seat the woman in position 2. There are 3 different ways to seat a man in position 3 since the person in position 3 is a man and there are 3 men left. Similarly, there are 3 choices for a woman in position 4. Proceeding in this fashion the generalized counting principle gives that there are $4 \cdot 4 \cdot 3 \cdot 3 \cdot 2 \cdot 2 \cdot 1 \cdot 1 = (4!)^2$ different ways to obtain $\{M\,W\,M\,W\,M\,W\,M\,W\}$. If we interchange men and women in the preceding discussion it is clear that there will also be this many ways to obtain $\{W\,M\,W\,M\,W\,M\,W\,M\}$. Therefore,

$$P(B) = \frac{(4!)^2}{8!} + \frac{(4!)^2}{8!} = \frac{1}{35}.$$

∎

Example 1.40   Consider the 4 married couples of the previous example. What is the probability that at least 1 man will sit next to his wife? Let $A = \{$at least one man sits next to his wife$\}$. This event can be broken into mutually exclusive events as follows.

$$A = \{\{\text{exactly 1 man sits next to his wife}\}$$

$$\cup \{\text{exactly 2 men sit next to their wives}\}$$

$$\cup \{\text{exactly 3 men sit next to their wives}\}$$

$$\cup \{\text{all 4 men sit next to their wives}\}\}$$

Because of this it might appear easier to deal with $A' = \{$no man sits next to his wife$\}$. A counting argument for A' might go something like the following. There are 8 ways to seat the first person. Once that person is chosen the next person is selected from the remaining 6 people who are not married to the first person. There are 6 people left, and the person in seat 3 must be chosen from the remaining 5 who are not married to the occupant of chair 2. Proceeding in this fashion there will be 4 ways to seat a person in seat 4, 3 ways to choose a person for seat 5, 2 for seat 6 and 1 each for the remaining 2 seats. This would give that there are $8 \cdot 6!$ ways for A' to occur, and

$$P(A) = 1 - P(A') = 1 - \frac{8 \cdot 6!}{8!} = \frac{6}{7}.$$

While there is an intuitive appeal to the counting argument just presented, it contains a serious flaw that the reader is invited to discover. Because of the complications this flaw introduces in determining how to count the number of outcomes in A', it turns out to be easier to compute $P(A)$ directly. In addition, the complications involved in computing the probability that no man sits next to his wife also make it very difficult to compute $P(\{\text{exactly } i \text{ couples sit next to each other}\}), i = 1, 2, 3, 4$. Therefore, consider

$$E_i = \{\text{married couple } i \text{ in adjacent chairs}\}, i = 1, 2, 3, 4.$$

When the events are worded in this fashion, the E's are no longer mutually exclusive. Stating that married couple number 1 are seated next to each other implies nothing about the seating arrangements for the other 3 couples. Now we see that $E_1 \cup E_2 \cup E_3 \cup E_4 = A$. Since A is the union of these events that are not mutually exclusive, we must use Theorem 1.9. Therefore,

$$P(A) = P(E_1 \cup E_2 \cup E_3 \cup E_4)$$
$$= P(E_1) + P(E_2) + P(E_3) + P(E_4) - P(E_1 \cap E_2) - P(E_1 \cap E_3)$$
$$- P(E_1 \cap E_4) - P(E_2 \cap E_3) - P(E_2 \cap E_4) - P(E_3 \cap E_4)$$
$$+ P(E_1 \cap E_2 \cap E_3) + P(E_1 \cap E_2 \cap E_4)$$
$$+ P(E_1 \cap E_3 \cap E_4) + P(E_2 \cap E_3 \cap E_4)$$
$$- P(E_1 \cap E_2 \cap E_3 \cap E_4).$$

To compute $P(E_i)$ we treat couples i that are required to sit next to each other as a unit. Within that unit there will be 2 ways to arrange the husband and wife. We then proceed in a manner similar to computing $P(B)$ in Example 1.39. Therefore,

$$P(E_i) = \frac{7! \cdot 2!}{8!}, \quad i = 1, 2, 3, 4.$$

In computing the probability of the 2-way intersections we treat married couples i and j as units. We are then seating 6 "people," and within each couple, seating each of them in 2 possible ways. Thus,

$$P(E_i \cap E_j) = \frac{6! \cdot (2!)^2}{8!}, \quad \begin{cases} i = 1, 2, 3 \quad \text{and} \\ j = i + 1, \ldots, 4. \end{cases}$$

By using similar arguments we obtain the following.

$$P(E_i \cap E_j \cap E_k) = \frac{5! \cdot (2!)^3}{8!}, \quad \begin{cases} i = 1, 2, \\ j = i + 1, \ldots, 3 \quad \text{and} \\ k = j + 1, \ldots, 4. \end{cases}$$

$$P(E_1 \cap E_2 \cap E_3 \cap E_4) = \frac{4! \cdot (2!)^4}{8!}$$

Therefore, we obtain

$$P(A) = 4 \cdot 2 \frac{7!}{8!} - 6 \cdot 2^2 \frac{6!}{8!} + 4 \cdot 2^3 \frac{5!}{8!} - 2^4 \frac{4!}{8!}$$
$$= \frac{54}{105}.$$

■

Example 1.41   *Acceptance Sampling*   In many manufacturing situations, contracts are written specifying a maximum allowable proportion of defective items in each lot. To determine if the output of the production process is of acceptable quality, the customer typically selects a specified number of items at random from a production run for inspection. If the number of defective items is above a certain predetermined number, then the lot is rejected by the customer. Determination of how many items to sample and the maximum number of defective units in an acceptable lot will be left to our discussion of statistical procedures in Chapter 12. Using probability, however, we can determine the probability that a lot with too many defective units will be accepted by the testing scheme employed. Similarly, we can determine the probability that the output of a production run with an acceptable proportion of defective items will be inadvertently rejected by the sampling process. To be specific, suppose that a production run contains 1000 units and 5% of the units will be randomly selected and tested. If the maximum allowable percentage for defective items in the entire run is 2% and a lot is rejected if more than 1 defective item is found, what is the probability that a production run with 2.5% defective items will be discovered and rejected? Sampling of this type is always done without replacement. In fact, many times the item selected is destroyed during the testing procedure. Since we do not care when a particular item is selected, counting will be done by using combinations. Therefore, there will be $\binom{1000}{50}$ elementary events in the sample space. If there are 2.5% defective items then the lot contains 25 bad units. Let $A = \{$more than 2 defective items in sample$\}$, and let $E_i = \{$exactly i defective items in sample$\}$, $i = 0, 1, \ldots, 50$. We see that the E's are mutually exclusive and $\bigcup_{i=1}^{50} E_i = A$. We shall compute $P(A')$ since $A' = E_0 \cup E_1$. Therefore,

$$P(A) = 1 - P(A') = 1 - [P(E_0) + P(E_1)].$$

A convenient way to count the number of elementary events in E_i is to view it as being a 2-stage experiment. In the first stage we select i items from the group of 25 defective units. This can be done in $\binom{25}{i}$ ways. In stage 2 we select $50 - i$ nondefective items from the $1000 - 25$ good units. This can be done in $\binom{1000-25}{50-i}$ ways. We then use the basic counting principle to find that the number of elementary events in E_i is $\binom{25}{i} \cdot \binom{1000-25}{50-i}$. Therefore,

$$P(A) = 1 - \frac{\binom{25}{0} \cdot \binom{1000-25}{50-0} + \binom{25}{1} \cdot \binom{1000-25}{50-1}}{\binom{1000}{50}} \doteq 0.356.$$

The previous example illustrates one practical problem in computing probabilities. Combinations involving large numbers, such as $\binom{1000}{50}$, are difficult to evaluate. The numbers will overflow most hand calculators. The task can be made easier by using an approximation to $n!$ known as *Stirling's formula*. It is pretty accurate when n is large.

$$n! \approx \sqrt{2\pi n}\, n^n e^{-n} \tag{1.21}$$

As we shall see later, probabilities such as those in the acceptance sampling example can be approximated by using other probability models that require less computational effort.   ■

Exercises 1.6

1. An urn contains 5 white balls, 3 black balls, and 2 blue balls. Two balls are drawn from the urn. Let A = {both balls the same color} and B = {second ball drawn white}.
   a. Find $P(A)$ and $P(B)$ if the balls are drawn with replacement.
   b. Find $P(A)$ and $P(B)$ if the balls are drawn without replacement.

2. Urn I contains 4 white balls and 2 red balls. Urn II contains 4 white balls and 4 red balls. A ball is selected from Urn I and placed in Urn II. Then a ball is drawn from Urn II. What is the probability that the ball selected from Urn II is white?

3. A group of three students is given a list of five books and asked to read one of them and write a report on it. If each student selects a book randomly, what is the probability that all three students will read the same book?

4. A restaurant menu has 5 entrees. In a party of 5 people, what is the probability that r of the entrees will be ordered for r = 1, 2, 3, 4, and 5? Assume that each member of the party selects randomly from the menu.

5. A fair die is rolled 10 times. What is the probability that at least 1 roll will yield a 1?

6. Four cards are chosen from a standard 52-card deck without replacement. What is the probability that at least 2 cards will be from the same suit?

7. Five cards are chosen at random from a standard 52-card deck without replacement. Find the probability of the following poker hands.
   a. Four of a kind. That is, four cards of the same denomination and any other card.
   b. A full house. That is, three cards of one denomination and two cards of another denomination.
   c. One pair. That is, two cards of one denomination and three cards from three other different denominations.

8. Compute the probability that Sam will win the game described in Problem 6 of Exercises 1.3. Compute the probability that Joe will win.

9. Explain why the first method for computing the probability that at least one married couple will sit next to each other in Example 1.40 is incorrect.

10. Do Example 1.40 when the seats are arranged in a circle.

11. Items of clothing are inspected after manufacture for defects. Those with no defects are sold under the manufacturer's label at full price, those with minor or serious (or both) defects are sold under the manufacturer's label as factory seconds, and those with any major defects are destroyed. In a production run, assume that defects occur in the following proportions.

Degree of Defects	%
Major defects	1.0%
Serious defects	1.5%
Minor defects	2.0%
Both major and serious defects	0.5%
Both major and minor defects	0.4%
Both serious and minor defects	0.6%
All three defects	0.1%

   a. What is the probability that a randomly selected item of clothing will be sold at full price?

b. What is the probability that a randomly selected item of clothing will be sold as a factory second?

12. Let there be five pairs of trousers with matching shirts. If trousers are randomly paired one at a time with a shirt find the probability that at least one pair of trousers will be assigned to its matching shirt if the pairing is done in each of two ways.

   a. A pair of trousers is matched with one of the five shirts, the next pair of trousers is matched with one of the four remaining shirts, and so forth. That is, pairing is done without replacement.

   b. A pair of trousers is matched with one of the five shirts. The second pair of trousers is matched with any of the five shirts and so on. That is, any pair of trousers is free to be matched with any shirt (pairing with replacement).

13. Use Stirling's formula to compute an approximation to $\binom{100}{20}$.

14. In Example 1.41 compute the probability that a production run with 2% defective items will be rejected.

15. A lottery machine contains balls numbered 0 through 9. In a particular game a single ball is drawn from each of n identical machines. Find the probability that the ball numbered 0 is selected at least once. How many machines must be used for this probability to exceed $\frac{1}{2}$?

16. A tub of water contains $n - 1$ red apples and 1 green apple. If apples are bobbed for until the green one occurs, compute the probability that at least k apples must be selected if apples are bobbed for

   a. with replacement, and

   b. without replacement.

17. A ball is drawn from an urn containing n white balls and m black balls. Find n and m if $P(\text{ball is white}) = p$ and $P(\text{ball is black}) = 2p^2$. Your answer will not be unique.

18. In an acceptance sampling scheme, items are sampled without replacement until the second defective item is found. If a production run contains N units, M of which are bad, compute the probability that the second defective unit will be the ith unit sampled.

19. A bus begins its route containing m passengers and makes n stops. If a random number of people (possibly none) get off at each stop, compute the probability that

   a. no people get off at at least one stop,

   b. no people get off at exactly one stop, and

   c. no people get off at exactly i stops.

   You may assume that $m > n$ and that, when the bus completes its route, all people have left the bus.

20. Show that any sample space that consists of equally likely outcomes can contain only a finite number of elementary events.

Chapter Summary

Set Theory

Set: A collection of outcomes of an experiment.

Set equality: $A = B$ if they contain exactly the same elements.

Subset: $A \subset B$ if for any $x \in A$ then $x \in B$.

If A, B, C are subsets in S, then we have the following.

$$\emptyset \subseteq A \subseteq S$$

$$A \subseteq A$$

If $A \subseteq B$ and $B \subseteq C$ then $A \subseteq C$.

$A = B$ if and only if $A \subseteq B$ and $B \subseteq A$.

Union: $A \cup B = \{x : x \in A \text{ or } x \in B\}$

Intersection: $A \cap B = \{x : x \in A \text{ and } x \in B\}$

Complement: $A' = \{x : x \in S, x \notin A\}$
           If $A \subset B$ then $A \cup B = B$ then $A \cap B = A$ then $A \subset B$

Algebra of sets: $A \cup B = B \cup A$

  Symmetry: $A \cap B = B \cap A$

   Associativity: $(A \cup B) \cup C = A \cup (B \cup C)$
                  $(A \cap B) \cap C = A \cap (B \cap C)$

   Distribution: $A \cup (B \cap C) = (A \cup B) \cap (A \cup C)$
                  $A \cap (B \cup C) = (A \cap B) \cup (A \cap C)$
                  $A \cup \emptyset = A$
                  $A \cup S = S$
                  $A \cap \emptyset = \emptyset$
                  $A \cap S = A$

Idempotent: $A \cup A = A$
                  $A \cap A = A$

Complementation: $(A')' = A$
                  $A \cup A' = S$
                  $S' = \emptyset$
                  $\emptyset' = S$

DeMorgan's Laws: $(A \cup B)' = A' \cap B'$
                  $(A \cap B)' = A' \cup B'$ **(Section 1.2)**

Sample Space: Set of all possible outcomes of an experiment.   **(Section 1.2)**

Random Experiments

1. An experiment that has a known sample space.
2. The particular outcome of any given run of the experiment cannot be known with certainty beforehand.   **(Section 1.3)**

Axioms of Probability

1. $P(E) \geq 0$
2. $P(S) = 1$
3. If $A \cap B = \emptyset$, then $P(A \cup B) = P(A) + P(B)$
4. If $A_1, A_2, \ldots$ are mutually exclusive, then

$$P\left(\bigcup_{i=1}^{\infty} A_i\right) = \sum_{I=1}^{\infty} P(A_i).$$

$P(A') = 1 - P(A)$

$P(\emptyset) = 0$

If $A_1, A_2, \ldots, A_n$ mutually exclusive, then

$$P\left(\bigcup_{i=1}^{n} A_i\right) = \sum_{I=1}^{n} P(A_i)$$

If $A \cup B$, then $P(A) \leq P(B)$

$P(A \cup B) = P(A) + P(B) - P(A \cap B)$

$$P\left(\bigcup_{i=1}^{n} A_i\right) = \sum_{i=1}^{n} P(A_i) - \sum_{i<j} P(A_i \cap A_j) + \sum_{i<j<k} P(A_i \cap A_j \cap A_k)$$
$$\sum_{i<j<k<l} P(A_i \cap A_j \cap A_k \cap A_l)$$
$$+ (-1)^{n+1} P\left(\bigcap_{i=1}^{n} A_i\right)$$

(Section 1.4)

Combinatorics

General Counting Principle: If subexperiment i has n_i outcomes for $i = 1, 2, \ldots, k$, then the experiment has $\prod_{i=1}^{k} n_i$ outcomes.

Permutations: $P(n, r) = \frac{n!}{(n-r)!}$.

There are n^r permutations of n objects taken r times with repetition.

Combinations: $\binom{n}{r} = \frac{n!}{r!(n-r)!}$

$$\binom{n}{r} = \binom{n-1}{r-1} + \binom{n-1}{r}$$

Binomial Theorem:

$$(x + y)^n = \sum_{i=1}^{n} \binom{n}{i} x^i y^{n-1}$$

Multinomial Theorem:

$$(x_1 + x_2 + \cdots + x_k)^n = \sum \binom{n}{n_1! \, n_2! \, \cdots \, n_k!} x_1^{n_1} x_2^{n_2} \cdots x_k^{n_k}$$

Distribution of balls in urns: The number of ways to distribute n identical balls in r different urns is $\binom{n+r-1}{n}$

If a sample space S consists of $N(S)$ equally likely outcomes, and $N(A)$ is the number of elementary events in A, then

$$P(A) = \frac{N(A)}{N(S)}.$$

(Section 1.5)

Review Exercises

1. Show that the method for computing probabilities for finite sample spaces with equally likely outcomes satisfies the axioms of probability.

2. Give an example of a sample with a countably infinite number of elementary events that assigns a positive probability to each elementary event. (Hint: Consider a geometric series.)

3. For each of the following, prove or find a counterexample.

   a. If A and B are mutually exclusive events, and B and C are mutually exclusive events, then A and C are mutually exclusive events.

   b. If $A_1, A_2, \ldots,$ is a sequence of events in the sample space S, then

$$\bigcup_{i=1}^{\infty} A_i = S$$

   c. If $P(A) < P(B)$, then $A \subset B$.

   d. $P(A \cap B) \geq 1 - P(A') - P(B')$.

4. In a procedure known as simple random sampling, members of a population are numbered and these numbers are randomly chosen without replacement to obtain the members of the sample. Suppose that a given population has n members that are numbered from 1 to n. We take a sample of r members using simple random sampling. Find the probabilities of the following events.

   a. The smallest number of any member selected is m.

   b. The ith smallest number of any member is m.

5. Two cards are selected at random from a standard deck of 52. A Blackjack occurs when 1 of the cards is an Ace and the other is either a 10, Jack, Queen, or a King. What is the probability that the 2 cards form a Blackjack?

6. A town has 5 plumbers. If 5 houses have plumbing problems on a given day, what is the probability that exactly i of the plumbers are called, for $i = 1, 2, 3, 4, 5$. Explain what assumptions, if any, you are making.

7. A basketball team consists of two guards, two forwards, and one center. A certain team's roster has five guards, four forwards, and two centers. Five people are selected at random from the roster. What is the probability that

   a. a proper team will be selected?

   b. at least two guards will be selected?

8. Four married couples attend a party. Men are randomly assigned to women for a dance. Find the probability that no man will dance with his wife.

9. A group of two men and two women is randomly divided into groups of two each. What is the probability that no group will be of the same sex?

10. A realtor is showing a house to a prospective buyer. He has n keys in his pocket, one of which will open the door to the house. The keys are tried at random.

   a. If the keys are thrown away after being tried, what is the probability that the correct key will be found on the ith try?

   b. If the keys are not thrown away after being tried, what is the probability that the correct key will be found on the ith try?

11. A club consists of four seniors, four juniors, three sophomores, and two freshmen. A committee of four is selected at random from the club. What is the probability that two of the members will be from the same class?

12. In a television game show, a bag contains seven chips, three of which are marked with an X and four with an O. A player is to select one chip at a time from the bag. If all four chips with an O are removed before receiving the first X, the player wins. Find the probability that a player will win the game.

13. One of many tests used to evaluate the performance of random number generators for computers is based on what are known as *runs*. For example, if an algorithm generates 0.4, 0.3, 0.2, 0.4, 0.5, 0.8, and 0.7, this sequence is said to have a *run-up* if a number is larger than its predecessor and to have a *run-down* if the next number is smaller. If a run-up is coded by a 1 and a run-down is coded by a 0, the example sequence would be coded as 001110. A run is said to be of length r if r consecutive runs are of the same type. The example sequence can be seen to have a run-down of length 2, followed by a run-up of length 3, followed by a run-down of length 1. The so-called *runs test* is based on the probability of observing certain distributions of runs in a supposedly random sequence of numbers. To illustrate the ideas involved, suppose that an algorithm generates random sequences of length n consisting of 0's and 1's. That is, 0 and 1 are equally likely outcomes. Note that this can also be viewed as an experiment where sampling is done with replacement from an urn containing 2 balls numbered 0 and 1.

   a. Find the probability that all runs have length 1 (0's and 1's alternate).

   b. Suppose that a sequence contains m 0's and $n - m$ 1's. Find the probability that there are r runs. Hint: Consider the 2 cases where r is even ($r = 2k$) and where r is odd ($r = 2k + 1$).

Conditional Probability and Independence

Chapter 1 presented us with a definition of probability and showed how to compute probabilities for certain kinds of experiments. In addition, we saw how to decompose complicated events into the union of a number of events whose probabilities were easier to compute by using the generalized counting principle and combinations. In this chapter we shall take a closer look at events that are the intersection of other events.

As an example, suppose we have a box that contains 3 red balls and 7 black balls. We select 2 balls without replacement from the box and wish to determine the probability that both balls will be black, {bb}. By using the basic counting principle we see that

$$P(\text{bb}) = \frac{7 \cdot 6}{10 \cdot 9} = \frac{21}{45}.$$

Suppose we are interested in the probability that the second ball will be black. We can decompose this event into the union of {rb} and {bb}, which are mutually exclusive. Using the basic counting principle on each we obtain

$$P(\text{second black}) = P(\text{rb}) + P(\text{bb})$$

$$= \frac{3 \cdot 7}{10 \cdot 9} + \frac{21}{45} = \frac{7}{10}.$$

Now, suppose we look at an experiment where a single ball is selected from a box containing 3 red balls and 6 black balls. Here the probability that the ball will be black is

$$P(\text{b}) = \frac{6}{9}.$$

The sample space for the second experiment is what would result from having the first ball drawn in the first experiment be black. This kind of idea leads us to the notion of conditional probability. If we consider only those outcomes in the sample space where the first ball drawn is black we say that we are using a *restricted sample space*. With conditional probabilities we confine attention to a subset of the sample space where a particular event has occurred. This is the restricted sample space. Then, in that restricted sample space, we compute the probability that another event occurs. In general, the probability of an event occurring in a restricted sample space will be different from what we would obtain for the whole sample space. The probability that an event occurs in such a restricted sample space is known as a *conditional probability*. The event that determines which elementary events will be in the restricted sample space is referred to as the *conditioning event*. Suppose a restricted sample space consists of those outcomes where an event B occurs. It is standard practice to refer to the probability that another event, A, occurs in the restricted sample space as the "probability of A given B" and denote it by

$$P(A \mid B).$$

This does not mean that $A \subset B$. The conditional probability is merely the proportion of outcomes in B which are also in A. The event A in the original sample space can have outcomes that are not in B. Thus, in our experiment with the balls in the box we would say that

$$P(\{\text{second ball black}\} \mid \{\text{first ball black}\}) = \frac{6}{9}.$$

Note that this value is different from the so-called *unconditional* probability of the second ball being black which was $\frac{7}{10}$.

Example 2.1   Suppose we have an inexpensive but relatively crude screening test for tuberculosis. We are primarily interested in whether a person who shows a positive result on this test is likely actually to have tuberculosis. Let $T = \{\text{person has tuberculosis}\}$, and $F = \{\text{person has positive result on screening test}\}$. We are interested only in those people where F has occurred. That is, F is the conditioning event. The probability that is of interest would be

$$P(T \mid F).$$

∎

As can be seen, conditional probabilities are important because they allow us to make use of additional information about the results of a random experiment. In addition, we shall see later that conditional probability can be a useful tool for computing the probability of more complicated events.

Let's return to the box containing 3 red balls and 7 black ones. This time we shall sample 2 balls with replacement. First, the probability that both balls will be black is given by

$$P(\text{bb}) = \frac{7 \cdot 7}{10 \cdot 10} = \frac{49}{100}.$$

In addition, since the first ball is returned to the box it is easy to see that the probability that the second ball will be black does not depend on whether or not the first ball was black. The second ball is being selected from a box containing 3 red balls and 7 black balls. Thus,

$$P(\{\text{second ball black}\} \mid \{\text{first ball black}\}) = \frac{7}{10}.$$

In addition, we see that

$$P(\{\text{second ball black}\} \mid \{\text{first ball not black}\}) = \frac{7}{10}.$$

Now the unconditional probability that the second ball will be black is found to be

$$P(\{\text{second ball black}\}) = P(\text{rb}) + P(\text{bb})$$
$$= \frac{3 \cdot 7}{10 \cdot 10} + \frac{7 \cdot 7}{10 \cdot 10} = \frac{7}{10}.$$

In this experiment we note that

$$P(\{\text{second ball black}\}) = P(\{\text{second ball black}\} \mid \{\text{first ball black}\})$$
$$= P(\{\text{second ball black}\} \mid \{\text{first ball not black}\}).$$

Pairs of events such as these where the conditional probability of one given the other is the same as the unconditional probability are said to be *stochastically independent* or simply *independent*. Independence is a central concept in much of probability and statistics.

2.2 Conditional Probability

In Example 1.34 we considered an experiment where 2 fair dice were rolled. In that experiment we noted that the sample space consisted of 36 equally likely ordered pairs. Let $T = \{\text{sum of both dice is ten}\}$, and $F = \{\text{first die is a four}\}$. We wish to compute the probability of T given F. The restricted sample space consists of those ordered pairs with a 4 in the first position. They are

$$(4, 1), \quad (4, 2), \quad (4, 3), \quad (4, 4), \quad (4, 5), \quad \text{and} \quad (4, 6).$$

Only one of these 6 elementary events is in T. Thus,

$$P(T \mid F) = \frac{1}{6}.$$

Now, in looking at the sample space for the experiment we see that the elementary events in T are

$$(4, 6), \quad (5, 5), \quad \text{and} \quad (6, 4).$$

We can then compute the following probabilities.

$$P(T \cap F) = \frac{1}{36}, \text{ and}$$

$$P(F) = \frac{6}{36}.$$

Note that

$$\frac{P(T \cap F)}{P(F)} = \frac{1}{6} = P(T \mid F).$$

This is not a coincidence. It is a general result regarding all conditional probabilities. We can see this by the following argument. $P(T \cap F)$ is the proportion of sample space that is in both T and F. Therefore, it is also a measure of how much of the sample space is in T when we restrict ourselves to those outcomes in F. $P(F)$ is the proportion of sample space which is in F. Now look at the Venn diagram of Figure 2.1. From this we see that $P(T \mid F)$ is the proportion of F that is in $T \cap F$. This proportion is $\frac{P(T \cap F)}{P(F)}$. To put it another way, knowing that F occurred means that we are restricted to only those outcomes in F. Out of those outcomes the ones that are also in T are precisely those in $T \cap F$.

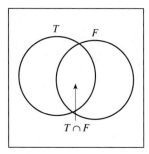

Figure 2.1

This leads to the following definition.

Definition 2.1

Given two events, A and B, with $P(B) > 0$, then the *conditional probability of A given the occurrence of B* is

$$P(A \mid B) = \frac{P(A \cap B)}{P(B)}. \qquad (2.1)$$

We leave it as an exercise to show that conditional probability satisfies the axioms of probability. This means that the results given in Chapter 1 can also be applied to conditional probabilities.

We shall now illustrate the use of this definition with a few examples.

Example 2.2

A fair coin is tossed 3 times. What is the probability that at least 2 heads occur given that the first toss is a head? Let $A = \{$at least two heads$\}$, and $B = \{$heads on first toss$\}$. We wish to find $P(A \mid B)$. Then $A \cap B$ is the event that at least 2 heads occur with the first toss being a head. The sample space in Example 1.6 (see page 13) shows that there are 3 elementary events out of the 8 in $A \cap B$. They are $\{HHT\}$, $\{HTH\}$, and $\{HHH\}$. In addition, there are 4 elementary events in B. They are $\{HTT\}$, $\{HHT\}$, $\{HTH\}$, and $\{HHH\}$. There are also

4 elementary events in A. They are $\{HHT\}$, $\{HTH\}$, $\{THH\}$, and $\{HHH\}$. Thus, Equation 2.1 gives that

$$P(A \mid B) = \frac{P(A \cap B)}{P(B)} = \frac{\frac{3}{8}}{\frac{4}{8}} = \frac{3}{4}.$$

Also note that

$$P(B \mid A) = \frac{P(A \cap B)}{P(A)} = \frac{\frac{3}{8}}{\frac{4}{8}} = \frac{3}{4}.$$

■

Example 2.3

An urn contains 10 black balls and 6 white balls. Let $W = \{$second ball white$\}$, and $B = \{$first ball black$\}$. We wish to compute $P(W \mid B)$ and $P(B \mid W)$. We note that the event of getting a white ball second, in terms of elementary events, really means that W occurs if we get a black ball first followed by a white ball or if we get a white ball first followed by another white ball. Similarly, the event B occurs if we get a black ball followed by a white ball or a black ball followed by a second black ball. That is,

$$W = W \cap (B \cup B') \quad \text{and} \quad B = B \cap (W \cup W').$$

By using this and the basic counting principle we see that

$$P(W \cap B) = \frac{10 \cdot 6}{16 \cdot 15} = \frac{1}{4},$$
$$P(B) = P[B \cap (W \cup W')]$$
$$= P[(B \cap W) \cup (B \cap W')] = P(B \cap W) + P(B \cap W')$$
$$= \frac{1}{4} + \frac{10 \cdot 9}{16 \cdot 15} = \frac{5}{8}, \quad \text{and}$$
$$P(W) = P[W \cap (B \cup B')] = P[(W \cap B) \cup (W \cap B')]$$
$$= \frac{1}{4} + \frac{6 \cdot 5}{16 \cdot 15} = \frac{3}{8}.$$

Now, Equation 2.1 gives

$$P(W \mid B) = \frac{P(W \cap B)}{P(B)} = \frac{\frac{1}{4}}{\frac{5}{8}} = \frac{2}{5}, \quad \text{and}$$

$$P(B \mid W) = \frac{P(W \cap B)}{P(W)} = \frac{\frac{1}{4}}{\frac{3}{8}} = \frac{2}{3}.$$

■

Here, the 2 conditional probabilities are different. This illustrates that conditional probabilities are not, in general, symmetric. That is, it is usually the case that $P(A \mid B) \neq P(B \mid A)$. The 2 will be equal if and only if $P(A) = P(B)$.

Example 2.4   A 5-card poker hand is selected from a standard deck of 52 cards. We wish to compute the probability that the hand will contain 4 aces given that there are at least 3 aces in it. Let $T = \{$at least 3 aces$\}$, and $F = \{4$ aces$\}$. In this case we note that if the hand had 4 aces it had to have had at least 3 aces. In other words, $(T \cap F) = F$, and $P(T \cap F) = P(F)$. To compute the various probabilities we shall treat this as a 2-stage experiment. In Stage 1 we shall select the aces, and in Stage 2 we shall choose the other cards from the remaining 48. As we saw in Example 1.37, there are $\binom{52}{5}$ elementary events in the sample space. When F occurs we select all 4 aces and a single card from the other 48. This means that there are $\binom{4}{4} \cdot \binom{48}{1}$ elementary events in F. Therefore,

$$P(T \cap F) = P(F) = \frac{\binom{4}{4} \cdot \binom{48}{1}}{\binom{52}{5}}.$$

Now T occurs if there are exactly 3 aces or exactly 4 aces in the hand. By using the same reasoning as before we see that

$$P(T) = P(\{\text{exactly 3 aces}\} \cup \{\text{exactly 4 aces}\})$$

$$= P(\{\text{exactly 3 aces}\}) + P(F)$$

$$= \frac{\binom{4}{3} \cdot \binom{48}{2}}{\binom{52}{5}} + \frac{\binom{4}{4} \cdot \binom{48}{1}}{\binom{52}{5}}.$$

Therefore, Equation 2.1 gives

$$P(F \mid T) = \frac{\binom{4}{4} \cdot \binom{48}{1}}{\binom{4}{3} \cdot \binom{48}{2} + \binom{4}{4} \cdot \binom{48}{1}} = \frac{1}{95}.$$

∎

As we mentioned in the introduction, conditional probability can be a useful tool for computing other probabilities. This is because Equation 2.1 can be rewritten to obtain what is sometimes called the *multiplication rule* for the probability of the intersection of two events.

$$P(A \cap B) = P(A \mid B)P(B) \tag{2.2}$$

The utility is that in many cases it may be difficult to compute $P(A \cap B)$ directly but relatively easier to find, say, $P(A \mid B)$ and $P(B)$. We begin by illustrating the use of Equation 2.2 with a simple example.

Example 2.5

A probability class consists of 10 men and 16 women. Two different students are selected at random from the class. We wish to find the probability that both will be men. Let $M_i = \{i$th student is male$\}$, $i = 1, 2$. We have a choice on whether to condition on the first male or the second. In this instance it seems most reasonable to condition on the first male. Therefore, we shall compute

$$P(\{\text{both students male}\}) = P(M_1 \cap M_2) = P(M_2 \mid M_1)P(M_1).$$

For the first student there are 10 males out of 26 students. Therefore,

$$P(M_1) = \frac{10}{26}.$$

After the first male has been removed from the class, there remain 9 men out of 25 students. Therefore,

$$P(M_2 \mid M_1) = \frac{9}{25}.$$

Thus, Equation 2.2 gives

$$P(M_1 \cap M_2) = \frac{9}{25} \cdot \frac{10}{26} = \frac{9}{65}.$$ ■

It may be argued that this is no less work than counting elementary events in the sample space. On the other hand, it does illustrate the basic technique and it is easier to follow what is going on. The next example is a little more complicated and, hopefully, illustrates the value of the multiplication rule.

Example 2.6

Bookshelf 1 contains 5 mystery novels and 3 science fiction novels. Bookshelf 2 contains 4 mystery novels and 6 science fiction novels. A bookshelf is selected at random, and then a book is taken from that shelf. We wish to determine the probability that a science fiction novel is chosen. Three events naturally arise from this description. Let $F_i = \{$science fiction novel taken from bookshelf $i\}$, $i = 1, 2$, and $B = \{$bookshelf 1 is picked$\}$. A science fiction novel is chosen by either taking it from bookshelf 1 or bookshelf 2. It is natural to condition on which bookshelf is picked. Thus,

$$P(\{\text{science fiction novel}\}) = P(\{\text{science fiction novel from shelf 1}\}$$
$$\cup \{\text{science fiction novel from shelf 2}\})$$
$$= P[(B \cap F_1) \cup (B' \cap F_2)].$$

Now, these 2 events are mutually exclusive. Since it is equally likely that we pick either bookshelf we see that

$$P(B) = P(B') = \frac{1}{2}.$$

If bookshelf 1 is picked then there are 3 ways out of 8 to choose a science fiction novel. If the book comes from bookshelf 2 then there are 6 out of 10 ways to get a science fiction novel. Therefore,

$$P(F_1 \mid B) = \frac{3}{8}, \quad \text{and} \quad P(F_2 \mid B') = \frac{6}{10}.$$

Combining all this we get

$$\begin{aligned}
P(\{\text{science fiction novel}\}) &= P(B \cap F_1) + P(B' \cap F_1) \\
&= P(F_1 \mid B)P(B) + P(F_2 \mid B')P(B') \\
&= \frac{3}{8} \cdot \frac{1}{2} + \frac{6}{10} \cdot \frac{1}{2} \\
&= \frac{39}{80}.
\end{aligned}$$

■

This last example illustrates an idea that is used quite often when computing probabilities. Notice that the events B and B' were mutually exclusive and their union is the sample space. The computation then became one of determining the probability that we would get a science fiction novel in each portion of the sample space determined by B and B'. The overall probability then turned out to be a weighted average of these probabilities. The partitioning of the sample space made it easier to compute the individual pieces. This is a perfectly general technique that is stated formally in the following theorem.

Theorem 2.1

Suppose that the sample space, S, can be partitioned into mutually exclusive events, $B_1, B_2, \ldots, B_n$. That is,

$$B_i \cap B_j = \varnothing, \quad \text{if } i \neq j, \quad \text{and} \quad \bigcup_{i=1}^{n} B_i = S.$$

Then, for any event, A,

$$P(A) = \sum_{i=1}^{n} P(A \mid B_i)P(B_i). \tag{2.3}$$

Proof   Figure 2.2 illustrates the theorem.

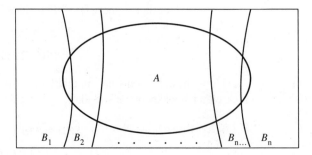

Figure 2.2

As the figure shows, the event A can be written as

$$A = A \cap S = A \cap \left(\bigcup_{i=1}^{n} B_i \right) = \bigcup_{i=1}^{n} (A \cap B_i).$$

Since the B_i are mutually exclusive events, so are the events $A \cap B_i$. Then we can use Theorem 1.6 and Equation 2.2 to finish the proof.

$$P(A) = \sum_{i=1}^{n} P(A \cap B_i) = \sum_{i=1}^{n} P(A \mid B_i)B(B_i)$$

$\blacksquare$

Note: A collection of events $B_1, B_2, \ldots, B_n$ such that

$$\bigcup_{i=1}^{n} B_i = S,$$

is said to be *collectively exhaustive*.

Example 2.7

An insurance company has found that 60% of all drivers who have had automobile accidents within the past 12 months will have another accident in the next 12 months. Only 10% of those who have not had an accident in the past year will have one in the next year. National statistics show that 30% of all drivers had an accident in the past year. We wish to determine the probability that a driver selected randomly will have an automobile accident in the next year. Here it makes sense to partition the population of drivers into those who had an accident in the past year and those who did not. Let $A = \{$driver had an accident in the past year$\}$, and $B = \{$driver will have an accident during next 12 months$\}$. In terms of these events the data given in the example indicate that

$$P(A) = 0.30, \quad P(A') = 0.70, \quad P(B \mid A) = 0.60, \quad \text{and} \quad P(B \mid A') = 0.10.$$

Since A and A' are mutually exclusive and collectively exhaustive, we can use Theorem 2.1 as follows.

$$P(B) = P(B \mid A)P(A) + P(B \mid A')P(A')$$
$$= (0.6)(0.3) + (0.1)(0.7) = 0.25$$

$\blacksquare$

The ideas we have been discussing can be extended to the computation of probabilities for the intersections of more than two events. Suppose we wish to find $P(A \cap B \cap C)$. The event $A \cap B \cap C$ can be viewed as $A \cap (B \cap C)$. If $P(A \mid B \cap C) > 0$, then we can use Equation 2.2 to get

$$P(A \cap B \cap C) = P(A \mid B \cap C)P(B \cap C).$$

If $P(C) > 0$, then Equation 2.2 can be used a second time to obtain

$$P(A \cap B \cap C) = P(A \mid B \cap C)P(B \mid C)P(C).$$

With this as a basis we can use induction to show that this can be extended to the intersection of any number of events. The general result is summarized in the following theorem. The proof is left as an exercise.

Theorem 2.2

Let $A_1, A_2, \ldots, A_n$, be a collection of events. If

$$P\left(\bigcap_{i=1}^{n} A_i\right) > 0,$$

then

$$P(A_1 \cap A_2 \cap \cdots \cap A_n) = P(A_n \mid A_1 \cap A_2 \cap \cdots \cap A_{n-1})$$
$$\cdot P(A_{n-1} \mid A_1 \cap A_2 \cap \cdots \cap A_{n-2}) \cdots \qquad (2.4)$$
$$\cdot P(A_3 \mid A_1 \cap A_2)P(A_2 \mid A_1)P(A_1)$$

Example 2.8

Five cards are selected without replacement from a standard deck of 52 cards. We wish to find the probability that all 5 cards will be hearts. Let $H_i = \{i\text{th card a heart}\}$, $i = 1, 2, 3, 4, 5$. The event of getting 5 hearts is the intersection of these 5 events. Then Equation 2.4 can be used as follows.

$$P(\{\text{five hearts}\}) = P(H_1 \cap H_2 \cap H_3 \cap H_4 \cap H_5)$$
$$= P(H_5 \mid H_1 \cap H_2 \cap H_3 \cap H_4)P(H_4 \mid H_1 \cap H_2 \cap H_3)$$
$$\cdot P(H_3 \mid H_1 \cap H_2)P(H_2 \mid H_1)P(H_1)$$

When the first card is drawn there are 13 hearts out of the 52 cards in the deck. On the second draw, if a heart has been removed there are now 12 hearts in the remaining 51 cards. On the third card, if 2 hearts have been taken out of the deck, there will be 10 hearts in the 50 cards that are left. Continuing in this fashion we see that

$$P(H_1) = \frac{13}{52} = \frac{1}{4},$$

$$P(H_2 \mid H_1) = \frac{12}{51},$$

$$P(H_3 \mid H_1 \cap H_2) = \frac{11}{50},$$

$$P(H_4 \mid H_1 \cap H_2 \cap H_3) = \frac{10}{49}, \quad \text{and}$$

$$P(H_5 \mid H_1 \cap H_2 \cap H_3 \cap H_4) = \frac{9}{48} = \frac{3}{16}.$$

Combining these we get

$$P(\{\text{five hearts}\}) = \frac{3}{16} \cdot \frac{10}{49} \cdot \frac{11}{50} \cdot \frac{12}{51} \cdot \frac{1}{4}$$

$$= \frac{33}{66640}.$$

Before leaving this example we note that the probability could have been obtained by using the ideas from Chapter 1 as follows.

$$P(\{\text{five hearts}\}) = \frac{\dbinom{13}{5}}{\dbinom{52}{5}}$$

∎

The next example requires that we use Theorem 2.2. It is a special case of what is known as an *urn model*. They have been used successfully to model such things as the spread of disease and the exchange of heat between two bodies. A more complete discussion of urn models can be found in *An Introduction to Probability Theory and Its Applications,* vol. 1, by William Feller.

Example 2.9   An urn originally contains 1 white ball and 99 black balls. A ball is selected from the urn. If the ball is white it is replaced along with 2 more white balls. If the ball is black it is simply replaced. This process is repeated 3 times. We wish to find the probability that all 3 balls selected will be white. This experiment could describe the spread of a disease. The removal of a white ball represents a member of a population becoming infected. The placing of 2 additional white balls in the urn models the heightened probability of infection for the rest of the population. Highly contagious diseases would be modeled by placing a large number of additional white balls in the urn. The act of removing balls can be used to represent units of time.

Let $W_i = \{\text{select a white ball on draw } i\}$, for $i = 1, 2$, and 3. From the description of the experiment we know that the probability of obtaining the first white ball is

$$P(W_1) = \frac{1}{100}.$$

After the first white ball is drawn the urn then contains 102 balls, 3 of which are white. Thus,

$$P(W_2 \mid W_1) = \frac{3}{102}.$$

Similarly, after 2 white balls have been chosen the urn then has 104 balls, 5 of which are white. This gives that

$$P(W_3 \mid W_1 \cap W_2) = \frac{5}{104}.$$

We wish to determine $P(W_1 \cap W_2 \cap W_3)$. An application of Theorem 2.2 gives

$$P(W_1 \cap W_2 \cap W_3) = P(W_3 \mid W_1 \cap W_2)P(W_2 \mid W_1)P(W_1)$$

$$= \frac{5}{104} \cdot \frac{3}{102} \cdot \frac{1}{100}$$

$$= \frac{1}{707020}.$$

■

Exercises 2.2

1. Let $A = \{\text{male}\}$, $B = \{\text{over 50 years old}\}$, and $C = \{\text{unemployed}\}$ be events describing a randomly selected person. Write the following in terms of A, B, and C.

   a. Person is over 50 and employed.

   b. A male is unemployed.

   c. An unemployed male is over 50.

   d. A woman is over 50 and employed.

2. Verify that conditional probability satisfies the axioms of probability. That is, show the following.

   a. If $P(B) > 0$, then $P(A \mid B) \geq 0$.

   b. If $P(B) > 0$, then $P(S \mid B) = 1$.

   c. If $A \cap B = \emptyset$, and $P(C) > 0$, then

   $$P(A \cup B \mid C) = P(A \mid C) + P(B \mid C).$$

   d. If $A_1, A_2, \ldots$ is a sequence of mutually exclusive events, and $P(B) > 0$, then

   $$P\left(\bigcup_{i=1}^{\infty} A_i \mid B\right) = \sum_{i=1}^{\infty} P(A_i \mid B).$$

3. Prove each of the following or give a counter example.

   a. $P(A \mid A) = 1$.

   b. $P(A \cup B \mid C) = P(A \mid C) + P(B \mid C) - P(A \cap B \mid C)$.

   c. If $P(A \mid B) = 0$, then $A = \emptyset$.

   d. If $A \subset B$, then $P(A \mid B) = 1$.

4. Prove Theorem 2.2.

5. An urn contains 5 white balls and 3 black balls. Two balls are selected from the urn. Find the probability that both balls will be the same color if sampling is done (a) without replacement and (b) with replacement.

6. A fair die is rolled five times. Given that at least one roll resulted in an even number, find the probability that two or more rolls will result in even numbers.

7. Five balls are randomly distributed in 5 urns.

   a. Find the probability that one of the urns will contain exactly 3 balls if exactly 2 urns are empty.

   b. Find the unconditional probability that one of the urns will contain exactly 3 balls.

8. This is another form of an urn model. An urn contains four balls numbered 1, 2, 3, and 4. A ball is selected from the urn. That ball is then returned to the urn along with as many balls of the same number. That is, if the ball numbered 3 was selected then the original ball plus three additional balls numbered 3 are put back in the urn. A ball is then drawn from the urn.

   a. What is the probability that the number on the second ball will be the same as that of the first ball?

   b. What is the probability that the number on the second ball will be even?

9. Five numbers are placed at random in an empty 5×5 matrix. What is the probability that no 2 numbers will be in the same row or column?

10. Meteorologist Ray N. Orshine has observed that when it is cloudy it rains 30% of the time. In his area it is cloudy 60% of the time. What is the probability that at a randomly selected time it will be raining?

11. Three barrels contain apples. Barrel I has 2 red apples and 5 green ones, barrel II has 4 red apples and 4 green ones, and barrel III has 3 red apples and 4 green ones. If an apple is selected at random from each barrel, find the probability that exactly 2 green apples were chosen.

12. A bowl contains 6 balls numbered 1, 2, 3, 4, 5, and 6. Two balls are selected with replacement from the bowl. Find the probability that at least 1 of the balls was a 3 given that the sum of the 2 balls was 6.

13. Three cards are selected at random without replacement from a standard deck of 52 playing cards. What is the probability that the first card was a queen given that the second and third cards were queens?

14. Two fair dice are rolled. Compute the probability that at least one 4 will be showing under the following conditions.

   a. The sum of the 2 dice is 8.

   b. The sum of the 2 dice is more than 8.

15. Five cards are drawn from a well-shuffled deck of 52 cards. The first card drawn is the 10 of diamonds.

   a. What is the probability that there will be at least one more 10 in the other 4 cards?

   b. What is the probability that there will be at least 3 diamonds in the hand?

16. In a certain company, 40% of the microcomputers are Macintoshes and 60% are IBM-compatible computers. 55% of Macintosh usage is for word processing, 30% goes to maintaining data bases, and 15% is spent on other tasks. On the IBM-compatible 40% is devoted to word processing, 40% to working with data bases, and 20% is used for other things. If a computer is chosen at random, what is the probability that it is being used for word processing?

2.3 ═══
Bayes' Rule

Bayes' rule is a straightforward extension of the ideas discussed in Section 2.2. It is named after the English philosopher Thomas Bayes who outlined the idea in a paper published posthumously in the middle of the 18th century. First proved by Laplace in the early 19th century, the rule has had a profound effect in the field of statistics. Statistical problems can be approached from a viewpoint that is based on Bayes' rule. This gives rise to a class of procedures that are collectively called

Bayesian statistics. The same problems can be approached from a direction that makes no use of Bayes' rule. These procedures are called *classical* or *non-Bayesian statistics*. There is still a lively debate going on regarding which view of statistics is most appropriate. We shall encounter these ideas when the discussion turns to statistical theory. For now, we will view Bayes' rule as a method to "turn around" conditional probabilities. That is, given $P(A \mid B_i)$, for $i = 1, 2, \ldots, n$, Bayes' rule computes $P(B_i \mid A)$ as follows.

Theorem 2.3

> **Bayes' Rule**   Let $B_1, B_2, \ldots, B_n$ be a set of mutually exclusive and collectively exhaustive events such that $P(B_i) > 0$, for $i = 1, 2, \ldots, n$. Let A be another event such that $P(A) > 0$. Then
>
> $$P(B_j \mid A) = \frac{P(A \mid B_j)P(B_j)}{\displaystyle\sum_{i=1}^{n} P(A \mid B_i)P(B_i)}, \quad \text{for } j = 1, 2, \ldots, n. \qquad (2.5)$$

Proof   Since $P(A) > 0$, $P(B_j \mid A)$ is defined by Equation 2.1 to be

$$P(B_j \mid A) = \frac{P(A \cap B_j)}{P(A)}.$$

Using Equation 2.2 in the numerator we obtain

$$P(B_j \mid A) = \frac{P(A \mid B_j)P(B_j)}{P(A)}.$$

Since the B's are mutually exclusive and collectively exhaustive, the conditions of Theorem 2.1 are satisfied, and we can write

$$P(A) = \sum_{i=1}^{n} P(A \mid B_i)P(B_i).$$

Substituting this for $P(A)$ in the denominator gives

$$P(B_j \mid A) = \frac{P(A \mid B_j)P(B_j)}{\displaystyle\sum_{i=1}^{n} P(A \mid B_i)P(B_i)}.$$

∎

We now give some examples that illustrate the use of Bayes' rule.

Example 2.10

Urn I contains 5 white balls and 3 red balls. Urn II contains 4 white balls and 6 red balls. An urn is selected at random and a ball is drawn. Find the probability that, if the ball selected is white, it came from Urn I. Let $U_i = \{$select urn $i\}$, $i = 1, 2$, and $W = \{$white ball$\}$. In terms of these events, we wish to compute

$P(U_1 \mid W)$. U_1 and U_2 are mutually exclusive and collectively exhaustive. In addition, from the information given we see that

$$P(U_1) = P(U_2) = \frac{1}{2},$$

$$P(W \mid U_1) = \frac{5}{8}, \quad \text{and}$$

$$P(W \mid U_2) = \frac{4}{10}.$$

All of the conditions of Theorem 2.3 are satisfied, and we can use Equation 2.5 to obtain the following.

$$P(U_1 \mid W) = \frac{P(W \mid U_1)P(U_1)}{P(W \mid U_1)P(U_1) + P(W \mid U_2)P(U_2)}$$

$$= \frac{\frac{5}{8} \cdot \frac{1}{2}}{\frac{5}{8} \cdot \frac{1}{2} + \frac{4}{10} \cdot \frac{1}{2}}$$

$$= \frac{25}{41} \qquad \blacksquare$$

Example 2.11

A screening test for a certain disease has been found to detect the presence of the disease 98% of the time when administered to an afflicted person. In 6% of the cases, a well person will be incorrectly diagnosed as having the disease by this test. Studies have shown that 4% of the population have the disease. A patient is given the test, and the result is positive. What is the probability that this person has the disease? Let $D = \{\text{patient has disease}\}$, and $T = \{\text{test positive}\}$. In terms of these events we wish to compute $P(D \mid T)$. From the information above we know that

$$P(D) = 0.04,$$

$$P(D') = 0.96,$$

$$P(T \mid D) = 0.98, \quad \text{and}$$

$$P(T \mid D') = 0.06.$$

Since the conditions of Bayes's rule are satisfied, we compute

$$P(D \mid T) = \frac{P(T \mid D)P(D)}{P(T \mid D)P(D) + P(T \mid D')P(D')}$$

$$= \frac{(0.98) \cdot (0.04)}{(0.98) \cdot (0.04) + (0.06) \cdot (0.96)}$$

$$= 0.405.$$

This result seems to go against intuition. The fact that the disease is fairly rare makes the low probability of giving an incorrect positive reading more important. Thus, mass screening tests can be a relatively poor indicator for the presence of disease. On the practical side, most physicians and patients would rather make use of screening tests which have an extremely high chance of detecting a disease when it is present and put up with a little inconvenience due to the test's giving false-positive results than have it the other way around. This presumes that the screening test has relatively low risk. It would be a different matter if it were something like a strong x-ray. ■

Thus far we have only partitioned the sample space into two mutually exclusive events. The next example shows the use of Bayes' rule when the partition consists of more events.

Example 2.12   An urn contains 2 white balls and 2 black balls. A number is randomly chosen from the set $\{1, 2, 3, 4\}$ and that many balls are removed from the urn. We wish to find the probability that the number i, $i = 1, 2, 3, 4$, was chosen if at least 1 black ball was removed from the urn. Let $A_i = \{$number i chosen$\}$, $i = 1, 2, 3, 4$, and $B = \{$at least 1 black ball removed from the urn$\}$. Using these events we wish to determine $P(A_i \mid B)$, for $i = 1, 2, 3, 4$. From the information given we know that

$$P(A_1) = P(A_2) = P(A_3) = P(A_4) = \frac{1}{4},$$

$$P(B \mid A_1) = \frac{1}{2},$$

$$P(B \mid A_2) = 1 - P(B' \mid A_2) = 1 - \frac{2 \cdot 1}{4 \cdot 3} = \frac{5}{6}, \quad \text{and}$$

$$P(B \mid A_3) = P(B \mid A_4) = 1.$$

Again the conditions of Bayes' rule are satisfied and we get

$$P(A_j \mid B) = \frac{P(B \mid A_j)P(A_j)}{\displaystyle\sum_{i=1}^{4} P(B \mid A_i)P(A_i)}$$

$$= \frac{P(B \mid A_j) \cdot \frac{1}{4}}{\frac{1}{2} \cdot \frac{1}{4} + \frac{5}{6} \cdot \frac{1}{4} + 1 \cdot \frac{1}{4} + 1 \cdot \frac{1}{4}}$$

$$= P(B \mid A_j) \cdot \frac{3}{10}$$

$$= \begin{cases} \frac{3}{20}, & \text{if } i = 1, \\ \frac{1}{4}, & \text{if } i = 2, \quad \text{and} \\ \frac{3}{10}, & \text{if } i = 3 \text{ or } 4. \end{cases}$$

■

The following example is a variation of a classical problem. The answer is relatively unintuitive. Although such behavior is not advocated, it has been used by some to make money from unsuspecting friends.

Example 2.13   There are 3 urns. Urn I contains 2 white balls, urn II contains 1 white ball and 1 black ball, and urn III contains 2 black balls. An urn is chosen at random and a ball is selected from that urn. If the ball removed from the urn is white, what is the probability that the ball remaining in the urn is also white? Pause for a minute and guess what the answer is before proceeding. Let $U_i = \{$urn i chosen$\}$, $i = 1, 2, 3$, and $W = \{$ball drawn is white$\}$. The desired probability is $P(U_1 \mid W)$. From the information we know that

$$P(U_1) = P(U_2) = P(U_3) = \frac{1}{3},$$

$$P(W \mid U_1) = 1,$$

$$P(W \mid U_2) = \frac{1}{2}, \quad \text{and}$$

$$P(W \mid U_3) = 0.$$

Bayes' rule gives

$$P(U_1 \mid W) = \frac{P(W \mid U_1)P(U_1)}{P(W \mid U_1)P(U_1) + P(W \mid U_2)P(U_2) + P(W \mid U_3)P(U_3)}$$

$$= \frac{1 \cdot \frac{1}{3}}{1 \cdot \frac{1}{3} + \frac{1}{2} \cdot \frac{1}{3} + 0 \cdot \frac{1}{3}}$$

$$= \frac{2}{3}.$$

Many times people will guess that the probability should be $\frac{1}{2}$. The reasoning usually goes as follows. Since we already know that a white ball was selected that means either urn I or urn II was selected. Thus, if we selected one of these two urns, the probability that it will be urn I must be $\frac{1}{2}$. The problem with this argument is that it implicitly assumes the events $(U_1 \mid W)$ and $(U_2 \mid W)$ are equally likely. While it is true that the overall experiment has a sample space consisting of 6 equally likely events, this does not necessarily carry over to outcomes affected by conditioning. The sample space is as follows.

	Ball 1	Ball 2
Urn I	$\boxed{W}$	$\boxed{W}$
Urn II	B	$\boxed{W}$
Urn III	B	B

The occurrence of W means that one of the 3 boxed equally likely events occurred. Of those, 2 are in the event U_1 and 1 in U_2. Hence, we get a probability of $\frac{2}{3}$. ∎

Exercises 2.3

1. Golfer A misses his 6-foot putts 60% of the time and golfer B misses her 6-foot putts 40% of the time. One of the golfers is chosen at random and proceeds to make a 6-foot putt. What is the probability that B was chosen?

2. A student takes a true-false examination containing 20 questions. On looking at the examination the student finds that he knows the answer to 8 of the questions, which he proceeds to answer correctly. He then randomly answers the remaining 12 questions. The instructor selects 1 of the questions at random and finds that this student answered the question correctly. What is the probability that the student knew the answer to this question?

3. In a production line there are 3 machines manufacturing a component at a rate of 100 items per hour. Machine I is running well, and 98% of its output is within specifications. Machine II is also running in control with 95% of its output within specifications. Machine III, however, is in need of adjustment, and only 50% of its output is within specifications. An equal number of components from a production run of all 3 machines is placed in a single bin, and a component is selected and found to be within specifications. What is the probability that the component selected was produced by machine III?

4. An insurance company categorizes its customers as to whether they are high risk clients, medium risk clients, or low risk clients. They find that 10% are high risk, 30% are medium risk, and 60% are low risk. The company has also noted that, over a given period of time, 5% of the high risk clients file a claim, 1% of the medium risk clients file a claim, and 0.1% of the low risk clients file a claim. What is the probability that a randomly selected claim will come from a high risk client?

5. In a certain county, 50% of the voters consider themselves to be conservative, 40% consider themselves to be liberal, and the remaining 10% indicate no preference. In a recent election, 75% of the voters with a conservative orientation voted, 85% of the liberals voted, and 50% of the unaligned voters voted. If a person who voted is selected at random, what is the probability that he or she is conservative?

6. A box contains 6 dice. Die i has i faces painted black and $6 - i$ faces painted white, for $i = 1, 2, \ldots, 6$. A die is selected from the box and is rolled. What is the probability that die 3 was chosen if a black face shows on the top face of the die after rolling?

7. There are 3 baskets of fruit. Basket I contains 3 apples, 2 oranges, and 4 bananas. Basket II contains 4 apples, 2 oranges, and 1 banana. Basket III contains 2 apples, 4 oranges, and 2 bananas. A basket is selected at random and a piece of fruit removed. What is the probability that basket II was chosen if the piece of fruit removed was an orange?

8. A college with 100 faculty members offers 3 different types of medical coverage: health maintenance coverage, local doctor coverage, and national coverage. Fifty-five of the faculty members have the national coverage and of those, 30 submitted claims in excess of $100 during the past year. Twenty-five of the faculty members belong to the health maintenance coverage group and 10 of those submitted claims in excess of $100 during the past year. Ten of those with local doctor coverage submitted claims during the past year that exceeded $100. If a randomly selected claim is in excess of $100, what is the probability that the faculty member subscribed to the health maintenance coverage?

9. A wine merchant notices that 50% of his sales are red wines, 40% are white wines, and 10% are rosés. Among those groups 75% of the red wine buyers serve the wine with

red meat, 50% of the white wine buyers serve the wine with red meat, and 80% of the rosé buyers serve it with red meat. If a randomly selected wine buyer indicates that the wine is to be served with red meat, what is the probability that he or she bought a white wine?

2.4 Independence

The examples given in the previous two sections show that there can be a difference between the unconditional and conditional probability of an event. That is, the restriction to a subset of the sample space corresponding to the conditioning event affects the probability of occurrence of another event. In this section we will consider events where $P(A \mid B) = P(A)$. If the conditional probability and the unconditional probability of an event are the same, then we can say that the value of the probability of occurrence of one event does not depend on whether the other event occurred or not. Two events that behave in this manner are said to be *independent*. Sometimes the term *stochastically independent* is used to distinguish this type of independence from, say, linear independence. We shall take the term "independent" to mean stochastically independent and, where necessary, refer to other types of independence by their full names. The formal definition of independence is based on this idea.

Definition 2.2

Two events, A and B, are said to be *independent* if

$$P(A \mid B) = P(A). \qquad (2.6)$$

If A and B are not independent, they are said to be *dependent*.

Suppose that $P(A \mid B) = P(A)$. From Equation 2.1 we see that this implies

$$P(A) = \frac{P(A \cap B)}{P(B)}.$$

It then follows that

$$P(A \cap B) = P(A)P(B). \qquad (2.7)$$

This is often used as the formal definition of independence although it does not give much insight into the idea of events being independent. Equation 2.7 gives us the means for checking if two events are independent or not. All we need to do is compute $P(A \cap B)$, $P(A)$, and $P(B)$ and see whether or not Equation 2.7 holds. If so, A and B are independent and, if not, they are dependent.

Example 2.14

A card is drawn at random from a standard deck of 52 cards. Let $K = \{$card is a king$\}$, and $H = \{$card is a heart$\}$. We shall use Equation 2.7 to determine if H and K are independent events. Since there are 4 kings in the deck from which to choose, we see that $P(K) = \frac{4}{52} = \frac{1}{13}$. Also, there are 13 hearts in a deck, which gives that $P(H) = \frac{1}{4}$. Only one card is the king of hearts, which means that $P(K \cap H) = \frac{1}{52}$. We therefore see

$$P(K)P(H) = \frac{1}{13} \cdot \frac{1}{4} = \frac{1}{52} = P(K \cap H).$$

Since Equation 2.6 holds we conclude that H and K are independent.   ■

Example 2.15   A coin is tossed 3 times. Let $O = \{$at least one head$\}$, $F = \{$first toss a head$\}$, and $T = \{$third toss a head$\}$. We wish to determine if O and F are independent and if T and F are independent. The sample space for this experiment is given in Example 1.6. To get out of the habit of relying on enumeration of the sample space to compute probabilities, we shall use the generalized counting principle. It is most convenient to view the experiment as consisting of 3 stages, one for each toss. Each toss has 2 possible outcomes which means that there will be $2 \cdot 2 \cdot 2 = 8$ elementary events in the sample space. We shall test the independence of O and F first. The complement of O is that no heads occur, which happens in only one way. Therefore,

$$P(O) = 1 - P(O') = 1 - \frac{1}{8} = \frac{7}{8}.$$

If the first toss is fixed at a head there still are 2 possibilities for the second and third tosses. Thus,

$$P(F) = \frac{1 \cdot 2 \cdot 2}{8} = \frac{1}{2}.$$

Now note that $F \subset O$ because if F occurs then O must. However, it is possible to get 1 or more heads without the first toss being a head. Therefore, we see that

$$\frac{1}{2} = P(O \cap F) \neq P(O)P(F) = \frac{7}{8} \cdot \frac{1}{2} = \frac{7}{16}.$$

This shows that O and F are dependent events.

To check the independence of T and F we need to compute $P(T)$ and $P(T \cap F)$. When T occurs the first 2 tosses have 2 possible outcomes and the third must be a head. Therefore,

$$P(T) = \frac{2 \cdot 2 \cdot 1}{8} = \frac{1}{2}.$$

When both T and F occur then the first and third tosses have a single outcome, but the second has 2 possibilities. This gives that

$$P(T \cap F) = \frac{1 \cdot 2 \cdot 1}{8} = \frac{1}{4} = P(T)P(F).$$

Therefore, T and F are independent events. This fact seems intuitively clear. The result of the first coin toss should not have any effect on the other tosses.   ■

The definition of independence has a few immediate consequences. These stem from the notion that when two events are independent, the probability of occurrence or nonoccurrence of one does not depend on the occurrence or nonoccurrence of the other. The following theorem makes this idea more definite.

Theorem 2.4

If A and B are independent events, then

1. A' and B' are independent,
2. A' and B are independent, and
3. A and B' are independent.

Proof   We shall prove the third part and leave the proof of the other two as exercises. Since A and B are independent, we know from Equation 2.7 that $P(A \cap B) = P(A)P(B)$. We need to show that this implies that $P(A \cap B') = P(A)P(B')$. We need somehow to rewrite $P(A \cap B')$ so that we can make use of the fact that A and B are independent. To do this we begin by noting that A can be written as the union of two mutually exclusive events as follows.

$$A = (A \cap B) \cup (A \cap B')$$

We can calculate $P(A)$ by using Axiom 3.

$$P(A) = P(A \cap B) + P(A \cap B')$$

We now can use the fact that A and B are independent to complete the proof.

$$P(A \cap B') = P(A) - P(A)P(B)$$
$$= P(A)[1 - P(B)]$$
$$= P(A)P(B')$$

The proofs of the other parts make use of this result.   ∎

So far we have been considering the independence of a pair of events. We now wish to extend this idea to collections of more than two events. To be consistent with the independence of two events any definition of the joint independence of more than two events should reflect the notion that the probability of occurrence or nonoccurrence of any single event is not affected by the occurrence or nonoccurrence of any combination of the other events. That is, each of the following pairs of events should be independent.

$$A, B \qquad A, C \qquad B, C$$
$$A, B \cap C \qquad B, A \cap C \qquad C, A \cap B$$

Since the notation $A \cap B \cap C$ is really shorthand for $A \cap (B \cap C)$ or $(A \cap B) \cap C$ or $(A \cap C) \cap B$, we can apply Equation 2.6 twice as follows.

$$P(A \cap B \cap C) = P[A \cap (B \cap C)]$$
$$= P(A)P(B \cap C)$$
$$= P(A)P(B)P(C)$$

Thus, we arrive at

$$P(A \cap B \cap C) = P(A)P(B)P(C). \tag{2.8}$$

This means that any pair of events should also be independent in accordance with Definition 2.2. Thus, a suitable definition for the joint independence of A, B, and C must ensure that Equation 2.8 holds and that Equation 2.7 holds for all possible pairs of the three events. There would be no problem with simply making this the definition, but it is a little complicated. We would like to determine whether or not we can use a simpler form and have it imply all of the above. Two possibilities come to mind. We check them both out.

First, is it enough to require that Equation 2.7 hold for every pair of events to be sure that Equation 2.8 will hold? The answer turns out to be that it is not enough. It is not too hard to construct events that are pairwise independent but fail to satisfy Equation 2.8.

Example 2.16   Suppose we select a point at random from

$$S = \{(0,0,0), (1,1,0), (0,1,1), (1,0,1)\}.$$

Define the event A_i to be that the ith coordinate of the element selected is 1, for $i = 1, 2, 3$. From these we see that

$$A_1 \cap A_2 = \{(1,1,0)\},$$
$$A_1 \cap A_3 = \{(1,0,1)\},$$
$$A_2 \cap A_3 = \{(0,1,1)\}, \quad \text{and}$$
$$A_1 \cap A_2 \cap A_3 = \varnothing.$$

Therefore, we obtain the following probabilities.

$$P(A_1) = P(A_2) = P(A_3) = \frac{1}{2},$$

$$P(A_1 \cap A_2) = P(A_1 \cap A_3) = P(A_2 \cap A_3) = \frac{1}{4} \quad \text{and}$$

$$P(A_1 \cap A_2 \cap A_3) = 0$$

Now,

$$P(A \cap B) = P(A)P(B) = \frac{1}{4},$$

$$P(A \cap C) = P(A)P(C) = \frac{1}{4}, \quad \text{and}$$

$$P(B \cap C) = P(B)P(C) = \frac{1}{4}, \quad \text{but}$$

$$P(A \cap B \cap C) = 0 \neq P(A)P(B)P(C).$$

Thus, pairwise independence does not necessarily imply Equation 2.8.   ■

Second, does the fact that Equation 2.8 holds imply that Equation 2.7 must hold for all pairs of events? Here, too, the answer is "no" as the following counterexample illustrates.

Example 2.17

Suppose that a number is chosen at random from $\{1,2,3,4,5,6,7,8\}$. Define the following events.

$$A = \{1,2,3,4\},$$
$$B = \{1,5,6,7\}, \quad \text{and}$$
$$C = \{1,2,3,8\}.$$

Therefore,

$$A \cap B = \{1\},$$
$$A \cap C = \{1,2,3\},$$
$$B \cap C = \{1\}, \quad \text{and}$$
$$A \cap B \cap C = \{1\}.$$

From this we see that

$$P(A) = P(B) = P(C) = \frac{1}{2},$$

$$P(A \cap B \cap C) = \frac{1}{8} = P(A)P(B)P(C), \quad \text{but}$$

$$P(A \cap C) = \frac{3}{8} \neq P(A)P(C).$$

■

This discussion shows that we are forced to define the joint independence of three or more events by requiring that all possible subsets obey equations that are similar to Equation 2.8. This means that checking the independence of a set of events requires verification that the probability of the intersection of any subset is equal to the product of the probabilities of the events in that intersection. We state this formally as follows.

Definition 2.3

The events $A_1, A_2, \ldots, A_n$ are said to be *independent* if for every subset, B, of $\{1, 2, \ldots, n\}$ then

$$P\left(\bigcap_{i \in B} A_i\right) = \prod_{i \in B} P(A_i). \tag{2.9}$$

The notation of this theorem is a little complicated. What it says, in essence, is the following. Suppose we have four events, A, B, C, and D. To declare that they are independent we must verify that each of the following is true.

$$P(A \cap B \cap C \cap D) = P(A)P(B)P(C)P(D),$$
$$P(A \cap B \cap C) = P(A)P(B)P(C),$$
$$P(A \cap B \cap D) = P(A)P(B)P(D),$$
$$P(A \cap C \cap D) = P(A)P(C)P(D),$$
$$P(B \cap C \cap D) = P(B)P(C)P(D),$$
$$P(A \cap B) = P(A)P(B),$$
$$P(A \cap C) = P(A)P(C),$$
$$P(A \cap D) = P(A)P(D),$$
$$P(B \cap C) = P(B)P(C),$$
$$P(B \cap D) = P(B)P(D), \quad \text{and}$$
$$P(C \cap D) = P(D)P(D).$$

For five events we need to check all two-way, three-way, four-way, and five-way intersections. Needless to say, this can get a little tedious. We shall illustrate how this process proceeds, and then begin to work on finding out if we can determine if events are independent by looking at the experiment.

Example 2.18   Three fair coins are tossed. Let $H_1 = \{\text{head on coin 1}\}$, $H_2 = \{\text{head on coin 2}\}$, and $T_3 = \{\text{tail on coin 3}\}$. We wish to determine if these 3 events are independent. That fact that each coin is fair implies that all elementary events are equally likely. Therefore, the generalized counting principle gives that

$$P(H_1) = \frac{1 \cdot 2 \cdot 2}{8} = \frac{1}{2},$$
$$P(H_2) = \frac{2 \cdot 1 \cdot 2}{8} = \frac{1}{2},$$
$$P(T_3) = \frac{2 \cdot 2 \cdot 1}{8} = \frac{1}{2},$$
$$P(H_1 \cap H_2) = \frac{1 \cdot 1 \cdot 2}{8} = \frac{1}{4} = P(H_1)P(H_2),$$
$$P(H_1 \cap T_3) = \frac{1 \cdot 2 \cdot 1}{8} = \frac{1}{4} = P(H_1)P(T_3),$$
$$P(H_2 \cap T_3) = \frac{2 \cdot 1 \cdot 1}{8} = \frac{1}{4} = P(H_2)P(T_3), \quad \text{and}$$
$$P(H_1 \cap H_2 \cap T_3) = \frac{1 \cdot 1 \cdot 1}{8} = \frac{1}{8} = P(H_1)P(H_2)P(T_3).$$

Since all the multiplication rules hold we conclude that H_1, H_2, and T_3 are independent events.

The coin tossing experiment of the preceding example can be viewed as consisting of three separate random experiments of one coin toss each. Clearly each individual coin toss satisfies the definition of being a random experiment. For this reason each toss is called a *subexperiment* of the overall experiment. Each coin toss has its own sample space, $\{H, T\}$. In addition, the sample space of the overall experiment is an ordered tuple where the ith entry is the elementary event in the sample space for the ith toss that occurred. In set theory terminology the overall sample space is the *Cartesian product* of the subexperiment sample spaces. For example, the Cartesian product of the sets

$$A = \{1, 2\}, \quad B = \{2, 3\}, \quad \text{and} \quad C = \{1, 4\}$$

would be denoted by $A \times B \times C$ and would consist of the following 8 ordered triples.

$$(1, 2, 1) \quad (1, 2, 4) \quad (1, 3, 1) \quad (1, 3, 4)$$
$$(2, 2, 1) \quad (2, 2, 4) \quad (2, 3, 1) \quad (2, 3, 4)$$

Viewed in this way the coin tossing experiment would be said to consist of 3 identical *subexperiments* because the sample space of each subexperiment is the same. An experiment where we draw 5 cards without replacement from a standard deck can be viewed as consisting of 5 subexperiments. Each subexperiment corresponds to the drawing of a card. In this case, however, the subexperiment sample spaces are not the same. The sample space for the first card has 52 elementary events while that of the second card only has 51 elementary events. Each subexperiment is commonly referred to as a *trial*. We give a formal definition.

Definition 2.4

> An experiment is said to consist of n *trials* if the experiment can be divided into n subexperiments, each of which satisfies Definition 1.7. In addition, the sample space for the experiment is the Cartesian product of the n subexperiment sample spaces.

Suppose we roll a fair die n times. This experiment consists of n identical trials. Clearly, the outcome of the first roll will not affect the probability of observing a specific result on the second or succeeding trials. In such a situation we would say that the trials would be independent. Computing the probability of occurrence of any elementary event would be greatly simplified in such a situation. Equation 2.9 could be used to compute probabilities. These ideas lead to the definition of *independent trials*.

Definition 2.5

> Let an experiment consist of n trials where S_i is the collection of possible outcomes for trial $i, i = 1, 2, \ldots n$. The experiment is said to consist of n *independent trials* if $S_1, S_2, \ldots, S_n$ are independent.

This means that if S_i is the collection of possible outcomes for subexperiment $i, i = 1, 2, \ldots, n$, then the sample space for the overall experiment will be

$$S = S_1 \times S_2 \times \ldots \times S_n.$$

S is said to be the *combinatorial product* of $S_1, S_2, \ldots, S_n$. A *combinatorial product event* in this sample space is the Cartesian product of events from each of the subexperiment sample spaces. That is, if $A_i \subset S_i, i = 1, 2, \ldots, n$, then we can define the combinatorial product event A to be

$$A = A_1 \times A_2 \times \ldots A_n.$$

It turns out that $P(A)$ can be calculated as follows.[*]

$$P(A) = P[(A_1, A_2, \ldots, A_n)] = P(A_1)P(A_2) \ldots P(A_n) \qquad (2.10)$$

The idea is that the probability of outcomes of individual trials should be easier to compute than that of the overall event of which they are a part. The notion of independent trials does for intersections what mutually exclusive events has done for unions.

One easy way to recognize a set of independent trials is to look at the sample spaces of the trials. If the sample space for each trial is unaffected by the outcomes of the other trials, then the trials are independent. Examples of experiments consisting of independent trials are sampling with replacement, tossing coins or dice, and removing balls from separate urns. Experiments that involve sampling without replacement result in trials that are dependent.

Example 2.19

We draw 5 cards from a standard deck with replacement. We wish to find the probability that all 5 cards will be face cards (that is, Jack, Queen, or King). Since we are sampling with replacement the experiment consists of 5 identical independent trials. Let $F_i = \{\text{face card on trial } i\}$, $i = 1, 2, 3, 4, 5$. The sample space for each trial contains 52 elementary events, one for each card in the deck. Since the sample spaces are identical the number of ways to obtain a face card on any trial is 12. Therefore,

$$P(F_i) = \frac{12}{52} = \frac{3}{13}, \quad i = 1, 2, 3, 4, 5.$$

Now Equation 2.10 gives

$$P(5 \text{ face cards}) = P(F_1 \cap F_2 \cap F_3 \cap F_4 \cap F_5)$$
$$= P(F_1)P(F_2)P(F_3)P(F_4)P(F_5)$$
$$= \left(\frac{3}{13}\right)^5.$$

∎

---

[*] For a development of this equation the reader is referred to *An Introduction to Probability Theory and Its Applications*, vol. 1, by William Feller.

Example 2.20

An urn contains w white balls and b black balls. A ball is sampled from the urn with replacement and the process is repeated. Since sampling is done with replacement the urn is always restored to its original configuration, and we have a set of identical and independent trials. We wish to determine the probability that the first white ball will be removed from the urn on the kth draw, the probability that the rth white ball will be removed from the urn on the kth draw, and the probability that a white ball will ever be drawn. Let $W_k = \{\text{white ball on } k\text{th draw}\}$, $k = 1, 2, \ldots$. Also let $F_k = \{\text{first white ball on } k\text{th draw}\}$, $R = \{r\text{th white ball on } k \text{ draw}\}$, and $E = \{\text{white ball ever drawn}\}$. Getting the first white ball on the kth draw means that the first $k - 1$ draws must be black and the kth must be white. Thus, in terms of the trial events, we see that F_k is

$$F_k = W_1' \cap W_2' \cap \cdots \cap W_{k-1}' \cap W_k.$$

Since the trials are identical we see that on any one there are w ways to draw a white ball, b ways to draw a black ball, and $w + b$ possible outcomes. Therefore,

$$P(W_i) = \frac{w}{w + b}, \quad \text{and} \quad P(W_i') = \frac{b}{w + b}, \quad i = 1, 2, \ldots.$$

We can now use Equation 2.10 to obtain the following.

$$P(F_k) = P(W_1' \cap W_2' \cap \cdots \cap W_{k-1}' \cap W_k)$$

$$= P(W_1')P(W_2') \cdots P(W_{k-1}')P(W_k)$$

$$= \left(\frac{b}{w + b}\right)^{k-1} \cdot \frac{w}{w + b}$$

To determine the probability of R we assume that $k \geq r$. Otherwise $P(R) = 0$. Now R occurs only if we draw $r - 1$ white balls and $k - r$ black balls in the first $k - 1$ draws and then get a white ball on the kth draw. One elementary event in R is

$$W_1 \cap W_2 \cap \cdots \cap W_{r-1} \cap W_r' \cap \cdots \cap W_{k-1}' \cap W_k.$$

If we use Equation 2.10 as we did above we get

$$P(W_1 \cap W_2 \cap \cdots \cap W_{r-1} \cap W_r' \cap \cdots \cap W_{k-1}' \cap W_k) = \left(\frac{w}{w + b}\right)^{r-1} \left(\frac{b}{w + b}\right)^{k-r} \frac{w}{w + b}.$$

Notice that, in the end, this calculation only depended on the number of black and white balls that are drawn in the first $k - 1$ draws and not the order in which they occurred. Therefore, any arrangement of $k - 1$ draws consisting of $r - 1$ white balls and $k - r$ black balls will have this same probability. Different arrangements of black and white balls are elementary events and, hence, are mutually exclusive. There are $\binom{k-1}{r-1}$ different arrangements of this sort. Therefore,

$$P(R) = \binom{k-1}{r-1}\left(\frac{w}{w + b}\right)^r \left(\frac{b}{w + b}\right)^{k-r}.$$

To compute $P(E)$ we assume that balls are drawn until a white ball is obtained and that $w > 0$. From this description we see that

$$P(E) = P\left(\bigcup_{k=1}^{\infty} F_k\right).$$

That is, E occurs if we get the first white ball on the first draw, the first white ball on the second draw, and so on. Since the F_k's are mutually exclusive events $P(E)$ would be the sum of the $P(F_k)$'s. Therefore,

$$P(E) = \sum_{k=1}^{\infty} \frac{w}{w+b}\left(\frac{b}{w+b}\right)^{k-1}.$$

Since $0 \le \frac{b}{w+b} < 1$, if we let $a = \frac{w}{w+b}$, $r = \frac{b}{w+b}$, and $n = k - 1$ we can write

$$P(E) = \sum_{n=0}^{\infty} ar^n.$$

This form we recognize from calculus as the sum of a geometric series whose value is

$$\sum_{n=0}^{\infty} ar^n = \frac{a}{1-r}.$$

Therefore, we arrive at

$$P(E) = \frac{w/(w+b)}{1 - b/(w+b)} = 1.$$

∎

The last part of the previous example points out something interesting about probabilities. The probability that we will ever draw a white ball was found to be one. This means that the probability that we will never draw a white ball is zero. There is, however, one elementary event in E', namely we can draw black balls forever. This shows that sample spaces can contain elementary events whose probability are zero. Thus, there are objects other than the empty set with zero probabilities. So $P(A) = 0$ does not necessarily mean that A never occurs.

We mentioned in Chapter 1 that modern probability theory began with the correspondence between Pascal and Fermat regarding a problem posed by the French gambler Chevalier de Mere. It has become known as the *problem of the points*. Since it concerns an experiment that consists of independent trials we consider its solution now.

Example 2.21   Two players, A and B, play a game. Each play of the game has 2 outcomes which we label "success" and "failure." On any play of the game $P(\text{success}) = p$ and $P(\text{failure}) = q = 1 - p$. Player A wins 1 point whenever a success occurs, and player B wins 1 point on each failure. We wish to calculate the probability that player A will win the match if he needs n more points to win and player B needs m more points to win. In terms of the problem posed by de Mere, player A has m coins, and player B has n coins. The game ends when one player has no coins. The question was how to split the pot among the 2 players if play is suspended at this point. Player A will get a proportion of the pot equal to his probability of winning all of the coins from that point of the match. Player B gets the remainder of the pot.

The key to the problem is to break up the event that Player A wins into the union of some more specific events. Note that Player A will win if he gets n points and player B gets $m - 1$ or fewer wins. Let

$$P_{n,m} = P(A \text{ wins when } A \text{ needs } n \text{ points and } B \text{ needs } m \text{ points}).$$

As indicated this can be decomposed as follows.

$$P_{n,m} = P(\{A \text{ wins}\} \cap \{B \text{ gets} \leq m - 1 \text{ points}\})$$

$$= P\left[\{A \text{ wins}\} \cap \left(\bigcup_{i=0}^{m-1} \{B \text{ gets } i \text{ points}\}\right)\right]$$

$$= P\left[\bigcup_{i=0}^{m-1} (\{A \text{ wins}\} \cap \{B \text{ gets } i \text{ points}\})\right]$$

Since the events in the union are mutually exclusive we can use Theorem 1.6 to get

$$P_{n,m} = \sum_{i=0}^{m-1} P(\{A \text{ wins}\} \cap \{B \text{ gets } i \text{ points}\}).$$

Now, when Player A wins the last play of the game must result in a "success." When seen in this manner we note that this problem has much in common with Example 2.20 where we considered the probability of drawing the rth white ball on the kth draw. Here each event in the sum is the occurrence of the nth "success" occurring on the $(n + i)$th play of the game. Thus, if we equate getting a white ball with "success" and getting a black ball with "failure" we can see from Example 2.20 that

$$P(\{A \text{ wins}\} \cap \{B \text{ gets } i \text{ points}\}) = \binom{n + i - 1}{i} p^n q^i, i = 0, 1, \ldots, m - 1.$$

Therefore, the solution to the problem of the points is

$$P_{n,m} = \sum_{i=0}^{m-1} \binom{n + i - 1}{i} p^n q^i.$$

◼

We conclude this section with an example from the field of genetics. Probability theory has been applied with great success to this area. We shall show a simple example that illustrates how the topics discussed in this chapter can be used.

Example 2.22   Before getting to the actual example we need to introduce some terminology. Characteristics that cells pass on to succeeding generations are determined by *chromosomes*. The number of chromosomes varies from organism to organism. The entity that controls a particular characteristic is called a *gene*. Genes combine in strings to make up chromosomes. The position of a gene within the chromosome is called its *locus*. We shall consider what is called the *diploid* case. Here the chromosomes come in pairs. Mammals fall into this category. The different types of genes that may occur at a particular locus are known as *alleles,* and the possible combinations of alleles within a chromosome are called the *genotypes*.

Consider a simple case where the alleles are denoted by A and a. In addition, the genotypes are assumed to be AA, Aa, and aa. In genetics terminology, AA and aa are called *homozygotes,* and Aa is called a *heterozygote*. The various physical manifestations of the allele combinations are called *phenotypes*. For example, if A is dominant and a recessive, then AA and Aa will give the same phenotype. Reproductive cells, known as *gametes,* are formed by splitting with each half receiving one of the two genes in a pair. The homozygotes produce only one type of gamete, all A or all a, while heterozygotes produce equal numbers of A and a gametes. New organisms will have the characteristics acquired from two parental gametes. At each generation a new organism receives a particular gamete from any parent with probability $\frac{1}{2}$. In addition, the mating of gametes to new organisms are independent trials. This means we can view n progeny from a particular generation as constituting n independent trials, each of which is the result of 2 coin tosses. That is, if we mate 2 Aa genotypes we get offspring that are AA, Aa, or aa with probability $\frac{1}{4}$, $\frac{1}{2}$, and $\frac{1}{4}$, respectively. If we mate AA with aa, type Aa offspring will occur with probability 1. Similarly, mating 2 AA or aa genotypes gives only AA or aa offspring. We shall assume that parents mate randomly. That is, it is equally likely that any male will mate with any female. This is obviously a simplification of the way things occur in nature.

In our model there are 3 possible genotypes for each parent. Thus, there are 9 possible pairs of parental genotypes. Random mating implies that we can view the process as that of independently and randomly selecting 1 male gamete and 1 female gamete from the parental population to form the genotype of 1 child. Suppose that both male and female genotypes AA, Aa, and aa occur with proportions u, v, and w, respectively, and that $u + v + w = 1$. Because mating is random the composition of the succeeding generation will be as shown below.

Parents	Offspring	Proportion
AA, AA	AA	u^2
AA, Aa	$\frac{1}{2}AA + \frac{1}{2}Aa$	$2uv$
AA, aa	Aa	$2uw$
Aa, Aa	$\frac{1}{4}AA + \frac{1}{2}Aa + \frac{1}{4}aa$	v^2
Aa, aa	$\frac{1}{2}Aa + \frac{1}{2}aa$	$2vw$
aa, aa	aa	w^2

Thus, in the next generation, the probability of occurrence of genotype AA will be $u^2 + uv + \frac{1}{4}v^2$, that of genotype Aa will be $uv + 2uw + \frac{1}{2}v^2 + vw$, and for genotype aa will be $\frac{1}{2}v^2 + vw + w^2$. If we let

$$p = u + \frac{1}{2}v, \quad \text{and}$$

$$q = \frac{1}{2}v + w$$

these become as follows.

AA	Aa	aa
p^2	$2pq$	q^2

Let's determine the probability of each genotype after another mating. If we repeat the above analysis we find that the probabilities for each genotype are as shown below. The details are left as an exercise.

AA	Aa	aa
$(p^2 + pq)^2$	$2(p^2 + pq)(q^2 + pq)$	$(pq + q^2)^2$

Using the fact that $p + q = 1$ we get

AA	Aa	aa
p^2	$2pq$	q^2

Thus, from the second generation onward the probability of each genotype is the same. Such a population is called *stable*. The fact that a stable population occurs after one mating cycle under random mating is known as the *Hardy-Weinberg law*. ∎

Exercises 2.4

1. In each of the following, two events are described. Explain whether or not it would be reasonable to assume that the events are independent. Give reasons for conclusion.
   a. $A = \{$a man is over 35 years old$\}$, $B = \{$a man weighs over 200 pounds$\}$.
   b. $A = \{$a person owns a telephone$\}$, $B = \{$a person is a Republican$\}$.
   c. $A = \{$a person is a college graduate$\}$, $B = \{$a person owns a personal computer$\}$.
   d. $A = \{$a student regularly attends class$\}$, $B = \{$a student gets goods grades$\}$.
   e. $A = \{$a monkey types *CanneryRow* on a typewriter$\}$, $B = \{$a monkey types *MobyDick* on a typewriter$\}$.

2. A number is selected from the set $\{1, 2, \ldots, 20\}$. Let

$$A = \{x : 1 \leq x \leq 10\},$$

$$B = \{x : 6 \leq x \leq 15\}, \quad \text{and}$$

$$C = \{x : 11 \leq x \leq 20\}.$$

Are A, B, and C independent? If not, which events, if any, are pairwise independent?

3. A fair coin is tossed. How many tosses are necessary for the events $A = \{$both heads and tails appears$\}$ and $B = \{$at most 1 tail$\}$ to be independent?

4. Prove Part a of Theorem 2.4. That is, if A and B are independent, show that A' and B' are independent.

5. An office building has a main entrance consisting of 4 doors. It has been observed that 75% of those leaving the building will use the center doors. Assume that people choose a door to leave independently. What is the probability that out of 10 people

   a. no one uses the center doors.

   b. at least one person uses the side doors.

6. Use Parts a and c of Theorem 2.4 to prove Part b.

7. Two-thirds of the employees at a company live within 10 miles of work.

   a. What is the probability that 3 randomly selected employees live within 10 miles of work?

   b. What is the probability that at least 1 of 3 randomly selected employees lives more than 10 miles from work?

   c. What is the probability that exactly 1 of 3 randomly selected employees lives more than 10 miles from work?

8. A fair die is rolled until an even number is obtained and this experiment is repeated for a total of three times. Find the probability that each run of the experiment will take the same number of rolls.

9. For the following, prove or give a counterexample.

   a. If A and B are independent, and A and C are independent, then B and C are independent.

   b. If A and B are mutually exclusive, then A and B are dependent.

   c. If A, B, and C are independent, then A is independent of $B \cup C$.

   d. If A is independent of B, and C is independent of $A \cap B$, then A, B, and C are independent.

10. Show that, A, B, and C are independent, then

$$P(A \cup B \cup C) = 1 - P(A')P(B')P(C').$$

11. Let A, B, and C be independent events. Express the probability of each of the following events in terms of $P(A)$, $P(B)$, and $P(C)$.

    a. None of the events A, B, or C occurs.

    b. Exactly two of the events A, B, or C occurs.

    c. More than one of the events A, B, or C occurs.

    d. At most two of the events A, B, or C occurs.

12. An urn contains 20 balls, 4 of which are white. Two players, A and B, remove balls alternately until someone gets a white ball and wins. If player A goes first, find the probability that he will win if balls are removed

    a. without replacement, and

    b. with replacement.

13. Suppose that A, B, and C are independent events and that $P(A) = \frac{1}{4}$, $P(B) = \frac{1}{5}$, and $P(C) = \frac{1}{2}$.

   a. Find the probability that none of these 3 events occurs.

   b. Find the probability that either A or B occurs.

   c. Find the probability that exactly 1 of the 3 events occurs.

14. In Example 22 show that the probability of each genotype in the third generation is the same as in the second generation.

15. The system diagrammed below will function if a path from A to B can be found through function components. If component i functions with probability $p_i, i = 1, 2, 3, 4, 5$, find the probability that the system will function properly.

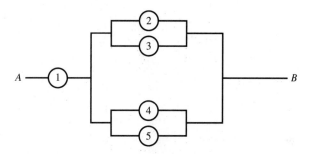

16. A very delicate, but vital, component in a ship's navigation system is known to fail 2.5% of the time. To increase system reliability, n of these components are installed in parallel. Find the smallest value for n to ensure that the navigation system will remain fully functional 99% of the time.

17. In the child's game odd man out players toss fair coins. If any player gets a result that is different from all of the other players then that player is the odd man and loses.

   a. If a game has four players, what is the probability that there will be an odd man on a given toss?

   b. How many players are needed before the probability of having an odd man is less than or equal to 0.10?

   c. If a game has 4 players, what is the probability that k tosses will be needed to get an odd man, for $k = 1, 2, 3, \ldots$?

18. In a particular family the probability of each child being a girl is 0.55. If this family has 4 children, what is the probability that

   a. there will be exactly 1 girl, and

   b. there will be at least 1 boy.

19. There are 3 boxes. Box I contains 1 red ball and 2 white balls. Box II contains 2 red balls and 3 white balls. Box III contains 3 red balls and 1 white ball. One ball is drawn from each box. Find the probability that

   a. there will be 2 white balls and 1 red ball,

   b. there will be at least 2 white balls, and

   c. there will be more red balls than white balls.

20. A fair die is tossed until an even number results. This process is repeated for a total of three times. What is the probability that each run of the experiment will require the same number of tosses to achieve that first even number?

Chapter Summary

Conditional Probability: If $P(B) > 0$

$$P(A \mid B) = \frac{P(A \cap B)}{P(B)}$$

Multiplication Rule: $P(A \cap B) = P(A \mid B)P(B)$

If $B_1, B_2, \ldots, B_k$ are mutually exclusive and $\bigcup_{i=1}^{k} B_i = S$ then

$$P(A) = \sum_{i=1}^{k} P(A \mid B_i)P(B_i).$$

If $P\left(\bigcap_{i=1}^{n} A_i\right) > 0$ then

$$P\left(\bigcap_{i=1}^{n} A_i\right) = P(A_n \mid A_1 \cap A_2 \cap \cdots \cap A_{n-1})$$

$$\cdot P(A_{n-1} \mid A_1 \cap A_2 \cap \cdots \cap A_{n-2}) \cdots$$

$$\cdot P(A_3 \mid A_1 \cap A_2)P(A_2 \mid A_1)P(A_1). \qquad \textbf{(Section 2.2)}$$

Bayes' Rule: If $B_1, B_2, \ldots, B_k$ are mutually exclusive and $\bigcup_{i=1}^{k} B_i = S$ and $P(A) > 0$ then

$$P(B_j \mid A) = \frac{P(A \mid B_j)P(B_j)}{\sum_{i=1}^{n} P(A \mid B_i)P(B_i)}, \qquad \text{for } i = 1, 2, \ldots, n \qquad \textbf{(Section 2.3)}$$

Independence:

A and B are independent events if $P(A \mid B) = P(A)$.

A and B are independent if and only if $P(A \cap B) = P(A)P(B)$.

If A and B are independent, then so are A' and B, A and B' as well as A' and B'. **(Section 2.4)**

Trials: An experiment consists of N trials if it can be divided into n subexperiments each of which is a random experiment and the sample space is the Cartesian product of the subexperiment sample spaces. If the subexperiment sample spaces are independent from one another then the trials are independent. **(Section 2.4)**

Review Exercises

1. In Example 2.22 suppose that, for a certain characteristic, type A genes are dominant and type a genes are recessive. For example, type A genes may lead to brown eyes and type a genes blue eyes.

   a. If all parents in a given population are Aa, find the probability that a randomly chosen child will be of type aa.

   b. If parents are evenly divided among all three genotypes, find the probability that a child selected at random will be of type AA or Aa.

2. (Breeding) Assume that there are three genotypes for a certain breed of dog, AA, Aa, and aa. Type aa offspring have certain traits that are considered undesirable. For this reason only dogs of type AA and Aa are permitted to mate. Show that such a policy will eventually result in the disappearance of type aa dogs. Assume that the original proportion of each genotype is $\frac{1}{3}$.

3. (Polya's Urn Model) This experiment has been used as a model for the spread of contagious disease. An urn initially contains w white balls and b black balls. A ball is drawn from the urn and replaced along with c $(c > 0)$ balls of the same color. The process is repeated. Let $W_i = \{$ball selected on the ith draw is white$\}$. Find the following.

   a. $P(W_2 \mid W_1')$

   b. $P(W_1 \mid W_2)$

   c. $P(W_2)$.

4. A particular system contains n identical components, k of which are necessary for the system to function satisfactorily. Assume that for any single component

$$P(\text{component fails in } \leq a \text{ hours}) = 1 - e^{-0.2a}, a > 0.$$

   a. Find the probability that the system is functioning after t hours if components fail independently.

   b. How many components are necessary for the probability that the system will function for at least 10 hours to be at least 0.95 $k = 2$?

5. A fair die is tossed three times. What is the probability that the outcome of each toss will exceed that of the preceding toss?

6. It is known that 80% of all politicians are liars, whereas only 40% of all lawyers are liars. At a particular gathering 75% of the attendees are lawyers and the rest are politicians. An attendee is selected at random.

   a. What is the probability that the person chosen is a liar?

   b. If the person chosen is a liar, what is the probability that he or she was a lawyer?

7. Let A, B, and C be three events and that the following are true.

$$P(A \mid C) \geq P(B \mid C), \quad \text{and} \quad P(A \mid C') \geq P(B \mid C')$$

Show that $P(A) \geq P(B)$.

8. Let A and B be independent events. Show that, if $P(B) > 0$, then

$$P(A \cup B \mid B') = P(A).$$

9. At a certain golf club 3 of the members have been known to cheat on their scores. Golfer A cheats on his score 30% of the time, Golfer B cheats on his score 50% of the time, and Golfer C cheats on his score 70% of the time. Assume that these golfers cheat independently of one another.

   a. Find the probability that at least one of them will cheat.

   b. Find the probability that exactly one of them will cheat.

10. There are 2 boxes. Box I contains 3 black balls and 6 white balls. Box II contains 6 black balls and 3 white balls. One of the boxes is chosen at random. From there on balls will be chosen with replacement from that box.

a. Find the probability that the first ball removed will be white.

b. If the first two balls are black, find the probability that the third one will also be black.

c. If the first two balls are black, find the probability that balls are being removed from Box II.

11. A box contains b black balls and w white balls. Balls are selected at random from the box.

a. What is the probability that all of the white balls will be removed before obtaining a black one?

b. What is the probability that all of the white balls will be removed before all of the black ones?

12. The system diagrammed below will function if a path from A to B can be found through functioning components. If component i functions with probability p_i, $i = 1, 2, 3, 4$, find the probability that the system will function properly.

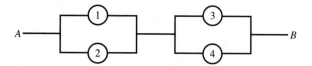

13. A box contains 7 dice. Die i has $7 - i$ faces painted black with the remainder being painted white, for $i = 1, 2, 3, 4, 5, 6, 7$. A die is selected at random from the box.

a. If the die is tossed once and a black face shows, what is the probability that die i was chosen, $i = 1, 2, 3, 4, 5, 6, 7$?

b. If the same die is tossed again, what is the probability that another black face will be obtained?

14. A fair die is thrown two times. Define the following events.

$$A = \{\text{first toss is even}\}$$

$$B = \{\text{second toss is even}\}$$

$$C = \{\text{both tosses are the same}\}$$

Are A, B, and C independent events? If not, are any of them pairwise independent?

15. A fair die is tossed five times. What is the probability that a face with fewer than three dots will show at least two times?

3

Random Variables

Introduction

Many times it is desirable to replace an event in a sample with a number that represents it. For example, if we toss a coin 10 times, we usually are interested in how many heads and tails occurred, not the outcome on each individual coin. So, if the outcome $\{HHHTHTTHHT\}$ occurs, the relevant information would be that 6 heads were seen. In other cases, it is natural for the sample space to contain events that are numerical. In the game spinner of Example 1.8, it is most practical to describe a particular outcome as being so many radians in a certain direction away from a reference point. When an experiment is repeated many times and the results are to be stored in a computer, it is common practice to use numerical codes to describe outcomes. For example, in the Mendel pea experiment, we could code the results of a crossing by a 1 for a wrinkled green seed, a 2 for a smooth green seed, a 3 for a wrinkled yellow seed, and a 4 for a smooth yellow seed. In each case listed, we have established a rule for pairing a unique real number with each possible outcome of a random experiment. Such a rule is called a *random variable (r.v.)*. This means that when using random variables the outcome of an experiment is, in effect, replaced by a real number, and the original sample space is replaced by a sample space of real numbers. Most applications use random variables to describe events. This motivates the following formal definition.

Definition 3.1

> A *random variable* is a random experiment whose sample space consists of real numbers.

Random variables are usually classified as being either *discrete* or *continuous*. We shall define these terms more formally later. For now, we shall describe a random variable as being discrete if it can assume as many values as there are

95

integers. Otherwise, it will be considered continuous. It is theoretically possible for a random variable to be neither strictly discrete nor continuous. Fortunately, such random variables rarely have any practical value.

Traditionally, the sample space for a random variable is denoted by uppercase letters such as X, Y or Θ. The elements in the sample space for a random variable are called the *values assigned to the random variable* and are denoted by lowercase letters such as x, y, and θ. Therefore, it is common practice simply to write $\{X = x\}$ to describe the event that a random experiment whose sample space is X takes the value x.

There is a computational convenience to the use of random variables. The values of a random variable are elementary events. This means that the events $\{X = a\}$ and $\{X = b\}$ will be mutually exclusive when $a \neq b$. This means that we can compute $P(\{X = a\} \cup \{X = b\})$ using Axiom 3 of Definition 1.8, rather than the more complicated Theorem 1.8.

Example 3.1

In tossing 3 fair coins, let the r.v. X be defined as X = number of tails. Thus, X can assume one of the values 0, 1, 2, or 3. This means that we would categorize X as being a discrete random variable. For various values of X, we compute

$$P(X = 0) = P(\{HHH\}) = \frac{1}{8},$$

$$P(X = 1) = P(\{HHT\} \cup \{HTH\} \cup \{THH\}) = \frac{3}{8},$$

$$P(X = 2) = P(\{TTH\} \cup \{THT\} \cup \{HTT\}) = \frac{3}{8},$$

$$P(X = 3) = P(\{TTT\}) = \frac{1}{8}, \quad \text{and}$$

$$P(X \leq 1) = P(\{X = 0\} \cup \{X = 1\}) = P(X = 0) + P(X = 1) = \frac{1}{2}.$$

In addition, we note that

$$P\left(\bigcup_{i=0}^{3}\{X = i\}\right) = \sum_{i=0}^{3} P(X = i) = 1.$$

This can serve as a useful check. Since every outcome of a random experiment results in X being assigned one of its possible values, the union of all possible values of X must be equivalent to the sample space, and that union must have probability one. If calculations do not verify this we should check to make sure we have accounted for all possible values of X and have properly computed the probability of occurrence for each one. ■

This example also highlights the following. Even though the sample space consisted of equally likely outcomes, the probability of $\{X = a\}$ does not necessarily equal the probability of $\{X = b\}$. Many beginning students mistakenly feel

that, since there are four possible values for X, then each.
$\frac{1}{4}$. This example shows why this is not the case. Many time.
is made up of a union of a number of these equally likely out.

Example 3.2

An urn contains 6 white balls and 4 black balls. A ball is drawn wi.
replacement from the urn and its color noted. This is repeated until a white ball
is found. Let X = number of trials until a white ball is removed. The possible
values for X are the integers 1, 2, 3, 4, and 5. Again, this is a discrete random
variable. We then compute

$$P(X = 1) = P(\{W\}) = \frac{6}{10},$$

$$P(X = 2) = P(\{BW\}) = P(W \mid B)P(B) = \frac{6}{9} \cdot \frac{4}{10} = \frac{4}{15},$$

$$P(X = 3) = P(\{BBW\}) = \frac{6}{8} \cdot \frac{3}{9} \cdot \frac{4}{10} = \frac{1}{10},$$

$$P(X = 4) = P(\{BBBW\}) = \frac{6}{7} \cdot \frac{2}{8} \cdot \frac{3}{9} \cdot \frac{4}{10} = \frac{1}{35}, \quad \text{and}$$

$$P(X = 5) = P(\{BBBBW\}) = \frac{6}{6} \cdot \frac{1}{7} \cdot \frac{2}{8} \cdot \frac{3}{9} \cdot \frac{4}{10} = \frac{1}{210}.$$

Again we note that

$$P(X \leq 5) = P\left(\bigcup_{i=1}^{5}\{X = i\}\right)$$

$$= \sum_{i=1}^{5} P(X = i) = 1.$$

Again this is due to the fact that the only values of X that can occur are the
integers 1, 2, 3, 4, and 5. Therefore, the sample space is represented by these
integers, and the probability that X will assume one of these values must be 1. ∎

Example 3.3

In a lottery, 3 numbers are selected with replacement from the set $\{0, 1, 2, 3, 4,
5, 6, 7, 8, 9\}$. Let X = largest number chosen. In a properly designed lottery, each
of the 10 numbers is an equally likely outcome in each of the 3 trials, and the
trials are independent. If the largest number chosen is i, then all 3 numbers must
be less than or equal to i, and at least 1 number must be equal to i. Probably the
easiest way to count in this experiment is by using permutations. $\{X = i\}$ occurs
when 2 numbers are chosen with replacement from the set $\{0, 1, 2, \ldots, i\}$, and the
other is equal to i. This can happen in 1 of 3 ways. First, only 1 number is equal
to i and the other 2 are less than i. This can happen in $3 \cdot 1 \cdot (i - 1)^2$ ways. Second,
2 numbers can be equal to i and 1 less than i. This can happen in $3 \cdot 1 \cdot (i - 1)$
ways. Finally, all 3 numbers can be equal to i. This can happen in only 1 way.
Thus, we get that $\{X = i\}$ occurs in $3i^2 - 6i + 3 + 3i - 3 + 1 = 3i^2 - 3i + 1$ ways.

The number of ways to draw 3 numbers with replacement is $10^3 = 1000$. Thus,

$$P(X = i) = \frac{3i^2 - 3i + 1}{1000}, \quad \text{for } i = 0, 1, 2, 3, 4, 5, 6, 7, 8, 9.$$

Again we compute that

$$\sum_{i=0}^{9} P(X = i) = 1.$$

∎

Example 3.4

All the examples thus far have been discrete r.v.'s. We now consider a continuous one—the game spinner of Example 1.8 (see page 14). Let Y = radians counterclockwise from the reference point. Y can take any value in $[0, 2\pi)$. With a fair spinner we can assume that the probability that it will stop in a given sector is proportional to the number of radians subtended by that sector. That is, we can say that, given $[a, b] \subset [0, 2\pi)$,

$$P(a \leq Y \leq b) = k \cdot (b - a), \quad \text{where } k \text{ is a real constant.}$$

The value of k is determined by noting that, since the observed value for Y must be in $[0, 2\pi)$, $\{Y = 2\pi\}$ is impossible and

$$P(0 \leq Y \leq 2\pi) = P(\{0 \leq Y < 2\pi\} \cup \{Y = 2\pi\})$$
$$= P(0 \leq Y < 2\pi) + 0$$
$$= 1 = k \cdot (2\pi - 0) = k \cdot 2\pi.$$

Therefore,

$$k = \frac{1}{2\pi}.$$

From this we get that

$$P(a \leq Y \leq b) = \frac{b - a}{2\pi}, \quad \text{for } 0 \leq a \leq b < 2\pi.$$

The fact that Y only assumes values in the interval $[0, 2\pi)$ implies that, if $b \geq 2\pi$, then

$$P(a \leq Y \leq b) = P(a \leq Y < 2\pi).$$

Similarly, if $a < 0$, the fact that 0 is the smallest possible value for Y implies that

$$P(a \leq Y \leq b) = P(0 \leq Y \leq b).$$

∎

Exercises 3.1

1. A fair die is rolled once. Suppose we lose \$1 for each 1 rolled; win \$2 for each 2, 3, or 4 rolled; and gain or lose \$0 for each 5 or 6 rolled. Let X = winnings in dollars.

    a. What are the possible values for X?

    b. Find the probability associated with each possible value for X.

    c. What is the probability that we will lose money in this game?

2. An urn contains 5 balls numbered 1, 2, 3, 4, and 5. Two balls are selected without replacement from the urn. Let X = sum of the 2 balls, and let Y = difference of the 2 balls.

    a. What values can X take on?

    b. Find the probability for each possible value of X.

    c. What values can Y take on?

    d. Find the probability for each possible value of Y.

3. An elevator begins a trip at the basement floor with 5 people. The building has 3 floors, each of which is a possible stop. Assume that people get off at these 3 floors randomly—that is, any person is equally likely to get off at any floor. Let X = number of people that get off on the first floor.

    a. What values can X take on?

    b. Find the probability for each possible value of X.

4. Five digit codes are selected at random from the set $\{0, 1, 2, 3, 4, 5, 6, 7, 8, 9\}$ with replacement. Let the random variable X be the number of zeroes in a randomly chosen code.

    a. What are the possible values for X?

    b. Find $P(X = i)$ for each possible value of i.

5. A continuous random variable, X, can take on any positive real number. Find the possible values for the following functions of X.

    a. $\sqrt{X}$

    b. X^2

    c. $\ln X$

    d. $1 - e^{-X}$

3.2
The Distribution Function

The *distribution function (d.f.)* plays a central role in the study of random variables. This is because a distribution function can be defined for any random variable. This has the benefit that all properties of random variables can be stated in terms of values of the distribution function. In particular, the probability of any event can always be found by using the distribution function. We begin with a definition.

Definition 3.2

> Given a random variable X, the *distribution function*, $F_X(c)$ is
>
> $$F_X(c) = P(X \le c). \qquad (3.1)$$

It is important to note that the distribution function is a *probability*. In particular, it is the probability of a special event, namely that the random variable takes a value less than or equal to the number in the argument of the function. We begin by looking at the distribution function for a discrete random variable.

Example 3.5   Consider the coin tossing experiment in Example 3.1. Using the results derived there, it is easy to see that the following are true.

$$F_X(0) = P(X \le 0) = P(X = 0) = \frac{1}{8},$$

$$F_X(1) = P(X \le 1) = P(X = 0) + P(X = 1) = \frac{1}{2},$$

$$F_X(2) = P(X \le 2) = P(X = 0) + P(X = 1) + P(X = 2) = \frac{7}{8}, \quad \text{and}$$

$$F_X(3) = P(X \le 3) = \sum_{i=0}^{3} P(X = i) = 1.$$

Therefore, we see that

$$F_X(c) = \begin{cases} 0, & \text{if } c < 0, \\ \dfrac{1}{8}, & \text{if } 0 \le c < 1, \\ \dfrac{1}{2}, & \text{if } 1 \le c < 2, \\ \dfrac{7}{8}, & \text{if } 2 \le c < 3, \quad \text{and} \\ 1, & \text{if } 3 \le c. \end{cases}$$

We can now use this function to compute any probability regarding the value of X. For example, we see immediately that, since $\frac{5}{2}$ is between 2 and 3,

$$P\left(X \le \frac{5}{2}\right) = F_X\left(\frac{5}{2}\right) = \frac{7}{8}.$$

The graph of this d.f. is shown in Figure 3.1.

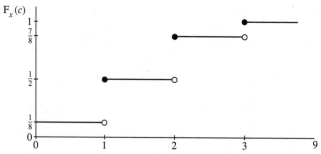

Figure 3.1

A look at the graph reveals the following. First, the graph is nondecreasing. It shows jumps at each point that is a possible value for X. In addition, the magnitude of the jump at the point $X = a$ is exactly equal to $P(X = a)$. These characteristics are common to the distribution function for all discrete random variables.   ■

We now look at the distribution for a continuous random variable. We are interested in two things. We are interested in what are the similarities with discrete random variables and, also, what are the differences.

Example 3.6   Recall the game spinner in Example 3.4. We found that

$$P(a \le Y \le b) = \frac{b - a}{2\pi}, \quad \text{for } 0 \le a \le b < 2\pi, \quad \text{and } 0 \text{ elsewhere.}$$

By using Definition 3.2 we find that the distribution function for Y is

$$F_Y(c) = P(Y \le c) = 0, \quad \text{if } c < 0,$$

$$F_Y(c) = P(Y \le c) = P(0 \le Y \le c) = \frac{c}{2\pi}, \quad \text{if } 0 \le c < 2\pi, \quad \text{and}$$

$$F_Y(c) = P(Y \le c) = P(0 \le Y < 2\pi) = 1, \quad \text{if } 2\pi \le c.$$

Therefore,

$$F_Y(c) = \begin{cases} 0, & \text{if } c < 0, \\ \dfrac{c}{2\pi}, & \text{if } 0 \le c < 2\pi, \quad \text{and} \\ 1, & \text{if } 2\pi \le c. \end{cases}$$

The graph of this distribution function is shown in Figure 3.2.

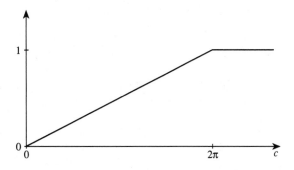

Figure 3.2

Note that this graph is also nondecreasing, but unlike Figure 3.1, it is continuous. This is where the term *continuous* comes from. It is characteristic of the d.f. for all continuous random variables.   ■

These two examples indicate that a distribution is a nondecreasing function that is defined for all real numbers. Also, since it is a probability it must range between

0 and 1. The following theorem gives some other properties of the distribution function. The proof is omitted.

Theorem 3.1

If $F_X(c)$ is the distribution function for the random variable X, then

1. $F_X(c)$ is a nondecreasing function. That is, if $a < b$, then $F_X(a) \le F_X(b)$.
2. $F_X(c)$ is right continuous. That is,

$$\lim_{a \to c^+} F_X(a) = F_X(c).$$

3. $\lim_{c \to -\infty} F_X(c) = 0.$
4. $\lim_{c \to \infty} F_X(c) = 1.$

We now wish to use the properties of the distribution function to calculate the probability of various events. Throughout this discussion we shall assume that $F_X(c)$ is available and we can find its value for any real number.

Case 1: $P(X \le a)$ It is clear from Definition 3.2 that we can calculate $P(X \le a)$ by simply evaluating it at $c = a$.

Case 2: $P(X < a)$ Using the distribution function to compute $P(X < a)$ requires more work. This is because Theorem 3.1 only guarantees that $F_X(c)$ is right continuous. This means that we cannot say that $P(X < a) = P(X \le a)$. The reader has seen ideas like this in calculus courses. Recall that it was not necessary that the left-hand limit at any point equal the right-hand limit. Therefore, to evaluate this probability we take the limit of the distribution from the left. That is,

$$P(X < a) = \lim_{c \to a^-} F_X(c). \tag{3.2}$$

Case 3: $P(X > a)$ By using the complement we note that $P(X > a) = 1 - P(X \le a)$. Then, by using Case 1 we see that

$$P(X > a) = 1 - F_X(a). \tag{3.3}$$

Case 4: $P(X \ge a)$ Again, using the complement we obtain that $P(X \ge a) = 1 - P(X < a)$. Then, from Case 2 we get

$$P(X \ge a) = 1 - \lim_{c \to a^-} F_X(c). \tag{3.4}$$

By using a combination of these four cases we can compute the probability associated with any interval of a random variable. The following theorem summarizes the results. The proof is left as an exercise.

Theorem 3.2

Let X be a random variable with distribution function $F_X(c)$. Then, for real numbers a and b,

1. $P(a < X \leq b) = F_X(b) - F_X(a)$, if $a < b$,
2. $P(a \leq X \leq b) = F_X(b) - \lim_{c \to a^-} F_X(c)$,
3. $P(a < X < b) = \lim_{c \to b^-} F_X(c) - F_X(a)$, and
4. $P(a \leq X < b) = \lim_{d \to b^-} F_X(d) - \lim_{c \to a^-} F_X(c)$.

Although these equations look imposing they are fairly easy to use in practice. First we note that if the endpoint of the interval is a point where the distribution function is continuous then $P(X < a) = P(X \leq a)$. This means that we can simply take the value of the distribution function at that point to obtain $P(X < a)$. If the point a where we need to take a limit is a point where the distribution function is discontinuous then the limit from the left will be the value of the distribution function evaluated at a^-. That is, if the distribution function is such that

$$F_X(c) = \begin{cases} g_1(c), & c < a, \quad \text{and} \\ g_2(c), & a \leq c, \end{cases}$$

then $\lim_{c \to a^-} F_X(c) = g_1(a)$. To make these ideas clear we now illustrate the use of the distribution to compute probabilities.

Example 3.7

Let the r.v. X have the following distribution function.

$$F_X(c) = \begin{cases} 0, & c < 0, \\ \dfrac{1}{4}, & 0 \leq c < 1, \\ \dfrac{1}{2}c, & 1 \leq c < 2, \quad \text{and} \\ 1, & 2 \leq c \end{cases}$$

The graph of the distribution function is shown in Figure 3.3.

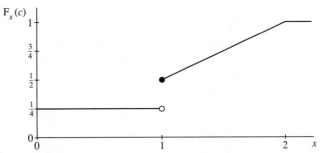

Figure 3.3

We wish to find the following probabilities.

1. $P(X < 1)$. Using Equation 3.2 we find that

$$P(X < 1) = \lim_{c \to 1^-} F_X(c) = \frac{1}{4}.$$

2. $P(X = 1)$. Note that $\{X \le 1\} = \{X < 1\} \cup \{X = 1\}$. Then using Part (1) of Theorem 3.2 we find

$$P(X = 1) = P(X \le 1) - P(X < 1)$$
$$= F_X(1) - \lim_{c \to 1^-} F_X(c)$$
$$= \frac{1}{2} - \frac{1}{4} = \frac{1}{4}.$$

3. $P(X > \frac{1}{2})$. In this case we use Equation 3.3 directly to get

$$P\left(X > \frac{1}{2}\right) = 1 - F_X\left(\frac{1}{2}\right) = \frac{3}{4}.$$

4. $P(X \le 3)$. We get this directly from Equation 3.1.

$$P(X \le 3) = F_X(3) = 1$$

5. $P\left(\frac{1}{2} < X \le \frac{3}{2}\right)$. From Part (1) of Theorem 3.2 we see that

$$P\left(\frac{1}{2} < X \le \frac{3}{2}\right) = F_X\left(\frac{3}{2}\right) - F_X\left(\frac{1}{2}\right)$$
$$= \frac{1}{2} \cdot \frac{3}{2} - \frac{1}{4} = \frac{1}{2}.$$

6. $P\left(X = \frac{3}{4}\right)$. By comparison with $P(X = 1)$ we find that

$$P\left(X = \frac{3}{4}\right) = F_X\left(\frac{3}{4}\right) - \lim_{c \to \frac{3}{4}^-} F_X\left(\frac{3}{4}\right)$$
$$= F_X\left(\frac{3}{4}\right) - F_X\left(\frac{3}{4}\right) = 0.$$

7. Find the value k such that $P(X \le k) = \frac{2}{3}$. This is different from the previous problems. Here we have been told the value of the distribution function and are asked to find the value of c that achieves it. If we look at Figure 3.3 we see that $F_X(c) = \frac{2}{3}$ occurs on that portion where $F_X(c)$ is of the form $F_X(c) = \frac{1}{2}c$. Thus, we know

$$\frac{1}{2}k = \frac{2}{3}, \quad \text{or}$$
$$k = \frac{4}{3}.$$

■

The answer to Part 6 of Example 3.7 illustrates a general property of all distribution functions. We state it as a theorem. The proof is left as an exercise.

Theorem 3.3

Let $F_X(c)$ be the distribution function of the random variable X. If $F_X(c)$ is continuous at $c = a$, then $P(X = a) = 0$.

Part 7 of Example 3.7 illustrates a type of question that is commonly asked of a distribution function. That is, given a probability of k find the value c of X such that

$$F_X(c) = k.$$

These values can be used to describe the shape of the distribution function. In practical applications, they can also be used to determine how rare it is to find outcomes in a given interval. These values for c are referred to as *quantiles*. We define them formally as follows.

Definition 3.3

The *kth quantile* of the distribution of the random variable X is denoted by γ_k and is that value of X such that

$$P(X < \gamma_k) \leq \frac{k}{100} \leq P(X \leq \gamma_k). \tag{3.5}$$

In terms of the distribution function, γ_k is that value of X such that

$$\lim_{c \to \gamma_k} F_X(c) \leq \frac{k}{100} \leq F_X(\gamma_k). \tag{3.6}$$

Thus, for example, the 20th quantile of the distribution of X would be denoted as γ_{20} and is that value such that the probability of being less than or equal to it is no more than 0.20 and the probability of being greater than or equal to it is no more than 0.80. Definition 3.3 also says that, in a very loose sense, the quantiles are values of the inverse of the distribution. That is, they are roughly those points where the value of the distribution function equals or "passes through" a given probability level.

Certain quantiles are used often enough to warrant receiving special names. γ_{50} divides the distribution in half and is known as the *median* of the distribution. It is used as a measure of the center of a distribution. γ_{25}, γ_{50}, and γ_{75} divide the distribution into quarters and are known as the *quartiles* of the distribution. γ_{25} is called the *first quartile*, and γ_{75} is called the *third quartile*. $\gamma_{10}, \gamma_{20}, \ldots, \gamma_{90}$ divide the distribution into tenths and are known as the *deciles* of the distribution. γ_{10} is called the *first decile*, γ_{20} is the *second decile*, and so on.

Example 3.8

For the distribution function of Example 3.7 we wish to find (1) the median, γ_{50}; (2) the first quartile, γ_{25}; and (3) the ninth decile, γ_{90}.

1. γ_{50}: By definition this is the value of X such that $F_X(\gamma_{50}) \leq 0.5$ and $1 - F_X(\gamma_{50}) \leq 0.5$. We note that $F_X(c) = 0.5$ when $c = 1$, $F_X(c) < 0.5$ when $c < 1$, and $F_X(c) > 0.5$ when $c > 1$. Thus, we see that $\gamma_{50} = 1$.

2. γ_{25}: This determination is less obvious because $F_X(c) = 0.25$ for $0 \leq c < 1$. Therefore, any value in $[0,1)$ can be the first quartile. This shows that the value of a quantile is not necessarily unique.

3. γ_{90}: The ninth decile is found in a straightforward manner. We see that $F_X(1.8) = 0.9$. In addition, the distribution function is continuous with a positive first derivative, which shows that it is increasing at $c = 1.8$. This implies that $F_X(c) < 0.9$ when $c < 1.8$ and $F_X(c) > 0.9$ when $c > 1.8$. Therefore, $\gamma_{90} = 1.8$.

∎

The last case shows a technique that can be used in general. Whenever the distribution has a positive first derivative in a neighborhood surrounding γ_k then that quantile will always occur at the point where the distribution function is such that $F_X(c) = \frac{k}{100}$.

Example 3.9

We wish to find the median value for the largest number drawn in the lottery of Example 3.3. From the values of $P(X = i)$ in that example, we find that the distribution function is

$$F_X(c) = \begin{cases} 0 & , c < 0, \\ \dfrac{1}{220} & , 0 \leq c < 1, \\ \dfrac{1}{55} & , 1 \leq c < 2, \\ \dfrac{1}{22} & , 2 \leq c < 3, \\ \dfrac{1}{11} & , 3 \leq c < 4, \\ \dfrac{7}{44} & , 4 \leq c < 5, \\ \dfrac{14}{55} & , 5 \leq c < 6, \\ \dfrac{21}{55} & , 6 \leq c < 7, \\ \dfrac{6}{11} & , 7 \leq c < 8, \\ \dfrac{3}{4} & , 8 \leq c < 9, \quad \text{and} \\ 1 & , 9 \leq c. \end{cases}$$

We note that $\frac{21}{55} < 0.5$ and $\frac{6}{11} > 0.5$. We also see that $\lim_{c \to 7^-} F_X(c) = \frac{21}{55} < 0.5$ and $F_X(7) = \frac{6}{11} > 0.5$. This means that the point $c = 7$ is the point that meets the specifications of the definition. Therefore, $\gamma_{50} = 7$. This illustrates why the definition needed to be so complicated. It is often necessary when dealing with discrete random variables. ■

Example 3.9 hints at an easy way to find any percentile. Note that the distribution function in that example was never exactly equal to a probability of 0.5. However, the median occurred at the value where the distribution function *passed through* a value of 0.5. As an illustration consider the first quartile for the distribution function of Example 3.9. If you follow the definition, it turns out that $\gamma_{25} = 5$. Note that this is precisely where the distribution function jumped from being less than 0.25 to being greater than or equal to 0.25.

Exercises 3.2

1. Find the distribution function for the random variable described in Exercise 1 of Section 3.1.

2. Find the distribution function for the random variables described in Exercise 2 of Section 3.1.

3. Find the distribution function for the random variable described in Exercise 3 of Section 3.1.

4. Find the distribution function for the random variable described in Exercise 4 of Section 3.1.

5. Prove Theorem 3.2.

6. Prove Theorem 3.3.

7. The distribution function of the random variable X is given below.

$$F_X(a) = \begin{cases} 0 & , a < 0, \\ \dfrac{a}{2} & , 0 \le a < 1, \\ \dfrac{3}{4} & , 1 \le a < 3, \\ \dfrac{3}{8} + \dfrac{a}{8} & , 3 \le a < 4, \quad \text{and} \\ 1 & , 4 \le a. \end{cases}$$

Find each of the following.
a. $P(X \le 1)$
b. $P(X > 3)$
c. $P\left(\frac{1}{2} \le X < 2\right)$
d. $P(5 < X \le 6)$
e. $P(X = 4)$
f. $P(X = 3)$

8. For each of the following either prove or give a counterexample.
a. $\lim_{a \to c} P(X \le a) = P(X \le c)$
b. $\lim_{a \to c} P(X < a) = P(X < c)$
c. $\lim_{a \to c^-} P(X > a) = P(X > c)$
d. $\lim_{a \to c^+} P(X > a) = P(X > c)$

9. Find the median and first and third quartiles for the distribution function derived in Exercise 1 of Section 3.1.

10. Find the median and first and third quartiles for the distribution function derived in Exercise 2 of Section 3.1.

11. Find the median and first and third quartiles for the distribution function derived in Exercise 3 of Section 3.1.

12. Find the median and first and third quartiles for the distribution function derived in Exercise 4 of Section 3.1.

13. Use the distribution function given in Exercise 6 of Section 3.1 to find the following percentiles.

    a. γ_{50}

    b. γ_{25}

    c. γ_{75}

    d. γ_{90}

3.3
Discrete Random Variables

In the introduction we mentioned that we will be categorizing random variables according to whether they are discrete or continuous. In the next two sections we shall discuss both types. We first begin with a formal definition of a discrete random variable.

Definition 3.4

> A random variable is said to be *discrete* if it assigns positive probability to, at most, a countable number of values.

This definition implies that experiments that are described by discrete random variables have sample spaces that contain, at most, a countable number of elementary events. In addition, it implies that a probability can be associated with each possible value of a discrete random variable. The values of these probabilities are not necessarily distinct. This probability is commonly referred to as the *probability function (p.f.)* of the random variable.

Definition 3.5

> If X is a discrete random variable with distribution function $F_X(c)$, the *probability function of X* is denoted by $p_X(c)$ and is
>
> $$p_X(c) = P(X = c). \tag{3.7}$$

The probability function is sometimes referred to as the *probability mass function*, the *discrete density function*, or the *probability density function*. We prefer the term probability function because it describes exactly what it is—a function whose value is the probability of occurrence of the event $\{X = c\}$. The probability function can be written in terms of the distribution function as follows.

$$p_X(c) = F_X(c) - \lim_{a \to c^-} F_X(a) \tag{3.8}$$

Suppose that the random variable X has positive probability at the points $c_1, c_2, \ldots$. That is, $p_X(c) > 0$, for $c = c_1, c_2, \ldots$, and $p_X(c) = 0$, otherwise. We can use the probability function to compute the distribution function by

$$F_X(a) = P(X \le a) = \sum_{c_i \le a} p_X(c_i). \tag{3.9}$$

The notation $\sum_{c_i \le a}$ represents the sum over the c_i's that are less than or equal to a. Since a random variable must assume one of its possible values,

$$\sum_{c_i} p_X(c_i) = 1. \tag{3.10}$$

Here, the notation $\sum_{c_i}$ represents summation over all c_i, where $p_x(c_i) > 0$. We have seen instances of Equation 3.10 in Example 3.1, Example 3.2, and Example 3.3.

We can use the probability function to compute the probability that a random variable will lie within a given interval as follows. Assume that a discrete random variable X can give a positive probability to a and b, where $a < b$. Then

$$P(a \le X \le b) = \sum_{c=a}^{b} p_X(c). \tag{3.11}$$

Furthermore, if X does not assign positive probability to any values in the open interval (a, b), then

$$p_X(b) = F_X(b) - F_X(a). \tag{3.12}$$

This is a consequence of the fact that $\lim_{x \to b^-} F_X(x) = F_X(a)$.

Example 3.10

Let X be a discrete r.v. that can assume any value in the set $\{1, 2, 3, 4, 5\}$, and let the p.f. be

$$p_x(c) = \begin{cases} \dfrac{1}{k}, & c = 1, 2, 3, 4, 5, \quad \text{and} \\ 0, & \text{otherwise.} \end{cases}$$

We wish to do the following.

1. Find the value of k to make $p_X(c)$ a probability function.
2. Find $P(X = 3)$.
3. Find $P(X > 3)$.
4. Find the d.f. $F_X(c)$.
5. Find the median of the distribution of X.

These will illustrate the ideas we have been discussing up to now. In computing the above, we have the following.

1. If $p_X(c)$ is a p.f., then we know from Equation 3.9 that

$$1 = \sum_{c=1}^{5} \frac{1}{k} = \frac{5}{k}.$$

Thus, $k = 5$ and the p.f. is

$$p_X(c) = \begin{cases} \dfrac{1}{5}, & c = 1, 2, 3, 4, 5, \quad \text{and} \\ 0, & \text{otherwise.} \end{cases}$$

2. Using the p.f. we find that

$$P(X = 3) = p_X(3) = \frac{1}{5}.$$

3. Since X assigns positive probability to only the values in the set $\{1, 2, 3, 4, 5\}$, the event $\{X > 3\}$ is equivalent to the event $\{4 \le X \le 5\}$. Therefore, we can use Equation 3.11 to get

$$P(X > 3) = P(4 \le X \le 5) = p_X(4) + P_X(5) = \frac{2}{5}.$$

4. To compute the distribution function we need only look at those points where the probability function is greater than zero. Therefore, Equation 3.9 gives

$$F_X(c) = \begin{cases} 0, & c < 1, \\ \dfrac{1}{5}, & 1 \le c < 2, \\ \dfrac{2}{5}, & 2 \le c < 3, \\ \dfrac{3}{5}, & 3 \le c < 4, \\ \dfrac{4}{5}, & 4 \le c < 5, \quad \text{and} \\ 1, & 5 \le c. \end{cases}$$

The graph of this d.f. is shown in Figure 3.4.

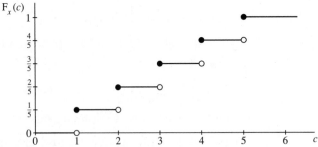

Figure 3.4

This graph is typical for the distribution function of any discrete random variable. It increases in a series of steps. The jumps occur at the values where the probability function is positive, and the size of each jump is the value of the probability function at that point.

5. By using the d.f. we see that $P(X < 3) < 0.5$ and $P(X \le 3) > 0.5$. Since the d.f. "passes through" 0.5 when $c = 3$ we find that $\gamma_{50} = 3$.

■

Example 3.11   Let the r.v. Y be defined on all the positive integers. Let p be a value such that $0 < p < 1$, and define the p.f. as follows.

$$p_Y(c) = \begin{cases} kp^c, & c = 1, 2, 3, \ldots, \quad \text{and} \\ 0, & \text{otherwise.} \end{cases}$$

Here Y can assume a countable number of values. We wish to do the following.

1. Find the value of k to make $p_Y(c)$ a valid probability function.
2. Find $P(2 \le Y \le 4)$.
3. Find $P(Y \ge 3)$.

   We shall take them in order.

1. To find k we begin with Equation 3.10.

$$1 = \sum_{c=1}^{\infty} kp^c = k \sum_{c=1}^{\infty} p^c$$

Since $0 < p < 1$ this would be the sum of a geometric series if the summation began at $c = 0$ (see Example 2.20). We can achieve this by adding and subtracting that term as follows.

$$\sum_{c=1}^{\infty} p^c = \sum_{c=0}^{\infty} p^c - p^0.$$

Therefore, we get

$$1 = k \left(\sum_{c=0}^{\infty} p^c - 1 \right)$$

$$= k \left(\frac{1}{1 - p} - 1 \right)$$

$$= k \frac{p}{1 - p}.$$

Therefore, $k = \frac{1-p}{p}$, and the p.f. is

$$p_Y(c) = \begin{cases} (1 - p)p^{c-1}, & c = 1, 2, 3, \ldots, \quad \text{and} \\ 0, & \text{otherwise.} \end{cases}$$

2. Equation 3.11 gives

$$P(2 \le Y \le 4) = p_X(2) + p_X(3) + p_X(4)$$

$$= (1 - p)(p + p^2 + p^3).$$

3. If we look at the complement of $\{Y \geq 3\}$ we find that

$$P(Y \geq 3) = 1 - P(Y < 3) = 1 - P(Y \leq 2).$$

This last equality is due to the fact that Y does not assign positive probability to any value in (2,3). We get

$$P(Y \geq 3) = 1 - (1 - p)(1 + p) = p^2.$$

∎

Exercises 3.3

1. A box contains 4 red chips and 3 black chips. Three chips are removed from the box without replacement. Let X = number of red chips selected.
   a. Find the probability function for X.
   b. Find $P(X = 0)$.
   c. Find $P(X > 1)$.
   d. Find the median value for X.

2. A bowl contains 10 balls numbered 1 through 10. Four balls are taken from the bowl without replacement. Let X = number of even numbers selected.
   a. Find the probability function for X.
   b. Find $P(X = 3)$.
   c. Find the distribution function for X.
   d. Find the median value for X.

3. A bowl contains 10 balls numbered 1 through 10. Three balls are taken from the bowl. Let X = the highest number drawn. Find the probability function for X for each of the following cases.
   a. Sampling is done without replacement.
   b. Sampling is done with replacement.

4. In a certain carnival game, a beanbag is thrown at a small opening. A player begins with 3 beanbags. Suppose that a player has a probability of 0.4 of hitting the opening on any single toss of a beanbag. The game stops when a player either hits the opening or runs out of beanbags. Let X be the number of tosses in a single play of the game. Find the probability function for X.

5. A random variable can assume any number from the set $\{0,1,2,3,4,5,6,7,8,9\}$. It has a probability function of the form

$$p_X(c) = \begin{cases} kc, & c = 0, 1, 2, 3, 4, 5, 6, 7, 8, 9, \quad \text{and} \\ 0, & \text{otherwise.} \end{cases}$$

   a. Find the value for k to make this a valid probability function for X.
   b. Find $P(X = 3)$.
   c. Find P when X is even.
   d. Find the distribution function for X.
   e. Find the median value for X.

6. A discrete random variable X can assume any value from the nonnegative integers $\{0, 1, 2, \ldots\}$. It has a p.f. of the form

$$p_X(c) = \begin{cases} \dfrac{k2^c}{c!}, & c = 0, 1, 2, 3, \ldots, \quad \text{and} \\ 0, & \text{otherwise.} \end{cases}$$

a. Use the Taylor series for e^x to find the value of k to make $p_X(c)$ a valid probability function.

b. Find $P(X = 0)$.

c. Find $P(X \geq 2)$.

d. For what value of c is $p_X(c)$ a maximum?

e. Find the median of the distribution for X.

f. Sketch the graph of the distribution function for $0 \leq X \leq 5$.

7. A word game uses 100 tiles. Forty-five tiles contain vowels and 55 tiles contain consonants. Five tiles are drawn without replacement from the 100 tiles. Let X be the difference between the number of vowels and the number of consonants.

a. Find the probability function for X.

b. What is the probability that the number of vowels selected will be greater than the number of consonants?

8. The squares on a tic-tac-toe board are numbered one through nine. Chips numbered one through nine are placed in a bowl. A chip is drawn from the bowl with replacement and the square on the tic-tac-toe board with that number is marked with an X. This process is repeated until a chip containing the number of an already marked square is removed. Let Y = number of times a chip is removed from the bowl.

a. Find the probability function for Y.

b. What is the probability that we will stop removing chips before one-half of the tic-tac-toe board has been marked?

9. Suppose that a random variable X has the following probability function.

a	-1	0	1	2	4	6	8
$p_X(a)$	0.10	0.15	0.20	0.15	0.10	0.15	0.05

a. Find the probability X will be odd.

b. Find the probability X will be positive.

c. Find $P(X \geq 2 \mid X > 0)$.

10. In a certain game of chance the probability of winning on any given play is 0.4. A gambler begins with a stake of \$127 and adopts the following strategy.

i. Begin by betting \$1. If he wins stop playing.

ii. Otherwise double the bet and play again.

iii. Proceed in this fashion until he either wins or runs out of money.
Let X be the number of times the gambler plays the game.

a. Find the probability function for X.

b. What is the probability that the gambler will not lose all of his money?

11. In the personnel committee of a government agency it takes a vote of 9 of the 12 members to promote an employee. Suppose that each committee member will vote to promote a qualified person with probability 0.8 and will vote to promote an unqualified person with probability 0.3. Suppose that each committee member acts independently and that 75% of the employees are qualified for promotion. What is the probability that the committee will correctly promote or fail to promote an employee?

3.4
Continuous Random Variables

As far as we are concerned a *continuous random variable* is one that is not discrete. For most continuous random variables of practical interest, this implies that the distribution function of a continuous random variable is continuous and nondecreasing. This is in contrast to the distribution function of a discrete random variable which has discontinuities at those values where the probability function is positive. In calculus we learned that the definite integral of a piecewise continuous function is continuous. This fact is the basis for the formal definition of a continuous random variable.

Definition 3.6

A random variable X is called a *continuous random variable* if there exists a function $f_X(x)$ such that

1. $f_X(x) \geq 0$ for $-\infty < x < \infty$, and
2. For any set A of real numbers,

$$P(X \in A) = \int_A f_X(x)\,dx. \tag{3.13}$$

The function $f_X(x)$ is called the *probability density function of X*.

It should be pointed out that, in applications, experimental observations are never truly continuous. This is because experimental data are collected by taking measurements with some kind of instrument (chronometers, scales, and so on). Any instrument has a limit to the accuracy of its measurements. This has the effect of making a set of measurements taken on an object, which produces continuous data, become discrete. When dealing with such data this fact is usually ignored, and the observations are treated as being continuous. This is because approximations of discrete data by continuous random variables have practical advantages. They are easier to work with, and the approximations are quite accurate.

The probability density function (p.d.f.) of a continuous random variable is the continuous counterpart of the probability function of a discrete random variable. The nonnegativity requirement ensures that probabilities calculated using Equation 3.13 will satisfy Axiom 1 of Definition 1.8. Another consequence of the definition is that, since Equation 3.13 must be true for any set A of real numbers, all improper integrals used for computing probabilities must converge. Since random variables map the sample space onto the real numbers it must be true that

$$\int_{-\infty}^{\infty} f_X(x)\,dx = 1. \tag{3.14}$$

Other properties of the probability density function are given in the following theorem.

Theorem 3.4

> Let X be a continuous random variable with distribution function $F_X(c)$ and probability density function $f_X(x)$. Then the following are true.
>
> 1. $P(a \le X \le b) = \int_a^b f_X(x)\,dx$.
> 2. $P(X = a) = 0$.
> 3. $F_X(c) = \int_{-\infty}^c f_X(x)\,dx$.
> 4. If $f_X(c)$ is continuous, then $\frac{d}{dc}F_X(c) = f_X(c)$.

Proof

1. This is an immediate consequence of the definition of a continuous random variable.

$$P(X \in [a,b]) = P(a \le X \le b)$$

$$= \int_a^b f_X(x)\,dx$$

2. If we let $a = b$ in Part 1 of this theorem and use the fact from calculus that $\int_a^a f(x)\,dx = 0$ we see that

$$P(X = a) = P(a \le X \le a) = \int_a^a f_X(x)\,dx = 0.$$

3. By combining Definition 3.2 and Definition 3.6 we find immediately that

$$F_X(c) = P(X \le c) = P(X \in (-\infty, c]) = \int_{-\infty}^c f_X(x)\,dx.$$

4. Recall from calculus that there were two parts to the Fundamental Theorem of Calculus. The more familiar part demonstrated the relation between definite integrals and antiderivatives. The other part talked about taking derivatives of definite integrals with respect to the upper limit of integration. The proof of this part is an immediate consequence of an application of this other part of the Fundamental theorem. From Part 1 we know that

$$F_X(c) = \int_{-\infty}^c f_X(x)\,dx.$$

We now differentiate $F_X(c)$ with respect to c and use the Fundamental Theorem of Calculus as follows.

$$\frac{d}{dc} F_X(c) = \frac{d}{dc} \int_{-\infty}^{c} f_X(x)\, dx = f_X(c)$$

■

A consequence of Part (2) of Theorem 4 is that, for any continuous random variable,

$$P(a < X \leq b) = P(a \leq X \leq b) = P(a \leq X < b) = P(a < X < b). \qquad \textbf{(3.15)}$$

We now illustrate the ideas discussed in this section with some examples.

Example 3.12   Let X be a continuous r.v. whose p.d.f. is of the form

$$f_X(x) = \begin{cases} kx & , 0 \leq x \leq 2, \\ k(4 - x), & 2 < x \leq 4, \quad \text{and} \\ 0 & , \text{otherwise.} \end{cases}$$

We wish to do the following.

1. Find the value of k to make $f_X(x)$ a valid probability density function.
2. Find $F_X(c)$.
3. Find $P(1 < X < 5)$.
4. Find $P(X \geq 3)$.
5. Find the median of the distribution of X.

Solutions include the following.

1. Using Equation 3.14 we get that

$$1 = \int_{-\infty}^{\infty} f_X(x)\, dx$$

$$= \int_{-\infty}^{0} 0\, dx + \int_{0}^{2} kx\, dx + \int_{2}^{4} k(4 - x)\, dx + \int_{4}^{\infty} 0\, dx$$

$$= \frac{1}{2} kx^2 \Big|_{0}^{2} + \left(4kx - \frac{1}{2}kx^2\right) \Big|_{2}^{4} = 4k.$$

Therefore, $k = \frac{1}{4}$, and the p.d.f. is

$$f_X(x) = \begin{cases} \dfrac{1}{4}x & , 0 \leq x \leq 2, \\ 1 - \dfrac{1}{4}x, & 2 < x \leq 4, \quad \text{and} \\ 0 & , \text{otherwise.} \end{cases}$$

2. We use Theorem 3.4 to obtain the d.f. as follows.

   a. For $c < 0$:

$$F_X(c) = \int_{-\infty}^{c} 0\,dx = 0.$$

   b. For $0 \le c \le 2$:

$$F_X(c) = P(X < 0) + P(0 \le X \le c)$$

$$= 0 + \int_0^c \frac{1}{4} x\,dx$$

$$= \frac{c^2}{8}.$$

   c. For $2 < c \le 4$:

$$F_X(c) = P(X \le 2) + P(2 < X \le c)$$

$$= \frac{1}{2} + \int_2^c (1 - \frac{1}{4}x)\,dx$$

$$= \frac{1}{2} + c - \frac{c^2}{8} - 2 + \frac{1}{2}$$

$$= -1 + c - \frac{c^2}{8}.$$

   d. For $4 < c$:

$$F_X(c) = P(X \le 4) + P(4 < X < c)$$

$$= 1 + \int_4^c 0\,dx = 1.$$

   In summary,

$$F_X(c) = \begin{cases} 0 & , c < 0, \\[2mm] \dfrac{c^2}{8} & , 0 \le c \le 2, \\[4mm] -1 + c - \dfrac{c^2}{8} & , 2 < c \le 4, \quad \text{and} \\[4mm] 1 & , 4 < c. \end{cases}$$

The graph of the d.f. is given in Figure 3.5.

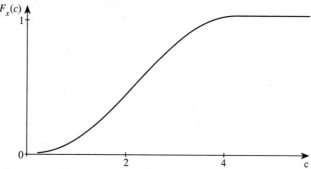

Figure 3.5

3. Equation 3.15 and Part 1 of Theorem 3.4 give that

$$P(1 < X < 5) = P(1 \le X \le 5) = \int_1^5 f_X(x)\,dx$$

$$= \int_1^2 \frac{1}{4}x\,dx + \int_2^4 (1 - \frac{1}{4}x)\,dx + \int_4^5 0\,dx$$

$$= \frac{3}{8} + \frac{1}{2} + 0 = \frac{7}{8}.$$

4. We can use the distribution function and Equation 3.15 as follows.

$$P(X \ge 3) = 1 - P(X < 3) = 1 - P(X \le 3)$$

$$= 1 - F_X(3) = 1 - \frac{7}{8} = \frac{1}{8}$$

5. Since the d.f. for X is continuous and strictly increasing in the region around where $F_X(c) = 0.5$ the median will occur uniquely at that point. By scanning the d.f. we find that $F_X(c) = 0.5$ occurs when $c = 2$. Therefore, $\gamma_{50} = 2$.   ∎

Example 3.13   It has been found that, until a component fails, the time, in hours, can be modeled by a continuous random variable T with a p.d.f. of the form

$$f_T(t) = \begin{cases} \lambda e^{-\lambda t}, & 0 < t, \quad \text{and} \\ 0, & \text{otherwise.} \end{cases}$$

A random variable with this p.d.f. is said to have an *exponential distribution*. Here, $\frac{1}{\lambda}$ is the mean time to failure for a component. Experience shows that, on average, half of all components of a certain type fail within 5 hours.

1. What is the probability that one of these components will fail within 10 hours?

2. If a machine has 4 of these components working independently, what is the probability that exactly 2 of them will fail within 10 hours?

Solutions include the following.

1. The first thing we need to do is to determine the value of λ. The fact that half of all components fail within 5 hours makes a statement about a long-term proportion that can be interpreted as a probability. Therefore, we know that $P(T \le 5) = \frac{1}{2}$. We then use this and Part 1 of Theorem 3.4 to solve for λ as follows.

$$\frac{1}{2} = \int_0^5 \lambda e^{-\lambda x}\,dx$$

$$= -e^{-\lambda x}\big|_0^5 = -e^{-5\lambda} - [-e^0]$$

$$= 1 - e^{-5\lambda}$$

Thus,

$$\lambda = \frac{\ln 2}{5}.$$

We now compute $P(T \le 10)$ by using Equation 3.13 as follows.

$$P(T \le 10) = \int_0^{10} \frac{\ln 2}{5} e^{-t\frac{\ln 2}{5}} \, dt$$

$$= 1 - e^{2\ln 2} = 1 - \frac{1}{4} = \frac{3}{4}$$

2. Since we are told that the 4 components function independently we can view them as constituting 4 identical independent trials. From Part 1 we know that the probability that any single component will fail within 10 hours is $\frac{3}{4}$. Let X = number of components failing within 10 hours, and

$$p = P(\text{single component fails within 10 hours})$$

$$= P(T \le 10) = \frac{3}{4}, \quad \text{and}$$

$$q = P(\text{single component does not fail within 10 hours})$$

$$= \frac{1}{4}.$$

We wish to compute $P(X = 2)$. If 2 components fail and 2 do not this can happen in $\binom{4}{2}$ different ways.

{fail, fail, not fail, not fail} {fail, not fail, fail, not fail}
{fail, not fail, not fail, fail} {not fail, fail, fail, not fail}
{not fail, fail, not fail, fail} {not fail, not fail, fail, fail}

Each of these 6 possible arrangements of 2 components failing and 2 components not failing will have a probability of p^2q^2. Therefore,

$$P(X = 2) = \binom{4}{2} p^2 q^2$$

$$= 6 \cdot \left(\frac{3}{4}\right)^2 \left(\frac{1}{4}\right)^2$$

$$= \frac{27}{128}.$$

Notice that this problem involved both a discrete and continuous random variable. We used the continuous random variable T to compute the probability that each of the 4 independent trials would show a failure within 10 hours. The discrete random variable X was then used to compute the probability of having this event occur in exactly 2 of the 4 trials. This kind of technique shows up quite often in studying system reliability. ∎

Exercises 3.4

1. Which of the following could be probability density functions?
   a. $f(x) = \frac{1}{2}$, for $1 < x < 3$, and 0, otherwise.
   b. $f(x) = \frac{2}{3}(x^2 - 3x + 2)$, for $0 < x < 3$, and 0, otherwise.

   c. $f(x) = xe^{-x}$, for $0 < x$, and 0, otherwise.

   d. $f(x) = \frac{1}{x}$, for $1 < x < e$, and 0, otherwise.

2. For each of the following probability density functions compute $F_X(c)$, $P(0 < X < 1)$, $P(X^2 < 4)$, and $P(-1 < 3X)$.

   a. $f_X(x) = \frac{1}{4}$, if $0 < x < 4$, and 0, otherwise.

   b. $f_X(x) = |x|$, if $-1 < x < 1$, and 0 otherwise.

   c. $f_X(x) = e^{-(x+2)}$, for $-2 < x$, and 0, otherwise.

3. A continuous r.v. is said to have a *Cauchy distribution* if it has a p.d.f. of the form

$$f_X(x) = \frac{1}{\pi(1 + x^2)}, \quad -\infty < x < \infty.$$

   a. Show that this is a valid probability density function.

   b. Find the distribution function for X.

   c. Find $P(|X| \leq 1)$.

   d. Find γ_{72}.

4. A continuous random variable X has a p.d.f. of the form

$$f_X(x) = \begin{cases} kx^3, & 0 \leq x \leq 1, \quad \text{and} \\ 0, & \text{otherwise.} \end{cases}$$

   a. Find k to make this a valid probability density function.

   b. Find and sketch the d.f. for X.

   c. Find $P(0.25 < X < 0.75)$.

   d. Find a so that $P(X \geq a) = 0.1$.

5. The p.d.f. of the continuous random variable X is of the form

$$f_X(x) = \begin{cases} \dfrac{k}{\sqrt{x^3}}, & 1 < x, \quad \text{and} \\ 0, & \text{otherwise.} \end{cases}$$

   a. Find the value of k to make $f_X(x)$ a valid probability density function.

   b. Sketch the graph of $F_X(c)$.

   c. Find $P(0 \leq X < 4)$.

   d. Find $P(X \geq 5)$.

6. The diameter of a circle D is a random variable whose p.d.f. is

$$f_D(d) = \begin{cases} 1, & 1 < d < 2, \quad \text{and} \\ 0, & \text{otherwise.} \end{cases}$$

Find the probability that the area of the circle will be more than 4.

7. It has been found that personal income (in tens of thousands of dollars) in a certain group can be modeled as a continuous r.v. with a p.d.f. of the form

$$f_X(x) = \begin{cases} \dfrac{2}{x^3}, & k \leq x, \quad \text{and} \\ 0, & \text{otherwise.} \end{cases}$$

A random variable with this p.d.f. is said to have a *Pareto distribution.* Note that X is income in tens of thousands of dollars. Wage earners are drawn from a population with replacement.

   a. What is the probability that a given wage earner in this group will earn more than $20,000?

   b. If 4 wage earners are selected from this group, what is the probability that at least one of them will earn more than $20,000?

8. If X is a continuous r.v., prove that

$$P(a < X \le b) = P(a \le X \le b) = P(a \le X < b) = P(a < X < b).$$

9. A point is chosen at random on a line of length L. Let X = the distance from the left end of the line to the point. Then the p.d.f. of X is

$$f_X(x) = \begin{cases} \dfrac{1}{L}, & 0 \le x \le L, \quad \text{and} \\ 0, & \text{otherwise.} \end{cases}$$

Find the probability that the ratio of the distance of X from the left end to the distance from the right end of the line segment is at least 2.

10. Find the distribution function for the random variable whose probability density function is given in Exercise 3.7.

11. The lifetime X of a certain component is a continuous random variable with a distribution function of

$$F_X(a) = \begin{cases} 0, & a < 0, \quad \text{and} \\ 1 - (1 + a)e^{-a}, & 0 \le a. \end{cases}$$

   a. Verify that this is a valid distribution function.

   b. Find the probability density function, $f_X(x)$.

   c. Find $P(1 < X < 2)$.

12. Let X be a random variable with a probability density function of

$$f_X(x) = \begin{cases} \dfrac{1}{4}, & 0 \le x \le 4, \quad \text{and} \\ 0, & \text{otherwise.} \end{cases}$$

Find the probability that the graph of the parabola $y^2 + 1 - X$ touches or crosses the horizontal axis.

3.5
Random
Vectors

There are times when an observer wishes to keep track of more than one quantity in a random experiment. In industrial quality control, for example, it may be necessary to measure both length and diameter of a product to determine if the output is within specifications. Here it would be appropriate to define two continuous random variables, X = diameter and Y = length. We would then be interested in events of the form $\{\{a \le X \le b\} \cap \{c \le Y \le d\}\}$. It is common practice to write an event such as this in the form

$$\{a \le X \le b,\ c \le Y \le d\}.$$

The comma is understood to be equivalent to "and." The values a and b would be tolerance limits on diameter and c and d would be tolerance limits for length.

As another example, a psychologist interested in analyzing the effects of violence in television programming might wish to measure the age and sex of a subject as well as reactions to viewing programs containing violence. Here we would use three random variables such as $S =$ sex, $A =$ age, and $R =$ reaction. S is a discrete random variable. Even though ages are usually rounded to the nearest year many researchers would treat A as a continuous random variable. If ages are grouped such as < 5, 5–9, 10–14, and so forth, then A would be used as a discrete random variable. Depending on the measurement scale used, R could be either discrete or continuous.

In each of these examples, we see that each experimental unit has more than one random variable defined on it. Thus, when the random experiment is conducted we are actually observing the value of a *random vector*. In the quality control example we would be observing a value for the random vector (X, Y), and in the psychology example each child would give an observation for the random vector (S, A, R).

Since random variables are functions from the sample space to the real numbers R^1 the natural extension to random vectors is the following.

Definition 3.7

> A *random vector with n components* is a random experiment whose sample space consists of points in n-dimensional Euclidean space, R^n.

Proceeding as we did with random variables the starting point for discussing random vectors is the n-dimensional version of the distribution function. It is called the *joint distribution function*. In order to keep the notation relatively simple we shall begin by considering vectors with two components.

Definition 3.8

> Let (X, Y) be a random vector defined on a sample space S. Given any two real numbers, a and b, the *joint distribution function* for (X, Y) is
>
> $$F_{X,Y}(a, b) = P(X \le a,\ Y \le b). \qquad (3.16)$$

This definition makes no requirement that both random variables be discrete or continuous. Mixtures are permitted, but we will not use them in this book. As before, the comma is shorthand notation for an intersection. The joint distribution function is the probability that an observation of the ordered pair (X, Y) will fall in the quarter plane $\{X \le a\} \cap \{Y \le b\}$. The properties of the joint distribution function are the natural extensions of those for univariate distribution functions given in Theorem 3.1. The proof is omitted.

Theorem 3.5

Let $F_{X,Y}(a,b)$ be the joint distribution function for the random vector (X,Y). Then the following are true.

1. If $a \leq c$ and $b \leq d$, then $F_{X,Y}(a,b) \leq F_{X,Y}(c,d)$.

2. $\lim_{\substack{a \to c^+ \\ b \to d^+}} F_{X,Y}(a,b) = F_{X,Y}(c,d)$.

3. $\lim_{a \to -\infty} F_{X,Y}(a,b) = \lim_{b \to -\infty} F_{X,Y}(a,b) = 0$.

4. $\lim_{\substack{a \to \infty \\ b \to \infty}} F_{X,Y}(a,b) = 1$.

5. $\lim_{a \to \infty} F_{X,Y}(a,b) = F_Y(b)$, and $\lim_{b \to \infty} F_{X,Y}(a,b) = F_X(a)$, where $F_X(a)$ and $F_Y(b)$ are the distribution functions of X and Y, respectively.

A common notation for $\lim_{b \to \infty} F_{X,Y}(a,b)$ is $F_{X,Y}(a,\infty)$. Similarly, $F_{X,Y}(\infty,b)$ denotes $\lim_{a \to \infty} F_{X,Y}(a,b)$.

When one obtains the distribution function of a single random variable from the joint distribution function by using Part 5 of Theorem 3.5, the term *marginal distribution* is commonly used. This sometimes confuses students by making them think there are two different distribution functions for a random variable—the ordinary distribution function of Section 3.2 and the marginal one of this section. This is not the case. The distribution function for, say, X from Part 5 is the same as we would have gotten in Section 3.2 if we took the original sample space and never defined Y on it. Remember the axioms of probability ensure that there is only one value for $P(X \leq a)$.

Any probabilities involving (X,Y) can be answered by using the joint distribution function. One useful result called the *Rectangle rule* is

$$P(a < X \leq b, c < Y \leq d) = F_{X,Y}(b,d) - F_{X,Y}(a,d) - F_{X,Y}(b,c) + F_{X,Y}(a,c). \tag{3.17}$$

The proof is left as an exercise.

Example 3.14

We toss a fair coin three times. Let X = number of tails, and Y = number of tails on the first toss. It is difficult to find a nice way to write the joint distribution function. We shall, therefore, evaluate it at those points that represent possible outcomes of the experiment. Those are $(0,0)$, $(1,0)$, $(2,0)$, $(1,1)$, $(2,1)$, and $(3,1)$. We evaluate the probability associated with each point as follows.

(a, b)	$P(X = a, Y = b)$
$(0,0)$	$\frac{1}{8}$
$(1,0)$	$\frac{1}{4}$
$(2,0)$	$\frac{1}{8}$
$(1,1)$	$\frac{1}{8}$
$(2,1)$	$\frac{1}{4}$
$(3,1)$	$\frac{1}{8}$

A graph of this is shown in Figure 3.6.

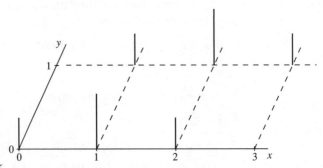

Figure 3.6

We than calculate the following values for the joint distribution function.

(a, b)	$F_{X, Y}(a, b)$
$(0,0)$	$\frac{1}{8}$
$(1,0)$	$\frac{3}{8}$
$(2,0)$	$\frac{1}{2}$
$(3,0)$	$\frac{1}{2}$
$(0,1)$	$\frac{1}{8}$
$(1,1)$	$\frac{1}{2}$
$(2,1)$	$\frac{7}{8}$
$(3,1)$	1

These were obtained as follows.

$$F_{X,Y}(0,0) = P(X = 0, Y = 0),$$

$$F_{X,Y}(1,0) = P(X = 0, Y = 0) + P(X = 1, Y = 0),$$

$$F_{X,Y}(2,0) = P(X = 0, Y = 0) + P(X = 1, Y = 0) + P(X = 2, Y = 0),$$

$$F_{X,Y}(3,0) = F_{X,Y}(2,0),$$

$$F_{X,Y}(0,1) = F_{X,Y}(0,0),$$

$$F_{X,Y}(1,1) = P(X = 0, Y = 0) + P(X = 1, Y = 0) + P(X = 1, Y = 1),$$

$$F_{X,Y}(2,1) = P(X = 0, Y = 0) + P(X = 1, Y = 0) + P(X = 2, Y = 0)$$
$$+ P(X = 1, Y = 1) + P(X = 2, Y = 1), \quad \text{and}$$

$$F_{X,Y}(3,1) = P(X = 0, Y = 0) + P(X = 1, Y = 0) + P(X = 2, Y = 0)$$
$$+ P(X = 1, Y = 1) + P(X = 2, Y = 1) + P(X = 3, Y = 1)$$

A graph of this distribution function is given in Figure 3.7.

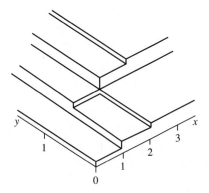

Figure 3.7

Note that the distribution function for X can be obtained by noting that

$$F_X(a) = F_{X,Y}(a, \infty) = F_{X,Y}(a, 1).$$

Therefore,

$$F_X(a) = \begin{cases} 0 \,, a < 0, \\ \dfrac{1}{8} \,, 0 \le a < 1, \\ \dfrac{1}{2} \,, 1 \le a < 2, \\ \dfrac{7}{8} \,, 2 \le a < 3, \quad \text{and} \\ 1 \,, 3 \le a. \end{cases}$$

This is just what we obtained in Example 3.5.

   To illustrate the Rectangle rule let's compute $P(0 < X \le 2, 0 < Y \le 1)$. By Equation 3.17 this is

$$P(0 < X \le 2, 0 < Y \le 1) = F_{X,Y}(2, 1) - F_{X,Y}(0, 1) - F_{X,Y}(2, 0) + F_{X,Y}(0, 0)$$

$$= \frac{7}{8} - \frac{1}{8} - \frac{1}{2} + \frac{1}{8}$$

$$= \frac{3}{8}. \qquad \blacksquare$$

   Although the joint distribution function is valuable when discussing the theoretical aspects of random vectors it can be difficult to use when calculating probabilities. Example 3.14 should give some idea of how cumbersome it is even in simple experiments. For this reason it is more convenient to compute probabilities using the vector counterparts of the probability function and the probability density function. We consider the discrete case first.

Definition 3.9

Let (X, Y) be a vector of discrete random variables. Then the *joint probability function* of (X, Y) is

$$p_{X,Y}(a, b) = P(X = a, Y = b). \tag{3.18}$$

It is clear that this is the natural generalization of the single variable probability function. In terms of the joint distribution function, Equation 3.18 is

$$p_{X,Y}(a, b) = F_{X,Y}(a, b) - \lim_{\substack{c \to a^- \\ d \to b^-}} F_{X,Y}(c, d).$$

Given the joint probability function we can find the probability function of one of the random variables by summing over the possible values of the other as follows.

$$p_X(a) = \sum_{b_i} p_{X,Y}(a, b_i)$$

$$p_Y(b) = \sum_{a_i} p_{X,Y}(a_i, b) \tag{3.19}$$

Since the random vector (X, Y) maps the sample space onto R^2, summing over all possible values of X and Y must equal the probability of the sample space. In other words,

$$\sum_{a_i} \sum_{b_j} p_{X,Y}(a_i, b_j) = 1.$$

In addition, given the joint probability function we can find the joint distribution function by

$$F_{X,Y}(c, d) = \sum_{a_i \leq c} \sum_{b_j \leq d} p_{X,Y}(a_i, b_j). \tag{3.20}$$

Example 3.15

An urn contains 3 balls numbered 1, 2, and 3. Two balls are chosen from the urn without replacement. Let X = sum of the 2 balls, and Y = number on the first ball. X can take the values 3, 4, or 5, while Y can take the values 1, 2, or 3. Then the random vector (X, Y) can take the values (3,1), (3,2), (4,1), (4,3), (5,2), or (5,3). The joint probability function is given below. Note that there are 6 equally likely elementary events in the sample space. Each possible value of the random vector corresponds to an elementary event.

a	b	$p_{X,Y}(a, b)$
3	1	$\frac{1}{6}$
3	2	$\frac{1}{6}$
4	1	$\frac{1}{6}$
4	3	$\frac{1}{6}$
5	2	$\frac{1}{6}$
5	3	$\frac{1}{6}$

A graph is this joint p.f. is given in Figure 3.8.

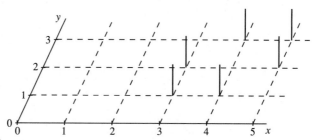

Figure 3.8

To get values for the p.f. for X, we use Equation 3.19 as follows.

$$p_X(3) = p_{X,Y}(3,1) + P_{X,Y}(3,2)$$

$$= \frac{1}{6} + \frac{1}{6} = \frac{1}{3}$$

Equation 3.20 can be used to find values for the joint distribution function. For example,

$$F_{X,Y}(3,0) = P(X \leq 3, Y \leq 0) = 0,$$

$$F_{X,Y}(4,2) = P(X \leq 4, Y \leq 2)$$

$$= p_{X,Y}(3,1) + p_{X,Y}(3,2) + P_{X,Y}(4,1)$$

$$= \frac{1}{2}, \quad \text{and}$$

$$F_{X,Y}(5,4) = P(X \leq 5, Y \leq 4) = 1.$$

■

Example 3.16

A deck contains 4 cards numbered 1, 2, 3, and 4. Three cards are chosen at random from the deck without replacement. Let X = sum of the first 2 cards, and Y = sum of the last 2 cards. Here we see that X and Y both have 3, 4, 5, 6, and 7 as possible values. By using elementary counting techniques we obtain the following joint probability function.

		Y					
		3	4	5	6	7	$p_X(a)$
X	3	0	$\frac{1}{24}$	$\frac{1}{12}$	$\frac{1}{24}$	0	$\frac{1}{6}$
	4	$\frac{1}{24}$	0	$\frac{1}{12}$	0	$\frac{1}{24}$	$\frac{1}{6}$
	5	$\frac{1}{12}$	$\frac{1}{12}$	0	$\frac{1}{12}$	$\frac{1}{12}$	$\frac{1}{3}$
	6	$\frac{1}{24}$	0	$\frac{1}{12}$	0	$\frac{1}{24}$	$\frac{1}{6}$
	7	0	$\frac{1}{24}$	$\frac{1}{12}$	$\frac{1}{24}$	0	$\frac{1}{6}$
$p_Y(b)$		$\frac{1}{6}$	$\frac{1}{6}$	$\frac{1}{3}$	$\frac{1}{6}$	$\frac{1}{6}$	

A graph of this joint p.f. is given in Figure 3.9.

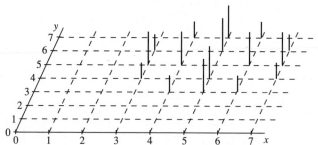

Figure 3.9

Notice that the probability functions are obtained by summing across the rows and down the columns of the table. This is the motivation for the term *marginal* that we mentioned earlier in this section. These probability functions appear in the margins of the table of the joint probability function. From the table it is easy to verify that

$$\sum_{a=3}^{7} \sum_{b=3}^{7} p_{X,Y}(a,b) = 1.$$

■

For the vector version of continuous random variables we extend Definition 3.6 to R^2 as follows.

Definition 3.10

A random vector (X, Y) is called a *continuous random vector* if there exists a function $f_{X,Y}(x, y)$ such that

1. $f_{X,Y}(x, y) \geq 0$ for all $(x, y) \in R^2$, and
2. For any set $A \subseteq R^2$,

$$P[(X, Y) \in A] = \iint_{\{(x,y) \in A\}} f_{X,Y}(x, y) \, dx \, dy.$$

The function $f_{X,Y}(x, y)$ is called the *joint probability density function* of (X, Y).

If X and Y are jointly continuous, then the individual random variables are also continuous. We can use the joint probability density function to find the probability density function of a single random variable by the following.

$$f_X(x) = \int_{-\infty}^{\infty} f_{X,Y}(x, y) \, dy, \quad \text{and}$$

$$f_Y(y) = \int_{-\infty}^{\infty} f_{X,Y}(x, y) \, dx$$

(3.21)

In addition, since some point in R^2 must occur we have

$$\int_{-\infty}^{\infty} \int_{-\infty}^{\infty} f_{X,Y}(x, y) \, dx \, dy = 1.$$

This is the vector version of Equation 14. Since the joint distribution function, $F_{X,Y}(a,b)$, is the probability of the event $\{X \le a, Y \le b\}$ it can be evaluated by integrating the joint probability density function as follows.

$$F_{X,Y}(a,b) = \int_{-\infty}^{a} \int_{-\infty}^{b} f_{X,Y}(x,y)\,dy\,dx \qquad \text{(3.22)}$$

This is analogous to Part 3 of Theorem 3.4. That theorem also showed that we can obtain the probability density function by differentiating the distribution function. If the joint p.d.f. is continuous, then this property is extended to continuous random vectors as follows. We state it without proof.

$$\frac{\partial^2}{\partial a\,\partial b} F_{X,Y}(a,b) = \frac{\partial^2}{\partial a\,\partial b} \int_{-\infty}^{a} \int_{-\infty}^{b} f_{X,Y}(x,y)\,dy\,dx$$

$$= f_{X,Y}(a,b) \qquad \text{(3.23)}$$

Example 3.17   Let (X,Y) have the joint p.d.f.

$$f_{X,Y}(x,y) = \begin{cases} 2e^{-2x-y}, & 0 < x,\ 0 < y, \quad \text{and} \\ 0, & \text{otherwise.} \end{cases}$$

1. Verify that $f_{X,Y}(x,y)$ is a valid joint probability density function.
2. Find $f_Y(y)$.
3. Find $P(X > 1, Y < 2)$.
4. Find $P(X > Y)$.

Solution: The graph of this joint p.d.f. is shown in Figure 3.10.

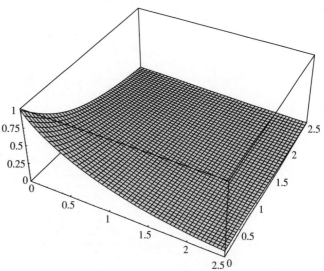

Figure 3.10

1. $f_{X,Y}(x, y)$ is clearly nonnegative, which satisfies Part 1 of Definition 3.10. We must also check that the integral over R^2 is equal to one. $f_{X,Y}(x, y)$ differs from zero only in the first quadrant as shown below.

$$
\begin{array}{c|c}
 & y \\
f_{X,Y}(x, y) = 0 & f_{X,Y}(x, y) > 0 \\
\hline
f_{X,Y}(x, y) = 0 & f_{X,Y}(x, y) = 0 \\
 & \quad\quad\quad x
\end{array}
$$

Thus, we get that

$$
\int_{-\infty}^{\infty} \int_{-\infty}^{\infty} f_{X,Y}(x, y)\, dx\, dy = \int_{0}^{\infty} \int_{0}^{\infty} 2e^{-2x-y}\, dx\, dy
$$

$$
= \int_{0}^{\infty} \left[\lim_{a \to \infty} -e^{-2x-y} \Big|_{0}^{a} \right] dy
$$

$$
= \int_{0}^{\infty} e^{-y}\, dy
$$

$$
= \lim_{b \to \infty} -e^{-y} \Big|_{0}^{b} = 1.
$$

2. We integrate over x to obtain $f_Y(y)$ as follows.

$$
f_Y(y) = \int_{-\infty}^{\infty} f_{X,Y}(x, y)\, dx
$$

$$
= \int_{0}^{\infty} e^{-2x-y}\, dx
$$

$$
= e^{-y}
$$

Therefore,

$$
f_Y(y) = \begin{cases} e^{-y}, & 0 < y, \quad \text{and} \\ 0, & \text{otherwise.} \end{cases}
$$

3. The region corresponding to the event $\{X > 1, Y < 2\}$ is shown in Figure 3.11.

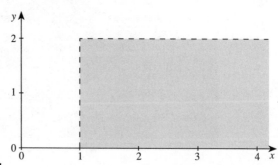

Figure 3.11

We use Part 2 of Definition 3.10 and this region to get

$$P(X > 1, Y < 2) = \int_1^\infty \int_{-\infty}^2 f_{X,Y}(x,y)\,dy\,dx$$

$$= \int_1^\infty \int_0^2 2e^{-2x-y}\,dy\,dx$$

$$= \int_1^\infty \left[-2e^{-2x-y}\Big|_0^2\right]dx$$

$$= \int_1^\infty 2e^{-2x}(1-e^{-2})\,dx$$

$$= -(1-e^{-2})\lim_{a\to\infty} e^{-2x}\Big|_1^a$$

$$= e^{-2} - e^{-4}.$$

4. The region described by the event $\{X > Y\}$ is triangular as shown in Figure 3.12.

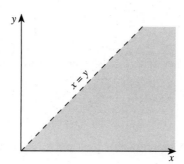

Figure 3.12

Another application of Part 2 of Definition 3.10 gives

$$P(X > Y) = P(-\infty < X < \infty, -\infty < Y < X)$$

$$= \int_{-\infty}^\infty \int_{-\infty}^x f_{X,Y}(x,y)\,dy\,dx$$

$$= \int_0^\infty \int_0^x 2e^{-2x-y}\,dy\,dx$$

$$= \int_0^\infty 2e^{-2x}\left(1 - e^{-x}\right)dx$$

$$= \lim_{a\to\infty}\left(-e^{-2x} + \frac{2}{3}e^{-3x}\right)\Big|_0^a$$

$$= \frac{1}{3}.$$

Example 3.18   A point is randomly chosen from within a circle of radius a. The position of the point is measured in polar coordinates. Thus, the circle is the set of points $\{0 \leq r \leq a, 0 \leq \theta < 2\pi\}$. Suppose that the point is selected in a manner such that the joint p.d.f. is

$$f_{R,\Theta}(r, \theta) = \begin{cases} kr, \ 0 \leq r \leq a, \ 0 \leq \theta < 2\pi, \quad \text{and} \\ 0 \ , \text{otherwise.} \end{cases}$$

1. Find the value of k to make this a valid joint probability density function.
2. Find the p.d.f. of R.
3. Find the p.d.f. of Θ.
4. Find the joint d.f. of R and Θ.

Solutions include the following.

1. We are looking for the value of k such that

$$\int_{-\infty}^{\infty} \int_{-\infty}^{\infty} f_{R,\Theta}(r, \theta) \, dr \, d\theta = 1.$$

So,

$$1 = \int_{0}^{a} \int_{0}^{2\pi} kr \, d\theta \, dr = \pi a^2 k.$$

Therefore, $k = \frac{1}{\pi a^2}$, and the joint p.d.f. is

$$f_{R,\Theta}(r, \theta) = \begin{cases} \dfrac{r}{\pi a^2} , \ 0 \leq r \leq a, \ 0 \leq \theta < 2\pi, \quad \text{and} \\ 0 \ , \text{otherwise.} \end{cases}$$

A graph of this joint p.d.f. is shown in Figure 3.13.

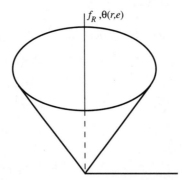

$f_R,_\Theta(r,e)$

Figure 3.13

2. The p.d.f. of R is found by integrating over Θ as follows.

$$f_R(r) = \int_0^{2\pi} \frac{r}{\pi a^2} \, d\theta = \frac{2r}{a^2}$$

Therefore,

$$f_R(r) = \begin{cases} \dfrac{2r}{a^2}, & 0 \le r \le a, \quad \text{and} \\ 0, & \text{otherwise.} \end{cases}$$

3. The p.d.f. of Θ is found by integrating over R.

$$f_\Theta(\theta) = \int_0^a \frac{r}{\pi a^2}\, dr = \frac{1}{2\pi}$$

Thus,

$$f_\Theta(\theta) = \begin{cases} \dfrac{1}{2\pi}, & 0 \le \theta < 2\pi, \quad \text{and} \\ 0, & \text{otherwise.} \end{cases}$$

4. To compute the d.f., $F_{R,\Theta}(c,d)$, there are 5 cases.
   a. $c < 0$ or $d < 0$: $F_{R,\Theta}(c,d) = 0$ since $f_{R,\Theta}(r,\theta) = 0$ in this region.
   b. $0 \le c \le a$ and $0 \le d < 2\pi$:

$$\begin{aligned} F_{R,\Theta}(c,d) &= \int_{-\infty}^c \int_{-\infty}^d f_{R,\Theta}(r,\theta)\, d\theta\, dr \\ &= \int_0^c \int_0^d \frac{r}{\pi a^2}\, d\theta\, dr \\ &= \frac{c^2 d}{2\pi a^2}. \end{aligned}$$

   c. $0 \le c \le a$ and $2\pi \le d$: Here we note that $f_{R,\Theta}(r,\theta) = 0$ when $2\pi \le d$. Therefore,

$$\begin{aligned} F_{R,\Theta}(c,d) &= \int_0^c \int_0^{2\pi} \frac{r}{\pi a^2}\, d\theta\, dr \\ &= \frac{r^2}{a^2}. \end{aligned}$$

   d. $a < c$ and $0 \le d < 2\pi$: Similar to the last case $f_{R,\Theta}(r,\theta) = 0$ when $a < c$. Thus,

$$\begin{aligned} F_{R,\Theta}(c,d) &= \int_0^a \int_0^d \frac{r}{\pi a^2}\, d\theta\, dr \\ &= \frac{d}{2\pi}. \end{aligned}$$

   e. $a < c$ and $2\pi \le d$:

$$\begin{aligned} F_{R,\Theta}(c,d) &= \int_0^a \int_0^{2\pi} \frac{r}{\pi a^2}\, d\theta\, dr \\ &= 1. \end{aligned}$$

In summary,

$$F_{R,\Theta}(c,d) = \begin{cases} 0 & , c < 0, \quad \text{or } d < 0, \\ \dfrac{c^2 d}{2\pi a^2} & , 0 \le c \le a, \, 0 \le d < 2\pi, \\ \dfrac{r^2}{a^2} & , 0 \le c \le a, \, 2\pi \le d, \\ \dfrac{d}{2\pi} & , a < c, \, 0 \le d < 2\pi, \quad \text{and} \\ 1 & , a < c, \, 2\pi \le d. \end{cases}$$

■

Up to now we have been concentrating on random vectors consisting of two components. We wish to extend these ideas to random vectors of other lengths. In fact, statistical procedures usually begin with a random vector of size n. A random vector of length n is the mathematical representation of a *random sample of size n* from some target population. The ideas we have developed for vectors of size 2 generalize in a straightforward manner to higher dimensions. We shall simply state the results without proof.

Suppose we define a random vector having n components on a sample space to be $(X_1, X_2, \ldots, X_n)$. This is a function whose domain is the sample space and whose range is n-dimensional Euclidean space, R^n. The joint distribution function would be

$$F_{X_1, X_2, \ldots, X_n}(a_1, a_2, \ldots, a_n) = P(X_1 \le a_1, X_2 \le a_2, \ldots, X_n \le a_n). \tag{3.24}$$

If $(X_1, X_2, \ldots, X_n)$ is a vector of *discrete* random variables, then we define the joint probability function by

$$p_{X_1, X_2, \ldots, X_n}(a_1, a_2, \ldots, a_n) = P(X_1 = a_1, X_2 = a_2, \ldots, X_n = a_n). \tag{3.25}$$

If A is any set in R^n, then

$$P[(X_1, X_2, \ldots, X_n) \in A] = \sum \sum \cdots \sum_{\{(a_1, a_2, \ldots, a_n) \in A\}} p_{X_1, X_2, \ldots, X_n}(a_1, a_2, \ldots, a_n). \tag{3.26}$$

The probability function of a single random variable is obtained by summing the other $n - 1$ random variables over their ranges. That is,

$$p_{X_i}(a_i) = \sum_{a_1} \cdots \sum_{a_{i-1}} \sum_{a_{i+1}} \cdots \sum_{a_n} p_{X_1, X_2, \ldots, X_n}(a_1, a_2, \ldots, a_n). \tag{3.27}$$

Example 3.19   Suppose we toss a fair coin 3 times. Let X = number of tails on the first 2 tosses, Y = number of tails on the last 2 tosses, and Z = number of tails on the second toss. Here X and Y can take on the values 0, 1, or 2. Z will be either 0 or 1.

By using elementary counting techniques we get the following joint probability function.

a	b	c	$P_{X,Y,Z}(a, b, c)$
0	0	0	$\frac{1}{8}$
0	0	1	0
0	1	0	$\frac{1}{8}$
0	1	1	0
0	2	0	0
0	2	1	0
1	0	0	$\frac{1}{8}$
1	0	1	0
1	1	0	$\frac{1}{8}$
1	1	1	$\frac{1}{8}$
1	2	0	0
1	2	1	$\frac{1}{8}$
2	0	0	0
2	0	1	0
2	1	0	0
2	1	1	$\frac{1}{8}$
2	2	0	0
2	2	1	$\frac{1}{8}$

From this we get the following probability functions.

$$p_X(a) = \begin{cases} \dfrac{1}{4}, & a = 0, 2, \\ \dfrac{1}{2}, & a = 1, \quad \text{and} \\ 0, & \text{otherwise} \end{cases}$$

$$p_Y(b) = \begin{cases} \dfrac{1}{4}, & b = 0, 2, \\ \dfrac{1}{2}, & b = 1, \quad \text{and} \\ 0, & \text{otherwise} \end{cases}$$

$$p_Z(c) = \begin{cases} \dfrac{1}{2}, & c = 0, 1, \quad \text{and} \\ 0, & \text{otherwise} \end{cases}$$

■

The vector $(X_1, X_2, \ldots, X_n)$ is said to be *continuous* if there exists a joint probability density function, $f_{X_1, X_2, \ldots, X_n}(x_1, x_2, \ldots, x_n)$, which is nonnegative, and, for any set A in R^n,

$$P[(X_1, X_2, \ldots, X_n) \in A] = \underset{\{x_1, x_2, \ldots, x_n\} \in A}{\int \cdots \int} f_{X_1, X_2, \ldots, X_n}(x_1, x_2, \ldots, x_n) \, dx_1 \, dx_2 \, \cdots \, dx_n.$$

(3.28)

The probability density function of any single random variable can be obtained by integrating over the other $n - 1$ random variables. In other words,

$$f_{X_i}(x_i) = \int_{-\infty}^{\infty} \cdots \int_{-\infty}^{\infty} f_{X_1, X_2, \ldots, X_n}(x_1, x_2, \ldots, x_n) \, dx_1 \cdots dx_{i-1} \, dx_{i+1} \cdots dx_n. \quad (3.29)$$

Finally, if the joint probability density function is continuous it can be obtained from the joint distribution function by differentiating as follows.

$$f_{X_1, X_2, \ldots, X_n}(a_1, a_2, \ldots, a_n) = \frac{\partial^n}{\partial a_1 \partial a_2 \cdots \partial a_n} F_{X_1, X_2, \ldots, X_n}(a_1, a_2, \ldots, a_n). \quad (3.30)$$

Example 3.20   Let the continuous random vector (X, Y, Z) have a joint p.d.f. of

$$f_{X,Y,Z}(x, y, z) = \begin{cases} e^{-x-y-z}, & 0 < x, \, 0 < y, \, 0 < z, \quad \text{and} \\ 0, & \text{otherwise.} \end{cases}$$

1. Verify that this is a valid joint probability density function.
2. Find the p.d.f. for Y.
3. Find $P(X < Y + Z)$.

Solutions include the following.

1. To verify that this is a valid joint p.d.f. we first note that it is a nonnegative function. We then integrate to observe that

$$\int_{-\infty}^{\infty} \int_{-\infty}^{\infty} \int_{-\infty}^{\infty} f_{X,Y,Z}(x, y, z) \, dx \, dy \, dz = \int_0^{\infty} \int_0^{\infty} \int_0^{\infty} e^{-x-y-z} \, dx \, dy \, dz = 1.$$

2. To find the p.d.f. for Y we must integrate over X and Z. Therefore, since the joint p.d.f. is positive only for positive values of X and Z,

$$f_Y(y) = \int_0^{\infty} \int_0^{\infty} e^{-x-y-z} \, dx \, dz$$
$$= e^{-y}.$$

Therefore,

$$f_Y(y) = \begin{cases} e^{-y}, & 0 < y, \quad \text{and} \\ 0, & \text{otherwise.} \end{cases}$$

3. The tricky part is determining the limits of integration that describe that portion of R^3 where $X < Y + Z$. Suppose we choose to integrate on X, then on Y, and finally on Z. With Y and Z held fixed we note that X can range between 0 and

$Y + Z$. Y and Z are free to take on any value. Therefore, we get that

$$P(X < Y + Z) = \int_0^\infty \int_0^\infty \int_0^{y+z} e^{-x-y-z} \, dx \, dy \, dz$$

$$= \int_0^\infty \int_0^\infty \left(e^{-y-z} - e^{-2y-2z}\right) \, dy \, dz$$

$$= \int_0^\infty \left(e^{-z} - \frac{1}{2}e^{-2z}\right) \, dz$$

$$= \frac{3}{4}.$$

■

Exercises 3.5

1. A college club consists of 4 men and 6 women. Two members are chosen at random without replacement. Let

$$X = \begin{cases} 0, & \text{if first person chosen is male, and} \\ 1, & \text{if first person chosen is female,} \end{cases}$$

$$Y = \begin{cases} 0, & \text{if second person chosen is male, and} \\ 1, & \text{if second person chosen is female.} \end{cases}$$

a. Find the joint p.f. of X and Y.

b. Find the p.f. of X.

c. Find the p.f. of Y.

d. Find the joint d.f. of X and Y.

2. Four men are available for assignment to 2 jobs. To make the selection process fair each is assigned a number from the set $\{1, 2, 3, 4\}$. Then 2 numbers are selected from this set. Let X = largest number selected, and Y = smallest number selected. Find the joint p.f. of X and Y and the p.f.'s of X and Y if selection is done

a. with replacement, and

b. without replacement.

3. A box contains 8 coins, 2 of which are counterfeit. Coins are removed from the box, one at a time, until both counterfeit coins are found. Let X = the number of coins removed to find the first counterfeit one, and let Y = the number of additional coins removed to find the second counterfeit one.

a. Find the joint probability function for X and Y.

b. What is the probability that not all of the coins will be removed from the box?

4. The random vector (X, Y) is continuous with a joint p.d.f. of

$$f_{X,Y}(x, y) = \begin{cases} k(xy - x), & 0 < x < 1, 0 < y < 1, \quad \text{and} \\ 0, & \text{otherwise.} \end{cases}$$

a. Find k to make $f_{X,Y}(x, y)$ a valid joint probability density function.

b. Find $f_X(x)$.

c. Find $F_Y(y)$.

d. Find $P(X \geq \frac{1}{2}, Y < \frac{3}{4})$.

e. Find $P(X + Y < 1)$.

f. Find $P(X \leq \frac{1}{2} \mid Y \geq \frac{1}{2})$.

5. The random vector (X, Y) is continuous with a joint p.d.f. of

$$f_{X,Y}(x, y) = \begin{cases} \dfrac{kx}{y^2}, y^{-1} < x < y, 1 < y < 2, & \text{and} \\ 0, \text{ otherwise.} \end{cases}$$

a. Find k to make $f_{X,Y}(x, y)$ a valid joint probability density function.
b. Find $f_X(x)$.
c. Find $f_Y(y)$.

6. The random vector (X, Y) is continuous with a joint p.d.f. of

$$f_{X,Y}(x, y) = \begin{cases} \dfrac{1}{4}, 0 \le x \le 2, 1 \le y \le 3, & \text{and} \\ 0, \text{ otherwise.} \end{cases}$$

a. Verify that this is a valid joint probability density function.
b. Find $P(X \le Y)$.
c. Find $P(|Y - X| \le 1)$.

7. Prove Part 4 of Theorem 3.5.

8. A company is conducting an auto safety campaign. An inspection of headlights and brakes resulted in the following joint probability function for X = number of defective brakes and Y = number of defective headlights.

		b		
$p_{X,Y}(a, b)$		0	1	2
	0	0.40	0.10	0.05
	1	0.15	0.10	0.03
a	2	0.05	0.02	0.02
	3	0.03	0.02	0.01
	4	0.02	0.01	0.00

a. What is the probability that a randomly selected car will have exactly one defective headlight and one defective brake?
b. What is the probability that a randomly selected car will have at most one defective headlight and one defective brake?
c. What is the probability that a randomly selected car will have no defective headlights?

9. Two components in a personal computer system have lifetimes that are distributed with a joint probability density function of

$$f_{X,Y}(x, y) = \begin{cases} xe^{-x(1+y)}, 0 \le x, 0 \le y, & \text{and} \\ 0, \text{ otherwise.} \end{cases}$$

a. Find the p.d.f. for X.
b. Find the p.d.f. for Y.
c. Find the probability that the lifetime of at least one component exceeds 2.

10. Let (X, Y) be a random vector. Show that

$$P(X \le a) = P(X \le a, Y < \infty).$$

11. Let $F_{X,Y}(a, b)$ be the joint d.f. of the random vector (X, Y).

a. Show that

$$P(a < X \le b, c < Y \le d) = F_{X,Y}(b, d) - F_{X,Y}(a, d) - F_{X,Y}(b, c) + F_{X,Y}(a, c).$$

b. If $F_X(a)$ and $F_Y(b)$ are the d.f.'s of X and Y, respectively, show that

$$P(X > a, Y > b) = 1 - F_X(a) - F_Y(b) + F_{X,Y}(a, b).$$

12. An urn contains 4 balls numbered 1, 2, 3, and 4. Two balls are drawn from the urn. Let X = largest numbered ball, Y = smallest numbered ball, and Z = sum of the 2 balls. Find the joint p.f. of (X, Y, Z) when sampling is done

a. with replacement, and

b. without replacement.

13. The continuous random vector has a joint p.d.f. of

$$f_{X,Y,Z}(x, y, z) = \begin{cases} kxyz, & 0 < x < y < z < 1, \quad \text{and} \\ 0, & \text{otherwise.} \end{cases}$$

a. Find k to make this a valid probability density function.

b. Find $f_X(x)$.

c. Find $f_Y(y)$.

d. Find $f_Z(z)$.

e. Find $P(X < YZ)$.

14. Let $F_{X,Y,Z}(a, b, c)$ be the joint d.f. of the random vector (X, Y, Z). Show that

a. $F_X(a) = F_{X,Y,Z}(a, \infty, \infty)$, and

b. $F_{X,Y}(a, b) = F_{X,Y,Z}(a, b, \infty)$.

15. Let (X, Y, Z) be a continuous random vector with a joint p.d.f. of

$$f_{X,Y,Z}(x, y, z) = \begin{cases} 1, & 0 \le x \le 1, 0 \le y \le 1, 0 \le z \le 1, \quad \text{and} \\ 0, & \text{otherwise.} \end{cases}$$

a. Find $P(X + Y \ge Z)$.

b. Find $P\left(\frac{X}{Y} < Z\right)$.

c. Find $P(X \le Y \le Z)$.

16. Prove Equation 3.30.

3.6
Independent Random Variables

Since random variables are numerical descriptions of events it makes sense to discuss independence. For example, given a random vector (X, Y), the events $\{X \in A\}$ and $\{Y \in B\}$ are, according to Equation 2.7, independent if and only if $P(X \in A, Y \in B) = P(X \in A)P(Y \in B)$. We would like to extend this notion to having the two random variables be declared independent. In order to say this, it is natural to require that this multiplication rule be true regardless of which sets A and B are used. This is the basis of the following definition.

Definition 3.11

The random variables X and Y are said to be *independent* if, for any pair of sets of real numbers, A and B,

$$P(X \in A, Y \in B) = P(X \in A)P(Y \in B). \qquad (3.31)$$

Some consequences of this definition are given in the following theorem. The proof is left as an exercise.

Theorem 3.6

Let (X, Y) be a random vector.

1. X and Y are independent if and only if for any real numbers, a and b,

$$F_{X,Y}(a,b) = F_X(a)F_Y(b). \qquad (3.32)$$

2. If (X, Y) is a vector of discrete random variables, then X and Y are independent if and only if for any real numbers, a and b,

$$p_{X,Y}(a,b) = p_X(a)p_Y(b). \qquad (3.33)$$

3. If (X, Y) is a continuous random vector, then X and Y are independent if and only if for any real numbers, x and y,

$$f_{X,Y}(x,y) = f_X(x)f_Y(y). \qquad (3.34)$$

This theorem is useful for checking the independence of random variables. We can conclude that X and Y are independent if the joint distribution function for (X, Y) factors into the product of the distribution functions for X and Y. Similarly, if the joint probability function of a discrete random vector factors into the product of the probability functions, then the random variables are independent. A similar check can be applied to the joint probability density function of a continuous random vector. These checks are usually more convenient than a direct application of Definition 3.11. It must be pointed out that the check for independence means that not only does the functional form factor but that the limits must also not show any dependency between the random variables. We illustrate these ideas with some examples.

Example 3.21

Two fair dice are rolled. Let X = number of dots showing on the first die, and Y = number of dots showing on the second die. Since the sample space consists of 36 equally likely events we use basic counting methods to see that the joint probability function is

$$p_{X,Y}(a,b) = \begin{cases} \dfrac{1}{36}, & a = 1, 2, \ldots, 6;\ b = 1, 2, \ldots, 6, \quad \text{and} \\ 0, & \text{otherwise.} \end{cases}$$

We first note that the limits on a do not depend on the value of b and likewise for those of b. Thus, we proceed to determine if the p.f.'s factor. It is easy to see that the p.f. for X is

$$p_X(a) = \sum_{a=1}^{6} p_{X,Y}(a,b) = \frac{1}{6}.$$

Therefore,

$$p_X(a) = \begin{cases} \dfrac{1}{6}, & a = 1, 2, \ldots, 6, \quad \text{and} \\ 0, & \text{otherwise.} \end{cases}$$

Proceeding similarly we find that

$$p_Y(b) = \begin{cases} \dfrac{1}{6}, & b = 1, 2, \ldots, 6, \quad \text{and} \\ 0, & \text{otherwise.} \end{cases}$$

Thus, we see that $p_{X,Y}(a,b) = p_X(a)p_Y(b)$ for all a and b and conclude that X and Y are independent. ∎

Example 3.22   Let the continuous random vector (X, Y) have a joint p.d.f. of

$$f_{x,Y}(x,y) = \begin{cases} e^{-x-y}, & 0 < x, \ 0 < y, \quad \text{and} \\ 0, & \text{otherwise.} \end{cases}$$

As in the previous example we first note that the limits of x and y do not depend on each other. So we obtain the p.d.f.'s for X and Y as follows. For X we integrate over y to find

$$f_X(x) = \int_{\infty}^{\infty} f_{X,Y}(x,y)\, dy = \int_0^{\infty} e^{-x-y}\, dy$$
$$= e^{-x}.$$

Therefore,

$$f_X(x) = \begin{cases} e^{-x}, & 0 < x, \quad \text{and} \\ 0, & \text{otherwise.} \end{cases}$$

In a similar manner we find that

$$f_Y(y) = \begin{cases} e^{-y}, & 0 < y, \quad \text{and} \\ 0, & \text{otherwise.} \end{cases}$$

From these we see that $f_{X,Y}(x,y) = f_X(x)f_Y(y)$ and conclude that X and Y are independent. ∎

Example 3.23   Let the continuous random vector (X, Y) have a joint p.d.f. of

$$f_{X,Y}(x,y) = \begin{cases} \dfrac{1}{y^2}, & 0 < x < y, \ 1 < y < e, \quad \text{and} \\ 0, & \text{otherwise.} \end{cases}$$

The region where the joint p.d.f. is positive is shown in Figure 3.14.

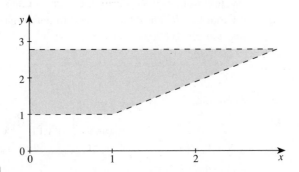

Figure 3.14

From the joint p.d.f. we see that the limits for X depend on the value of y and vice versa. Therefore, we immediately conclude that X and Y are not independent. ■

When considering random vectors of more than two variables the notion of independence generalizes in a straightforward manner.

Definition 3.12

The random variables $X_1, X_2, \ldots, X_n$ are said to be *independent* if, for any sets $A_1, A_2, \ldots, A_n$ of real numbers,

$$P(X_1 \in A_1, X_2 \in A_2, \ldots, X_n \in A_n) = P(X_1 \in A_1)P(X_2 \in A_2) \cdots P(X_n \in A_n).$$
$$(3.35)$$

We give a version of Theorem 3.6 for random vectors of arbitrary size. The proof is omitted.

Theorem 3.7

Let $(X_1, X_2, \ldots, X_n)$ be a random vector.

1. $X_1, X_2, \ldots, X_n$ are independent if and only if for any real vector $(a_1, a_2, \ldots, a_n)$

$$F_{X_1, X_2, \ldots, X_n}(a_1, a_2, \ldots, a_n) = F_{X_1}(a_1)F_{X_2}(a_2) \cdots F_{X_n}(a_n).$$
$$(3.36)$$

2. If $X_1, X_2, \ldots, X_n$ are discrete random variables, then they are independent if and only if for any real vector $(a_1, a_2, \ldots, a_n)$

$$p_{X_1, X_2, \ldots, X_n}(a_1, a_2, \ldots, a_n) = p_{X_1}(a_1)p_{X_2}(a_2) \cdots p_{X_n}(a_n).$$
$$(3.37)$$

3. If $(X_1, X_2, \ldots, X_n)$ is a continuous random vector, then the component random variables are independent if and only if, for any real vector $(x_1, x_2, \ldots, x_n)$,

$$f_{X_1, X_2, \ldots, X_n}(x_1, x_2, \ldots, x_n) = f_{X_1}(x_1)f_{X_2}(x_2) \cdots f_{X_n}(x_n).$$
$$(3.38)$$

An important application of this theorem comes from experiments that consist of independent trials. In such cases, it is usually easy to derive the probability function or probability density function of each trial. Theorem 3.7 then shows that we can obtain the joint probability function or joint probability density function by forming the appropriate product of the probability functions or probability density functions of the individual trials. An important special case of this idea occurs when the trials are identical as well as independent. In this instance, each trial will then have the same probability function or probability density function. Thus, if $X_i, i = 1, 2, \ldots, n$, is the random variable associated with the ith trial then we can make the following definition.

Definition 3.13

> If the random variables $X_1, X_2, \ldots, X_n$ are independent and $F_{X_i}(a) = F(a), i = 1, 2, \ldots, n$, then $X_1, X_2, \ldots, X_n$ are said to be *independent and identically distributed (i.i.d.) random variables.*

This means that, if we have n independent and identically distributed discrete random variables, each with

$$p_{X_i}(a) = p_X(a), i = 1, 2, \ldots, n,$$

then the joint probability function will be

$$p_{X_1, X_2, \ldots, X_n}(a_1, a_2, \ldots, a_n) = \prod_{i=1}^{n} p_X(a_i). \tag{3.39}$$

Similarly, if we have n i.i.d. continuous random variables, each with

$$f_{X_i}(x) = f_X(x), i = 1, 2, \ldots, n,$$

then the joint probability density function will be

$$f_{X_1, X_2, \ldots, X_n}(x_1, x_2, \ldots, x_n) = \prod_{i=1}^{n} f_X(x_i). \tag{3.40}$$

Example 3.24

Let X, Y, and Z be i.i.d. continuous r.v.'s, each having a p.d.f. of

$$f(t) = \begin{cases} e^{-t}, & 0 < t, \quad \text{and} \\ 0, & \text{otherwise.} \end{cases}$$

We wish to find $P(X < Y - Z)$. Since the r.v.'s are i.i.d. we use Equation 3.40 to find the joint p.d.f. as follows.

$$\begin{aligned} f_{X,Y,Z}(x, y, z) &= f_X(x) f_Y(y) f_Z(z) \\ &= f(x) f(y) f(z) \\ &= \begin{cases} e^{-x-y-z}, & 0 < x, 0 < y, 0 < z, \quad \text{and} \\ 0, & \text{otherwise} \end{cases} \end{aligned}$$

Now we complete the solution as follows.

$$P(X < Y - Z) = P(X + Z < Y)$$

$$= \int_0^\infty \int_0^\infty \int_{x+z}^\infty e^{-x-y-z} \, dy \, dx \, dz$$

$$= \int_0^\infty \int_0^\infty e^{-2x-2z} \, dx \, dz$$

$$= \int_0^\infty \frac{1}{2} e^{-2z} \, dz$$

$$= \frac{1}{4}$$

∎

Exercises 3.6

1. The joint p.f. of (X, Y) is given in the table below. Determine if X and Y are independent.

		Y		
		0	1	2
X	0	$\frac{1}{6}$	$\frac{1}{6}$	$\frac{1}{6}$
	1	$\frac{1}{6}$	$\frac{1}{6}$	$\frac{1}{6}$

2. A book has r pages numbered 1, 2, ..., r. n pages are chosen from the book with replacement. Let X = largest page number selected, and Y = smallest page number selected. Determine if X and Y are independent.

3. Determine if the two random variables X and Y of Example 3.15 are independent.

4. Determine if the two random variables X and Y of Example 3.16 are independent.

5. Determine if the two random variables R and Θ of Example 3.18 are independent.

6. Determine if the three random variables X, Y, and Z of Example 3.19 are independent.

7. Determine if the three random variables X, Y, and Z of Example 3.20 are independent.

8. Determine if the two random variables X and Y of Exercise 1 of Section 3.5 are independent.

9. Determine if the two random variables X and Y of Exercise 2 of Section 3.5 are independent.

10. Determine if the two random variables X and Y of Exercise 3 of Section 3.5 are independent.

11. Determine if the two random variables X and Y of Exercise 4 of Section 3.5 are independent.

12. Determine if the three random variables X, Y, and Z of Exercise 9 of Section 3.5 are independent.

13. Points are selected at random on a line segment of length L. Each point selected is modeled by a continuous r.v. with a p.d.f. of

$$f(x) = \begin{cases} \dfrac{1}{L}, & 0 < x < L, \quad \text{and} \\ 0, & \text{otherwise.} \end{cases}$$

   a. How many points must be selected so that the probability that all points will fall in the interval $\left(\dfrac{L}{2}, L\right)$ will be less than $\dfrac{1}{2}$?

   b. How many points must be selected so that the probability that at least one point will fall in the interval $\left(\dfrac{3}{4}L, L\right)$ will be at least $\dfrac{1}{2}$?

14. Prove Theorem 3.6.

15. Let X and Y be independent and identically distributed random variables each having a p.d.f. of

$$f(t) = \begin{cases} e^{-t}, & 0 \le t, \quad \text{and} \\ 0, & \text{otherwise.} \end{cases}$$

   Find $P(X + Y \ge 2)$.

16. Let X and Y be independent and identically distributed random variables each having a p.f. of

$$p(a) = \begin{cases} (a!\, e)^{-1}, & a = 0, 1, 2, \ldots, \quad \text{and} \\ 0, & \text{otherwise.} \end{cases}$$

   Find $P(X + Y \le 5)$.

3.7
Conditional Distributions

In Chapter 2 we defined the conditional probability of an event A given the occurrence of the event B to be

$$P(A \mid B) = \frac{P(A \cap B)}{P(B)}, \quad \text{if } P(B) > 0.$$

We then introduced the notion of independence of events. Since we have discussed the independence of random variables it is appropriate to investigate *conditional distributions*. In the case of discrete random variables we give the following definition. It is essentially the same idea used to define conditional probability.

Definition 3.14

> Let X and Y be discrete random variables. The *conditional probability function of X given $Y = y$* is,
>
> $$p_{X|Y}(x \mid y) = \frac{p_{X,Y}(x, y)}{p_Y(y)}, \qquad (3.41)$$
>
> for all y such that $p_Y(y) > 0$.

Example 3.25   An urn contains 5 white balls, 3 red balls, and 2 black balls. Two balls are drawn from the urn without replacement. Let X = number of white balls drawn, and Y = number of red balls drawn. We wish to determine $p_{X|Y}(x \mid y)$. We first need the joint p.f. of X and Y. We leave it as an exercise to show that $p_{X,Y}(x, y)$ is

$$p_{X,Y}(x, y) = \begin{cases} \dfrac{\dbinom{5}{x}\dbinom{3}{y}\dbinom{2}{2-x-y}}{\dbinom{10}{2}} , & 0 \le x + y \le 2, \quad \text{and} \\[2em] 0 & , \text{otherwise.} \end{cases}$$

The p.f. for Y is found by summing this over x as follows.

$$p_Y(y) = \sum_{x=0}^{2-y} p_{X,Y}(x, y)$$

$$= \frac{\dbinom{3}{y}\dbinom{7}{2-y}}{\dbinom{10}{2}}$$

Therefore,

$$p_Y(y) = \begin{cases} \dfrac{\dbinom{3}{y}\dbinom{7}{2-y}}{\dbinom{10}{2}} , & y = 0, 1, 2, \quad \text{and} \\[2em] 0 & , \text{otherwise.} \end{cases}$$

An alternative way to derive $p_Y(y)$ without using the joint p.f. is to view the experiment as drawing red and nonred balls and compute it directly from the description of the experiment. In any event we now use Equation 3.41 as follows.

$$p_{X|Y}(x \mid y) = \frac{p_{X,Y}(x, y)}{p_Y(y)}$$

$$= \frac{\dbinom{5}{x}\dbinom{3}{y}\dbinom{2}{2-x-y}}{\dbinom{10}{2}} \cdot \frac{\dbinom{10}{2}}{\dbinom{3}{y}\dbinom{7}{2-y}}$$

$$= \frac{\dbinom{5}{x}\dbinom{2}{2-x-y}}{\dbinom{7}{2-y}}$$

Therefore,

$$p_{X|Y}(x \mid y) = \begin{cases} \dfrac{\dbinom{5}{x}\dbinom{2}{2-x-y}}{\dbinom{7}{2-y}}, & 0 \le x \le 2 - y, \quad \text{and} \\ 0, & \text{otherwise.} \end{cases}$$

In the case of two continuous random variables there is a definition for the conditional distribution based on probability density functions. ∎

Definition 3.15

Let X and Y be continuous random variables. Then, for any y such that $f_Y(y) > 0$, the *conditional probability density function of X given Y = y* is

$$f_{X|Y}(x \mid y) = \frac{f_{X,Y}(x, y)}{f_y(y)}. \tag{3.42}$$

Example 3.26

Let X and Y be two continuous random variables with a joint p.d.f. of

$$f_{X,Y}(x, y) = \begin{cases} 12(x^2 - xy), & 0 < x < 1, \, 0 < y < 1, \quad \text{and} \\ 0, & \text{otherwise.} \end{cases}$$

We wish to find $f_{X|Y}(x|y)$. The p.d.f. for Y is found by integrating over X as follows.

$$f_Y(y) = \int_0^1 12(x^2 - xy)\, dx = 4 - 6y$$

Therefore,

$$f_Y(y) = \begin{cases} 4 - 6y, & 0 < y < 1, \quad \text{and} \\ 0, & \text{otherwise.} \end{cases}$$

From these we use Equation 3.42 to find

$$f_{X|Y}(x \mid y) = \frac{f_{X,Y}(x, y)}{f_Y(y)}$$

$$= \frac{6(x^2 - xy)}{2 - 3y}.$$

Therefore,

$$f_{X|Y}(x \mid y) = \begin{cases} \dfrac{6(x^2 - xy)}{2 - 3y}, & 0 < x < 1, \quad \text{and} \\ 0, & \text{otherwise.} \end{cases}$$

We can also define the *conditional distribution function*. To be consistent with our previous definitions of distribution functions the conditional distribution function will be

$$F_{X|Y}(a \mid y) = P(X \le a \mid Y = y). \tag{3.43}$$

The following definition implements this idea for discrete and continuous random variables.   ∎

Definition 3.16

Let X and Y be two random variables. If X and Y are discrete random variables, then for all y such that $p_Y(y) > 0$, the *conditional distribution function of X given Y = y* is

$$F_{X|Y}(a \mid y) = \sum_{\{x \le a\}} p_{X|Y}(x \mid y). \tag{3.44}$$

If X and Y are continuous random variables, then for all y such that $f_Y(y) > 0$, the *conditional distribution function of X given Y = y* is

$$F_{X|Y}(a \mid y) = \int_{-\infty}^{a} f_{X|Y}(x \mid y)\, dx. \tag{3.45}$$

Example 3.27

We wish to find the conditional distribution function for the conditional probability function found in Example 3.25. There are 3 cases to be considered here. We shall compute the conditional d.f. for the case when $y = 1$ and leave the other 2 as exercises. Since these were discrete random variables we use Equation 3.44 to obtain the following.

$$F_{X|Y}(a \mid y) = \begin{cases} 0\ , a < 0, \\ \dfrac{2}{7}\ , 0 \le a < 1, \quad \text{and} \\ 1\ , 1 \le a \end{cases}$$

∎

Example 3.28

We wish to find the conditional distribution function for the conditional p.d.f. found in Example 3.26. Since these were continuous random variables we use Equation 3.45 to obtain the following.

$$F_{X|Y}(a \mid y) = \begin{cases} 0 \qquad\ , a \le 0, \\ \dfrac{2a^3 - 3a^2 y}{2 - 3y}\ , 0 < a < 1, \quad \text{and} \\ 1 \qquad\ , 1 \le a \end{cases}$$

∎

Exercises 3.7

1. Let the random vector (X, Y) have a joint p.f. of

$$p_{X,Y}(x, y) = \begin{cases} \dfrac{2x - y}{15}\ , x = 1, 2, 3; \quad y = 1, 2; \quad \text{and} \\ 0 \quad , \text{otherwise.} \end{cases}$$

a. Find $p_{X|Y}(x \mid y)$.
b. Find $p_{Y|X}(y \mid x)$.

2. The random vector (X, Y) has a joint p.f. of

$$p_{X,Y}(x, y) = \begin{cases} \dfrac{xy}{18}, & x = 1, 2; \quad y = 1, 2, 3; \quad \text{and} \\ 0, & \text{otherwise.} \end{cases}$$

   a. Find $p_{X|Y}(x \mid y)$.
   b. Find $p_{Y|X}(y \mid x)$.

3. The random vector (X, Y) has a joint p.d.f. of

$$f_{X,Y}(x, y) = \begin{cases} 6(y^2 - x^2), & 0 < x < y < 1, \quad \text{and} \\ 0, & \text{otherwise.} \end{cases}$$

   a. Find $f_{X|Y}(x \mid y)$.
   b. Find $f_{Y|X}(y \mid x)$.

4. Show that if X and Y are independent discrete random variables, then

$$p_{X|Y}(x \mid y) = p_X(x).$$

5. Show that if X and Y are independent continuous random variables, then

$$f_{X|Y}(x \mid y) = f_X(x).$$

6. A box contains 3 black balls and 3 white balls. Three balls are drawn at random without replacement. Let X = number of black balls drawn on the first 2 draws, and let Y = total number of black balls removed.

   a. Find $p_{X|Y}(x \mid y)$.
   b. Find $p_{Y|X}(y \mid x)$.

7. The random vector (X, Y) has a joint p.d.f. of

$$f_{X,Y}(x, y) = \begin{cases} k, & 0 \leq x \leq 20, \ 10 \leq y \leq 30, \quad \text{and} \\ 0, & \text{otherwise.} \end{cases}$$

   a. Find the value of k to make this a valid joint probability density function.
   b. Find $f_{X|Y}(x \mid y)$.

8. The random vector (X, Y) has a joint p.d.f. of

$$f_{X,Y}(x, y) = \begin{cases} 2x^3 + 2y^3, & 0 \leq x \leq 1, \ 0 \leq y \leq 1, \quad \text{and} \\ 0, & \text{otherwise.} \end{cases}$$

   a. Find $f_{X|Y}(x \mid y)$.
   b. Find $f_{Y|X}(y \mid x)$.

Chapter Summary

Random Variable: A random experiment whose sample space consists of real numbers.   **(Section 3.1)**

Distribution Function: $F_X(a) = P(A \leq a)$.

1. $F_X(a)$ is nondecreasing.

2. $F_X(a)$ is right continuous.
3. $\lim_{a \to -\infty} F_X(a) = 0$
4. $\lim_{a \to \infty} F_X(a) = 1$

$$P(a < X \le b) = F_X(b) - F_X(a)$$

$$P(a \le X \le b) = F_X(b) - \lim_{c \to a^-} F_X(c)$$

$$P(a < X < b) = \lim_{c \to b^-} F_X(c) - F_X(a)$$

$$P(a \le X < b) = \lim_{d \to b^-} F_X(d) - \lim_{c \to a^-} F_X(c)$$

If $F_X(a)$ is continuous at $a = c$ then $P(X = c) = 0$.   **(Section 3.2)**

Quantiles: The kth quantile, γ_k is the value of X such that

$$P(X < \gamma_k) \le \frac{k}{100} \le P(X \le \gamma_k)$$

Median = γ_{50}
1st quartile = γ_{25}
3rd quartile = γ_{75} **(Section 3.2)**

Discrete Random Variable: Assigns positive probability to at most a countable number of values.

Probability Function: $p_x(a) = P(X = a)$.

$$F_X(a) = \sum_{c_i \le a} p_X(c_i)$$

$$p_X(a) = F_X(a) - \lim_{c \to a^-} F_X(c)$$

$$\sum_{c_i} p_X(c_i) = 1$$ **(Section 3.3)**

Continuous Random Variable: There exists a nonnegative probability density function, $f_X(x)$ such that

$$P(X \in A) = \int_A f_X(x)\, dx.$$

$$\int_{-\infty}^{\infty} f_X(x)\, dx = 1$$

$$P(a \le X \le b) = \int_a^b f_X(x)\, dx$$

$$P(X = a) = 0$$

$$F_X(a) = \int_{-\infty}^{a} f_X(x)\, dx$$

If $f_X(c)$ is continuous then $\frac{d}{dc}F_X(c) = f_X(c)$ **(Section 3.4)**

Random Vector: A random experiment whose sample space consists of points in R^n.

Joint Distribution Function: $F_{X,Y}(a,b) = P(X \le a, Y \le b)$.

$F_X(a) = \lim_{b \to \infty} F_{X,Y}(a,b)$

Discrete Random Vector has a joint probability function of

$$p_{X,Y}(a,b) = P(X = a, Y = b).$$

$$p_X(a) = \sum_{b_i} p_{X,Y}(a,b_i)$$

Continuous random vector has a joint probability density function that is non-negative and

$$P[(X,Y) \in A] = \int\int_A f_{X,Y}(x,y)\,dx\,dy.$$

$f_X(x) = \int_{-\infty}^{\infty} f_{X,Y}(x,y)\,dy$

$f_{X,Y}(a,b) = \frac{\partial^2}{\partial a \partial b} F_{X,Y}(a,b)$ **(Section 3.5)**

Independent Random Variables: X and Y are independent if for any sets A and B

$$P(X \in A, Y \in B) = P(X \in A)P(Y \in B).$$

X and Y are independent if and only if $F_{X,Y}(a,b) = F_X(a)F_Y(b)$.

Discrete random variables X and Y are independent if and only if

$$p_{X,Y}(a,b) = p_X(a)p_Y(b).$$

Continuous random variables X and Y are independent if and only if

$$f_{X,Y}(x,y) = f_X(x)f_Y(y).$$

$X_1, X_2, \ldots, X_n$ are independent and identically distributed (i.i.d.) if they are jointly independent and have the same distribution function.   **(Section 3.6)**

Conditional Probability Function: If X and Y are discrete then

$$p_{X|Y}(x \mid y) = \frac{p_{X,Y}(x,y)}{p_Y(y)}.$$

$F_{X|Y}(a \mid y) = \sum_{x \le a} p_{X|Y}(x \mid y)$ **(Section 3.7)**

Conditional Probability Density Function: If X and Y are continuous then

$$f_{X|Y}(x \mid y) = \frac{f_{X,Y}(x,y)}{f_Y(y)}.$$

$F_{X|Y}(a \mid y) = \int_{-\infty}^{a} f_{X|Y}(x \mid y)\,dx$ **(Section 3.7)**

**Review
Exercises**

1. A box contains 10 balls, 3 of which are red. Two balls are selected from the box. Let Y = number of red balls. Find the probability function and distribution function for Y if sampling is done

   a. with replacement, and

   b. without replacement.

2. A probability class consists of 15 juniors and 20 seniors. Each meeting 5 are selected at random to present problems to the rest of the class. Let X be the number of juniors that are selected.

   a. Find $P(X = 0)$.

   b. Find $P(4 \leq X \leq 5)$.

   c. Find $p_X(a)$.

3. For each of the following find the constant k that make each a valid probability function.

   a. $p_X(a) = \frac{x}{k}$, $a = 1, 2, 3, 4, 5, 6$.

   b. $p_x(a) = kx^3$, $a = 1, 2, \ldots, 10$.

   c. $p_X(a) = k \left(\frac{1}{4}\right)^a$, $a = 1, 2, \ldots$.

4. Which of the following could be probability density functions?

   a. $f(x) = \frac{1}{2}$, for $-1 < x < 1$, and 0, otherwise.

   b. $f(x) = 2x$, for $0 < x < 2$, and 0, otherwise.

   c. $f(x) = \frac{1}{2}e^{-|x|}$, for $-\infty < x < \infty$.

   d. $f(x) = \frac{1}{x^2}$, for $1 < x < \infty$, and 0, otherwise.

5. An urn contains 5 white balls, 3 red balls and 2 black balls. Two balls are drawn from the box without replacement. Let X = number of white balls, and Y = number of red balls.

   a. Show that the joint p.f. of X and Y is

   $$p_{X,Y}(x, y) = \begin{cases} \dfrac{\dbinom{5}{x}\dbinom{3}{y}\dbinom{2}{2-x-y}}{\dbinom{10}{2}}, & 0 \leq x + y \leq 2, \quad \text{and} \\ 0, & \text{otherwise.} \end{cases}$$

   b. Are X and Y independent random variables?

6. Determine if the random variables X and Y from the following exercises are independent.

   a. Exercise 1, Section 3.7.

   b. Exercise 2, Section 3.7.

   c. Exercise 3, Section 3.7.

7. Let X and Y be independent random variables with the following p.d.f.'s.

   $$f_X(x) = \begin{cases} 1, 0 \leq x \leq 1, & \text{and} \\ 0, \text{otherwise.} \end{cases}$$

   $$f_Y(y) = \begin{cases} \dfrac{3}{8}y^2, 0 \leq y \leq 2, & \text{and} \\ 0, \text{otherwise.} \end{cases}$$

   Find $P(Y > X)$.

8. Let (X, Y, Z) be a random vector with a joint p.d.f. of

$$f_{X,Y,Z}(x, y, z) = \begin{cases} 6, 0 \leq x < y < z \leq 1, & \text{and} \\ 0, \text{ otherwise.} \end{cases}$$

a. Find $f_X(x)$, $f_Y(y)$, and $f_Z(z)$.
b. Find $P(X + Y > Z)$.

9. Let the random vector (X, Y) have a joint p.d.f. of

$$f_{X,Y}(x, y) = \begin{cases} \dfrac{1}{\pi}, x^2 + y^2 \leq 1, & \text{and} \\ 0, \text{ otherwise.} \end{cases}$$

a. Show that this is a valid joint probability density function.
b. Find $f_X(x)$.
c. Find $P(X > Y)$.

10. The conditional p.d.f. of X given $Y = y$ is

$$f_{X|Y}(x \mid y) = \begin{cases} \dfrac{k_1 x}{y^2}, 1 < x < y, & \text{and} \\ 0, \text{ otherwise.} \end{cases}$$

The p.d.f. for Y is

$$f_Y(y) = \begin{cases} \dfrac{k_2}{y^3}, 1 < y, & \text{and} \\ 0, \text{ otherwise.} \end{cases}$$

a. Find the values of k_1 and k_2.
b. Find $f_{X,Y}(x, y)$.
c. Find $P(X < 2 \mid Y = 3)$.

11. An experiment consists of n identical independent trials. X_i is the r.v. associated with the ith trial. Each X_i has a p.f. of

$$p_X(a) = \begin{cases} 1 - p, a = 0, \\ p, a = 1, & \text{and} \\ 0, \text{ otherwise.} \end{cases}$$

Let $Y = \sum_{i=1}^{n} X_i$ = number of trials whose outcome is 1.
a. Show that the p.f. of Y is

$$p_y(a) = \begin{cases} \binom{n}{a} p^a (a - p)^{n-a}, a = 0, 1, \ldots, n, & \text{and} \\ 0, \text{ otherwise.} \end{cases}$$

b. Verify that $\sum_{a=0}^{n} p_Y(a) = 1$.

4

Expectation

4.1
Introduction

At this point we have a general idea of what random variables are and how to use them to compute probabilities for random experiments. We now turn to the question of describing properties of probability distributions. One of the most important concepts in this regard is that of *expected value*. To give some meaning to this idea consider the following situation. Suppose we are asked to play a game of chance. The rules of the game are as follows.

1. A fair die is tossed.
2. If the die shows 1 dot we win $10.
3. If the die shows 6 dots we win $2.
4. On any other outcome we lose $3.

Is this a fair game? By the term *fair* we mean that the game does not, in the long run, result in one player winning more than the other. One way to answer this would be to play the game a large number of times. When this game was simulated on a computer and run one million times it turned out that, on the average, the reader would win about $0.01, per play. This is pretty strong evidence that the game is a fair one. However, we can calculate the game's fairness with a lot less technological support. Let X = the amount that we win, in dollars, on a play of the game. Negative numbers denote losses. From the rules of the game we see that $\frac{1}{6}$ of the time X will be 10, $\frac{1}{6}$ of the time X will be 2, and $\frac{2}{3}$ of the time X will be -3. Thus, if we compute a *weighted average* of the possible outcomes with the weightings being the proportion of the time a particular result occurs, we find that

$$\text{Avg. outcome} = \frac{1}{6} \cdot 10 + \frac{1}{6} \cdot 2 + \frac{2}{3} \cdot (-3).$$
$$= 0.$$

According to this we would expect neither to win nor lose any money on an *average* play of this game. In other words, our *expected winnings* per play is zero. This is the idea underlying expected value. The expected value of a random variable measures the "typical" outcome of a random experiment.

Another important use of expected values is to describe the shape of the distribution of a random variable. In Chapter 3 we saw that the median, γ_{50}, could be used to describe, in some sense, the "middle" of the distribution. It also turns out that we can use the distance between the first and third quartiles to determine if one probability distribution is more "spread out" than another. That is, if a probability distribution has a value of $\gamma_{75} - \gamma_{25}$, which is large compared to that of another, then the values of that random variable are said to show more dispersion than the other. In this chapter we will investigate other quantities called *moments* that are used to describe the distribution of a random variable. These quantities are special types of expected values.

In calculus we studied the behavior of a function by looking at its graph. The area under the graph could be found by using definite integrals. In addition, we could tell whether the function was increasing or decreasing, concave up or concave down by looking at the values of its derivatives. Expected values play a similar role in summarizing the attributes of the distribution of a random variable. These expected values are often the quantities of interest in applications of probability. For example, management executives often must make decisions when the exact result cannot be predicted. In such cases, it is common practice to base decisions on what the average outcome would be if the same conditions were to occur a large number of times. If the uncertainty is modeled by a probability distribution, then this average outcome will turn out to be an expected value.

4.2
Expected Value

As we saw in the introduction, a weighted average can be used to determine the average outcome of a random experiment. The formal definition of *expected value* can be motivated by viewing our game of chance as follows. X, the amount we win per play, is a discrete random variable with a probability function of

$$p_X(a) = \begin{cases} \dfrac{1}{6}, & \text{if } a = 2 \text{ or } 10, \\ \dfrac{2}{3}, & \text{if } a = -3, \text{ and} \\ 0, & \text{otherwise.} \end{cases}$$

If we let $E[X]$ denote our expected winnings we then computed $E[X]$ as follows.

$$E[X] = \frac{1}{6} \cdot 10 + \frac{1}{6} \cdot 2 + \frac{2}{3} \cdot (-3)$$

In other words, we take those values of the random variable that are assigned positive probability and computed a weighted sum where the weightings are the probabilities associated with each one. Based on this we have the following formal definition.

Definition 4.1

Let X be a random variable. The *expected value of X*, denoted by $E[X]$, is

$$E[X] = \sum_{a_i} a_i \, p_X(a_i), \qquad (4.1)$$

if X is a discrete random variable with a probability function of $p_X(a)$. If X is a continuous random variable with a probability density function of $f_X(x)$, then

$$E[X] = \int_{-\infty}^{\infty} x f_X(x) \, dx. \qquad (4.2)$$

If $\sum_{a_i} a_i \, p_X(a_i)$ or $\int_{-\infty}^{\infty} x f_X(x) \, dx$ does not converge absolutely, then the expected value is said not to exist.

Note: By absolute convergence we mean that $\sum_{a_i} |a_i| \, p_X(a_i)$ or $\int_{-\infty}^{\infty} |x| f_X(x) \, dx$, whichever is appropriate, must converge to a finite value.

If X is a discrete random variable, Equation 4.1 shows that $E[X]$ is a weighted average of the possible values of X. Each value is weighted by the value of the probability function at that point. If X is a continuous random variable, Equation 4.2 shows that $E[X]$ is the centroid of the probability density function. In either case, $E[X]$ measures the location of the "center" of the distribution. To use an analogy from physics, $E[X]$ is the "center of gravity" of the distribution. Along with the median we now have two different measures of the "center" of the distribution. We will compare them in some detail in Chapter 8. We illustrate Definition 4.1 with some examples.

Example 4.1

Let the random variable X have a p.f. of

$$p_X(a) = \begin{cases} 0.2, & a = 0, \\ 0.3, & a = 1, \\ 0.1, & a = 3, \\ 0.4, & a = 5, \quad \text{and} \\ 0, & \text{otherwise.} \end{cases}$$

Since this is a discrete random variable $E[X]$ is computed using Equation 4.1 as follows.

$$E[X] = \sum_{a_i} a_i \, p_X(a_i)$$

$$= 0 \cdot (0.2) + 1 \cdot (0.3) + 3 \cdot (0.1) + 5 \cdot (0.4)$$

$$= 2.6 \qquad \blacksquare$$

Example 4.2

The random variable X has a p.f. of

$$p_X(a) = \begin{cases} \dfrac{\lambda^a e^{-\lambda}}{a!}, & a = 0, 1, 2, \ldots, \quad \text{and} \\ 0, & \text{otherwise, where } \lambda > 0. \end{cases}$$

A random variable with this p.f. is said to have a *Poisson distribution*. Since it is discrete, we use Equation 4.1 and the Taylor series for e^x to compute $E[X]$.

$$E[X] = \sum_{a=0}^{\infty} a \frac{\lambda^a e^{-\lambda}}{a!}$$

$$= \sum_{a=1}^{\infty} \frac{\lambda^a e^{-\lambda}}{(a-1)!}$$

Now let $b = a - 1$.

$$E[X] = \sum_{b=0}^{\infty} \frac{\lambda^{b+1} e^{-\lambda}}{b!}$$

$$= \lambda e^{-\lambda} \sum_{b=0}^{\infty} \frac{\lambda^b}{b!}$$

The summation is the Taylor series for e^{λ}. Thus,

$$E[X] = \lambda e^{-\lambda} e^{\lambda}$$

$$= \lambda.$$

■

Example 4.3   The random variable X has a p.f. of

$$p_X(a) = \begin{cases} \binom{n}{a} p^a (1-p)^{n-a}, & a = 0, 1, \ldots, n, \quad \text{and} \\ 0 & , \text{otherwise, where } 0 \le p \le 1. \end{cases}$$

A random variable with this p.f. is said to have a *binomial distribution* because each value of $p_X(a)$ is a term in the binomial expansion of $[p + (1-p)]^n$. We use Equation 4.1 to find $E[X]$.

$$E[X] = \sum_{a=0}^{n} a \binom{n}{a} p^a (1-p)^{n-a}$$

$$= \sum_{a=1}^{n} \frac{n!}{(n-a)!(a-1)!} p^a (1-p)^{n-a}$$

$$= np \sum_{a=1}^{n} \frac{(n-1)!}{(n-a)!(a-1)!} p^{a-1} (1-p)^{n-a}$$

Now let $b = a - 1$.

$$E[X] = np \sum_{b=0}^{n-1} \binom{n-1}{b} p^b (1-p)^{n-1-b}$$

This is the binomial expansion of $[p + (1 - p)]^{n-1}$. Therefore,

$$E[X] = np[p + (1 - p)]^{n-1}$$
$$= np.$$

∎

Example 4.4

The random variable X has a p.f. of

$$p_X(a) = \begin{cases} \dfrac{1}{a(a + 1)} , & a = 1, 2, \ldots, \text{ and} \\ 0 , & \text{otherwise.} \end{cases}$$

It may not be clear that this is a probability function. It is clearly nonnegative, and the sum of the p.f. is

$$\sum_{a=1}^{\infty} \frac{1}{a(a + 1)} = \sum_{a=1}^{\infty} \left(\frac{1}{a} - \frac{1}{a + 1} \right) = 1 - \lim_{a \to \infty} \frac{1}{a + 1} = 1.$$

Using Equation 4.1 we get

$$E[X] = \sum_{a=1}^{\infty} a \frac{1}{a(a + 1)}$$
$$= \sum_{a=1}^{\infty} \frac{1}{a + 1},$$

which can be seen to diverge by comparison with the harmonic series

$$\sum_{n=1}^{\infty} \frac{1}{n}.$$

Therefore, $E[X]$ does not exist.

∎

Example 4.5

Let X be a r.v. with a p.d.f. of

$$f_X(x) = \begin{cases} 1, & 0 \le x \le 1, \text{ and} \\ 0, & \text{otherwise.} \end{cases}$$

A random variable whose p.d.f. is constant over an interval and zero everywhere else is said to have a *uniform distribution*. Since this is a continuous r.v. we use Equation 4.2 as follows.

$$E[X] = \int_{-\infty}^{\infty} x f_X(x) \, dx$$
$$= \int_{0}^{1} x \cdot 1 \, dx$$
$$= \frac{1}{2}$$

∎

Example 4.6   Let the r.v. X have a p.d.f. of

$$f_X(x) = \begin{cases} \lambda e^{-\lambda x}, & 0 < x, \quad \text{and} \\ 0, & \text{otherwise, where } 0 < \lambda. \end{cases}$$

A random variable with this p.d.f. is said to have an *exponential distribution*. We use Equation 4.2, integration by parts, and L'Hopital's rule to compute $E[X]$ as follows.

$$E[X] = \int_0^\infty x \lambda e^{-\lambda x} \, dx$$

$$= -x e^{-\lambda x} \Big|_0^\infty + \int_0^\infty e^{\lambda x} \, dx$$

$$= -\lim_{a \to \infty} a e^{-\lambda a} - \frac{1}{\lambda} \left(\lim_{b \to \infty} e^{-\lambda b} - 1 \right)$$

$$= \frac{1}{\lambda}$$

■

Example 4.7   Let the r.v. X have a p.d.f. of

$$f_X(x) = \frac{1}{2} e^{-|x|}, \quad -\infty < x < \infty.$$

A random variable with this p.d.f. is said to have a *double exponential* distribution. Equation 4.2 gives

$$E[X] = \int_{-\infty}^\infty x \frac{1}{2} e^{-|x|} \, dx = \int_{-\infty}^0 x \frac{1}{2} e^x \, dx + \int_0^\infty x \frac{1}{2} e^{-x} \, dx.$$

Now if we let $y = -x$ in the first integral, we obtain

$$E[X] = -\int_0^\infty \frac{1}{2} y e^{-y} \, dy + \int_0^\infty \frac{1}{2} x e^{-x} \, dx = 0.$$

Here we exploited the fact that $x e^{-|x|}$ is an odd function.   ■

Example 4.8   Let X have a p.d.f. of

$$f_X(x) = \begin{cases} \dfrac{1}{x^2}, & 1 \le x, \quad \text{and} \\ 0, & \text{otherwise.} \end{cases}$$

Here Equation 4.2 gives

$$E[X] = \int_1^\infty x \frac{1}{x^2} \, dx = \int_0^\infty \frac{1}{x} \, dx = \lim_{a \to \infty} \ln(a) - \ln(1) = \infty.$$

Therefore, $E[X]$ does not exist.   ■

Exercises 4.2

1. Find $E[X]$, if it exists, for the following distributions.

   a.
   $$p_X(a) = \begin{cases} \dfrac{1}{5}, & a = 1, 2, 3, 4, 5, \quad \text{and} \\ 0, & \text{otherwise} \end{cases}$$

   b.
   $$p_X(a) = \begin{cases} \dfrac{6}{\pi^2 a^2}, & a = 1, 2, \ldots, \quad \text{and} \\ 0, & \text{otherwise} \end{cases}$$

   c.
   $$p_X(a) = \begin{cases} \dfrac{2a}{n(n+1)}, & a = 1, 2, \ldots, n, \quad \text{and} \\ 0, & \text{otherwise} \end{cases}$$

2. Find $E[X]$, if it exists, for each of the following distributions.

   a.
   $$f_X(x) = \begin{cases} 6x(1-x), & 0 < x < 1, \quad \text{and} \\ 0, & \text{otherwise} \end{cases}$$

   b.
   $$f_X(x) = \begin{cases} \dfrac{2}{x^3}, & 1 < x, \quad \text{and} \\ 0, & \text{otherwise} \end{cases}$$

   c.
   $$f_X(x) = \dfrac{1}{\sqrt{2\pi}} e^{-(1/2)x^2}, \quad -\infty < x < \infty$$

3. Let X be a r.v. with a p.f. of

   $$p_X(a) = \begin{cases} p(1-p)^{a-1}, & a = 1, 2, \ldots, \quad \text{and} \\ 0, & \text{otherwise, where } 0 < p < 1. \end{cases}$$

   Show that $E[X] = \frac{1}{p}$. (Hint: You can view $E[X]$ as being the derivative of a geometric series.)

4. Let the r.v. X have a p.d.f. of

   $$f_X(x) = \dfrac{e^{-x}}{(1 + e^{-x})^2}, \quad -\infty < x < \infty.$$

   Find $E[X]$.

5. A box contains 6 white balls and 4 black balls. Balls are removed from the box without replacement until either a white ball is removed or 3 balls have been drawn. Find the expected number of balls that will be removed from the box.

6. A target consists of 3 concentric circles of radii $\frac{1}{2}$, 1, and 2 feet, respectively. A shot hitting the inner circle scores 4 points. A hit in the next ring gets 3 points. A shot in the outer ring gets 2 points, and any shot outside all 3 circles gets no points. Let R be the distance, in feet, from the center of the circles, and let it have a p.d.f. of

   $$f_R(r) = \begin{cases} \dfrac{2}{\pi} \dfrac{1}{1 + r^2}, & 0 \le r, \quad \text{and} \\ 0, & \text{otherwise.} \end{cases}$$

   Find the expected score for a shot at this target.

7. The sentence "The quick brown fox jumped over the lazy dog" is a common typing drill. Suppose that a word is selected at random from this sentence. Find the expected length of the word selected.

8. In a roulette game, the wheel has 18 red numbers, 18 black numbers, and 2 green numbers. Players are allowed to bet that the wheel will stop on either a red number or a black number. A player who bets that a black number will occur receives the amount he bet and an amount equal to his bet if a black number occurs. In other words, if a dollar is bet and is successful, then I get my dollar back plus another dollar. A similar payoff occurs for a red number. If a particular color is bet on and another color occurs, then the amount of the bet is lost. If one dollar is bet on this game, what is the expected payoff for one play of the game?

9. Suppose in the roulette game described in Exercise 8 you follow the following betting strategy. Begin by betting one dollar on red. If you lose, double your bet on red for the next play. If you lose again, double your previous bet and play red again. This continues until a red occurs, whereupon you quit. What are your expected winnings?

10. An urn contains 6 white balls and 4 black balls. Balls are drawn without replacement until a black ball occurs. Find the expected number of balls that will be drawn.

11. Let X be a continuous r.v. whose p.d.f. is symmetric about $x = a$. Show that $E[X] = a$.

12. Let X be a r.v. with p.d.f. $f_X(x)$ and d.f. $F_X(a)$ such that $E[X]$ exists. Show that

$$E[X] = \int_0^\infty [1 - F_X(x)]\, dx - \int_{-\infty}^0 F_X(-x)\, dx.$$

Hint: Begin with

$$\int_0^\infty [1 - F_X(x)]\, dx = \int_0^\infty \int_x^\infty f_X(y)\, dy\, dx, \quad \text{and}$$

$$\int_{-\infty}^0 F_X(-x)\, dx = \int_{-\infty}^0 \int_{-\infty}^{-x} f_X(y)\, dy\, dx.$$

Then interchange the order of integration.

4.3
Expectation of a Function of a Random Variable

Oftentimes we are interested in the expected value of some function of a random variable X rather than $E[X]$. For example, suppose the game described in the introduction is altered as follows. We are asked to pay $5 per roll to play the game. The operator of the game tells us that each 1 rolled will give a payoff of $12. In addition, each 3 or 5 rolled will pay us $1, but any even roll loses. We are interested in whether the expected payoff equals or exceeds $5. If we let X be the number of dots showing after a roll, it is clear that the payoff will be a function of X, say $g(X)$. What we wish to compute is $E[g(X)]$. In addition to this kind of question, the expectation of other functions of a random variable provide additional information about the distribution of that random variable. We will look at these in more detail in Section 4.4. In the second half of this book we shall also see that it is common practice to compare statistical procedures by looking at the expectation of functions of random variables.

One method for finding $E[g(X)]$ is to find the probability function or probability density function for $Y = g(X)$ using methods which will be considered in Chapter 6. Given the distribution for $Y = g(X)$ we would then use Equation 4.1 or Equation 4.2 to compute $E[Y]$. We illustrate with some examples.

Example 4.9   Consider the game described in the introduction to this section. Since the die is fair we know that X will have a p.f. of

$$p_X(a) = \begin{cases} \dfrac{1}{6}, & a = 1, 2, 3, 4, 5, 6, \quad \text{and} \\ 0, & \text{otherwise.} \end{cases}$$

If we let $Y = g(X)$ be the payoff upon getting X, we see that

$$g(a) = \begin{cases} 12, & a = 1, \\ 1, & a = 3, 5, \quad \text{and} \\ -5, & a = 2, 4, 6. \end{cases}$$

From the description of the game we find that the p.f. of Y is

$$p_Y(a) = \begin{cases} \dfrac{1}{6}, & a = 12, \\ \dfrac{1}{3}, & a = 1, \\ \dfrac{1}{2}, & a = -5, \quad \text{and} \\ 0, & \text{otherwise.} \end{cases}$$

Then, Equation 4.1 gives that

$$E[Y] = 12 \cdot \frac{1}{6} + 1 \cdot \frac{1}{3} + (-5) \cdot \frac{1}{2} = -\frac{1}{6}.$$

This indicates that, on the average, we will lose money by playing this game. It is interesting to note that

$$\sum_{a=1}^{6} g(a) p_X(a) = 12 \cdot \frac{1}{6} + (-5) \cdot \frac{1}{6} + 1 \cdot \frac{1}{6} + (-5) \cdot \frac{1}{6} + 1 \cdot \frac{1}{6} + (-5) \cdot \frac{1}{6}$$

$$= -\frac{1}{6}.$$

■

Example 4.10   Let the r.v. X have a p.d.f. of

$$f_X(x) = \begin{cases} 2x, & 0 < x < 1, \quad \text{and} \\ 0, & \text{otherwise.} \end{cases}$$

We wish to find the expected value of $g(X) = 1 - X$. It turns out that the p.d.f. for $Y = 1 - X$ is

$$f_Y(y) = \begin{cases} 2(1 - y), & 0 < y < 1, \quad \text{and} \\ 0, & \text{otherwise.} \end{cases}$$

We now use Equation 4.2 to get

$$E[Y] = \int_0^1 y\,2(1-y)\,dy = \frac{1}{3}.$$

Here, too, we note the following.

$$\int_{-\infty}^{\infty} g(x)f_X(x)\,dx = \int_0^1 (1-x)\,2x\,dx = \frac{1}{3} \qquad \blacksquare$$

At the end of each of these two examples we indicated an interesting fact about $E[\,g(X)]$. It turns out that this is not merely a coincidence. We can always find $E[\,g(X)]$ without first having to determine the probability function or probability density function of $g(X)$. We state this result as a theorem.

Theorem 4.1

Let X be a random variable, and let $g(X)$ be a function X. Then the following are true.

a. When X is a discrete random variable with probability function $p_X(a)$,

$$E[\,g(X)] = \sum_{a_i} g(a_i)\,p_X(a_i). \qquad (4.3)$$

b. When X is a continuous random variable with probability density function $f_X(x)$,

$$E[\,g(X)] = \int_{-\infty}^{\infty} g(x)f_X(x)\,dx. \qquad (4.4)$$

Proof We shall prove the theorem for the case when X is a discrete random variable. The continuous case requires that we use the result of Exercise 10 of Section 4.2 and is left as an exercise. We begin by letting $Y = g(X)$. Let $b_1, b_2, \ldots$ be the possible values of Y, and $a_1, a_2, \ldots$ be the possible values of X. Note that $g(X)$ is not necessarily a one-to-one function. Therefore, there may be many a_i's that are transformed to a particular value b_j. Let $A_j = \{a_i : g(a_i) = b_j\}$. That is A_j is the set of values of X that are mapped to b_j. This means that there is a relation between the p.f. for X and the p.f. for Y as follows.

$$p_Y(b_j) = P(Y = b_j) = P(X \in A_j) = \sum_{a \in A_j} p_X(a)$$

Using this we compute $E[Y]$ as follows.

$$E[\,g(X)] = E[Y]$$

$$= \sum_j b_j\,p_Y(b_j)$$

$$= \sum_j b_j \cdot \left(\sum a \in A_j p_X(a) \right)$$

$$= \sum_j \sum_{a \in A_j} b_j p_X(a)$$

We now note that $b_j = g(a)$ and finish the proof as follows.

$$E[g(X)] = \sum_j \sum_{a \in A_j} g(a)p_X(a)$$

$$= \sum_{\text{all } a} g(a)p_X(a) \qquad \blacksquare$$

Example 4.11   Let X be a r.v. with a p.f. of

$$p_X(a) = \begin{cases} \dfrac{\lambda^a e^{-\lambda}}{a!} , & a = 0, 1, 2, \ldots, \quad \text{and} \\ 0 , & \text{otherwise, where } \lambda > 0. \end{cases}$$

We wish to find $E[X^2]$. Since X is discrete we use Equation 4.3 to get

$$E[X^2] = \sum_{a=0}^{\infty} a^2 \frac{\lambda^a e^{-\lambda}}{a!}$$

$$= \sum_{a=0}^{\infty} a^2 \frac{\lambda^a e^{-\lambda}}{a!} - \sum_{a=0}^{\infty} a \frac{\lambda^a e^{-\lambda}}{a!} + \sum_{a=0}^{\infty} a \frac{\lambda^a e^{-\lambda}}{a!}$$

$$= \sum_{a=2}^{\infty} \frac{\lambda^a e^{-\lambda}}{(a-2)!} + E[X]$$

$$= \lambda^2 \sum_{a=2}^{\infty} \frac{\lambda^{a-2} e^{-\lambda}}{(a-2)!} + E[X]$$

$$= \lambda^2 \sum_{b=0}^{\infty} \frac{\lambda^b e^{-\lambda}}{b!} + E[X]$$

$$= \lambda^2 + E[X].$$

From Example 4.2 we know that $E[X] = \lambda$. Therefore,

$$E[X^2] = \lambda^2 + \lambda.$$

This would be a hard problem without Theorem 4.1. $\qquad \blacksquare$

Example 4.12   The volume (in cm^3), V, of ball bearings manufactured by a certain process has a p.d.f. of

$$f_V(v) = \begin{cases} \dfrac{1}{v}, & 1 < v < e, \quad \text{and} \\ 0, & \text{otherwise.} \end{cases}$$

We wish to find the expected value of the radius of the ball bearings. Let R be the radius of the sphere. The volume of a sphere of radius r is

$$V = \frac{4}{3} \pi R^3.$$

Therefore, we can use Equation 4.4 to find

$$E[R] = E\left[\left(\frac{3V}{4\pi}\right)^{1/3}\right]$$

$$= \int_1^e \left(\frac{3v}{4\pi}\right)^{1/3} \frac{1}{v}\, dv$$

$$= \left(\frac{3}{4\pi}\right)^{1/3} 3(e^{1/3} - 1) \quad \text{cm.}$$

∎

Example 4.13   A game is said to be *fair* if a player antes an amount equal to the expected winnings. A card is drawn from a standard deck of 52 cards. The payoffs are as follows.

Card	Result
Ace	Win 20 cents
Face card	Win 10 cents
Other card	Lose 5 cents

We wish to determine how much a player should ante to make this game fair. Let X be the number of the card drawn. That is, $X = 2$ if a deuce is drawn, 3 for a trey, and so forth. For the cards without numbers, let $X = 11$ for a jack, 12 for a queen, 13 for a king, and 1 for an ace. The function $g(X)$ for the payoffs would be as follows.

X	$g(X)$
1	0.20
$2, 3, \ldots, 10$	-0.05
$11, 12, 13$	0.10

The p.f. for X is

$$p_X(x) = \begin{cases} \dfrac{1}{13}, & x = 1, 2, \ldots, 13, \quad \text{and} \\ 0, & \text{otherwise.} \end{cases}$$

For the game to be fair,

$$\text{Amount of ante} = E[g(X)] = \sum_{x=1}^{13} g(x) p_X(x)$$

$$= \frac{1}{13}[1 \cdot (0.2) + 9 \cdot (-0.05) + 3 \cdot (0.1)]$$

$$= \frac{5}{13} \quad \text{cents.}$$

∎

A useful consequence of Theorem 4.1 is the following.

Theorem 4.2

Let X be a random variable. If a and b are constants, then

$$E[aX + b] = a E[X] + b. \qquad (4.5)$$

Proof We give the proof for the continuous case. If X is discrete, the proof is the same with sums replacing integrals. Let $f_X(x)$ be the probability density function of X. By Theorem 4.1

$$E[aX + b] = \int_{-\infty}^{\infty} (ax + b) f_X(x)\, dx$$

$$= a \int_{-\infty}^{\infty} x f_X(x)\, dx + b \int_{-\infty}^{\infty} f_X(x)\, dx$$

$$= a E[X] + b.$$

■

Note: If $a = 0$, then Theorem 4.2 gives that $E[b] = b$. In other words, the expected value of a constant is that constant.

Example 4.14

Ships mooring in the harbor of a particular country must reserve fresh water in advance. To prompt ships to reserve sufficient water, the local port authority charges for water according to the following rules. The basic price for reserved water is a dollars per gallon. If a ship does not reserve enough water, a fee of $\frac{a}{2}$ dollars per extra gallon is levied. A ship's water consumption X is a random variable with a p.d.f. of $f_X(x)$. We wish to know how many gallons the ship should reserve in order to minimize expected costs. Let $y =$ gallons of water reserved. Then water costs $C(y)$ are

$$C(y) = \begin{cases} ay & , X \le y, \quad \text{and} \\ ay + \dfrac{3}{2}a(X - y), & y < X. \end{cases}$$

Then the expected water costs are

$$E[C(y)] = \int_0^y ay f_X(x)\, dx + \int_y^{\infty} [ay + \frac{3}{2}a(x - y)] f_X(x)\, dx$$

$$= E[ay] + \frac{3}{2}a \int_y^{\infty} (x - y) f_X(x)\, dx.$$

In terms of integration on x, ay is a constant. Thus,

$$E[C(y)] = ay + \frac{3}{2}a \int_y^{\infty} (x - y) f_X(x)\, dx$$

$$= ay + \frac{3}{2}a \left(\int_y^{\infty} x f_X(x)\, dx - y[1 - F_X(y)] \right).$$

Note that $F_X(y)$ is the d.f. for X evaluated at y. We now differentiate with respect to y to find the minimum value. Recall that the Fundamental Theorem of Calculus gives that

$$\frac{d}{dy}\int_y^\infty f_X(x)\,dx = -\frac{d}{dy}\int_\infty^y f_X(x)\,dx = -f_X(y).$$

Therefore,

$$\frac{d}{dy}E[C(y)] = a + \frac{3}{2}a\frac{d}{dy}\int_y^\infty xf_X(x)\,dx - \frac{3}{2}a\frac{d}{dy}y[1 - F_X(y)]$$

$$= a + \frac{3}{2}a\left(yf_X(y) - yf_X(y) - [1 - F_X(y)]\right)$$

$$= -\frac{1}{2}a + \frac{3}{2}aF_X(y).$$

Setting this derivative equal to zero gives that

$$F_X(y) = \frac{1}{3}.$$

It is easy to check that the second derivative is positive everywhere. Thus, expected costs are minimized if the ship orders water equal to the value of the distribution of X where $F_X(y) = \frac{1}{3}$. ∎

In the case of a random vector $(X_1, X_2, \ldots, X_n)$ we can use Theorem 4.1 as a basis for calculating $E[g(X_1, X_2, \ldots, X_n)]$. We state the result as a theorem without proof.

Theorem 4.3

Let $(X_1, X_2, \ldots, X_n)$ be a random vector, and $g(X_1, X_2, \ldots, X_n)$ be a function of $(X_1, X_2, \ldots, X_n)$.

a. If $(X_1, X_2, \ldots, X_n)$ is a vector of discrete random variables with joint probability function $p_{X_1, X_2, \ldots, X_n}(a_1, a_2, \ldots, a_n)$, then

$$E[g(X_1, \ldots, X_n)] = \sum_{a_1} \cdots \sum_{a_n} g(a_1, \ldots, a_n)p_{X_1, \ldots, X_n}(a_1, \ldots, a_n). \qquad (4.6)$$

b. If $(X_1, X_2, \ldots, X_n)$ is a vector of continuous random variables with joint probability density function $f_{X_1, X_2, \ldots, X_n}(x_1, x_2, \ldots, x_n)$, then

$$E[g(X_1, \ldots, X_n)] = \int_{-\infty}^\infty \cdots \int_{-\infty}^\infty g(x_1, \ldots, x_n)f_{X_1, \ldots, X_n}(x_1, \ldots, x_n)\,dx_1 \cdots dx_n.$$
$$(4.7)$$

Example 4.15   Let (X, Y) have a joint p.d.f. of

$$f_{X,Y}(x, y) = \begin{cases} 6xy, & 0 < x < y < 1, \quad \text{and} \\ 0, & \text{otherwise.} \end{cases}$$

We wish to find $E[XY]$. Since X and Y are continuous r.v.'s Equation 4.7 gives

$$E[XY] = \int_{-\infty}^{\infty} \int_{-\infty}^{\infty} xy f_{X,Y}(x,y)\, dx\, dy.$$

$$= \int_{0}^{1} \int_{0}^{y} xy\, 6xy\, dx\, dy$$

$$= \int_{0}^{1} 2y^5\, dy$$

$$= \frac{1}{3}.$$

■

It is common for applications of probability theory to form linear combinations of random variables. The expected value of such a combination can be found by using Theorem 4.3. We give the result as a theorem below.

Theorem 4.4

> Let X and Y be random variables. If $E[X]$ and $E[Y]$ exist, then for constants a and b
>
> $$E[aX + bY] = aE[X] + bE[Y]. \tag{4.8}$$

Proof   We give the proof when X and Y are continuous. The proof of the discrete case is identical with summation replacing integration. Equation 4.7 gives

$$E[aX + bY] = \int_{-\infty}^{\infty} \int_{-\infty}^{\infty} (ax + by) f_{X,Y}(x,y)\, dx\, dy$$

$$= a \int_{-\infty}^{\infty} \int_{-\infty}^{\infty} x f_{X,Y}(x,y)\, dx\, dy + b \int_{-\infty}^{\infty} \int_{-\infty}^{\infty} x f_{X,Y}(x,y)\, dx\, dy$$

$$= a \int_{-\infty}^{\infty} x \left(\int_{-\infty}^{\infty} f_{X,Y}(x,y)\, dy \right) dx + b \int_{-\infty}^{\infty} y \left(\int_{-\infty}^{\infty} f_{X,Y}(x,y)\, dx \right) dy.$$

The interior integrals are the p.d.f.'s of X and Y, respectively. Thus,

$$E[aX + bY] = a \int_{-\infty}^{\infty} x f_X(x)\, dx + b \int_{-\infty}^{\infty} y f_Y(y)\, dy.$$

These integrals are $E[X]$ and $E[Y]$, respectively. Therefore,

$$E[aX + bY] = aE[X] + bE[Y].$$

■

This theorem shows that expectations share a linearity property with derivatives and integrals. That is, the expected value of a linear combination is the linear combination of the expected values. This result can be extended to a linear combination of any number of random variables by using induction. The proof is left as an exercise.

Theorem 4.5

> Let $X_1, X_2, \ldots, X_n$ be random variables such that $E[X_i]$ exists for $i = 1, 2, \ldots, n$. Then for constants $a_1, a_2, \ldots, a_n$
>
> $$E[a_1 X_1 + a_2 X_2 + \ldots + a_n X_n] = a_1 E[X_1] + a_2 E[X_2] + \cdots + a_n E[X_n]. \quad \textbf{(4.9)}$$

Example 4.16

A random number generator produces a series of digits in the range 0 through 9. If X_i is ith digit produced, for $i = 1, 2, \ldots, n$, then $X_1, X_2, \ldots, X_n$ are i.i.d. random variables. Let each random variable have a p.f. of

$$p_X(a) = \begin{cases} \dfrac{1}{10}, & a = 0, 1, 2, 3, 4, 5, 6, 7, 8, 9, \quad \text{and} \\ 0, & \text{otherwise.} \end{cases}$$

We wish to find the expected value of the sum of the n digits. Since these are discrete random variables we use Equation 4.9 to show that

$$E[X_1 + X_2 + \cdots + X_n] = E[X_1] + E[X_2] + \cdots + E[X_n].$$

Since the r.v.'s are i.i.d., $E[X_1] = E[X_2] = \ldots = E[X_n]$. This means that

$$E[X_1 + X_2 + \cdots + X_n] = nE[X_1].$$

We compute $E[X_1]$ from Equation 4.1 as follows.

$$E[X_1] = \sum_{a=0}^{9} a \frac{1}{10} = \frac{9}{2}$$

Therefore,

$$E[X_1 + X_2 + \cdots + X_n] = \frac{9}{2} n. \qquad \blacksquare$$

If a set of random variables is independent, then we can use Theorem 3.6 and Theorem 4.3 to obtain the following result.

Theorem 4.6

> Let X and Y be independent random variables. Then, provided $E[g(X)]$ and $E[h(Y)]$ exist,
> $$E[g(X)h(Y)] = E[g(X)]E[h(Y)]. \qquad \textbf{(4.10)}$$

Proof We give the proof for the case that (X, Y) is a continuous random vector. The proof for the discrete case is identical with sums replacing integrals. Let $f_{X,Y}(x, y)$ be the joint p.d.f. of (X, Y). Then Theorem 4.3 gives

$$E[g(X)h(Y)] = \int_{-\infty}^{\infty} \int_{-\infty}^{\infty} g(x)h(y)f_{X,Y}(x, y)\, dx\, dy.$$

Since X and Y are independent Theorem 3.6 shows that the joint p.d.f. factors into the product of the p.d.f.'s. We then complete the proof as follows.

$$E[g(X)h(Y)] = \int_{-\infty}^{\infty}\int_{-\infty}^{\infty} g(x)h(y)f_X(x)f_Y(y)\,dx\,dy$$

$$= \left(\int_{-\infty}^{\infty} g(x)f_X(x)\,dx\right)\left(\int_{-\infty}^{\infty} h(y)f_Y(y)\,dy\right)$$

$$= E[g(X)]E[h(Y)]$$

■

Example 4.17   Let X and Y be independent r.v.'s. The p.d.f. of X is

$$f_X(x) = \begin{cases} 2x, 0 < x < 1, & \text{and} \\ 0, \text{ otherwise.} \end{cases}$$

The p.d.f. of Y is

$$f_Y(y) = \begin{cases} 2(1-y), 0 < y < 1, & \text{and} \\ 0, \text{ otherwise.} \end{cases}$$

We wish to find $E[XY]$. Since X and Y are independent we can use Equation 4.10 with $g(X) = X$ and $h(Y) = Y$. Therefore,

$$E[XY] = E[X]E[Y].$$

The individual expected values are as follows.

$$E[X] = \int_0^1 x\,2x\,dx = \frac{2}{3}, \quad \text{and}$$

$$E[Y] = \int_0^1 y\,2(1-y)\,dy = \frac{1}{3}$$

Thus,

$$E[XY] = \frac{2}{3}\cdot\frac{1}{3} = \frac{2}{9}.$$

■

As might be suspected, Theorem 4.6 can be extended to random vectors of arbitrary length. The proof is by induction and is left as an exercise.

Theorem 4.7

Let $(X_1, X_2, \ldots, X_n)$ be a vector of independent random variables. Then, if $E[g_i(X_i)]$ exists, for $i = 1, 2, \ldots, n$,

$$E[g_1(X_1)g_2(X_2)\cdots g_n(X_n)] = E[g_1(X_1)]E[g_2(X_2)]\cdots E[g_n(X_n)]. \tag{4.11}$$

Exercises 4.3

1. Prove Theorem 4.1 for the case of a continuous random variable. Use the result of Exercise 12 of Section 4.2.

2. Grades in a statistics course for social science majors follow a distribution with a p.d.f. of

$$f_X(x) = \begin{cases} \dfrac{3}{32}x(4 - x), & 0 < x < 4, \quad \text{and} \\ 0, & \text{otherwise.} \end{cases}$$

The father of a prospective student offers to pay either \$5 or two times the grade received in the range [0,4]. Which offer should the student take?

3. In Exercise 6 of Section 4.2, find the expected score after five shots at the target.

4. Consider the following game. A bowl contains 15 numbered balls. There are i balls bearing the number i for $i = 1, 2, 3, 4,$ and 5. Player A draws a ball from the bowl. Player B is blindfolded and guesses the number on the ball drawn by Player A. Player B pays Player A an amount equal to the square of the difference between his guess and the actual number drawn. What number should Player B guess to minimize his expected losses?

5. In Exercise 4, what strategy should Player B use if he pays an amount equal to the absolute value of the difference between his score and the actual number drawn?

6. A drug store purchases a certain magazine for 50 cents which it sells for one dollar. The store must purchase all magazines at the beginning of the month and dispose of any unsold copies. If magazine sales are a random variable whose distribution can be approximated by a p.d.f. of

$$f_X(x) = \begin{cases} \dfrac{x}{5000}, & 0 \le x \le 100, \quad \text{and} \\ 0, & \text{otherwise.} \end{cases}$$

How many magazines should the drug store purchase per month in order to maximize expected profit?

7. Two evenly matched tennis players compete in a "best of seven sets" match. Each player wins \$D from his opponent for each set won. On average what will the winning player earn per set won? Assume that each player tries as hard as possible to win each set.

8. A certain laboratory procedure for testing blood for a rare disease costs \$100 per test. To cut costs to patients it is decided that the blood from 10 samples will be combined and tested. If the test reveals the presence of the disease then the 10 samples will be tested individually for the disease. What is the maximum proportion of the population that can have this disease in order to make this strategy less expensive than testing all samples individually? Assume sampling with replacement.

9. A right triangle has one side, not the hypotenuse, fixed at length 1. The angle between it and the hypotenuse is a random variable X having a probability density function of

$$f_X(x) = \begin{cases} \dfrac{4}{\pi}, & 0 \le x \le \frac{\pi}{4}, \quad \text{and} \\ 0, & \text{otherwise.} \end{cases}$$

Find the expected area of the triangle.

10. In Example 14 how much water should a ship request if there is also a fee imposed of $\frac{q}{4}$ dollars per gallon of water that is requested but not used?

11. In a transmission system the data are encoded and sent as eight-digit binary numbers. Each *binary digit* is called a *bit,* and the eight bits that constitute a number are called a *byte.* On a certain noisy circuit each bit has a probability p of being in error. The system is designed to detect and correct, at most, one erroneous bit in each byte. If more than one bit in a byte is in error the system will not detect that the byte is wrong. What is the expected number of incorrect bytes that will go undetected if n bytes are sent?

12. Show that $E[(X - a)^2]$ is a minimum when $a = E[X]$.

13. Show that $E\left[|X - a|\right]$ is a minimum when a is the median of the distribution of X.

14. Prove Theorem 4.5.

15. Prove Theorem 4.7.

16. Let X, Y, and Z be independent random variables such that $E[X] = E[Y] = E[Z] = 1$. Find $E[X(2Y - Z)]$.

17. Let X, Y, and Z be i.i.d. random variables each having a probability function of

$$p(a) = \begin{cases} \dfrac{1}{4}, & a = 1, 2, 3, 4, \quad \text{and} \\ 0, & \text{otherwise.} \end{cases}$$

Find $E[X - (Y + Z)^2]$.

4.4
Moments

The expected value of certain functions of a random variable are of particular interest. These are *moments* of distribution. Moments give useful information about the properties of a distribution. These numerical summaries make it easier to compare distributions. In addition, in statistical analyses it is typically the case that we are interested in using sample data to discover information about the values of certain moments or functions of them. For instance, it is common for investigators to wish to determine the value of $E[X]$. Other properties of probability distributions are also of interest. One often wishes to compare two distributions to determine which is more widely dispersed about its mean. The degree to which the graph of a distribution is nonsymmetric is important in statistics. For these kinds of questions it is useful to be able to describe the properties of the distribution of a random variable by using numerical summaries. This idea should be familiar from Calculus courses where, for instance, we calculated the curvature of a function in order to compare the graph with a standard curve such as a circle. In probability theory, moments provide useful numerical summaries of the properties of the distribution of a random variable. As an illustration, consider the graphs of the two p.d.f.'s, $f_X(x)$ and $f_Y(y)$, shown in Figure 4.1.

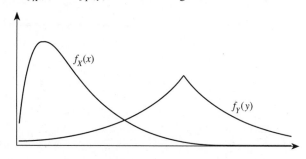

Figure 4.1

It can be seen that most of the values of Y are larger than those of X. However, the values of X are less dispersed than are those of Y. In addition, the graph of $f_X(x)$ is somewhat skewed, while the graph of $f_Y(y)$ appears to be symmetric. These properties can all be measured by using moments of distribution.

Definition 4.2

> Let X be a random variable. The *rth moment of X* is denoted μ'_r and is
>
> $$\mu'_r = E[X^r], \quad \text{for } r = 1, 2, \ldots. \tag{4.12}$$
>
> In addition, the *rth central moment of X* is denoted μ_r and is
>
> $$\mu_r = E\left[(X - E[X])^r\right], \quad \text{for } r = 2, 3, \ldots. \tag{4.13}$$

Note that moments are numbers computed by sums or integrals. Therefore, they are constants. There is no point in considering the first central moment because if $E[X]$ exists we would find that

$$\mu_1 = E[(X - E[X])]$$
$$= E[X] - E[E[X]]$$
$$= E[X] - E[X]$$
$$= 0.$$

The first moment of X is the expected value of X which we discussed in Section 4.2. There we noted that $E[X]$ can be used as a measure of the location of the "center" of the distribution. The term *measure of central tendency* is often used. $E[X]$ is referred to as the *mean* of the distribution and is denoted by μ_X. Using this notation central moments become

$$\mu_r = E[(X - \mu_X)^r].$$

As we have seen the moments of a distribution may not exist. Example 4.4 and Example 4.8 are cases where the distribution does not have a mean. It should be clear from Definition 4.2 that, when the mean of a distribution does not exist, none of the central moments exist. The existence of a mean does not, however, guarantee the existence of other moments as the following example shows.

Example 4.18   Let X be a random variable with a p.d.f. of

$$f_X(x) = \begin{cases} \dfrac{2}{x^3}, & 1 < x, \quad \text{and} \\ 0, & \text{otherwise.} \end{cases}$$

Note that the mean exists by the following calculation.

$$\mu_X = \int_1^\infty x \, \frac{2}{x^3} \, dx$$

$$= -\lim_{a \to \infty} \frac{2}{a} + 2$$

$$= 2$$

However, computing μ_2' gives

$$\mu_2' = E[X^2] = \int_1^\infty x^2 \, \frac{2}{x^3} \, dx$$

$$= -\lim_{a \to \infty} 2 \ln(a) + 0$$

$$= -\infty.$$

Thus, we see that the second moment does not exist. In fact, it turns out that no moment besides the mean exists for this random variable. ∎

The second central moment, μ_2, is used to measure the dispersion of the distribution about its mean. It is commonly referred to as the *variance of the distribution* and is denoted by Var(X). One often sees the symbol σ_X^2 used instead of Var(X). We shall use whichever one seems to be most convenient at the time. The units of Var(X) are the square of the units of X. For this reason the positive square root of the variance, called the *standard deviation of X,* is often used in practical applications. The standard deviation of X is usually denoted by σ_X. A formula that is often useful for computing Var(X) is given by the following theorem.

Theorem 4.8

> Let X be a random variable such that Var(X) exists. Then
>
> $$\text{Var}(X) = E[X^2] - \mu_X^2. \tag{4.14}$$

Proof   The idea is to use Definition 4.2, expand the square, and then use the linearity properties of expectation to get the result. These give the following.

$$\text{Var}(X) = E[(X - \mu_X)^2]$$

$$= E[X^2 - 2\mu_X X + \mu_X^2]$$

$$= E[X^2] - 2\mu_X E[X] + \mu_X^2$$

$$= E[X^2] - 2\mu_X^2 + \mu_X^2$$

$$= E[X^2] - \mu_X^2$$

∎

In certain cases it turns out to be computationally more convenient to note the following. We leave the verification as an exercise.

$$\text{Var}(X) = E[X(X - 1)] + \mu_X - \mu_X^2 \tag{4.15}$$

It is usually easier to work with a random variable if it is discrete and if the p.f. involves factorial terms.

Example 4.19   Let X be a Poisson r.v. with a p.f. of

$$p_X(a) = \begin{cases} \dfrac{\lambda^a e^{-\lambda}}{a!} & , a = 0, 1, 2, \ldots, \quad \text{and} \\ 0 & , \text{otherwise, where } \lambda > 0. \end{cases}$$

In Example 4.2 we found that $E[X] = \mu_X = \lambda$. In Example 4.9 we found that $E[X^2] = \lambda^2 + \lambda$. Therefore, using Equation 4.14 we find that

$$\text{Var}(X) = \lambda^2 + \lambda - \lambda^2 = \lambda.$$

∎

Example 4.20   Let X be a binomial r.v. with a p.f. of

$$p_X(a) = \begin{cases} \dbinom{n}{a} p^2(1 - p)^{n-a} & , a = 0, 1, 2, \ldots, n, \quad \text{and} \\ 0 & , \text{otherwise, where } 0 < p < 1. \end{cases}$$

In this case it turns out to be easier to compute the variance by using Equation 4.15. This proceeds as follows. We first evaluate $E[X(X - 1)]$.

$$E[X(X - 1)] = \sum_{a=0}^{n} a(a - 1) \binom{n}{a} p^a (1 - p)^{n-a}$$

$$= \sum_{a=0}^{n} a(a - 1) \binom{n}{a} p^a (1 - p)^{n-a}$$

$$= \sum_{a=2}^{n} \frac{n!}{(a - 2)!(n - a)!} p^a (1 - p)^{n-a}$$

$$= n(n - 1)p^2 \sum_{a=2}^{n} \binom{n - 2}{a - 2} p^{a-2} (1 - p)^{n-a}$$

Now let $b = a - 2$. Thus,

$$E[X^2] = n(n - 1)p^2 \sum_{b=0}^{n-2} \binom{n - 2}{b} p^b (1 - p)^{n-2-b}.$$

The summation is the binomial expansion of $[p + (1 - p)]^{n-2}$. Therefore,

$$E[X(X - 1)] = n(n - 1)p^2.$$

Then Equation 4.15 and the fact that $E[X] = np$ give

$$\text{Var}(X) = n^2p^2 - np^2 + np - n^2p^2 = np(1 - p).$$

∎

Example 4.21   Let X be a uniformly distributed r.v. with a p.d.f. of

$$f_X(x) = \begin{cases} 1, 0 < x < 1, & \text{and} \\ 0, \text{otherwise.} \end{cases}$$

In Example 4.5 we found that $\mu_X = \frac{1}{2}$. We now find $E[X^2]$ to be

$$E[X^2] = \int_0^1 x^2 \, dx = \frac{1}{3}.$$

Therefore, by Equation 4.14 we get

$$\text{Var}(X) = \frac{1}{3} - \left(\frac{1}{2}\right)^2 = \frac{1}{12}.$$

∎

Many applications make linear transformations on random variables. The effect of such a transformation on the mean was given in Theorem 4.2. We now look at the variance.

Theorem 4.9

Let X be a random variable whose variance exists. Then

$$\text{Var}(aX + b) = a^2 \, \text{Var}(X). \tag{4.16}$$

Proof   We begin by using Theorem 4.8 as follows.

$$\text{Var}(aX + b) = E[(aX+b)^2] - (E[aX + b])^2$$
$$= E[a^2X^2 + 2abX + b^2] - (aE[X] + b)^2$$

We now use the linearity properties of expectation to continue as follows.

$$= a^2E[X^2] + 2abE[X] + b^2 - a^2 \, (E[X])^2 - 2abE[X] - b^2$$
$$= a^2E[X^2] - a^2 \, (E[X])^2$$
$$= a^2 \, \text{Var}(X)$$

∎

This theorem shows that variance is affected by changes in scale but not by changes in location.

When computing the variance of a linear combination of random variables we need to be a little careful. Recall that the mean of a linear combination was simply the linear combination of the means. Unfortunately, this is not the case when dealing with variance. Suppose that we have a random vector (X, Y) and wish to find $\text{Var}(aX + bY)$. Let $E[X] = \mu_x$, and $E[Y] = \mu_Y$. Definition 4.2 gives that

$$\text{Var}(aX + bY) = E[(aX + bY - E[aX + bY])^2].$$

By combining terms involving the same random variable, expanding the square, and making use of the fact that the expectation of a sum is the sum of the expectations, we obtain

$$\text{Var}(aX + bY) = E[a^2(X - \mu_X)^2] + E[b^2(Y - \mu_Y)^2] + 2E[ab(X - \mu_X)(Y - \mu_Y)]$$
$$= a^2\, \text{Var}(X) + b^2\, \text{Var}(Y) + 2abE[(X - \mu_X)(Y - \mu_Y)].$$

The last expectation is called the *covariance of X and Y*, and it is denoted by $\text{Cov}(X, Y)$. As we shall see later covariance is a measure of the degree to which X and Y are linearly related. We give a formal definition of covariance.

Definition 4.3

Let (X, Y) be a random vector with $E[X] = \mu_X$, and $E[Y] = \mu_Y$. The *covariance of X and Y* is

$$\text{Cov}(X, Y) = E[(X - \mu_X)(Y - \mu_Y)]. \qquad (4.17)$$

There is a version of Equation 4.17 that is often useful in computing covariances. We state it as a theorem and leave the proof as an exercise.

Theorem 4.10

Let X and Y be random variables whose variances exist. Then

$$\text{Cov}(X, Y) = E[XY] - \mu_X \mu_Y. \qquad (4.18)$$

Example 4.22

Let X and Y be random variables whose variances exist. Further suppose that $Y = aX + b$. We wish to compute the covariance of X and Y. Let $E[X] = \mu_X$, and $E[Y] = \mu_Y$. We wish to use Equation 4.18. Theorem 4.2 gives

$$\mu_Y = E[aX + b] = a\mu_X + b.$$

Furthermore,

$$E[XY] = E[X(aX + b)] = aE[X^2] + b\mu_X.$$

Then Equation 4.18 gives

$$\text{Cov}(X, Y) = aE[X^2] + b\mu_X - \mu_X(a\mu_X + b)$$
$$= a(E[X^2] - \mu_X^2)$$
$$= a\, \text{Var}(X).$$

An important special case occurs when X and Y are independent random variables. We state it as a theorem. ■

Theorem 4.11

Let X and Y be independent random variables whose variances exist. Then

$$\text{Cov}(X, Y) = 0.$$

Proof We shall compute the covariance using Theorem 4.10 and then conclude by applying Theorem 4.6.

$$
\begin{aligned}
\text{Cov}(X, Y) &= E[XY] - \mu_X \mu_Y \quad \text{(by Theorem 4.10)} \\
&= E[X]E[Y] - \mu_X \mu_Y \quad \text{(by Theorem 4.6)} \\
&= \mu_X \mu_Y - \mu_X \mu_Y = 0
\end{aligned}
$$

■

We must point out that this theorem only states that when two random variables are independent it follows that their covariance will be zero. The converse, however, is not necessarily true. It happens that there exist pairs of dependent random variables possessing a covariance of zero. This means that we cannot use covariance as a test for independence.

Example 4.23

Let X and Y have the following joint probability function.

		Y		
		0	1	2
X	0	$\frac{1}{4}$	0	$\frac{1}{4}$
	1	0	$\frac{1}{2}$	0

From this we see that the distributions are as follows.

$$
p_X(a) = \begin{cases} \dfrac{1}{2}, & a = 0, 1, \quad \text{and} \\ 0, & \text{otherwise} \end{cases}
$$

$$
p_Y(b) = \begin{cases} \dfrac{1}{4}, & b = 0, 2, \\ \dfrac{1}{2}, & b = 1, \quad \text{and} \\ 0, & \text{otherwise} \end{cases}
$$

We note that

$$
p_{X,Y}(0, 1) = 0 \neq p_X(0) p_Y(1) = \frac{1}{4}.
$$

Therefore, X and Y are dependent random variables. We will use Equation 4.18 to compute the covariance. Thus, the necessary expectations are as follows.

$$E[XY] = 0 \cdot 0 \cdot \frac{1}{4} + 0 \cdot 1 \cdot 0 + 0 \cdot 2 \cdot \frac{1}{4} + 1 \cdot 0 \cdot 0 + 1 \cdot 1 \cdot \frac{1}{2} + 1 \cdot 2 \cdot 0$$

$$= \frac{1}{2}$$

$$E[X] = 0 \cdot \frac{1}{2} + 1 \cdot \frac{1}{2}$$

$$= \frac{1}{2}$$

$$E[Y] = 0 \cdot \frac{1}{4} + 1 \cdot \frac{1}{2} + 2 \cdot \frac{1}{4}$$

$$= 1$$

Then Equation 4.18 gives

$$\text{Cov}(X, Y) = \frac{1}{2} - \frac{1}{2} \cdot 1 = 0.$$

Therefore, dependent random variables can have a zero covariance. ■

We can now combine all of these results into a useful statement about the variance of a linear combination of random variables. The foregoing discussion constitutes a proof.

Theorem 4.12

Let X and Y be random variables such that $\text{Var}(X)$ and $\text{Var}(Y)$ exist. Then

$$\text{Var}(aX + bY) = a^2 \text{Var}(X) + b^2 \text{Var}(Y) + 2ab \text{Cov}(X, Y). \qquad \textbf{(4.19)}$$

Furthermore, if X and Y are independent, then

$$\text{Var}(aX + bY) = a^2 \text{Var}(X) + b^2 \text{Var}(Y). \qquad \textbf{(4.20)}$$

Example 4.24

Let (X, Y) be a continuous random vector with a joint p.d.f. of

$$f_{X,Y}(x, y) = \begin{cases} 4xy, & 0 < x < 1, \ 0 < y < 1, \quad \text{and} \\ 0, & \text{otherwise} \end{cases}$$

To find $\text{Var}(aX + bY)$ we note that the joint p.d.f. factors into a product of the p.d.f.'s for X and Y. Thus, X and Y are independent. Their p.d.f.'s are given below.

$$f_X(x) = \begin{cases} 2x, & 0 < x < 1, \quad \text{and} \\ 0, & \text{otherwise} \end{cases}$$

$$f_Y(y) = \begin{cases} 2y, & 0 < y < 1, \quad \text{and} \\ 0, & \text{otherwise} \end{cases}$$

This also shows that X and Y i.i.d. random variables. Therefore, $\text{Var}(X) = \text{Var}(Y)$, and Equation 4.20 gives

$$\text{Var}(aX + bY) = a^2\,\text{Var}(X) + b^2\,\text{Var}(Y)$$
$$= (a^2 + b^2)\,\text{Var}(X).$$

We then compute $\text{Var}(X)$ by using Theorem 4.8 as follows.

$$E[X] = \int_0^1 x\,2x\,dx = \frac{2}{3},$$

$$E[X^2] = \int_0^1 x^2\,2x\,dx = \frac{1}{2}, \quad \text{and}$$

$$\text{Var}(X) = \frac{1}{2} - \left(\frac{2}{3}\right)^2 = \frac{5}{18}$$

Therefore,

$$\text{Var}(aX + bY) = \frac{5(a^2 + b^2)}{18}.$$ ∎

Example 4.25   Let (X, Y) be a continuous random vector with a joint p.d.f. of

$$f_{X,Y}(x, y) = \begin{cases} 8xy, & 0 < x < y < 1, \quad \text{and} \\ 0, & \text{otherwise.} \end{cases}$$

We note that the limits of X depend on the value of Y. Therefore, X and Y are not independent. This means that we must use Equation 4.19 to compute $\text{Var}(aX + bY)$. We find that the p.d.f.'s for X and Y are as follows.

$$f_X(x) = \begin{cases} 4x(1 - x^2), & 0 < x < 1, \quad \text{and} \\ 0, & \text{otherwise} \end{cases}$$

$$f_Y(y) = \begin{cases} 4y^3, & 0 < y < 1, \quad \text{and} \\ 0, & \text{otherwise} \end{cases}$$

Using these and the joint p.d.f. we obtain the necessary expectations as shown below.

$$E[X] = \int_0^1 x\,4x(1 - x^2)\,dx$$

$$= \frac{8}{15}$$

$$E[X^2] = \int_0^1 x^2\,4x(1 - x^2)\,dx$$

$$= \frac{1}{3}$$

$$E[Y] = \int_0^1 y\,4y^3\,dy$$

$$= \frac{4}{5}$$

$$E[Y^2] = \int_0^1 y^2\, 4y^3\, dx$$

$$= \frac{2}{3}$$

$$E[XY] = \int_0^1 \int_0^y xy\, 8xy\, dx\, dy$$

$$= \frac{4}{9}$$

Therefore, we use Equation 4.14 to compute the variances and Equation 4.18 to compute the covariance.

$$\mathrm{Var}(X) = \frac{1}{3} - \left(\frac{8}{15}\right)^2 = \frac{11}{225},$$

$$\mathrm{Var}(Y) = \frac{2}{3} - \left(\frac{4}{5}\right)^2 = \frac{2}{75}, \quad \text{and}$$

$$\mathrm{Cov}(X,Y) = \frac{4}{9} - \frac{8}{15} \cdot \frac{4}{5} = \frac{4}{225}$$

Then Equation 4.19 gives that

$$\mathrm{Var}(aX + bY) = a^2\, \mathrm{Var}(X) + b^2\, \mathrm{Var}(Y) + 2ab\, \mathrm{Cov}(X,Y)$$

$$= a^2 \frac{11}{225} + b^2 \frac{2}{75} + 2ab \frac{4}{225}$$

$$= \frac{11a^2 + 6b^2 + 8ab}{225}.$$ ∎

Theorem 4.12 can be extended to the linear combination of any number of random variables. We state the result as a theorem and leave the proof as an exercise.

Theorem 4.13

Let $X_1, X_2, \ldots, X_n$ be a set of random variables whose variances exist. Let $a_1, a_2, \ldots, a_n$ be constants. Then

$$\mathrm{Var}\left(\sum_{i=1}^n a_i X_i\right) = \sum_{i=1}^n a_i^2\, \mathrm{Var}(X_i) + 2 \sum_{j<k} a_j a_k\, \mathrm{Cov}(X_j, X_k). \qquad (4.21)$$

Furthermore, if $X_1, X_2, \ldots, X_n$ are independent, then

$$\mathrm{Var}\left(\sum a_i X_i\right) = \sum_{i=1}^n a_i^2\, \mathrm{Var}(X_i). \qquad (4.22)$$

As we stated earlier the covariance can be used to measure the relationship between two random variables. The units of covariance are the product of the units of the two random variables. This makes it difficult to determine the strength of a relationship. For this reason, a unitless measure called *correlation* is usually used. We define it as follows.

Definition 4.4

Let X and Y be random variables whose variances exist and are not equal to zero. The *correlation between X and Y* is denoted by $\rho_{X,Y}$ and computed by

$$\rho_{X,Y} = \frac{\text{Cov}(X,Y)}{\sqrt{\text{Var}(X)\ \text{Var}(Y)}}. \qquad (4.23)$$

We now wish to establish the range of values for correlation. In order to do this we first give a special case of what is known as the *Cauchy-Schwarz Inequality*. It will also be useful in later discussions.

Theorem 4.14

Cauchy-Schwarz Inequality   Let X and Y be random variables such that $E[X^2]$ and $E[Y^2]$ exist. Then

$$(E[XY])^2 \leq E[X^2]\,E[Y^2]. \qquad (4.24)$$

Furthermore, equality holds if and only if $X = kY$ for some real constant k.

Proof   Before proceeding we point out that the proof is highly unintuitive. For those with a background in linear algebra this result is a special case of the more general result for inner product spaces. We begin by considering

$$E[(kX + Y)^2].$$

This quantity must be positive unless there is some k such that $Y = -kX$. We now expand the expectation to obtain.

$$0 < E[(kX + Y)^2] = k^2 E[X^2] + 2kE[XY] + E[Y^2]$$

This is a quadratic equation in k. Since it has no real roots the discriminant from the quadratic formula must be negative. In other words,

$$4\,(E[XY])^2 - 4E[X^2]\,E[Y^2] < 0,$$

or

$$(E[XY])^2 < E[X^2]\,E[Y^2].$$

If $Y = -kX$, then we get

$$E[XY] = kE[X^2] = \sqrt{k^2\,\left(E[x^2]\right)^2}$$

$$= \sqrt{E[X^2]\,E[k^2 X^2]}$$

$$= \sqrt{E[X^2]\,E[Y^2]},$$

or
$$(E[XY])^2 = E[X^2]E[Y^2].$$

Combining these completes the proof.   ∎

We can now use the Cauchy-Schwarz Inequality to establish the range of possible values for correlation.

Theorem 4.15

> Let X and Y be random variables whose variances exist and are not equal to 0. Then
> $$-1 \le \rho_{X,Y} \le 1.$$

Proof From the Cauchy-Schwarz Inequality we see that

$$\mathrm{Cov}^2(X,Y) = E[(X-\mu_X)(Y-\mu_Y)]^2 \le E[(X-\mu_X)^2]E[(Y-\mu_Y)^2] = \mathrm{Var}(X)\,\mathrm{Var}(Y).$$

Thus,
$$\mathrm{Cov}^2(X,Y) \le \mathrm{Var}(X)\,\mathrm{Var}(Y),$$

and
$$\rho_{X,Y}^2 = \frac{\mathrm{Cov}^2(X,Y)}{\mathrm{Var}(X)\,\mathrm{Var}(Y)} \le 1.$$

Therefore,
$$-1 \le \rho_{X,Y} \le 1.   ∎$$

Note that when X and Y are independent $\rho_{X,Y} = 0$ because $\mathrm{Cov}(X,Y) = 0$. If $\rho_{X,Y} = 0$, then X and Y are said to be *uncorrelated*. We cannot conclude that uncorrelated random variables are necessarily independent because of the fact that a 0 covariance does not always imply independence.

Example 4.26   In Example 4.25 we found that

$$\mathrm{Var}(X) = \frac{11}{225},$$

$$\mathrm{Var}(Y) = \frac{2}{75}, \quad \text{and}$$

$$\mathrm{Cov}(X,Y) = \frac{4}{225}.$$

Using these data, Equation 4.23 gives that the correlation between X and Y is

$$\rho_{X,Y} = \frac{\dfrac{4}{225}}{\sqrt{\dfrac{11}{225}\dfrac{2}{75}}} = -0.492.$$

∎

The correlation between X and Y gives a numerical measure of the degree to which X and Y are linearly related. The following example shows that $\rho_{X,Y}$ achieves ± 1 when Y is a linear function of X.

Example 4.27

Let X and Y be random variables such that $Y = aX + b$. From Theorem 4.9 we see that $\mathrm{Var}(Y) = a^2\,\mathrm{Var}(X)$. In Example 4.22 we found that $\mathrm{Cov}(X,Y) = a\,\mathrm{Var}(X)$. Therefore, we find that the correlation between X and Y is

$$\rho_{X,Y} = \frac{a\,\mathrm{Var}(X)}{\sqrt{\mathrm{Var}(X)\,a^2\,\mathrm{Var}(Y)}}$$

$$= \frac{a\,\mathrm{Var}(X)}{|x|\,\mathrm{Var}(X)}$$

$$= \begin{cases} 1, & \text{if } a > 0, \quad \text{and} \\ -1, & \text{if } 0 < a. \end{cases}$$

■

We have seen that the mean gives the location of the distribution, and the variance measures the amount of dispersion about the mean. There are two other measures that are used to gain additional information about the shape of the distribution of a random variable. The mean and variance give no information regarding whether the distribution is symmetric about the mean or skewed. For that, a measure based on the third central moment is sometimes used.

Definition 4.5

> Let X be a random variable whose third moment exists. Then the *skewness* is denoted by γ_1 and defined to be
>
> $$\gamma_1 = \frac{E[(X - \mu_X)^3]}{[\mathrm{Var}(X)]^{3/2}}. \tag{4.25}$$

Skewness is a unitless quantity. If the distribution is symmetric about its mean, then $\gamma_1 = 0$. If the distribution is skewed with a long tail to the right, it is said to be *positively skewed* or *skewed to the right*. If the distribution is skewed with a long tail to the left, it is said to be *negatively skewed* or *skewed to the left*. If a distribution is positively skewed, then $\gamma_1 > 0$, and if it is negatively skewed, then $\gamma_1 < 0$. The drawback is that γ_1 is not a definitive indicator of skewness. It is possible to construct random variables with skewed distributions that have $\gamma_1 = 0$. We leave it as an exercise.

A descriptive measure based on the fourth central moment is also used. We define it as follows.

Definition 4.6

Let X be a random variable whose fourth moment exists. Then the *kurtosis* is

$$\gamma_2 = \frac{E[(X - \mu_X)^4]}{[\text{Var}(X)]^2}. \tag{4.26}$$

Like skewness, kurtosis is unitless. There is some disagreement regarding precisely what property of a distribution is measured by the kurtosis. Kurtosis is often referred to as *peakedness*. This term indicates that kurtosis can be thought of as a measure of whether the distribution has a distinct peak or is relatively flat. This idea is illustrated in Figure 4.2.

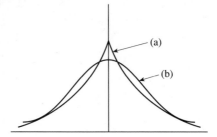

Figure 4.2

The p.d.f. denoted by (a) has a more pronounced peak than that of (b). It turns out that the kurtosis of the p.d.f. marked (a) is higher than that of (b). Others interpret kurtosis as a measure of how much of the distribution is concentrated in the center versus the tails. Still others use kurtosis as a measure of how far a distribution deviates from a reference distribution called the *normal* distribution. In the case of the normal distribution, $\gamma_2 = 3$. The normal distribution will be discussed in Chapter 5.

Example 4.28

Let X be a r.v. whose p.d.f. is

$$f_X(x) = \begin{cases} x e^{-x}, & x > 0, \quad \text{and} \\ 0, & \text{otherwise.} \end{cases}$$

The graph of this p.d.f. is shown in Figure 4.3.

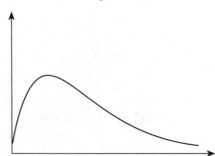

Figure 4.3

It is clear from the graph that $f_X(x)$ is skewed to the right. Therefore, we would expect that skewness would be positive. The heavy tail to the right indicates that we would expect kurtosis to be relatively high. We shall find both γ_1 and γ_2 to illustrate how the computations go. Each of the integrals relies heavily on integration by parts. The details of finding the antiderivatives are omitted.

$$E[X] = \int_0^\infty x\,xe^{-x}\,dx = 2$$

$$E[X^2] = \int_0^\infty x^2\,xe^{-x}\,dx = 6$$

$$E[(X - \mu_X)^3] = \int_0^\infty (x - 2)^3\,xe^{-x}\,dx$$

$$= \int_0^\infty (x^4 - 6x^3 + 12x^2 - 8x)\,e^{-x}\,dx$$

$$= 24 - 36 + 24 - 8 = 4$$

$$E[(X - \mu_X)^4] = \int_0^\infty (x - 2)^4\,xe^{-x}\,dx$$

$$= \int_0^\infty (x^5 - 8x^4 + 24x^3 - 32x^2 + 16x)\,e^{-x}\,dx$$

$$= 120 - 192 + 144 - 64 + 14 = 24$$

Armed with these expectations we find the following.

$$\text{Var}(X) = E[X^2] - \mu_X^2 = 6 - (2)^2 = 2$$

$$\gamma_1 = \frac{E[(X - \mu_X)^3]}{[\text{Var}(X)]^{3/2}} = \frac{4}{2^{3/2}} = \sqrt{2}$$

$$\gamma_2 = \frac{E[(X - \mu_X)^4]}{[\text{Var}(X)]^2} = \frac{24}{2^2} = 6$$

The skewness is verified to be positive as was suspected. It is hard to give an interpretation for the value of kurtosis since the distribution is skewed to the right. ∎

Determining the properties of a distribution through the use of moments is an imprecise business. This is because we have been vague in defining the attributes they measure. It is difficult, for example, to give a precise meaning to a value of $\gamma_1 = 3$ other than to say that it indicates that the distribution is skewed to the right. That is, it is impossible to state that such a value for skewness indicates that the distribution is slightly skewed, moderately skewed, or heavily skewed. On the other hand, it is possible to determine if a distribution is more skewed than another by comparing their values for γ_1. For this reason these measures are useful for comparing one distribution with another.

Exercises 4.4

1. Find Var(X), if it exists, for the distributions given in Exercise 1 of Section 4.2.

2. Find Var(X), if it exists, for the distributions given in Exercise 2 of Section 4.2.

3. Find Var(X) for the p.f. given in Exercise 3 of Section 4.2.

4. Let a random variable, X, have a p.d.f. of

$$f_X(x) = \begin{cases} a + bx + cx^2, & 0 < x < 1, \quad \text{and} \\ 0 & , \text{ otherwise.} \end{cases}$$

If $E[X] = \frac{1}{3}$, and $\text{Var}(X) = \frac{1}{18}$, find a, b, and c.

5. A point is chosen in a circle of radius a. Let (R, Θ), the coordinates of the point in polar coordinates, have a joint p.d.f. of

$$f_{R,\Theta}(r, \theta) = \begin{cases} kr, & 0 \le r \le a, \ 0 \le \theta < 2\pi, \quad \text{and} \\ 0, & \text{otherwise.} \end{cases}$$

Find the mean and variance of the distance of the point from the center of the circle.

6. Show that

$$\text{Var}(X) = E[X(X - 1)] + \mu_X - \mu_X^2.$$

7. An insurance company receives claims each day, the number of which have a distribution whose mean is μ_N and whose variance is σ_N^2. The amount of each claim has a distribution whose mean is μ_A and whose variance is σ_A^2. Find the mean and variance of the amount of money the company must pay each day. Assume that the number of claims and the amount per claim are independent.

8. Let X have a p.d.f. of

$$f_X(x) = \begin{cases} \dfrac{1}{2\pi}, & 0 \le x < 2\pi, \quad \text{and} \\ 0, & \text{otherwise.} \end{cases}$$

Let $Y = \cos x$, and $Z = \sin x$. Find $\text{Cov}(Y, Z)$.

9. An urn contains 5 balls numbered 1, 2, 3, 4, and 5. Two balls are selected with replacement. Let X and Y be the sum and difference of the balls drawn, respectively. Compute $\rho_{X,Y}$. Are X and Y independent?

10. Let X, Y, and Z be random variables such that each has a mean of 0 and a variance of 1. In addition, any pair of random variables has a correlation of $\frac{1}{2}$. Compute
    a. $\text{Var}(X + Y + Z)$, and
    b. $\text{Cov}(X + Y, Z)$.

11. Let $X_1, X_2, \ldots, X_n$ be i.i.d. random variables each with a mean of μ and a variance of σ^2. Let $\bar{X} = \frac{1}{n}\sum_{i=1}^{n} X_i$. Show that
    a. $E[\bar{X}] = \mu$,
    b. $\text{Var}(\bar{X}) = \frac{\sigma^2}{n}$, and
    c. $E\left[\frac{1}{n-1}\sum_{i=1}^{n}(X_i - \bar{X})^2\right] = \sigma^2$.

12. Prove Theorem 4.10.

13. Let X and Y be random variables with a correlation of $\rho_{X,Y}$. Find the correlation between $T = aX + b$ and $U = cY + d$.

14. Let X and Y be random variables such that
$$\mu_X = 1, \quad \mu_Y = 2, \quad \sigma_X^2 = 16, \quad \sigma_Y^2 = 9, \quad \text{and } \rho_{X,Y} = \frac{2}{5}.$$

   a. Find $E[2X - Y]$.
   b. Find $\text{Var}(2X - Y)$.

15. Let X be a r.v. with a p.d.f. of
$$f_X(x) = \begin{cases} 1, 0 < x < 1, & \text{and} \\ 0, \text{otherwise.} \end{cases}$$

   Let and $Y = X^2$, and find $\rho_{X,Y}$.

16. Show that, if $E[X^2]$ and $E[Y^2]$ exist, then $E[XY]$ exists.

17. Prove Theorem 4.13.

4.5
Moment Generating Functions

Another useful expectation, when it exists, is $E\left[e^{tX}\right]$, where t is a real number. It is called the *moment generating function (m.g.f.) of X*. As its name suggests, it can be used to compute the moments of the distribution of a random variable. As we shall see in Chapter 6, it can be a valuable tool for deriving the distribution of a function of random variables. From a theoretical standpoint, it can also be invaluable for proving the so-called limit theorems that will be investigated in Chapter 7. A formal definition is given below.

Definition 4.7

> Let X be a random variable. The *moment generating function of X* is
> $$M_X(t) = E\left[e^{tX}\right], \tag{4.27}$$
> provided there is some $h > 0$ such that $M_X(t)$ exists for all $|t| < h$.

The term "moment generating function" arises from the fact that $E[X^r]$ can be computed from $M_X(t)$ in the following manner. Using the Taylor series for e^x we see that

$$M_X(t) = E\left[e^{tX}\right] = E\left(\sum_{i=0}^{\infty} \frac{(tX)^i}{i!}\right) = \sum_{i=0}^{\infty} \frac{t^i}{i!} E[X^i]$$

$$= 1 + tE[X] + \frac{1}{2}t^2 E[X^2] + \frac{t^3}{3!}E[X^3] + \frac{t^4}{4!}E[X^4] + \cdots.$$

If we evaluate $M_X(t)$ at $t = 0$, we note that

$$M_X(0) = 1.$$

If we take the first derivative of $M_X(t)$ with respect to t, we get

$$M_X'(t) = E[X] + tE[X^2] + \frac{t^2}{2!}E[X^3] + \frac{t^3}{3!}E[X^4] + \cdots.$$

If we set $t = 0$, we get
$$M'_X(0) = E[X].$$

The second derivative of $M_X(t)$ with respect to t is
$$M''_X(t) = E[X^2] + tE[X^3] + \frac{t^2}{2!}E[X^4] + \cdots .$$

Setting t to zero gives
$$M''_X(0) = E[X^2].$$

If we proceed in this fashion we find that
$$M_X^{(r)}(0) = E[X^r]. \tag{4.28}$$

Thus, the moment generating function, when it exists, is a power series in t whose terms involve the moments of the distribution. We illustrate with some examples.

Example 4.29   Let X be an exponential r.v. with a p.d.f. of
$$f_X(x) = \begin{cases} e^{-x}, & 0 < x, \quad \text{and} \\ 0, & \text{otherwise.} \end{cases}$$

Using Equation 4.27 we find that
$$M_X(t) = E\left[e^{tX}\right] = \int_{-\infty}^{\infty} e^{tx} f_X(x) \, dx$$
$$= \int_0^{\infty} e^{tx} e^{-x} \, dx = \int_0^{\infty} e^{-x(1-t)} \, dx$$
$$= -\frac{1}{1-t}\left[\lim_{a \to \infty} e^{-a(1-t)} - 1\right]$$
$$= \frac{1}{1-t}, \quad \text{if } t < 1.$$

Thus, the m.g.f. exists with $h = 1$. The first and second moments of X are found as follows.
$$M'_X(t) = \frac{1}{(1-t)^2}, \quad \text{and}$$
$$E[X] = M'_X(0) = 1$$
$$M''_X(t) = \frac{2}{(1-t)^3}, \quad \text{and}$$
$$E[X^2] = M''_X(0) = 2 \qquad\qquad \blacksquare$$

Example 4.30   Let X be a Poisson r.v. with a p.f. of
$$p_X(a) = \begin{cases} \dfrac{\lambda^a e^{-\lambda}}{a!}, & a = 0, 1, 2, \ldots, \quad \text{and} \\ 0, & \text{otherwise.} \end{cases}$$

Equation 4.27 gives

$$M_X(t) = \sum_{a=0}^{\infty} e^{ta} \frac{\lambda^a e^{-\lambda}}{a!}$$

$$= \sum_{a=0}^{\infty} \frac{(\lambda e^t)^a e^{-\lambda}}{a!}$$

$$= e^{-\lambda} \sum_{a=0}^{\infty} \frac{(\lambda e^t)^a}{a!}.$$

The quantity in the sum is the Taylor series for $e^{\lambda e^t}$. Therefore,

$$M_X(t) = e^{-\lambda} e^{\lambda e^t}$$

$$= e^{\lambda(e^t - 1)}.$$

Since the Taylor series for e^x converges for all values of x, $M_X(t)$ exists with h being infinite. We shall use this to verify previous results.

$$M_X'(t) = \lambda e^t e^{\lambda(e^t - 1)}, \quad \text{and}$$

$$E[X] = M_X(0) = \lambda$$

$$M_X''(t) = \lambda e^t e^{\lambda(e^t - 1)} + \lambda^2 e^{2t} e^{\lambda(e^t - 1)}, \quad \text{and}$$

$$E[X^2] = M_X''(0) = \lambda + \lambda^2$$

These are identical to the values obtained in Examples 4.2 and 4.11.   ■

Example 4.31   Let X be a discrete r.v. with a p.f. of

$$p_X(a) \begin{cases} \dfrac{1}{a(a+1)}, & a = 1, 2, \ldots, \quad \text{and} \\ 0 & , \text{ otherwise.} \end{cases}$$

This is the r.v. discussed in Example 4.4. There we found that $E[X]$ did not exist, which would lead us to guess that the moment generating function would also not exist. We verify this as follows.

$$M_X(t) = \sum_{a=1}^{\infty} e^{ta} \frac{1}{a(a+1)}$$

The ratio test applied to this sum gives that

$$\lim_{n \to \infty} \left| \frac{a_{n+1}}{a_n} \right| = \lim_{n \to \infty} \frac{n\, e^{t(n+1)}}{(n+1)\, e^{tn}}$$

$$= e^t \lim_{n \to \infty} \frac{n}{n+1}$$

$$= e^t,$$

which is greater than 1 if $t > 0$. Thus, there is no value of $h > 0$ for which the summation converges for all $|t| < h$ which means that the moment generating function does not exist.   ■

When the moment generating function of a random variable exists, it is unique. This is due to the fact that a power series representation of a function is unique in its domain of convergence. This means that the distribution of a random variable can be identified by its moment generating function when it exists. Some useful facts regarding moment generating functions are given in the following theorems.

Theorem 4.16

Let $M_X(t)$ be the moment generating function for the random variable X. If $Y = aX + b$, where a and b are real constants, then

$$M_Y(t) = e^{tb} M_X(at). \qquad (4.29)$$

Proof   We use Equation 4.27 and the properties of expectations as follows.

$$M_Y(t) = E\left[e^{tY}\right]$$
$$= E\left[e^{t(aX+b)}\right]$$
$$= E\left[e^{(at)X} e^{tb}\right]$$

Since e^{tb} is a constant with regard to an expectation on X we get

$$M_Y(t) = e^{tb} E\left[e^{(at)X}\right].$$

Now the expectation is the moment generating function of X evaluated at at. Thus, we complete the proof.

$$M_Y(t) = e^{tb} M_X(at)$$

■

Theorem 4.17

Let X and Y be independent random variables with moment generating functions $M_X(t)$ and $M_Y(t)$, respectively. Let $Z = aX + bY$, where a and b are real constants. Then

$$M_Z(t) = M_X(at)M_Y(bt). \qquad (4.30)$$

Proof   For this proof we use Equation 4.27 and Theorem 4.6 as follows.

$$M_Z(t) = E\left[e^{tZ}\right]$$
$$= E\left[e^{t(aX+bY)}\right]$$
$$= E\left[e^{(at)X} e^{(bt)Y}\right]$$

Now the independence of X and Y gives

$$M_Z(t) = E\left[e^{(at)X}\right] E\left[e^{(bt)Y}\right]$$
$$= M_X(at) M_Y(bt).$$

■

This can be extended to the linear combination of any number of independent random variables. The proof is left as an exercise.

Theorem 4.18

Let $X_1, X_2, \ldots, X_n$ be independent random variables with moment generating functions $M_{X_i}(t)$, for $i = 1, 2, \ldots, n$. Let $Z = \sum_{i=1}^{n} a_i X_i$, where $a_1, a_2, \ldots, a_n$, are real constants. Then

$$M_Z(t) = \prod_{i=1}^{n} M_{X_i}(a_i t). \tag{4.31}$$

Example 4.32

Let X and Y be i.i.d. random variables, each with a Poisson p.f. of

$$p_X(a) = \begin{cases} \dfrac{\lambda^a e^{-\lambda}}{a!}, & a = 0, 1, 2, \ldots, \quad \text{and} \\ 0, & \text{otherwise, where } \lambda > 0. \end{cases}$$

We wish to find the m.g.f. of $Z = aX + bY$. We know from Theorem 4.8 that, since X and Y are independent,

$$M_Z(t) = M_X(at) M_Y(bt).$$

Since X and Y are identically distributed they will have the same moment generating function. In Example 4.30 we found the m.g.f. for this probability function. Therefore,

$$M_X(t) = M_Y(t) = e^{\lambda(e^t - 1)}.$$

Then we combine these facts to find

$$M_X(at) = e^{\lambda(e^{at} - 1)},$$
$$M_Y(bt) = e^{\lambda(e^{bt} - 1)}, \quad \text{and}$$
$$M_Z(t) = e^{\lambda(e^{at} - 1)} e^{\lambda(e^{bt} - 1)}$$
$$= e^{\lambda(e^{at} + e^{bt} - 2)}.$$

Note that if $a = b = 1$, then

$$M_Z(t) = e^{2\lambda(e^t - 1)}.$$

This is of the same form as $M_X(t)$ with λ replaced by 2λ. Since the m.g.f. is unique, when it exists, this shows that if X and Y are i.i.d. Poisson random variables, then $Z = X + Y$ is also a Poisson random variable.   ■

The moment generating function for jointly distributed random variables is a straightforward extension of the single variable case. The notation becomes more complicated.

Definition 4.8

Let $X_1, X_2, \ldots, X_n$ be random variables. The *joint moment generating function of* $X_1, X_2, \ldots, X_n$ is

$$M_{X_1, X_2, \ldots, X_n}(t_1, t_2, \ldots, t_n) = E\left[e^{t_1 X_1 + t_2 X_2 + \ldots + t_n X_n}\right], \qquad (4.32)$$

provided there is an $h > 0$ such that the expectation exists for all $|t_i| < h, i = 1, 2, \ldots, n$.

Note that if each t, except for t_i, is set to zero then

$$M_{X_1, X_2, \ldots, X_n}(0, 0, \ldots, 0, t_i, 0, \ldots, t_n) = E\left[e^{t_i X_i}\right]$$

$$= M_{X_i(t_i)}. \qquad (4.33)$$

Thus, the moment generating function of a single random variable can be obtained from the joint moment generating in this fashion.

In addition, if $X_1, X_2, \ldots, X_n$ are independent random variables, we can use Theorem 4.7 as follows.

$$M_{X_1, X_2, \ldots, X_n}(t_1, t_2, \ldots, t_n) = E\left[e^{t_1 X_2 + t_2 X_2 + \cdots + t_n X_n}\right]$$

$$= E\left[e^{t_1 X_1} e^{t_2 X_2} \cdots e^{t_n X_n}\right]$$

$$= E\left[e^{t_1 X_1}\right] E\left[e^{t_2 X_2}\right] \ldots E\left[e^{t_n X_n}\right]$$

$$= \prod_{i=1}^{n} M_{X_i}(t_i) \qquad (4.34)$$

This gives another way to check whether random variables are independent. If the joint moment generating function factors into a product of their individual moment generating functions, then they are independent. If not, then they are dependent.

Example 4.33

Let X and Y be random variables with a joint p.d.f. of

$$f_{X,Y}(x, y) = \begin{cases} 2e^{-x-y}, & 0 < x < y, \quad \text{and} \\ 0, & \text{otherwise.} \end{cases}$$

We note that X and Y are not independent because the joint p.d.f. does not factor. The limits depend on each other. We wish to compute the joint moment generating function. Equation 4.32 gives

$$M_{X,Y}(t,u) = E\left[e^{tX+uY}\right]$$

$$= \int_0^\infty \int_0^y e^{tx+uy}\, 2e^{-x-y}\, dx\, dy$$

$$= \int_0^\infty \frac{2}{1-t}\left[1 - e^{-y(1-t)}\right] e^{-y(1-u)}\, dy$$

$$= \frac{2}{(1-u)(2-t-u)},$$

as long as $u \neq 1$ and $t + u < 2$. Therefore, we can let $h = 1$ in Definition 4.8.

Next we wish to find the individual moment generating functions for X and Y. To do this we use Equation 4.33 as follows.

$$M_X(t) = M_{X,Y}(t,0) = \frac{2}{2-t}, \quad \text{and}$$

$$M_Y(u) = M_{X,Y}(0,u) = \frac{2}{u^2 - 3u + 2}$$

■

The joint moment generating function can be used to calculate such expectations as $E[XY]$. We state the general result as a theorem without proof.

Theorem 4.19

> Let $X_1, X_2, \ldots, X_n$ be random variables having a joint moment generating function of $M_{X_1, X_2, \ldots, X_n}(t_1, t_2, \ldots, t_n)$. Then
>
> $$E[X_1^{r_1}, X_2^{r_2}, \ldots, X_n^{r_n}] = \frac{\partial^{r_1+r_2+\cdots+r_n}}{\partial t_1^{r_1} \partial t_2^{r_2} \cdots \partial t_n^{r_n}} M_{X_1, X_2, \ldots, X_n}(0, 0, \ldots, 0). \qquad (4.35)$$

Example 4.34

We wish to use Theorem 4.19 to compute $E[XY]$ for the joint distribution discussed in Example 4.33. The joint moment generating function was

$$M_{X,Y}(t,u) = \frac{2}{(1-u)(2-t-u)},$$

for $u \neq 1$ and $t + u < 2$. To compute $E[XY]$ we proceed as follows.

$$\frac{\partial^2}{\partial u\, \partial t} M_{X,Y}(t,u) = \frac{8 - 2t - 4u}{(1-u)^2(2-t-u)^3}$$

Setting $t = u = 0$ gives

$$E[XY] = 1.$$

■

Exercises 4.5

1. Let the random variable X have a p.f. of

$$p_X(a) = \begin{cases} \binom{n}{a} p^a (1-p)^{n-a}, & a = 0, 1, 2, \ldots, n, \quad \text{and} \\ 0, & \text{otherwise.} \end{cases}$$

a. Show that
$$M_X(t) = \left[(1-p) + pe^t \right]^n, \quad \text{for all } t.$$

b. Use this m.g.f. to compute $E[X]$ for this distribution.

c. Use this m.g.f. to compute $\text{Var}(X)$ for this distribution.

2. Find the moment generating function, if it exists, of random variables with the following distributions.

a.
$$p_X(a) = \begin{cases} p(1-p)^a, & a = 0, 1, 2, \ldots, \quad \text{and} \\ 0, & \text{otherwise} \end{cases}$$

b.
$$f_X(x) = \begin{cases} \dfrac{1}{b-a}, & a \le x \le b, \quad \text{and} \\ 0, & \text{otherwise} \end{cases}$$

c.
$$f_X(x) = \begin{cases} \dfrac{1}{2} e^{-|x|}, & -\infty < x < \infty \end{cases}$$

3. Let the random variable X have a moment generating function of
$$M_X(t) = e^{(1/2)t^2}.$$

Compute the mean, variance, skewness, and kurtosis of X.

4. Let X and Y be i.i.d. random variables. The distribution of each has a moment generating function of
$$M_X(t) = \frac{\alpha}{\alpha - t}, \quad \text{where } \alpha > 0.$$

a. Find the moment generating function of $T = X + Y$.

b. Find the moment generating function of $U = 2X$.

c. It turns out that
$$f_X(x) = \begin{cases} \alpha e^{-\alpha x}, & 0 < x, \quad \text{and} \\ 0, & \text{otherwise.} \end{cases}$$

What did your answers in Parts a and b tell you about the p.d.f.'s of T and U?

5. Let (X, Y) have a joint p.f. of

$$p_{X,Y}(a, b) = \begin{cases} \dfrac{n!}{a!\, b!\, (n-a-b)!} p^a q^b r^{n-a-b}, & 0 \le a + b \le n, \text{ and} \\ 0, & \text{otherwise,} \end{cases}$$

where a and b are integers, and p, q, and r are nonnegative real numbers such that $p + q + r = 1$. This is called a *trinomial distribution* because each value of the p.f. is a term in the trinomial expansion of $(p + q + r)^n$.

a. Find the joint moment generating function of X and Y.

b. Find the moment generating function of X.

6. Let (X, Y) have a joint p.d.f. of

$$f_{X,Y}(x, y) = \begin{cases} e^{-y}, & 0 < x < y, \quad \text{and} \\ 0, & \text{otherwise.} \end{cases}$$

a. Find the joint moment generating function of X and Y.

b. Find the individual moment generating functions of X and Y.

7. Let the random vector (X, Y) have a joint moment generating function of $M_{X,Y}(t_1, t_2)$. Show how you would use this moment generating function to compute $E[X]$, $E[Y]$, $E[X^2]$, and $E[XY]$.

8. Prove Theorem 4.18.

9. A box contains 6 white balls and 4 black balls. Balls are drawn without replacement from the box until either a white ball occurs or 3 balls have been removed. Let X be the number of balls removed.

a. Find the moment generating function for X.

b. Use the moment generating function to compute the mean and variance for X.

10. A random variable X has a moment generating function of

$$M_X(t) = \frac{\pi t e^t}{\sin \pi t}, |t| < \frac{1}{2}.$$

a. Find $E[X]$ and $\text{Var}(X)$.

b. Find $E[5X + 2]$.

4.6
Conditional
Expectation

Some applications of probability require that we investigate the expected value of random variables in conditional distributions. For example, in what is known as *regression analysis,* we desire to estimate the mean of a random variable, say Y, given that we know the value of another random variable, say X. Values of (X, Y) are observed, and from these the goal is to make a good estimate of the value of $E[Y \mid X = x]$. As might be suspected the *conditional expectation* is based on the conditional distribution of a random variable. We define it following.

Definition 4.9

Let X and Y be random variables. The *conditional expectation of Y given $X = x$* is found as follows.

a. If X and Y are discrete random variables with a conditional probability function of $p_{Y|X}(a \mid x)$, then

$$E[Y \mid X = x] = \sum_{a_i} a_i \, p_{Y|X}(a_i \mid x), \qquad (4.36)$$

provided the sum converges absolutely.

b. If X and Y are continuous random variables with a conditional probability density function of $f_{Y|X}(y \mid x)$, then

$$E[Y \mid X = x] = \int_{-\infty}^{\infty} y f_{Y|X}(y \mid x) \, dy, \qquad (4.37)$$

provided the integral converges absolutely.

Conditional expectation can be extended to functions of random variables in a manner similar to that of our discussion of unconditional expectations in Section 4.3. We state the result as a theorem without proof.

Theorem 4.20

Let X and Y be random variables, and let $g(Y)$ be a function of Y.

a. If X and Y are discrete random variables with a conditional probability function of $p_{Y|X}(a \mid x)$, then

$$E[g(Y) \mid X = x] = \sum_{a_i} g(a_i) p_{Y|X}(a_i \mid x), \qquad (4.38)$$

provided the sum converges absolutely.

b. If X and Y are continuous random variables with a conditional probability density function of $f_{Y|X}(y \mid x)$, then

$$E[g(Y) \mid X = x] = \int_{-\infty}^{\infty} g(y) f_{Y|X}(y \mid x) \, dy, \qquad (4.39)$$

provided the integral converges absolutely.

An important application of this theorem is the computation of the conditional variance of Y given $X = x$. The conditional mean is often denoted by $\mu_{Y|X}$, and the conditional variance is denoted by $\sigma^2_{Y|X}$. $E[Y \mid X = x]$ is often referred to as the *regression of Y on X*.

Example 4.35

Let X and Y have a joint p.d.f. of

$$f_{X,Y}(x, y) = \begin{cases} 8xy, & 0 < x < y < 1, \quad \text{and} \\ 0, & \text{otherwise.} \end{cases}$$

We wish to compute the conditional mean and variance of X given $Y = y$. First we must determine $f_{X|Y}(x \mid y)$. The p.d.f. for Y is found as follows.

$$f_Y(y) = \int_0^y 8xy \, dx = 4y^3$$

Thus,

$$f_Y(y) = \begin{cases} 4y^3 , 0 < y < 1, & \text{and} \\ 0 , \text{otherwise.} \end{cases}$$

Then Equation 3.42 gives that

$$f_{X|Y}(x \mid y) = \frac{f_{X,Y}(x, y)}{f_Y(y)}$$

$$= \begin{cases} \dfrac{2x}{y^2} , 0 < x < y, & \text{and} \\ 0 , \text{otherwise.} \end{cases}$$

To obtain $\mu_{X|Y}$ we use Equation 4.37.

$$\mu_{X|Y} = \int_0^y x \frac{2x}{y^2} dx$$

$$= \frac{2}{3} y$$

We use Equation 4.39 to compute the conditional variance.

$$\sigma^2_{X|Y} = E[(X - \mu_{X|Y})^2 \mid Y = y]$$

$$= \int_0^y \left(x - \frac{2}{3} y \right)^2 \frac{2}{3} y \, dy$$

$$= \frac{y^2}{18}$$

■

We give another basic and extremely useful result. As we shall see it can serve as a valuable tool in computing unconditional expectations.

Theorem 4.21

Let $E[Y \mid X]$ represent that function of the random variable X whose value at $X = x$ is $E[Y \mid X = x]$. Then

$$E[Y] = E[E[Y \mid X]]. \tag{4.40}$$

In particular, this means that

$$E[Y] = \sum_a E[Y \mid X = x] p_X(x), \tag{4.41}$$

if X and Y are discrete, and

$$E[Y] = \int_{-\infty}^{\infty} E[Y \mid X = x] f_X(x) \, dx \tag{4.42}$$

if X and Y are continuous.

Proof We shall give a proof for the case when X and Y are continuous. The proof for the discrete case is similar and is left as an exercise. We use Equation 3.42, Equation 4.37, and Equation 4.42 and proceed as follows.

$$E[E[Y \mid X]] = \int_{-\infty}^{\infty} E[Y \mid X = x] f_X(x)\, dx$$

$$= \int_{-\infty}^{\infty} \int_{-\infty}^{\infty} y f_{Y\mid X}(y \mid x) f_X(x)\, dy\, dx$$

$$= \int_{-\infty}^{\infty} \int_{-\infty}^{\infty} y \frac{f_{X,Y}(x, y)}{f_X(x)} f_X(x)\, dy\, dx$$

$$= \int_{-\infty}^{\infty} \int_{-\infty}^{\infty} y f_{X,Y}(x, y)\, dx\, dy$$

$$= \int_{-\infty}^{\infty} y f_Y(y)\, dy$$

$$= E[Y]$$

∎

In effect, this theorem states that an alternative way to compute $E[Y]$ is to take a weighted average of each value of $E[Y \mid X = x]$. The weightings are, either the probability that $X = x$, in the discrete case, or the value of the p.d.f., $f_X(x)$, in the continuous case. This should sound strangely similar to our initial discussion of expected value in Section 4.1.

We shall indicate how this result can simplify the determination of expected values.

Example 4.36   A morning radio show conducts a game according to the following rules. A number between 1 and 100 is selected at random, and the disc jockey announces what number is chosen. Call it n. The announcer then randomly selects a number between 1 and n and asks that the first caller guess the value of the second number chosen. We assume that caller's guesses are equally likely to be any of the numbers between 1 and n and wish to determine the expected value of the number guessed by the caller. Let X = the first number chosen, and Y = the number guessed by the caller from the audience. We wish to determine $E[Y]$. We use Equation 4.40 as follows. If $X = n$ then

$$p_X(x) = \frac{1}{100}, \qquad p_{Y\mid X}(y \mid x = n) = \frac{1}{n},$$

and

$$E[Y \mid X] = \sum_{a=1}^{n} a \frac{1}{n} = \frac{n(n+1)}{2} \frac{1}{n} = \frac{n+1}{2}.$$

Thus, Equation 4.41 gives that

$$E[Y] = \sum_{n=1}^{100} \frac{n+1}{2} \frac{1}{100}$$

$$= \left[\frac{100(101)}{2} + 100 \right] \frac{1}{200}$$

$$= \frac{103}{4} = 25.75.$$

■

We introduced the idea of the *regression of Y on X* at the beginning of this section. We wish to investigate this idea when it is assumed that this regression has a particular functional form. In statistical applications, it is common practice to assume that there is a linear relationship between X and Y. The following example illustrates an important aspect of this idea.

Example 4.37 Let X and Y be r.v.'s with a joint p.d.f. of $f_{X,Y}(x, y)$ and a conditional p.d.f. of Y given $X = x$ of $f_{Y|X}(y \mid x)$. Assume that $E[Y \mid X] = a + bX$. We wish to find a and b in terms of μ_X, μ_Y, σ_X^2, σ_Y^2, and $\rho_{X,Y}$. From Equation 3.42 we know that

$$f_{Y|X}(y \mid x) = \frac{f_{X,Y}(x, y)}{f_X(x)}.$$

By our assumption the conditional mean is

$$E[Y \mid X] = a + bX.$$

Therefore we use Theorem 4.21 to note that

$$\mu_Y = E[Y] = E[E[Y \mid X]] = E[a + bX] = a + bE[X] = a + \mu_X.$$

We also note that Theorem 4.21 gives us that

$$E[XY] = E[E[XY \mid X]] = E[XE[Y \mid X]].$$

We can bring X out of the interior expectation because it is a constant when we condition on $X = x$. We then use the fact that $E[X^2] = \sigma_X^2 + \mu_X^2$ to obtain the following.

$$E[XY] = E[XE[Y \mid X]] = E[X(a + bX)]$$

$$= aE[X] + bE[X^2]$$

$$= a\mu_X + b(\sigma_X^2 + \mu_X^2)$$

Combining this with the fact that $E[XY] = \text{Cov}(X, Y) + \mu_X\mu_Y$, and $\text{Cov}(X, Y) = \rho_{X,Y}\sigma_X\sigma_Y$ we find

$$\rho_{X,Y}\sigma_X\sigma_Y = a\mu_X + b(\sigma_X^2 + \mu_X^2).$$

This gives us a system of two equations involving the unknowns a and b which we solve to obtain that

$$a = \mu_Y - \rho_{X,Y} \frac{\sigma_Y}{\sigma_X} \mu_X, \quad \text{and}$$

$$b = \rho_{X,Y} \frac{\sigma_Y}{\sigma_X}.$$

Therefore,

$$E[Y \mid X = x] = \mu_Y + \rho_{X,Y} \frac{\sigma_Y}{\sigma_X}(x - \mu_X).$$

We leave it as an exercise to show that

$$\text{Var}(Y \mid X = x) = \sigma_Y^2(1 - \rho_{X,Y}^2).$$

■

Exercises 4.6

1. An urn contains 5 balls numbered 1, 2, 3, 4, and 5. Balls are drawn without replacement. Let X and Y be the number of draws necessary to obtain the 1 and 2, respectively.
   a. Find $E[X \mid Y = 2]$.
   b. Find $E[Y \mid X = 2]$.

2. As punishment for committing a particularly heinous crime you have been sentenced to jail under the following terms. Upon entering jail you draw a ball from a box containing 3 balls numbered 0, 1, and 3, respectively. If you draw the ball numbered 0 you get out of jail immediately. If you draw the 1 or 3 you replace the ball in the box and stay that many years in jail, at which time you draw again. This is repeated until you draw the zero and go free. How long do you expect to be in jail? Condition on the first ball drawn.

3. Urn I contains 3 white balls and 2 black balls. Urn II contains 2 white balls and 1 black ball. Two balls are drawn from Urn I and placed in Urn II, and then 2 balls are drawn from Urn II. Let X and Y be the number of white balls drawn from Urns I and II, respectively. Compute $E[Y \mid X = i]$, for $i = 0$, 1, and 2.

4. Let (X, Y) have a joint p.d.f. of

$$f_{X,Y}(x, y) = \begin{cases} e^{-y}, & 0 < x < y, \quad \text{and} \\ 0, & \text{otherwise.} \end{cases}$$

   a. Compute $E[Y \mid X = x]$.
   b. Compute $E[X \mid Y = y]$.
   c. Compute $\text{Var}(Y \mid X = x)$.

5. Let (X, Y) have a joint p.d.f. of

$$f_{X,Y}(x, y) = \begin{cases} 2 - x - y, & 0 < x < 1, \, 0 < y < 1, \quad \text{and} \\ 0, & \text{otherwise.} \end{cases}$$

   a. Find $E[Y \mid X = x]$.
   b. Find $E[Y^2 \mid X = x]$.
   c. Find $\text{Var}(Y \mid X = x)$.

6. If X and Y are independent random variables, show that $E[Y \mid X = x]$ does not depend on x.

7. Let $E[Y \mid X = x] = a + bx$. Show that

$$\text{Var}(Y \mid X = x) = \sigma_Y^2(1 - \rho_{X,Y}^2).$$

8. Let $E[Y \mid X = x] = a + bx$. Given values for a, b, and x we can *predict* the value of $E[Y \mid X = x]$. The values of a and b that minimize $E[(Y - (a + bx))^2]$ give what is known as the *best linear predictor* of X.

a. Show that the best linear predictor of $E[Y \mid X = x]$ occurs when

$$a = \mu_Y - \rho_{X,Y}\frac{\sigma_Y}{\sigma_X}\mu_X, \quad \text{and}$$

$$b = \rho_{X,Y}\frac{\sigma_Y}{\sigma_X}.$$

Chapter Summary

Expected Value:

If X is discrete,

1. $E[X] = \displaystyle\sum_{a_i} a_i p_X(a_i),$ **(Section 4.2)**

2. $E[g(X)] = \displaystyle\sum_{a_i} g(a_i) p_X(a_i).$ **(Section 4.2)**

If X is continuous,

1. $E[X] = \displaystyle\int_{-\infty}^{\infty} x f_X(x)\, dx,$ **(Section 4.2)**

2. $E[g(X)] = \displaystyle\int_{-\infty}^{\infty} g(x) f_X(x)\, dx.$ **(Section 4.2)**

$E[aX + bY] = aE[X] + bE[Y].$

If X and Y are independent, then

$$E[g(X)h(Y)] = E[g(X)]\,E[h(Y)].$$

The rth moment: $\mu_r' = E[X^r].$

Mean of X is $\mu_X = E[X].$ **(Section 4.4)**

The rth central moment: $\mu_r = E[(X - E[X])^r].$

Variance of X is $\text{Var}(X) = \sigma_X^2 = E[(X - \mu_X)^2].$

$\text{Var}(X) = E[X^2] - \mu_X^2.$ **(Section 4.4)**

Cauchy-Schwarz Inequality: If $E[X^2]$ and $E[Y^2]$ exist, then

$$E[XY] \leq E[X^2]\,E[Y^2]. \qquad \qquad \textbf{(Section 4.4)}$$

Covariance: $\text{Cov}(X, Y) = E[(X - \mu_X)(Y - \mu_Y)]$.   (Section 4.4)

Correlation: $\rho_{X,Y} = \dfrac{\text{Cov}(X, Y)}{\sqrt{\text{Var}(X)\,\text{Var}(Y)}}$.

If X and Y are independent, then $\text{Cov}(X, Y) = \rho_{X,Y} = 0$.   (Section 4.4)

$\text{Var}(aX + bY) = a^2\,\text{Var}(X) + b^2\,\text{Var}(Y) + 2ab\,\text{Cov}(X, Y)$.   (Section 4.4)

Skewness: $\gamma_1 = \dfrac{E[(X - \mu_X)^3]}{[\text{Var}(X)]^{3/2}}$.   (Section 4.4)

Kurtosis: $\gamma_2 = \dfrac{e[(X - \mu_X)^4]}{[\text{Var}(X)]^2}$.   (Section 4.4)

Moment Generating Function: $M_X(t) = E[e^{tX}]$.

$E[X^r] = M_X^{(r)}(0)$.

$M_{aX+b}(t) = e^{tb} M_X(at)$.

If X and Y are independent, $M_{aX+bY}(t) = M_X(at)M_Y(bt)$.

If (X, Y) is a random vector, then $M_{X,y}(t_1, t_2) = E[e^{t_1 X + t_2 Y}]$.

$M_X(t_1) = M_{X,Y}(t_1, 0)$.

$E[X^r Y^s] = \dfrac{\partial^{r+s}}{\partial t_1^r \partial t_2^s} M_{X,Y}(0, 0)$.   (Section 4.5)

Conditional Expectation:

If X and Y are discrete,

$$E[X \mid Y = y] = \sum_{a_i} p_{X|Y}(a_i \mid y)$$

$$E[g(X) \mid Y = y] = \sum_{a_i} g(a_i) p_{X|Y}(a_i \mid y).$$

If X and Y are continuous,

$$E[X \mid Y = y] = \int_{-\infty}^{\infty} x f_{X|Y}(x \mid y)\, dx,$$

$$E[g(X) \mid Y = y] = \int_{-\infty}^{\infty} g(x) f_{X|Y}(x \mid y)\, dx.$$
                    (Section 4.6)

$E[Y] = E[E[Y \mid X]]$.   (Section 4.6)

Review Exercises

1. Prove or give a counterexample for each of the following.
   a. $E[X^2] \geq (E[X])^2$.
   b. $\text{Cov}(R + S, T + U) = \text{Cov}(R, T) + \text{Cov}(R, U) + \text{Cov}(S, T) + \text{Cov}(S, U)$.
   c. If X and Y are independent, then $E[X \mid Y = y] = E[X]$.
   d. If X and Y are uncorrelated, then $E[X \mid Y = y] = E[X]$.

2. Let X and Y be i.i.d. random variables each with a p.f. of

$$p(a) = \begin{cases} \binom{n}{a} p^a (1-p)^{n-a} , & a = 0, 1, 2, \ldots, n, \quad \text{and} \\ 0 & , \text{otherwise.} \end{cases}$$

a. Find the moment generating function of X.

b. Find the moment generating function of $X + Y$.

c. What is the p.f. of $X + Y$?

3. Let the random vector (X, Y) have a joint moment generating function of $M_{X,Y}(t_1, t_2)$. Show how you would use this moment generating function to compute

a. $E[X^2]$,

b. $E[Y^3]$, and

c. $E[X^2 Y^2]$.

4. For some distributions calculating $E[X^r]$ and $E[(X-E[X])^r]$ can be quite difficult. Sometimes the job can be made easier by calculating what are called *factorial moments*. The rth factorial moment is defined to be

$$\mu^*_r = E[X(X-1)(X-2)\cdots(X-r+1)].$$

a. Write the mean and variance of a random variable X in terms of the factorial moments.

b. Use your answer to Part a to compute the variance of a random variable X whose p.d.f. is

$$p_X(a) = \begin{cases} \dfrac{\binom{M}{a}\binom{N-M}{n-a}}{\binom{N}{n}} , & a = 0, 1, 2, \ldots, n, \quad \text{and} \\ 0 & , \text{otherwise, where } n < M. \end{cases}$$

5. Let X have a p.d.f. of

$$f_X(x) = \begin{cases} \dfrac{1}{2\pi} , & 0 \le x < 2\pi, \quad \text{and} \\ 0 , & \text{otherwise.} \end{cases}$$

Let $Y = \cos X$, and $Z = \sin X$. Find $\text{Cov}(Y, Z)$.

6. Let X and Y be random variables such that

$$\mu_X = 3, \quad \mu_Y = 1, \quad \sigma_X^2 = 16, \quad \sigma_Y^2 = 9, \quad \text{and} \quad \rho_{X,Y} = 0.4.$$

a. Find $E[2X - Y]$.

b. Find $\text{Var}(2X - Y)$.

7. Suppose that the skewness and kurtosis of a random variable X exist. Let $Y = a + bX$. Find the skewness and kurtosis of Y in terms of the skewness and kurtosis of X.

8. Let the i.i.d. random variables X and Y have a p.d.f. of

$$f(x) = \begin{cases} 1, & 0 \leq x \leq 1, \quad \text{and} \\ 0, & \text{otherwise.} \end{cases}$$

   a. Find μ_X, σ_X^2 and $M_X(t)$.
   b. Find the moment generating function for $X + Y$.

9. Let X be a random variable whose p.d.f. is

$$f_X(x) = \begin{cases} x & , \ 0 \leq x \leq 1, \\ 2 - x, & 1 < x \leq 2, \quad \text{and} \\ 0 & , \ \text{otherwise.} \end{cases}$$

   a. Find $E[X]$, $\text{Var}(X)$, and $M_X(t)$.
   b. Compare $M_X(t)$ with $M_{X+Y}(t)$ from Part b of Exercise 8. What does this say about the p.d.f. of $X + Y$?

10. Let X have a p.d.f. of

$$f_X(x) = \begin{cases} k\left(1 - \dfrac{x^2}{a^2}\right), & -a \leq x \leq a, \quad \text{and} \\ 0 & , \ \text{otherwise.} \end{cases}$$

   Express k and a in terms of the variance and kurtosis of X.

11. Show that if a random variable is symmetric then all the the odd central moments will be zero.

12. Let (X, Y) have a joint probability density function of

$$f_{X,Y}(x, y) = \begin{cases} \dfrac{1}{2}, & 0 \leq x \leq y \leq 1, \quad \text{and} \\ 0, & \text{otherwise.} \end{cases}$$

   Find $E[Y \mid X]$.

13. Let X be a continuous random variable having a probability density function of

$$f_X(x) = \begin{cases} \alpha e^{-\alpha x}, & 0 \leq x, \quad \text{and} \\ 0 & , \ \text{otherwise.} \end{cases}$$

   Show that $E[X^r] = \frac{r!}{\alpha^r}$.

14. Suppose that the random variable X has $E[X^r] = 0$, for r an odd integer, and $\text{Var}(X) = 1$. Let $Y = aX^2 + b$ and compute $\rho_{X,Y}$.

15. Show that, if $E[X \mid Y = y] = E[X]$ for all y, then $\text{Cov}(X, Y) = 0$.

5

Probability Distributions

5.1
Introduction

Various probability distributions have been presented in the preceding chapters. Some of these are of particular interest for either practical or theoretical reasons. Certain random variables arise naturally in random experiments. Others are excellent models for observed phenomena. Still others are valuable because they are convenient and accurate approximations to more complicated distributions. The ability to recognize a probability function or probability density function can make a problem easier to solve by exploiting the properties of such functions.

This chapter addresses a small number of the distributions that are used in practice. For an additional source that either expounds on the distributions presented in the text or presents distributions not discussed here, see the four volume series *Distributions in Statistics* by Norman Johnson and Samuel Kotz.

Example 5.1

We wish to give a simple example of how random variables can sometimes help in evaluating definite integrals. A random variable that has a p.d.f. of the form

$$f_X(x) = \begin{cases} \dfrac{(n+m)!}{n!\,m!} x^n(1-x)^m, & 0 \le x \le 1, \quad \text{and} \\ 0, & \text{otherwise,} \end{cases}$$

is said to have a *beta* distribution. We shall investigate this distribution in more detail in Section 5.9. For now, suppose we wish to evaluate the following definite integral.

$$\int_0^1 x^4(1-x)^6 \, dx$$

This would be a tedious job using calculus techniques. Expanding $(1 - x)^6$ and integrating each component separately would take some time. However, we can recognize that the integrand is similar in form to the beta p.d.f. with the leading constant missing. Thus, if we multiply and divide by $\frac{10!}{4!\,6!}$ we obtain

$$\frac{1}{210} \int_0^1 \frac{10!}{4!\,6!}\, x^4 (1 - x)^6 \, dx.$$

The integral must have a value of 1 since it is that of a beta p.d.f. over the range where it is positive. Thus, the value of the original integral must be $\frac{1}{210}$. ∎

5.2
Hypergeometric Distribution

The hypergeometric distribution arises when we sample without replacement from a finite group of objects. For example, suppose we have an urn containing N balls, M of which are black and the remaining $N - M$ are white. A sample of n is drawn without replacement and the number of black balls noted. If we let X be the number of black balls drawn, then the probability function for X is

$$p_X(a) = \begin{cases} \dfrac{\dbinom{M}{a}\dbinom{N-M}{n-a}}{\dbinom{N}{n}} & , a = 0, 1, \ldots, \min(n, M), \quad \text{and} \\[2em] 0 & , \text{otherwise}. \end{cases} \qquad (5.1)$$

A random variable with this probability function is said to have a *hypergeometric distribution with parameters N, M, and n*. We leave it as an exercise to verify that it has the properties of a probability function. It is common practice to say that X is a hypergeometric random variable and to use the notation $X \sim H(n, M, N)$ as shorthand to state this. Thus, $X \sim H(5, 15, 50)$ would mean that X is a hypergeometric random variable with $n = 5$, $M = 15$, and $N = 50$. The parameters are fixed constants that are determined by the random experiment. The set of all hypergeometric distributions is commonly referred to as the *family* of hypergeometric distributions. The hypergeometric distribution describes exactly those experiments where sampling is done without replacement and we are counting the number of units in the sample possessing a certain attribute. If an object selected is one of the M objects with the desired attribute, it is commonly referred to as a "success." If not, it is called a "failure." Thus, a hypergeometric random variable counts the number of successes in a sample of size n drawn without replacement from a collection of N objects, M of which are successes.

The term "hypergeometric" arises from the fact that the right-hand side of Equation 5.1 can be written using a special mathematical function known as the hypergeometric function. This function is the solution to a differential equation, due to Gauss, called the hypergeometric equation. The equation has the form

$$x(1 - x)y'' + [c - (a + b + 1)x]y' - aby = 0.$$

For additional detail the interested reader can consult any good differential equations text.

The complicated nature of the probability function for the hypergeometric distribution makes it difficult to compute moments. It turns out to be more convenient to use factorial moments, which were introduced in Exercise 4 of Section 4.7. The rth factorial moment is given by

$$\mu_r^* = E[X(X-1)(X-2)\cdots(X-r+1)]. \tag{5.2}$$

Thus, $E[X] = \mu_1^*$, and $E[X^2] = \mu_2^* + \mu_1^*$. We then use these to compute the mean of the hypergeometric distribution as follows.

Example 5.2

Let $X \sim H(n,M,N)$. Assume that $n \leq M$. Then the mean is the first factorial moment.

$$E[X] = \sum_{a=0}^{n} a \frac{\dbinom{M}{a}\dbinom{N-M}{n-a}}{\dbinom{N}{n}}$$

$$= \sum_{a=1}^{n} \frac{M\dbinom{M-1}{a-1}\dbinom{N-M}{n-a}}{\dbinom{N}{n}}.$$

$$= \frac{nM}{N} \sum_{a=1}^{n} \frac{\dbinom{M-1}{a-1}\dbinom{N-M}{n-a}}{\dbinom{N-1}{n-1}}$$

We now let $b = a - 1$.

$$E[X] = \frac{nM}{N} \sum_{b=0}^{n-1} \frac{\dbinom{M-1}{b}\dbinom{N-M}{n-1-b}}{\dbinom{N-1}{n-1}}$$

The terms in the summation are those of the probability function of a random variable with a $H(n-1, M-1, N-1)$ distribution. Since the sum is from 0 to $n-1$ it must have a value of 1. Therefore,

$$E[X] = \frac{nM}{N}. \tag{5.3}$$

This is another illustration of the idea presented in Example 5.1. It is a very powerful tool that we will use quite often. We leave it as an exercise to show that

$$\text{Var}(X) = \frac{N-n}{N-1} \frac{nM}{N}\left(1 - \frac{M}{N}\right). \tag{5.4}$$

The moment generating function for the hypergeometric distribution exists. It is written in terms of the hypergeometric function, and we omit it. It can be found in the first volume of Johnson and Kotz.   ∎

Example 5.3   A lot consisting of 100 electrical resistors is inspected for acceptance by the following procedure. Five resistors are selected at random and the resistance of each is measured. If the resistance of each is within 10% of the rated value then the lot is accepted. If any deviate by more than 10% then the lot is rejected. We wish to determine the probability that a lot containing 20 defective resistors will be rejected.

Let X = number of defective resistors. We are sampling without replacement from a group of 100 objects, 20 of which are successes. Since X is counting the number of successes that occur we see that $X \sim H(5, 20, 100)$. Thus,

$$P(\text{lot rejected}) = P(1 \le X \le 5) = 1 - P(X = 0)$$

$$= 1 - \frac{\binom{20}{5}\binom{100-20}{5-0}}{\binom{100}{5}}$$

$$= 1 - \frac{24040016}{75287520}$$

$$= 0.681.$$

Exercises 5.2   1. Let X have a $H(n, M, N)$ distribution.

   a. Show that
   $$\text{Var}(X) = \frac{N-n}{N-1} \frac{nM}{N}\left(1 - \frac{M}{N}\right).$$

   b. Assume N becomes infinitely large while M stays a fixed proportion of N. That is, as N grows the ratio $\frac{M}{N}$ stays at a fixed value, p. Show that

   $$\lim_{\substack{N \to \infty \\ p=\frac{M}{N}}} \text{Var}(X) = np(1-p).$$

2. A particular species of fish is considered to be near extinction in a particular lake if fewer than 20 are present. To test if this is the case, 5 fish of this type are caught, tagged, and returned to the lake. Five more fish of this type are caught at a later date and the number of tagged ones noted. If there are only 20 fish of this species left in the lake, what is the probability that more than 2 of the fish in the second catch will be tagged?

3. Show that the probability function for the hypergeometric distribution sums to one. Assume that $n \le M$.

4. A company is trying to fill a managerial vacancy. In the applicant pool, 20 people are considered to be equally qualified. Eight of these 20 people belong to minorities. Budgetary constraints dictate that only 5 of these people can be called in for an interview. What is the probability that no more than 1 minority applicant will be interviewed if the people are chosen randomly?

5. In a class of 10 students, 4 of them are left-handed. What is the probability that, if 3 students are chosen at random without replacement, all of them will be left-handed?

6. A lake in a certain city receives processed waste from 20 local industrial plants. In an attempt to enforce pollution laws, a state official randomly tests the effluent from 5 of these companies. If any effluent has illegally high amounts of pollutants, then all 20 plants are tested. If half of the plants are in violation of state pollution laws, what is the probability that the state inspector will not find this out?

7. A game show asks that contestants determine a phrase by selecting letters until it is possible to guess the entire phrase. A success occurs if the contestant guesses the correct letter in the correct position in the phrase. Suppose that the current phrase is "She sells seashells by the seashore," and further suppose that a contestant has been tipped off that the letter *s* is in the puzzle. What is the probability that if the contestant selects a position at random and guesses the letter *s*, that he will not be successful?

5.3
Distributions Based on Independent Bernoulli Trials

A number of random experiments can be interpreted as sampling with replacement from a finite population. If each unit sampled is inspected to determine whether or not it has a certain attribute, then we have what is known as a sequence of *Bernoulli trials*.

Definition 5.1

> An experiment is said to consist of *independent Bernoulli trials* if it satisfies the following.
>
> 1. The outcome of each trial can be interpreted as being either a *success* or a *failure*.
> 2. For each trial $P(success) = p$, and $P(failure) = q = 1 - p$.
> 3. Each trial is independent.

For an experiment consisting of a single Bernoulli trial, we say that the random variable X is a *Bernoulli random variable with parameter p* if $X = 1$ for a success, and $X = 0$ for a failure. With this definition it should be clear that the probability function for X is

$$p_X(a) = \begin{cases} p , & a = 1, \\ q , & a = 0, \quad \text{and} \\ 0 , & \text{otherwise.} \end{cases} \tag{5.5}$$

It is easy to verify that, for a Bernoulli random variable,

$$E[X] = p,$$

$$\text{Var}(X) = p q, \quad \text{and}$$

$$M_X(t) = q + p e^t, -\infty < t < \infty.$$

Suppose now that we perform n independent Bernoulli trials and let

$$X_i = \begin{cases} 1 , & \text{if trial } i \text{ is a success, and} \\ 0 , & \text{if trial } i \text{ is a failure, } i = 1, 2, \ldots, n. \end{cases}$$

If we define

$$Y = \sum_{i=1}^{n} X_i,$$

then Y is the number of successes in the n independent Bernoulli trials. The random variable Y is said to have a *binomial distribution with parameters n and p,*where n is the number of trials and p is the probability of success on any given trial. The notation is $Y \sim B(n, p)$. Y can assume the values 0, 1, 2, ..., n. To obtain the probability function $p_Y(a)$, consider a single outcome having a successes, such as the one where the first a trials are successes and the remaining $n - a$ trials are failures. The probability of this outcome is

$$P(\underbrace{S \cap S \cap \cdots \cap S}_{a \text{ times}} \cap \underbrace{F \cap F \cap \cdots \cap F}_{n-a \text{ times}}) = \underbrace{p \cdot p \cdots p}_{a \text{ times}} \underbrace{q \cdot q \cdots q}_{n-a \text{ times}}$$

$$= p^a q^{n-a}.$$

Any other outcome with a successes and $n - a$ failures will only be a rearrangement of the above. Thus, the probability of obtaining a successes in n trials will be $p^a q^{n-a}$ added together for each of the $\binom{n}{a}$ possible arrangements of a successes and $n - a$ failures. Therefore, the probability function for a binomial random variable with parameters n and p is

$$p_X(a) = \begin{cases} \binom{n}{a} p^a q^{n-a} & , a = 0, 1, \ldots, n, \quad \text{and} \\ 0 & , \text{otherwise, where } 0 \le p \le 1, \quad \text{and, } q = 1 - p. \end{cases} \tag{5.6}$$

The distribution function for various values of n and p is tabulated in Table 1 of Appendix A.

We can make use of the fact that a binomial random variable is the sum of i.i.d. Bernoulli random variables to obtain the mean, variance, and moment generating function.

$$E[Y] = E\left[\sum_{i=1}^{n} X_i\right]$$

$$= \sum_{i=1}^{n} E[X_i]$$

$$= \sum_{i=1}^{n} p = np$$

$$\text{Var}(Y) = \text{Var}\left(\sum_{i=1}^{n} X_i\right)$$

$$= \sum_{i=1}^{n} \text{Var}(X_i)$$

$$= \sum_{i=1}^{n} pq = npq$$

$$M_Y(t) = E\left[e^{tY}\right]$$

$$= E\left[e^{t(X_1 + X_2 + \cdots + X_n)}\right]$$

$$= \prod_{i=1}^{n} E\left[e^{tX_i}\right]$$

$$= \left[M_X(t)\right]^n$$

$$= (q + pe^t)^n \quad -\infty < t < \infty$$

Example 5.4   Let X be a random variable with a m.g.f. of

$$M_X(t) = \left(\frac{3}{4} + \frac{1}{4}e^t\right)^{10}.$$

This is the m.g.f. of a r.v. whose distribution is $B\left(10, \frac{1}{4}\right)$. Therefore,

$$p_X(a) = \begin{cases} \binom{10}{a}\left(\frac{1}{4}\right)^a \left(\frac{3}{4}\right)^{10-a} & , a = 0, 1, 2, \ldots, 10, \quad \text{and} \\ 0 & , \text{otherwise.} \end{cases}$$

We wish to find $P(2 \leq X < 4)$. Now, by using Table 1 we find that

$$P(2 \leq X < 4) = F_X(3) - F_X(1)$$

$$= 0.7759 - 0.2440$$

$$= 0.5319.$$

■

Example 5.5   In a manufacturing process, it is desired to devise an acceptance sampling scheme that will sample with replacement from a bin containing the output of a production run and accept the run only if no defective items are found. We wish to determine the smallest sample size that will give a probability of no more than 0.05 of accepting a lot with 10% defective items.

Let X = number of defective items in a sample of n items. Then $X \sim B(n, 0.10)$, and we are looking for the smallest n so that $P(X = 0) \leq 0.05$. Now Equation 5.6 gives

$$P(X = 0) = (0.9)^n, \quad \text{which gives}$$

$$(0.9)^n \leq 0.05, \quad \text{or}$$

$$n \geq \frac{\ln(0.05)}{\ln(0.9)}$$

$$= 28.43.$$

Therefore, the minimum sample size is 29 items.   ■

The hypergeometric and binomial random variables differ only in the fact that sampling is done without replacement in the hypergeometric case and with replacement for the binomial. This means that the trials are dependent for a hypergeometric random variable and independent in the case of the binomial. Now suppose we sample 3 balls without replacement from a box containing 1000 balls, 400 of which are white. The exact probability that all 3 balls will be white is

$$\left(\frac{400}{1000}\right)\left(\frac{399}{999}\right)\left(\frac{398}{998}\right) = 0.0637.$$

If the sample is done with replacement, the exact probability that all 3 balls will be white is

$$\left(\frac{400}{1000}\right)^3 = 0.064.$$

This indicates that if we sample without replacement from a large number of objects the trials are not independent, but they are "almost independent." In other words, probabilities for hypergeometric random variables can be closely approximated by using the binomial probability function. This idea is made formal in the following theorem. The proof is left as an exercise.

Theorem 5.1

Let X be a random variable having a hypergeometric distribution with parameters n, $M = N_p$, and N, where $0 \leq p \leq 1$. Then

$$\lim_{N \to \infty} p_X(a) = \begin{cases} \binom{n}{a} p^a (1-p)^{n-a}, & a = 0, 1, 2, \ldots, n, \quad \text{and} \\ 0 & , \text{ otherwise.} \end{cases}$$

Experimentation has shown that this approximation of the hypergeometric distribution by the binomial distribution is useful if $n < 0.05N$.

Another useful result concerns the sums of binomial random variables. It turns out that, under the right conditions, the sum of two binomial random variables also has a binomial distribution. This is made explicit in the following theorem. We defer the proof until Chapter 6.

Theorem 5.2

Let X be a $B(n, p)$ random variable and Y be an independent $B(m, p)$ random variable. Then $S = X + Y$ will be a $B(n + m, p)$ random variable.

Now suppose that independent Bernoulli trials are conducted until a success occurs. If we let X be the number of trials then X has what is known as a *geometric distribution*. Notice here that the number of successes is fixed and the number of trials is random. This differs from the binomial case where the number of trials was fixed with the number of successes being random. In this case, the range of possible values for X is the positive integers. We can derive the probability function as follows. If X equals a, then the first $a - 1$ trials must be failures and the ath trial a success. If the probability of success on each trial is p and the probability of failure is $q = 1 - p$ we see that

$$P(X = a) = \underbrace{q \cdot q \cdots q}_{a-1 \text{ times}} p$$

$$= q^{a-1}p.$$

Thus, the probability function for X is

$$p_X(a) = \begin{cases} q^{a-1}p, & a = 1, 2, \ldots, \quad \text{and} \\ 0, & \text{otherwise, where } 0 \le p \le 1, \quad \text{and } q = 1 - p. \end{cases} \tag{5.7}$$

The geometric random variable gets its name from the fact that $p_X(a)$ is the ath term of the geometric series that is studied in calculus courses. We leave it as an exercise to verify that the mean, variance, and moment generating function are as follows.

$$E[X] = \frac{1}{p}.$$

$$\text{Var}(X) = \frac{q}{p^2}.$$

$$M_X(t) = \frac{p e^t}{1 - q e^t}, \quad q e^t < 1$$

Example 5.6

An urn contains 1 white ball and 1 black ball. A ball is sampled with replacement until the white ball is drawn. We let X be the number of trials and compute the probability that X will be an odd number. Clearly X has a geometric distribution with $p = q = \frac{1}{2}$, and

$$P(X \text{ is odd}) = \sum_{i=1}^{\infty} q^{2(i-1)} p$$

$$= \sum_{i=1}^{\infty} \left(q^2\right)^{i-1} p$$

$$= \frac{p}{1 - q^2}$$

$$= \frac{2}{3}.$$

∎

If independent Bernoulli trials are conducted until the rth success is obtained, and we let X be the number of trials, we say that X has a *negative binomial distribution*. The parameters of this distribution are p, the probability of success on any given trial, and r, the number of successes. The notation for this is $X \sim NB(r, p)$. This is an extension of the geometric distribution just as the binomial distribution was for the Bernoulli case. The probability function is obtained as follows. If the

rth success occurs on trial a, then $r - 1$ successes occurred in the first $a - 1$ trials followed by a success on the ath trial. One such outcome has probability

$$\underbrace{pp\cdots p}_{r-1 \text{ terms}}\ \underbrace{qq\cdots q}_{a-r \text{ terms}}\ p = q^{a-r}p^r.$$

Since there are $\binom{a-1}{r-1}$ possible arrangements of $r - 1$ successes and $a - r$ failures, we obtain that

$$P(X = a) = \binom{a-1}{r-1}q^{a-r}p^r.$$

Thus, the probability function for a negative binomial random variable is

$$p_X(a) = \begin{cases} \binom{a-1}{r-1}q^{a-r}p^r, & a = r, r+1, r+2, \ldots, \quad \text{and} \\ 0 & , \text{ otherwise, where } 0 \le p \le 1, \quad \text{and, } q = 1 - p. \end{cases}$$
$$(5.8)$$

The negative binomial distribution derives its name from the negative binomial series, which is also discussed in some calculus texts. The probability function at $X = a$ is the ath term in a negative binomial series for $(p + q)^{-r}$. Note that the geometric distribution is $NB(1, p)$. We leave it as an exercise that the mean, variance, and moment generating function for the negative binomial distribution are as follows.

$$E[X] = \frac{r}{p}.$$

$$\text{Var}(X) = \frac{rq}{p^2}.$$

$$M_X(t) = \left(\frac{pe^t}{1 - qe^t}\right), \quad qe^t < 1$$

Example 5.7

An urn contains b black balls and w white ones. Balls are selected one at a time with replacement. We wish to find the probability that r black balls will be withdrawn before s white ones are drawn. Let X be the number of balls drawn. We note that the event occurs only if rth black ball is drawn on or before the $(r + s - 1)$th trial. Now, $X \sim NB(r, p)$, where $p = \frac{b}{w+b}$. Therefore,

$$P(r \text{ black balls drawn before } s \text{ white balls}) = \sum_{a=r}^{r+s+1}\binom{a-1}{r-1}\left(\frac{w}{w+b}\right)^{a-r}\left(\frac{b}{w+b}\right)^r.$$

■

There is a result regarding the sum of independent negative binomial random variables that is analogous to Theorem 5.2. We give it in the following theorem and again defer the proof until Chapter 6.

Theorem 5.3

> Let X be a $NB(r,p)$ random variable and Y be an independent $NB(s,p)$ random variable. Then the random variable $S = X + Y$ will be a $NB(r+s,p)$ random variable.

Exercises 5.3

1. Let X be a Bernoulli random variable with parameter p. Show that
   a. $E[x] = p$,
   b. $\text{Var}(X) = pq$, and
   c. $M_X(t) = q + pe^t$.

2. Prove Theorem 5.1.

3. Let X have a geometric distribution with parameter p. Show that
   a. $E[X] = \dfrac{1}{p}$,
   b. $\text{Var}(X) = \dfrac{q}{p^2}$, and
   c. $M_X(t) = \dfrac{pe^t}{1 - qe^t}, qe^t < 1$.

4. Let X have a negative binomial distribution with parameters r and p. Show that
   a. $E[X] = \dfrac{r}{p}$,
   b. $\text{Var}(X) = \dfrac{rq}{p^2}$, and
   c. $M_X(t) = \left[\dfrac{pe^t}{1 - qe^t}\right]^r, qe^t < 1$.

5. An urn contains 6 white balls and 2 black balls. A ball is drawn and then replaced. This process is repeated.
   a. What is the probability that in 4 draws, exactly 2 black balls will occur?
   b. What is the probability that the second black ball will occur on the fourth draw?
   c. How few draws are required so that the probability of obtaining at least 1 black ball is 0.90?

6. An examination consists of 10 multiple-choice questions, each with 4 possible answers. At least 7 correct answers are needed to pass. What is the probability that a student who is completely unprepared and guesses at each answer will pass the examination?

7. A lawn mower manufacturer advertises that its product will start on the first attempt 90% of the time. What is the probability that a mower that starts only 80% of the time on the first try will be considered acceptable if 10 starts are attempted?

8. In an attempt to distinguish between 2 methods for training workers in using a computer spreadsheet, trainees are divided into 2 groups. Then 1 member of a group is paired with a member of the other group. If both methods are equally effective in training personnel, what is the probability that in training 12 pairs, one method will be judged more effective 8 or more times?

9. In baseball, the World Series lasts until one team wins 4 games. Suppose the San Francisco Giants meet the Cleveland Indians in the World Series. The Giants are

considered the superior team with a probability of 0.6 of winning any single game between themselves and the Indians.

   a. What is the probability that the Giants will win the World Series in 5 games?

   b. What is the probability that the Giants will win the World Series?

10. In an acceptance sample scheme for a manufacturing process, items are drawn with replacement from the output of a production run until either 3 defective items have been found or until 10 items have been selected. If 3 defective items are found, the process is declared to be out of control. If a production run of 100 items contains 10 defective ones, what is the probability that it will not be declared to be out of control?

11. Dr. U. N. Scrupulus is determined to prove that everyone has some ability in extrasensory perception (e.s.p.). To do this he takes a subject with absolutely no e.s.p. and has this person tell which of 4 colors is behind a screen. This is repeated 10 times. The good doctor will conclude that a person has e.s.p. if more than half of the guesses are correct. If the subject does not get more than half of the guesses correct, the test is repeated until a positive result is obtained. What is the probability that it will take more than 5 runs of the experiment to produce a positive result?

12. A student is asked to select 15 cards without replacement from a deck containing 500 cards numbered 1 to 500. What is the (approximate) probability that at least half of the numbers selected will be even?

13. A fair die is being rolled.

   a. Find the probability that at least five rolls are needed to obtain the first one.

   b. Find the probability that it will take fewer than seven rolls to obtain the second one if the first one occurs on the third roll.

14. One hundred of the houses in a development containing 500 homes are white. If 5 houses are selected at random, without replacement, what is the probability that fewer than 2 of them will be white?

15. Five 4-digit numbers are selected at random with replacement. What is the probability that at least $\frac{1}{2}$ of them will be divisible by 3?

16. Forty percent of the registered voters in a certain large city are Democrats.

   a. Find the probability that more than half of a sample of 10 voters will be Democrats.

   b. Find the probability that at least 4 voters must be sampled in order to obtain a Democrat.

17. In the first draft of a manuscript, 750 pages of a total of 1000 pages have at least 1 typographical error. Find the probability that out of 5 randomly selected pages fewer than 2 pages have any typographical errors.

5.4
Multinomial
Distribution

Many experiments consisting of a fixed number of repeated independent trials have results where the outcome of each trial can be assigned to one of more than two distinct categories. For example, in sampling registered voters, we might want to indicate whether a given voter is a Democrat, a Republican, or neither. Such situations are similar to a binomial experiment except that, instead of having only two possible outcomes on each trial, we have many. In this case the experiment is said to consist of independent *multinomial trials*. We give a formal definition.

Definition 5.2

> An experiment is said to consist of independent *multinomial trials* if the following are true.
>
> 1. The outcome of each trial can be assigned to one of k mutually exclusive and collectively exhaustive categories $C_1, C_2, \ldots, C_k$.
> 2. On each trial the probability that an observation x will fall in category C_i is $P(x \in C_i) = p_i$, for $i = 1, 2, \ldots, k$, such that $p_1 + p_2 + \cdots + p_k = 1$.
> 3. The trials are independent.

Note that a Bernoulli trial is a special case of a multinomial trial where $k = 2$. In addition, if $k = 3$ the trials are said to be *trinomial*.

Suppose we have a random experiment consisting of n independent trinomial trials. If we let X_i be the number of trials whose outcome falls in category C_i, for $i = 1$ and 2, then the number of outcomes in category C_3 is $n - X_1 - X_2$. We seek the joint probability function for the random vector (X_1, X_2). The line of reasoning is similar to that used to derive the binomial probability function. Consider the event $\{X_1 = a_1, X_2 = a_2\}$, and let $a_3 = n - a_1 - a_2$, where a_1, a_2, and a_3 are nonnegative integers such that $a_1 + a_2 + a_3 = n$. One such occurrence is where the outcomes of the first a_1 trials are in category C_1, the next a_2 trials are in category C_2, and the last a_3 trials are in category C_3. This has a probability of

$$\underbrace{p_1 p_1 \cdots p_1}_{a_1 \text{ in } C_1} \underbrace{p_2 p_2 \cdots p_2}_{a_2 \text{ in } C_2} \underbrace{p_3 p_3 \cdots p_3}_{a_3 \text{ in } C_3} = p_1^{a_1} p_2^{a_2} p_3^{a_3}.$$

Now, recalling the multinomial coefficients described in Section 1.5 we see that there are

$$\binom{n}{a_1, a_2, a_3} = \frac{n!}{a_1! \, a_2! \, a_3!}$$

possible arrangements of these outcomes, each of which has the probability just computed. Therefore,

$$P(X_1 = a_1, X_2 = a_2) = \binom{n}{a_1, a_2, a_3} p_1^{a_1} p_2^{a_2} p_3^{a_3}.$$

This is a value of the joint probability function for (X_1, X_2). Thus, the joint probability function for a random vector having a *trinomial distribution with parameters n, p_1, p_2, and p_3* is

$$p_{X_1, X_2}(a_1, a_2) = \begin{cases} \binom{n}{a_1, a_2, n - a_1 - a_2} p_1^{a_1} p_2^{a_2} p_3^{n - a_1 - a_2}, & \begin{array}{l} a_1, a_2, \text{ nonnegative} \\ \text{integers such} \\ \text{that } a_1 + a_2 \leq n \end{array} \\ 0 & , \text{ otherwise,} \end{cases}$$

$$(5.9)$$

where p_1, p_2, and p_3 are nonnegative real numbers such that $p_1 + p_2 + p_3 = 1$.

The distribution gets its name from the fact that $p_{X_1, X_2}(a_1, a_2)$ is a term in the multinomial expansion of $(p_1 + p_2 + p_3)^n$. We leave it as an exercise to show that the joint moment generating function for (X_1, X_2) is

$$M_{X_1, X_2}(t_1, t_2) = \left(p_1 e^{t_1} + p_2 e^{t_2} + p_3\right)^n \tag{5.10}$$

for all values of t_1 and t_2. Note that

$$M_{X_1}(t_1) = M_{X_1, X_2}(t_1, 0)$$

$$= \left(p_1 e^{t_1} + p_2 + p_3\right)^n$$

which, since $p_2 + p_3 = 1 - p_1$, is the moment generating function for a $B(n, p_1)$ random variable. Thus, the marginal probability function for any single random variable in a trinomial distribution is binomial. This means that

$$E[X_i] = np_i, \quad \text{and}$$

$$\text{Var}(X_i) = np_i(1 - p_i), \quad \text{for } i = 1, 2.$$

Example 5.8

Let (X, Y) have a trinomial distribution with parameters n, p_1, p_2, and p_3. We wish to determine the correlation between X and Y. We shall use the joint moment generating function and Theorem 4.19 to determine the moments. Recall that

$$\rho_{X,Y} = \frac{\text{Cov}(X, Y)}{\sqrt{\text{Var}(X)\ \text{Var}(Y)}}, \quad \text{where}$$

$$\text{Cov}(X, Y) = E[XY] - E[X]E[Y].$$

We know that

$$E[X] = np_1,$$

$$E[Y] = np_2,$$

$$\text{Var}(X) = np_1(1 - p_1), \quad \text{and}$$

$$\text{Var}(Y) = np_2(1 - p_2).$$

We use Theorem 4.19 to determine $E[XY]$ as follows. The cross partial derivative of the joint moment generating function is

$$\frac{\partial^2}{\partial t_1\, \partial t_2} M_{X,Y}(t_1, t_2) = n(n - 1)p_1 p_2 e^{t_1} e^{t_2} \left(p_1 e^{t_1} + p_2 e^{t_2} + p_3\right)^{n-2}.$$

Setting $t_1 = t_2 = 0$ we find that

$$E[XY] = n(n - 1)p_1 p_2, \quad \text{and}$$

$$\text{Cov}(X, Y) = n(n - 1)p_1 p_2 - n^2 p_1 p_2$$

$$= -np_1 p_2.$$

Therefore,

$$\rho_{X,Y} = \frac{-np_1p_2}{\sqrt{np_1(1-p_1)\,np_2(1-p_2)}}$$

$$= -\sqrt{\frac{p_1p_2}{(1-p_1)(1-p_2)}}.$$

This is valid as long as $0 < p_1 < 1$ and $0 < p_2 < 1$. ∎

Example 5.9   Suppose that (X,Y) has a trinomial distribution with parameters n, p_1, p_2, and p_3. We wish to find the conditional probability function for X given that $Y = k$. From Equation 3.42 we note that

$$p_{X|Y}(a \mid k) = \frac{p_{X,Y}(a,k)}{p_Y(k)}.$$

Therefore, since Y has a distribution that is $B(n,p_2)$, we find that

$$p_{X|Y}(a \mid k) = \frac{\dfrac{n!}{a!\,k!\,(n-a-k)!}\,p_1^a p_2^k p_3^{n-a-k}}{\dfrac{n!}{k!(n-k)!}\,p_2^k(1-p_2)^{n-k}}$$

$$= \frac{(n-k)!}{a!\,(n-a-k)!}\left(\frac{p_1}{1-p_2}\right)^a\left(\frac{1-p_1-p_2}{1-p_2}\right)^{n-a-k}, \quad \text{or}$$

$$p_{X|Y}(a \mid k) = \begin{cases} \dbinom{n-k}{a}\left(\dfrac{p_1}{1-p_2}\right)^a\left(\dfrac{p_3}{1-p_2}\right)^{n-k-a} &, a = 0,1,\ldots,n-k, \quad \text{and} \\ 0 &, \text{otherwise.} \end{cases}$$

Thus, the conditional distribution of X, given $Y = k$ is $B\left(n-k, \frac{p_1}{1-p_2}\right)$. In addition, the conditional expectation of X given $Y = k$ is

$$E[X \mid Y = k] = (n-k)\frac{p_1}{1-p_2}.$$

∎

The multinomial distribution for $k > 3$ is a straightforward generalization of the trinomial distribution. Given categories $C_1, C_2, \ldots, C_k$, we define the random variable X_i to be the number of trials whose outcome is in category i, for $i = 1, 2, \ldots, k$. Using the probability function for the trinomial distribution for a model we can arrive at the fact that the probability function for the *multinomial distribution with parameters* n, p_1, p_2, $\ldots$, p_k is

$$p_{X_1,X_2,\ldots,X_k}(a_1,a_2,\ldots,a_k) = \binom{n}{a_1,a_2,\ldots,a_k}p_1^{a_1}p_2^{a_2}\cdots p_k^{a_k}, \tag{5.11}$$

for $a_1, a_2, \ldots, a_k$ nonnegative integers such that $a_1 + a_2 + \cdots + a_k = n$, and zero, otherwise, with the understanding that $p_1, p_2, \ldots, p_k$ are nonnegative real numbers such that $p_1 + p_2 + \cdots + p_k = 1$.

It can also be shown that the joint moment generating function for X_1, X_2, $\ldots$, X_k is

$$M_{X_1, X_2, \ldots, X_k}(t_1, t_2, \ldots, t_k) = \left(p_1 e^{t_1} + p_2 e^{t_2} + \cdots + p_k e^{t_k} \right)^n, \qquad (5.12)$$

for all values of $t_1, t_2, \ldots, t_k$.

Example 5.10

We wish to compute $\text{Cov}(X_i, X_j)$ for a general multinomial distribution. To do this we shall use the joint moment generating function. The means are computed as follows.

$$\frac{\partial}{\partial t_i} M_{X_1, X_2, \ldots, X_k}(t_1, t_2, \ldots, t_k) = n p_i e^{t_i} \left(p_1 e^{t_1} + p_2 e^{t_2} + \cdots + p_k e^{t_k} \right)^{n-1}$$

Therefore,

$$E[X_i] = np_i, \quad \text{and} \quad E[X_j] = np_j.$$

To evaluate $E[XY]$, we take the second partial derivative as follows.

$$\frac{\partial^2}{\partial t_1 \partial t_2} M_{X_1, X_2, \ldots, X_k}(t_1, t_2, \ldots, t_k) = n(n-1)p_i e^{t_i} p_j e^{t_j} \left(p_1 e^{t_1} + p_2 e^{t_2} + \cdots + p_k e^{t_k} \right)^{n-2}$$

Setting $t_1 = t_2 = \cdots t_k = 0$ we find that

$$E[XY] = n(n-1)p_i p_j,$$

and

$$\text{Cov}(X_i, X_j) = n(n-1)p_i p_j - n^2 p_i p_j = -np_i p_j.$$

Note that this is the same result we found for the trinomial distribution. ■

Exercises 5.4

1. Let (X, Y) have a trinomial distribution with parameters n, p_1, p_2, and p_3. Show that

$$M_{X,Y}(t_1, t_2) = \left(p_1 e^{t_1} + p_2 e^{t_2} + p_3 \right)^n,$$

for all values of t_1, t_2, and t_3.

2. A fair die is tossed 4 times. Let X be the number of times a toss results in a 1 or a 2. Let Y be the number of times a toss is either a 3, 4, or 5.

   a. Compute $P(X = 0, Y = 2)$.

   b. Compute $P(X = Y)$.

3. Five cards are drawn with replacement from a standard deck of 52 cards. Let X be the number of face cards drawn, and let Y be the number of cards with even numbers drawn.

   a. Give the joint probability function for X and Y.

   b. Find $P(X = 3, Y = 1)$.

   c. Find $P(Y = 2 \mid X = 1)$.

4. Four fair dice are tossed 5 times. Let X be the number of times all dice show the same amount, and let Y be the number of times exactly two of the dice show the same number.
   a. Find the joint probability function for (X,Y).
   b. Compute $E[XY]$.

5. Over the course of the season, soccer team A will play soccer team B a total of 5 times. In any given game, team A will win with probability 0.4, team B will win with probability 0.5, and the rest of the time the game will result in a tie. Assume that the games are independent.
   a. What is the probability that team A will win 2 games and lose the other 3?
   b. What is the probability that team B will win more games than team A?

6. The Washington Capitals and the New York Rangers of the National Hockey League are in the same division and play each other many times in the course of a season. It is generally felt that, in any single game, the Capitals should win 65% of the time and the Rangers should win 30% of the time with a tie resulting the remainder of the time.
   a. What is the probability that the Capitals and Rangers would each win 3 out of 6 randomly selected games?
   b. What is the probability that the Capitals would win if the 2 teams met in a best 3-out-of-5 playoff series?

7. A popular arcade game requires that players maneuver randomly generated shapes in order to fill a rectangular area. There are 5 distinct shapes in the game. Call these shapes A, B, C, D, and E. The probabilities that any given shape will be generated are denoted by p_A, p_B, p_C, p_D, and p_E and they have values 0.25, 0.25, 0.15, 0.15, and 0.20, respectively. Ten shapes are generated during the game.
   a. What is the probability that there will be 3 A's, 3 B's, 3 C's, 1 D, and 0 E's?
   b. What is the probability that there will be one of each of the shapes A, B, C, D, with the remainder being shape E?
   c. What is the probability that there will be more E's than the other shapes put together?

5.5
Poisson Distribution

Suppose that a random variable X has a $B(n,p)$ distribution where n is large and p is small. It can be difficult in a practical sense to compute $P(X = a)$. The binomial coefficient $\binom{n}{a}$ can easily result in integers that exceed the storage capacity of most hand calculators and computers. One possibility is to use approximations to $n!$ such as Stirling's formula, but there can be some computational problems in raising the small number p to large powers. Roundoff and truncation errors become significant. In such cases, however, there is a useful and accurate approximation to this probability. We state it as a theorem and leave the proof as an exercise.

Theorem 5.4

Let the random variable X have a binomial distribution with parameters n and p, and for any n let $\lambda = np$. Then

$$\lim_{\substack{n \to \infty \\ np = \lambda}} P(X = a) = \begin{cases} \dfrac{\lambda^a e^{-\lambda}}{a!}, & a = 0, 1, 2 \ldots, \text{ and} \\ 0, & \text{otherwise.} \end{cases} \qquad (5.13)$$

Example 5.11

Let X have a $B(100, 0.001)$ distribution. We wish to compute $P(X = 1)$. A hand calculator gives that, to 4 decimal places, the exact probability is

$$P(X = 1) = \binom{100}{1}(0.001)^1(0.999)^{99} = 0.0906.$$

Using $\lambda = np = (100)(0.001) = 0.1$ in Equation 5.13 gives

$$P(X = 1) \approx \frac{(0.1)^2\, e^{-0.1}}{1!} = 0.0905.$$

The results agree to within 3 decimal places.   ∎

The right-hand side of Equation 5.13 can be shown to constitute a valid probability function. Since $\lambda = np > 0$, we see that Equation 5.13 is nonnegative for any real number a. Now, using the Taylor series for e^x we get

$$\sum_0^\infty \frac{\lambda^a e^{-\lambda}}{a!} = e^{-\lambda} \sum_0^\infty \frac{\lambda^a}{a!} = e^{-\lambda} e^\lambda = 1.$$

Thus, in addition to being an approximation for the binomial distribution, the right-hand side of Equation 5.13 is the probability function for some random variable. The distribution of this random variable is named after Simeon D. Poisson who first discussed it in his 1837 book *Recherches sur la Probabilité des Jugements en Matière Criminelle et en Matière Civili, Précédées des Regeles Générales du Calculudes Probabilitiés*. Poisson approached the distribution as an approximation to the binomial distribution.

To be formal, a random variable X is said to have a *Poisson distribution with parameter* λ, denoted $X \sim Po(\lambda)$, if it has a probability function of the form

$$p_X(a) = \begin{cases} \dfrac{\lambda^a e^{-\lambda}}{a!}, & a = 0, 1, 2, \ldots, \text{ and} \\ 0, & \text{otherwise, where } \lambda > 0. \end{cases} \tag{5.14}$$

The Poisson distribution function for selected values of λ is given in Table 2 of Appendix A. In Examples 4.11, 4.19, and 4.30, we showed that the mean, variance, and moment generating function for the Poisson distribution are as follows.

$$E[X] = \lambda,$$

$$\text{Var}(X) = \lambda, \quad \text{and}$$

$$M_X(t) = e^{\lambda(e^t - 1)}, \quad -\infty < t < \infty$$

In addition to being a useful approximation to the binomial distribution, there are a number of random phenomena that have been successfully modeled by the Poisson distribution. Some examples include:

1. the number of deaths per year due to the kick of a horse in the Prussian Army Corps during the period 1875–1894.

2. the number of α particles emitted by a source in a fixed period of time.
3. the number of bombs striking a city block in London over a fixed period of time during World War II.
4. the number of customers arriving at a bank each minute.
5. the number of raisins in a slice of raisin bread.
6. the number of cars passing through an intersection each minute.
7. the number of jobs submitted to a computer's operating system each minute.

These examples have certain things in common. Each example considered events taking place over a fixed interval of time or within a fixed quantity of material. In that interval the random variable counted the number of times a particular event occurred. This is no coincidence. It is possible to state conditions that characterize the types of experiments that give rise to random variables having a Poisson distribution. We give an informal verbal description of them.

If a random variable that counts the number of times a particular event occurs in a given interval of length h satisfies the following assumptions, it will have a Poisson distribution. Let X be the number of events occurring in the interval h.

1. $P(X = 1)$ is proportional to the length of the interval when h is small.
2. $P(X \geq 2)$ is essentially 0 when h is small.
3. The random variables counting the number of events occurring in any collection of nonoverlapping intervals are independent.

The proof that these assumptions give rise to a Poisson distribution can be found in many calculus texts and in more advanced books on probability.

Example 5.12   A seed company sells packages containing 100 marigold seeds. It is known that, due to uncontrollable circumstances, any batch of seeds will contain 1% seeds of a different variety. Let X be the number of variant seeds in a package of marigold seeds. Then $X \sim B(100, 0.01)$. We wish to use the Poisson approximation to compute the probability that a given package of seeds will contain no more than one variant seed. We note that $\lambda = np = (100)(0.01) = 1$ and use Equation 5.13 to get

$$P(X \leq 1) \approx \frac{\lambda^0 e^{-\lambda}}{0!} + \frac{\lambda^1 e^{-\lambda}}{1!}$$

$$= e^{-1} + e^{-1}$$

$$= 0.736.$$

This result is the same, to 3 decimal places, as the exact value obtained using the binomial distribution. Thus, as long as p is small enough, n need not be very large. ∎

It is fairly common that investigators make many observations of a random experiment having a Poisson distribution. For example, we may be interested in the number of raisins in an entire loaf of raisin bread when it is known that the number of raisins per slice follows a Poisson distribution. The number of raisins in the loaf would be the sum of independent Poisson random variables. The distribution of the sum of two independent Poisson random variables is given by the following theorem. We defer the proof until Chapter 6.

Theorem 5.5

> Let X and Y be independent Poisson random variables with parameters λ and η, respectively. Then the distribution of $S = X + Y$ will be Poisson with parameter $\lambda + \eta$.

Example 5.13

The number of $\frac{1}{8}$ flaws in a square foot of a certain type of high-quality steel is known to follow a Poisson distribution with parameter $\frac{1}{4}$. We wish to find the probability that a 1 ft by 8 ft piece of this type of steel will contain no more than a single $\frac{1}{8}$-in. flaw. We partition the sheet into 8 1-ft by 1-ft pieces, and let X_i, $i = 1, 2, \ldots, 8$, be the number of flaws in each square foot section. Then $X_1, X_2, \ldots, X_8$ are i.i.d. $Po\left(\frac{1}{4}\right)$ random variables. An application of Theorem 5.5 gives that $Y = X_1 + X_2 + \cdots + X_8$ will have a Poisson distribution with parameter $\lambda = 8 \cdot \frac{1}{4} = 2$. Then we obtain that

$$P(Y \leq 1) = \frac{2^0 e^{-2}}{0!} + \frac{2^1 e^{-2}}{1!} = 0.406.$$

Now suppose we wish to determine the probability that 3 randomly selected 1-ft squares taken from a large number of squares will contain no flaws. Let Z be the number of squares with flaws. We are sampling without replacement from a group of squares. Therefore, the exact distribution of Z will be hypergeometric. However, because the number of squares from which the sample is taken is large we shall use the binomial distribution to approximate $p_Z(a)$. Thus, approximately $Z \sim B(3, p)$, where p is the probability that a square contains flaws. We obtain this probability from the fact that the number of flaws in a square has a $Po\left(\frac{1}{4}\right)$ distribution. Therefore,

$$p = 1 - \frac{\left(\frac{1}{4}\right)^0 e^{-1/4}}{0!} = 1 - 0.7788 = 0.2212,$$

and

$$P(Z = 0) = \binom{3}{0}(0.2212)^0(1 - 0.2212)^3 = 0.4723. \qquad \blacksquare$$

Exercises 5.5

1. Let $X \sim B(500, 0.002)$ find an approximate value for $P(X \geq 2)$.

2. Prove Theorem 5.4.

3. Let X have a Poisson distribution with $\lambda = 3$. Find each of the following.
   a. $P(X \leq 4)$
   b. $P(X \geq 2)$
   c. $P(2 \leq X \leq 6)$
   d. The median of X.

4. For what value of λ does the maximum value of $p_X(a)$ at $a = 6$?

5. Let X have a Poisson distribution with parameter λ. Show that the probability that X is even is $\frac{1}{2}(1 + e^{-2\lambda})$.

6. A bin contains 200 items, one of which is defective. You sample a single item from the bin with replacement and repeat the process 100 times. What is the (approximate) probability that you will

   a. obtain the defective item at least once,

   b. obtain the defective item exactly once.

7. The first draft of a calculus text consisting of 1000 pages has only 100 pages with no errors. The number of errors on each page follows a Poisson distribution. How many pages should contain exactly 1 error?

8. A particularly dangerous section of an interstate highway is known to average 1.1 traffic accidents per day. The number of accidents per day follows a Poisson distribution. What is the probability that, on a random selected day, there will be more than 2 traffic accidents on this section of highway?

9. Refer to the section of highway discussed in Exercise 8. What is the probability that, in a randomly selected week, there will be fewer than 5 traffic accidents?

10. A particular brand of chocolate chip cookies is known to average 5 chocolate chips per cookie. The number of chips per cookie follows a Poisson distribution. What is the probability that you will find a cookie with no chocolate chips in it? How many cookies would you expect to have to sample before finding a chipless cookie?

11. The number of orders every 10 minutes for tickets at a telephone box office facility follows a Poisson distribution with $\lambda = 2$.

    a. Find the probability that 8 orders will be received in a 30-minute time period.

    b. If the phones are down for 20 minutes, what is the probability that no orders will be missed?

    c. How many orders would be expected in a 20-minute time period?

12. A software company has discovered a "bug" in a little used feature of one of its products. The decision is made not to announce it publicly, but rather to give a corrected version of the product to any user who calls the company after having encountered the problem. Assume that only 0.4% of people running the program use the affected feature. A random sample of 1000 users is made.

    a. What is the expected number of users that have encountered the problem?

    b. What is the probability that 5 or more users have encountered the problem?

    c. What is the probability that no users have encountered the problem?

13. During a severe thunderstorm, lightning occurs on the average of every 6 seconds. Assume a Poisson distribution for the number of flashes per minute.

    a. What is the probability that no flashes will occur during a given minute?

    b. What is the probability that between 6 and 12 flashes, inclusive, will occur in a minute?

14. While two computers communicate with each other, tests, called *parity checks,* are performed by the receiving computer to determine if an error has occurred during transmission of a packet of data. Suppose that a particularly noisy telephone line gives an average of five errors per second.

    a. What is the probability of detecting no errors in a minute of transmission?

    b. If at least one error is detected in a given minute, what is the probability that at least five errors occurred during that minute?

15. In a certain part of west Texas, there are an average of 10 armadillos per square mile. How large of a circular region should be selected to ensure that the probability of finding at least 5 armadillos is at least 0.95?

5.6
Uniform
Distribution

In some random experiments, a random variable can assume any value in an interval of finite length. The experiment with the game spinner introduced in Example 1.3 is such a case. If the probability that a random variable will take a value in a subinterval is proportional to the length of the subinterval, then the random variable is said to have a *uniform distribution*. We can use this idea to develop the probability density function. Suppose that X is a random variable that is uniformly distributed over the interval $[a, b]$. If $[c, d] \subset [a, b]$, then the statement that the probability is proportional to the length of the subinterval means that

$$P(c \leq X \leq d) = k(d - c),$$

where k is the constant of proportionality. Since the interval $[a, b]$ is the entire sample space we know that

$$k(b - a) = 1, \quad \text{or } k = \frac{1}{b - a}.$$

From this we can obtain the distribution function for X. For $c \in [a, b]$

$$F_X(c) = P(X \leq c) = P(a \leq X \leq c) = \frac{c - a}{b - a}.$$

Now, differentiating with respect to c gives the probability density function.

$$\frac{d}{dc} F_X(c) = \frac{1}{b - a}$$

Therefore, we say that X has a *uniform distribution with parameters a and b*, denoted by $X \sim U(a, b)$, if it has a probability density function of the form

$$f_X(x) = \begin{cases} \dfrac{1}{b - a}, & a \leq x \leq b, \quad \text{and} \\ 0, & \text{otherwise.} \end{cases} \tag{5.15}$$

The graph of the probability density function is shown in Figure 5.1.

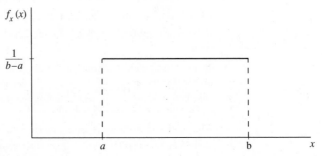

Figure 5.1

We leave it as an exercise to show that the mean, variance, and moment generating function are as follows.

$$E[X] = \frac{b + a}{2},$$

$$\text{Var}(X) = \frac{(b - a)^2}{12}, \quad \text{and}$$

$$M_X(t) = \frac{e^{tb} - e^{ta}}{t(b - a)}, \quad -\infty < t < \infty$$

Example 5.14   A point is chosen at random from a line segment of length L. It is reasonable to interpret this as meaning that, if X is the distance from one end of the line segment, then $X \sim U(0, L)$. Therefore,

$$f_X(x) \begin{cases} \dfrac{1}{L}, & 0 \le x \le L, \quad \text{and} \\ 0, & \text{otherwise.} \end{cases}$$

Thus,

$$P\left(\frac{1}{3}L \le X \le \frac{1}{2}L\right) = \int_{L/3}^{L/2} \frac{1}{L} \, dx$$

$$= \frac{1}{6}.$$

In addition, if we let A be the event that the point will divide the line into pieces such that the shorter segment is no more than $\frac{1}{3}$ as long as the longer piece, then

$$P(A) = P\left(\left\{0 \le X \le \frac{L}{4}\right\} \cup \left\{\frac{3L}{4} \le X \le L\right\}\right)$$

$$= P\left(0 \le X \le \frac{L}{4}\right) + P\left(\frac{3L}{4} \le X \le L\right)$$

$$= \int_0^{L/4} \frac{1}{L} \, dx + \int_{(3L/4)L}^{L} \frac{1}{L} \, dx$$

$$= \frac{1}{2}.$$

The next two examples are famous problems involving uniform random variables. They are part of an area sometimes referred to as geometric probability.

Example 5.15   *Buffon's Needle Problem*   A flat surface is covered with parallel lines that are a distance D apart. A needle of length L is randomly dropped on the surface. We wish to determine the probability that the needle will intersect a line. We shall consider the case where $L \le D$ and leave the solution for $L > D$ as an exercise.

The position of the needle can be determined by measuring the distance X from the center of the needle to the nearest line and the angle Θ measured clockwise from a perpendicular dropped from the center of the needle to the nearest line. This is shown in Figure 5.2.

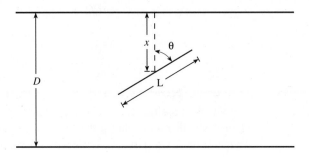

Figure 5.2

It is reasonable to assume that $X \sim U\left(0, \frac{D}{2}\right)$, $\Theta \sim U\left(-\frac{\pi}{2}, \frac{\pi}{2}\right)$, and that X and Θ are independent. The needle lies on the hypotenuse of a right triangle whose two other sides are the nearest line and the perpendicular dropped from the center of the needle to that line. The needle will intersect the nearest line whenever

$$\frac{X}{\frac{1}{2}L} \le \cos\Theta, \quad \text{or } X \le \frac{L}{2}\cos\Theta.$$

Then,

$$P\left(X \le \frac{L}{2}\cos\Theta\right) = \iint\limits_{\{x \le (L/2)\cos\theta\}} f_X(x) f_\Theta(\theta)\,dx\,d\theta$$

$$= \int_{-\pi/2}^{\pi/2} \int_0^{(L/2)\cos\theta} \frac{1}{\frac{1}{2}D}\frac{1}{\pi}\,dx\,d\theta$$

$$= \frac{2}{\pi D} \int_{-\pi/2}^{\pi/2} \frac{L}{2}\cos\theta\,d\theta$$

$$= \frac{2L}{\pi D}.$$

$\blacksquare$

Example 5.16   *Bertrand's Paradox*   A chord of a circle of radius r is chosen at random. We wish to determine the probability that the length of the chord will be greater than the side of the equilateral triangle inscribed inside the circle.

First, assume that a line tangent to the circle at some point is moved at a uniform rate of speed along a path through the circle such that it remains perpendicular to a line through the center of the circle and the original point of tangency. The line is stopped at a randomly selected time to obtain the chord. This is illustrated in Figure 5.3a.

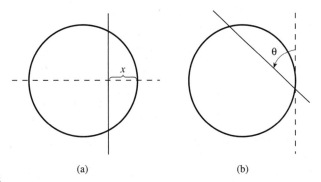

(a)   (b)

Figure 5.3

Let X be the distance from the point of tangency to the line, then $X \sim U(0, 2r)$. When X is in $\left(\frac{r}{2}, \frac{3r}{2}\right)$, the line will create a chord that is longer than the side of the equilateral triangle. Therefore, the desired probability is

$$P\left(\frac{r}{2} < X < \frac{3r}{2}\right) = \int_{r/2}^{3r/2} \frac{1}{2r}\,dx = \frac{1}{2}.$$

Now suppose that the chord is chosen by anchoring a line at a point on the circle and rotating the line at a constant angular rate through the circle and stopping it at a randomly selected time. This is illustrated in Figure 5.3b. If we let Θ be as shown, then it is reasonable to assume that $\Theta \sim U(0, \pi)$. The chord will be longer than the side of the equilateral triangle when $\Theta \in \left(\frac{\pi}{3}, \frac{2\pi}{3}\right)$. Therefore, the desired probability is

$$P\left(\frac{\pi}{3} < \Theta < \frac{2\pi}{3}\right) = \int_{\pi/3}^{2\pi/3} \frac{1}{\pi}\,d\theta = \frac{1}{3}.$$

This is Bertrand's paradox. The same event computed two different ways gives different values for the probability. There is a third way that gives yet another answer which is left as an exercise. The reader is invited to determine why this happens and show why there is no paradox here. ∎

Exercises 5.6

1. Suppose that the difference, X, between the scheduled arrival time and actual arrival time, in minutes, of a bus follows a $U(-4, 4)$ distribution.
   a. Find $P(X \le 0)$.
   b. Find $P(-1 < X < 1)$.
   c. Find $P(-2 \le X \le 3)$.

2. Let X have a uniform distribution with parameters a and b. Show that
   a. $E[X] = \dfrac{b+a}{2}$,
   b. $\text{Var}(X) = \dfrac{(b-a)^2}{12}$, and
   c. $M_X(t) = \dfrac{e^{tb} - e^{ta}}{t(b-a)}$.

3. Compute the skewness and kurtosis for a random variable with a $U(a, b)$ distribution.

4. Let $X \sim U(a, b)$. Find $P[|X - E[X]| \geq \text{Var}(X)]$.

5. Solve Buffon's needle problem, Example 5.15, for the case where the length of the needle exceeds the distance between the parallel lines. That is, for $L > D$.

6. Reconsider Bertrand's paradox in Example 5.16. Select a random chord by choosing the midpoint of the chord at random from any point inside the circle of radius r. Compute the probability that such a chord will be longer than the length of the inscribed equilateral triangle. (Note: This problem does not use the uniform distribution.)

7. Why is Bertrand's paradox not a paradox?

8. Two points are selected at random on a line segment of length L. What is the probability that the three line segments thus formed will create a triangle?

9. Two friends agree to meet at a diner at "noonish." Since it is not certain that either can keep the date it is agreed that neither should wait any longer than 5 minutes after arriving. Person A arrives at a time that is uniformly distributed between 11:50 and 12:05. Person B arrives at a time that is uniformly distributed between 11:55 and 12:10. What is the probability that they will miss each other?

10. A certain component in a manufacturing process fails quite rarely. Because of this spares are not kept on hand but must be ordered from the manufacturer. Arrival time for the new component is uniformly distributed over a period of 2 to 7 days. While the manufacturing process is shut down work must be subcontracted. The cost of doing this consists of a fixed overhead cost, c_o, and a per day cost of c_d. Thus, the cost is

$$C = c_o + c_d X.$$

   a. Find the probability that the process will be shut down for more than 3 days when a component fails.

   b. Find the expected cost for the failure of a component in terms of c_o and c_d.

11. A satellite orbits the Earth at a uniform rate of 3 radians per hour. A second satellite orbits the Earth at a uniform rate of 1 radian per hour. The orbits of both satellites lie in the same plane. Find the probability that the 2 satellites will be within 1 radian of each other as viewed from the Earth.

12. Suppose that the thermostat for an oven regulator maintains the actual temperature, X, within 5 degrees of the desired temperature.

   a. What is the probability that the actual temperature will be within 2 degrees of the desired temperature?

   b. What is the probability that the actual temperature will be more than 1 degree away from the desired temperature?

5.7
Normal Distribution

A random variable X is said to have a *normal distribution with parameters μ and σ^2*, denoted by $X \sim N(\mu, \sigma^2)$, if it has a probability density function of the form

$$f_X(x) = \frac{1}{\sqrt{2\pi\sigma^2}} e^{-(x-\mu)^2/2\sigma^2}, \quad \begin{cases} -\infty < x < \infty, \\ -\infty < \mu < \infty, \quad \text{and} \\ 0 < \sigma^2. \end{cases} \tag{5.16}$$

Figure 5.4 shows the graph of the probability density function when $\mu = 0$ and various values of σ^2. Note that the probability density function is symmetric about μ. It also turns out that the inflection points occur at $\mu \pm \sigma$. The $N(0,1)$ distribution is called the *standard normal distribution*. The standard normal distribution is almost universally referred to by the letter Z. For the rest of this book we shall use Z exclusively to refer to a standard normal random variable.

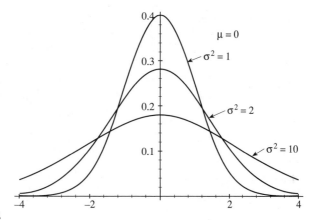

Figure 5.4

The normal distribution is one of the most important in probability and statistics. The earliest known derivation of the distribution was published in 1733 by de Moivre. There he used the normal distribution as an approximation for a binomial random variable. Many of the methods that are commonly referred to as "classical statistics" are based on the normal distribution. One of the main theoretical justifications for this is a result known as the *Central Limit theorem*. We shall study this theorem in some detail in Chapter 7. Roughly speaking, the Central Limit theorem states that if we form certain sums of random variables, the distribution of these sums approaches a normal distribution as the number of items being summed becomes infinitely large.

It is far from obvious that Equation 5.16 is a valid probability density function. It is clearly nonnegative, but we need to show that

$$\int_{-\infty}^{\infty} \frac{1}{\sqrt{2\pi\sigma^2}} e^{-(x-\mu)^2/2\sigma^2}\, dx = 1.$$

First, let $z = \frac{x-\mu}{\sigma}$ and make the substitution to get

$$I = \int_{-\infty}^{\infty} \frac{1}{\sqrt{2\pi}} e^{-(x-\mu)^2/2\sigma^2}\, dx$$

$$= \int_{-\infty}^{\infty} \frac{1}{\sqrt{2\pi}} e^{-z^2/2}\, dz.$$

The problem now is that there is no closed form antiderivative for $e^{-z^2/2}$. To get around this, we have to pull something of a rabbit out of a hat. We note that if $I = 1$, then so does I^2. The idea is to consider I^2 and change from rectangular to polar coordinates to evaluate I^2. Thus, we begin with

$$I^2 = \left(\int_{-\infty}^{\infty} \frac{1}{\sqrt{2\pi}} e^{-z^2/2} \, dz \right) \left(\int_{-\infty}^{\infty} \frac{1}{\sqrt{2\pi}} e^{-y^2/2} \, dy \right)$$

$$= \int_{-\infty}^{\infty} \int_{-\infty}^{\infty} \frac{1}{2\pi} e^{-(z^2+y^2)/2} \, dz \, dy.$$

Recall that in transforming from rectangular to polar coordinates we use

$$y = r \cos \theta, \quad z = r \sin \theta, \quad \text{and} \quad dz \, dy = r \, dr \, d\theta.$$

This gives

$$I^2 = \int_0^{\infty} \int_0^{2\pi} \frac{1}{2\pi} e^{-r^2/2} r \, d\theta \, dr$$

$$= \lim_{a \to \infty} \int_0^a r e^{-r^2/2} \, dr$$

$$= \lim_{a \to \infty} -e^{-r^2/2} \Big|_0^a$$

$$= 1.$$

Now that Equation 5.16 is known to be a probability density function we can find the moment generating function for a normal random variable and use it to determine the mean and variance. Let $X \sim N(\mu, \sigma^2)$. Then

$$M_X(t) = E\left[e^{tX} \right]$$

$$= \int_{-\infty}^{\infty} e^{tx} \frac{1}{\sqrt{2\pi\sigma^2}} e^{-(x-\mu)^2/2\sigma^2} \, dx$$

$$= \int_{-\infty}^{\infty} \frac{1}{\sqrt{2\pi\sigma^2}} e^{[x^2 - 2(\mu+\sigma^2 t)x + \mu^2]/2\sigma^2} \, dx.$$

We now complete the square to obtain

$$M_X(t) = \int_{-\infty}^{\infty} \frac{1}{\sqrt{2\pi\sigma^2}} e^{\mu t + (\sigma^2 t^2)/2} e^{-[x-(\mu-\sigma^2 t)]^2/2\sigma^2} \, dx$$

$$= e^{\mu t + (\sigma^2 t^2)/2} \int_{-\infty}^{\infty} \frac{1}{\sqrt{2\pi\sigma^2}} e^{-[x-(\mu+\sigma^2 t)]^2/2\sigma^2} \, dx.$$

The integrand is the probability density function of a $N(\mu + \sigma^2 t, \sigma^2)$ random variable. Thus, the value of the integral must be 1 for all values of t. Therefore,

$$M_X(t) = e^{\mu t + (\sigma^2 t^2)/2}, \quad -\infty < t < \infty. \tag{5.17}$$

We now differentiate the moment generating function with respect to t to determine the mean and variance of a normal random variable.

$$M_X'(t) = (\mu + \sigma^2 t) e^{\mu t + (\sigma^2 t^2)/2}, \quad \text{and}$$

$$E[X] = M_X'(0) = \mu$$

Also,

$$M_X''(t) = (\mu + \sigma^2 t)^2 e^{\mu + (\sigma^2 t^2)/2} + \sigma^2 e^{\mu + (\sigma^2 t^2)/2},$$

$$E[X^2] = M_X''(0) = \mu^2 + \sigma^2, \quad \text{and}$$

$$\text{Var}(X) = E[X^2] - \mu^2 = \sigma^2.$$

Therefore, we see that the two parameters of the normal distribution are its mean and variance. It is common practice to state that "X has a normal distribution with mean μ and variance σ^2."

In showing that Equation 5.16 is a probability density function, we simplified our notation by making the linear transformation $z = \frac{x-\mu}{\sigma}$. Note that this resulted in the probability density for a standard normal random variable. The fact that linear transformations of normal random variables always result in other normal distributions is shown in the theorem below. We defer the proof until Chapter 6.

Theorem 5.6

> Let X be a random variable with a $N(\mu, \sigma^2)$ distribution. Then the distribution of $Y = aX + b$ is $N(a\mu + b, a^2\sigma^2)$.

An important application of this theorem is in the computation of probabilities involving normal distributions. Many calculus texts use a function of the form e^{-x^2} to illustrate that not all continuous functions have closed form antiderivatives. This means that ordinary calculus techniques cannot help us find

$$P(a < X \le b) = \int_a^b \frac{1}{\sqrt{2\pi\sigma^2}} e^{-(x-\mu)^2/2\sigma^2} \, dx.$$

Because of this, values of the distribution function for normal random variables must be found using numerical techniques such as Simpson's rule or Gaussian quadrature. Values of the distribution function for the standard normal random variable are given in Table 3 of Appendix A. A consequence of Theorem 5.6 is that this one table is sufficient for calculating probabilities involving any normal distribution. All we need to do is transform a given problem into an equivalent one involving the standard normal distribution. Let $\Phi(z)$ denote the value of the distribution function for the standard normal random variable, Z, at z. That is, $\Phi(z) = P(Z \le z)$. If $X \sim N(\mu, \sigma^2)$, then we use $X = Z\sigma + \mu$ to get

$$F_X(a) = P(X \le a)$$

$$= P(Z\sigma + \mu \le a)$$

$$= P\left(Z \le \frac{a - \mu}{\sigma}\right)$$

$$= \Phi\left(\frac{a - \mu}{\sigma}\right). \tag{5.18}$$

The reader will notice that Table 3 only gives values of $\Phi(z)$ for $z \ge 0$. If z is negative we can exploit the symmetry of the normal distribution to obtain

$$\Phi(-z) = 1 - \Phi(z). \tag{5.19}$$

The proof of this is left as an exercise.

Example 5.17   Let $X \sim N(10, 25)$. We use Equation 5.18 and Equation 5.19 to compute the following probabilities.

$$P(X \leq 20) = \Phi\left(\frac{20 - 10}{5}\right) = \Phi(2) = 0.9772$$

$$P(5 > X) = 1 - P(X \leq 5) = 1 - \Phi\left(\frac{5 - 10}{5}\right) = 1 - \Phi(-1)$$

$$= 1 - [1 - \Phi(1)] = \Phi(1) = 0.8413$$

$$P(12 \leq X \leq 15) = \Phi\left(\frac{15 - 10}{5}\right) - \Phi\left(\frac{12 - 10}{5}\right)$$

$$= \Phi(1) - \Phi(0.4) = 0.1859$$

$$P(|X - 10| \leq 15) = P(-5 \leq X \leq 25)$$

$$= \Phi\left(\frac{25 - 10}{5}\right) - \Phi\left(\frac{-5 - 10}{5}\right)$$

$$= \Phi(3) - \Phi(-3) = \Phi(3) - [1 - \Phi(3)]$$

$$= 2\Phi(3) - 1 = 0.9974 \qquad \blacksquare$$

Example 5.18   The system for assigning examination grades known as "grading on the curve" is based on the normal distribution. If scores approximately follow a normal curve with mean μ and variance σ^2 then grades are determined according to the following rule.

Grade	Score (x)
A	$\mu + \sigma < x$
B	$\mu < x \leq \mu + \sigma$
C	$\mu - \sigma < x \leq \mu$
D	$\mu - 2\sigma < x \leq \mu - \sigma$
F	$x < \mu - 2\sigma$

We wish to compute the probability of receiving each possible grade.

$$P(A) = P(\mu + \sigma < X) = 1 - \Phi(1) = 0.1587,$$

$$P(B) = P(\mu < X \leq \mu + \sigma) = \Phi(1) - \Phi(0) = 0.3413,$$

$$P(C) = P(\mu - \sigma < X \leq \mu) = \Phi(0) - \Phi(-1) = 0.3413,$$

$$P(D) = P(\mu - 2\sigma < X \leq \mu - \sigma) = \Phi(-1) - \Phi(-2) = 0.1358, \quad \text{and}$$

$$P(F) = P(X \leq \mu - 2\sigma) = \Phi(-2) = 0.0228.$$

Therefore, under this scheme about 16% of the grades will be A's, 34% will be B's, 34% will be C's, 14% will be D's, and 2% will be F's.   ■

We said that the normal distribution was first derived by de Moivre as an approximation to the binomial distribution. The result given by him was for the case when $p = \frac{1}{2}$. In 1812, Laplace generalized this to any value p. For this reason the theorem regarding using the normal distribution to approximate the binomial bears both names. It is a special case of the *Central Limit theorem* which will be discussed in Chapter 7. We defer a proof until that time.

Theorem 5.7

(de Moivre-Laplace Theorem)   Let X_n have a $B(n,p)$ distribution for $n = 1, 2, \ldots$. Then

$$\lim_{np(1-p) \to \infty} P\left(a \le \frac{X_n - np}{\sqrt{np(1-p)}} \le b\right) = \Phi(b) - \Phi(a). \qquad (5.20)$$

This gives us a second approximation for computing binomial probabilities when n gets large. Recall that we can use the Poisson distribution with parameter $\lambda = np$ when n gets large and p is small. The normal approximation can be used for any value of p when n is large. Experience has shown that n can be considered to be large enough when both $np \ge 10$ and $n(1-p) \ge 10$.

Example 5.19

A particular form of cancer has been found to be curable by ordinary means in only 15% of the cases. A new drug treatment has been proposed for this type of cancer. Researchers will consider this treatment to be an improvement over existing methods if it results in cures for at least 20% of the patients treated. We wish to determine how many patients should be treated to give a probability of no more than 0.025 that the new treatment will result in at least 20% cures when it is, in fact, no improvement. We assume that n will turn out to be large enough to use the de Moivre-Laplace theorem. Let X_n be the number of cures out of n patients. If the treatment is not more effective, then $X_n \sim B(n, 0.15)$. We wish to find the smallest n such that

$$P(X_n \ge 0.20n) \le 0.025.$$

Using the de Moivre-Laplace theorem we find that

$$P(X_n \ge 0.20n) = P\left(\frac{0.20n - 0.15n}{\sqrt{0.15(0.85)n}} \le \frac{X_n - 0.15n}{\sqrt{0.15(0.85)n}}\right)$$

$$= 1 - \Phi\left(\frac{0.05\sqrt{n}}{\sqrt{0.1275}}\right)$$

$$= 1 - \Phi(0.14\sqrt{n})$$

$$\le 0.025.$$

Thus, we want n such that

$$\Phi(0.14\sqrt{n}) \geq 0.975.$$

From Table 3 of Appendix A we find that $\Phi(19.6) = 0.975$. Therefore, we find that $\sqrt{n} \geq 13.997$ or $n \geq 195.9$. This means that we should use the new treatment on 196 patients. To ensure that the use of the de Moivre-Laplace theorem was valid we note that $np = 196(0.15) = 29.4 > 10$, and $n(1 - p) = 196(0.85) = 166.6 > 10$.   ■

Exercises 5.7

1. Let Z have a standard normal distribution. Show that

$$\Phi(-z) = 1 - \Phi(z).$$

2. Compute the skewness and kurtosis for a $N(0, 1)$ random variable.

3. Show that the inflection points of the p.d.f. of a normal distribution occur at $\mu \pm \sigma$.

4. Let Z have a standard normal distribution. Find each of the following.
    a. $P(Z \leq 1.1)$
    b. $P(Z \leq -0.75)$
    c. $P(Z > -0.4)$
    d. $P(0.2 \leq Z < 0.3)$
    e. $P(|Z| \leq 0.6)$
    f. k such that $P(Z \leq k) = 0.2$.

5. Let X have a normal distribution with mean 20 and variance 16. Find each of the following.
    a. $P(X < 22)$
    b. $P(X > 19)$
    c. $P(23 < X < 24)$
    d. $P(X < 12)$
    e. k such $P(X > k) = 0.1$.

6. Let X have a normal distribution with $P(X \leq 5) = 0.1150$ and $P(X > 10) = 0.5793$. Find $P(X \leq 15)$.

7. The average rainfall during the month of April follows a normal distribution with a mean of 5 in. and a standard deviation of 0.6 in. What is the probability that the rainfall in April will differ from the normal amount by more than 1 in.?

8. In the design of a new class of aircraft carrier, the Navy needs to determine the length of bunks for berthing the crew. It is known that the male population is approximately normally distributed with a mean height of 72 inches and a standard deviation of 2 inches. How long should a bunk be in order to accommodate at least 90% of all naval personnel? Assume that naval personnel are randomly selected from the population at large.

9. A fair coin is tossed 400 times. What is the probability of obtaining at least 230 heads? Would you consider investigating the premise that the coin is fair? Explain.

10. Twenty items are selected without replacement from a bin containing 1000 items. If 10% of the items are defective, what is the (approximate) probability of drawing 2 or more defective items?

11. Suppose that 60% of all students seeking assistance at a university mathematics center ask for help in basic statistics. Let X be the number of students seeking statistics help in a sample of 100 students using the mathematics center.

   a. Find $P(50 \le X \le 70)$.

   b. Find $P(X \le 65)$.

   c. Find $P(X \ge 70)$.

12. Let X have a binomial distribution with $n = 36$ and $p = 0.4$. Find the following (approximate) probabilities.

   a. $P(X \le 15)$

   b. $P(X \ge 20)$

   c. $P(X = 14)$

13. The weight of the contents of 20-lb bags of dog food follows a normal distribution with a mean of 20 lb and a standard deviation of 4 oz. What is the probability that the weight of a randomly selected bag of dog food will be within $\frac{1}{2}$ lb of the advertised weight?

14. The *mode* of the distribution of a continuous random variable is the point at which the p.d.f. achieves a maximum. For a normal random variable with mean μ and variance σ^2 show that the mean, median, and mode are all equal.

15. A certain brand of radial tire has advertised that at least 95% of all drivers get at least 30,000 miles of tread life. Suppose that tread life follows a normal distribution with a mean of 40,000 miles and a standard deviation of 4000 miles. Is the manufacturer's claim justified?

5.8
Gamma
Distribution

A random variable X is said to have a *gamma distribution with parameters α and β*, denoted by $X \sim G(\alpha, \beta)$, if it has a probability density function of the form

$$f_X(x) = \begin{cases} \dfrac{x^{\alpha-1}\, e^{-x/\beta}}{\Gamma(\alpha)\, \beta^\alpha}, & 0 < x, \quad \text{and} \\[2mm] 0, & \text{otherwise, where } 0 < \alpha,\, 0 < \beta. \end{cases} \qquad (5.21)$$

Some example of the curves generated by these probability density functions are given in Figures 5.5 and 5.6.

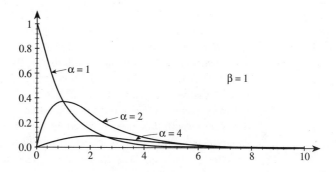

Figure 5.5

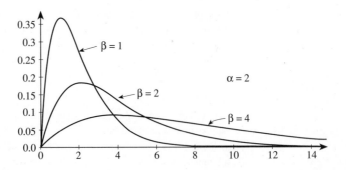

Figure 5.6

$\Gamma(\alpha)$ is the *gamma function* evaluated at α. It occurs frequently in science and engineering work and is defined by

$$\Gamma(\alpha) = \int_0^\infty x^{\alpha-1} e^{-x} \, dx. \tag{5.22}$$

It is a common problem in calculus courses to show that

$$\Gamma(1) = \int_0^\infty e^{-x} \, dx = 1, \quad \text{and}$$

$$\Gamma(2) = \int_0^\infty x e^{-x} \, dx = 1.$$

In general, if $\alpha > 1$, we can use integration by parts to show that

$$\Gamma(\alpha) = (\alpha - 1)\Gamma(\alpha - 1). \tag{5.23}$$

This means that if α is a positive integer, then $\Gamma(\alpha) = (\alpha - 1)!$. It is left as an exercise to show that

$$\Gamma\left(\frac{1}{2}\right) = \sqrt{\pi}. \tag{5.24}$$

This means that

$$\Gamma\left(\frac{3}{2}\right) = \frac{1}{2}\sqrt{\pi},$$

$$\Gamma\left(\frac{5}{2}\right) = \frac{3}{2}\Gamma\left(\frac{3}{2}\right) = \frac{3}{4}\sqrt{\pi}, \quad \text{and so on.}$$

There are some important special cases of the gamma distribution. When X is distributed as a $G(1, \beta)$ random variable, it is said to have an *exponential distribution with parameter* β. Another special case occurs when $\alpha = \frac{n}{2}$, where n is a positive integer, and $\beta = 2$. The $G\left(\frac{n}{2}, 2\right)$ distribution is called the χ^2 *distribution* (read as chi-squared) *distribution with n degrees of freedom*. The χ^2 distribution with n degrees of freedom is commonly denoted by $\chi^2(n)$. The χ^2 distribution will be discussed at length in Chapter 9.

We leave it as an exercise to show that if $X \sim G(\alpha, \beta)$, then the mean, variance, and moment generating function are as follows.

$$E[X] = \alpha\beta,$$

$$\text{Var}(X) = \alpha\beta^2, \quad \text{and}$$

$$M_X(t) = \left(\frac{1}{1 - \beta t} \right)^{\alpha}, t < \frac{1}{\beta}$$

The gamma distribution has found application in a variety of situations. One of the earliest was to use the χ^2 distribution to approximate the distribution of certain statistics based on samples from multinomial distributions. These give rise to a variety of procedures that come under the collective heading of *chi-squared statistics*. More recently, the gamma distribution has been used to model the distribution of the time to failure of components. The exponential distribution was first used in this situation to model the lifetime of a single component. Suppose we have n identical components whose lifetimes follow an exponential distribution with parameter β. We allow one of these to fail, then replace it, allow that one to fail, replace it, and so forth. It turns out that the total lifetime for all n components will have a $G(n, \beta)$ distribution. In addition, the gamma distribution has, in some cases, been used to model single component lifetimes. In recent years, the Weibull distribution, which is discussed later in this chapter, has replaced the gamma distribution as the popular model for these problems. Gamma distributions have also been used to approximate normal distributions by choosing α to be large. The justification for this is, again, the Central Limit theorem, which will be discussed in Chapter 7. This approximation is used because in some situations it is desirable to ensure that a random variable will be nonnegative.

The following theorems cite an important situation that gives rise to chi-square distributions. The proofs are deferred until Chapter 6.

Theorem 5.8

> Let Z have a standard normal distribution. Then $Z^2 \sim \chi^2(1)$.

Theorem 5.9

> Let X have a $G(\alpha, \beta)$ distribution and Y have an independent $G(\eta, \beta)$ distribution. Then $X + Y$ has a $G(\alpha + \eta, \beta)$ distribution.

Example 5.20

The number of hundreds of hours that a light bulb burns is exponentially distributed with $\beta = 4$. We wish to determine the probability that, if the bulb burns for 300 hours, it will burn for at least an additional 200 hours. Let X be the burning time in hundreds of hours. Then $X \sim G(1, 4)$, and we wish to determine

$$P(X \geq 5 \mid X \geq 3).$$

Since this is a conditional probability we use Equation 2.1 as follows.

$$P(X \geq 5 \mid X \geq 3) = \frac{P(\{X \geq 5\} \cap \{X \geq 3\})}{P(X \geq 3)}$$

$$= \frac{P(X \geq 5)}{P(X \geq 3)}$$

$$= \frac{\displaystyle\int_5^\infty \frac{1}{4} e^{-x/4}\, dx}{\displaystyle\int_3^\infty \frac{1}{4} e^{-x/4}\, dx}$$

$$= e^{-1/2}.$$

Note that

$$P(X \geq 2) = \int_2^\infty \frac{1}{4} e^{-x/4}\, dx = e^{-1/2}.$$

This demonstrates the fact that the exponential distribution is what is called *memoryless*. That is, the probability that a component will survive an additional r hours given that it has lasted for s does not depend on s. It is just as though we begin with a new component and compute the probability that it will last r hours.   ∎

There is an interesting relationship between the gamma and Poisson distributions. If we compare the gamma p.d.f. with the Poisson p.f. it is clear that they are very similar in appearance. Keep in mind that $\Gamma(a) = (a - 1)!$ when a is a positive integer.

$$f(x) = \frac{x^{\alpha-1}\, e^{-x/\beta}}{\Gamma(\alpha)\, \beta^\alpha}$$

$$p(a) = \frac{\lambda^a\, e^{-\lambda}}{a!}$$

They are very similar with the roles of parameter and variable interchanged. This connection can be exploited when calculating $P(X \leq x)$, where $X \sim G(n, \beta)$. If n is an integer and $n \geq 2$, we integrate by parts to obtain

$$P(X \leq x) = \int_0^x \frac{t^{n-1}\, e^{-t/\beta}}{(n-1)!\, \beta^n}\, dt$$

$$= \frac{-t^{n-1}\, e^{-t/b}}{(n-1)!\, \beta^n}\bigg|_0^x + \int_0^x \frac{t^{n-2}\, e^{-t/\beta}}{(n-2)!\, \beta^{n-1}}\, dt$$

$$= \int_0^x \frac{t^{n-2}\, e^{-t/\beta}}{(n-2)!\, \beta^{n-1}}\, dt - \frac{(x/\beta)^{n-1}\, e^{-x/\beta}}{(n-1)!}.$$

If we keep integrating by parts for a total of $n - 1$ times and note that

$$\int_0^x \frac{1}{\beta} e^{-t/\beta}\, dt = 1 - e^{-x/\beta},$$

then we obtain that

$$\int_0^x \frac{t^{n-1} e^{-t/\beta}}{(n-1)! \, \beta^n} \, dt = 1 - \sum_{i=0}^{n-1} \frac{(x/\beta)^i}{i!} \, e^{-x/\beta}, \quad \text{for } 0 < x. \tag{5.25}$$

This means that we can use tables of the $Po\left(\frac{x}{\beta}\right)$ distribution to evaluate the distribution function of a $G(\alpha, \beta)$ random variable whenever α is an integer. In addition, it also means that we can approximate values for the distribution function of a $Po(\lambda)$ random variable by numerically integrating to find the value of the distribution function of a $G(\alpha, 1)$ random variable at $x = \lambda$. This is, in fact, how we derived Table 2 in Appendix A.

There is another practical side to Equation 5.25 from random experiments. Suppose that n identical components with a mean lifetime of β hours are operated such that as one fails another is put in its place. If the lifetime of each component follows a $G(1, \beta)$ distribution, then the lifetime of the entire system will have a $G(n, \beta)$ distribution. The integral on the left-hand side of Equation 5.25 is the probability that the system will survive less than or equal to x hours. If the average lifetime of a component is β hours, we would expect to see $\frac{x}{\beta}$ components fail in x hours. Therefore, if the number of components that fail in x hours should follow a Poisson distribution with parameter $\lambda = \frac{x}{\beta}$. Therefore, the probability that the system survives x hours or less is the same as observing n or more component failures in x hours. This is precisely what the right-hand side of Equation 5.25 computes.

Exercises 5.8

1. Prove that $\Gamma(\alpha) = (\alpha - 1)\Gamma(\alpha - 1)$.

2. Show that $\Gamma\left(\frac{1}{2}\right) = \sqrt{\pi}$.

3. Let X have a gamma distribution with parameters α and β. Show that
   a. $E[X] = \alpha\beta$.
   b. $\text{Var}(X) = \alpha\beta^2$.
   c. $M_X(t) = \left(\dfrac{1}{1 - \beta t}\right)^\alpha, \; t < \dfrac{1}{\beta}$.

4. Compute the skewness for a $G(\alpha, \beta)$ random variable.

5. Let X have a $G(\alpha, \beta)$ distribution. Determine where the probability density function for X reaches a maximum.

6. Let X have a $G(1, \beta)$ distribution. If $P(X \le 1) = 0.5$, find $P(X \le 2)$.

7. It has been found that the time, in minutes, between the arrival of parcels at a next-day delivery service has an exponential distribution with parameter β. Let X be the number of parcels arriving in t minutes. Show that

$$P(X = a) = \frac{\left(\dfrac{t}{\beta}\right)^a e^{-t/\beta}}{a!}, a = 0, 1, 2, \ldots.$$

8. Fill in the details of the derivation of Equation 5.25.

9. Let X have a gamma distribution with $\alpha = 5$ and $\beta = \frac{1}{3}$. Determine $P(X \leq 1)$.

10. The number of minutes that airline flights are behind schedule follow a gamma distribution with parameters $\alpha = 2$ and $\beta = 5$. Find the following.

   a. The probability that a randomly selected flight will be late by no more than 5 minutes.

   b. The probability that a randomly selected flight will be late by more than 10 minutes.

   c. The median number of minutes that a flight will be late.

11. The time, X, in seconds, between the arrival of successive customers at a drive-in teller follows an exponential distribution with a mean time of 15 seconds.

   a. What is the standard deviation of the time between the successive arrival of 2 customers?

   b. Find $P(5 \leq X \leq 20)$.

12. Let X have a $G(1, \beta)$ distribution. Find the median of X.

13. The *mode* of the distribution of a continuous random variable is that point at which the p.d.f. achieves a maximum. Let X have a $G(\alpha, \beta)$ distribution and show that for X mode $\leq$ median $\leq$ mean.

5.9 Other Continuous Distributions

In this section we introduce certain other distributions that are of either theoretical or practical interest. We shall give the probability density functions and show some typical graphs for them. Applications will be given in subsequent chapters.

5.9.1 Weibull Distribution

A random variable X is said to have a *Weibull distribution with parameters α, β, and θ* if it has a probability density function of the form

$$f_X(x) = \begin{cases} \left(\dfrac{\alpha}{\beta}\right)\left(\dfrac{x-\theta}{\beta}\right)^{\alpha-1} e^{-(x-\theta/\beta)^\alpha}, & \theta < x, \quad \text{and} \\ 0, & \text{otherwise}, \\ \text{where } -\infty < \theta < \infty, 0 < \alpha, 0 < \beta. \end{cases} \tag{5.26}$$

Figure 5.7 gives the graph of this probability density function for selected values of α with $\beta = 1$ and $\theta = 0$.

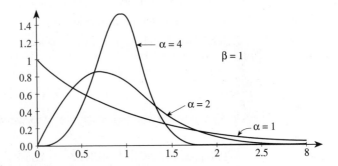

Figure 5.7

It turns out that, if X has a Weibull distribution with parameters α, β, and θ, then

$$E[X] = \Gamma\left(\frac{\alpha + 1}{\alpha}\right), \quad \text{and}$$

$$\text{Var}(X) = \Gamma\left(\frac{2 + \alpha}{\alpha}\right) - \left[\Gamma\left(\frac{\alpha + 1}{\alpha}\right)\right]^2.$$

This distribution was first introduced by the Swedish physicist Waloddi Weibull in 1939 to model the tensile strength of materials. Since that time it has become popular for many situations, such as life testing of components. Much of its popularity is due to the wide variety of shapes the distribution can assume by varying the parameters. There are no particular theoretical reasons that make its use preferable to, say, the gamma distribution. Mostly empirical evidence can be described quite easily by choosing a Weibull distribution with the appropriate values for its parameters.

5.9.2 Beta Distribution

A random variable X is said to have a *beta distribution with parameters α and β,* denoted by $X \sim Be(\alpha, \beta)$, if it has a probability density function of the form

$$f_X(x) = \begin{cases} \dfrac{\Gamma(\alpha + \beta)}{\Gamma(\alpha)\,\Gamma(\beta)} x^{\alpha - 1}(1 - x)^{\beta - 1}, & 0 \le x \le 1, \quad \text{and} \\[2mm] 0 & , \text{ otherwise, where } 0 < \alpha, 0 < \beta. \end{cases} \tag{5.27}$$

Figure 5.8 shows the probability density function for selected values of α and β.

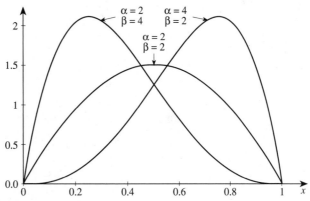

Figure 5.8

It turns out that if X has a $Be(\alpha, \beta)$ distribution, then

$$E[X] = \frac{\alpha}{\alpha + \beta}, \quad \text{and}$$

$$\text{Var}(X) = \frac{\alpha\beta}{(\alpha + \beta)^2(\alpha + \beta + 1)}.$$

Note that when $\alpha = \beta = 1$ we have a $U(0,1)$ distribution. So the uniform distribution is, in some sense, a special case of the beta distribution.

The distribution of certain functions of normal and gamma random variables turn out to have a beta distribution. In addition, the parameters α and β give the beta distribution the ability to serve as a useful model for many situations where the range of a random variable is a known, finite interval.

A relationship similar to that between the gamma and Poisson distributions exists between the beta and binomial distributions. Suppose α and β are positive integers and $\beta = n - \alpha + 1$. We leave it as an exercise to verify that

$$\int_0^x \frac{n!}{(\alpha - 1)!\,(n - \alpha)!}\, t^{\alpha-1}(1 - t)^{n-\alpha}\, dt = \sum_{i=\alpha}^n \binom{n}{i} x^i (1 - x)^{n-i}. \tag{5.28}$$

Thus, there is an intimate relation between $P(X \le x)$ for a $Be(\alpha, n - \alpha + 1)$ random variable and the distribution function of a $B(n, x)$ random variable.

5.9.3 Cauchy Distribution

A random variable X is said to have a *Cauchy distribution with parameter θ* if it has a probability density function of the form

$$f_X(x) = \frac{1}{\pi[1 + (x - \theta)^2]}, \quad \begin{cases} -\infty < x < \infty, \quad \text{and} \\ -\infty < \theta < \infty. \end{cases} \tag{5.29}$$

As can be seen in Figure 5.9, the graph of this p.d.f. is similar in appearance to that of the normal distribution, but it is flatter and has heavier tails.

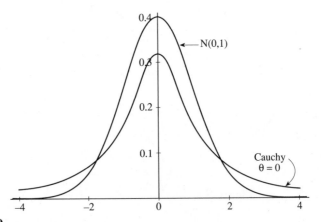

Figure 5.9

The Cauchy distribution can be used to describe certain physical problems. For example, suppose a source emits particles from the point S in Figure 5.10, which are uniformly distributed between $\frac{-\pi}{2}$ and $\frac{\pi}{2}$. The distribution of the particles arriving at the plate X units away from the point directly opposite the source will have a Cauchy distribution. The proof of this is left to Exercise 2 in Section 5.11.

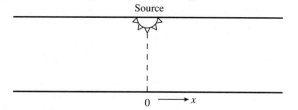

Figure 5.10

In addition, the Cauchy distribution is often used to evaluate the performance of statistical procedures when dealing with a distribution having extremely heavy tails. It is also interesting that none of the moments of the Cauchy distribution exist. This distribution is also a special case of Student's t distribution, which will be discussed in Chapter 9.

5.9.4   Double Exponential Distribution

A random variable X is said to have a *double exponential distribution with parameters* θ *and* σ if it has a probability density function of the form

$$f_X(x) = \frac{1}{2\sigma} e^{-|x-\theta|/\sigma}, \quad \begin{cases} -\infty < x < \infty, \\ -\infty < \theta < \infty, \quad \text{and} \\ 0 \ < \sigma. \end{cases} \tag{5.30}$$

Laplace discovered this distribution in 1774. For this reason it is sometimes called the *Laplace distribution*. Figure 5.11 shows the graph of this probability density function when $\theta = 0$ and $\sigma = 1$.

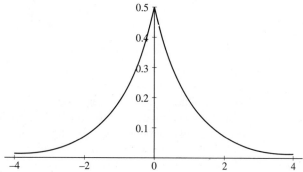

Figure 5.11

As can be seen, the graph looks like a combination of two exponential distributions. Therefore, it should not be too surprising to find that the double exponential distribution can be used to model the difference in the failure time of two identical components, each of which has an exponential distribution. Other than that, the double exponential distribution is mainly of theoretical interest. It is useful for determining the behavior of statistical procedures in the presence of heavily tailed, symmetric distributions. We leave it as an exercise to show that, if X has a double exponential distribution with parameters θ and σ, then

$$E[X] = \theta,$$

$$\text{Var}(X) = 2\sigma^2, \quad \text{and}$$

$$M_X(t) = \frac{e^{t\theta}}{1 - \sigma^2 t^2}, \quad \text{for } t < \frac{1}{\sigma}.$$

5.9.5 Logistic Distribution

A random variable X is said to have a *logistic distribution with parameters θ and σ* if it has a probability density function of the form

$$f_X(x) = \frac{e^{-(x-\theta)/\sigma}}{\sigma \left[1 + e^{-(x-\theta)/\sigma}\right]^2}, \quad \begin{cases} -\infty < x < \infty, \\ -\infty < \theta < \infty, \quad \text{and} \\ 0 < \sigma. \end{cases} \qquad \textbf{(5.31)}$$

As Figure 5.12 shows, the graph of this probability density function is quite similar to that of the normal distribution. It does, however, have slightly heavier tails.

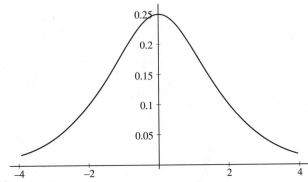

Figure 5.12

The distribution has been very popular for demographic modeling as well as for describing economic growth. In addition, analysts have noted that many naturally occurring distributions tend to be slightly more heavily tailed than the normal. This makes the logistic a reasonable alternative. We leave it as an exercise to show that

$$E[X] = \theta,$$

$$\text{Var}(X) = \frac{\sigma^2 \pi^2}{3}, \quad \text{and}$$

$$M_X(t) = e^{t\theta} \, \pi \sigma t \, \csc(\pi \sigma t), \quad \text{for } |t| < \frac{1}{2\sigma}.$$

Exercises 5.9

1. Let X have a Weibull distribution with parameters α, β, and θ. Derive the distribution function for X.

2. Let X have a probability density function of

$$f_X(x) = \begin{cases} kx^2 e^{-x^3/64}, & 0 < x, \quad \text{and} \\ 0, & \text{otherwise.} \end{cases}$$

Find k.

3. The time to failure, in hours, of a component in a particular system has been found to follow a Weibull distribution with $\alpha = 2$, $\beta = \frac{1}{2}$, and $\theta = 0$. What is the probability that out of 5 such components selected at random 1 will fail within 1 hour?

4. A company has found that the time, in days, it takes for customers to respond to a questionnaire follows a Weibull distribution with $\alpha = 1.5$, $\beta = 2$, and $\theta = 4$.

   a. What is the mean response time? What is the standard deviation?

   b. What is the probability that a randomly selected questionnaire will be returned in between 3 and 7 days, inclusive?

5. Let X have a $Be(\alpha, \beta)$ distribution. Show that

   a. $E[X] = \dfrac{\alpha}{\alpha + \beta}$.

   b. $\text{Var}(X) = \dfrac{\alpha\beta}{(\alpha + \beta)^2(\alpha + \beta + 1)}$.

6. Derive Equation 5.28.

7. Let X have a beta distribution with $\alpha = 4$ and $\beta = 7$. Find $P(X \leq 0.6)$.

8. The staff of a congressional committee has determined that the expenditure of funds for scientific space projects during their development period follows a beta distribution with parameters $\alpha = 1$ and $\beta = 2$. If a particular program has spent $\frac{1}{2}$ of its funds, how far along, in percent, is it in its development period?

9. Let X have a Cauchy distribution with parameter θ. Derive the distribution function for X.

10. Let X have a double exponential distribution with parameters θ and σ.

   a. Show that the moment generating function is

$$M_X(t) = \frac{e^{t\theta}}{1 - \sigma^2 t^2}, \quad t < \frac{1}{\sigma}.$$

   b. Use this moment generating function to derive $E[X]$ and $\text{Var}(X)$.

11. Let X have a double exponential distribution with $\theta = 0$ and $\sigma = 1$. Find a general formula for computing the pth quantile of X for any value of p.

12. The time it takes to complete a particular task in a production line has been found to follow a double exponential distribution with a mean of 10 minutes and a standard deviation of 2 minutes. What is the probability that it will take between 7 and 12 minutes to complete the task?

13. Compute the kurtosis for a random variable having a double exponential distribution with parameters $\theta = 0$ and $\sigma = 1$.

14. Let X have a logistic distribution with parameters θ and σ. Find the distribution function for X.

15. The random variable X has a logistic distribution with parameters $\theta = 5$ and $\sigma = 2$. Find $P(X \geq 3)$.

16. Let X have a logistic distribution with parameters θ and σ. Show that $E[X] = \theta$.

17. Compute the kurtosis for a logistic random variable with parameters $\theta = 0$ and $\sigma = 1$.

18. Let X have a logistic distribution with $\theta = 0$ and $\sigma = 1$. Derive a general formula for determining the pth quantile of X.

19. Let X have a logistic distribution with $\theta = 100$ and $\sigma = 5$. Find each of the following probabilities.

    a. $P(X \leq 80)$

    b. $P(95 \leq X \leq 110)$

    c. $P(|X - 100| \geq 10)$

5.10
Bivariate Normal Distribution

One of the most important distributions for a continuous random vector is the *multivariate normal*. It is the generalization to random vectors of the normal distribution. Much of what is known in statistics as *multivariate analysis* is based on this distribution. In this section we consider the special case where the random vector has two components. The random vector (X, Y) is said to have a *bivariate normal distribution with parameters* μ_X, μ_Y, σ_X^2, σ_Y^2, and ρ if it has a joint probability density function of the form

$$f_{X,Y}(x, y) = \frac{1}{2\pi\sqrt{\sigma_X^2 \sigma_Y^2 (1 - \rho^2)}} e^{-Q/2}, \quad \begin{cases} -\infty < x < \infty, \\ -\infty < y < \infty, \end{cases} \tag{5.32}$$

where

$$-\infty < \mu_x < \infty, \quad -\infty < \mu_Y < \infty, \quad 0 < \sigma_X^2, \quad 0 < \sigma_Y^2, \quad -1 < \rho < 1, \quad \text{and}$$

$$Q = \frac{1}{1 - \rho^2} \left[\left(\frac{x - \mu_X}{\sigma_X} \right)^2 - 2\rho \left(\frac{x - \mu_X}{\sigma_X} \right) \left(\frac{y - \mu_Y}{\sigma_Y} \right) + \left(\frac{y - \mu_Y}{\sigma_Y} \right)^2 \right].$$

Figure 5.13 shows a typical graph of this joint probability density function. It is a bell-shaped surface with a maximum at (μ_X, μ_Y). In addition, any plane that cuts the surface parallel to the x-y plane will graph as an ellipse. Also, the intersection of any plane perpendicular to the x-y plane that passes through (μ_X, μ_Y) will graph as the curve of a univariate normal distribution.

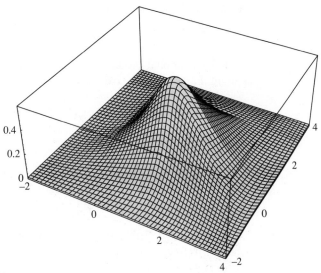

Figure 5.13

It is a tedious calculation to obtain the joint moment generating function for the bivariate normal distribution. It is derived by methods similar to those used in the univariate case. We, therefore, state it as a theorem without proof.

Theorem 5.10

> Let (X, Y) have a bivariate normal distribution with parameters μ_X, μ_Y, σ_X^2, σ_Y^2, and ρ. Then the joint moment generating function is
>
> $$M_{X,Y}(t_1, t_2) = e^{t_1 \mu_X + t_2 \mu_Y + [(t_1 \sigma_X)^2 + 2 t_1 t_2 \rho \sigma_X \sigma_Y + (t_2 \sigma_Y)^2]/2}, \qquad (5.33)$$
>
> for all values of t_1 and t_2.

The moment generating functions for X and Y are shown below.

$$M_X(t_1) = M_{X,Y}(t_1, 0) = e^{t_1 \mu_X + \sigma_X^2 t_1^2 / 2}, \quad \text{and}$$

$$M_Y(t_2) = M_{X,Y}(0, t_2) = e^{t_2 \mu_Y + \sigma_Y^2 t_2^2 / 2}.$$

These are the moment generating functions of univariate normal distributions. Therefore, $X \sim N(\mu_X, \sigma_X^2)$ and $Y \sim N(\mu_Y, \sigma_Y^2)$. This means that

$$E[X] = \mu_X,$$

$$E[Y] = \mu_Y,$$

$$\text{Var}(X) = \sigma_X^2, \quad \text{and}$$

$$\text{Var}(Y) = \sigma_Y^2.$$

We leave it as an exercise to show that the correlation

$$\rho_{X,Y} = \rho.$$

The conditional probability density function of Y, given $X = x$, is found by applying Equation 3.42. We leave it as an exercise to show that, by starting with

$$f_{Y|X}(y \mid x) = \frac{f_{X,Y}(x, y)}{f_X(x)},$$

we obtain, after some simplification,

$$f_{Y|X}(y \mid x) = \frac{1}{\sqrt{2\pi\sigma_Y^2(1 - \rho^2)}} e^{-\frac{1}{2\sigma_Y^2(1-\rho^2)}\left[y - \left(\mu_Y + \frac{\rho\sigma_Y}{\sigma_X}[x - \mu_X]\right)\right]^2}.$$

The conditional probability density of X, given $Y = y$ is of the same form with X and Y interchanged. This gives us the result that the distribution of Y, given $X = x$, is normal with

$$\mu_{Y|X} = \mu_Y + \frac{\rho\sigma_Y}{\sigma_X}(x - \mu_X), \quad \text{and}$$

$$\sigma_{Y|X}^2 = \sigma_Y^2(1 - \rho^2).$$

Also the distribution of X, given $Y = y$, is normal with

$$\mu_{X|Y} = \mu_X + \frac{\rho\sigma_X}{\sigma_Y}(y - \mu_Y), \quad \text{and}$$

$$\sigma_{X|Y}^2 = \sigma_X^2(1 - \rho^2).$$

Recall that in our discussion of correlation we made the statement that independent random variables will be uncorrelated, but that the converse is not necessarily true. We provided an example to demonstrate this fact. An important exception occurs in the case of jointly distributed normal random variables. Here, it turns out that uncorrelated normal random variables are always independent.

Theorem 5.11

> If (X, Y) has a bivariate normal distribution with $\rho = 0$, then X and Y are independent.

Proof We shall use the joint probability density function for (X, Y) and show that it factors into the probability density functions X and Y when $\rho = 0$. Setting ρ to 0 in Equation 5.32 gives that

$$Q = \left(\frac{x - \mu_X}{\sigma_X}\right)^2 + \left(\frac{y - mu_y}{\sigma_Y}\right)^2,$$

and thus

$$f_{X,Y}(x, y) = \frac{1}{2\pi\sqrt{\sigma_X^2 \sigma_Y^2}} e^{-\left[(x-\mu_X/\sigma_X)^2 + (y-mu_y/\sigma_Y)^2\right]/2}$$

$$= \frac{1}{\sqrt{2\pi\sigma_X^2}} e^{-(x-\mu_X/\sigma_X)^2/2} \frac{1}{\sqrt{2\pi\sigma_Y^2}} e^{-(y-\mu_Y/\sigma_Y)^2/2}$$

$$= f_X(x) f_Y(y).$$

Thus the joint p.d.f. is the product of the individual p.d.f.'s of X and Y which means that they are independent. ∎

Exercises 5.10

1. Let (X, Y) have a bivariate normal distribution. Use the joint moment generating function to show that $\rho_{X,Y} = \rho$.

2. Let (X, Y) have a bivariate normal distribution with parameters μ_X, μ_Y, σ_X^2, σ_Y^2, and ρ. Fill in the details to show that the conditional distribution of Y, given $X = x$, is normal with

$$\mu_{Y|X} = \mu_Y + \frac{\rho\sigma_Y}{\sigma_X}(x - \mu_X), \quad \text{and}$$

$$\sigma_{Y|X}^2 = \sigma_Y^2(1 - \rho^2).$$

3. Let (X, Y) have a bivariate normal distribution with parameters $\mu_X = 10$, $\mu_Y = 20$, $\sigma_X^2 = 4$, $\sigma_Y^2 = 25$, and $\rho = 0.8$. Find $P(Y \geq 22 \mid X = 12)$.

4. Let (X, Y) have a bivariate normal distribution with $\mu_X = 4$ and $\mu_Y = 8$. Find ρ if

$$P(X > 3) = P(Y \leq 10) = P(Y \leq 9 \mid X = 5) = 0.8413.$$

5. Let (X, Y) have a bivariate normal distribution with

$$Q = x^2 + 2y^2 - 2xy - 2x + 2y + 1.$$

Determine μ_X, μ_Y, σ_X^2, σ_Y^2, and ρ.

Chapter Summary

The probability functions or probability density functions, means, variances, and moment generating functions of the random variables discussed in this chapter are listed in Appendix B.

Bernoulli Trials:

1. Each outcome is a success or a failure.
2. $P(\text{success}) = p$.
3. Trials are independent.

If X is hypergeometric with $M = Np$, then

$$\lim_{N \to \infty} p_X(a) = \binom{n}{a} p^a (1-p)^{n-a}.$$

If $X \sim B(n,p)$ and $Y \sim B(m,p)$ and X, Y are independent, then
$X + Y \sim B(n + m, p)$. **(Section 5.3)**

Negative Binomial Distribution: If $X \sim NB(r,p)$ and $Y \sim NB(s,p)$ and X, Y are independent, then $X + Y \sim NB(r + s, p)$. **(Section 5.3)**

Multinomial Trials:

1. k distinct outcomes.
2. $P(\text{category } i) = p_i, \; p_1 + \cdots + p_k = 1$.
3. Trials are independent. **(Section 5.4)**

Poisson Distribution: If $X \sim B(n,p)$, then

$$\lim_{\substack{N \to \infty \\ np = \lambda}} p_X(a) = \frac{\lambda^a e^{-\lambda}}{a!}.$$ **(Section 5.5)**

If $X \sim Po(\lambda)$ and $Y \sim Po(\eta)$ with X, Y independent, then
$X + Y \sim Po(\lambda + \eta)$. **(Section 5.5)**

Normal Distribution: If $X \sim N(\mu, \sigma^2)$, then $aX + b \sim N(a\mu + b, b^2\sigma^2)$.
(Section 5.7)

de Moivre–Laplace Theorem: If $X_n \sim B(n,p)$, then

$$\lim_{np(1-p) \to \infty} P\left(a \le \frac{X_n - np}{\sqrt{np(1-p)}} \le b\right) = \Phi(b) - \Phi(a).$$ **(Section 5.7)**

Gamma Distribution: If $Z \sim N(0,1)$, then $Z^2 \sim \chi^2(1)$.   **(Section 5.8)**

If $X \sim G(\alpha, \beta)$ and $Y \sim G(\eta, \beta)$ with X, Y independent,
then $X + Y \sim G(\alpha + \eta, \beta)$. **(Section 5.8)**

Bivariate Normal Distribution: If (X, Y) bivariate normal with $\rho = 0$,
then X and Y are independent. **(Section 5.10)**

Review Exercises

1. An investigator wishes to determine whether the outcomes from a random experiment indicate the presence of a normal distribution with a mean of 10 and a variance of 4 or a logistic distribution with a mean of 10 and a variance of 4. Would the investigator expect to see more values below 8 with the normal distribution or with the logistic distribution?

2. Consider the diagram below. The angle Θ is uniformly distributed over $\left(-\frac{\pi}{2}, \frac{\pi}{2}\right)$. Show that X has a Cauchy distribution.

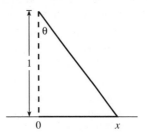

3. Show that a plane parallel to the x-y plane cutting the joint probability density function of a bivariate normal distribution will show an ellipse. That is, fix $f_{X,Y}(X, Y)$ at some value and show that the (x, y) values form an ellipse.

4. A bin contains 20 light bulbs, 5 of which are bad. Four bulbs are selected from the bin. Let X be the number of defective light bulbs drawn. Find $P(X > 1)$ if
   a. sampling is done with replacement, and
   b. sampling is done without replacement.

5. Consider the bin described in Exercise 4. Light bulbs are drawn with replacement.
   a. What is the probability that the first defective bulb will be drawn by the time the third bulb is selected?
   b. What is the probability that the second defective bulb will be drawn by the time the fourth bulb is selected?
   c. If the first light bulb is found by the time the second bulb is drawn, what is the probability that it will take fewer than three additional draws to get the second defective bulb?
   d. What would your answer to Part a be if sampling is done without replacement?

6. An average of two incoming telephone calls arrive at a college switchboard each minute. What is the probability that more than half of five randomly chosen one-minute periods will have more than two phone calls in them? Assume a Poisson distribution.

7. Let X have a normal distribution with $P(X \le 5) = 0.1587$ and $P(X > 4) = 0.9332$. Find $P(X \ge 7.5)$.

8. Use p.d.f.'s to evaluate each of the following integrals.
   a. $\int_0^\infty e^{-x^2}\, dx$
   b. $\int_0^1 x^6 \sqrt{1 - x}\, dx$
   c. $\int_0^\infty x^4 e^{-3x}\, dx$
   d. $\int_0^{0.5} x^3 (1 - x)^7\, dx$
   e. $\int_0^5 x^4 e{-x}\, dx$

9. Let X have a geometric distribution with parameter p. What is the probability that X will be even?

10. A Coast Guard radio station monitors various radio frequencies for evidence of drug-related activity offshore. A particular frequency is monitored on a schedule of 5 minutes

on and 10 minutes off on a regular cycle beginning at the start of each hour. A drug runner who does not know this pattern uses this frequency at a time randomly chosen from a time in the period from 1:10 a.m. to 1:55 a.m. What is the probability that his communication will be detected by the Coast Guard?

11. A component in a system has a lifetime that follows an exponential distribution with a mean lifetime of 100 hours. System operators replace components immediately upon failure. If there is only 1 spare available, what is the probability that the system will function for at least 250 hours?

12. Five independent observations are taken from a population having a normal distribution with a mean of 25 and a standard deviation of 4. What is the probability that more than half of the observations will lie within 1 standard deviation of the mean?

13. The number of sparrows that sit on a 1-foot section of fence has a Poisson distribution with $\lambda = 1.5$. Find the probability that on 3 randomly selected 1-foot sections of fence, there will be more than 5 sparrows.

14. Let X have a double exponential distribution with $\theta = 0$ and $\sigma = 1$, and let Y have an independent logistic distribution with $\theta = 1$. Find $P(X > Y)$.

15. Let X be a geometric random variable with a probability of success of p. Show that

$$P(X = a - 1 + b \mid X > a - 1) = P(X = b).$$

6 Distributions of Functions of Random Variables

6.1 Introduction

There are times when we know the distribution of a given random variable, say X, but the object of interest is actually some function of this random variable such as $Y = f(X)$. Suppose, for example, that we have a radar used for air traffic control near an airport. Assume that the radar gives the distance to an aircraft accurately but the angle of elevation is a random variable that follows a uniform distribution. If we are interested in determining the altitude of the aircraft, this will be a random variable. More precisely, if we let X be the angle of elevation of the radar, then the altitude would be the random variable $Y = D \sin X$, where D is the distance to the plane. For another example, suppose that a leaf spring for a truck is composed of three leaves, and the thickness of each leaf follows a certain distribution. We might be interested in the distribution of the total spring thickness given that we know the distribution of the thickness of each spring. That is, if we let X_1, X_2, and X_3 be the individual spring thicknesses, we want to determine the distribution of the random variable $Y = X_1 + X_2 + X_3$.

In this chapter we give some of the basic techniques for determining the distribution of functions of random variables along with examples of their application. In addition, we shall show how to find the distribution of functions of a random vector.

6.2 Transformations of Discrete Random Variables

Suppose that X is a discrete random variable with a probability function of $p_X(a)$. If $Y = g(X)$, then it is a relatively simple matter to obtain the probability function for Y. Let $x_1, x_2, \ldots,$ be the possible values for X, and $y_1, y_2, \ldots,$ be the possible values of $y_j = g(x_i)$. Note that $g(X)$ is not necessarily a one-to-one function. Now let

$$A_j = \{x_i : y_j = g(x_i)\}.$$

That is, A_j contains all the values of X that get mapped to y_j. Then it is obvious that

$$p_Y(y_j) = \sum_{A_j} p_X(x). \qquad (6.1)$$

Example 6.1

Suppose X is a discrete random variable with a probability function of

$$p_X(a) = \begin{cases} \dfrac{1}{n}, & a = 1, 2, \ldots, n, \quad \text{and} \\ 0, & \text{otherwise.} \end{cases}$$

We wish to determine the probability function for $Y = 2X - 1$. The possible values for Y are $1, 3, \ldots, 2n - 1$. Here, $g(X) = 2X - 1$ is a one-to-one function. This means that each A_j contains a single point. Therefore, we find that

$$p_Y(c) = \begin{cases} \dfrac{1}{n}, & c = 1, 3, \ldots, n, \quad \text{and} \\ 0, & \text{otherwise.} \end{cases}$$

Example 6.2

Suppose $X \sim Po(\lambda)$. We wish to find the probability function of $Y = X^2$. Since the possible values of X are the nonnegative integers X^2 is again a one-to-one function. Therefore, we find immediately that

$$p_Y(c) = p_X(\sqrt{c})$$
$$= \frac{\lambda^{\sqrt{c}} e^{-\lambda}}{(\sqrt{c})!},$$

or, in summary

$$p_Y(c) = \begin{cases} \dfrac{\lambda^{\sqrt{c}} e^{-\lambda}}{(\sqrt{c})!}, & c = 0, 1, 4, \ldots, \quad \text{and} \\ 0, & \text{otherwise.} \end{cases}$$

It may not be immediately clear why we would ever need to perform transformations such as those in the above examples. If the discrete random variables described in Chapter 5 are counting the number of successes or the number of trials, what interpretation could we give to the quantity $\sqrt{\text{number of successes}}$? The answer is generally "not much." In statistical applications there are, however, times when we are interested in functions of discrete random variables. We illustrate this in the following.

Example 6.3

In statistics, a common problem is to try to use the results of a random experiment (commonly called a *random sample*) to estimate the value of a parameter of a probability function. When trying to determine the value of p in a $B(n, p)$ distribution, we use the quantity $Y = \frac{X}{n}$, where X is the number of successes in

n trials. In such problems, the distribution of Y is of interest. We shall see why in later chapters. In the context of this discussion, we have that $X \sim B(n,p)$ and $g(X) = \frac{X}{n}$. This is again a one-to-one mapping. Therefore, we find

$$p_Y(c) = p_X(nc),$$

or

$$p_Y(c) = \begin{cases} \binom{n}{nc} p^{nc}(1-p)^{n-nc} & , c = 0, \frac{1}{n}, \frac{2}{n}, \ldots, 1, \quad \text{and} \\ 0 & , \text{otherwise.} \end{cases}$$

■

We now give an example when $g(X)$ is not one to one.

Example 6.4   Suppose X has a probability function of

$$p_X(a) = \begin{cases} \dfrac{1}{2n+1} & , a = 0, \pm 1, \pm 2, \ldots, \pm n, \quad \text{and} \\ 0 & , \text{otherwise.} \end{cases}$$

We wish to determine the probability function for $Y = X^2$. Since X assumes both positive and negative values X^2 is not a one-to-one transformation. In this case we note that, except for $y = 0$, two values of X are mapped to each Y. Therefore, we find that

$$p_Y(a) = p_X(\sqrt{a}) + p_X(-\sqrt{a}), \text{ if } a \neq 0, \quad \text{and}$$

$$p_Y(0) = p_X(0).$$

Since the possible values for Y are $0, 1, 4, \ldots, n^2$ we find that the probability function for Y is

$$p_Y(a) = \begin{cases} \dfrac{1}{2n+1} & , a = 0 \\ \dfrac{2}{2n+1} & , a = 1, 4, \ldots, n^2, \quad \text{and} \\ 0 & , \text{otherwise.} \end{cases}$$

■

If (X,Y) is a discrete random vector, the procedure for finding the distribution of $T = f(X,Y)$ is a generalization of what we have just done. In other words, suppose that $(x_1, y_1), (x_2, y_2), \ldots,$ are the possible values of (X,Y) and $t_1, t_2, \ldots,$ are the possible values of T. As before, let

$$A_j = \{(x_i, y_i) : t_j = f(x_i, y_i)\}.$$

Then we find that

$$p_T(t_j) = \sum_{A_j} p_{X,Y}(x, y). \tag{6.2}$$

An example should help to clarify the general procedure.

Example 6.5

Suppose that the number of jobs being submitted to a computer system at a large government agency per minute follows a Poisson distribution with parameter $\lambda = 1$. We wish to determine the probability that there will be more than 5 jobs submitted in a 2-minute period. We can safely assume that the number of jobs submitted in separate minutes are independent. Let X be the number of jobs submitted in the first minute and Y be the number of jobs submitted in the second minute. Then X and Y are i.i.d. $Po(1)$ random variables, and we wish to determine the probability function for $T = X + Y$. Theorem 5.5 can be used in this case to show that $S \sim Po(2)$, but we wish to illustrate the general procedure with this example. Thus, we shall generalize the current example and give a proof for Theorem 5.5. Let X and Y be independent random variables such that $X \sim Po(\lambda)$ and $Y \sim Po(\eta)$. We wish to determine the probability function of $T = X + Y$. Since X and Y are independent, the joint probability function for (X, Y) is

$$p_{X,Y}(a,b) = p_X(a)p_X(b) = \begin{cases} \dfrac{\lambda^a \, \eta^b \, e^{-\lambda - \eta}}{a! \, b!} & \begin{array}{l} a = 0, 1, 2 \ldots, \\ b = 0, 1, 2, \ldots, \end{array} \quad \text{and} \\ 0 & \text{, otherwise.} \end{cases}$$

We now use Equation 6.2 as follows. Suppose $t_j = c$.

$$p_T(c) = \sum_{x+y=c} \frac{\lambda^x \, \eta^y \, e^{-\lambda - \eta}}{x! \, y!}$$

$$= \sum_{x=0}^{c} \frac{\lambda^x \, \eta^{c-x} \, e^{-\lambda - \eta}}{x! \, (c - x)!}$$

$$= \frac{e^{-\lambda - \eta}}{c!} \sum_{x=0}^{c} \frac{c!}{x! \, (c - x)!} \lambda^x \, \eta^{c-x}$$

The sum is the binomial expansion of $(\lambda + \eta)^c$ which leaves us with

$$p_T(c) = \begin{cases} \dfrac{(\lambda + \eta)^c \, e^{-(\lambda + \eta)}}{c!} & , c = 0, 1, 2, \ldots, \quad \text{and} \\ 0 & \text{, otherwise.} \end{cases}$$

This is the probability function of a $Po(\lambda + \eta)$ random variable.

To complete the problem we set $\lambda = \eta = 1$, note that $S \sim Po(2)$ and use Table 2 of Appendix A to compute

$$P(S > 5) = 1 - P(X \le 5) = 1 - 0.9834 = 0.0166.$$ ∎

Exercises 6.2

1. In a game of chance, a box contains 1 ball numbered 1, 2 balls numbered 2, 3 balls numbered 3, 4 balls numbered 4, and 5 balls numbered 5. A single ball is removed

from the box and a payoff is made that is equal to 4 times the reciprocal of the number drawn. Find the probability function for the amount of the payoff.

2. Let X be a geometric random variable with parameter p. Find the probability function for $Y = \sqrt{X}$, the positive square root of X.

3. Let X have a probability function of

$$p_X(a) = \begin{cases} \dfrac{1}{3n+1} & , a = -n, -n+1, \ldots, 2n, \quad \text{and} \\ 0 & , \text{otherwise.} \end{cases}$$

Find the probability function for $Y = X^2$.

4. Let T and U be i.i.d. random variables each with a probability function of

$$p(a) = \begin{cases} \dfrac{1}{10} & , a = 0, 1, \ldots, 9, \quad \text{and} \\ 0 & , \text{otherwise.} \end{cases}$$

Find the probability function for $10T + U$.

5. Numbers are selected at random from the interval $[0, 1]$ until a value greater than 0.75 is obtained. This is repeated twice. Let X and Y represent the number of trials in each of the 2 runs. Thus, X and Y will be i.i.d. geometric random variables with parameter 0.25. Find the distribution of $T = X + Y$, the total number of trials for both runs of the experiment.

6. Show that

$$\sum_{r=0}^{a} \binom{a}{r}\binom{b}{s-r} = \binom{a+b}{s}.$$

Hint: Consider terms in $(1 + x)^a(1 + x)^b = (1 + x)^{a+b}$.

7. Use the result in Exercise 6 and Equation 6.2 to show that, if X and Y are independent random variables with $X \sim B(n,p)$ and $Y \sim B(m,p)$, then $X + Y \sim B(n + m, p)$. That is, prove Theorem 5.2.

8. Let (X, Y) have a trinomial distribution with a joint p.f. of

$$p_{X,Y}(a,b) = \begin{cases} \left(\dfrac{2}{a!b!(2-a-b)!}\right)\left(\dfrac{1}{3}\right)^2 & , (a,b) = \begin{cases} (0,0), & (1,0), & (0,1), \\ (1,1), & (2,0), & (0,2), \end{cases} \quad \text{and} \\ 0 & , \text{otherwise.} \end{cases}$$

Find the joint p.f. for $T = X + Y$, and $U = X - Y$.

9. Let X have a p.f. of

$$p_X(a) = \begin{cases} \dfrac{1}{5} & , a = 1, 3, 5, 7, 9, \quad \text{and} \\ 0 & , \text{otherwise.} \end{cases}$$

Find the p.f. for $Y = \ln X$.

10. Let X and Y be i.i.d. $B\left(5, \dfrac{1}{4}\right)$ random variables. Find the p.f. for $T = XY$.

6.3
The Distribution Function Method

Suppose that X is a continuous random variable with a known distribution function, $F_X(x)$, and we wish to find the probability density function of $Y = g(X)$. The method based on distribution functions can be used whenever we can express the event $\{Y \le y\}$ in terms of some equivalent set of values of X. Once we can do this, then the distribution function for Y can be expressed in terms of the distribution function for X. This is then differentiated with respect to y to get the probability density function for Y. We illustrate the method with an example.

Example 6.6

Let $X \sim U(0, \theta)$. We wish to find the probability density function of $Y = aX + b$. The distribution function for X is

$$F_X(x) = \begin{cases} 0 & , x \le 0, \\ \dfrac{x}{\theta} & , 0 < x \le \theta, \quad \text{and} \\ 1 & , \theta < x. \end{cases}$$

Using the distribution function for Y we find

$$\begin{aligned} F_Y(y) &= P(Y \le y) \\ &= P(aX + b \le y) \\ &= P\left(X \le \frac{y - b}{a}\right) \\ &= F_X\left(\frac{y - b}{a}\right) \\ &= \begin{cases} 0 & , y \le b, \\ \dfrac{y - b}{a\theta} & , b < y \le a\theta + b, \quad \text{and} \\ 1 & , a\theta + b < y. \end{cases} \end{aligned}$$

To get the probability density function we differentiate $F_Y(y)$ with respect to y to get

$$f_Y(y) = \begin{cases} \dfrac{1}{a\theta} & , b \le y \le a\theta + b, \quad \text{and} \\ 0 & , \text{otherwise.} \end{cases}$$

This is the probability density function of a $U(b, a\theta + b)$ random variable. ∎

The preceding example can be generalized to a linear transformation of any continuous random variable. We leave the proof as an exercise.

Theorem 6.1

Let X be a continuous random variable with probability density function $f_X(x)$. Then the probability density function for $Y = aX + b$ is

$$f_Y(y) = \frac{1}{|a|} f_X\left(\frac{y - b}{a}\right). \tag{6.3}$$

In some cases, we can use the distribution function method for transformations that are not one to one. The key is the ability to write the distribution function for the new random variable in terms of the distribution function of the known random variable. As was the case with discrete random variables the specifics of the derivation depend on the transformation.

Example 6.7

Let the continuous r.v. X have a p.d.f. of $f_X(x)$. We wish to find the p.d.f. of $Y = X^2$. This is not a linear transformation. We use the distribution function as follows.

$$F_Y(y) = P(Y \leq y)$$

$$= P(X^2 \leq y)$$

$$= P(-\sqrt{y} \leq X \leq \sqrt{y})$$

$$= F_X(\sqrt{y}) - F_X(-\sqrt{y})$$

We have been able to express the distribution function of Y in terms of that for X. We now differentiate with respect to y to obtain the probability density function. This gives

$$f_Y(y) = \frac{1}{2\sqrt{y}} \left[f_X(\sqrt{y}) + f_X(-\sqrt{y}) \right].$$

∎

It is easier to see why transformations of continuous random variables can be useful. For example, we may wish to convert measurements made in units of feet to meters. Theorem 6.1 would give the probability density function of the transformed measurements. Other uses can be found that bear a resemblance to the so-called related rates problems encountered in a calculus course. An example of this is given in the following.

Example 6.8

A device known as a *planimeter* computes the area of a figure by tracing its perimeter. Consider an experiment where a planimeter is used to measure the area A of a square in cm^2. This value is then used to compute the length L of the sides in cm. Assume that measurements of A follow a $U(24, 26)$ distribution. We wish to determine the distribution of the computed values for L. Note that $L = \sqrt{A}$. The distribution function for A is

$$F_A(a) = \begin{cases} 0 & , a < 24, \\ \dfrac{a - 24}{2} & , 24 \leq a \leq 26, \quad \text{and} \\ 1 & , 26 < a. \end{cases}$$

Beginning with the distribution function for L we find that

$$F_L(l) = P(L \le l)$$

$$= P(\sqrt{A} \le l)$$

$$= P(A \le l^2)$$

$$= F_A(l^2)$$

$$= \begin{cases} 0 & , l < \sqrt{24}, \\ \dfrac{l^2 - 24}{2} & , \sqrt{24} \le l \le \sqrt{26}, \quad \text{and} \\ 1 & , \sqrt{26} < l. \end{cases}$$

Differentiating with respect to l and using the Fundamental Theorem of Calculus gives

$$f_L(l) = \begin{cases} l , \sqrt{24} \le l \le \sqrt{26}, \quad \text{and} \\ 0 , \text{otherwise.} \end{cases}$$

■

Exercises 6.3

1. The time X it takes for n identical components to fail follows a $\chi^2(n)$ distribution. Find the probability density function for the average failure time, $Y = \frac{X}{n}$.

2. The length X of a side of a cube is selected at random in such a manner that it has a probability density function of

$$f_X(x) = \begin{cases} \dfrac{1}{x^2} , 1 \le x, \quad \text{and} \\ 0 , \text{otherwise.} \end{cases}$$

Find the probability density function for the volume of the cube.

3. A common method for generating random observations from a specified distribution is to use a generator that returns random observations from a $U(0, 1)$ distribution. Suppose you observe X from a $U(0, 1)$ distribution and compute $Y = -\ln X$. What is the distribution of Y?

4. Let $X \sim G(2, \beta)$. Find the probability density function for $Y = X^2$.

5. Let X be a continuous random variable with a distribution function of $F_X(x)$. Find the probability density function for $Y = F_X(x)$.

6. Let X be a continuous random variable with a probability density function of $f_X(x)$ and a distribution function of $F_X(x)$. Show that the probability density function for $Y = |X|$ is

$$f_Y(y) = \begin{cases} f_X(y) + f_X(-y) , 0 \le y, \quad \text{and} \\ 0 , \text{otherwise.} \end{cases}$$

7. Use the result of Exercise 6 to determine the distribution of $Y = |X|$, where X has a double exponential distribution with $\theta = 0$ and $\sigma = 1$.

8. Prove Theorem 6.1.

9. Use Theorem 6.1 to prove that if $X \sim N(\mu, \sigma^2)$ then $aX + b \sim N(a\mu + b, a^2\sigma^2)$.

10. Let X and Y be i.i.d. $U(0, 1)$ random variables. Show that the p.d.f. for $T = |X - Y|$ is

$$f_T(t) = \begin{cases} 2(1 - t), 0 \leq t \leq 1, & \text{and} \\ 0 & , \text{otherwise.} \end{cases}$$

6.4
The Transformation of Variables Method

The next method is also based, in some sense, on the distribution function of the new random variable. This method avoids the intermediate step of finding the distribution function which must then be differentiated to obtain the probability density function. It can be viewed as being based on the distribution function because we are essentially using the substitution method for finding antiderivatives to obtain the integrand in a distribution function calculation. We state the method as a theorem.

Theorem 6.2

If X is a continuous random variable and $Y = g(X)$ is a one-to-one, differentiable transformation, then the probability density function for Y is

$$f_Y(y) = f_X[g^{-1}(y)] \left| \frac{d}{dy} g^{-1}(y) \right|, \qquad (6.4)$$

where $g^{-1}(Y)$ is the inverse function for $g(X)$.

Proof We shall prove the theorem for the case of an increasing function. The basic procedure is to begin with the distribution for Y, $F_Y(y)$, and rewrite it in terms of the distribution function for X, $F_X(x)$ which we know. Now that we have the distribution function for Y we differentiate with respect to y to obtain the probability density function. The details are as follows.

$$F_Y(y) = P(Y \leq y)$$
$$= P(g(X) \leq y)$$
$$= P\left[X \leq g^{-1}(y)\right]$$
$$= F_X\left[g^{-1}(y)\right]$$

Then

$$f_Y(y) = \frac{d}{dy} F_X\left[g^{-1}(y)\right]$$
$$= f_X\left[g^{-1}(y)\right] \frac{d}{dy} g^{-1}(y).$$

Example 6.9   Let $X \sim U(1,2)$. We wish to find the p.d.f for $Y = \frac{1}{X}$. Since $g(X) = \frac{1}{X}$ is one to one in the interval $[1,2]$ it has an inverse that is $g^{-1}(Y) = \frac{1}{Y}$. Now

$$f_X(x) = \begin{cases} 1 , 1 \le x \le 2, & \text{and} \\ 0 , \text{otherwise}, & \text{and} \end{cases}$$

$$\frac{d}{dy} g^{-1}(y) = -\frac{1}{y^2}.$$

Therefore, Equation 6.4 gives

$$f_Y(y) = \begin{cases} \dfrac{1}{y^2} , \frac{1}{2} \le y \le 1, & \text{and} \\ 0 , \text{otherwise}. \end{cases}$$ ∎

Example 6.10   Let $Z \sim N(0,1)$. We wish to find the p.d.f. for $Y = e^Z$. Here, $g(z) = e^z$ which gives $g^{-1}(y) = \ln(y)$. Now,

$$f_Z(z) = \frac{1}{\sqrt{2\pi}} e^{-z^2/2}, -\infty < z < \infty, \quad \text{and}$$

$$\frac{d}{dy} g^{-1}(y) = \frac{1}{y}.$$

Thus, Equation 6.4 gives

$$f_Y(y) = \begin{cases} \dfrac{1}{y\sqrt{2\pi}} e^{-[\ln(y)]^2/2} , 0 < y, & \text{and} \\ 0 , \text{otherwise}. \end{cases}$$

This last example is a special case of what is known as the *lognormal distribution*. More generally, a random variable X such that $\ln(X)$ has a normal distribution is said to have a lognormal distribution. ∎

Example 6.11   We return to one of the problems posed in the introduction to this chapter. We have a radar used for air traffic control near an airport. Assume that the radar gives the distance to an aircraft accurately but the angle X, measured in radians, of elevation is a random variable that follows a $U(\theta - 0.1, \theta + 0.1)$, where θ is the actual angle of elevation. We are interested in determining the distribution of the altitude Y of the aircraft when it is D miles from the radar. From Figure 6.1 we see that $Y = D \sin X$.

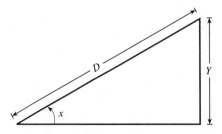

Figure 6.1

In the range of interest $g(X) = D \sin X$ will be a one-to-one transformation with an inverse of $g^{-1}(Y) = \sin^{-1}(\frac{Y}{D})$. Now,

$$f_X(x) = \begin{cases} 5, & \theta - 0.1 \le x \le \theta + 0.1, \quad \text{and} \\ 0, & \text{otherwise,} \quad \text{and} \end{cases}$$

$$\frac{d}{dy} g^{-1}(y) = \frac{1}{\sqrt{D^2 - y^2}}.$$

Thus, Equation 6.4 gives

$$f_Y(y) = \begin{cases} \dfrac{5}{\sqrt{D^2 - y^2}}, & D \sin(\theta - 0.1) \le y \le D \sin(\theta + 0.1), \quad \text{and} \\ 0, & \text{otherwise.} \end{cases}$$

∎

The transformation method can be extended to cover functions that are not one to one. It is a little bit complicated. We shall introduce the general procedure and then illustrate it with a couple of examples.

Assume that a continuous random variable X, having a p.d.f. of $f_X(x)$, can assume values in an interval (not necessarily finite) I, and we wish to determine the p.d.f. for $Y = g(X)$ which is an arbitrary function from I to some interval I'.

1. Fix y arbitrarily. Then determine which values of X get mapped to this value of y. Let

$$A_y = \{x : y = g(x)\}.$$

2. Apply Equation 6.4 to transform $f_X(x)$ at each value of X.

3. Obtain the p.d.f. for Y by summing the transformed p.d.f.'s for all values of X that are mapped to the same value of Y. That is,

$$f_Y(y) = \sum_{A_y} f_X \left[g^{-1}(y) \right] \left| \frac{d}{dy} g^{-1}(y) \right|. \tag{6.5}$$

Admittedly this seems rather involved. Hopefully, the next two examples will clarify what is going on. We shall not attempt to prove that the scheme just given works.

Example 6.12   Let X be a $U(-1, 2)$ random variable with a p.d.f. of

$$f_X(x) = \begin{cases} \dfrac{1}{3}, & -1 \le x \le 2, \quad \text{and} \\ 0, & \text{otherwise.} \end{cases}$$

We wish to find the p.d.f. for $Y = X^2$. From Figure 6.2 we see that this is not a one-to-one transformation in the interval $[-1, 2]$.

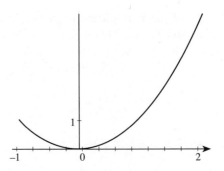

Figure 6.2

However, if we partition $[-1, 2]$ into the subintervals $[-1, 0]$, $[0, 2]$ and $Y = X^2$ is one to one within each subinterval. Since $Y = X^2$ transforms X to the same range of values of Y in $[-1, 0]$ and $[0, 1]$, the idea is to transform the p.d.f. separately on each subinterval and for those that map X into the same value of Y add the corresponding transformed p.d.f.'s. We consider each of the cases separately.

Case 1: $X \in [-1, 0]$.

$$g^{-1}(y) = -\sqrt{y}, \quad \text{and}$$

$$\frac{d}{dy} g^{-1}(y) = -\frac{1}{2\sqrt{y}}$$

Thus, the transformed p.d.f. is

$$f_X(-\sqrt{y}) \frac{1}{2\sqrt{y}} = \frac{1}{6\sqrt{y}}.$$

Case 2: $X \in [0, 2]$.

$$g^{-1}(y) = \sqrt{y}, \quad \text{and}$$

$$\frac{d}{dy} g^{-1}(y) = \frac{1}{2\sqrt{y}}$$

The transformed p.d.f. is

$$f_X(\sqrt{y}) \frac{1}{2\sqrt{y}} = \frac{1}{6\sqrt{y}}.$$

Thus, for $Y \in [0, 1]$, we add the results from Case 1 and Case 2 to obtain

$$f_Y(y) = \frac{1}{6\sqrt{y}} + \frac{1}{6\sqrt{y}} = \frac{1}{3\sqrt{y}}.$$

For $Y \in [1, 4]$ we simply have the transformed p.d.f. from Case 2. So, in summary

$$f_Y(y) = \begin{cases} \dfrac{1}{3\sqrt{y}} & , \ 0 \le y \le 1, \\[2mm] \dfrac{1}{6\sqrt{y}} & , \ 1 < y \le 4, \quad \text{and} \\[2mm] 0 & , \ \text{otherwise.} \end{cases}$$

■

Example 6.13   We would like to use this method to give a proof that the square of a standard normal random variable has a $\chi^2(1)$ distribution. This is Theorem 5.8. Since a standard normal random variable ranges over all of the real numbers, the function $Y = Z^2$ will map to the nonnegative real numbers. Thus, we can partition $I = (-\infty, \infty)$ into $I_1 = (-\infty, 0]$ and $I_2 = [0, \infty)$. Here both x and $-x$ get mapped to the same value of y. This means that the p.d.f. for each y will be the sum of the transformations for $x = \sqrt{y}$ and $x = -\sqrt{y}$.

For $x = \sqrt{y}$

$$g^{-1}(y) = \sqrt{y}, \quad \text{and}$$

$$\frac{d}{dy}g^{-1}(y) = \frac{1}{2\sqrt{y}}.$$

The transformed p.d.f. is

$$f_X\left(\sqrt{y}\right)\frac{1}{2\sqrt{y}} = \frac{1}{2\sqrt{y}\sqrt{2\pi}}e^{-y/2}.$$

For $x = -\sqrt{y}$

$$g^{-1}(y) = -\sqrt{y}, \quad \text{and}$$

$$\frac{d}{dy}g^{-1}(y) = -\frac{1}{2\sqrt{y}}.$$

Therefore, the transformed p.d.f. is

$$f_X\left(\sqrt{y}\right)\frac{1}{2\sqrt{y}} = \frac{1}{2\sqrt{y}\sqrt{2\pi}}e^{-y/2}.$$

Adding these together gives

$$f_Y(y) = \begin{cases} \dfrac{y^{-1/2}}{\sqrt{\pi}\,2^{1/2}}e^{-y/2}, & 0 \le y, \quad \text{and} \\ \\ 0 & , \text{otherwise.} \end{cases}$$

This is the p.d.f. of a $G\left(\frac{1}{2}, 2\right)$ random variable.   ∎

Exercises 6.4

1. Do Exercise 1 of Section 6.3 using the transformation of variables method.

2. Do Exercise 2 of Section 6.3 using the transformation of variables method.

3. Do Exercise 3 of Section 6.3 using the transformation of variables method.

4. Do Exercise 4 of Section 6.3 using the transformation of variables method.

5. Let Z be a standard normal random variable. Find the probability density function for $Y = |Z|$.

6. Let X be a $U(-1, 2)$ random variable with a p.d.f. of

$$f_X(x) = \begin{cases} \dfrac{1}{3}, & -1 \le x \le 2, \quad \text{and} \\ \\ 0, & \text{otherwise.} \end{cases}$$

Find the p.d.f. for $Y = |X|$.

7. The radius of a circle R is a random variable whose p.d.f. is

$$f_R(r) = \begin{cases} 2r \, , \, 0 < r < 1, & \text{and} \\ 0 \, , \, \text{otherwise.} \end{cases}$$

Find the p.d.f. for the area of the circle.

8. Let X have a $G(1, 1)$ distribution. Find the p.d.f. for $Y = \sqrt{X}$, the positive square root of X.

9. Prove Theorem 6.2 for the case where $Y = g(X)$ is a decreasing function.

10. Let $X \sim U\left(-\frac{\pi}{2}, \frac{\pi}{2}\right)$. Show that the p.d.f. for $Y = \tan(X)$ is

$$f_Y(y) = \frac{1}{\pi(1 + y^2)}, \, -\infty < y < \infty.$$

6.5
Moment Generating Function Method

In some cases, it is possible to exploit the intimate relation between random variables and their moment generating functions to determine the distribution of functions involving them. In those cases where it works, the use of moment generating functions usually requires far less effort than the methods discussed thus far. It can be particularly useful for deriving the distribution of sums of independent random variables.

The idea behind the method is as follows. Suppose we know the distribution of the random variable X. Therefore, we also know the moment generating function for X, if it exists. Suppose that $Y = g(X)$. If Y is continuous and its moment generating function exists, we would compute it by

$$M_Y(t) = E\left[e^{tY}\right]$$

$$= \int_{-\infty}^{\infty} e^{ty} f_Y(y)\, dy.$$

This would seem to require that we know the probability density function for Y. However, we can use Theorem 4.1 and the known probability density function for X as follows.

$$M_Y(t) = E\left[e^{tY}\right]$$

$$= E\left[e^{tg(X)}\right]$$

$$= \int_{-\infty}^{\infty} e^{tg(x)} f_X(x)\, dx \tag{6.6}$$

If the resulting $M_Y(t)$ can be recognized as being the moment generating function of a known random variable, we have found the distribution of Y. The shortcoming is that we may not be able to identify which family of distributions the moment generating function is from. The development we have given is the same for discrete random variables if integrals are replaced by sums. That is, if $Y = g(X)$, then

$$M_Y(t) = \sum_a e^{tg(a)} p_X(a). \tag{6.7}$$

In the case where $(X_1, X_2, \ldots, X_n)$ is a random vector with a known joint distribution and we wish to determine the distribution of $Y = g(X_1, X_2, \ldots, X_n)$, we can determine $M_Y(t)$ as follows. In the case of a discrete random vector,

$$M_Y(t) = \sum_{x_1} \sum_{x_2} \cdots \sum_{x_n} e^{tg(x_1, x_2, \ldots, x_n)} p_{X_1, X_2, \ldots, X_n}(x_1, x_2, \ldots, x_n). \qquad (6.8)$$

If the random vector is continuous

$$M_Y(t) = \int_{-\infty}^{\infty} \int_{-\infty}^{\infty} \cdots \int_{-\infty}^{\infty} e^{tg(x_1, x_2, \ldots, x_n)} f_{X_1, X_2, \ldots, X_n}(x_1, x_2, \ldots, x_n) \, dx_1 \, dx_2 \cdots dx_n.$$
$$(6.9)$$

Example 6.14   Let X be a discrete random variable whose probability function is

$$p_X(a) = \begin{cases} \dfrac{a}{10} & , a = 1, 2, 3, 4, \quad \text{and} \\ 0 & , \text{otherwise.} \end{cases}$$

We wish to determine the probability function for $Y = 2X - 1$ by using moment generating functions. We compute the moment generating function directly by using Equation 6.7 as follows.

$$M_Y(t) = \sum_{a=1}^{4} e^{2a-1} \frac{a}{10}$$

$$= \frac{1}{10} e^t + \frac{2}{10} e^{3t} + \frac{3}{10} e^{5t} + \frac{4}{10} e^{7t}$$

Since this is of the form

$$\sum_a e^{ta} p(a)$$

we see that the probability function for Y is

$$p_Y(a) = \begin{cases} \dfrac{2a - 1}{10} & , a = 1, 3, 5, 7, \quad \text{and} \\ 0 & , \text{otherwise.} \end{cases}$$

■

Many times this method does not require that we compute the moment generating function by directly evaluating sums or integrals. It is possible in some cases to begin with the moment generating function for the new random variable and to find some way to write it in terms of the moment generating function of the known random variable. This is like the approach taken with the distribution function method and is the basis for the claim that this method can require less work than the others.

Example 6.15   Suppose that $X \sim B(n, p)$. We wish to determine the distribution of $Y = n - X$. We begin with the moment generating function for Y, make use of Theorem 4.16, and proceed as follows.

$$M_Y(t) = E\left[e^{tY}\right]$$

$$= E\left[e^{t(n-X)}\right]$$

$$= e^{tn} E\left[e^{-tX}\right]$$

$$= e^{tn} M_X(-t)$$

$$= e^{tn}\left(q + pe^{-t}\right)^n$$

$$= \left(qe^t + p\right)^n$$

This is the moment generating of a $B(n, q)$ random variable. Therefore, $Y \sim B(n, q)$. ∎

We mentioned that the use of moment generating functions can be particularly useful in determining the distribution of sums of random variables. We give an example of how this works.

Example 6.16   Suppose $X \sim G(\alpha, \beta)$ and $Y \sim G(\eta, \beta)$ and that X and Y are independent. We wish to determine the distribution of $S = X + Y$. This will prove Theorem 5.9. We use moment generating functions and Theorem 4.16 as follows.

$$M_S(t) = E\left[e^{tS}\right]$$

$$= E\left[e^{t(X+Y)}\right]$$

$$= E\left[e^{tX}\right] E\left[e^{tY}\right]$$

$$= M_X(t) M_Y(t)$$

$$= \left(\frac{1}{1 - \beta t}\right)^\alpha \left(\frac{1}{1 - \beta t}\right)^\eta$$

$$= \left(\frac{1}{1 - \beta t}\right)^{\alpha + \eta}$$

This is the moment generating function of a $G(\alpha + \eta, \beta)$ random variable. Therefore, $S \sim G(\alpha + \eta, \beta)$. ∎

Example 6.17   Suppose that $X \sim N(\alpha, \beta)$, $Y \sim N(\gamma, \delta)$, and that X and Y are independent. We wish to find the distribution of $V = aX + bY$. As in the previous example we use moment generating functions and Theorem 4.17 as follows.

$$M_V(t) = E\left[e^{tV}\right]$$

$$= E\left[e^{t(aX+bY)}\right]$$

$$= E\left[e^{atX}\right] E\left[e^{btY}\right]$$

$$= M_X(at)\, M_Y(bt)$$

$$= e^{at\alpha + a^2 t^2 \beta/2}\, e^{bt\gamma + b^2 t^2 \delta/2}$$

$$= e^{(a\alpha + b\gamma)t + t^2(a^2\beta + b^2\delta)/2}$$

This is the moment generating function of a $N(a\alpha + b\beta, a^2\beta + b^2\delta)$ random variable. Thus, a linear combination of independent normal random variables has a normal distribution. In fact, linear combinations of normal random variables have normal distributions even when the original random variables are not independent. We leave this case as an exercise. ∎

Exercises 6.5

1. Let X be a discrete random variable with a probability function of

$$p_X(a) = \begin{cases} \dfrac{1}{5} & , a = 0, \pm 1, \pm 2, \quad \text{and} \\ 0 & , \text{otherwise.} \end{cases}$$

Use moment generating functions to find the probability function of the following.

a. $Y = X^3 + 8$

b. $Y = |X|$

c. $Y = \cos \dfrac{\pi X}{2}$

2. Let X have a $G(1, 1)$ distribution. Use moment generating functions to find the probability density function of $Y = e^{-X}$.

3. Use moment generating functions to solve Exercise 5 of Section 6.2.

4. Show that, if $X_1, X_2, \ldots, X_n$ are i.i.d. $Po(\lambda)$ random variables, then $T = X_1 + X_2 + \cdots + X_n$ has a $Po(n\lambda)$ distribution.

5. Let $X \sim N(\mu, \sigma^2)$ and $Y \sim N(\eta, \tau^2)$, where X and Y are dependent with correlation ρ. Find the distribution of $T = aX + bY$, where a and b are real constants.

6. Let X and Y be i.i.d. $N(0, 1)$ random variables. In many statistical applications, linear combinations of i.i.d. normal random variables are used to estimate parameter values. Find the joint moment generating function for $U = aX + bY$ and $V = cX + dY$. What conditions must be satisfied by a, b, c, and d for U and V to be independent? Use moment generating functions.

7. Let X have a double exponential distribution with parameters $\theta = 0$ and $\sigma = 0$. Use moment generating functions to determine the distribution of $Y = |X|$.

8. Suppose that $X \sim N(\mu, \sigma^2)$. Use moment generating functions to show that the distribution of $Y = aX + b$ is $N(a\mu + b, a^2\sigma^2)$. That is, prove Theorem 5.6.

9. Suppose that $X \sim Po(\lambda)$ and $Y \sim Po(\eta)$ and that X and Y are independent. Use moment generating functions to show that $X + Y \sim Po(\lambda + \eta)$.

10. Suppose that $X \sim B(n, p)$ and $Y \sim B(m, p)$ and that X and Y are independent. Use moment generating functions to show that $X + Y \sim B(n + m, p)$.

11. Use moment generating functions to show that the square of a standard normal random variable has a chi-squared distribution with 1 degree of freedom.

12. Suppose that X and Y are i.i.d. $G(1, \beta)$ random variables. Use moment generating functions to show that $T = X - Y$ has a double exponential distribution.

6.6 Sums of Independent Random Variables

Random variables that are constructed by computing the sum of some independent random variables play a central role in many of the applications of probability. For this reason the ability to determine the distribution of such sums is of special importance. We have seen that, in some cases, moment generating functions can be an effective tool. In other cases, however, moment generating functions come up short.

Example 6.18

Let X_1, X_2, and X_3 be i.i.d. $U(0, 1)$ random variables. We wish to compute the moment generating function of $T = X_1 + X_2 + X_3$. Following the ideas of Section 6.5 we proceed as follows.

$$
\begin{aligned}
M_T(t) &= E\left[e^{tT}\right] \\
&= E\left[e^{t(X_1 + X_2 + X_3)}\right] \\
&= E\left[e^{tX_1}\right] E\left[e^{tX_2}\right] E\left[e^{tX_3}\right] \\
&= M_{X_1}(t) M_{X_2}(t) M_{X_3}(t) \\
&= \frac{e^t - 1}{t} \frac{e^t - 1}{t} \frac{e^t - 1}{t} \\
&= \left(\frac{e^t - 1}{t}\right)^3
\end{aligned}
$$

We have found the moment generating function for the sum of 3 i.i.d. uniform random variables. Unfortunately, it is a moment generating function that we do not recognize. So having found the moment generating function is useless to us. ∎

In this section we introduce an alternative approach based on the distribution function method that is useful in cases where the moment generating function of the sum cannot be recognized as being a member of any particular family of distributions. Assume that X and Y are independent continuous random variables.

Let $T = X + Y$ and compute the distribution function for T as follows.

$$F_T(t) = P(T \leq t)$$

$$= P(X + Y \leq t)$$

$$= \iint\limits_{\{x+y\leq t\}} f_X(x)f_Y(y)\,dx\,dy$$

$$= \int_{-\infty}^{\infty} \int_{-\infty}^{t-y} f_X(x)f_Y(y)\,dx\,dy$$

$$= \int_{-\infty}^{\infty} F_X(t - y)f_Y(y)\,dy. \tag{6.10}$$

$F_T(t)$ is known as the *convolution of the distributions $F_X(x)$ and $F_Y(y)$*. We now differentiate the convolution with respect to t to obtain the probability density function for $T = X + Y$.

$$f_T(t) = \frac{d}{dt} \int_{-\infty}^{\infty} F_X(t - y)f_Y(y)\,dy$$

$$= \int_{-\infty}^{\infty} \frac{d}{dt} F_X(t - y)f_Y(y)\,dy$$

$$= \int_{-\infty}^{\infty} f_X(t - y)f_Y(y)\,dy \tag{6.11}$$

We can determine the distribution of a sum of any number of independent random variables by repeated use of Equation 6.11. It is certainly not as convenient as moment generating functions, but when they fail it is the only alternative. We illustrate how to use Equation 6.11 with some examples.

Example 6.19   Let X and Y be i.i.d. $U(0, 1)$ random variables. We wish to determine the p.d.f. for $T = X + Y$. Now

$$f_X(u) = f_Y(u) = \begin{cases} 1, & 0 \leq u \leq 1, \quad \text{and} \\ 0, & \text{otherwise.} \end{cases}$$

Equation 6.11 becomes

$$f_T(t) = \int_{-\infty}^{\infty} f_X(t - y)f_Y(y)\,dy = \int_0^1 f_X(t - y)\,dy.$$

We must consider two cases.

Case 1: $0 \leq t \leq 1$. In this region $f_X(t - y) = 1$ is positive as long as y is between zero and t, and it is zero, otherwise. Therefore,

$$f_T(t) = \int_0^t dy = t.$$

Case 2: $1 < t \le 2$. Here $f_X(t - y) = 1$ when y is between $t - 1$ and 1, and zero, otherwise. Thus,

$$f_T(t) = \int_{t-1}^{1} dy = 2 - t.$$

Combining these results gives

$$f_T(t) = \begin{cases} t & , 0 \le t \le 1, \\ 2 - t & , 1 < t \le 2, \quad \text{and} \\ 0 & , \text{otherwise.} \end{cases}$$

The graph of this p.d.f. is shown in Figure 6.3. It is clear why this is called a *triangular distribution*.

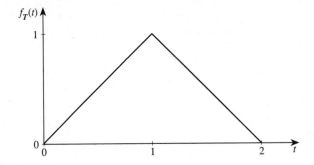

Figure 6.3

Example 6.20

Suppose that X and Y are i.i.d. $G(\alpha, \beta)$ random variables. We wish to determine the p.d.f. for $T = X + Y$. Moment generating functions could be used here, but we wish to illustrate the convolution method.

$$f_X(u) = f_Y(u) = \begin{cases} \dfrac{u^{\alpha-1}}{\Gamma(\alpha)\,\beta^{\alpha}} e^{-u/\beta} & , 0 < u, \quad \text{and} \\ 0 & , \text{otherwise} \end{cases}$$

Since y can range from 0 to t Equation 6.11 gives

$$f_T(t) = \int_0^t \frac{(t-y)^{\alpha-1} y^{\alpha-1}}{\left[\Gamma(\alpha)\right]^2 \beta^{2\alpha}} e^{-(t-y)/\beta} e^{-y/\beta} \, dy$$

$$= \frac{e^{-t/\beta}}{\beta^{2\alpha}} \int_0^t \frac{1}{\left[\Gamma(\alpha)\right]^2} (t-y)^{\alpha-1} y^{\alpha-1} \, dy.$$

The integrand is similar in form to the p.d.f. for a beta distribution with both parameters equal to α. We need to multiply and divide by $\Gamma(2\alpha)$ and $t^{2\alpha-2}$. Then we must make a substitution to change the range of integration from 0 to t to 0 to 1.

$$f_T(t) = \frac{t^{2\alpha-2}}{\Gamma(2\alpha)\,\beta^{2\alpha}} e^{-t/\beta} \int_0^t \frac{\Gamma(2\alpha)}{\left[\Gamma(\alpha)\right]^2} \left(1 - \frac{y}{t}\right)^{\alpha-1} \left(\frac{y}{t}\right)^{\alpha-1} dy$$

Let $x = \frac{y}{t}$. Therefore, $dy = t\,dx$, and we get

$$f_T(t) = \frac{t^{2\alpha-1}}{\Gamma(2\alpha)\,\beta^{2\alpha}} e^{-t/\beta} \int_0^1 \frac{\Gamma(2\alpha)}{[\Gamma(\alpha)]^2} (1-x)^{\alpha-1} x^{\alpha-1}\,dx.$$

The integral is now that of a beta p.d.f. over its range. Therefore, it has a value of 1, and we get

$$f_T(t) = \begin{cases} \dfrac{t^{2\alpha-1}}{\Gamma(2\alpha)\,\beta^{2\alpha}} e^{-t/\beta}, & 0 < t, \quad \text{and} \\[2ex] 0 & , \text{ otherwise.} \end{cases}$$

This is the p.d.f. of a $G(2\alpha, \beta)$ random variable. This is far more work than would be involved when using moment generating functions, but it does illustrate some interesting computational techniques. ∎

Exercises 6.6

1. Let $X \sim U(0, 1)$ and $Y \sim U(0, 2)$, where X and Y are independent. Find the probability density function for $T = X + Y$.

2. Two components operate in parallel in a system. Let X be the time to failure for one component and Y be the time to failure for the other. When one component fails, the other one is placed in operation. Suppose that $X \sim G(1, \beta)$ and $Y \sim G(1, \gamma)$, where X and Y are independent. Find the probability density function for $T = X + Y$, the time to failure for this system. Is it a gamma distribution?

3. Let U, V, and W be i.i.d. $U(0, 1)$ random variables. Find the probability density function for $T = U + V + W$. Use the results of Example 6.19.

4. Use convolutions to show that the sum of 2 i.i.d. $N(0, 1)$ random variables has a $N(0, 2)$ distribution.

5. Let X and Y be independent random variables whose p.d.f.'s are as follows.

$$f_X(x) = \begin{cases} 3x^2, & 0 \le x \le 1, \quad \text{and} \\ 0, & \text{otherwise} \end{cases}$$

$$f_Y(y) = \begin{cases} 3(1-y)^2, & 0 \le y \le 1, \quad \text{and} \\ 0, & \text{otherwise} \end{cases}$$

Use convolutions to determine the p.d.f. for $T = X + Y$.

6.7
Distribution of Functions of Several Random Variables

In Section 6.4, we discussed how to find the probability density function of a function of a single continuous random variable based on the substitution method for finding antiderivatives. In this section we consider the extension of this approach to the case of several functions of several random variables. Not surprisingly, this method is based on that for transforming multiple integrals. We present the method and leave its proof to more advanced texts.

We begin by considering the case of two functions of two random variables. Suppose that (X, Y) is a continuous random vector having a joint probability density function of $f_{X,Y}(x, y)$. We wish to find the joint probability density function for the random vector (U, V), where $U = g_1(X, Y)$ and $V = g_2(X, Y)$. The method is applicable if the functions g_1 and g_2 satisfy the following conditions.

1. If $f_{X,Y}(x, y) > 0$ for $(x, y) \in A$, then $u = g_1(x, y)$ and $v = g_2(x, y)$ form a one-to-one transformation from $A \subseteq \mathbb{R}^2$ to $B \subseteq \mathbb{R}^2$. That is, there exist unique functions h_1 and h_2 such that $x = h_1(u, v)$ and $y = h_2(u, v)$ for all $(u, v) \in B$.

2. The functions h_1 and h_2 have continuous first partial derivatives at all points $(u, v) \in B$, and the determinant

$$J = \begin{vmatrix} \dfrac{\partial h_1}{\partial u} & \dfrac{\partial h_1}{\partial v} \\[2ex] \dfrac{\partial h_2}{\partial u} & \dfrac{\partial h_2}{\partial v} \end{vmatrix} = \frac{\partial h_1}{\partial u}\frac{\partial h_2}{\partial v} - \frac{\partial h_1}{\partial v}\frac{\partial h_2}{\partial u} \neq 0.$$

The determinant J is known as the *Jacobian* of the transformation. Hence, this method is often referred to as the *Jacobian method*. If these conditions are satisfied it turns out that the joint probability density function for (U, V) is given by

$$f_{U,V}(u, v) = \begin{cases} f_{X,Y}[h_1(u, v), h_2(u, v)] \, |J| & , \text{ for all } (u, v) \in B, \text{ and} \\ 0 & , \text{ otherwise.} \end{cases} \tag{6.12}$$

To illustrate these conditions we recall a transformation that should be familiar to the reader. We wish to convert from rectangular to polar coordinates. For each $(x, y) \in \mathbb{R}^2$ we wish to convert to (r, θ), where $r = g_1(x, y) = \sqrt{x^2 + y^2}$ and $\theta = g_2(x, y) = \tan^{-1}\frac{y}{x}$. Note that the set $A = \mathbb{R}^2$, and the set $B = \{(r, \theta): -\infty < r < \infty, 0 \le \theta < 2\pi\}$. From calculus courses we know that $x = h_1(r, \theta) = r \cos \theta$ and $y = h_2(r, \theta) = r \sin \theta$ for all $(r, \theta) \in B$. Therefore, there is a one-to-one transformation from A to B which means Condition 1 is satisfied. In addition,

$$\frac{\partial h_1}{\partial r} = \cos \theta, \quad \frac{\partial h_1}{\partial \theta} = -r \sin \theta, \quad \frac{\partial h_2}{\partial r} = \sin \theta, \quad \text{and} \quad \frac{\partial h_2}{\partial \theta} = r \cos \theta.$$

Each of these partial derivatives is continuous. Therefore, the Jacobian is

$$J = \begin{vmatrix} \cos \theta & -r \sin \theta \\ \sin \theta & r \cos \theta \end{vmatrix} = r \cos^2 \theta + r \sin^2 \theta = r.$$

Therefore, as long as $r \neq 0$ Condition 2 is satisfied. Thus, if a random vector in rectangular coordinates is converted to polar coordinates, we would have

$$f_{R,\Theta}(r, \theta) = f_{X,Y}(r \cos \theta, r \sin \theta) \, |r|.$$

We now illustrate the application of the Jacobian method with some examples.

Example 6.21   Suppose that X and Y are i.i.d. $G(\alpha, \beta)$ random variables. We wish to find the joint p.d.f. for $U = \frac{X}{X+Y}$ and $V = X + Y$. In this case $A = \{(x, y): 0 < x, 0 < y\}$ and $B = \{(u, v): 0 < u < 1, 0 < v\}$. Since $U = g_1(X, Y) = \frac{X}{X+Y}$ and $V = g_2(X, Y) =$

$X + Y$ we find that $X = h_1(U, V) = UV$ and $Y = h_2(U, V) = V - UV$. Therefore,

$$J = \begin{vmatrix} v & u \\ -v & 1 - u \end{vmatrix} = v(1 - u) + uv = v = |J|.$$

Since X and Y are i.i.d. $G(1, \beta)$ random variables, we have that

$$f_X(t) = f_Y(t) = \begin{cases} \dfrac{1}{\beta} e^{-t/\beta}, & 0 < t, \quad \text{and} \\ 0, & \text{otherwise.} \end{cases}$$

Thus, the joint p.d.f. is

$$f_{X,Y}(x, y) = \frac{1}{\beta^2} e^{-(x+y)/\beta}.$$

Therefore, Equation 6.12 gives that

$$f_{U,V}(u, v) = \begin{cases} \dfrac{v}{\beta^2} e^{-v/\beta}, & 0 < u < 1, 0 < v, \text{ and} \\ 0, & \text{otherwise.} \end{cases}$$

Note that

$$f_{U,V}(u, v) = f_U(u)f_V(v), \quad \text{where}$$

$$f_U(u) = \begin{cases} 1, & 0 < u < 1, \quad \text{and} \\ 0, & \text{otherwise,} \quad \text{and} \end{cases}$$

$$f_V(v) = \begin{cases} \dfrac{v}{\beta^2} e^{-v/\beta}, & 0 < v, \quad \text{and} \\ 0, & \text{otherwise.} \end{cases}$$

This means that $U \sim U(0, 1)$, $V \sim G(2, \beta)$, and that U and V are independent. ∎

Example 6.22   Suppose that X and Y are i.i.d. $N(0, 1)$ random variables. We wish to find the joint p.d.f. for $U = X + Y$ and $V = X + 2Y$. Here $A = B = \mathbb{R}^2$. Since $u = g_1(x, y) = x + y$ and $v = g_2(x, y) = x + 2y$ we find that the inverse functions are $x = h_1(u, v) = 2u - v$ and $y = h_2(u, v) = v - u$. Thus,

$$J = \begin{vmatrix} 2 & -1 \\ -1 & 1 \end{vmatrix} = 1,$$

and $|J| = 1$. The fact that X and Y are i.i.d. standard normal random variables means that

$$f_X(z) = f_Y(z) = \frac{1}{\sqrt{2\pi}} e^{-z^2/2}, \quad -\infty < z < \infty.$$

This gives a joint probability density function of

$$f_{X,Y}(x, y) = \frac{1}{2\pi} e^{-(x^2+y^2)/2}.$$

Then Equation 6.12 gives

$$f_{U,V}(u, v) = \frac{1}{2\pi} e^{-[(2u-v)^2+(v-u)^2]/2}$$

$$= \frac{1}{2\pi} e^{-[5u^2-6uv+2v^2]/2}, \quad \text{for all } (u, v) \in \mathbb{R}^2.$$

Now this is the p.d.f. for a bivariate normal distribution with

$$\mu_U = \mu_V = 0, \quad \sigma_U^2 = 2, \quad \sigma_V^2 = 5, \quad \text{and} \quad \rho = \sqrt{\frac{9}{10}}.$$

Since ρ is not zero U and V are not independent. This is also clear from the fact that the joint p.d.f. does not factor.   ∎

Suppose we are given the joint probability density function of $(X_1, X_2, \ldots, X_n)$ and wish to determine the joint probability density function of $(Y_1, Y_2, \ldots, Y_n)$, where

$$Y_1 = g_1(X_1, X_2, \ldots, X_n),$$
$$Y_2 = g_2(X_1, X_2, \ldots, X_n),$$
$$\vdots$$
$$Y_n = g_n(X_1, X_2, \ldots, X_n).$$

The method is the natural extension of that for random vectors with two components. That is, we require that $g_1, g_2, \ldots, g_n$ satisfy the following conditions.

1. If $f_{X_1, X_2, \ldots, X_n}(x_1, x_2, \ldots, x_n) > 0$ for $(x_1, x_2, \ldots, x_n) \in A$, then $g_1, g_2, \ldots, g_n$ form a one-to-one transformation from $A \subseteq \mathbb{R}^n$ to $B \subseteq \mathbb{R}^n$. That is, there exist unique functions $h_1, h_2, \ldots, h_n$ such that

$$x_1 = h_1(y_1, y_2, \ldots, y_n),$$
$$x_2 = h_2(y_1, y_2, \ldots, y_n),$$
$$\vdots$$
$$x_n = h_n(y_1, y_2, \ldots, y_n)$$

for each $(y_1, y_2, \ldots, y_n) \in B$.

2. Each of the functions $h_1, h_2, \ldots, h_n$ possesses a continuous partial derivative for every $(y_1, y_2, \ldots, y_n) \in B$, and the Jacobian

$$J = \begin{vmatrix} \dfrac{\partial h_1}{\partial y_1} & \dfrac{\partial h_1}{\partial y_2} & \cdots & \dfrac{\partial h_1}{\partial y_n} \\ \dfrac{\partial h_2}{\partial y_1} & \dfrac{\partial h_2}{\partial y_2} & \cdots & \dfrac{\partial h_2}{\partial y_n} \\ \vdots & \vdots & \ddots & \vdots \\ \dfrac{\partial h_n}{\partial y_1} & \dfrac{\partial h_n}{\partial y_2} & \cdots & \dfrac{\partial h_n}{\partial y_n} \end{vmatrix} \neq 0.$$

Under these conditions it can be shown that

$$f_{Y_1,\ldots,Y_n}(y_1,\ldots,y_n) = f_{X_1,\ldots,X_n}[h_1(y_1,\ldots,y_n),\ldots,h_n(y_1,\ldots,y_n)]\,|J|, \qquad \textbf{(6.13)}$$

for all $(y_1,\ldots,y_n) \in B$, and 0, otherwise.

The notation is somewhat messy. Hopefully, the following example will clarify how to apply this method.

Example 6.23

Suppose that U, V, and W are i.i.d. $G(1,\beta)$ random variables. We wish to compute the joint p.d.f. for $R = U$, $S = U + V$, and $T = U + V + W$. We also wish to compute the p.d.f. for S. Note that

$$A = \{(u,v,w) : 0 < u, 0 < v, 0 < w\}, \quad \text{and}$$

$$B = \{(r,s,t) : 0 < r < s < t\}.$$

Here

$$r = g_1(u,v,w) = u, \quad s = g_2(u,v,w) = u+v, \quad \text{and} \quad t = g_3(u,v,w) = u+v+w.$$

This gives that the inverse functions are

$$u = h_1(r,s,t) = r, \quad v = h_2(r,s,t) = s-r, \quad \text{and} \quad w = h_3(r,s,t) = t-s.$$

Therefore, the Jacobian is

$$J = \begin{vmatrix} 1 & 0 & 0 \\ -1 & 1 & 0 \\ 0 & -1 & 1 \end{vmatrix} = 1, \quad \text{and} \quad |J| = 1.$$

Now,

$$f_U(x) = f_V(x) = f_W(x) = \begin{cases} \dfrac{1}{\beta}e^{-x/\beta}, & 0 < x, \quad \text{and} \\ 0, & \text{otherwise.} \end{cases}$$

This gives that the joint probability density function for (U,V,W) is

$$f_{U,V,W}(u,v,w) = \begin{cases} \dfrac{1}{\beta^3}e^{-(u+v+w)/\beta}, & 0 < u,\ 0 < v,\ 0 < w, \quad \text{and} \\ 0, & \text{otherwise.} \end{cases}$$

Combining all this in Equation 6.13 gives

$$f_{R,S,T}(r,s,t) = \begin{cases} \dfrac{1}{\beta^3}e^{-t/\beta}, & 0 < r < s < t, \quad \text{and} \\ 0, & \text{otherwise.} \end{cases}$$

Note that R, S, and T are not independent.

In order to determine the p.d.f. for S we must integrate over R and T as follows.

$$f_S(s) = \int_s^\infty \int_0^s \frac{1}{\beta^3} e^{-t/\beta} \, dr \, dt$$

$$= \int_s^\infty \frac{1}{\beta^3} s e^{-t/\beta} \, dt$$

$$= \lim_{a \to \infty} -\frac{1}{\beta^2} s e^{t/\beta} \Big|_s^a$$

$$= \begin{cases} \dfrac{1}{\beta^2} s e^{-s/\beta}, & 0 < s, \quad \text{and} \\[2mm] 0 & , \text{ otherwise} \end{cases}$$

Thus, $S \sim G(2, \beta)$.   ∎

It is common in applications that we begin with a random vector, and we wish to determine the distribution of a function of that random vector. In order to apply this method, we need to define as many new random variables as we begin with. For example, suppose we begin with $(X_1, X_2, \ldots, X_n)$, and we want the distribution of $(Y_1, Y_2, \ldots, Y_m)$, where $m < n$. In order to use the Jacobian method, we must, in addition to $(Y_1, Y_2, \ldots, Y_m)$, define $n - m$ additional random variables. We can now use the Jacobian transformation to obtain the joint p.d.f. for $(Y_1, Y_2, \ldots, Y_n)$. We then integrate over $Y_{m+1}, Y_{m+2}, \ldots, Y_n$ to obtain the desired joint p.d.f. for $(Y_1, Y_2, \ldots, Y_m)$.

Suppose, for example, that in Example 6.23 we were only interested in determining the distribution of $T = U + V + W$. We then would have created the two other random variables $R = U$ and $S = U + V$, performed the Jacobian transformation, and then integrated with respect to R and S to obtain the desired probability density function.

Example 6.24

Suppose that X and Y are i.i.d. $G(1, \beta)$ random variables and that the distribution of $U = \frac{X}{X+Y}$ is desired. Since we have only one function of X and Y we need to come up with a second one in order to use the Jacobian method. $V = X + Y$ would seem to be a natural choice since it appears in the definition of U and also it appears in the exponent of the joint p.d.f. for (X, Y). Note here that

$$A = \{(x, y) : 0 < x, 0 < y\}, \quad \text{and}$$

$$B = \{(u, v) : 0 < u < 1, 0 < v\}.$$

Since

$$u = g_1(x, y) = \frac{x}{x + y}, \quad \text{and} \quad v = g_2(x, y) = x + y$$

we find that the inverse transformation is

$$x = h_1(u, v) = uv, \quad \text{and} \quad y = v(1 - u).$$

Therefore, the Jacobian is

$$J = \begin{vmatrix} v & u \\ -v & 1-u \end{vmatrix} = v, \quad \text{and} \quad |J| = v.$$

Now

$$f_X(t) = f_Y(t) = \begin{cases} \dfrac{1}{\beta} e^{-t/\beta}, & 0 < t, \quad \text{and} \\[2mm] 0 & , \text{ otherwise.} \end{cases}$$

Therefore, the joint p.d.f. of (X, Y) is

$$f_{X,Y}(x, y) = \begin{cases} \dfrac{1}{\beta^2} e^{-(x+y)/\beta}, & 0 < x, 0 < y, \quad \text{and} \\[2mm] 0 & , \text{ otherwise.} \end{cases}$$

An application of Equation 6.13 gives that

$$f_{U,V}(u, v) = \begin{cases} \dfrac{v}{\beta^2} e^{-v/\beta}, & 0 < u < 1, \, 0 < v, \quad \text{and} \\[2mm] 0 & , \text{ otherwise.} \end{cases}$$

From Example 6.23 we know that $V \sim G(2, \beta)$. Therefore, we see that

$$f_{U,V}(u, v) = f_U(u) f_V(v),$$

where

$$f_U(u) = \begin{cases} 1, & 0 < u < 1, \quad \text{and} \\[2mm] 0, & \text{otherwise,} \quad \text{and} \end{cases}$$

$$f_V(v) = \begin{cases} \dfrac{v}{\beta^2} e^{-v/\beta}, & 0 < u < 1, \, 0 < v, \quad \text{and} \\[2mm] 0 & , \text{ otherwise.} \end{cases}$$

Therefore, $U \sim U(0, 1)$. ■

In Chapter 9 we shall see additional examples of the application of the Jacobian method to develop the distributions of random variables that are important in statistical applications.

Before leaving this section we wish to point out that it is possible to extend this method to cover multivariate transformations that do not possess a unique inverse. The ideas are similar to those shown in Section 6.4. We shall show how this works for the case when we begin with a random vector of size 2. The generalization to vectors of arbitrary length is straightforward.

Assume that a continuous random vector (X, Y), having a joint p.d.f. of $f_{X,Y}(x, y)$, can assume values in a region (not necessarily finite) R and we wish to determine the joint p.d.f. for $(T, U) = (g_1(X, Y), g_2(X, Y))$, which is an arbitrary function from R to some region R'.

1. Fix (u, v) arbitrarily. Then determine which values of (X, Y) get mapped to this value of (u, v). Let

$$A_{u,v} = \{(x, y) : (u, v) = (g_1(x, y), g_2(x, y))\}.$$

2. Apply Equation 6.12 to transform $f_{X,Y}(x, y)$ at each value of (X, Y).
3. Obtain the joint p.d.f. for (U, V) by summing the transformed joint p.d.f.'s for all values of (X, Y) that are mapped to the same value of (U, V). That is,

$$f_{U,V}(u, v) = \sum_{A_{u,v}} f_{X,Y}\left(g_1^{-1}(u, v), g_2^{-1}(u, v)\right) |J|. \tag{6.14}$$

Example 6.25   Suppose that X and Y are i.i.d. $U(0, \theta)$ random variables. We wish to determine the joint p.d.f. of

$$T = X + Y, \quad \text{and} \quad U = \max(X, Y).$$

This is not a one-to-one transformation because it is possible for either X or Y to be the largest observed random variable. Therefore, the point (a, b) gets mapped to the same value of (u, v) as does the point (b, a). The joint p.d.f. of (X, Y) is

$$f_{X,Y}(x, y) = \frac{1}{\theta^2}, 0 \le x \le \theta, 0 \le y \le \theta.$$

From the fact that

$$t = g_1(x, y) = x + y, \quad \text{and} \quad u = g_2(x, y) = \max(x, y).$$

We have that
$$A = \{(x, y) : 0 \le x \le \theta, 0 \le y \le \theta\}, \quad \text{and}$$
$$B = \{(u, v) : u \le t \le 2u, 0 \le u \le \theta\}.$$

There are two cases.

Case 1: $x = u$, and $y = t - u$.

$$J = \begin{vmatrix} 0 & 1 \\ 1 & 0 \end{vmatrix} = -1, \quad \text{and} \quad |J| = 1.$$

Thus,

$$f_{T,U}(t, u) = \frac{1}{\theta^2}.$$

Case 2: $x = t - u$, $y = u$.

$$J = \begin{vmatrix} 1 & -1 \\ 0 & 1 \end{vmatrix} = 1, \quad \text{and} \quad |J| = 1.$$

Thus,

$$f_{T,U}(t, u) = \frac{1}{\theta^2}$$

We now sum these two to obtain that

$$f_{T,U}(t,u) = \begin{cases} \dfrac{2}{\theta^2}, & u \le t \le 2u, \ 0 \le u \le \theta, \quad \text{and} \\ 0, & \text{otherwise.} \end{cases}$$

To obtain the p.d.f. for U we integrate with respect to T as follows.

$$f_U(u) = \int_u^{2u} \frac{2}{\theta^2}\, dt$$

$$= \left. \frac{2t}{\theta^2} \right|_u^{2u}$$

$$= \frac{2u}{\theta^2}$$

Thus, in summary,

$$f_U(u) = \begin{cases} \dfrac{2u}{\theta^2}, & 0 \le u \le \theta, \quad \text{and} \\ 0, & \text{otherwise.} \end{cases}$$

∎

Exercises 6.7

1. Let X and Y be i.i.d. $U(0,1)$ random variables. Find the probability density function for
   a. $X - Y$,
   b. XY, and
   c. $\dfrac{X}{Y}$.

2. Repeat Exercise 1, parts (a) and (c) with X and Y being i.i.d. $G(1,1)$ random variables.

3. In target practice, the distance from the center of the target is measured in polar coordinates and then converted to rectangular coordinates. Let R and Θ be independent random variables with $\Theta \sim U(-\pi, \pi)$, and

$$f_R(r) = \begin{cases} r e^{-r^2/2}, & 0 \le r, \quad \text{and} \\ 0, & \text{otherwise.} \end{cases}$$

   Find the joint probability density function for $X = R \cos \Theta$, and $Y = R \sin \Theta$. Are X and Y independent?

4. A popular method, due to Box and Muller, for generating random observations from a normal distribution is to use 2 independent observations, U and V, from a $U(0,1)$ distribution and then compute

$$Z_1 = \sqrt{-2 \ln U} \, \cos(2\pi V), \quad \text{and } Z_2 = \sqrt{-2 \ln U} \, \sin(2\pi V).$$

   Show that Z_1 and Z_2 are i.i.d. $N(0,1)$ random variables.

5. Let U, V, and W be i.i.d. $G(1,1)$ random variables. Find the joint probability density function for

$$R = U + V + W, \quad S = \frac{U}{U+V}, \quad \text{and } T = \frac{U+V}{U+V+W}.$$

   Are R, S, and T independent?

6. Let U, V, and W be i.i.d. random variables each with a probability density function of

$$f(x) = \begin{cases} \dfrac{2x}{3} , 1 < x < 2, & \text{and} \\ 0 , \text{otherwise.} \end{cases}$$

Find the probability density function for $T = UVW$. (*Hint:* Let $R = V$, $S = W$, and find the joint probability density function for R, S, and T.)

7. Let X and Y be i.i.d. $N(0, 1)$ random variables. Find the joint p.d.f. of $U = \frac{X}{Y}$ and $V = |Y|$. In addition, show that U has a Cauchy distribution.

8. Let X and Y have a joint p.d.f. of

$$f_{X,Y}(x, y) = \begin{cases} 2e^{-x-y} , 0 < x < y < \infty, & \text{and} \\ 0 , \text{otherwise.} \end{cases}$$

   a. Find the joint p.d.f. for $T = X + Y$ and $U = Y$.
   b. Find the p.d.f. for T.
   c. Find the p.d.f. for U.

9. Let X_1, X_2, and X_3 have a joint p.d.f. of

$$f_{X_1,X_2,X_3}(x_1, x_2, x_3) = \begin{cases} 6e^{-x_1-x_2-x_3} , 0 < x_1 < x_2 < x_3, & \text{and} \\ 0 , \text{otherwise.} \end{cases}$$

   a. Find the joint p.d.f. for $T = X_1$, $U = X_1 + X_2$, and $V = X_1 + X_2 + X_3$.
   b. Find the p.d.f. for U.

10. Let X and Y have a joint p.d.f. of

$$f_{X,Y}(x, y) = \begin{cases} \dfrac{2}{3}(2x + y) , 0 < x < y < 1, & \text{and} \\ 0 , \text{otherwise.} \end{cases}$$

   a. Find the joint p.d.f. for $T = XY$ and $U = X$.
   b. Find the p.d.f. for T.

Chapter Summary

Discrete Transformations: Let $Y = g(X)$. Then

$$A_j = \{x_i : y_j = g(x_i)\}, \quad \text{and}$$

$$p_Y(y_j) = \sum_{A_j} p_X(x).$$

(Section 6.2)

Continuous Transformations: Let $Y = g(X)$.   (Section 6.3)

Distribution Function Method: $F_X(x)$ known. Begin with

$$F_Y(y) = P(Y \le y).$$

Rewrite in terms of an equivalent set of values of X. Evaluate this using the known $F_X(x)$. Differentiate to obtain $f_X(x)$.   **(Section 6.3)**

Transformation of Variables Method: $f_X(x)$ known. If $Y = g(X)$ is one to one, then

$$f_Y(y) = f_X\left[g^{-1}(y)\right] \left|\frac{d}{dy} g^{-1}(y)\right|.$$

If $Y = g(X)$ is not one to one, let

$$A_y = \{x : y = g(x)\}, \quad \text{and}$$

$$f_Y(y) = \sum_{A_y} f_X\left[g^{-1}(y)\right] \left|\frac{d}{dy} g^{-1}(y)\right|.$$ **(Section 6.4)**

Moment Generating Function Method: $M_X(t)$ exists and is known. Begin with

$$M_Y(t) = E\left[e^{tY}\right].$$

Either evaluate the moment generating function directly, or rewrite it in terms of the known moment generating function for X.   **(Section 6.5)**

Sums of Independent Random Variables: X and Y are independent random variables, and $T = X + Y$. Then

$$f_T(t) = \int_{-\infty}^{\infty} f_X(t - y) f_Y(y)\, dy.$$ **(Section 6.6)**

Functions of Several Random Variables: $f_{X_1,\dots,X_n}(x_1,\dots,x_n)$ known. Let

$$Y_1 = g_1(X_1, X_2, \dots, X_n),$$
$$Y_2 = g_2(X_1, X_2, \dots, X_n),$$
$$\vdots$$
$$Y_n = g_n(X_1, X_2, \dots, X_n)$$

be such that we have the following.

1. If $f_{X_1, X_2, \dots, X_n}(x_1, x_2, \dots, x_n) > 0$ for $(x_1, x_2, \dots, x_n) \in A$, then $g_1, g_2, \dots, g_n$ form a one-to-one transformation from $A \subseteq \mathbb{R}^n$ to $B \subseteq \mathbb{R}^n$. That is, there exist unique functions $h_1, h_2, \dots, h_n$ such that

$$x_1 = h_1(y_1, y_2, \dots, y_n),$$
$$x_2 = h_2(y_1, y_2, \dots, y_n),$$
$$\vdots$$
$$x_n = h_n(y_1, y_2, \dots, y_n)$$

for each $(y_1, y_2, \dots, y_n) \in B$.

2. Each of the functions $h_1, h_2, \ldots, h_n$ possesses a continuous partial derivative for every $(y_1, y_2, \ldots, y_n) \in B$ and the Jacobian

$$
J = \begin{vmatrix}
\dfrac{\partial h_1}{\partial y_1} & \dfrac{\partial h_1}{\partial y_2} & \cdots & \dfrac{\partial h_1}{\partial y_n} \\
\dfrac{\partial h_2}{\partial y_1} & \dfrac{\partial h_2}{\partial y_2} & \cdots & \dfrac{\partial h_2}{\partial y_n} \\
\vdots & \vdots & \ddots & \vdots \\
\dfrac{\partial h_n}{\partial y_1} & \dfrac{\partial h_n}{\partial y_2} & \cdots & \dfrac{\partial h_n}{\partial y_n}
\end{vmatrix} \neq 0.
$$

Then

$$
f_{Y_1,\ldots,Y_n}(y_1,\ldots,y_n) = f_{X_1,\ldots,X_n}[h_1(y_1,\ldots,y_n),\ldots,h_n(y_1,\ldots,y_n)]\,|J|,
$$

for all $(y_1,\ldots,y_n) \in B$, and 0, otherwise.   **(Section 6.7)**

Review Exercises

1. A consultant has determined that the number of requests per day for an item follows a Poisson distribution and that a supplier should consider opening an additional distribution facility if the mean number of requests moves from its present amount of 200 per day to 500 per day. To determine the mean number of requests, the consultant suggests that the company measure the average number of requests for 10 randomly selected days. From these data the company should conclude that the mean number of requests is at 500 per day if the computed average exceeds 400. What is the probability that, if the actual mean number of requests is still at 200 per day, the company will conclude it has reached 500 per day using this scheme? Customers must order the item in quantities of 1000 each.

2. A class in analysis of measurements is conducting an experiment where it is trying to determine the height of a building by measuring the angle subtended by the building when standing exactly 100 feet from the base of the building. If the angle measurements taken by members of the class are uniformly distributed between 29 and 31 degrees, find the probability density function for the computed height of the building.

3. In an aquarium, it has been observed that a species of seal in a 10 meter by 10 meter tank swims around the tank in a random pattern that seems to have the following joint probability density function.

$$
f_{X,Y}(x,y) = \begin{cases} \dfrac{(10-x)(10-y)}{2500} & , 0 < x < 10, \ 0 < y < 10, \quad \text{and} \\ 0 & , \text{otherwise} \end{cases}
$$

The feeding area is located at the corner with coordinates $(0,0)$. What is the probability that a seal will be more than 5 meters from the feeding area?

4. Let X and Y be i.i.d. $N(0,1)$ random variables. Find the probability density function for $T = \dfrac{|X|}{|Y|}$.

5. Let X and Y be i.i.d. $N(0,1)$ random variables. Does the random variable $T = XY$ have a chi-squared distribution?

6. Let X and Y be independent continuous random variables, and let $T = \frac{X}{Y}$. Show that the probability density function for T is

$$f_T(t) = \int_{-\infty}^{\infty} x f_X(x) f_Y(xt) \, dx.$$

$\Bigg($ *Hint:* Consider an approach similar to that used to derive the convolution of X and Y in Section 6.6. Note that $\left\{ \frac{x}{y} \le t \right\} = \{x < 0, y \ge xt\} \cup \{x > 0, y \le xt\}.\Bigg)$

7. Let X have a geometric distribution with parameter p. Find the probability function for
   a. $5X + 1$, and
   b. $(X - 1)^2$.

8. Let X have a Poisson distribution with parameter $\lambda = 2$. Let $Y = X$, if $X \le 4$, and $Y = 4$, if $X > 4$. Find the probability function for Y.

9. Let X be a positive continuous random variable with a probability density function of $f_X(x)$. Let $Y = \frac{1}{X}$, and write the probability density function for Y in terms of that for X.

10. Suppose that X and Y are i.i.d. $N(0, 1)$ random variables. Find the joint distribution of $U = X$ and $V = X^2 + Y^2$. (*Note:* This is not a one-to-one transformation.)

11. Let X and Y be i.i.d. random variables each having a p.d.f. of

$$f(t) = \begin{cases} 2(1 - t), \ 0 < t < 1, & \text{and} \\ 0 & , \text{otherwise.} \end{cases}$$

   a. Find the joint p.d.f. of $T = \max(X, Y)$ and $U = \min(X, Y)$.
   b. Find the p.d.f. for $R = T - U$.

12. Let $X_1, X_2, \ldots, X_n$ be i.i.d. $G(1, \beta)$ random variables. Show the distribution of

$$\bar{X} = \frac{1}{n} \sum_{i=1}^{n} X_i \text{ is } G(n, \beta/n)$$

   as follows:
   a. by using moment generating functions,
   b. by using convolutions and the transformation method.

13. Let X have a probability function of

$$p_X(a) = \begin{cases} \dfrac{1}{6}, \ a = -2, -1, 0, 1, 2, 3, & \text{and} \\ 0, \text{otherwise.} \end{cases}$$

   Find the probability function for $Y = X(X - 1)$.

14. Suppose that the circumference of a circle C is a random variable having a p.d.f. of

$$f_C(c) = \begin{cases} \dfrac{2}{7}(c + 1), \ 2 \le c \le 3, & \text{and} \\ 0 & , \text{otherwise.} \end{cases}$$

    a. Find the p.d.f. for the radius of the circle.

    b. Find the p.d.f. for the area of the circle.

15. Let X and Y have a joint p.d.f. of

$$f_{X,Y}(x, y) = \begin{cases} \dfrac{1}{2}xy, & 0 < x < y < 2, \quad \text{and} \\ 0, & \text{otherwise.} \end{cases}$$

Find the joint p.d.f. of $T = \dfrac{X}{Y}$ and $U = Y$.

7

Limit Theorems

Introduction

In Chapter 1 we indicated that we wanted our concept of probability to reflect the notion of long run proportions. That is, if we performed a random experiment a large number of times, then the proportion of the times that an event, say A, occurs would approach $P(A)$. This concept can be interpreted as a sequence of random variables. Let X_i take the value 1 if A occurs on trial i and 0 if not. Define Y_n, for $n = 1, 2, \ldots$, to be

$$Y_n = \frac{1}{n} \sum_{i=1}^{n} X_i,$$

which is the proportion of times that the event A occurs in n trials. If $p = P(A)$ then the idea of long run proportion implies that

$$\lim_{n \to \infty} Y_n = p.$$

Based on experience from calculus we would be inclined to say that the sequence of random variables $\{Y_1, Y_2, Y_3, \ldots\}$ *converges to p*. Following this idea we would require that the sequence of Y_n's converge to p for all points in the sample space. Unfortunately, this will not work. Suppose we toss a fair coin a large number of times and observe only tails (a 0). Then each Y_n will be equal to 0. This can happen with random variables even though the probability of such an occurrence is quite small when n gets large. Therefore, we cannot require convergence in such a strict sense. It is more appropriate to permit Y_n to deviate from p but require that the probability that Y_n differs from p by a small amount tends toward 0 in the limit as n gets large. It is our purpose in this chapter to be more precise about exactly what such a statement might mean when dealing with a sequence of random variables. We shall investigate 2 different convergence concepts. There are other types of convergence, but the 2 we will discuss are the ones that will interest us in the statistical applications to come later.

The first type of convergence we shall consider is known as *convergence in probability* or *stochastic convergence*. Historically, this type of convergence was studied first. Results involving this type of convergence are usually referred to as *weak laws of large numbers*. The term *law of large numbers* comes from the fact that we are interested in behavior as *n* becomes large. The term *weak* comes from the fact that there is a hierarchy in the convergence concepts we are considering. One type of convergence is said to be *stronger* than another if the fact that a sequence converges in one sense implies that convergence occurs in the other, but the converse is not necessarily true.

The other type of convergence that we consider is *convergence in distribution*. Here we are interested in conditions under which the distribution functions of the random variables in a sequence converge to another distribution function. These ideas give rise to the *Central Limit theorems*. We introduced some of these theorems in Chapter 5. One was the de Moivre-Laplace theorem. There, a sequence of binomial random variables converged to a normal distribution. Another was the fact that the probability function of a hypergeometric random variable approached that of a binomial random variable as the number of objects in the population became large. A final one was the fact that the probability function of a binomial random variable approached that of a Poisson random variable as *n* became large subject to the condition that $np = \lambda$. The Central Limit theorems are of practical value in that we can approximate the probability of an event by using the distribution to which the sequence converges, the so-called *limiting distribution*. In many cases, calculations using the limiting distribution are much easier to perform than are those using the exact distribution. For example, compare using the hypergeometric probability function with using the binomial probability function.

The material in this chapter is, by far, the most mathematically sophisticated of anything discussed thus far. A detailed discussion of these topics is well beyond the background level assumed for this book. The interested reader who has some training in real analysis and complex variables can consult more advanced works such as *An Introduction to Modern Probability Theory and Its Applications,* vol. 2, by William Feller.

7.2
Convergence in Probability

The notion of convergence in probability is based on the probability of a sequence of random variables becoming arbitrarily near some constant value. If this probability has a limiting value of one, then the sequence of random variables is said to *converge in probability*. A formal statement of this concept follows.

Definition 7.1

> Suppose $X_1, X_2, \ldots$ is a sequence of random variables defined on a sample space S. The sequence *converges in probability to the random variable X* if, for any $\varepsilon > 0$,
>
> $$\lim_{n \to \infty} P(|X_n - X| < \varepsilon) = 1. \qquad (7.1)$$

Notice that we are saying that a sequence of random variables converges to another random variable. If we wish to show that a sequence converges in probability to a constant, say a, we would define the random variable X such that $P(X = a) = 1$. An alternative to Equation 7.1 that is often used is to state that a sequence of random variables converges in probability to X if

$$\lim_{n \to \infty} P(|X_n - X| \geq \varepsilon) = 0. \tag{7.2}$$

This says, in effect, that if a sequence of random variables converges in probability to a random variable X, then the probability that the X_n's deviate from X by a small amount tends toward zero. It does not prohibit X_n from deviating wildly from X; it just says that the probability of that happening becomes quite small. We illustrate the definition with an example.

Example 7.1   Let $X_1, X_2, \ldots$ be a sequence of discrete random variables such that

$$p_{X_n}(a) = \begin{cases} \dfrac{1}{2n} & , a = \pm n, \\[2mm] 1 - \dfrac{1}{n} & , a = 0, \quad \text{and} \\[2mm] 0 & , \text{otherwise.} \end{cases}$$

In addition, let X be a random variable where $P(X = 0) = 1$. In order to show that the sequence converges in probability to 0, we need to verify that for any $\varepsilon > 0$

$$\lim_{n \to \infty} P(|X_n| < \varepsilon) = 1,$$

or equivalently that

$$\lim_{n \to \infty} P(|X_n| \geq \varepsilon) = 0.$$

Now, for a given value of n and any $\varepsilon < n$ we have

$$P(|X_n| \geq \varepsilon) = P(X_n = -n) + P(X_n = n) = \frac{1}{n}.$$

Now, in the limit, all values of ε will be less than n. Therefore, we have

$$\lim_{n \to \infty} P(|X_n| \geq \varepsilon) = \lim_{n \to \infty} \frac{1}{n} = 0.$$

Therefore, by Equation 7.2 we see that the sequence converges in probability to 0. This example also illustrates what was said earlier. Notice that as n increases X_n can assume values of $\pm n$, but the probability that it will do so becomes arbitrarily small. This is the idea behind convergence in probability.   ∎

Using Definition 7.1 to demonstrate or refute convergence in probability can be a difficult task. For this reason we wish to illustrate some other methods that can be used in many cases. One such technique determines if

$$\lim_{n \to \infty} E[(X_n - a)^2] = 0.$$

The net result will be that, if this expectation has a limiting value of zero, then the sequence will converge in probability to a. The method is based on a famous result known as *Chebyshev's Inequality*. We shall derive it by first considering Markov's Inequality.

Theorem 7.1

> **Markov's Inequality** Let X be random variable whose mean exists and is nonnegative (i.e., $P(X \geq 0) = 1$). Then for any $a > 0$
>
> $$P(X \geq a) \leq \frac{E[X]}{a}. \tag{7.3}$$

Proof We give a proof for the case when X is continuous with a probability density function of $f_X(x)$. The argument for the discrete case is identical if integrals are replaced by sums. The idea is to begin by computing $E[X]$, which was assumed to exist. We then use some basic facts concerning definite integrals.

$$E[X] = \int_{-\infty}^{\infty} x f_X(x)\, dx$$

$$= \int_{0}^{\infty} x f_X(x)\, dx$$

This is because of the nonnegativity of X. We now use the fact that an integral from a to b can be found by integrating from a to c and then from c to b.

$$E[X] = \int_{0}^{a} x f_X(x)\, dx + \int_{a}^{\infty} x f_X(x)\, dx$$

Since the first integrand is nonnegative between 0 and a, that integral will be greater than or equal to 0. Therefore,

$$E[X] \geq \int_{a}^{\infty} x f_X(x)\, dx.$$

Now, in the interval $[a, \infty)$ we have that $x f_X(x) \geq a f_X(x)$. Thus,

$$E[X] \geq \int_{a}^{\infty} a f_X(x)\, dx$$

$$= a \int_{a}^{\infty} f_X(x)\, dx$$

$$= a P(X \geq a).$$

Thus,

$$E[X] \geq a P(X \geq a), \quad \text{or}$$

$$P(X \geq a) \leq \frac{E[X]}{a}. \qquad \blacksquare$$

Given Markov's Inequality it becomes relatively straightforward to obtain Chebyshev's Inequality.

Theorem 7.2

> **Chebyshev's Inequality** If X is a random variable such that $E[(X - a)^2]$ exists, then for any $k > 0$
>
> $$P(|X - a| \geq k) \leq \frac{E[(X - a)^2]}{k^2}. \qquad (7.4)$$

Proof First note that the random variable $|X - a|$ is nonnegative. The other key observation is that the values of X that are in the event $\{|X - a| \geq k\}$ are precisely those that are in the event $\{(X - a)^2 \geq k^2\}$. Therefore,

$$P(|X - a| \geq k) = P[(X - a)^2 \geq k^2].$$

Now the random variable $(X - a)^2$ is also nonnegative. Therefore, in Markov's Inequality, we let X be replaced by $(X - a)^2$ and we let a be replaced by k^2. Then we have that

$$P(|X - a| \geq k) = P[(X - a)^2 \geq k^2] \leq \frac{E[(X - a)^2}{k^2}.$$

∎

An important special case of Chebyshev's Inequality occurs when $a = E[X]$. Then $E[(X - a)^2] = \text{Var}(X)$ and

$$P(|X - E[X]| \geq k) \leq \frac{\text{Var}(X)}{k^2}. \qquad (7.5)$$

In this instance, we can put an upper bound on the probability that a random variable will deviate from its mean when we know only the mean and variance. This bound may not necessarily be very close to the actual probability, but sometimes this may be the best that we can do.

Example 7.2

Suppose that $Z \sim N(0, 1)$. We wish to compare the results from Chebyshev's Inequality with the actual probabilities. Chebyshev's Inequality with $a = 0$ becomes

$$P(|Z| \geq k) \leq \frac{1}{k^2}.$$

We shall use Table 3 in Appendix A to determine the actual probabilities. For various values of k we find the following.

$$P(|Z| \geq k)$$

k	Chebyshev	Actual
1	1.000	0.3174
2	0.250	0.0456
3	0.111	0.0026

Thus, in the case of the normal distribution, the Chebyshev bounds are quite crude. ■

Example 7.3

A bowl contains 10 balls numbered $1, 2, \ldots, 10$. One ball is selected with replacement, and the process is repeated. We wish to use Chebyshev's Inequality to estimate how many draws will be needed to ensure that the proportion of 2's drawn is between 0.05 and 0.15 with probability at least 0.9. Let X be the number of 2's selected. We know that $X \sim B(n, 0.1)$. Thus, $E[X] = 0.1n$, and $\text{Var}(X) = n(0.1)(0.9) = 0.09n$. Therefore, $E\left[\frac{X}{n}\right] = 0.1$, and $\text{Var}\left(\frac{X}{n}\right) = \frac{1}{n^2}\text{Var}(X) = \frac{0.09}{n}$. We wish to determine a value of n so that

$$P\left(0.05 < \frac{X}{n} < 0.15\right) = P\left(\left|\frac{X}{n} - 0.1\right| < 0.05\right) > 0.9.$$

This is equivalent to determining n such that

$$P\left(\left|\frac{X}{n} - 0.1\right| \geq 0.05\right) \leq 0.10.$$

This is a Chebyshev's Inequality statement. Therefore, we can use Equation 7.5 with $k = 0.05$ and $\text{Var}\left(\frac{X}{n}\right) = \frac{0.09}{n}$. From this we get

$$P\left(\left|\frac{X}{n} - 0.1\right| \geq 0.05\right) \leq \frac{0.09}{0.0025n} = 0.10.$$

Therefore,

$$n = \frac{0.09}{0.00025} = 360.$$

■

We now use Chebyshev's Inequality to prove a result regarding convergence in probability that will be useful in proving a form of the Weak Law of Large Numbers as well as in later discussions. It gives a sufficient condition for convergence in probability.

Theorem 7.3

> Let $X_1, X_2, \ldots$ be a sequence of random variables such that $E[X_i] = \mu$ exists and is finite, for $i = 1, 2, \ldots$. If
>
> $$\lim_{n \to \infty} \text{Var}(X_n) = 0,$$
>
> then, for any $\varepsilon > 0$,
>
> $$\lim_{n \to \infty} P(|X_n - \mu| < \varepsilon) = 1.$$

Proof We shall use Chebyshev's Inequality to show convergence in probability. Since the members of the sequence have the same mean Chebyshev's Inequality gives that

$$P(|X_n - \mu| \geq \varepsilon) \leq \frac{\mathrm{Var}(X_n)}{\varepsilon^2}, \quad \text{or}$$

$$P(|X_n - \mu| < \varepsilon) \geq 1 - \frac{\mathrm{Var}(X_n)}{\varepsilon^2}.$$

Taking the limit as n becomes infinitely large and using the fact that the limiting variance is zero gives

$$\lim_{n \to \infty} P(|X_n - \mu| < \varepsilon) \geq 1 - \lim_{n \to \infty} \frac{\mathrm{Var}(X_n)}{\varepsilon^2} = 1.$$

∎

We now are in a position to prove one form of the *Weak Law of Large Numbers*. The idea is that an experiment is performed n times with the value of a random variable X being observed each time. The arithmetic average (called the *sample mean* in statistics) of the observed values is computed. The Weak Law of Large Numbers states that if n is large, the sample mean rarely deviates from the mean of the distribution of X (called the *population mean* in statistics) by very much. In other words, the sample mean converges in probability to the population mean.

Theorem 7.4

Weak Law of Large Numbers   Let $X_1, X_2, \ldots$ be a sequence of independent and identically distributed random variables such that $E[X_i] = \mu$ and $\mathrm{Var}(X_i) = \sigma^2$ exist and are finite. Define

$$Y_n = \frac{1}{n} \sum_{i=1}^{n} X_i.$$

Then, for any $\varepsilon > 0$,

$$\lim_{n \to \infty} P(|Y_n - \mu| < \varepsilon) = 1.$$

Proof   The idea is to use Theorem 7.3 to show convergence in probability. Since the X_i's are independent and identically distributed we see that

$$E[Y_n] = E\left[\sum_{i=1}^{n} \frac{X_i}{n}\right] = \frac{1}{n} \sum_{i=1}^{n} E[X_i] = \mu, \quad \text{and}$$

$$\mathrm{Var}(Y_n) = \mathrm{Var}\left(\sum_{i=1}^{n} \frac{X_i}{n}\right) = \frac{1}{n^2} \sum_{i=1}^{n} \mathrm{Var}(X_i) = \frac{\sigma^2}{n}.$$

Now, we note that

$$\lim_{n \to \infty} \mathrm{Var}(Y_n) = \lim_{n \to \infty} \frac{\sigma^2}{n} = 0.$$

Thus, Theorem 7.3 gives

$$\lim_{n \to \infty} P(|Y_n - \mu| < \varepsilon) \geq 1 - \lim_{n \to \infty} \frac{\sigma^2}{n \varepsilon^2} = 1.$$

∎

Remark: It is possible to prove a version of the Weak Law of Large Numbers which assumes only that $E[X]$ exists and is finite, but it is a more involved task.

It is important to note that Theorem 7.3 does not imply that, if the limiting variance is not zero, then the sequence does not converge in probability. In fact, it is possible for the limiting variance to be infinitely large and still have convergence in probability.

Example 7.4

Consider the sequence of random variables from Example 7.1. Recall that X_n had a probability function of

$$p_{X_n}(a) = \begin{cases} \dfrac{1}{2n} & , \ a = \pm n, \\ 1 - \dfrac{1}{n} & , \ a = 0, \quad \text{and} \\ 0 & , \ \text{otherwise.} \end{cases}$$

We showed in Example 7.1 that this sequence converged in probability to 0. Now,

$$E[X_n] = -n \frac{1}{2n} + 0 \left(1 - \frac{1}{n} \right) + n \frac{1}{2n} = 0.$$

Thus,

$$\text{Var}(X_n) = E[X_n^2] = (-n)^2 \frac{1}{2n} + (0)^2 \left(1 - \frac{1}{n} \right) + n^2 \frac{1}{2n} = n, \quad \text{and}$$

$$\lim_{n \to \infty} \text{Var}(X_n) = \lim_{n \to \infty} n = \infty.$$

This result is reminiscent of the so-called nth term test for infinite series. There you knew a series diverged if the limiting value of the nth term was not 0, but the converse was not necessarily true. ∎

We now give what is often called the *Bernoulli Law of Large Numbers*. It is really just a special case of the weak law of large numbers. We leave the proof as an exercise.

Theorem 7.5

Let $X_1, X_2, \ldots$ be a sequence of independent and identically distributed binomial random variables with $n = 1$ and $P(\text{success}) = p$. Define

$$Y_n = \sum_{i=1}^{n} \frac{X_i}{n}.$$

Then, for any $\varepsilon > 0$,

$$\lim_{n \to \infty} P(|Y_n - p| < \varepsilon) = 1.$$

Note that the Bernoulli Law of Large Numbers, in effect, says the following. Consider the event A in a random experiment to be a success with $P(A) = p$. Conduct the experiment many times. Then, in the long run, the proportion of the time A occurs converges in probability to $P(A)$. This shows that the long run proportion interpretation for probability is valid. This result is also what people commonly refer to as the *law of averages*.

In some cases we can determine convergence in probability of a sequence of random variables to a constant by investigating the sequence of distribution functions. This approach is justified by the following theorem. We omit the proof.

Theorem 7.6

Let $X_1, X_2, \ldots$ be a sequence of random variables with distribution functions $F_{X_1}(a), F_{X_2}(a), \ldots$, respectively. Let X be a random variable such that $P(X = c) = 1$. If

$$\lim_{n \to \infty} F_{X_n}(a) = \begin{cases} 0, & a < c, \quad \text{and} \\ 1, & c \le a, \end{cases}$$

then, for any $\varepsilon > 0$,

$$\lim_{n \to \infty} P(|X_n - X| < \varepsilon) = 1.$$

Example 7.5

Let $X_1, X_2, \ldots$ be a sequence of i.i.d. $G(1,1)$ random variables. Let $Y_n = \min(X_1, X_2, \ldots, X_n)$, for $n = 1, 2, \ldots$. We wish to use Theorem 7.6 to show that the sequence $\{Y_1, Y_2, \ldots\}$ converges in probability to 0. The d.f. for Y_n is obtained by observing that if the smallest random variable is larger than y, then they all must be. This gives

$$P(Y_n > y) = P(X_1 > y, X_2 > y, \ldots, X_n > y).$$

Since the X's are independent we have that

$$P(Y_n > y) = P(X_1 > y) P(X_2 > y) \cdots P(X_n > y).$$

Using the fact that the X's are identically distributed gives

$$P(Y_n > y) = \left(\int_y^\infty e^{-x} \, dx \right)^n = \left(e^{-y} \right)^n$$

$$= e^{-ny}$$

$$= 1 - F_{Y_n}(y).$$

Thus,

$$F_{Y_n}(y) = \begin{cases} 0 & , y < 0, \quad \text{and} \\ 1 - e^{-ny} & , 0 \leq y. \end{cases}$$

Now $\lim_{n \to \infty} e^{-ny} = 0$ for any $0 < y$. Thus, we get

$$\lim_{n \to \infty} F_{Y_n}(y) = \begin{cases} 0 & , y < 0, \quad \text{and} \\ 1 & , 0 \leq y. \end{cases}$$

Then by Theorem 7.6 the sequence $\{Y_1, Y_2, \ldots\}$ converges in probability to 0. This seems intuitively reasonable. In effect, this example shows that, if we run an experiment with a random variable that has a gamma distribution often enough, the smallest value observed becomes, on average, arbitrarily close to the smallest observable value.   ■

Another tool that is often useful in determining convergence in probability is the moment generating function. Recall that we stated that there is a one-to-one correspondence between the distribution of a random variable and its moment generating function. It also turns out to be true that, if a sequence of random variables converges in probability to another random variable and the sequence of moment generating functions converges to a moment generating function, that limiting moment generating function is the moment generating function of the limiting random variable. We hope to make this more clear after presenting a formal statement. This theorem, which we present without proof, will also form the basis of our proof of the Central Limit theorem in Section 7.3.

Theorem 7.7

Let $X_1, X_2, \ldots$ be a sequence of random variables with distribution functions $F_{X_1}(a), F_{X_2}(a), \ldots$ and moment generating functions $M_{X_1}(t), M_{X_2}(t), \ldots$, that exist for all n in an interval of the form $-h < t < h$, where $h > 0$. If there exists a distribution function $F_X(a)$ with corresponding moment generating function $M_X(t)$ that exists for $-h < -r < t < r < h$, such that

$$\lim_{n \to \infty} M_{X_n}(t) = M_X(t),$$

then

$$\lim_{n \to \infty} F_{X_n}(a) = F_X(a).$$

Example 7.6

We wish to use Theorem 7.7 to prove the relationship between the binomial and Poisson distributions we cited in Chapter 5. Let $X \sim B(N, p)$ with $\lambda = np$. The moment generating function for X_n is

$$M_{X_n}(t) = \left(1 - p + pe^t\right)^n$$

$$= \left(1 - \frac{\lambda}{n} + \frac{\lambda}{n}e^t\right)^n$$

$$= \left(1 + \frac{\lambda(e^t - 1)}{n}\right)^n$$

Recall from calculus that

$$\lim_{n \to \infty} \left(1 + \frac{x}{n}\right)^n = e^x. \tag{7.6}$$

Now taking the limit of the moment generating function as n becomes infinitely large gives

$$\lim_{n \to \infty} M_{X_n}(t) = \lim_{n \to \infty} \left(1 + \frac{\lambda(e^t - 1)}{n}\right)^n.$$

$$= e^{\lambda(e^t - 1)}.$$

This is the moment generating function of a $Po(\lambda)$ random variable. Thus, by Theorem 7.7, this sequence of binomial random variables converges to a Poisson random variable. ∎

In order to show that a sequence of random variables converges in probability to a constant c, Theorem 7.6 says that we can show that the sequence of random variables converges in probability to a random variable X such that $P(X = c) = 1$. Now Theorem 7.7 states that when the moment generating functions exist we need only show that the sequence of moment generating functions converges to $M_X(t) = e^{tc}$. When dealing with the Weak Law of Large Numbers we often begin by considering a sequence of independent and identically distributed random variables, say $X_1, X_2, \ldots$, and form another sequence $Y_1, Y_2, \ldots$, where

$$Y_n = \frac{1}{n} \sum_{i=1}^{n} X_i.$$

In applying moment generating functions, we begin by finding the moment generating function for Y_n. This is done by using Theorem 4.18 as follows.

$$M_{Y_n}(t) = E\left[e^{tY_n}\right] = E\left[e^{(t/n)\sum_{i=1}^{n} X_i}\right]$$

$$= \prod_{i=1}^{n} E\left[e^{(t/n)X_i}\right] \tag{7.7}$$

$$= \left(E\left[e^{(t/n)X_1}\right]\right)^n$$

Example 7.7   Let $X_1, X_2, \ldots$ be a sequence of i.i.d. $N(\mu, \sigma^2)$ random variables, and let

$$Y_n = \frac{1}{n} \sum_{i=1}^{n} X_i.$$

We wish to show that $Y_1, Y_2, \ldots$ converges in probability to μ. We begin by using

Equation 7.6 to obtain the moment generating function for Y_n as follows.

$$M_{Y_n}(t) = \left(E\left[e^{(t/n)X_1}\right]\right)^n$$

$$= \left(e^{(t\mu/n)+\sigma^2 t^2/2n^2}\right)^n$$

$$= e^{t\mu + \sigma^2 t^2/2n}$$

We now take the limit as n becomes infinitely large to get

$$\lim_{n\to\infty} M_{Y_n}(t) = \lim_{n\to\infty} e^{t\mu + \sigma^2 t^2/2n}$$

$$= e^{t\mu}.$$

Therefore, by Theorem 7.7, $Y_1, Y_2, \ldots$ converges in probability to μ.   ■

Sometimes it is necessary to use a slightly different form for Equation 7.5. It is usually proved in advanced calculus courses that, if $g(n)$ is a function of n such that,

$$\lim_{n\to\infty} ng(n) = 0,$$

then

$$\lim_{n\to\infty}\left[1 + \frac{x}{n} + g(n)\right]^n = e^x. \tag{7.8}$$

An example of such a function is $g(n) = n^{-2}$. We put this to work in the next example.

Example 7.8

We wish to use moment generating functions to prove the Bernoulli Law of Large Numbers. Recall that $X_1, X_2, \ldots$ is a sequence of i.i.d. B(1,p) random variables, and

$$Y_n = \frac{1}{n}\sum_{i=1}^{n} X_i.$$

Now, Equation 7.6 gives that the moment generating function for Y_n is

$$M_{Y_n}(t) = \left(E\left[e^{(t/n)X_1}\right]\right)^n$$

$$= \left(1 - p + p\,e^{t/n}\right)^n.$$

A Taylor series expansion for $e^{t/n}$ is

$$e^{t/n} = 1 + \frac{t}{n} + \sum_{i=2}^{\infty} \frac{1}{i!}\left(\frac{t}{n}\right)^i$$

$$= 1 + \frac{t}{n} + g(n),$$

where $\lim_{n\to\infty} ng(n) = 0$. If we combine these, take the limit as n becomes infinitely large, and use Equation 7.7 we find that

$$\lim_{n\to\infty} M_{Y_n}(t) = \lim_{n\to\infty} \left[1 - p + p + \frac{pt}{n} + p\, g(n) \right]^n$$
$$= e^{pt}.$$

This is the moment generating function for a random variable such that $P(Y = p) = 1$. Therefore, by Theorem 7.7, $Y_1, Y_2, \ldots$ converges in probability to p. ∎

We now present some theorems that summarize a number of properties of convergence in probability. These facts will be useful when we discuss the large sample properties of statistical procedures.

Theorem 7.8

Let $X_1, X_2, \ldots$ be a sequence of random variables that converges in probability to c, and let $g(x)$ be a function that is continuous at $x = c$. Then the sequence of random variables $g(X_1), g(X_2), \ldots$ converges in probability to $g(c)$.

Proof From the definition of continuity we know that if $g(x)$ is continuous at $x = c$, then $g(c)$ exists and $\lim_{x\to c} g(x) = g(c)$. From the definition of the limit this implies that, for any $\varepsilon > 0$, there exists a $\delta > 0$ such that

$$|g(x) - g(c)| < \varepsilon, \quad \text{whenever} \quad |x - c| < \delta.$$

From this it follows that, if $a \in \{x : |x - c| < \delta\}$, then $a \in \{x : |g(x) - g(c)| < \varepsilon\}$. Therefore,

$$\{x : |X_n - c| < \delta\} \subseteq \{x : g(X_n) - g(c)| < \varepsilon\}.$$

Therefore, by Theorem 1.7 this implies that for any $\varepsilon > 0$ there exists a $\delta > 0$ such that

$$P(|g(X_n) - g(c)| < \varepsilon) \geq P(|X_n - c| < \delta).$$

Since $X_1, X_2, \ldots$ converges in probability to c we know that for any $\delta > 0$

$$\lim_{n\to\infty} P(|X_n - c| < \delta) = 1.$$

Then, combining these two facts we see immediately that, for any $\varepsilon > 0$,

$$\lim_{n\to\infty} P(|g(X_n) - g(c)| < \varepsilon) \geq \lim_{n\to\infty} P(|X_n - c| < \delta) = 1.$$ ∎

The next theorem is a direct consequence of Theorem 7.8. We leave the proof as an exercise.

Theorem 7.9

If $X_1, X_2, \ldots$ is a sequence of random variables that converges in probability to a nonzero constant c, then the sequence of random variables $\frac{X_1}{c}, \frac{X_2}{c}, \ldots$ converges in probability to one.

Example 7.9   Let $X_1, X_2, \ldots$ be a sequence of i.i.d. $N(0, \sigma^2)$ random variables, and let

$$Y_n = \frac{1}{n} \sum_{i=1}^{n} X_i^2.$$

We wish to show that $\sqrt{Y_1}, \sqrt{Y_2}, \ldots$ converges in probability to σ. In order to do this, we will first show that $Y_1, Y_2, \ldots$ converges in probability to σ^2 and then use Theorem 7.8. These are the kinds of things that we will be doing later in Chapter 10. We use the results of Example 6.7 to obtain the distribution of X_i^2 as follows. Recall that if $Y = X^2$, then

$$f_Y(y) = \frac{1}{2\sqrt{y}} \left[f_X(\sqrt{y}) + f_X(-\sqrt{y}) \right].$$

Therefore,

$$f_{X_i^2}(t) = \frac{1}{2\sqrt{t}} \left[f_{X_i}(\sqrt{t}) + f_{X_i}(-\sqrt{t}) \right]$$

$$= \frac{1}{\sqrt{2\pi\sigma^2}\sqrt{t}} e^{-t/2\sigma^2}, 0 \le t.$$

Therefore, $X_i^2 \sim G(\frac{1}{2}, 2\sigma^2)$. We now use Theorem 4.16 and Theorem 4.18 to determine the distribution of Y_n as follows.

$$M_{Y_n}(t) = \left(E\left[e^{(t/n)X_i^2} \right] \right)^n$$

$$= \left(\frac{1}{1 - 2\sigma^2 \dfrac{t}{n}} \right)^{n/2}$$

This is the moment generating function of a $G\left(\frac{1}{n}, 2\sigma^2\right)$ random variable. We use Theorem 4.18 to obtain the moment generating function for Y_n as follows.

$$M_{Y_n}(t) = \prod_{i=1}^{n} M_{X_i^2}\left(\frac{t}{n} \right) = \left(\frac{1}{1 - \dfrac{2\sigma^2 t}{n}} \right)^{n/2}$$

Thus, Y_n has a $G\left(\frac{n}{2}, \frac{2\sigma^2}{n}\right)$ distribution. Therefore,

$$E[Y_n] = \sigma^2, \quad \text{and} \quad \text{Var}(Y_n) = \frac{2\sigma^4}{n}.$$

Since $\lim_{n \to \infty} \text{Var}(Y_n) = 0$ we use Theorem 7.3 to observe that $Y_1, Y_2, \ldots$ converges in probability to σ^2.

The fact that σ^2 is positive means that $\sqrt{z}$ is continuous at $z = \sigma^2$. Therefore, Theorem 7.8 establishes that $\sqrt{Y_1}, \sqrt{Y_2}, \ldots$ converges in probability to σ. It is not hard to see that theorems like these can save a considerable amount of work. ∎

The next theorem states, in essence, that convergence in probability has properties similar to those of ordinary limits. The proofs of the respective parts are somewhat involved and, on the whole, uninteresting. For this reason they are omitted.

Theorem 7.10

> Let $X_1, X_2, \ldots$ be a sequence of random variables that converges in probability to c and $Y_1, Y_2, \ldots$ be a sequence of random variables that converges in probability to d. Then the following are true.
>
> 1. If $c = d$, then the sequence $X_1 - Y_1, X_2 - Y_2, \ldots$ converges in probability to 0.
> 2. The sequence $X_1 + Y_1, X_2 + Y_2, \ldots$ converges in probability to $c + d$.
> 3. If a is a constant, then the sequence $aX_1, aX_2, \ldots$ converges in probability to ac.
> 4. The sequence $X_1 Y_1, X_2 Y_2, \ldots$ converges in probability to cd.
> 5. If $d \neq 0$, then the sequence $\frac{X_1}{Y_1}, \frac{X_2}{Y_2}, \ldots$ converges in probability to $\frac{c}{d}$.

Note: Part 3 is a more general version of Theorem 7.9.

Example 7.10

Suppose $X_1, X_2, \ldots$ is a sequence of i.i.d. $N(\mu, 1)$ random variables and $Y_1, Y_2, \ldots$ is a sequence of i.i.d. $N(\tau, 1)$ random variables that are independent of the X's. Let

$$\bar{X}_n = \frac{1}{n} \sum_{i=1}^{n} X_i, \quad \text{and} \quad \bar{Y}_n = \frac{1}{n} \sum_{i=1}^{n} Y_i, \quad \text{for } n = 1, 2, \ldots.$$

We wish to show that the sequence $\bar{X}_1 \bar{Y}_1, \bar{X}_2 \bar{Y}_2, \ldots$ converges in probability to $\mu\tau$. Theorem 7.10 makes this a relatively simple task. To begin we use Theorem 7.3 to see that the sequence $\bar{X}_1, \bar{X}_2, \ldots$ converges in probability to μ and the sequence $\bar{Y}_1, \bar{Y}_2, \ldots$ converges in probability to τ. Now, part 4 of Theorem 7.10 tells us immediately that the sequence $\bar{X}_1 \bar{Y}_1, \bar{X}_2 \bar{Y}_2, \ldots$ converges in probability $\mu\tau$. This would be a hard problem to complete without the aid of this theorem. ∎

Exercises 7.2

1. The number of students enrolling in the freshman class at a small college is a random variable with a mean of 450. Using Markov's Inequality what can be said about the probability that the next class of freshmen will contain at least 475 students?

2. The number of transactions handled by a bank teller in a day is a random variable with a mean of 80 and a standard deviation of 5.

   a. What can be said about the probability that the teller will handle at least 100 transactions in a day?

   b. What can be said about the probability that the teller will handle between 70 and 90 transactions in a day?

3. Let $X \sim Po(\lambda)$. Use Chebyshev's Inequality to verify the following.

a. $P\left(\dfrac{\lambda}{2} < X < \dfrac{3\lambda}{2}\right) \geq \dfrac{\lambda - 4}{\lambda}$

b. $P(X \geq 2\lambda) \leq \dfrac{1}{4}\left(1 + \dfrac{1}{\lambda}\right)$

4. Determine whether or not the following sequences of independent random variables converges in probability.

a.
$$p_{X_n}(a) = \begin{cases} \dfrac{1}{n} & , a = 1, \\ \dfrac{n-1}{n} & , a = 0, \quad \text{and} \\ 0 & , \text{otherwise} \end{cases}$$

b.
$$p_{X_n}(a) = \begin{cases} \dfrac{1}{3} & , a = -2^n, 0, 2^n, \quad \text{and} \\ 0 & , \text{otherwise} \end{cases}$$

c.
$$p_{X_n}(a) = \begin{cases} \dfrac{1}{\sqrt{n}} & , a = -n, n, \\ 1 - \dfrac{2}{\sqrt{n}} & , a = 0, \quad \text{and} \\ 0 & , \text{otherwise} \end{cases}$$

d.
$$f_{X_n}(x) = \begin{cases} n, & 0 < x < \dfrac{1}{n}, \quad \text{and} \\ 0, & \text{otherwise} \end{cases}$$

5. Give an example of a sequence of random variables that converges in probability to zero but whose limiting variance is finite but not equal to zero.

6. Prove that if a random variable X is such that $\text{Var}(X) = 0$, then $P(X = E[X]) = 1$.

7. Use Theorem 7.4 to prove Theorem 7.5.

8. Let $X_1, X_2, \ldots$ be a sequence of i.i.d. $G(1, \beta)$ random variables. Let

$$Y_n = \frac{1}{n}\sum_{i=1}^{n} X_i, \quad \text{for } n = 1, 2, \ldots.$$

Show that $Y_1, Y_2, \ldots$ converges in probability to β by using

a. Chebyshev's Inequality, and

b. moment generating functions.

9. Let $X_1, X_2, \ldots$ be a sequence of i.i.d. geometric random variables with parameter p. Let

$$Y_n = \frac{n}{\sum_{i=1}^{n} X_i}, \quad \text{for } n = 1, 2, \ldots.$$

Show that $Y_1, Y_2, \ldots$ converges in probability to p.

10. Let $X_1, X_2, \ldots$ be a sequence of i.i.d. $Po(\lambda)$ random variables. Let

$$Y_n = \frac{1}{n} \sum_{i=1}^{n} X_i, \quad \text{for } n = 1, 2, \ldots.$$

Show that $Y_1, Y_2, \ldots$ converges in probability to λ by using

a. Chebyshev's Inequality, and

b. moment generating functions.

11. Prove Theorem 7.9.

12. Let $X_1, X_2, \ldots$ be a sequence of independent random variables such that $E[X_i] = \mu_i$, and $\text{Var}(X_i) = \sigma_i^2$, such that μ_i and σ_i^2 are finite for $i = 1, 2, \ldots$. Let

$$Y_n = \frac{1}{n} \sum_{i=1}^{n} X_i, \quad \text{for } n = 1, 2, \ldots.$$

Show that, if

$$\lim_{n \to \infty} \sum_{i=1}^{n} \frac{\sigma_i^2}{n^2} = 0,$$

then, for any $\varepsilon > 0$,

$$\lim_{n \to \infty} P \left(\left| Y_n - \frac{\mu_1 + \mu_2 + \cdots + \mu_n}{n} \right| \geq \varepsilon \right) = 0.$$

13. Let $X_1, X_2, \ldots$ be i.i.d. $U(0, 1)$ random variables. Let $Y_n = \max(X_1, X_2, \ldots, X_n)$.

a. Show that

$$F_{Y_n}(y) = \begin{cases} 0, & y < 0, \\ y^n, & 0 \leq y < 1, \quad \text{and} \\ 1, & 1 \leq y. \end{cases}$$

b. Show that $Y_1, Y_2, \ldots$ converges in probability to 1 by using distribution functions.

c. Show that $Y_1, Y_2, \ldots$ converges in probability to 1 by using Chebyshev's Inequality.

14. Show that the sequence of random variables in Example 1 converges in probability to zero by using distribution functions.

15. Let $X_1, X_2, \ldots$ be i.i.d. random variables each having a p.d.f. of

$$f(t) = \begin{cases} 3t^2, & 0 \leq t \leq 1, \quad \text{and} \\ 0, & \text{otherwise,} \end{cases}$$

and let $Y_1, Y_2, \ldots$ be such that

$$Y_n = \frac{1}{n} \sum_{i=1}^{n} X_i.$$

Determine whether or not $Y_1, Y_2, \ldots$ converges in probability to anything and, if so, to what.

7.3
Central Limit
Theorem

In Chapter 5 we discussed a number of results where, in the limit, a sequence of distribution functions tended toward another distribution function. Recall that as the total number of objects from which one sampled without replacement became infinitely large, the hypergeometric probability function tended toward a binomial probability function. Also, in an experiment with a binomial random variable, as n became infinitely large with np being held constant the probability function approached that of a Poisson random variable with $\lambda = np$. We also stated without proof the *de Moivre-Laplace theorem* which showed that, as the number of trials becomes infinitely large, the distribution function of a binomial random variable can be approximated by using the distribution function of a normal random variable. These are all examples of *convergence in distribution* of a sequence of random variables. The main idea is that the distribution functions of the random variables in the sequence converge to the distribution function of another random variable. This motivates the following definition.

Definition 7.2

> Let $X_1, X_2, \ldots$ be a sequence of random variables having distribution functions $F_{X_1}(a), F_{X_2}(a), \ldots$, respectively, and let X be a random variable with a distribution function $F_X(a)$. The sequence of random variables is said to *converge in distribution to* X if,
>
> $$\lim_{n \to \infty} F_{X_n}(a) = F_X(a), \qquad (7.9)$$
>
> for all a such that $F_X(x)$ is continuous at $x = a$.

Sometimes the term *converges in law* is used. Convergence in distribution is the weaker of the two types of convergence under discussion. This means that, if a sequence of random variables converges in probability to another random variable, it will also converge in distribution to that same random variable. However, a sequence of random variables can converge in distribution to a random variable and not converge in probability to anything. Convergence in probability has to do with the values of the random variables in the sequence while convergence in distribution means that the sequence of probabilities of events converges to those of a certain random variable.

One of the most important results regarding convergence in distribution is the so-called *Central Limit theorem*. In general terms, this theorem states that the sum of many independent random variables has a distribution that is approximately normal.

Example 7.11

We wish to demonstrate the central limit effect. Consider a set of i.i.d. $G(1, 1)$ random variables, $X_1, X_2, \ldots$. We wish to find the distribution of

$$Y_n = \sum_{i=1}^{n} X_i.$$

Using moment generating functions we see that the moment generating function of Y_n is

$$M_{Y_n}(t) = \left[M_{X_1}(t)\right]^n$$

$$= \left(\frac{1}{1-t}\right)^n.$$

Therefore, Y_n has a $G(n, 1)$ distribution. This means that the p.d.f. for Y_n is

$$f_{Y_n}(y) = \begin{cases} \dfrac{y^{n-1}}{(n-1)!}e^{-y}, & 0 < y, \quad \text{and} \\[2mm] 0 & , \text{ otherwise.} \end{cases}$$

Figure 7.1 shows the graph of this p.d.f. when $n = 1$, 5, and 10.

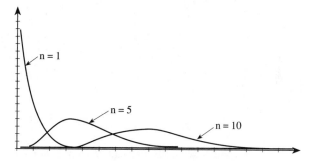

Figure 7.1

Note that as n increases the graph of the p.d.f. is "stretched out" and becomes more bell-shaped in appearance. ∎

For a long time experimenters had noted that their observations seemed to behave in a manner similar to that shown in Example 7.11. That is, as the number of observations increased the distribution of their sums more closely approximated the bell-shaped pattern of a normal distribution. The Central Limit theorem says, in essence, that the experimental evidence is not coincidental. We give one of many versions of the Central Limit theorem.

Theorem 7.11

Central Limit Theorem   Let $X_1, X_2, \ldots$ be a sequence of independent and identically distributed random variables. Let $E[X_i] = \mu$ and $\text{Var}(X_i) = \sigma^2$ exist and be finite. Define

$$Y_n = \frac{\sum_{i=1}^n X_i - n\mu}{\sigma\sqrt{n}} = \sqrt{n}\,\frac{\bar{X} - \mu}{\sigma}.$$

Let $F_{Y_n}(a)$ be the distribution function for Y_n. Then, for every a,

$$\lim_{n\to\infty} F_{Y_n}(a) = \int_{-\infty}^{a} \frac{1}{\sqrt{2\pi}}e^{-z^2/2}\,dz. \tag{7.10}$$

Proof We shall prove the theorem for the special case that $M_{X_i}(t)$ exists. Note that this requires that every moment of X_i exist. We begin by computing the moment generating function for Y_n. Again we use Theorem 4.18.

$$M_{Y_n}(t) = E\left[e^{tY_n}\right]$$

$$= E\left[e^{t(\Sigma_{i=1}^n X_i - n\mu)/\sigma\sqrt{n}}\right]$$

$$= e^{-\sqrt{n}t\mu/\sigma}\left(E\left[e^{tX_i/\sigma\sqrt{n}}\right]\right)^n$$

$$= \left(E\left[e^{t(X_i-\mu)/\sigma\sqrt{n}}\right]\right)^n$$

We now expand this by using a MacLaurin series and collect terms that go to 0 faster than does $\frac{1}{n}$. Call these $g(n)$.

$$M_{Y_n}(t) = \left(1 + \frac{t}{\sigma\sqrt{n}}E[X_i - \mu] + \frac{t^2}{2\sigma^2 n}E[(X_i - \mu)^2]\right.$$

$$\left. + \frac{t^3}{3!\sigma^3(\sqrt{n})^3}E[(X_i - \mu)^3] + \cdots\right)^n$$

$$= \left[1 + \frac{t^2}{2n} + g(n)\right]^n$$

We now use Equation 7.7 to find that

$$\lim_{n\to\infty} M_{Y_n}(t) = \lim_{n\to\infty}\left[1 + \frac{t^2}{2n} + g(n)\right]^n = e^{t^2/2}.$$

Since the limiting moment generating function for Y_n is that of a $N(0, 1)$ random variable we have shown that

$$\lim_{n\to\infty} F_{Y_n}(a) = \Phi(a).$$

∎

The proof of this theorem for the general case is done using the *characteristic function*, $E\left[e^{itX}\right]$, where $i^2 = -1$. The characteristic function exists for any random variable. A complete discussion of this function unfortunately requires some knowledge of complex variables, which is beyond the level assumed for this book.

Sequences of random variables that obey the Central Limit theorem are said to be *asymptotically normal*, and it is common practice to say that the sequence $X_1, \frac{1}{2}(X_1 + X_2), \frac{1}{3}(X_1 + X_2 + X_3), \ldots$ is *asymptotically $N(\mu, \frac{\sigma^2}{n})$*. We should point out that there are many forms of the Central Limit theorem. They each have certain advantages and disadvantages. By this we mean that each version of the theorem places restrictions on certain aspects of the distribution of the members of the sequence while relaxing others. For example, the theorem we are using requires that the members of the sequence be independent and identically distributed with each having a finite mean and variance. Another theorem due to Liapunov does not require that the members of the sequence be identically distributed, but in return places restrictions on the third moments of the distribution. Recall that the de Moivre-Laplace theorem requires not only that the members of the sequence be independent and identically distributed, but it further requires that the distribution be binomial. For the remainder of this book we shall use the term "Central Limit theorem" to refer to Theorem 7.11.

In Section 5.7 we stated without proof the de Moivre-Laplace theorem. We will now show that it is a special case of the Central Limit theorem. This theorem can be proven directly by using moment generating functions. The proof in that manner is left as an exercise.

Corollary 7.1

> **de Moivre-Laplace Theorem**   Let X_n have a binomial distribution with parameters n and p. Then
>
> $$\lim_{n \to \infty} P\left(\frac{X_n - np}{\sqrt{np(1-p)}} \leq a\right) = \Phi(a).$$

Proof   In section 5.3 we showed that if $Y_1, Y_2, \ldots, Y_n$ are independent and identically distributed $B(1, p)$ random variables, then

$$X_n = \sum_{i=1}^{n} X_i$$

will have a $B(n, p)$ distribution. Therefore, $E[X_n] = np < \infty$, and $\text{Var}(X) = np(1-p) < \infty$. This means that the sequence $Z_1, Z_2, \ldots$, where

$$Z_n = \frac{X_n - np}{\sqrt{np(1-p)}},$$

satisfies the requirements of the Central Limit theorem. Therefore, Theorem 7.11 shows that

$$\lim_{n \to \infty} P\left(\frac{X_n - np}{\sqrt{np(1-p)}} \leq a\right) = \lim_{n \to \infty} F_{Z_n}(a) = \Phi(a).$$

■

The Central Limit theorem can be of enormous practical value. It allows us to calculate probabilities for sums of a large number of independent and identically distributed random variables without having to know anything about the random variables except their mean and variance. In addition, computations are reduced to consulting tables of the normal distribution function. This avoids problems of having to evaluate possibly complicated sums or integrals. From Example 7.11 recall that the distribution function of a $G(10, 1)$ random variable requires that we integrate by parts *10* times.

An obvious question is how large n needs to be before the Central Limit theorem gives useful approximations. We have already mentioned that the de Moivre-Laplace theorem has been found to be appropriate when both np and $n(1 - p)$ are at least 5. In addition, researchers have found that a good rule of thumb for using the normal approximation in most cases is that n should be at least 30. This is not very large. It has also been discovered that convergence is faster if the distribution of the X's is relatively symmetric. For example, sums of uniformly distributed random variables may be approximated by a normal distribution for values of n as small as 6. In fact, an early computer algorithm for generating normally distributed random observations did exactly that. A uniform random number generator was run 6 times and the 6 values were summed and then transformed to have a mean of 0 and a variance of 1. The rule of thumb of 30 is safe for all but the most extremely skewed distributions.

Example 7.12

In golf it is a common practice to pace off the distance from a landmark to one's ball in order to determine the distance that the ball must be hit on the next shot. Suppose that a golfer's stride is randomly distributed with a mean length of 0.98 yards and a standard deviation of 0.1 yards. This golfer counts 70 paces from a marker to her ball. We wish to find the probability that the distance she measured is within 2 yards of 70 yards.

With no information regarding the shape of the distribution of the golfer's stride we are unable to use the methods of Chapter 6 to compute the exact probability of this event. However, we can view the 70 paces as being a set of i.i.d. continuous random variables $X_1, X_2, \ldots, X_{70}$ each having $\mu = 0.98$ and $\sigma = 0.1$. The total distance paced will be

$$Y_{70} = \sum_{i=1}^{70} X_i.$$

We wish to find $P(68 \leq Y_{70} \leq 72)$. Since $n = 70$, which is well above 30, we shall use the Central Limit theorem to approximate the probability as follows.

$$P(68 \leq Y_n \leq 72) = P\left(\frac{68 - (70)(0.98)}{(0.1)\sqrt{70}} \leq \frac{Y_n - n\mu}{\sigma\sqrt{n}} \leq \frac{72 - (70)(0.98)}{(0.1)\sqrt{70}}\right)$$

$$= P(-0.717 \leq Z_n \leq 4.064)$$

$$\approx \Phi(4.063) - \Phi(-0.717)$$

$$= \Phi(4.063) - [1 - \Phi(0.717)]$$

$$= 0.7642$$

Example 7.13

A portable radio uses a type of battery that, after installation, has a lifetime with a mean of 8 hours and a standard deviation of 8 hours. The radio is tested by running it continuously for 30 days with batteries being replaced immediately after they run down. We wish to find the probability that more than 100 batteries will be needed to test the radio.

Let X_i be the lifetime, in hours, of the ith battery. This means that, for the 100 batteries, $X_1, X_2, \ldots, X_{100}$ are i.i.d. random variables. If we define

$$Y_{100} = \sum_{i=1}^{100} X_i,$$

then we wish to determine $P(Y_{100} < 720)$. It might be reasonable to assume that the lifetimes of the batteries follow an exponential distribution with a mean of 8 hours. If we were to do this it would mean that Y_{100} would have a $G(100, 8)$ distribution. In such a case, the exact probability would be computed by

$$P(Y_{100} < 720) = \int_0^{720} \frac{x^{99}}{\Gamma(100)\, 8^{100}} e^{-x/8} \, dx.$$

Integrating by parts as many times as this would require is not a very appealing prospect. However, since Y_{100} is the sum of 100 i.i.d random variables each with $\mu = \sigma = 8$ we can approximate the desired probability by using the Central Limit theorem. Thus,

$$P(Y_{100} < 720) = P\left(\frac{Y_{100} - 100(8)}{8\sqrt{100}} < \frac{720 - 100(8)}{8\sqrt{100}}\right)$$

$$= P(Z_{100} < -1.00)$$

$$\approx \Phi(-1.00)$$

$$= 1 - \Phi(1.00)$$

$$= 0.1587.$$

As an illustration of how accurate the normal approximation can be, see Figure 7.2, which shows the plotted probability density functions for a $G(100, 8)$ random variable and for a $N(800, 6400)$ random variable. As can be seen the 2 curves are virtually indistinguishable.

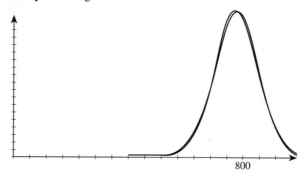

Figure 7.2

Exercises 7.3

1. A dairy ships cheddar cheese in bulk units of approximately 100 lbs apiece. Each unit is a random variable with a mean weight of 100 lbs and a standard deviation of 5 pounds. A shipment consists of 50 units of cheese. Use the Central Limit theorem to approximate the probability that the total weight of the shipment will be between 4950 pounds and 5050 pounds.

2. Economic data for use in a report are rounded to the nearest dollar before being tabulated. These rounded data are then summed to give an overall total. Assume that rounding follows a $U(-\frac{1}{2}, \frac{1}{2})$ distribution. If we add 100 rounded numbers, what is the probability that the rounding errors will be no more than \$4 in either direction?

3. Let $X_1, X_2, \ldots$ be a sequence of i.i.d. random variables each having a p.d.f. of

$$f(x) = \begin{cases} 3x^2, & 0 \le x \le 1, \quad \text{and} \\ 0, & \text{otherwise.} \end{cases}$$

Let $T_{30} = X_1 + X_2 + \cdots + X_{30}$. Use the Central Limit theorem to approximate $P(20 \le T_{30})$.

4. A fair die is rolled repeatedly. Let X_n be the sum of the number of dots in n rolls. Use the Central Limit theorem to determine the number of rolls necessary to ensure that

$$P\left(3.4 \le \frac{X_n}{n} \le 3.6\right) \ge 0.95.$$

5. A drug for controlling high blood pressure has been found to be effective in 75 out of 100 cases. A pharmaceutical company claims to have a more effective drug by citing results from a test wherein high blood pressure was effectively controlled for 80 out of 100 patients receiving the new drug. What do you think of this claim? (*Hint:* What would be the probability of the new drug being effective 80 or more times out of 100 if, in fact, it was no better than the old drug?).

6. A bowl contains 5 chips numbered 1, 2, 3, 4, and 5, respectively. Chips are drawn 100 times with replacement and the number noted.

   a. What is the approximate probability that the sum of these numbers will be at least 350?

   b. What is the approximate probability that the sum will be exactly 350?

7. Let X_n be a chi-squared random variable with n degrees of freedom, for $n = 1, 2, \ldots$. Show that, if

$$Y_n = \frac{X_n - n}{\sqrt{2n}},$$

then

$$\lim_{n \to \infty} F_{Y_n}(a) = \Phi(a).$$

8. Use moment generating functions directly to show that, if $X_1, X_2, \ldots$ is a sequence of i.i.d. $Po(\lambda)$ random variables and $Y_1, Y_2, \ldots$ is defined by

$$Y_n = \sqrt{n} \, \frac{\frac{1}{n}\sum_{i=1}^{n} X_i - \lambda}{\sqrt{\lambda}},$$

then

$$\lim_{n \to \infty} M_{Y_n}(t) = e^{t^2/2}.$$

9. Let $X_1, X_2, \ldots$ be a sequence of Poisson random variables where $X_n \sim Po(n)$. Show that

$$\lim_{n \to \infty} P\left(\frac{X_n - n}{\sqrt{n}} \leq a\right) = \Phi(a).$$

10. Let $X_1, X_2, \ldots, X_{100}$ be i.i.d. random variables each having a p.d.f. of

$$f(t) = \begin{cases} 3(1 - t)^2, & 0 \leq t \leq 1, \quad \text{and} \\ 0 & , \text{ otherwise.} \end{cases}$$

Find the (approximate) probability that $X_1 + X_2 + \cdots + X_{100}$ will be less than 40.

11. The time that a customer spends being serviced at a state lottery window follows an exponential distribution with a mean time of 2 minutes. What is the probability that the average time spent at the lottery window by 64 customers exceeds 3 minutes?

12. The number of calls arriving at a particular 800 number follows a Poisson distribution with an average of 3 calls per minute. What is the probability that there will be more than 200 calls arriving in an hour?

13. Use moment generating functions directly to prove the de Moivre-Laplace theorem.

14. A common rule of thumb for measuring one yard is to use the distance from your nose to your outstretched hand. Suppose that a particular individual measures distance in this way and that the measurements are a random variable with a mean distance of 37 inches with a standard deviation of 0.5 inch. Fifty yards of fishing line are measured using this method. Find the probability that the amount of line measured will be within 1 yard of 50 yards.

15. A certain medication has been found to produce unpleasant side effects in 0.1% of the patients who use it. What is the probability that, in a sample of 1000 patients on the drug, more than 0.15% will have had unpleasant side effects?

7.4 ═══
More on Convergence in Distribution

We conclude this discussion of convergence in distribution with a result due to Cramér that will be useful in developing some statistical procedures in large sample cases. The proof is outlined in Cramér's book, *Mathematical Methods of Statistics*. It serves as a useful way to tie together the two types of convergence we have been discussing.

Theorem 7.12

Let $X_1, X_2, \ldots$ be a sequence of random variables that converges in distribution of the random variable X. Let $Y_1, Y_2, \ldots$ be another sequence of random variables that converges in probability to a constant c. Define the following sequences of random variables.

1. $T_1, T_2, \ldots$, where $T_n = X_n + Y_n$.
2. $U_1, U_2, \ldots$, where $U_n = X_n Y_n$.
3. $V_1, V_2, \ldots$, where $V_n = \frac{X_n}{Y_n}$.

Then, the following are true.

1.

$$\lim_{n \to \infty} F_{T_n}(a) = F_X(a - c)$$

2.

$$\lim_{n \to \infty} F_{U_n}(a) = \begin{cases} F_X\left(\dfrac{a}{c}\right) & , c > 0, \quad \text{and} \\ 1 - F_X\left(\dfrac{a}{c}\right), & 0 < c \end{cases}$$

3.

$$\lim_{n \to \infty} F_{V_n}(a) = \begin{cases} F_X(ac) & , c > 0, \quad \text{and} \\ 1 - F_X(ac), & 0 < c \end{cases}$$

This theorem is not as imposing as it first appears. In essence, it states the following. We have a situation where $X_1, X_2, \ldots$ converges in distribution to X, and $Y_1, Y_2, \ldots$ converges in probability to c. Then $X_1 + Y_1, X_2 + Y_2, \ldots$ converges in distribution to the random variable $X + c$, $X_1 Y_1, X_2 Y_2, \ldots$ converges in distribution to the random variable cX if $c \neq 0$, and $\frac{X_1}{Y_1}, \frac{X_2}{Y_2}, \ldots$ converges to the random variable $\frac{X}{c}$ if $c \neq 0$. We illustrate the application of Theorem 7.12 with some examples.

Example 7.14   Let $X_1, X_2, \ldots$ be a sequence of i.i.d. random variables such that $E[X_i] = \mu$, $\text{Var}(X_i) = \sigma^2$, and $E\left[(X_i - \mu)^4\right]$ exists. Define $\bar{X}_1, \bar{X}_2, \ldots$ to be a sequence of *sample means*, where

$$\bar{X}_n = \frac{1}{n} \sum_{i=1}^{n} X_i,$$

and $S_1^2, S_2^2, \ldots$ to be a sequence consisting of terms of the form

$$S_n^2 = \frac{1}{n} \sum_{i=1}^{n} (X_i - \mu)^2.$$

Let $Z_1, Z_2, \ldots$ be a sequence where

$$Z_n = \sqrt{n} \frac{\bar{X} - \mu}{S_n}.$$

We wish to determine the limiting distribution of this sequence. First, note that we can rewrite Z_n as follows.

$$Z_n = \sqrt{n} \frac{\bar{X} - \mu}{S_n}$$

$$= \frac{\sum_{i=1}^{n} X_i - n\mu}{\sqrt{n S_n^2}}$$

$$= \frac{\frac{\sum_{i=1}^{n} X_i - n\mu}{\sigma \sqrt{n}}}{\sqrt{S_n^2 / \sigma^2}}$$

Let

$$T_n = \frac{\sum_{i=1}^{n} X_i - n\mu}{\sigma \sqrt{n}}.$$

Since $X_1, X_2, \ldots$ is a sequence of i.i.d. random variables with finite mean and variance, we can use the Central Limit theorem to see that $T_1, T_2, \ldots$ converges in distribution to a $N(0, 1)$ random variable. That is,

$$\lim_{n \to \infty} F_{T_n}(a) = \Phi(a).$$

If we can show that the sequence $S_1^2, S_2^2, \ldots$ converges in probability to a nonzero constant, then Theorem 7.12 can be used. To determine what, if anything, this sequence converges to in probability, proceed as follows. First, note that

$$E[S_n^2] = \frac{1}{n} \sum_{i=1}^{n} E\left[(X_i - \mu)^2\right] = \sigma^2, \quad \text{and}$$

$$\mathrm{Var}(S_n^2) = \frac{1}{n} E\left[(X_i - \mu)^4\right].$$

Now it is an immediate consequence of the Weak Law of Large Numbers that $S_1^2, S_2^2, \ldots$ converges in probability to $E\left[(X_i - \mu)^2\right] = \sigma^2$. Now the function $\sqrt{x}$ is continuous at $x = \sigma^2$. Therefore, Theorems 7.8 and 7.9 show that the sequence $\sqrt{\frac{S_1^2}{\sigma^2}}, \sqrt{\frac{S_2^2}{\sigma^2}}, \ldots$ converges in probability to 1. Thus, the sequence $Z_1, Z_2, \ldots$ is of the form

$$Z_n = \frac{T_n}{\sqrt{\frac{S_n^2}{\sigma^2}}},$$

where the numerator sequence converges in distribution to a $N(0, 1)$ random variable and the denominator sequence converges in probability to 1. Then Part 3 of Theorem 7.12 gives that

$$\lim_{n \to \infty} F_{Z_n}(a) = \Phi(1 \cdot a) = \Phi(a).$$

Thus, $Z_1, Z_2, \ldots$ converges in distribution to a $N(0, 1)$ random variable. As we said, this example does an excellent job of making use of most of the ideas covered in this chapter. ∎

Example 7.15   Suppose the sequence $X_1, X_2, \ldots$ is asymptotically $N(\mu, \sigma^2 / n)$ and the sequence $Y_1, Y_2, \ldots$ is asymptotically $N(\eta, \tau / n)$. We wish to determine the asymptotic distribution of $Z_1, Z_2, \ldots$, where

$$Z_n = \sqrt{n} \frac{X_n - \mu}{\sqrt{Y_n}}.$$

We know from the asymptotic normality of $X_1, X_2, \ldots$ that

$$\lim_{n \to \infty} P\left(\sqrt{n}\,\frac{X_n - \mu}{\sigma} \leq a\right) = \Phi(a).$$

Therefore, the sequence $(X_1 - \mu), \sqrt{2}\,(X_2 - \mu), \sqrt{3}\,(X_3 - \mu), \ldots$ is asymptotically $N(0, \sigma^2)$. We can use Theorem 7.4 to see that, since

$$\lim_{n \to \infty} \mathrm{Var}(Y_n) = \lim_{n \to \infty} \frac{\tau^2}{n} = 0,$$

$Y_1, Y_2, \ldots$ converges in probability to η. Then Part 3 of Theorem 7.12 gives that $Z_1, Z_2, \ldots$ converges in distribution to a random variable whose distribution function is that of a $N(0, \sigma^2)$ random variable evaluated at ηa. In other words,

$$\lim_{n \to \infty} F_{Z_n}(a) = \int_{-\infty}^{\eta a} \frac{1}{\sqrt{2\pi\sigma^2}}\, e^{-x^2/2\sigma^2}\, dx.$$

Let $y = \eta x$.

$$= \int_{-\infty}^{a} \sqrt{\frac{\eta^2}{2\pi\sigma^2}}\, e^{-\eta^2 y^2/2\sigma^2}\, dy$$

Since this is the d.f. of a $N(0, \sigma^2 / \eta^2)$ random variable, we find that $Z_1, Z_2, \ldots$ has an asymptotic $N(0, \sigma^2 / \eta^2)$ distribution. ∎

Exercises 7.4

1. Let $X_1, X_2, \ldots$ be a sequence of binomial random variables, where $X_n \sim B(n, p)$, and define $Y_1, Y_2, \ldots$ by

$$Y_n = \sqrt{n}\,\frac{\frac{X_n}{n} - p}{\sqrt{\frac{X_n}{n}\left(1 - \frac{X_n}{n}\right)}}.$$

Show that

$$\lim_{n \to \infty} P(Y_n \leq a) = \Phi(a).$$

2. Let $X_1, X_2, \ldots$ be a sequence of i.i.d. random variables each of which has a mean of 0, a variance of 1, $E[X^3] = 0$, and $E[X^4] = 2$. Let $Y_1, Y_2, \ldots$ be defined by

$$Y_n = \frac{X_1 + X_2 + \cdots + X_n}{\sqrt{X_1^2 + X_2^2 + \cdots + X_n^2}}.$$

Show that the limiting distribution of $Y_1, Y_2, \ldots$ is $N(0, 1)$.

3. Let $X_1, X_2, \ldots$ be a sequence of i.i.f. $G(1, \beta)$ random variables, and define $Y_1, Y_2, \ldots$ by

$$Y_n = \sqrt{n}\,\frac{\bar{X}_n - \beta}{\sqrt{\bar{X}_n^2}}, \quad \text{where } \bar{X}_n = \frac{1}{n}\sum_{i=1}^{n} X_i.$$

Show that

$$\lim_{n \to \infty} P(Y_n \leq a) = \Phi(a).$$

4. Let $X_1, X_2, \ldots$ be a sequence of i.i.d. random variables each having a p.d.f. of

$$f(t) = \begin{cases} 6t(1-t)^2, & 0 \le t \le 1, \quad \text{and} \\ 0 & , \text{ otherwise.} \end{cases}$$

Let $Y_1, Y_2, \ldots$ be defined by

$$Y_n = \sqrt{n} \, \frac{\bar{X}_n - \dfrac{1}{2}}{\sqrt{\dfrac{\bar{X}_n}{10}}}, \quad \text{where} \quad \bar{X}_n = \frac{1}{n} \sum_{i=1}^{n} X_i.$$

Show that

$$\lim_{n \to \infty} P(Y_n \le a) = \Phi(a).$$

Chapter Summary

Convergence in Probability: $X_1, X_2, \ldots$ converges in probability to X if, for any $\varepsilon > 0$,

$$\lim_{n \to \infty} P(|X_n - X| < \varepsilon) = 1. \qquad \textbf{(Section 7.2)}$$

Markov's Inequality: If X is a nonnegative random variable whose mean exists then, for any $a > 0$,

$$P(X \ge a) \le \frac{E[X]}{a}. \qquad \textbf{(Section 7.2)}$$

Chebyshev's Inequality: If X is a random variable such that $E[(X-a)^2]$ exists, then, for any $k > 0$,

$$P(|X - a| \ge k) \le \frac{E[(X-a)^2]}{k^2}. \qquad \textbf{(Section 7.2)}$$

If $X_1, X_2, \ldots$ is a sequence of random variables such that $E[X_i] = \mu < \infty$, for $i = 1, 2, \ldots$. If

$$\lim_{n \to \infty} \text{Var}(X_n) = 0,$$

then, for any $\varepsilon > 0$,

$$\lim_{n \to \infty} P(|X_n - \mu| < \varepsilon) = 1. \qquad \textbf{(Section 7.2)}$$

Weak Law of Large Numbers: Let $X_1, X_2, \ldots$ be a sequence of i.i.d. random variables such that $E[X_i] = \mu$ and $\text{Var}(X_i) = \sigma^2$ exist and are finite. Let

$$Y_n = \frac{1}{n} \sum_{i=1}^{n} X_i.$$

Then, for any $\varepsilon > 0$,

$$\lim_{n \to \infty} P(|Y_n - \mu| < \varepsilon) = 1. \qquad \textbf{(Section 7.2)}$$

Bernoulli Law of Large Numbers: Let X_n be a $B(n,p)$ random variable, and let $Y_n = \frac{X_n}{n}$. Then, for any $\varepsilon > 0$,

$$\lim_{n \to \infty} P(|Y_n - p| < \varepsilon) = 1. \qquad \text{(Section 7.2)}$$

Let $X_1, X_2, \ldots$ be a sequence of random variables with d.f.'s $F_{X_1}(a), F_{X_2}(a), \ldots,$ respectively. Let X be a random variable such that $P(X = c) = 1$. If

$$\lim_{n \to \infty} F_{X_n}(a) = \begin{cases} 0, & a < c, \quad \text{and} \\ 1, & c \le a, \end{cases}$$

then, for any $\varepsilon > 0$,

$$\lim_{n \to \infty} P(|X_n - X| < \varepsilon) = 1. \qquad \text{(Section 7.2)}$$

Let $X_1, X_2, \ldots$ be a sequence of random variables with d.f.'s $F_{X_1}(a), F_{X_2}(a), \ldots$ and moment generating functions $M_{X_1}(t), M_{X_2}(t), \ldots,$ which exist in an interval of the form $|t| < h$, where $h > 0$. If there exists a distribution function $F_X(a)$ with corresponding moment generating function $M_X(t)$ that exists for $|t| < r$, where $r < h$, such that

$$\lim_{n \to \infty} M_{X_n}(t) = M_X(t),$$

then

$$\lim_{n \to \infty} F_{X_n}(a) = F_X(a). \qquad \text{(Section 7.2)}$$

Let $X_1, X_2, \ldots,$ be a sequence of random variables that converges in probability to c, and let $g(x)$ be a function that is continuous at $x = c$. Then the sequence of random variables $g(X_1), g(X_2), \ldots$ converges in probability to $g(c)$. **(Section 7.2)**

Let $X_1, X_2, \ldots,$ be a sequence of random variables that converges in probability to a nonzero constant c. Then the sequence of variables $\frac{X_1}{c}, \frac{X_2}{c}, \ldots$ converges in probability to 1.

Let $X_1, X_2, \ldots$ be a sequence of random variables that converges in probability to c and $Y_1, Y_2, \ldots$ be a sequence of random variables that converges in probability to d. Then

1. if $c = d$, then $X_1 - Y_2, X_2 - Y_2, \ldots$ converges in probability to 0.
2. $X_1 + Y_1, X_2 + Y_2, \ldots$ converges in probability to $c + d$.
3. if a is a constant, then $aX_1, aX_2, \ldots$ converges in probability to ac.
4. $X_1 Y_1, X_2 Y_2, \ldots$ converges in probability to cd.
5. if $d \neq 0$, then $\frac{X_1}{Y_1}, \frac{X_2}{Y_2}, \ldots$ converges in probability to $\frac{c}{d}$. **(Section 7.2)**

Convergence in Distribution: Let $X_1, X_2, \ldots$ be a sequence of random variables having d.f.'s $F_{X_1}(a), F_{X_2}(a), \ldots,$ respectively, and let X be a random variable with a distribution function $F_X(a)$. The sequence of random variables is said to *converge in distribution to* X if,

$$\lim_{n \to \infty} F_{X_n}(a) = F_X(a),$$

for all a such that $F_X(x)$ is continuous at $x = a$. **(Section 7.3)**

Central Limit Theorem: Let $X_1, X_2, \ldots$ be a sequence of independent and identically distributed random variables. Let $E[X_i] = \mu$ and $\mathrm{Var}(X_i) = \sigma^2$ exist and be finite. Define

$$Y_n = \frac{\sum\limits_{i=1}^{n} X_i - n\mu}{\sigma\sqrt{n}} = \sqrt{n}\,\frac{\bar{X} - \mu}{\sigma}.$$

Let $F_{Y_n}(a)$ be the distribution function for Y_n. Then, for every a,

$$\lim_{n \to \infty} F_{Y_n}(a) = \int_{-\infty}^{a} \frac{1}{\sqrt{2\pi}} e^{-z^2/2}\, dz. \qquad \textbf{(Section 7.3)}$$

de Moivre-Laplace Theorem: Let X_n have a binomial distribution with parameters n and p. Then

$$\lim_{n \to \infty} P\left(\frac{X_n - np}{\sqrt{np(1-p)}} \leq a\right) = \Phi(a).$$

Let $X_1, X_2, \ldots$ be a sequence of random variables that converges in distribution of the random variable X. Let $Y_1, Y_2, \ldots$ be another sequence of random variables that converges in probability to a constant c. Define the following sequences of random variables.

1. $T_1, T_2, \ldots$, where $T_n = X_n + Y_n$.
2. $U_1, U_2, \ldots$, where $U_n = X_n Y_n$.
3. $V_1, V_2, \ldots$, where $V_n = \frac{X_n}{Y_n}$.

Then, the following are true.

1.

$$\lim_{n \to \infty} F_{T_n}(a) = F_X(a - c)$$

2.

$$\lim_{n \to \infty} F_{U_n}(a) = \begin{cases} F_X\left(\dfrac{a}{c}\right) & , c > 0, \quad \text{and} \\ 1 - F_X\left(\dfrac{a}{c}\right), & 0 < c \end{cases}$$

3.

$$\lim_{n \to \infty} F_{V_n}(a) = \begin{cases} F_X(ac) & , c > 0, \quad \text{and} \\ 1 - F_X(ac), & 0 < c \end{cases} \qquad \textbf{(Section 7.3)}$$

Review Exercises

1. A company contracts to supply ball bearings to a roller skate wheel manufacturer such that, at most, 2% of the bearings will be out of specification. The supplier agrees that the manufacturer can sample and measure some ball bearings before accepting delivery. How many bearings should the manufacturer sample and measure to be sure that a shipment containing 2.5% out of specification bearings will be detected 99% of the time?

2. Let $X_1, X_2, \ldots$ be a sequence of independent and identically distributed $U(0, \theta)$ random variables. Define $Y_n = \max(X_1, X_2, \ldots, X_n)$, for $n = 1, 2, \ldots$.

   a. Show that

   $$f_{Y_n}(y) = \begin{cases} \dfrac{n\, y^{n-1}}{\theta^n}, & 0 \le y \le \theta, \quad \text{and} \\ 0, & \text{otherwise.} \end{cases}$$

   b. Show that $Y_1, Y_2, \ldots$ converges in probability to θ.

3. Let $X_1, X_2, \ldots$ be a sequence of i.i.d. $G(1, \beta)$ random variables. Define the sequence $Y_1, Y_2, \ldots$ such that

   $$Y_n = \frac{1}{n} \sum_{i=1}^{n} (X_i - \beta)^2.$$

   Show that $Y_1, Y_2, \ldots$ converges in probability to β^2 by using Chebyshev's Inequality.

4. Let $X_1, X_2, \ldots$ be a sequence of i.i.d. $Po(\lambda)$ random variables, and define $Y_1, Y_2, \ldots$ by

   $$Y_n = \sqrt{n}\, \frac{\bar{X}_n - \lambda}{\sqrt{\bar{X}_n}}, \quad \text{where} \quad \bar{X}_n = \frac{1}{n} \sum_{i=1}^{n} X_i.$$

   Show that

   $$\lim_{n \to \infty} P(Y_n \le a) = \Phi(a).$$

5. A system uses a particular component whose lifetime follows a uniform distribution with a mean time to failure of 10 hours. The component is replaced upon failure.

   a. Use Chebyshev's Inequality to put a bound on the probability that 75 components replaced in the manner described will be enough to keep the system operating at least 1000 hours.

   b. Use the Central Limit theorem to approximate the probability that 75 components replaced in the described manner will be enough to keep the system operating at least 1000 hours.

6. In a grocery store checkout line, it has been found that the average time to process a customer is a random variable with a mean of 5 minutes and a standard deviation of 2 minutes. Use the Central Limit theorem to approximate the probability that 50 customers can go through the checkout line in less than 4 hours.

7. Suppose $X_1, X_2, \ldots$ is a sequence of i.i.d. random variables each having a probability function of

   $$p(a) = \begin{cases} \dfrac{a}{15}, & a = 1, 2, 3, 4, 5, \quad \text{and} \\ 0, & \text{otherwise.} \end{cases}$$

   Let $Y_{20} = X_1 + X_2 + \cdots + X_{20}$. Use the Central Limit theorem to approximate $P(Y_{20} \ge 80)$.

8. A sextant is a device used in navigation to determine the angle above the horizon of stars. In previous experiments, it has been found that errors in sextant measurements are a random variable with a mean of $0°$ and a standard deviation of $0.5°$.

  a. Use the Central Limit theorem to approximate the probability that 15 measurements will have an average error that is less than $\pm 0.1°$.

  b. How many observations are necessary to ensure that the average error of sextant measurements will be less than $\pm 0.01°$ with probability 0.95. Use the Central Limit theorem.

9. In order to certify that stainless steel is sufficiently free of flaws to be used in nuclear reactor components, an extensive series of screening tests are used to measure the number and size of imperfections. At one manufacturer it has been noted that 20% of the stainless steel produced is of sufficient quality to pass these rigid tests. If 100 randomly selected sheets of stainless steel from this manufacturer are tested, what is the probability that fewer than 15 will pass the screening tests?

10. Let $X_1, X_2, \ldots$ be a sequence of uncorrelated, but not necessarily independent, random variables such that $E[X_i] = 0$, $\mathrm{Var}(X_i) = 1$ for $i = 1, 2, \ldots$. Define

$$Y_n = \frac{1}{n} \sum_{i=1}^{n} X_i.$$

Show that $Y_1, Y_2, \ldots$ converges in probability to 0.

11. Let $X_1, X_2, \ldots$ be a sequence of i.i.d. random variables each having a p.d.f. of

$$f(t) = \begin{cases} e^{-(t-2)}, & 2 \le t, \quad \text{and} \\ 0, & \text{otherwise.} \end{cases}$$

Let $Y_n = \min(X_1, X_2, \ldots, X_n)$.

  a. Show that the p.d.f. for Y_n is

$$f_{Y_n}(y) = \begin{cases} e^{-n(y-2)}, & 2 \le y, \quad \text{and} \\ 0, & \text{otherwise.} \end{cases}$$

  b. Show that $Y_1, Y_2, \ldots$ converges in probability to 2.

  c. Let T_n be defined by

$$T_n = \frac{1}{n} \sum_{i=1}^{n} (X_i - Y_n).$$

  Show that $T_1, T_2, \ldots$ converges in probability to a constant and determine that constant.

12. A certain variety of tree grows an average of 5 feet per year with a standard deviation of 1 foot. What is the probability that the average growth in a sample of 81 such trees will be less than 4.5 feet in a given year?

13. Let X_n and Y_n be i.i.d. random variable each having a $B(n, p)$ distribution. Let

$$T_n = \frac{\frac{X_n}{n} - \frac{Y_n}{n}}{\sqrt{p(1-p)\left[\frac{2}{n}\right]}}.$$

Show that

$$\lim_{n\to\infty} P(T_n \le a) = \Phi(a).$$

14. The population of registered voters is known to show that 55% are in favor of a certain governmental policy. A sample of 100 voters is taken from each of two separate states. Use the result of Exercise 13 to approximate the probability that the observed proportion of voters in one state favoring the policy will differ from that of the other state by more than 0.5%.

15. Let X have a $G(\alpha, \beta)$ distribution. Use moment generating functions to show that

$$\lim_{\alpha\to\infty} P\left(\frac{X - \alpha\beta}{\sqrt{\alpha\beta^2}} \le a\right) = \Phi(a).$$

8 Statistical Models

Introduction

In virtually every phase of modern life, we are called upon to make decisions based on whatever, usually incomplete, information is available to us. People in business must make decisions regarding product lines, plant operations, and the like; politicians need to formulate and enact public policy; medical researchers need to determine the effectiveness of new treatments and medicines; and the list goes on. Because of the increased ease with which we can collect, store, and process data on high speed computers, policy makers now require that any decision be supported by evidence that the data have been analyzed properly. In a large number of cases, the data collected must be viewed as though they were generated by some random process. In those instances, it is the role of statistics to determine how best to process the data. In such cases, specific questions posed by investigators become questions regarding the distribution of a random variable. Perhaps a few examples would be in order.

Example 8.1

An ecologist knows that, if the smallmouth bass population in a certain lake drops below a known lower limit, the bass are endangered and must be restocked from a local hatchery. In order to determine how many bass are currently present, the ecologist goes out one day and catches and tags m smallmouth bass and returns them to the lake. A week later the ecologist returns and catches n smallmouth bass, r of which turn out to be tagged. How can these data best be used to obtain an estimate of the number of smallmouth bass in the lake? ∎

Example 8.2

During the nineteenth century Michelson and Morley conducted a series of experiments to determine the speed of light. Each time an observation of the speed of light was obtained, uncontrollable and unpredictable fluctuations occurred in the measurement process which gave rise to slightly different values. How can the experimental data best be used to determine the true speed of light?   ■

Example 8.3   A regional manager is trying to decide whether or not to provide low level background music in stenographic areas of all offices in her region. The company providing the service claims that such music will increase the productivity of clerical employees. To assess the vendor's claim the manager decided to take advantage of a 30-day free trial offer. One office in the region is selected at random (we shall see later how this can be done). The office is then observed for 30 days without background music and the average daily productivity for each clerical employee for that period is determined. The office is then outfitted with the music system, and the process is repeated for another 30-day period. How can these 2 sets of data best be used to determine if the background music system does, indeed, increase clerical productivity?   ■

Example 8.4   An investigator studying sexual discrimination in the hiring practices of accounting firms partitioned companies according to whether their staff sizes were small, medium, or large. Companies were randomly selected and data regarding company size as well as the number and sex of recent accounting graduates each had hired was obtained. How can these data best be used to determine if hiring patterns differ according to firm size?   ■

In each case it should be clear that the manner in which the data are collected, either by design as in Examples 8.3 and 8.4 or out of necessity as in Examples 8.1 and 8.2, injects a random component into the situation. The data can be viewed as being outcomes of a random experiment with an underlying probability distribution. Therefore, one goal of the statistician is to determine, as well as possible, what type of probability distribution (normal, gamma, Poisson, and so on) is appropriate for a given experimental situation. If this is successful, then the investigator's questions are restated as questions regarding the values of the parameters for that type of distribution. Then a procedure is selected that will "best" utilize the information contained in data. This chapter addresses the question of determining the underlying distribution, while the remainder of the book is devoted to developing and assessing techniques that answer questions about parameters of probability distributions. As we shall see, the determination of the proper probability distribution is something of an art. Thus, in subsequent chapters we will also look at what can be done if the proposed underlying distribution is either incorrect or indeterminate.

Before going much further it should be pointed out that statistical experiments are carefully designed entities. One does not simply go out, collect some information, and then try to figure out what to do with it. All of these things are planned before the first item of data is obtained. The sampling procedure is carefully devised and the statistical questions that will be posed are determined. The situation we are discussing in this chapter is one where the particular probability distribution is not known beforehand. For example, we may decide we want to determine the mean of the population distribution, but we need to know, if possible, if we are looking for the mean of a normal distribution or the mean of a gamma distribution.

8.2
Populations and Random Sampling

All statistical discussions assume the existence of a population, a population distribution, and a random sample from that population. Most readers probably have an intuitive notion of what we mean by these terms. In order to proceed mathematically, it is helpful to be a little more formal.

Definition 8.1

> A *population* is a set of objects about which information is desired.

In Example 8.1 the population consists of all of the smallmouth bass in the lake. The population in Example 8.2 is the collection of all possible values that could be observed using Michelson and Morley's apparatus. Example 8.3 has as its population all clerical employees of that company. Finally, the population for Example 8.4 is all accounting firms in the United States.

We will be studying situations where a representative subset of the population, which we will call a *random sample,* is selected and a measurement X_i is made on each. The underlying distribution of the X_i's is known as the *population distribution.* For example, it could turn out that the population of measurements of the speed of light in Example 8.2 follows a normal distribution centered at the actual speed of light. In such a case, the population distribution would be normal, and we would be interested in estimating the mean of that normal distribution. Keep in mind that the population is a collection of units while the population distribution is the probability distribution that describes the pattern of measurements taken on those units. Sometimes the units and the measurements are one and the same as in the speed of light example. On the other hand, in Example 8.3 the population is the set of clerical employees, but the population distribution is the probability distribution of employee productivities. Arriving at a statistical model for a given experimental situation requires that we determine the appropriate population distribution.

In some cases, the manner in which data are collected fixes the form of the underlying probability distribution. In Example 8.1 we make the simplifying, and not unreasonable, assumption that, at any given time, all smallmouth bass in the lake are equally likely to be caught. Then X, the number of smallmouth bass caught the second day, will have a hypergeometric distribution with parameters N = number of smallmouth bass in the lake, M = number of tagged smallmouth bass, and n = number of smallmouth bass caught the second day. In the example, a value of $X = r$ is observed. The value of a random variable that occurs after completion of a random experiment is called a *realization* of the random variable. Thus, in our case, we would say that the realization of X was r. In this experiment we wish to use the realization of X and the known values of the parameters M and n to estimate the value of the parameter N.

Example 8.1 is the only one of the four examples where the population distribution is fixed by the experimental procedure. In Example 8.2, the population distribution is determined by the experimental apparatus as well as the procedure for taking measurements, but we have little idea of what that distribution may be. We hinted earlier that it might be normal, but that was merely a conjecture. Example 8.3 has a similar problem in that there is nothing in the experimental procedure that determines what the distributions of productivities with and without music will be. In Example 8.4, we are confronted with a situation where we have a joint distribution of two discrete random variables, company size and sex of new employees, but again there is no hint as to the exact form of that joint distribution.

It should be noted that, in most real-life settings, the probability model that is used is an idealization of the true model. The key to making these models give usable results is making assumptions that

1. are reasonable,
2. do not severely distort what is actually happening, and
3. result in a distribution that is reasonably easy to deal with.

To illustrate, if the ecologist in Example 8.1 fished in only one section of the lake the equally likely assumption would make little sense and a hypergeometric model would not be appropriate. Thus, decisions based on the hypergeometric distribution would probably cause the lake to be restocked too frequently or not often enough.

Often experimental situations can be formulated in terms of Bernoulli trials. Recall that any selection procedure that results in repeated independent trials with two possible outcomes can result in independent Bernoulli trials. Sampling with replacement or sampling a relatively small number of items without replacement are the primary methods used. In those cases, binomial, geometric, and negative binomial distributions may be used depending on whether the data collected are a number of successes or a number of trials. In a similar vein, multinomial trials occur frequently. A sampling situation need only consist of repeated, independent trials with a fixed number of possible outcomes on each trial. An example of this might be an opinion survey where the possible responses are "for," "against," and "no opinion." In other cases, the experiment may not fix the population distribution, but we may be able to appeal to past experience to arrive at a good model.

Example 8.5

A political science professor is attempting to determine the amount of local support for a particular federal policy. He has his class randomly interview 1000 adults in the community and then he determines the number that are in favor of the policy. In this case the professor is interested in determining the proportion of adults that favor the policy. Hopefully, it is clear that a hypergeometric distribution fits the experimental situation exactly and the quantity of interest is $\frac{M}{N}$, where M = number of adults in the community favoring the policy and N = number of adults in the community. Recall in Chapter 5 we saw that if n is less than $0.05N$, then probabilities regarding a hypergeometric random variable can be approximated by using a binomial distribution. Thus, if N is known to be greater than about 20,000, then the simpler binomial distribution with $n = 1000$ and p = proportion of adults in favor of the policy may be used instead. In this instance, an estimate of p is desired. ■

Example 8.6

A traffic engineer has been asked to determine whether there is sufficient traffic volume at the intersection of a local street and a state highway to warrant the installation of a stop light. If the proportion of the time during daylight hours on weekdays when traffic entering the intersection from the local street exceeds 5 cars per minute is at least 0.20, then highway department policy calls for the installation of a stop light. If 1-minute periods are randomly selected and the number of cars passing through the intersection from the side street are counted, then, as we have already noted in Chapter 5, a Poisson distribution has been found to be a good model. If the random variable X = number of cars per minute, we are interested in whether or not $P(X > 5) \geq 0.20$. ■

Observations of continuous random variables never occur in practice. Any measuring device has limits on its accuracy and inherently rounds the data. Thus, technically all data are discrete. For practical purposes this discretization of continuous data by the measurement process can be safely ignored unless the values observed are close to the limits of accuracy. For example, suppose we wish to use a scale to measure the weights of newborn babies. If the scale is accurate to within 0.01 ounces, we may treat the measurements as continuous data. If, on the other hand, the scale is accurate only to the nearest pound, the observations should be treated as discrete data. In the latter case, serious consideration should be given to purchasing a better scale.

When dealing with continuous data it is rare that the experimental procedure can be devised to fix the probability distribution to some desired form as is often the case with discrete data. Sometimes we can appeal to past experience as was done in Example 8.6. There we used the fact that, over the years, researchers have discovered that Poisson distributions describe the probability structure of these kinds of observations quite well. In Chapter 5 we pointed out that experience has shown that some types of continuous data seem to follow certain types of distributions in practice. We mentioned that the time to failure of components often seems to follow an exponential, gamma, or Weibull distribution. In addition, whenever it is reasonable to assume that an observation is equally likely to assume any value within an interval, a uniform distribution is a good model. In other cases we may hope that studies of similar situations have been conducted and that we can make use of those results to arrive at a probability model. Otherwise, we will have to collect our own data and use them to postulate the underlying distribution. Examples 8.2 and 8.3 give rise to continuous data and, without additional information, it is by no means clear what type of distribution should be assumed in either situation. The next two sections are devoted to discussing methods that summarize the important characteristics of data and how these can be used to arrive at a reasonable population distribution.

Throughout this discussion we have used terms like "randomly selected" without being very precise about just what we mean or how we might do such a thing. We all have an intuitive feel for what is meant by the term "random sample." It somehow connotes a sense that, in a random sample from a given group, each member of the group has an equal chance of being in the sample. To be more precise, let's assume that we have a collection of N objects from which we wish to take a random sample of n objects without replacement. Clearly there are $\binom{N}{n}$ possible samples, and a random sampling procedure should make it equally likely to choose any one of them. This motivates the following definition.

Definition 8.2

> A *random sample of n objects* taken from a population of N objects is a procedure where each of the $\binom{N}{n}$ possible samples has a probability of $\binom{N}{n}^{-1}$ of being selected. The quantity n is called the *sample size*, and the quantity N is called the *population size*.

From the viewpoint of probability theory we give an alternative, but equivalent, definition. It will be more useful in our development of statistical theory.

Definition 8.3

> A *random sample of n objects* from a population is a set of n independent and identically distributed random variables $X_1, X_2, \ldots, X_n$. The quantity n is called the *sample size*.

When the experiment is performed, each of the random variables will be assigned specific numerical values. That is, we have the realizations of $X_1 = x_2$, $X_2 = x_2$, ..., $X_n = x_n$. At this point we say that the sample has been taken. We can then proceed to use the observed values to determine the attribute of the population distribution that was of interest.

It is logical to ask just how a random sample may be obtained in practice. If the population is small we might apply brute force to the problem by simply writing each possible sample of n objects on separate slips of paper, placing them in a hat, mixing them thoroughly, and selecting one piece of paper. Obviously such a procedure becomes unmanageable for realistic population sizes. In those cases, one method for drawing what is known as a *simple random sample* makes use of what are called *random number tables*. A small example of such a table is shown in Table 8.1.

Table 8.1   Random Numbers

16048	81899	04153	53381	79401	21438	83035	92350	36693
31238	18629	81953	05520	91962	04739	13092	97662	24822
94730	06496	73115	35101	47498	87637	99016	71060	88824
71013	18735	20286	57491	16703	23167	49323	45021	33132
12544	41035	80780	45393	30405	83946	23792	14422	15059
45799	22716	19792	09983	74353	16631	35006	85900	98275
23288	52390	16815	69298	82732	38480	96773	20206	42559
78985	05300	22164	24369	54224	35083	19687	38935	64202
14349	82674	66523	44133	00697	35552	35970	19124	31264
76384	17403	53363	44167	64486	64758	75366	76554	31601
78919	19474	23632	27889	47914	02584	37680	20801	72152
39339	03931	33309	57047	74211	63445	17361	62825	39908
05607	91284	74426	33278	43972	10119	89917	15665	52872
73823	73144	88662	09066	00903	20795	95452	92648	45454
09552	88815	16553	51125	42238	12426	87025	14267	20979
04508	64535	31355	86064	29472	16153	08002	26504	41744
81959	65642	74240	56302	00033	67107	21457	40742	29820
96783	29400	21840	15035	34537	33310	06116	21581	57802
02050	89728	17937	37621	47075	42080	97403	48626	55612
78095	83197	33732	05810	24813	86902	60397	16489	03264

A simple random sample can be achieved in the following manner. For a sample of size n from a population of size N, number the members from 1 to N. Then choose an arbitrary starting place in the random number table—for example, row 3, column 4. Here, we assume that a column consists of a 5-digit number. The entry with these coordinates would be 35101. Now proceed through the table in some predetermined manner selecting unused numbers that are between 1 and N. Stop when n distinct numbers have been chosen. The units with the numbers chosen constitute the members of the sample. Note that this is sampling without replacement.

Example 8.7

We want to select a random sample of 8 letters from the alphabet. Assign a number to each letter by letting $a = 1, b = 2, \ldots, z = 26$. Let's start at row 3 column 4 and proceed down that column in the random number table using the first 2 digits in each 5-digit block. If we reach the bottom of column 4 before getting 8 distinct, 2-digit numbers between 1 and 26, start at the bottom of column 5, selecting the first 2 digits and move up the column. If necessary, proceed on to column 6 taking the first 2 digits and go from top to bottom. Using this scheme we get the numbers 09, 24, 15, 05, 16, 21, 04, and 23. Thus, the simple random sample of 8 letters is i, x, o, e, p, u, d, and w. ∎

Sampling procedures such as these work well provided it is possible to identify and number all members of the population beforehand. In cases such as Example 8.1, where the population size is unknown, or Example 8.2, where the measurements do not yet exist and are theoretically infinite in number, it is not possible to ensure randomness through such a scheme. In those cases the experimenter must use great care to not inject nonrandom elements into the measurement process. It is not easy.

For the remainder of the book it will be assumed that somehow a random sample has been obtained. With finite populations, a procedure such as that involving the use of random number tables and, in other cases, good experimental technique has ensured that data were randomly collected. It should be noted that there are a wide variety of sampling methods. For further discussion of sampling theory consult a book on the topic, such as *Sampling Techniques* by William G. Cochran.

Exercises 8.2

1. The manager of a chain of supermarkets is planning to build a new store in an area of the city. Some of the existing stores in the chain have a a section devoted to gourmet foods, such as exotic coffees, teas, and cheeses. Some of the stores do not. The manager is trying to decide whether or not to build such a section in the new store. Devise a random experiment to help answer this question. Be sure to specify exactly what data should be collected and how to collect it. Can you make any *a priori* comments about the expected distribution of your population?

2. A medical researcher believes that the present death rate from typhoid in the United States has fallen below 15%. Death rate measures the percentage of people having the disease that die from it. Devise a random experiment to assess this belief. Be sure to specify exactly what data should be collected and how to collect it.

3. Advocates of the consumption of large doses of vitamin C claim that the daily intake of at least 1500 mg of the vitamin results in an overall reduction in the frequency and

severity of illness. A researcher for the Food and Drug Administration has been asked to assess this claim. Devise a random experiment to test the claim. Be sure to specify exactly what data should be collected and how to collect it.

4. A committee of the House of Representatives that is studying the issue of gun control is trying to determine the number of a certain brand of cheaply made handgun that are in circulation in the United States. The Justice Department has in its possession a large quantity of these handguns that have been confiscated in arrests throughout the country. Explain how these handguns might be used to answer the question. Explain whether or not these handguns can be considered to be a random sample.

5. Use the random number tables to take a sample of size six from the following list of last names.
Baumgartner, Miller, Wright, Choi, Krishnapriyan, Zucco, Barnett, Datta, Papadopol, Smith, Frank, Waterman, Chapman, Kwiatkowski, Schramm, O'Brien, Crawford, Simoson, Burkhead, Munz, Hsu, and DiCaprio.

6. Show that the method for taking a random sample by using random number tables satisfies Definition 8.2 for a random sample.

7. Suppose that you are using an instrument, such as a sextant, to take celestial navigation measurements. A sextant is a hand-held device consisting, roughly, of a protractor and a mirror that is used to measure the angle above the horizon, at sea, of astronomical objects, such as the sun, moon, and certain selected stars. Explain the potential sources of systematic error in using such an instrument. For each one explain briefly how to eliminate each one. In addition, explain what things might be expected to generate a random error in measurements.

8.3
Numerical Summaries

Recall that the type of situation being considered is one in which a random sample of observations is taken and the proper population is not predetermined. The first task for the statistician is to make an intelligent guess at the general form of the population distribution. In other words, we must first determine whether or not some type of, say, normal distribution appears to be reasonable before we make the effort to determine which particular normal distribution we have.

The techniques we will discuss in this section are numerical summaries of the data. Graphical methods will be discussed in the next section. Numerical summaries are referred to as *statistics*. This is another term that everyone has seen in a variety of contexts. For this reason it is necessary to be precise about what we shall mean when we call something a statistic.

Definition 8.4

> Given a set of random variables $X_1, X_2, \ldots, X_n$, a *statistic* is a function of those random variables that does not use any unknown parameter values.

For example, $\sum_{i=1}^{n}(X_i - \mu)$ would be a statistic if and only if the value of μ were known. Note that in our usage a statistic is a random variable.

Statistics may be used to describe the shape of the distribution of the data. If the sample is at all typical of the shape of the data it should provide a reasonable picture of the general shape of the population distribution. Thus, if the sample displays the attributes one would expect for some member of the family of gamma distributions it would seem reasonable to assume that the population distribution is a member of the gamma family. In Chapter 12 we will discuss the so-called *goodness-of-fit tests,* which are formal methods for assessing such assumptions. A wide variety of descriptive statistics have been proposed and are in varying degrees of use. We shall discuss only a small number of them. The statistics to be covered provide information about the location of the "center" of the data, the degree of dispersion of the data, the symmetry of the data, and the so-called tail weight of the data. The concept of a distribution being symmetric is as is defined in a standard calculus course. That is, a function $f(x)$ is symmetric about the point $x = a$ if $f(a - x) = f(a + x)$ for all x. Here we will be interested in whether or not there is a point of symmetry for the population density function. As far as the idea of tail weight is concerned we beg the indulgence of the reader. We will be a little more precise later, but for now we rely on an appeal to intuition regarding the concept that one distribution has "heavier" (thicker) tails than another distribution. Throughout the discussion that follows we assume that our random sample is $X_1 = x_1, X_2 = x_2, \ldots, X_n = x_n$.

8.3.1 Sample Mean

The sample mean is denoted $\bar{X}$ and is evaluated by

$$\bar{X} = \frac{1}{n} \sum_{i=1}^{n} X_i. \tag{8.1}$$

As we shall see later it is used as an estimator for the population mean. Since it is an arithmetic average it gives the location of the "center of gravity" of the data. That is, if a ruler were taken and a one-pound weight hung at each value of x_i, the ruler would balance at $\bar{X}$. The reader may have already seen the center of gravity or center of mass in physics or calculus courses.

Example 8.8   Suppose we have 5 observations of 1, 2, 3, 6, and 8. Using Equation 8.1 we find that $\bar{X} = 4$. Figure 8.1 illustrates the analogy with center of gravity.

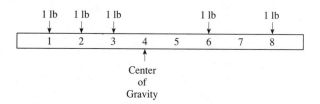

Figure 8.1

8.3.2 Sample Median

The sample median is sometimes denoted by $\widetilde{X}$. It is found by first placing the data in ascending order and finding the value that has as many observations smaller than it as are larger than it. If n is odd, then the median is the $\frac{n+1}{2}$th smallest observation. If n is even, then the median is the average of the $\frac{n}{2}$th and $(\frac{n}{2}+1)$st observations. By the way, ordered observations are commonly referred to as *order statistics*. We shall discuss order statistics in more detail in Chapter 9. As you can see the sample median measures the "center" of the data in a different sense from that of the sample mean. It is far less affected by extremely large or small sample values, and it is usually a more reliable measure in the presence of skewed data or data with heavy tails. For symmetric data with moderate tail weight the mean has been found to be a better measure of location. We shall pursue these ideas further in Chapter 10.

Example 8.9   Consider a sample consisting of $x_1 = 1$, $x_2 = 2$, $x_3 = 3$, and $x_4 = 4$. It is easy to check that both $\bar{X}$ and $\widetilde{X}$ have a value of 2.5. If we move $x_4 = 4$ to $x_4 = 5$, then $\widetilde{X}$ remains at 2.5 while $\bar{X}$ changes from 2.5 to 2.75. This illustrates the effect that extreme observations have on the sample mean compared to the sample median. ■

A comparison of the mean and the median can give an indication of whether the data are skewed. With symmetric data the mean and the median should be fairly close to each other. Data that are skewed such that there is a long tail to the right (positively skewed) will result in a mean, which is affected more by such a tail, that is larger than the median. If there is a long tail to the left (negatively skewed), then the mean will be less than the median. Keep in mind that, since we are dealing only with sample data, it is a very subjective decision whether a difference between the mean and the median is small enough to be consistent with samples from symmetric populations or is an indication of a skewed population distribution.

8.3.3 Sample Variance

The sample variance is denoted S^2 and defined by the following formula.

$$S^2 = \frac{1}{n-1} \sum_{i=1}^{n} (X_i - \bar{X})^2 \tag{8.2}$$

In some cases, a divisor of n instead of $n-1$ is used, and the quantity is still referred to as the sample variance. We shall always use a sample variance with a divisor of $n-1$ and use the symbol s^2. Another symbol will be used when the divisor is n. A more useful formula from a computational standpoint is the so-called *machine formula*.

$$S^2 = \frac{1}{n-1} \left(\sum_{i=1}^{n} X_i^2 - n\bar{X}^2 \right) \tag{8.3}$$

The sample variance is a measure of the degree of dispersion about the sample mean and is an estimator for the population variance. Because it uses squared deviations from the sample mean its use is not recommended in the presence of heavily tailed distributions. The square root of the sample variance, denoted by s, is referred to as the *sample standard deviation*. It is a popular measure of dispersion because it has the same units as the sample data. To give some interpretation to variance and standard deviation, recall that, in an analogy from physics, the mean was the center of gravity of the data. In a similar vein, variance and standard deviation are analogous to the moment of inertia of the data. In many elementary courses, students are presented with the rule of thumb, based on a normal distribution, that approximately $\frac{2}{3}$ of the sample data will lie within 1 standard deviation of the sample mean.

8.3.4 Median Absolute Deviation

The median absolute deviation is denoted *MAD* and is found by using

$$MAD = \text{median} \, | \, X_i - \widetilde{X} \, | \, . \tag{8.4}$$

It is a measure of dispersion that is far less sensitive to extremely large or small observations and is recommended to measure spread in skewed or heavily tailed samples.

8.3.5 Percentile-Based Measures of Dispersion

We are all familiar with the concept of percentiles. Most standard aptitude tests, such as the Scholastic Aptitude Tests (SAT's), report a numerical score and a ranking indicating the percentage of those taking the examination that receive a score equal to or less than that amount. This percentage is referred to as the *percentile rank* or *percentile score*.

Definition 8.5

> Given a sample of numerical observations, the *kth percentile*, denoted P_k, is that value, not necessarily observed, that has no more than $k\%$ of the observations less than it and no more than $(100 - k)\%$ of the observations larger than it.

This is the sample analog of the distribution quantiles that were discussed in Chapter 3. Note that the sample median is the $50th$ percentile, P_{50}, of the sample.

Certain sample percentiles are used frequently enough to be assigned special names. The median is one example. In addition, percentiles that are multiples of 25 are referred to as sample *quartiles* and denoted by Q_i, $i = 1, 2$, and 3. Thus, $Q_1 = P_{25}$, $Q_2 = \widetilde{X}$, and $Q_3 = P_{75}$. Percentiles that are multiples of 10 are referred to as sample *deciles* and denoted by D_i, $i = 1, 2, \ldots, 9$. Thus, $D_1 = P_{10}$, $D_2 = P_{20}$, and so forth.

There are two popular measures of dispersion that use quartiles—the *interquartile range (IQR)* and the *semi-interquartile range (SIR)*. They are defined as follows.

$$IQR = Q_3 - Q_1, \quad \text{and} \tag{8.5}$$

$$SIR = \frac{1}{2}IQR \tag{8.6}$$

In addition, one sometimes encounters the *decile deviation* which is defined to be $D_9 - D_1$.

Like the median absolute deviation, the percentile-based measures are relatively insensitive to extreme observations. Thus, they outperform the variance and standard deviation in the presence of skewed or heavily tailed samples. When the data are symmetric an indication of tail weight may be obtained by comparing the values of the median absolute deviation, the semi-interquartile range, and the standard deviation. In the presence of heavier tails, the greater number of extreme observations will tend to inflate the values of the variance and standard deviation. To illustrate we shall compare these quantities for the medium tailed normal distribution and the heavier tailed logistic distribution. Sample statistics should compare similarly. The uniform and double exponential cases are left as exercises.

Example 8.10   Consider a standard normal population. Clearly the standard deviation, σ, is 1. Due to symmetry, the mean and the median are both 0. Also symmetry implies that the distance from the median, $\tilde{\mu}$, to the 75th quantile, γ_{75}, is the same as the distance from the median to the 25th quantile, γ_{25}. Therefore, the semi-interquartile range will be the distance from $\tilde{\mu}$ to γ_{75}. Since $\tilde{\mu} = 0$, the value of γ_{75} will be the semi-interquartile range. γ_{75} is that value of Z such that $\Phi(z) = 0.75$. Table 3 of Appendix A shows that this value is $z = 0.67$. Thus, for the population $SIR = 0.67$. To determine the population median absolute deviation, we could proceed to determine the distribution of $|Z|$ and then find the median of that distribution. Fortunately, we are spared this task, again because of symmetry. We can argue that if $MAD = a$, then the median of the distribution of $|Z|$ is that value $z = a$ such that $P(|Z| \le a) = 0.5$. Thus,

$$P(|Z| \le a) = P(-a \le Z \le a) = 0.5.$$

Using Table 3 in Appendix A gives that $MAD = a = 0.67$. Note that, for a standard normal distribution,

$$MAD = SIR = 0.67, \quad \text{and}$$

$$\frac{\sigma}{SIR} = 1.49.$$

We hope the reader has not been thrown off by the use of SIR and MAD, which were defined as sample statistics, to denote population quantities. In the rare instances when this is done in the future, the context should make the use unambiguous. ∎

Example 8.11   Consider a logistic population with parameters $\theta = 0$ and $\sigma = 1$. The standard deviation is $\frac{\sigma\pi}{\sqrt{3}} = 1.81$. Note that this distribution is symmetric about 0.

Therefore, as was the case in the previous example, the semi-interquartile range is the distance between the median, which is $\theta = 0$, and the 75th quantile. γ_{75} is that value $x = a$ such that $P(X \le a) = 0.75$, which is found as follows.

$$0.75 = \int_{-\infty}^{a} \frac{e^{-x}}{(1 + e^{-x})^2} \, dx$$

$$= \frac{1}{1 + e^{-x}} \Bigg|_{-\infty}^{a}$$

$$= \frac{1}{1 + e^{-a}}$$

$$= 0.75$$

Solving for a gives that $SIR = \gamma_{75} = a = \ln(3) = 1.10$. By the same argument we used in the previous example, this is also the value of the median absolute deviation. (Is this true for any continuous, symmetric distribution?) Thus, we find that

$$MAD = SIR = 1.10, \quad \text{and}$$

$$\frac{\sigma}{SIR} = 1.65.$$

■

The logistic distribution has slightly heavier tails than the normal distribution. Thus, these examples support the claim that standard deviation grows more quickly with increasing tail weight than the other measures of dispersion. Calculations based on random samples from these populations should give similar results. That is, a random sample from a normal population should show $\frac{s}{SIR}$ to be about 1.5, while a sample from a logistic population would give $\frac{s}{SIR}$ near 1.65.

8.3.6 Sample Skewness

The sample skewness is denoted b_3 and is found by evaluating

$$b_3 = (s^2)^{-3/2} \frac{1}{n} \sum_{i=1}^{n} (x_i - \bar{X})^3. \tag{8.7}$$

It is a measure of the degree to which a distribution has unbalanced tails. Skewness is positive in the presence of a large tail to the right, and it is negative if the predominant tail is to the left. It is the sample version of the distribution skewness, γ_1, which was defined in Chapter 4. The division by the sample variance raised to the $\frac{3}{2}$ power serves as a standardizing factor and permits a meaningful comparison of the skewness values based on samples from different distributions. Symmetric samples have a skewness of 0, but it should be noted that skewed distributions may also have a skewness of 0. Thus, it is not a definite indicator of symmetrically distributed data. To adequately determine if a sample indicates a symmetric population, skewness should be used in conjunction with the graphical techniques to be discussed later. Also a comparison of the values of the sample mean and sample median should be made.

8.3.7 Sample Kurtosis

The sample kurtosis is denoted b_4 and is found by evaluating

$$b_4 = (s^2)^{-2} \frac{1}{n} \sum_{i=1}^{n} (x_i - \bar{X})^4. \tag{8.8}$$

The term *peakedness* is also used. This is the sample version of the distribution kurtosis, γ_2, which was discussed in Chapter 4. It is a measure of the degree to which a symmetric distribution has light, moderate, or heavy tail weight. Taking the fourth power of the deviations from the mean places greater emphasis on the contribution of observations in the tail regions of the sample. Dividing by the square of the sample variance serves to standardize the values of kurtosis. In this way, kurtosis measures the degree to which a distribution has values more than 1 standard deviation from the mean. The standard of comparison is the normal distribution, which is considered to have medium tail weight. The kurtosis, γ_2, of a normal distribution is 3. Based on this, distributions with kurtosis less than 3 are said to be *light tailed* (sometimes the term *platykurtic* is used), those with kurtosis near 3 are termed *medium tailed* (or *mesokurtic*), and those with kurtosis greater than 3 are called *heavy tailed* (or *leptokurtic*). The sample kurtosis has little meaning if the distribution is skewed.

8.3.8 An Example

To illustrate the use of these numerical summaries consider the following problem. Some investigators were interested in determining the average weight gain for cattle that had been placed on a new type of diet. To determine if the diet was superior to other diets, in terms of weight gain in pounds, 100 steers were placed on the diet with the results shown in Table 8.2. Each value is the pounds gained by a particular steer from having been on the new diet for a period of 20 days.

Table 8.2   Weight Gain of Cattle (in pounds)

13	34	40	47	17	34	40	47	21	34
40	48	48	40	35	22	23	35	40	49
49	41	35	24	25	36	41	49	50	41
36	26	27	36	41	50	51	42	36	27
28	37	42	51	51	43	37	28	28	37
43	52	52	43	38	29	29	38	43	52
53	43	38	29	30	39	44	53	54	44
39	30	31	39	44	55	56	45	39	31
31	40	45	57	58	45	40	32	32	40
46	45	56	46	40	33	33	40	46	67

The values of the numerical summaries are given in Table 8.3. The reader is encouraged to duplicate these results. Note that the mean and median have virtually the same value indicating that the sample is symmetric about that common value.

This conjecture is further supported by the fact that the skewness differs only slightly from 0. Thus, it seems safe to assume that the population distribution is one of the continuous, symmetric distributions. The kurtosis is almost exactly 3, $\frac{s}{SIR} = 1.56$ and $\frac{s}{MAD} = 1.62$. Based on the kurtosis, a normal distribution seems to be a reasonable model for this experiment. The other 2 quantities are somewhere between the values expected from a normal distribution and those for a logistic distribution. This leaves it somewhat unclear as to what to decide regarding tail weight. This illustrates some of the uncertainty in drawing definitive conclusions from sample data. We hope that the graphical methods to be discussed next will help with this situation.

Table 8.3   Descriptive Statistics for Cattle Data

Statistic	Value
Mean	39.79
Median	40.00
Variance	94.81
Standard deviation	9.74
Median absolute deviation	6.00
Interquartile range	12.75
Semi-interquartile range	6.38
Skewness	−0.10
Kurtosis	2.96

Exercises 8.3

1. Derive "machine formulas" for computing

   a. $\displaystyle\sum_{i=1}^{n} (x_i - \bar{X})^3$, and

   b. $\displaystyle\sum_{i=1}^{n} (x_i - \bar{X})^4$.

   These formulas should use only linear combinations of $\bar{X}$, $\sum_{i=1}^{n} x_i^2$, $\sum_{i=1}^{n} x_i^3$, and $\sum_{i=1}^{n} x_i^4$.

2. Give an example of a random variable that has a skewness of zero but is not symmetric about its mean. (*Hint:* Let X be discrete, have a mean of zero and take on only three values.)

3. The diameter of a heavy gauge metal wire was measured by 16 different people using a micrometer. Compute the mean, median, variance, median absolute deviation, semi-interquartile range, skewness, and kurtosis. Interpret your results.

$$
\begin{array}{cccc}
0.165 & 0.165 & 0.167 & 0.166 \\
0.168 & 0.165 & 0.170 & 0.167 \\
0.168 & 0.167 & 0.167 & 0.164 \\
0.166 & 0.168 & 0.166 & 0.165
\end{array}
$$

4. The following data are the weights of 45 randomly selected boys at a junior high school. Compute the mean, median, variance, median absolute deviation, semi-interquartile range, skewness, and kurtosis. Interpret your results.

117	119	131	125	130	117
122	112	120	125	120	121
121	128	122	133	125	134
122	127	123	122	123	134
123	125	153	120	113	124
135	181	112	118	135	131
120	124	164	131	146	114
141	172	128	119	124	125
147	151	131	124	133	141

5. The time, in hours, to failure of 40 randomly selected electrical components under heavy load are given below. Compute the mean, median, variance, median absolute deviation, semi-interquartile range, skewness, and kurtosis. Interpret your results.

72	18	16	4	28	82	75	9	4	30
35	11	34	13	63	11	74	93	94	42
7	53	94	57	99	47	48	16	15	33
37	82	67	84	9	77	18	84	33	96

6. The times, in seconds, to complete a task for a sample of 35 people are given below. Compute the mean, median, variance, median absolute deviation, semi-interquartile range, skewness, and kurtosis. Interpret your results.

24	26	24	27	33	23	27
25	38	22	28	25	25	24
23	22	26	20	21	23	20
21	26	21	39	22	25	21
23	31	21	20	22	23	33

7. The number of lines of computer code per hour written by a sample of 40 students in a Pascal programming course is given below. Compute the mean, median, variance, median absolute deviation, semi-interquartile range, skewness, and kurtosis. Interpret your results.

37	44	44	43	47	40	42	38	44	40
43	39	39	37	43	35	37	44	36	39
35	36	45	42	41	39	43	44	41	41
40	40	38	40	38	38	39	37	39	35

8. Conduct an experiment where you measure the ability of students in your school to guess the number of objects, such as jelly beans, in a jar. Take a random sample of 50 students and ask them to guess the number of items in the jar. Compute the mean, median, variance, median absolute deviation, semi-interquartile range, skewness, and kurtosis. Interpret your results.

9. Conduct an experiment where you measure the ability of students to estimate the area of an irregularly shaped geometric object. Take a random sample of 50 students and ask

them to estimate the area of a scalene triangle. Compute the mean, median, variance, median absolute deviation, semi-interquartile range, skewness, and kurtosis. Interpret your results.

8.4
Graphical
Summaries

There are many instances where the old adage of a picture being worth a thousand words has merit. This is particularly true in the determination of statistical models. After all, if one wishes to determine the shape of the probability distribution that gives rise to a particular random sample, why not draw a graph of the distribution of the sample and compare it with the shapes of known distributions? For instance, in the last section, the numerical summaries indicated that the pattern of weight gains for the 100 cattle fit well with either a normal or a logistic distribution. This section will describe some of the graphical methods that can be used to help distinguish between these two distributions. We shall assume that the random sample comes from a continuous population distribution. Samples from continuous populations usually present the most problems in model identification.

Before we begin it is appropriate to introduce the sample version of the distribution function. Recall that the distribution function of the random variable X, $F_X(a)$, represents the proportion of population values that are less than or equal to the value a. With this in mind it is perfectly logical to define a similar quantity for sample data.

Definition 8.6

Let $a_1, a_2, \ldots, a_n$ be the observed values of the random variables X_1, X_2, $\ldots$, X_n constituting a random sample from a population having density $f_X(x)$. Let $N(x)$ be the number of observed values that are less than or equal to x. Then the *empirical distribution function*, denoted $F_n(x)$, is

$$F_n(x) = \frac{N(x)}{n}. \tag{8.9}$$

Example 8.12

We wish to find and graph the empirical distribution for a sample of 5 observations whose values are 3, 7, 5, 4, and 5. Using Equation 8.9 the empirical distribution function is as shown in Table 8.4. A graph of this function is shown in Figure 8.2.

Table 8.4

x	$F_5(x)$
≤ 3	0.0
3	0.2
4	0.4
5	0.8
≥ 7	1.0

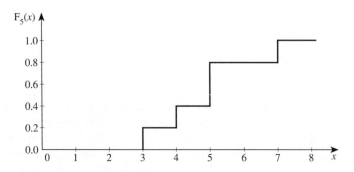

Figure 8.2

The empirical distribution has the same properties as the distribution function of a random variable. In addition, we can use the Bernoulli Law of Large Numbers to show that $F_n(a)$ converges in probability to $F_X(a)$. For practical purposes this implies that as sample sizes grow the empirical distribution function should give an accurate picture of the distribution function of the population. It is left as an exercise to graph the empirical distribution function of the cattle data in Section 8.3.8.   ■

From this discussion it follows that the empirical distribution function or its graph may be used to determine the general form of the population distribution. One method that can be used with symmetric data exploits this idea. More usually, however, it is easier to recognize a distributional shape by looking at the density function. We will consider two graphical displays that exploit this idea. Some of these techniques are part of a collection of procedures that has become known in recent years as *exploratory data analysis (EDA)*. This approach to analyzing data has been promoted by investigators such as Tukey. These methods have the advantage over traditional graphical methods in being easier to perform while being equally informative.

8.4.1 Quantile–Quantile Plots

The quantile–quantile plot compares the values of sample percentiles (quantiles) with those of the distribution of some random variable. This graph is the one we mentioned that compares an empirical distribution function with the distribution function of a random variable. It is really only useful for symmetric populations. One does not need to know or guess at the values of parameters of the density function to use this technique. First, a set of percentiles are computed using the hypothesized distribution function. Let's suppose for the sake of argument that these are the deciles $\gamma_{10}, \gamma_{20}, \ldots, \gamma_{90}$. Then the same percentiles are computed from the sample. In this case, these would be $P_{10}, P_{20}, \ldots, P_{90}$. Then, on a two-dimensional graph with values of the random variable on the vertical axis and sample values on the horizontal axis, plot the points (P_i, γ_i). This is a quantile–quantile plot. If the random sample was drawn from a population with a distribution of the same type as the density of the random variable on the vertical axis, the points should ideally lie on a straight line. If the plot is not straight, then the deviations can give an indication of how the data differ from the proposed distribution. For example, if the plot is curved in the tails, then this is evidence for the population from which the data were taken having lighter or heavier tails than the proposed distribution. A quantile–quantile plot like that shown in Figure 8.3 would indicate that the population distribution has heavier tails than the proposed one. We leave it to the exercises to determine what other typical quantile–quantile plots should look like.

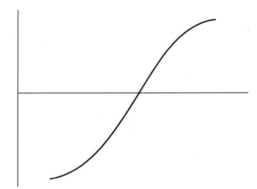

Figure 8.3

Example 8.13   We wish to compare the sample deciles of the cattle data X with the theoretical deciles of the standard normal random variable Z. The normal quantiles are those values of Z where the distribution function has values of $0.1, 0.2, \ldots, 0.9$. We find these values from Table 3 of Appendix A. The cattle data are placed in ascending order to determine the sample deciles. The results are shown in Table 8.5. The reader is encouraged to duplicate these results.

Table 8.5

Percentile Rank	Standard Normal (z)	Data (x)
10	-1.28	27.5
20	-0.84	31.0
30	-0.52	35.0
40	-0.25	38.0
50	0.00	40.0
60	0.25	42.0
70	0.52	45.0
80	0.84	48.5
90	1.28	52.0

The resulting quantile–quantile plot is shown in Figure 8.4. The points on this graph closely follow a straight line. This gives additional support to a conjecture that the population of cattle weight gains follows a normal distribution.

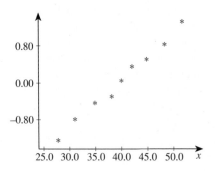

Figure 8.4

Example 8.14   As another example a sample of 40 observations was taken from a population. The data are given below.

$$
\begin{array}{cccccccccc}
72 & 18 & 16 & 4 & 28 & 82 & 9 & 4 & 30 & 35 \\
11 & 34 & 13 & 63 & 11 & 74 & 93 & 94 & 42 & 7 \\
53 & 94 & 57 & 99 & 47 & 48 & 16 & 15 & 33 & 37 \\
82 & 67 & 84 & 9 & 77 & 18 & 33 & 96 & 75 & 84
\end{array}
$$

We shall construct a quantile–quantile plot for the data against a normal distribution. We shall use the deciles as was the case in Example 8.12. The results are shown in Table 8.6.

Table 8.6

Percentile Rank	Standard Normal (z)	Data (x)
10	−1.28	9.0
20	−0.84	14.0
30	−0.52	18.0
40	−0.25	33.0
50	0.00	39.5
60	0.25	55.0
70	0.52	73.0
80	0.84	82.0
90	1.28	93.5

The resulting quantile–quantile plot is shown in Figure 8.5. Note that this time the points do not follow a straight line. The tails of the plot distinctly show that the quantiles of the data are closer together than are those for a normal distribution. This indicates that the sample data have lighter tails than would be expected for a sample from a normal population. The points in the center do not follow a straight line, but these could be due to sampling error. These data indicate that some lighter tailed model, such as a uniform distribution, might be more appropriate.

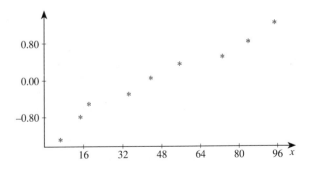

Figure 8.5

8.4.2 Histograms

The histogram is intended to give a visual description of the probability function or probability density function of the data. To construct a histogram we begin by dividing the range of observed values into equal width subdivisions called *classes*. The number of observations in each class is then plotted on a two-dimensional graph with the horizontal coordinates being the midpoint of each class and the vertical coordinates being proportional to the number of observations in each class divided by the width of that class. This has the effect that the areas of the bars of a histogram are proportional to the observed frequencies. If the classes happen to be of equal width, then the bar heights will represent frequencies.

Classes are created using a few simple guidelines. First, it is recommended that there be at least 5 but no more than 15 classes. It has been found that a more accurate picture of the data is obtained if there are neither too few nor too many classes. It is also common practice when constructing a histogram to make the class boundaries occur at points that are not observed values. This avoids having to devise a special rule for assigning observations with equal boundary values.

We shall construct a histogram for the cattle data. From the data we find that the smallest and largest observations are 13 pounds and 67 pounds, respectively. We shall divide the data into 6 classes with each having a width of 10 pounds. The class boundaries will be at 9.5, 19.5, 29.5, 39.5, 49.5, 59.5, and 69.5 pounds. Therefore, the class midpoints will be 14.5, 24.5, 34.5, 44.5, 54.5, and 64.5 pounds. We now go through the data and count the number of observations in each class. The results are shown in Table 8.7. We then use this information to draw the histogram as shown in Figure 8.6. Note that this shows that the data are fairly symmetric. It does little to answer questions regarding tail weight. This supports the other information from the numerical summaries and the quantile–quantile plot.

Table 8.7

Class Midpoint	Frequency
14.5	2
24.5	14
34.5	28
44.5	39
54.5	16
64.5	1

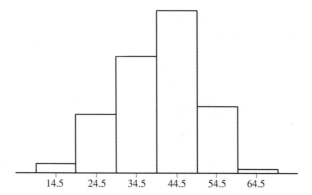

Figure 8.6

8.4.3 Stem-and-Leaf Plots

A stem-and-leaf plot is a simplified version of a histogram. It has the additional advantage of obtaining a diagram that retains the sample values. To do this we make use of the manner in which numbers are represented in base 10. That is, any number b can be written in the form

$$b = a_n 10^n + a_{n-1} 10^{n-1} + \cdots + a_1 10 + a_0.$$

What this means is that, for a particular power of 10, the observed data may be divided into units of that power of 10. That is, if data are assigned to groups whose values are 0–9, 10–19, 20–29, and so forth, we can identify an observation according to the group to which it belongs and the particular value it attains within that group. For example, an observation of 125 can be assigned to the group whose leading digits are 12 and that have, within that group, a value of 5. It could also be assigned to a group whose leading digit is 1 and that have, within that group, a value of 25. The numbers that are used to determine the groups are called the *stems*. The value each observation has within each group is called a *leaf*. For the steer data the logical way to divide the data is by 10s of pounds. Thus, each multiple of 10 pounds will be a *stem*. All lower order digits constitute *leaves*. A stem-and-leaf plot for the cattle data would look like that given in Figure 8.7. A vertical line is used to separate the stems from the leaves.

```
1 | 37
2 | 12345677888999
3 | 00111223344455666677778889999
4 | 000000000011112233333444555566677788999
5 | 0011122233456678
6 | 7
```

Figure 8.7

It is evident that there would be some advantage to being able to break up the data into finer units than the nearest 10s unit. Some conventions have been developed to assist in this. For example, the weight gains between 10 and 19 pounds might be subdivided into weight gains in the 10–14 pound and 15–19 pound range. A stem-and-leaf would then be as shown in Figure 8.8. This new subdivision gives a much clearer picture of the pattern for the underlying population. It should be pointed out that there are conventions for breaking the data into finer or coarser subdivisions than those shown here. The important thing to point out is that the stem-and-leaf plot can, for sufficiently large samples, give an accurate picture of the shape of the population probability density function. Figure 8.8 shows that the cattle weight gains are symmetric. This further supports the conjecture from Section 8.3.8 that the underlying population is symmetric. The stem-and-leaf plot, however, sheds little light on the question of the tail weight of the data.

```
 1  | 3
 1* | 7
 2  | 1234
 2* | 5677888999
 3  | 001112233444
 3* | 5566667778889999
 4  | 00000000001111223333344 4
 4* | 555566677788999
 5  | 00111222334
 5* | 56678
 6  |
 6* | 7
```

Figure 8.8

Exercises 8.4

1. Sketch the graph of the probability density function for a logistic random variable with $\theta = 0$ and $\sigma = 1$. Compare it with a sketch of the graph of a double exponential random variable with $\theta = 0$ and $\sigma = 1$. Sketch the graphs of both density functions on the same graph to facilitate making meaningful comparisons.

2. Show that a quantile–quantile plot for a $N(\mu_1, \sigma_1^2)$ versus a $N(\mu_2, \sigma_2^2)$ will graph as a straight line. That is, if γ_i is the ith quantile of a $N(\mu_1, \sigma_1^2)$ distribution, and v_i is the ith quantile of a $N(\mu_2, \sigma_2^2)$ distribution, show that $\gamma_i = a v_i + b$, where a and b are real constants.

3. Show that a quantile–quantile plot of a $G(1, \beta_1)$ random variable versus a $G(1, \beta_2)$ random variable will graph as a straight line when $\beta_1 \neq \beta_2$. That is, show that if X is the $G(1, \beta_1)$ random variable and Y is the $G(1, \beta_2)$ random variable, then $Y = aX + b$, where a and b are real constants.

4. Draw a typical quantile–quantile plot for a sample from a distribution that has lighter tails than the proposed distribution.

5. Draw a typical quantile–quantile plot for a sample from a gamma distribution when compared to a proposed normal distribution.

6. Compute and graph the empirical distribution of the cattle data. Compare your graph with that of the distribution function of a $N(40, 95)$ random variable.

7. Construct a quantile–quantile plot for the data in Exercise 3 of Section 8.3 to determine if the data follow a logistic distribution. If you conclude not, indicate whether the plot indicates a skewed distribution or one that is symmetric but with heavier or lighter tails.

8. Construct a quantile–quantile plot for the data in Exercise 4 of Section 8.3 to determine if the data follow a double exponential distribution. If you conclude not, indicate whether the plot indicates a skewed distribution or one that is symmetric but with heavier or lighter tails.

9. Construct a quantile–quantile plot for the data in Exercise 5 of Section 8.3 to determine if the data follow a uniform distribution. If you conclude not, indicate whether the plot indicates a skewed distribution or one that is symmetric but with heavier tails.

10. Construct a quantile–quantile plot for the data in Exercise 6 of Section 8.3 to determine if the data follow a normal distribution. If you conclude not, indicate whether the plot indicates a skewed distribution or one that is symmetric but with heavier or lighter tails.

11. Construct a quantile–quantile plot for the data in Exercise 7 of Section 8.3 to determine if the data follow a normal distribution. If you conclude not, indicate whether the plot indicates a skewed distribution or one that is symmetric but with heavier or lighter tails.

12. Construct a quantile–quantile plot for the data collected in Exercise 8 of Section 8.3 to determine if the data follow a normal distribution. If you conclude not, indicate whether the plot indicates a skewed distribution or one that is symmetric but with heavier or lighter tails.

13. Construct a quantile–quantile plot for the data collected in Exercise 9 of Section 8.3 to determine if the data follow a normal distribution. If you conclude not, indicate whether the plot indicates a skewed distribution or one that is symmetric but with heavier or lighter tails.

14. Construct a histogram and a stem-and-leaf plot for the data in Exercise 3 of Section 8.3. Interpret your results.

15. Construct a histogram and a stem-and-leaf plot for the data in Exercise 4 of Section 8.3. Interpret your results.

16. Construct a histogram and a stem-and-leaf plot for the data collected in Exercise 5 of Section 8.3. Interpret your results.

17. Construct a histogram and a stem-and-leaf plot for the data collected in Exercise 6 of Section 8.3. Interpret your results.

$$\frac{n-1}{n} s^2$$

18. Construct a histogram and a stem-and-leaf plot for the data collected in Exercise 7 of Section 8.3. Interpret your results.

Chapter Summary

Population: The set of objects about which information is desired. **(Section 8.2)**

Random Sample (1): Given a population of N objects a random sample of size n is a procedure where each of the $\binom{N}{n}$ possible samples has a probability of $\binom{N}{n}^{-1}$ of being selected. **(Section 8.2)**

Random Sample (2): A random sample of size n is a set of i.i.d. random variables $X_1, \ldots, X_n$.
Use random number tables to obtain a random sample from a finite population. Use good experimental design to obtain random samples in other cases. **(Section 8.2)**

Statistic: A random variable that is a function of variables in a random sample but no other unknown quantities. **(Section 8.3)**

Sample Mean: $\bar{X} = \frac{1}{n}\sum_{i=1}^{n} X_i.$ **(Section 8.3.1)**

Sample Median: $\tilde{X} = $ 50th percentile of sample data. **(Section 8.3.2)**

Sample Variance: $s^2 = \frac{1}{n-1}\sum_{i=1}^{n}(X_i - \bar{X})^2.$ **(Section 8.3.3)**

Machine Formula: $s^2 = \frac{1}{n-1}\left(\sum_{i=1}^{n} X_i^2 - n\bar{X}^2\right).$ **(Section 8.3.3)**

Median Absolute Deviation: $MAD = \text{median} \,|X_i - \tilde{X}|.$ **(Section 8.3.4)**

Semi-Interquartile Range: $SIR = \frac{1}{2}(Q_3 - Q_1).$ **(Section 8.3.5)**

Sample Skewness: $b_3 = (s^2)^{-3/2}\frac{1}{n}\sum_{i=1}^{n}(X_i - \bar{X})^3.$ **(Section 8.3.6)**

Sample Kurtosis: $b_4 = (s^2)^{-2}\frac{1}{n}\sum_{i=1}^{n}(X_i - \bar{X})^4.$ **(Section 8.3.7)**

Empirical Distribution Function: $F_n(x) = \dfrac{N(x)}{n}$, where $N(x)$ is the number of observations less than or equal to x. **(Section 8.4)**

Quantile–quantile Plot: Vertical axis are quantiles of hypothesized population distribution. Horizontal axis are corresponding percentiles from data. If plot is a straight line, then data came from hypothesized popuation. **(Section 8.4.1)**

Histogram: A plot where areas are proportional to sample frequencies. Gives a general picture of p.d.f. or p.f. of population from which the sample is taken. **(Section 8.4.2)**

Stem-and-Leaf Plot: Gives same information as the histogram but also retains and partially orders the observed values. **(Section 8.4.3)**

**Review
Exercises**

1. A local bank has installed automatic teller machines at each branch and curtailed Saturday business hours. There is some indication that customers are dissatisfied with this new policy. Devise a random experiment that would find out the degree to which customers are happy or unhappy with the lack of Saturday hours. Be sure to specify exactly what data should be collected and how it should be collected.

2. Show that, if the probability density function of a random variable X which depends on two parameters μ and σ is of the form

$$f_X(x) = f\left(\frac{x - \mu}{\sigma}\right),$$

and the graph of the probability density function is symmetric about the point $x = \mu$, then the quantile–quantile plot of this random variable with $\mu = a$ and $\sigma = b$ versus this random variable with $\mu = c$ and $\sigma = d$ will always be a straight line.

3. A random variable X has a Cauchy distribution with a p.d.f. of

$$f_X(x) = \frac{1}{\pi(1 + x^2)}, \quad -\infty < x < \infty.$$

 a. Show that the distribution of X is symmetric.

 b. Determine whether the distribution of X has light, medium, or heavy tails. It is only fair to point out that the kurtosis does not exist for this random variable.

4. Explain in detail how you would take a random sample of 50 from the student body at your school.

5. Conduct an experiment where you measure the ability of students in your school to estimate the height of an object such as a wooden stick. Take a random sample of 30 students and ask them to estimate the height of the object. Use the techniques discussed in this chapter to formulate an appropriate population distribution for these observations.

6. Use the techniques of this chapter to formulate an appropriate population model for the following data.

1.10	1.65	0.04	0.61	0.84	0.07	0.05	0.07	0.29	0.69
1.18	0.98	0.76	0.39	0.90	0.71	2.10	0.17	0.53	0.11
0.77	1.77	0.15	0.04	0.07	0.32	0.32	0.09	2.45	1.55
0.21	0.12	0.22	0.80	0.15	0.09	0.04	0.38	0.19	0.15

7. Use the techniques of this chapter to formulate an appropriate population model for the following data.

50	55	55	46	50	51	51	51	53	48
48	52	52	51	50	49	51	47	50	53
47	52	52	49	50	49	53	48	51	44
44	48	54	47	47	48	55	47	48	53

Sampling Distributions

Introduction

In this chapter, we are interested in determining the distributions of statistics computed from a random sample. In Chapter 6, we discussed methods to determine the distribution of functions of random variables. We will now use those techniques to find the distribution of certain functions that are frequently encountered in statistical applications. As we shall see later, the ability to determine the distribution of a statistic will be a critical part of the construction and evaluation of statistical procedures. In some cases, this will be a relatively easy task, while in others the best that can be done is to take a large sample and use limiting distributions. The term *sampling distribution* is used to emphasize that there is a difference between the distribution of the population from which the sample was taken and the distribution of some function of that sample. For example, we have seen that functions of normal random variables are not necessarily normally distributed. Populations have population distributions while statistics have sampling distributions. They are usually different from each other.

Distributions Based on Samples from Normal Populations

The normal distribution plays an important role in statistical theory for both historical reasons and because of the Central Limit theorem. Therefore, we shall begin by considering some probability distributions that are closely related to random samples from normal populations. Throughout this section we shall assume that we have a random sample $X_1, X_2, \ldots, X_n$ from a population that has a normal distribution with mean μ and variance σ^2. That is, $X_i \sim N(\mu, \sigma^2)$, for $i = 1, 2, \ldots, n$.

We first consider the distribution of an arbitrary linear combination of normal random variables. Recall in Chapter 6 we used the moment generating function method to show that the distribution of $\sum_{i=1}^{n} X_i$ is normal but with mean $n\mu$ and variance $n\sigma^2$. We state the result as a theorem. The proof is left as an exercise.

Theorem 9.1

Let $X_1, X_2, \ldots, X_n$ be independent random variables such that X_i has a $N(\mu_i, \sigma_i^2)$ distribution, and let $a_1, a_2, \ldots, a_n$ be real constants. Then the distribution of

$$Y = \sum_{i=1}^{n} a_i X_i \quad \text{is normal with} \quad \begin{cases} \mu_Y = \displaystyle\sum_{i=1}^{n} a_i \mu_i, \quad \text{and} \\[2mm] \sigma_Y^2 = \displaystyle\sum_{i=1}^{n} a_i^2 \sigma_i^2. \end{cases}$$

This fact will be quite useful in a number of later investigations. One important use of Theorem 9.1 is to determine the distribution of the sample mean.

Corollary 9.1

Let $X_1, X_2, \ldots, X_n$ be a random sample from a population whose distribution is $N(\mu, \sigma^2)$. Then the distribution of the sample mean,

$$\bar{X} = \frac{1}{n} \sum_{i=1}^{n} X_i,$$

is $N\left(\mu, \frac{\sigma^2}{n}\right)$.

Proof The proof is obtained by letting $a_i = \frac{1}{n}$, $\mu_i = \mu$, and $\sigma_i^2 = \sigma^2$, for $i = 1, 2, \ldots, n$ in Theorem 9.1. Thus,

$$\mu_{\bar{X}} = \sum_{i=1}^{n} \frac{1}{n} \mu$$

$$= \mu \frac{n}{n} = \mu, \quad \text{and}$$

$$\sigma_{\bar{X}}^2 = \sum_{i=1}^{n} \frac{1}{n^2} \sigma^2$$

$$= \sigma^2 \frac{n}{n^2} = \frac{\sigma^2}{n}.$$

∎

In Chapter 5 we saw that if X has a $N(\mu, \sigma^2)$ distribution, then $Z = \frac{X - \mu}{\sigma}$ has a standard normal distribution. We also made the statement in Chapter 5 that a random variable having a gamma distribution with parameters $\alpha = n/2$ and $\beta = 2$, where n is a positive integer, was said to have a χ^2 distribution with n degrees of freedom. We also showed in Chapter 6 that the square of a standard normal random variable, Z^2, has a chi-squared distribution with 1 degree of freedom. We can combine these ideas to cover the sum of the squares of a number of i.i.d.

standard normal random variables. This is a consequence of the fact that the sum of n i.i.d. $G(\alpha, \beta)$ random variables will have a $G(n\alpha, \beta)$ distribution. We state it as a definition.

Definition 9.1

> Let $Z_1, Z_2, \ldots, Z_n$ be independent and identically distributed $N(0, 1)$ random variables. Then the random variable
>
> $$U = \sum_{i=1}^{n} Z_i^2 \qquad \qquad (9.1)$$
>
> is said to have a *chi-squared distribution with n degrees of freedom* and is denoted by $\chi^2(n)$.

We now state a result that gives a common way in which chi-squared random variables occur in statistical applications.

Theorem 9.2

> Let $X_1, X_2, \ldots, X_n$ be independent and identically distributed $N(\mu, \sigma^2)$ random variables. Then
>
> $$Y = \sum_{i=1}^{n} \frac{(X_i - \mu)^2}{\sigma^2}$$
>
> has a chi-squared distribution with n degrees of freedom.

Proof   The proof follows directly from Definition 9.1. First let

$$Z_i = \frac{X_i - \mu}{\sigma}, i = 1, 2, \ldots, n.$$

Now each Z_i has a $N(0, 1)$ distribution. Therefore, each Z_i^2 has a chi-squared distribution with 1 degree of freedom. The Z^2's are independent random variables because they are functions of the X's, which are independent. We then note that

$$Y = \sum_{i=1}^{n} Z_i^2.$$

The result is then immediate from Definition 9.1.   ∎

Calculating probabilities using the probability density function of the chi-squared distribution is difficult. To help in this, approximate values of the distribution function

$$F_{\chi^2(n)}(a) = \int_0^a \frac{x^{n/2-1}}{\Gamma(\frac{n}{2}) \, 2^{n/2}} \, e^{-x/2} \, dx$$

are given in Table 4 of Appendix A. The column headings are values of the distribution function, and the column entries are the value of a that give that distribution function value with the degrees of freedom for that row.

Example 9.1

Suppose that V is a $\chi^2(9)$ random variable. Using Table 4 in Appendix A in the row labeled with $n = 9$, we find the following.

1. $P(V \leq 16.92) = 0.95$
2. $P(V \geq 2.70) = 1 - P(V < 2.70) = 1 - 0.025 = 0.975$

■

There are two other distributions that are important in statistical applications. They deal with ratios of random variables that occur frequently. The first of these was originally developed by W. S. Gosset in 1908. He analyzed this distribution while he was employed by the Guiness brewery. His employers would not permit him to publish his work under his own name so he used the pseudonym "Student." Hence, the distribution of the statistic is known as *Student's t distribution*.

Definition 9.2

Let Z be a standard normal random variable, and let U be a random variable that is independent of Z and possesses a chi-squared distribution with n degrees of freedom. Then the random variable

$$T = \frac{Z}{\sqrt{\dfrac{U}{n}}} \tag{9.2}$$

is said to have *Student's t distribution with n degrees of freedom* and is denoted by $T(n)$.

Note that the degrees of freedom is the same as that of the chi-squared random variable in the denominator. Often the name "Student" is omitted, and $T(n)$ is simply referred to as having a *t distribution with n degrees of freedom*.

A major use of the t distribution is in making inferences concerning the mean of a normal distribution. The obvious question concerning the p.d.f. for $T(n)$ is answered in the next theorem. The proof is a relatively straightforward application of the Jacobian technique covered in Chapter 6 and is left as an exercise.

Theorem 9.3

The probability density function for a random variable that has a t distribution with n degrees of freedom is

$$f_{T(n)}(t) = \frac{\Gamma\left(\dfrac{n+1}{2}\right)}{\sqrt{\pi n}\,\Gamma\left(\dfrac{n}{2}\right)\left(1 + \dfrac{t^2}{n}\right)^{(n+1)/2}}, \quad -\infty < t < \infty. \tag{9.3}$$

The graph of $f_{t(n)}(t)$ for various values of n is shown in Figure 9.1. Note that the probability density function is symmetric about $t = 0$. It also is similar in shape to the standard normal probability density function that is drawn for purposes of comparison. The graph of $f_{t(n)}$ has heavier tails than does the standard normal, but the tail weight decreases as n increases. In fact, it can be shown that the standard normal distribution is the limiting distribution as n becomes infinitely large. In addition, when $n = 1$, $T(n)$ has a Cauchy distribution.

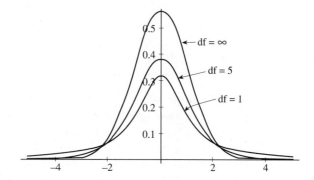

Figure 9.1

Calculations of probabilities using $f_{T(n)}(t)$ are difficult. Consequently, some approximate values for the distribution function

$$F_{T(n)}(a) = \int_{-\infty}^{a} f_{T(n)}(t)\,dt$$

are tabulated in Table 5 of Appendix A. The column headings are the values of the distribution function, and the column entries are the values of t that give that distribution function value for the degrees of freedom for that row. Note that the values in the row labeled as infinity are the same as those that would be calculated for the standard normal random variable. Also note that only distribution function values greater than 0.5 are given. For values less than 0.5 we can use the symmetry of the t distribution to show that

$$F_{T(n)}(-a) = 1 - F_{T(n)}(a). \tag{9.4}$$

Example 9.2   Let T have a Student's t distribution with 5 degrees of freedom. From Table 5 in Appendix A we find the following.

1. $P(T(5) \le 2.571) = 0.975$
2. $P(T(5) \le -2.015) = 1 - P(T(5) \le 2.015) = 1 - 0.95 = 0.05$
3. The value of b such that $P(-b < T(5) < b) = 0.99$ is, due to symmetry, the value b such that $P(T(5) \le b) = 0.995$. Thus, $b = 4.032$.

∎

The other commonly used ratio has what is known as an *F distribution*, which is defined as follows.

Definition 9.3

Let U be a random variable that has a chi-squared distribution with m degrees of freedom. Let V be a random variable that has a chi-squared distribution with n degrees of freedom that is independent of U. Then the random variable

$$F = \frac{\frac{U}{m}}{\frac{V}{n}} \tag{9.5}$$

is said to have an *F distribution with m and n degrees of freedom* and is denoted by $F(m, n)$.

Note that the degrees of freedom are an ordered pair. The first entry is the degrees of freedom for the numerator chi-squared random variable, and the second entry is the degrees of freedom for the denominator chi-squared random variable. This random variable has applications in making inferences concerning the variances of independent normal populations and in the set of procedures that are collectively known as *analysis of variance*. Those procedures are introduced in Chapter 14. The p.d.f. for a random variable with an $F(m, n)$ distribution can be obtained by using the Jacobian method of Chapter 6. The details are left as an exercise.

Theorem 9.4

The probability density function for a random variable having an F distribution with m and n degrees of freedom is

$$g_{F(m,n)}(f) = \begin{cases} \dfrac{\Gamma\left(\dfrac{m+n}{2}\right)}{\Gamma\left(\dfrac{m}{2}\right)\Gamma\left(\dfrac{n}{2}\right)} \dfrac{\left(\dfrac{m}{n}\right)^{m/2} f^{m/2-1}}{\left(1 + \dfrac{mf}{n}\right)^{(m+n)/2}}, & 0 < f, \quad \text{and} \\ \\ 0 & , \text{otherwise.} \end{cases} \tag{9.6}$$

The graph of $g_{F(m,n)}(f)$ for various values of m and n is shown in Figure 9.2. Note that the graph is skewed to the right with a shape reminiscent of a gamma random variable. Like the t distribution, integrals involving $g_{F(m,n)}(f)$ are hard to evaluate. Therefore, Table 6 in Appendix A gives approximate values for the distribution function for selected values of m and n. This table is in four parts. Each part is a particular value for the distribution function. In each part, the column headings are the numerator degrees of freedom and the row headings are the denominator degrees of freedom. The entry in the main body of the table is the value of F that achieves the distribution function value for that part of the table. The distribution function values given are all well above 0.50, which leads to the question of how to use the table to find a point such as that where the distribution function is equal to 0.05. The following theorem gives the answer.

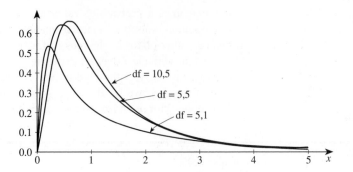

Figure 9.2

Theorem 9.5

Let F have an $F(m, n)$ distribution. Then

$$\int_0^a g_{F(m,n)}(x)\, dx = 1 - \int_0^{a^{-1}} g_{F(n,m)}(x)\, dx. \qquad (9.7)$$

Proof   The proof becomes simple once we recognize that, if F has an $F(m, n)$ distribution, then $H = \frac{1}{F}$ has an $F(n, m)$ distribution. We can then write out the integral

$$\int_0^a g_{F(m,n)}(f)\, df$$

and make a substitution to obtain an integral on $H = \frac{1}{F}$. The details are left as an exercise.   ∎

Example 9.3

Let f have an $F(9, 12)$ distribution. From Table 6 of Appendix A we find the following.

1. $P(F \le 2.80) = 0.95$
2. $P(F \le 0.196) = 1 - P(F^{-1} \le 5.11) = 1 - 0.99 = 0.01$
3. $P(a < F(9, 12) < b) = 0.95$ and $(F > b) = 0.025$ gives that one possible set of solutions is $b = 3.44$ and $a = \frac{1}{3.87} = 0.26$.   ∎

Statistics uses the normal, chi-square, t and F distributions in special ways. In many instances, we are interested in the value of a random variable such that the probability of being equal to or larger than it is a specified value. This usage has given rise to a special notation called the α *subscripting notation*. Consider a standard normal random variable, Z. The notation z_α is the value of the random variable Z such that

$$P(Z \ge z_\alpha) = \alpha.$$

The number α is a probability value while z_α is a value of the random variable Z. This notation is used in a similar manner for other random variables. The only difference is that often the degrees of freedom are included in the notation. Thus, if the random variable T has a t distribution with n degrees of freedom we would write $t(n)_\alpha$. Similarly, if the random variable has a chi-squared distribution with n degrees of freedom we would write $\chi^2(n)_\alpha$, and if the random variable F has an F distribution with m and n degrees of freedom we would use $F(m, n)_\alpha$.

Example 9.4

Let $\alpha = 0.05$.

1. Using Table 3 of Appendix A we see that

$$P(Z \geq z_{0.05}) = 1 - P(Z < z_{0.05}) = 0.95.$$

Thus, $z_{0.05} = 1.64$ or 1.65.

2. Suppose that T has a t distribution with 8 degrees of freedom. Use the table for the t distribution to see that $t(8)_{0.05} = 1.860$.

3. Suppose that V has a chi-squared distribution with 12 degrees of freedom. The chi-squared distribution with 12 degrees of freedom gives that $\chi(12)^2_{0.05} = 21.03$.

4. If F has an F distribution with 5 and 7 degrees of freedom we find that $F(5, 7)_{0.05} = 3.97$.

Exercises 9.2

1. Prove Theorem 9.1.

2. Let Z be a $N(0, 1)$ random variable and U be an independent chi-squared random variable with n degrees of freedom. Let

$$T = \frac{Z}{\sqrt{\dfrac{U}{n}}}, \quad \text{and}$$

$$S = U.$$

Use the Jacobian transformation to derive the joint probability density function of T and S. Then find the marginal density of T to derive the p.d.f. for a random variable having Student's t distribution with n degrees of freedom.

3. Show that the p.d.f. of random variable having a t distribution with 1 degree of freedom is a special case of the Cauchy distribution.

4. Let U be a chi-squared random variable with m degrees of freedom and let V be an independent chi-squared random variable with n degrees of freedom. Let

$$F = \frac{\dfrac{U}{m}}{\dfrac{V}{n}}, \quad \text{and}$$

$$G = V.$$

Use the Jacobian transformation to derive the joint probability density function of F and G. Then find the marginal density of F to derive the p.d.f. for a random variable having an F distribution with m and n degrees of freedom.

5. Fill in the details of the proof of Theorem 9.5.

6. Let the random variable F have an F distribution with m and n degrees of freedom. Show that as n becomes large the distribution of mF tends to that of a chi-squared random variable with m degrees of freedom.

7. Let $Z_1, Z_2, \ldots, Z_n$ be independent $N(0, 1)$ random variables, and let $X = \sum_{i=1}^{n} a_i Z_i$ and $Y = \sum_{i=1}^{n} b_i Z_i$. What conditions must be placed on the a's and b's so that X and Y will be independent?

8. Let T have a t distribution with 12 degrees of freedom. Find the following.
   a. a such that $P(-a < T < a) = 0.99$.
   b. b such that $P(|T| > b) = 0.10$.
   c. $P(T < -3.055)$
   d. $P(1.782 < T < 2.681)$
   e. $t(12)_{0.01}$

9. Let V have a chi-squared distribution with 18 degrees of freedom. Find the following.
   a. $P(V > 9.39)$
   b. a and b such that $P(a < V < b) = 0.90$.
   c. $\chi^2(18)_{0.025}$

10. Let F have an F distribution with 5 and 8 degrees of freedom. Find the following.
   a. $F(5, 8)_{0.025}$
   b. a and b such that $P(a < F < b) = 0.98$.
   c. $P(F > 6.63)$

9.3
Independence of $\bar{X}$ and s^2

We shall finish our discussion of sampling from normal populations with a result that will prove to be quite useful in later sections.

Theorem 9.6

Let $X_1, X_2, \ldots, X_n$ be a random sample from a $N(\mu, \sigma^2)$ population. Let

$$\bar{X} = \frac{1}{n} \sum_{i=1}^{n} X_i, \quad \text{and} \quad s^2 = \frac{1}{n-1} \sum_{i=1}^{n} (X_i - \bar{X})^2.$$

Then

1. $\bar{X}$ has a $N(\mu, \frac{\sigma^2}{n})$ distribution,
2. $\frac{(n-1)s^2}{\sigma^2}$ has a $\chi^2(n-1)$ distribution, and
3. $\bar{X}$ and s^2 are independent.

At first glance it probably seems a little odd that these two random variables are independent since s^2 contains $\bar{X}$ in it. Later in the book we will see how this result can be obtained almost trivially after we develop more theory regarding statistics and their distributions. At this point, however, it will take some effort to use what we have learned to prove this theorem. There are at least four ways to approach the proof. We could attempt to derive the joint distribution of $\bar{X}$ and $\frac{(n-1)s^2}{\sigma^2}$ and show that it factors into the individual distributions of each random variable. Another approach might be to derive the joint moment generating function and show that it factors into the desired marginal moment generating functions. A third approach might be to find the conditional p.d.f. for $\frac{(n-1)s^2}{\sigma^2}$ given $\bar{X}$ and show that it does not depend on $\bar{X}$ and has the desired form. The fourth method uses a proof by induction and the properties of the bivariate normal distribution. We shall pursue the fourth idea. It is not the most elegant approach, but it has the benefit of not demanding much more than we have presented thus far. We only need assume that the reader is familiar with inductive proofs. The proof that follows is given by Stephen Stigler in "Kruskal's Proof of the Joint Distribution of $\bar{X}$ and s^2," *The American Statistician*, vol. 38, no. 2 (May 1984).

Proof   First define

$$\bar{X}_n = \frac{1}{n} \sum_{i=1}^{n} X_i, \quad \text{and}$$

$$s_n^2 = \frac{1}{n-1} \sum_{i=1}^{n} (X_i - \bar{X}_n)^2.$$

We begin with the base case where $n = 2$. Here

$$\bar{X}_2 = \frac{X_1 + X_2}{2}, \quad \text{and}$$

$$s_2^2 = (X_1 - \bar{X}_2)^2 + (X_2 - \bar{X}_2)^2$$

$$= \frac{1}{4}(X_1 - X_2)^2 + \frac{1}{4}(X_2 - X_1)^2$$

$$= \frac{(X_1 - X_2)^2}{2}.$$

From Theorem 9.1 and Corollary 9.1 we know that $\bar{X}_2 \sim N(\mu, \frac{\sigma^2}{2})$, and $(X_2 - X_1)/\sqrt{2\sigma^2} \sim N(0, 1)$. Therefore, $\frac{s_2^2}{\sigma^2} \sim \chi^2(1)$. If we can show that $\bar{X}_2$ and s_2 are uncorrelated, we can use the fact that uncorrelated normal random variables are independent to see that the theorem is true for the base case. We can do this easily by computing the covariance between $X_1 + X_2$ and $X_1 - X_2$ as follows.

$$\text{Cov}(X_1 + X_2, X_1 - X_2) = \text{Cov}(X_1, X_1) - \text{Cov}(X_1, X_2) + \text{Cov}(X_2, X_1) - \text{Cov}(X_2, X_2)$$

$$= 0$$

We now assume that the theorem is true for a sample of size n and show that this implies that it holds for a sample of size $n + 1$. To do this we need to establish relationships that relate $\bar{X}_n$ and s_n^2 to $\bar{X}_{n+1}$ and s_{n+1}^2, respectively. We leave it as an exercise to show that

$$\bar{X}_{n+1} = \frac{n\bar{X}_n + X_{n+1}}{n + 1}, \quad \text{and}$$

$$ns^2_{n+1} = (n - 1)s^2_n + \frac{n}{n + 1}(X_{n+1} - \bar{X}_n)^2.$$

The inductive hypothesis assumes that

$$\bar{X}_n \sim N(\mu, \frac{\sigma^2}{n}),$$

$$\frac{(n - 1)s^2_n}{\sigma^2} \sim \chi^2(n - 1), \quad \text{and}$$

$\bar{X}_n$ and s^2_n are independent.

Now, Corollary 9.1 gives that

$$\bar{X}_{n+1} \sim N(\mu, \frac{\sigma^2}{(n + 1)}).$$

In addition, Theorem 9.1 gives that

$$X_{n+1} - \bar{X}_n \sim N(0, \frac{n + 1}{n}\sigma^2).$$

Therefore,

$$\sqrt{\frac{n}{n + 1}} \frac{(X_{n+1} - \bar{X}_n)}{\sigma} \sim N(0, 1) \quad \text{and}$$

$$\frac{n}{n + 1} \frac{(X_{n+1} - \bar{X}_n)^2}{\sigma} \sim \chi^2(1).$$

Therefore, $\frac{ns^2_{n+1}}{\sigma^2}$ is the sum of a $\chi^2(n - 1)$ random variable and an independent $\chi^2(1)$ random variable. Therefore, $\frac{ns^2_{n+1}}{\sigma^2} \sim \chi^2(n + 1)$.

The last item is to show that $\bar{X}_{n+1}$ and ns^2_{n+1} are independent. Since $\bar{X}_n$ is independent of s^2_n by the inductive hypothesis and X_{n+1} is indepdendent of s^2_n by the theorems for independent random variables, the demonstration becomes one of showing that

$$\text{Cov}(n\bar{X}_n + X_{n+1}, X_{n+1} - \bar{X}_n) = 0.$$

This is shown as follows.

$$\text{Cov}(n\bar{X}_n + X_{n+1}, X_{n+1} - \bar{X}_n) = \text{Cov}(n\bar{X}_n, X_{n+1}) - n\,\text{Cov}(\bar{X}_n, \bar{X}_n)$$

$$+ \text{Cov}(X_{n+1}, X_{n+1}) - \text{Cov}(X_{n+1}, \bar{X}_n)$$

$$= 0 - n\frac{\sigma^2}{n} + \sigma^2 - 0$$

$$= 0$$

Therefore, $\bar{X}_{n+1}$ and s^2_{n+1} are independent.

Exercises 9.3

1. Show the following.

    a.
    $$\bar{X}_{n+1} = \frac{n\bar{X}_n + X_{n+1}}{n+1}$$

    b.
    $$ns_{n+1}^2 = (n-1)s_n^2 + \frac{n}{n+1}\left(X_{n+1} - \bar{X}_n\right)^2$$

2. Suppose we have a sample of size n from a population with a $N(\mu, \sigma^2)$ distribution, and we compute the sample mean $\bar{X}$ and the sample variance s^2. Show that

    $$T = \sqrt{n}\frac{\bar{X} - \mu}{\sqrt{s^2}}$$

    has a t distribution with $n - 1$ degrees of freedom.

3. Let $X_1, X_2, \ldots, X_n$ be a random sample from a population with a $N(\mu_X, \sigma^2)$ distribution, and $Y_1, Y_2, \ldots, Y_m$ be a random sample from an independent having a $N(\mu_Y, \sigma^2)$ distribution. Show that

    $$T = \frac{\left(\bar{X} - \bar{Y}\right) - (\mu_X - \mu_Y)}{\sqrt{\dfrac{(n-1)s_X^2 + (m-1)s_Y^2}{n+m-2}\left(\dfrac{1}{n} + \dfrac{1}{m}\right)}}$$

    has a t distribution with $n + m - 2$ degrees of freedom.

4. Use Theorem 9.6 to find the mean and variance of s^2.

5. Let s^2 be the sample variance for a sample of 10 objects from a $N(5, 16)$ distribution. Find $P(3.712 < s^2 < 30.078)$.

6. Let $X_1, X_2, \ldots, X_5$ be a random sample from a $N(10, 25)$ distribution. Find $P(7 < \bar{X} < 12, s^2 \geq 60)$.

9.4

Order Statistics

A number of problems in probability can be solved more easily if we take a set of unordered independent random variables and place them in order before determining the probability of an event. Note that the percentile-based measures discussed in Chapter 8 are based on this idea.

Example 9.5

Consider a situation where 3 points are selected at random from an interval of length D. Suppose we wish to determine the probability that no point is closer than a distance of a units from the left end of the interval. Let the positions of the points be the random variables X_1, X_2, and X_3. It is easy to see that if the smallest X_i has position greater than or equal to a, then all 3 points will have a position of more than a. Therefore, if we let $Y = \min(X_1, X_2, X_3)$ we can write

$$P(X_1 \geq a, X_2 \geq a, X_3 \geq a) = P(Y \geq a).$$

The event has been reduced to a statement involving only a single random variable. The reader might say that this does not help much since we do not know the distribution of Y. That is the issue we shall address in this section.   ■

Some of the statistical procedures we shall discuss in subsequent chapters will result in statistics that will be functions of the largest observation, the smallest observation, the sample median, and the like. For samples of any size it is impractical to keep computing probabilities involving, say, the kth largest observation by using unordered random variables.

Example 9.6

Suppose we have a random sample of 15 from a $G(1, 1)$ distribution. Let $\widetilde{X}$ be the sample median which, in this case, is the 8th smallest observation, and suppose we wish to compute $P(\widetilde{X} \leq a)$. Since

$$P(\widetilde{X} \geq a) = P(8 \text{ or more observations out of } 15 \geq a)$$

it is, hopefully, easy to see that such a computation would be quite time consuming to carry out using the distribution of the unordered observations. If we had the probability density function of the eighth smallest observation, then we would need to compute only a single integral rather than a multiple integral involving 15 variables. The single integral should be easier to compute.   ■

These examples give some motivation as to why we would want to consider the transformation of a set of unordered random variables to a set of ordered ones. If the unordered random variables are independent and identically distributed, then the ordered random variables are called the *order statistics*. As far as notation is concerned, if we let $X_1, X_2, \ldots, X_n$ be a set of independent random variables, each of which has a distribution function of $F_X(x)$, then the set of random variables $Y_1, Y_2, \ldots, Y_n$, where

$$Y_1 = \min(X_1, X_2, \ldots, X_n)$$
$$Y_2 = \text{second smallest of } (X_1, X_2, \ldots, X_n)$$
$$\vdots$$
$$Y_n = \max(X_1, X_2, \ldots, X_n)$$

will be the order statistics and have the property that

$$Y_1 \leq Y_2 \leq \cdots \leq Y_n.$$

It is our goal in this section to determine the distribution of the order statistics when the original random variables are continuous. We begin with some important special cases.

Theorem 9.7

> Let $X_1, X_2, \ldots, X_n$ be independent and identically distributed continuous random variables each having a probability density function of $f_X(x)$ and a distribution function of $F_X(x)$. If $Y_1 \leq Y_2 \leq \cdots \leq Y_n$ are the order statistics then
>
> 1.
> $$f_{Y_n}(y) = nf_X(y)[F_X(y)]^{n-1}, \quad -\infty < y < \infty, \quad \text{and} \qquad (9.8)$$
>
> 2.
> $$f_{Y_1}(y) = nf_X(y)[1 - F_X(y)]^{n-1}, \quad -\infty < y < \infty. \qquad (9.9)$$

Proof   We shall use the distribution function method to prove Part 1. The proof of Part 2 is left as an exercise. The key is to note that if Y_n is the largest observation, then the event $\{Y_n \leq y\}$ means that all n observations are less than or equal to y. Thus,

$$F_{Y_n}(y) = P(Y_n \leq y)$$
$$= P(X_1 \leq y, X_2 \leq y, \ldots, Y_n \leq y).$$

We then use the fact that the X's are i.i.d. as follows.

$$F_{Y_n}(y) = P(X_1 \leq y, X_2 \leq y, \ldots, X_n \leq y)$$
$$= F_{X_1}(y) F_{X_2}(y) \cdots F_{X_n}(y) \quad \text{(independence)}$$
$$= [F_X(y)]^n \quad \text{(identical distributions)}$$

Differentiating the distribution function with respect to y gives

$$f_{Y_n}(y) = \frac{d}{dy} F_{Y_n}(y)$$
$$= n[F_X(y)]^{n-1} \frac{d}{dy} F_X(y)$$
$$= n[F_X(y)]^{n-1} f_X(y).$$

■

Example 9.7

This example is a continuation of Example 9.5. Let's use Equation 9.9 to obtain the probability that, if 3 points are selected at random from an interval of length D, no point will be within a units of the left end. If we call the 3 points X_1, X_2, and X_3, then it is reasonable to assume that they are i.i.d. $U(0, D)$ random variables. Thus, in the notation of Theorem 9.7,

$$f_X(x) = \begin{cases} \dfrac{1}{D}, & 0 \leq x \leq D, \quad \text{and} \\ 0, & \text{otherwise. Also} \end{cases}$$

$$F_X(x) = \begin{cases} 0 & , 0 < x, \\ \dfrac{x}{D} & , 0 \le x \le D, \quad \text{and} \\ 1 & , D < x. \end{cases}$$

If we let $Y_1 \le Y_2 \le Y_3$ be the order statistics, then we wish to compute $P(Y_1 \ge a)$. Using Equation 9.9 we find that

$$f_{Y_1}(y) = \begin{cases} 3 \left[1 - \dfrac{y}{D} \right]^2 \dfrac{1}{D} & , 0 \le y \le D, \quad \text{and} \\ 0 & , \text{otherwise.} \end{cases}$$

We then integrate to find

$$P(Y_1 \ge a) = \int_a^D 3 \left[1 - \frac{y}{D} \right]^2 \frac{1}{D} \, dy$$

$$= - \left[1 - \frac{y}{D} \right]^3 \Big|_a^D$$

$$= \left[1 - \frac{a}{D} \right]^3.$$

It is interesting to note that if n points are selected at random from this interval we would get that

$$P(Y_1 \ge a) = \left[1 - \frac{a}{D} \right]^n.$$

Now,

$$\lim_{n \to \infty} P(Y_1 \ge a) = \lim_{n \to \infty} \left[1 - \frac{a}{D} \right]^n$$

$$= \begin{cases} 0, & \text{if } a > 0, \quad \text{and} \\ 1, & \text{if } a = 0. \end{cases}$$

This implies that Y_1 converges in probability to 0. ∎

Determining the probability density function of the other order statistics is a little more involved. The general approach is as follows. Theorem 9.7 is a special case of the following general result.

Theorem 9.8

Let $X_1, X_2, \ldots, X_n$ be independent and identically distributed continuous random variables that each have a probability density function $f_X(x)$ and a distribution function of $F_X(x)$. If $Y_1 \le Y_2 \le \cdots \le Y_n$ are the order statistics, then, for $k = 1, 2, \ldots, n$,

$$f_{Y_k} = \frac{n!}{(k-1)!(n-k)!} \left[F_X(y) \right]^{k-1} \left[1 - F_X(y) \right]^{n-k}, \quad -\infty < y < \infty. \quad \textbf{(9.10)}$$

Proof Again we shall use the distribution function method. We begin by noting that, if the event $\{Y_k \le y\}$ occurs, at least k of the observations are less than or equal to y. The probability that any single observation will be less than or equal to y is the value of the distribution function $F_X(y)$. We can view the n observations as constituting a set of n repeated and independent trials. If we consider a success to be having an observation be less than or equal to y, the number of observations that are less than or equal to y is a binomial random variable with parameters n and $p = F_X(y)$. Thus, we get

$$F_{Y_k}(y) = P(Y_k \leq y)$$
$$= P(\text{at least } k \text{ observations} \leq y)$$
$$= \sum_{i=k}^{n} \binom{n}{i}[F_X(y)]^i[1 - F_X(y)]^{n-i}.$$

We now use Equation 5.28 to get

$$F_{Y_k}(y) = \int_0^{F_X(y)} \frac{n!}{(k-1)!(n-k)!} t^{k-1}(1-t)^{n-k}\,dt.$$

Now we differentiate with respect to y and use the Fundamental Theorem of Calculus to finish the proof.

$$f_{Y_k}(y) = \frac{d}{dy} F_{Y_k}(y)$$
$$= \frac{d}{dy} \int_0^{F_X(y)} \frac{n!}{(k-1)!(n-k)!} t^{k-1}(1-t)^{n-k}\,dt$$
$$= \frac{n!}{(k-1)!(n-k)!} [F_X(y)]^{k-1}[1 - F_X(y)]^{n-k} \frac{d}{dy} F_X(y)$$
$$= \frac{n!}{(k-1)!(n-k)!} [F_X(y)]^{k-1}[1 - F_X(y)]^{n-k} f_X(y)$$

■

Example 9.8

This example is a continuation of Example 9.6. Given a sample of 15 observations from a $G(1,1)$ distribution we wish to compute $P(\widetilde{X} \geq a)$. For 15 observations the sample median will be the 8th order statistic. In the notation of Theorem 9.9, we have

$$f_X(x) = \begin{cases} e^{-x}, & 0 < x, \quad \text{and} \\ 0, & \text{otherwise. Also} \end{cases}$$

$$F_X(x) = \begin{cases} 0, & 0 \leq x, \quad \text{and} \\ 1 - e^{-x}, & 0 < x. \end{cases}$$

Thus, Equation 9.10 gives that

$$f_{Y_8}(y) = \begin{cases} \dfrac{15!}{7!\,7!}[e^{-y}]^7\,[1 - e^{-y}]^7\,e^{-y}, & 0 < y, \quad \text{and} \\ 0, & \text{otherwise.} \end{cases}$$

Therefore we integrate to get

$$P(Y_8 \geq a) = \int_a^\infty \frac{15!}{7!\,7!} e^{-8y} [1 - e^{-y}]^7 \, dy$$

$$= \frac{15!}{7!\,7!} \left[\frac{1}{8} e^{-8a} - \frac{7}{9} e^{-9a} + \frac{21}{10} e^{-10a} - \frac{35}{11} e^{-11a} \right.$$

$$\left. + \frac{35}{12} e^{-12a} - \frac{21}{13} e^{-13a} + \frac{7}{14} e^{-14a} - \frac{1}{15} e^{-15a} \right].$$

This looks complicated and like a lot of work. It is, but to appreciate the savings in using this approach try doing it without using order statistics. ∎

Statistics that are functions of a pair of order statistics occur frequently in applications. Some examples include the following.

1. Sample range $= R = Y_n - Y_1$.
2. Midrange $= \frac{1}{2}(Y_1 + Y_n)$.
3. Sample median $= \tilde{X} = \frac{1}{2}(Y_{n/2} + Y_{n/2+1})$, for n even.

In order to determine the distribution of statistics such as these it is necessary to find the joint distribution of a pair of order statistics. The derivation is based on the same approach used in Theorem 9.8. The details are, however, more complicated. We suggest that the proof be skipped on a first reading of this material.

Theorem 9.9

> Let $X_1, X_2, \ldots, X_n$ be independent and identically distributed continuous random variables each having a probability density function of $f_X(x)$ and a distribution function of $F_X(x)$. Let $Y_1 \leq Y_2 \leq \cdots \leq Y_n$ be the order statistics. Then for any $1 \leq i < j \leq n$ and $x \leq y$
>
> $$f_{Y_i, Y_j}(x, y) = \frac{n!}{(i-1)!\,(j-i-1)!\,(n-j)!} [F_X(x)]^{i-1} [F_X(y) - F_X(x)]^{j-i-1}$$
>
> $$\cdot [1 - F_X(y)]^{n-j} f_X(x) f_X(y), \quad -\infty < x \leq y < \infty.$$
>
> $$(9.11)$$

Proof   We will use the distribution method again. First note that the event $\{Y_j \leq y\}$ can be written as the union of two mutually exclusive events as follows.

$$\{Y_j \leq y\} = \{Y_i \leq x, Y_j \leq y\} \cup \{Y_i > x, Y_j \leq y\}$$

Therefore, we can write

$$F_{Y_i, Y_j}(x, y) = P(Y_i \leq x, Y_j \leq y)$$

$$= P(Y_j \leq y) - P(Y_i > x, Y_j \leq y).$$

From the proof of Theorem 9.8 we have

$$P(Y_j \le y) = F_{Y_j}(y)$$

$$= \sum_{k=j}^{n} \binom{n}{k} [F_X(y)]^k [1 - F_X(y)]^{n-k}.$$

Recall that this was based on using a binomial distribution. The second probability can be evaluated using the multinomial distribution that was discussed in Section 5.4. For any single observation X_i, we find the following.

$$P(X_i \le x) = F_X(x),$$

$$P(x < X_i \le y) = F_X(y) - F_X(x), \quad \text{and}$$

$$P(X_i > y) = 1 - F_X(y)$$

Let L be the number of observations that fall in the interval $(-\infty, x]$, K be the number of observations that fall in $(x, y]$, and $M = n - K - L$ be the remaining observations that fall in (y, ∞). Then (K, L, M) is a trinomial random vector with parameters n, $p_1 = F_X(x)$, $p_2 = F_X(y) - F_X(x)$, and $p_3 = 1 - F_X(y)$. Then

$$P(Y_i > x, Y_j \le y) = P(\{\text{at least } i \text{ observations} > x\}$$

$$\cap \{\text{at least } j \text{ observations} \le y\})$$

$$= \sum_{l=0}^{i-1} \sum_{k=j-l}^{n-l} \frac{n!}{l!\, k!\, (n-k-l)!} p_1^l p_2^k p_3^{n-k-l}.$$

Combining these gives

$$F_{Y_i, Y_j}(x, y) = \sum_{k=j}^{n} \binom{n}{k} [F_X(y)]^k [1 - F_X(y)]^{n-k}$$

$$- \sum_{l=0}^{i-1} \sum_{k=j-l}^{n-l} \frac{n!}{l!\, k!\, (n-k-l)!} [F_X(x)]^l [F_X(y) - F_X(x)]^k [1 - F_X(y)]^{n-k-l}.$$

To obtain the joint probability density function, we use Equation 3.23 and differentiate first with respect to x and then with respect to y. We leave the details as an exercise. ∎

Example 9.9   Let $X_1, X_2, \ldots, X_n$ be a random sample from a $U(0, \theta)$ distribution. We wish to determine the distribution of the sample range $R = Y_n - Y_1$, where Y_1 and Y_n are the first and nth order statistics, respectively. To do this we will use Theorem 9.10 to obtain the joint p.d.f. of Y_1 and Y_n. We will then perform a Jacobian transformation to obtain the joint p.d.f. of R and $S = Y_n$. Then we shall integrate with respect to S to obtain the p.d.f. of R.

In the notation of Theorem 9.9, we have the following.

$$f_X(x) = \begin{cases} \dfrac{1}{\theta}, & 0 \le x \le \theta, \quad \text{and} \\[2mm] 0, & \text{otherwise} \end{cases}$$

$$F_X(x) = \begin{cases} 0, & 0 < x, \\[2mm] \dfrac{x}{\theta}, & 0 \le x \le \theta, \quad \text{and} \\[2mm] 1, & \theta < x \end{cases}$$

Using Equation 9.11 we find that

$$f_{Y_1, Y_n}(x, y) = \begin{cases} n(n-1) \left[\dfrac{y - x}{\theta} \right]^{n-2} \left(\dfrac{1}{\theta} \right)^2, & 0 \le x \le y \le \theta, \quad \text{and} \\[2mm] 0, & \text{otherwise.} \end{cases}$$

With $R = Y_n - Y_1$ and $S = Y_n$ we get $Y_1 = S - R$ and $Y_n = S$. Thus, the Jacobian is

$$J = \begin{vmatrix} 1 & -1 \\ 1 & 0 \end{vmatrix} = 1.$$

Therefore, the joint p.d.f. of R and S is

$$f_{R,S}(r, s) = \begin{cases} \dfrac{n(n-1) r^{n-2}}{\theta^n}, & 0 \le r \le s \le \theta, \quad \text{and} \\[2mm] 0, & \text{otherwise.} \end{cases}$$

We now integrate with respect to s to obtain the p.d.f. for R.

$$\begin{aligned} f_R(r) &= \int_r^\theta \frac{n(n-1) r^{n-2}}{\theta^n} \, ds \\[2mm] &= \frac{n(n-1) r^{n-2}(\theta - r)}{\theta^n} \end{aligned}$$

In summary we get

$$f_R(r) = \begin{cases} \dfrac{n(n-1) r^{n-2}(\theta - r)}{\theta^n}, & 0 \le r \le \theta, \quad \text{and} \\[2mm] 0, & \text{otherwise.} \end{cases}$$

■

There are occasions when we need to work with all of the order statistics. The joint probability density function for $Y_1, Y_2, \ldots, Y_n$ can be obtained in a number of ways, including the approach we have been using thus far in this discussion. As might be suspected the prospect of differentiating the distribution function n times to obtain the probability density function is somewhat arduous. For this reason we shall use the Jacobian method this time.

Theorem 9.10

> Let $X_1, X_2, \ldots, X_n$ be independent and identically distributed continuous random variables that each have a probability density function $f_X(x)$, and let $Y_1 \leq Y_2 \leq \cdots \leq Y_n$ be the order statistics. Then the joint probability density function of $Y_1, Y_2, \ldots, Y_n$ is
>
> $$f_{Y_1, Y_2, \ldots, Y_n}(y_1, y_2, \ldots, y_n) = n! f_X(y_1) f_X(y_2) \cdots f_X(y_n),$$
>
> $$-\infty < y_1 \leq y_2 \leq \cdots \leq y_n < \infty. \tag{9.12}$$

Proof   We shall use the Jacobian method for the case where the transformation is not one to one. First note that the joint p.d.f. for $X_1, X_2, \ldots, X_n$ is

$$f_{X_1, X_2, \ldots, X_n}(x_1, x_2, \ldots, x_n) = f_X(x_1) f_X(x_2) \cdots f_X(x_n).$$

Now for the event $\{a < y_1 < y_2 < \cdots < y_n < b\}$ we note that there are $n!$ permutations of the set of x's. For example, in the case when $n = 3$, each of the events

$$A_1 = \{a < x_1 < x_2 < x_3 < b\},$$
$$A_2 = \{a < x_1 < x_3 < x_2 < b\},$$
$$A_3 = \{a < x_2 < x_1 < x_3 < b\},$$
$$A_4 = \{a < x_2 < x_3 < x_1 < b\},$$
$$A_5 = \{a < x_3 < x_1 < x_2 < b\}, \quad \text{and}$$
$$A_6 = \{a < x_3 < x_2 < x_1 < b\}$$

gets mapped to the event

$$B = \{a < y_1 < y_2 < y_3 < b\}.$$

Each of these is a one-to-one transformation. The transformation from A_1 to B would be $Y_1 = X_1, Y_2 = X_2, Y_3 = X_3$, the transformation from A_2 to B would be $Y_1 = X_1, Y_2 = X_3, Y_3 = X_2$, and so forth. The Jacobian of the first transformation is

$$J_1 = \begin{vmatrix} 1 & 0 & 0 \\ 0 & 1 & 0 \\ 0 & 0 & 1 \end{vmatrix} = 1,$$

the Jacobian of the second transformation is

$$J_2 = \begin{vmatrix} 1 & 0 & 0 \\ 0 & 0 & 1 \\ 0 & 1 & 0 \end{vmatrix} = -1,$$

and so forth. Each Jacobian is the identity matrix with the rows permuted in the same way as the x's are permuted in being mapped to the order statistics. Since the determinant of a matrix undergoes a sign change when two rows are interchanged, we see that all 6 Jacobians will have a value of either $+1$ or -1.

Therefore, when $n = 3$,

$$f_{Y_1, Y_2, Y_3}(y_1, y_2, y_3) = f_X(y_1)f_X(y_2)f_X(y_3) + f_X(y_1)f_X(y_3)f_X(y_2)$$
$$+ \cdots + f_X(y_3)f_X(y_2)f_X(y_1)$$
$$= 3!f_X(y_1)f_X(y_2)f_X(y_3).$$

It is clear that this line of reasoning works for any value of n.   ∎

Example 9.10   Let $X_1, X_2, \ldots, X_n$ be i.i.d. random variables. A measure of dispersion known as *Gini's mean difference* is computed according to the following formula.

$$G = \sum_{j=2}^{n} \sum_{i=1}^{j-1} \frac{|X_j - X_i|}{\binom{n}{2}}.$$

Suppose we have a sample of size $n = 4$ from a $U(0,1)$ distribution and we wish to compute $E[G]$. This task can be simplified by the use of order statistics. Note that if $Y_1 \le Y_2 \le Y_3 \le Y_4$ are the order statistics, then G can be written as follows.

$$G = \frac{1}{6} \left[|X_2 - X_1| + |X_3 - X_1| + |X_3 - X_2| + |X_4 - X_1| + |X_4 - X_2| + |X_4 - X_3| \right]$$
$$= \frac{1}{6}[Y_2 - Y_1 + Y_3 - Y_1 + Y_3 - Y_2 + Y_4 - Y_1 + Y_4 - Y_2 + Y_4 - Y_3]$$
$$= \frac{1}{6}[3Y_4 + Y_3 - Y_2 - 3Y_1]$$

In the notation of Theorem 9.10 we have the following

$$f_X(x) = \begin{cases} 1, & 0 \le x \le 1, \quad \text{and} \\ 0, & \text{otherwise.} \end{cases}$$

Therefore, Equation 9.12 gives that

$$f_{Y_1, Y_2, Y_3, Y_4}(y_1, y_2, y_3, y_4) = \begin{cases} 4!, & 0 \le y_1 \le y_2 \le y_3 \le y_4 \le 1, \quad \text{and} \\ 0, & \text{otherwise.} \end{cases}$$

Therefore, to find $E[G]$ we proceed as follows.

$$E[G] = E[\frac{1}{6}(3Y_4 + Y_3 - Y_2 - 3Y_1)]$$
$$= 4 \int_0^1 \int_0^{y_4} \int_0^{y_3} \int_0^{y_2} (3y_4 + y_3 - y_2 - 3y_1) \, dy_1 \, dy_2 \, dy_3 \, dy_4$$
$$= \frac{1}{3}$$ ∎

As one might suspect the probability density functions for the order statistics can become quite complicated. For example, given a sample of size n from a

$N(0, 1)$ distribution, Equation 9.10 gives

$$f_{Y_k}(y) = \frac{n!}{(k-1)!\,(n-k)!} \left[\int_{-\infty}^{y} \frac{1}{\sqrt{2\pi}} e^{-z^2/2}\, dz \right]^{k-1} \left[1 - \int_{-\infty}^{y} \frac{1}{\sqrt{2\pi}} e^{-z^2/2}\, dz \right]^{n-k}$$

$$\cdot \frac{1}{\sqrt{2\pi}} e^{-y^2/2}, \quad -\infty < y < \infty.$$

The calculation of probabilities or expected values for Y_k is quite difficult. We would probably have to use numerical integration methods in cases like this. For this reason it is frequently useful to use limiting distributions when they exist.

In Chapter 4 we defined the kth quantile, γ_k, of the distribution of a random variable X to be the value such that

$$P(X \le \gamma_k) \le \frac{k}{100}, \quad \text{and} \quad P(X \ge \gamma_k) \le 1 - \frac{k}{100}.$$

Let Y_{k_n} be the kth order statistic from a sample of size n. This order statistic corresponds to a sample percentile. Since the distributions of the sample percentiles are complicated we wish to look at the large sample properties of them. The proofs of the results are beyond the level of this book and are omitted. The interested reader should consult a text such as *Mathematical Statistics,* by S. S. Wilks, or *Order Statistics,* by H. A. David. The first theorem discusses convergence in probability for order statistics. Note here that $[x]$ denotes the greatest integer that is less than or equal to x.

Theorem 9.11

Let $X_1, X_2, \ldots, X_n$ be independent and identically distributed continuous random variables each with distribution function $F_X(x)$. Let $\{Y_{[np]}, n = 1, 2, \ldots\}$, be a sequence of sample kth percentiles where $p = \frac{k}{100}$. If γ_k, the kth quantile of $F_X(x)$, is unique, then $Y_{[np]}$ converges in probability to γ_k.

This theorem says, in essence, that for large samples the kth sample percentile approaches the kth population quantile. Thus, the sample median converges to the population median, the sample semi-interquartile range converges to the population semi-interquartile range, and so forth. This also implies that graphs such as quantile–quantile plots, which are based on sample quantiles, will be fairly accurate representations of the population when sample sizes are large.

The next theorem discusses convergence in distribution for order statistics.

Theorem 9.12

If, in addition to the assumptions of Theorem 9.11, there is an $h > 0$, such that $F_X(x)$ has a derivative $f_X(x)$ at all $x \in (\gamma_k - h, \gamma_k + h)$, and $f_X(\gamma_k) > 0$, then

$$\lim_{n \to \infty} P(Y_{k_n} \le a) = \Phi \left(\frac{a - \gamma_k}{\sqrt{p(1-p)/n f_X^2(\gamma_k)}} \right). \tag{9.13}$$

Thus, for large samples, the distribution of the kth sample quantile, Y_{k_n}, is approximately normally distributed with a mean of γ_k and a variance of $\frac{p(1-p)}{nf_X^2(\gamma_k)}$.

Example 9.11

Let $p = \frac{1}{2}$ and consider the sequence of sample medians, $\tilde{X}_1, \tilde{X}_2, \ldots$, from a population that has a $N(\mu, \sigma^2)$ distribution. We know that the median of a normal distribution is unique and equal to μ. Furthermore, the p.d.f. of the normal distribution is differentiable everywhere and positive everywhere. Thus, all of the conditions of Theorem 9.11 and Theorem 9.12 are satisfied. In addition,

$$f_X(\gamma_{50}) = f_X(\mu) = \frac{1}{\sqrt{2\pi\sigma^2}}.$$

Thus, Theorem 9.11 shows that the sequence of sample medians converges in probability to μ. In addition, Theorem 9.12 gives that $\tilde{X}_1, \tilde{X}_2, \ldots$ converges in distribution to a random variable and has a normal distribution with a mean of μ and a variance of $\frac{\pi\sigma^2}{2n}$. ■

As might be expected, there is a result like Theorem 9.12 for the asymptotic joint distribution of a pair of order statistics. Let $Y_{[np_i]}$ and $Y_{[np_j]}$ be the ith and jth sample percentiles, respectively, from a sample of size n with $i < j$. In addition, let γ_i and γ_j be the ith and jth population quantiles, respectively. Also let $p_i = \frac{i}{100}$ and $p_j = \frac{j}{100}$. With this notation we give the following theorem without proof.

Theorem 9.13

If, in addition to the assumptions of Theorem 11, there is an $h > 0$ such that $F_X(x)$ has a derivative $f_X(x)$ at all points x such that $x \in (\gamma_i - h, \gamma_i + h) \cup (\gamma_j - h, \gamma_j + h)$, and $f_X(\gamma_i) > 0$ and $f_X(\gamma_j) > 0$, then the joint distribution of $(Y_{[np_i]}, Y_{[np_j]})$ is asymptotically bivariate normal with parameters

$$\mu_i = \gamma_i,$$

$$\mu_j = \gamma_j,$$

$$\sigma_i^2 = \frac{p_i(1 - p_i)}{nf_X^2(\gamma_i)},$$

$$\sigma_j^2 = \frac{p_j(1 - p_j)}{nf_X^2(\gamma_j)}, \quad \text{and} \tag{9.14}$$

$$\rho_{i,j} = \sqrt{\frac{p_i(1 - p_j)}{p_j(1 - p_i)}}, \quad \text{where } 0 < p_i < p_j < 1.$$

Example 9.12

We wish to use Theorem 9.13 to obtain the asymptotic distribution of the sample semi-interquartile range for a sample taken from a $U(0, 1)$ distribution. Theorem

9.11 states that Q_1, the sample first quartile, converges in probability to γ_{25}, and Q_3, the sample third quartile, converges in probability to γ_{75}. Thus, $i = 25$ and $j = 75$, which implies that $p_i = 0.25$ and $p_j = 0.75$. Also,

$$f_X(x) = \begin{cases} 1 \,, 0 \le x \le 1, & \text{and} \\ 0 \,, \text{otherwise}. \end{cases}$$

Then from Theorem 9.13 we find that (Q_1, Q_3) has an asymptotic distribution that is bivariate normal with parameters

$$\mu_{Q_1} = \gamma_{25} = 0.25,$$

$$\mu_{Q_3} = \gamma_{75} = 0.75,$$

$$\sigma^2_{Q_1} = \frac{(0.25)(0.75)}{n},$$

$$\sigma^2_{Q_3} = \frac{(0.75)(0.25)}{n}, \quad \text{and}$$

$$\rho_{Q_1,Q_3} = \sqrt{\frac{(0.25)(0.25)}{(0.75)(0.75)}} = \frac{1}{3}.$$

The semi-interquartile range is $\frac{1}{2}(Q_3 - Q_1) = \frac{1}{2}Q_3 - \frac{1}{2}Q_1$. This is a linear combination of random variables that are asymptotically normally distributed. Thus, the asymptotic distribution will also be normal with a mean of

$$\mu_{\frac{1}{2}(Q_3 - Q_1)} = \frac{1}{2}\mu_{Q_3} - \frac{1}{2}\mu_{Q_1}$$

$$= \frac{1}{2}(0.75) - \frac{1}{2}(0.25)$$

$$= 0.25,$$

and a variance of

$$\sigma^2_{\frac{1}{2}(Q_3 - Q_1)} = \frac{1}{4}\sigma^2_{Q_3} + \frac{1}{4}\sigma^2_{Q_1} - 2 \cdot \frac{1}{4} \cdot \rho_{Q_1,Q_3}\sigma_{Q_3}\sigma_{Q_1}$$

$$= \frac{3}{64\,n} + \frac{3}{64\,n} - \frac{1}{32\,n}$$

$$= \frac{1}{16\,n}.$$

This was obtained using the fact that

$$\text{Var}(aX + bY) = a^2 \, \text{Var}(X) + b^2 \, \text{Var}(Y) + 2\,a\,b \, \text{Cov}(X, Y), \quad \text{and}$$

$$\text{Cov}(X, Y) = \rho_{X,Y} \sqrt{\text{Var}(X) \, \text{Var}(Y)}.$$

■

Exercises 9.4

1. Prove Part 2 of Theorem 9.7.

2. Fill in the details of the proof of Theorem 9.9.

3. Two points are randomly selected on a line of length L. Let X and Y represent the distance from the left end of the line to each of the points. Show that

$$E\left[|X - Y|^n\right] = \frac{2L^n}{(n+1)(n+2)}.$$

4. Find the probability that three points selected at random from the circumference of a circle lie in the same semicircle.

5. The opening market prices per bushel for corn and wheat are known to independently follow a probability density function of

$$f(x) \begin{cases} e^{-(x-5)}, & 5 \le x, \quad \text{and} \\ 0, & \text{otherwise.} \end{cases}$$

An investor decides she will buy whichever commodity has the lower opening price. Find the probability density function for the price per bushel that the investor will have to pay. Find the expected price per bushel.

6. Show that the distribution of the kth order statistic based on a sample of size n from a $U(0, 1)$ distribution has a beta distribution and give its parameters.

7. Let $X_1, X_2, \ldots, X_5$ be i.i.d. random variables each having a p.d.f. of

$$f(t) = \begin{cases} 3t^2, & 0 \le t \le 1, \quad \text{and} \\ 0, & \text{otherwise.} \end{cases}$$

   a. Find the p.d.f. of Y_1, the smallest order statistic.
   b. Find the p.d.f. of Y_3, the sample median.

8. Consider the random sample described in Exercise 7.
   a. Find the joint p.d.f. of Y_1 and Y_5.
   b. Find the p.d.f. of the sample range $R = Y_5 - Y_1$.

9. Let $X_1, X_2, \ldots, X_6$ be i.i.d. random variables each having a p.d.f. of

$$f(t) = \begin{cases} \dfrac{1}{t^2}, & 1 \ge t, \quad \text{and} \\ 0, & \text{otherwise.} \end{cases}$$

   a. Find the p.d.f. of Y_6, the largest order statistic.
   b. Find the p.d.f. of Y_3.

10. Consider the random sample described in Exercise 9.
    a. Find the joint p.d.f. of Y_3 and Y_4.
    b. Use the results of Part a to find the p.d.f. of the sample median, $\frac{1}{2}(Y_3 + Y_4)$.

11. Another way to derive the p.d.f. of a single order statistic, say Y_k, is to begin with Theorem 9.10 and integrate the joint p.d.f. with respect to $Y_1, Y_2, \ldots, Y_{k-1}, Y_{k+1}, \ldots, Y_n$. Carry out the details of this derivation.

12. Find the asymptotic distribution of the sample median for a sample from a $G(1, \beta)$ distribution.

13. Find the asymptotic distribution of the sample median for a sample from a double exponential distribution with probability density function

$$f_X(x) = \frac{1}{2} e^{-|x|}, -\infty < x < \infty.$$

14. Suppose we have a random sample of size n from a population having a double exponential distribution with probability density function

$$f_X(x) = \frac{1}{2} e^{-|x|}, -\infty < x < \infty.$$

a. Find the asymptotic joint distribution of the first and third quartiles.

b. Use Part a to find the asymptotic distribution of the semi-interquartile range.

15. A system has 4 identical components operating in parallel for reliability. The system will function as long as at least one of these components is operating. The time to failure for each of these components follows a $G(1, 1)$ distribution. What is the expected time for the system to remain in operation?

16. A system is diagrammed below. Component A has a lifetime that follows a $G(1, 5)$ distribution. Components B and C are identical, each having a lifetime that follows a $G(1, 2)$ distribution. The system will function as long as a path through functioning components can be found. Find the expected lifetime for this system.

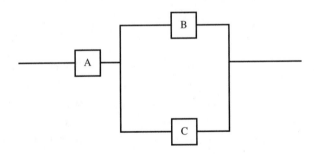

17. Let $Y_1 \leq Y_n$ be the first and nth order statistics from a sample of size n from a population having a distribution function of $F_X(x)$ and a probability density function of $f_X(x)$. Show that the joint distribution function of Y_1 and Y_n is, for $y_1 \leq y_n$,

$$F_{Y_1, Y_n}(y_1, y_n) = F_X^n(y_n) - [F_X(y_n) - F_X(y_1)]^n.$$

18. Let X be a discrete random variable with a probability function of

$$p(a) = \begin{cases} \frac{1}{5}, & a = 1, 2, 3, 4, 5, \quad \text{and} \\ 0, & \text{otherwise.} \end{cases}$$

Find the probability function for the largest observation based on a random sample of size 4 taken from a population which has this distribution. Why can't you use Equation 9.8?

Chapter Summary

If $X_1, \ldots, X_n$ are independent $N(\mu_i, \sigma_i^2)$ random variables, then

$$Y = \sum_{i=1}^{n} a_i X_i \quad \text{is normal with} \quad \begin{cases} \mu_Y = \displaystyle\sum_{i=1}^{n} a_i \mu_i, \quad \text{and} \\[2ex] \sigma_Y^2 = \displaystyle\sum_{i=1}^{n} a_i^2 \sigma_i^2. \end{cases}$$

(Section 9.2)

If $Z_1, \ldots, Z_n$ are i.i.d. $N(0, 1)$ random variables, then $\sum_{i=1}^{n} Z_i^2$ has a chi-squared distribution with n degrees of freedom. **(Section 9.2)**

If $X_1, \ldots, X_n$ are i.i.d. $N(\mu, \sigma^2)$ random variables, then $\sum_{i=1}^{n} \frac{(X_i - \mu)^2}{\sigma^2}$ has a chi-squared distribution with n degrees of freedom. **(Section 9.2)**

If Z has a $N(0, 1)$ distribution and U has an independent chi-squared distribution with n degrees of freedom, then $T = \frac{Z}{\sqrt{U/n}}$ has a Student's t distribution with n degrees of freedom. **(Section 9.2)**

If U has a chi-squared distribution with m degrees of freedom and V has an independent chi-squared distribution with n degrees of freedom, then $F = \frac{U/m}{V/n}$ has an F distribution with m and n degrees of freedom. **(Section 9.2)**

If F has an $F(m, n)$ distribution and G has an $F(m, n)$ distribution, then

$$P(F \le a) = 1 - P\left(G \le \frac{1}{a}\right).$$

(Section 9.2)

Let $X_1, X_2, \ldots, X_n$ be i.i.d. $N(\mu, \sigma^2)$ random variables. Let

$$\bar{X} = \frac{1}{n} \sum_{i=1}^{n} X_i, \quad \text{and} \quad s^2 = \frac{1}{n-1} \sum_{i=1}^{n} (X_i - \bar{X})^2.$$

Then
1. $\bar{X}$ has a $N(\mu, \frac{\sigma^2}{n})$ distribution,
2. $\frac{(n-1)s^2}{\sigma^2}$ has a $\chi^2(n-1)$ distribution, and
3. $\bar{X}$ and s^2 are independent. **(Section 9.3)**

Order Statistics: Let $X_1, X_2, \ldots, X_n$ be i.i.d. continuous random variables. Let $Y_1, Y_2, \ldots, Y_n$ be the X's placed in ascending order. Then $Y_1 \le Y_2 \le \cdots \le Y_n$ are the order statistics.

$$f_{Y_1}(y) = n f_X(y) [1 - F_X(y)]^{n-1}$$
$$f_{Y_n}(y) = n f_X(y) [F_X(y)]^{n-1}$$

$$f_{Y_k}(y) = nf_X(y)\,[F_X(y)]^{k-1}[1 - F_X(y)]^{n-k}$$

$$f_{Y_i,Y_j}(x, y) = \frac{n!}{(i-1)!\,(j-i-1)!\,(n-j)!}[F_X(x)]^{i-1}$$
$$\cdot\,[F_X(y) - F_X(x)]^{j-i-1}[1 - F_X(y)]^{n-j}$$
$$\cdot\,f_X(x)f_X(y), \text{ for } i < j$$

$$f_{Y_1,\dots,Y_n}(y_1,\dots,y_n) = n!\,f_X(y_1)\dots f_X(y_n). \qquad \textbf{(Section 9.4)}$$

Let $X_1, X_2, \dots, X_n$ be independent and identically distributed continuous random variables each with distribution function $F_X(x)$. Let $\{Y_{[np]}, n = 1, 2, \dots\}$, be a sequence of sample kth percentiles where $p = \frac{k}{100}$. If γ_k, the kth quantile of $F_X(x)$, is unique then $Y_{[np]}$ converges in probability to γ_k.   **(Section 9.4)**

If, in addition to the assumptions above, there is an $h > 0$, such that $F_X(x)$ has a derivative $f_X(x)$ at all $x \in (\gamma_k - h, \gamma_k + h)$, and $f_X(\gamma_k) > 0$, then

$$\lim_{n \to \infty} P(Y_{k_n} \le a) = \Phi\left(\frac{a - \gamma_k}{\sqrt{p(1-p)/nf_X^2(\gamma_k)}}\right). \qquad \textbf{(Section 9.4)}$$

If, also, there is an $h > 0$ such that $F_X(x)$ has a derivative $f_X(x)$ at all points x, such that $x \in (\gamma_i - h, \gamma_i + h) \cup (\gamma_j - h, \gamma_j + h)$, and $f_X(\gamma_i) > 0$ and $f_X(\gamma_j) > 0$, then the joint distribution of $(Y_{[np_i]}, Y_{[np_j]})$ is asymptotically bivariate normal with parameters

$$\mu_i = \gamma_i,$$

$$\mu_j = \gamma_j,$$

$$\sigma_i^2 = \frac{p_i(1 - p_i)}{nf_X^2(\gamma_i)},$$

$$\sigma_j^2 = \frac{p_j(1 - p_j)}{nf_X^2(\gamma_j)}, \quad \text{and}$$

$$\rho_{i,j} = \sqrt{\frac{p_i(1 - p_j)}{p_j(1 - p_i)}}, \quad \text{where } 0 < p_i < p_j < 1. \qquad \textbf{(Section 9.4)}$$

Review Exercises

1. If T has a Student's t distribution with ν degrees of freedom, show that T^2 has an F distribution and give the degrees of freedom.

2. Show that Student's t distribution tends to a standard normal distribution as the degrees of freedom tend to infinity.

3. Let $X_1, X_2, \dots, X_n$ be i.i.d. $N(\mu, \sigma^2)$ random variables, and let $Y_i = a + bX_i$, for $i = 1, 2, \dots, n$. Find $E[Y_i]$ and $\text{Var}(Y_i)$. In addition, give

$$\bar{Y} = \frac{1}{n}\sum_{i=1}^{n} Y_i, \quad \text{and}$$

$$s^2 = \frac{1}{n-1} \sum_{i=1}^{n} (Y_i - \bar{Y})^2$$

in terms of $X_1, X_2, \ldots, X_n$.

4. Show that the F distribution and the beta distribution are related by

$$B = \frac{\dfrac{mF}{n}}{1 + \dfrac{mF}{n}},$$

where F has an F distribution with m and n degrees of freedom and B has a beta distribution. What are the parameters of the resulting beta distribution?

5. Suppose that X has a $G(1, \beta)$ distribution.

a. Show that $Y = \dfrac{2X}{\beta}$ has a chi-squared distribution.

b. A plant manager has determined that a certain type of machine has a history of weekly oil consumption that seems to follow an exponential distribution with an average weekly oil consumption of 5 quarts. The plant has 4 of these machines. The manager wants to maintain enough oil on hand so that the probability that oil consumption exceeds the amount on hand is 0.05. How many quarts of oil should the manager have in stock at the beginning of each week? Use the result you obtained in Part a.

6. Let U be a $N(\mu_U, \sigma_U^2)$ random variable, V be a $N(\mu_V, \sigma_V^2)$ random variable, and W be a $N(\mu_W, \sigma_W^2)$ random variable. Assume U, V, and W are independent. Using these random variables form

a. a random variable with a $N(0, 1)$ distribution,

b. a random variable with a chi-squared distribution with 1 degree of freedom,

c. a random variable with a t distribution with 1 degree of freedom, and

d. a random variable with an F distribution with 1 and 2 degrees of freedom.

7. Compute the mean and variance of a random variable having a t distribution with n degrees of freedom and verify that

$$E[T] = 0, \quad \text{and}$$

$$\text{Var}(T) = \frac{n}{n-2}, \quad n > 2.$$

(Hint: $E[T^2] = E\left[Z^2 \dfrac{n}{V}\right] = E[Z^2] E\left[\dfrac{n}{V}\right]$, where $Z \sim N(0, 1)$ and $V \sim \chi^2(n)$).

8. Suppose that 4 numbers are chosen at random from the the interval $[0, 1]$. What is the probability that they will all lie in the interval $[0.25, 0.5]$?

9. Suppose that the random variable X has a chi-squared distribution with 20 degrees of freedom.

a. Find $P(X \le 12.443)$.

b. Find $P(X > 31.41)$.

c. Find 2 number a and b such that $P(a < X < b) = 0.94$.

d. Find $\chi^2(20)_{0.01}$.

10. Suppose that the random variable X has a t distribution with 25 degrees of freedom.
    a. Find $P(X \leq -1.708)$.
    b. Find a number k such that $P(X > k) = 0.95$.
    c. Find $t(25)_{0.05}$.

11. Suppose that the random variable X has an F distribution with 12 and 20 degrees of freedom.
    a. Find $P(X \leq 2.68)$.
    b. Find k such that $P(X > k) = 0.90$.

12. Let $X_1, X_2, \ldots, X_4$ be i.i.d. random variables each having a p.d.f. of

$$f(t) = \begin{cases} t & , 0 \leq t \leq 1, \\ 2 - t, & 1 < t \leq 2, \quad \text{and} \\ 0 & , \text{otherwise.} \end{cases}$$

    a. Find the p.d.f. of Y_4, the largest order statistic.
    b. Find the joint p.d.f. of Y_1 and Y_4.
    c. Find the p.d.f. of the sample range $R = Y_4 - Y_1$.

13. Let $X_1, X_2, \ldots, X_n$ be a random sample from a population having a p.d.f. of

$$f(t) = \begin{cases} 2t, & 0 \leq t \leq 1, \quad \text{and} \\ 0, & \text{otherwise.} \end{cases}$$

    a. Find the limiting distribution of the sample median.
    b. Find the joint limiting distribution of the first and third sample quartiles, Q_1 and Q_3.
    b. Find the limiting distribution of the semi-interquartile range, $\frac{1}{2}(Q_3 - Q_1)$.

10

Point Estimation

Introduction

In this chapter, we investigate the problem of using sample data to estimate the true values of parameters of probability density functions and probability functions. We assume that the experimenter has determined, either through the application of the techniques discussed in Chapter 8 or through special knowledge of the data, which family of distributions is an appropriate model for the data. By the term "family" we mean that the distribution of the data is known to be a member of the group of possible distributions of a certain type, such as normal distributions. However, we do not know which particular member of that family it is. The *estimation problem* is one of determining how to use the observations in the "best" fashion to estimate the values of the parameters of the distribution. For example, in Chapter 8, we analyzed a sample of cattle weight gains and concluded that it was reasonable to presume that the data follow a normal distribution. The question now becomes one of determining the mean, μ, and the variance, σ^2, in order to ascertain the particular normal distribution that best describes the data.

To be a little more precise, assume that $X_1, X_2, \ldots, X_n$ are i.i.d. random variables with a p.d.f. of $f_X(x \mid \theta_1, \theta_2, \ldots, \theta_k)$ (or, in the case of discrete random variables, a p.f. of $p_X(a \mid \theta_1, \theta_2, \ldots, \theta_k)$). The notation $f_X(x \mid \theta_1, \theta_2, \ldots, \theta_k)$ is used to indicate that the distribution depends on the values of the parameters $\theta_1, \theta_2, \ldots, \theta_k$. We wish to determine statistics $g_i(X_1, X_2, \ldots, X_n)$, $i = 1, 2, \ldots, k$, which can be used to estimate the value of each of the parameters. We would use g_1 to estimate the value of θ_1, g_2 to estimate the value of θ_2, and so on. These statistics are called *estimators* for the parameters, and the values computed for these statistics using sample data are called *estimates* of the parameters. We will denote the estimator for θ_i, which is $g_i(X_1, X_2, \ldots, X_n)$, by

$$\widehat{\theta}_i = g_i(X_1, X_2, \ldots, X_n), i = 1, 2, \ldots, k.$$

Thus, in the case of a normal distribution, we typically want to determine estimators $\hat{\mu}$ and $\widehat{\sigma^2}$ for μ and σ^2, respectively. One idea that takes some getting used to with this notation is that θ is a distribution parameter. It is a constant whose value we don't know. $\hat{\theta}$ is a random variable whose value is computed from sample data to estimate the value of θ.

We shall begin by looking at two methods of deriving estimators—the method of maximum likelihood and the method of moments. These methods do not necessarily derive the same estimators. This means that we will need some means for deciding between competing estimators. This will lead to a discussion of the properties of estimators and criteria for determining which estimator is "better." If the investigator has some prior knowledge of what the value of a parameter is likely to be, it would be desirable to use this information. This can be accomplished by the use of Bayes estimators, which will be discussed next. Finally, we look at the problems associated with assuming the wrong family of population distributions. This will lead to a discussion of the idea of robust estimation. There we will consider the idea of finding estimators that are less sensitive to departures from the assumed population distribution. That is, estimators whose behavior is good when the assumed distribution is correct but with the added benefit that their performance remains good over a range of distribution families.

10.2
Maximum Likelihood Estimation

The method of maximum likelihood is a popular technique for deriving estimators for distribution parameters. The general form for the procedure for computing these estimators was introduced by R. A. Fisher around 1922. We assume that we have a random sample $X_1, X_2, \ldots, X_n$ from a population with a probability density function of $f_X(x \mid \theta_1, \theta_2, \ldots, \theta_k)$ or a probability function of $p_X(a \mid \theta_1, \theta_2, \ldots, \theta_k)$, where the values of $\theta_1, \theta_2, \ldots,$ and θ_k are unknown. We will denote the p.d.f. or p.f. simply by $f(x \mid \theta)$ with the context indicating whether $f(x \mid \theta)$ is a p.d.f. or a p.f. and whether θ is a scalar or a vector of parameters. The joint p.d.f. or p.f. of $X_1, X_2, \ldots, X_n$ is

$$f_{X_1, X_2, \ldots, X_n}(x_1, x_2, \ldots, x_n \mid \theta) = f(x_1 \mid \theta) f(x_2 \mid \theta) \cdots f(x \mid \theta).$$

In this method it is assumed that we have the sample observations in hand. Thus, $(X_1, X_2, \ldots, X_n)$ is considered to be fixed at $(x_1, x_2, \ldots, x_n)$. If we evaluate the joint p.d.f. or p.f. at these values, we no longer have a joint distribution but rather a function of θ. This function of θ is called the *likelihood function*, which we shall denote by $L(\theta)$. Thus, the likelihood function is

$$L(\theta) = f(x_1 \mid \theta) f(x_2 \mid \theta) \cdots f(x_n \mid \theta). \tag{10.1}$$

Example 10.1   Let $X_1, X_2, \ldots, X_n$ be a random sample from a $N(\mu, \sigma^2)$ distribution. If we assume that both μ and σ^2 are not known, then Equation 10.1 gives that the likelihood function is

$$L(\mu, \sigma^2) = \prod_{i=1}^{n} \frac{1}{\sqrt{2\pi\sigma^2}} e^{-(x_i-\mu)^2/2\sigma^2}$$

$$= (2\pi\sigma^2)^{-n/2} e^{-(1/2\sigma^2)\sum_{i=1}^{n}(x_i-\mu)^2}. \qquad \blacksquare$$

Example 10.2   Consider a set of n Bernoulli trials each with a probability of success of p, where $0 < p < 1$. We can view the experiment as a sample of size n from a $B(1,p)$ distribution. Let $X_i = 0$ if the ith trial results in a failure, and $X_i = 1$ if the ith trial results in a success. Then $X_1, X_2, \ldots, X_n$ are i.i.d. $B(1,p)$ random variables. Let $Y = \sum_{i=1}^{n} X_i$. Then the likelihood function for this experiment is

$$L(p) = \prod_{i=1}^{n} p^{x_i}(1-p)^{1-x_i}$$

$$= p^{\sum_{i=1}^{n} x_i}(1-p)^{n-\sum_{i=1}^{n} x_i}$$

$$= p^y(1-p)^{n-y}.$$

It is also possible, however, to view the n Bernoulli trials as being a single observation from a $B(n,p)$ distribution. Let X be the number of successes in the n trials. In this case, the likelihood function would be

$$L(p) = \binom{n}{x} p^x(1-p)^{n-x}.$$

These are both legitimate likelihood functions for this problem. $\blacksquare$

Example 10.3   Let $X_1, X_2, \ldots, X_n$ be a random sample from a $U(0, \theta)$ distribution. The likelihood function is

$$L(\theta) = \begin{cases} \prod_{i=1}^{n} \dfrac{1}{\theta} & \text{, if } 0 \leq x_i \leq \theta, \quad \text{for } i = 1, 2, \ldots, n, \quad \text{and} \\ 0 & \text{, otherwise.} \end{cases}$$

$$= \begin{cases} \dfrac{1}{\theta^n} & \text{, if } 0 \leq x_i \leq \theta, \quad \text{for } i = 1, 2, \ldots, n, \quad \text{and} \\ 0 & \text{, otherwise.} \end{cases}$$

$\blacksquare$

For the sake of exposition assume that the parameter θ is a scalar. The method of maximum likelihood attempts to determine a value of θ that maximizes the value of $L(\theta)$. This value of θ, if it exists, will be some function of $x_1, x_2, \ldots, x_n$, say $t(x_1, x_2, \ldots, x_n)$. We then declare that $\widehat{\theta} = t(X_1, X_2, \ldots, X_n)$ is the maximum likelihood estimator for θ. Thus, in some sense, maximum likelihood estimators are those quantities that make it most likely for us to observe the sample values that occurred. In the case where more than one parameter is being estimated this method will derive a separate estimator for each parameter. In many cases of practical interest, there will be a unique maximum likelihood estimator that can be found by standard calculus techniques.

In a number of cases, it is possible to eliminate much of the algebra involved in finding the location of the maximum value for $L(\theta)$ by finding the value of θ that maximizes $\ln L(\theta)$. Recall from calculus that $\ln x$ is an increasing function of x that possesses an inverse function for all $x > 0$. Thus, $\ln f(x)$ has a derivative with the same sign as the derivative of $f(x)$. This means that the maximum of $\ln f(x)$ occurs at the same value of x as does the maximum of $f(x)$.

Example 10.4   Let X have a $B(n, p)$. In Example 10.2 we found a likelihood function to be

$$L(p) = \binom{n}{x} p^x (1 - p)^{n-x}.$$

We shall derive the maximum likelihood estimator for p by maximizing $L(p)$ directly and then by maximizing $\ln L(p)$. $L(p)$ is differentiable for all values of p. Therefore, differentiating $L(p)$ with respect to p gives

$$\frac{d}{dp} L(p) = \binom{n}{x} \left[x p^{x-1}(1 - p)^{n-x} - (n - x)p^x(1 - p)^{n-x-1} \right]$$

$$= \binom{n}{x} p^{x-1}(1 - p)^{n-x-1}[x(1 - p) - (n - x)p]$$

$$= \binom{n}{x} p^{x-1}(1 - p)^{n-x-1}[x - np].$$

Setting this equal to 0 and solving for p gives

$$p = \frac{x}{n}.$$

Therefore, the maximum likelihood estimator for p is

$$\hat{p} = \frac{X}{n}.$$

To illustrate the approach using logarithms we begin with

$$\ln L(p) = \ln \binom{n}{x} + x \ln p + (n - x) \ln(1 - p).$$

Differentiating this with respect to p gives

$$\frac{d}{dx} \ln L(p) = \frac{x}{p} - \frac{n - x}{1 - p}$$

$$= \frac{x(1 - p) - (n - x)p}{p(1 - p)}$$

$$= \frac{x - np}{p(1 - p)}.$$

Setting this equal to 0 and solving for p again gives

$$p = \frac{x}{n}.$$

It is easy to check that $p = \frac{x}{n}$ is the global maximum of both $L(p)$ and $\ln L(p)$. Note that the use of logarithms in this case resulted in less work than was the case in maximizing $L(p)$ directly. ∎

Example 10.5 Let $X_1, X_2, \ldots, X_n$ be a random sample from a $N(\mu, \sigma^2)$ distribution and assume that both μ and σ^2 are unknown. Here we need values of μ and σ^2 that *jointly* make $L(\mu, \sigma^2)$ a maximum. Recall from multivariable calculus that this involves the use of partial derivatives. In Example 10.1, we found the likelihood function to be

$$L(\mu, \sigma^2) = (2\pi\sigma^2)^{-n/2} e^{-(1/2\sigma^2)\sum_{i=1}^{n}(x_i - \mu)^2}.$$

Taking logarithms gives

$$\ln L(\mu, \sigma^2) = -\frac{n}{2}\ln(2\pi) - \frac{n}{2}\ln\sigma^2 - \frac{1}{2\sigma^2}\sum_{i=1}^{n}(x_i - \mu)^2.$$

Since there are 2 parameters to estimate we take partial derivatives with respect to μ and σ^2, respectively, and equate both to 0. This will give 2 equations that can be solved to obtain the maximum likelihood estimators as follows.

$$\frac{\partial}{\partial\mu}\ln L(\mu, \sigma^2) = \frac{1}{\sigma^2}\sum_{i=1}^{n}(x_i - \mu) = 0.$$

This can be solved immediately for μ to give

$$\mu = \frac{1}{n}\sum_{i=1}^{n} x_i = \bar{x}.$$

Therefore,

$$\hat{\mu} = \bar{X}.$$

Proceeding to the other partial derivative we find the following.

$$\frac{\partial}{\partial\sigma^2}L(\mu, \sigma^2) = -\frac{n}{2\sigma^2} + \frac{1}{2(\sigma^2)^2}\sum_{i=1}^{n}(x_i - \mu)^2 = 0$$

Solving for σ^2 gives

$$\sigma^2 = \frac{1}{n}\sum_{i=1}^{n}(x_i - \mu)^2.$$

Upon substituting $\hat{\mu}$ for μ we find that the maximum likelihood estimator for σ^2 is

$$\widehat{\sigma^2} = \frac{1}{n}\sum_{i=1}^{n}(X_i - \bar{X})^2.$$

∎

Example 10.6   Let $X_1, X_2, \ldots, X_n$ be a random sample from a $U(0, \theta)$ distribution. In Example 10.3 we found that the likelihood function is

$$L(\theta) = \begin{cases} \dfrac{1}{\theta^n}, & \text{if } 0 \le x_i \le \theta, \quad \text{for } i = 1, 2, \ldots, n, \quad \text{and} \\ 0, & \text{otherwise.} \end{cases}$$

The likelihood function has a simple form so we do not take logarithms. Differentiation with respect to θ gives

$$\frac{d}{d\theta} L(\theta) = \frac{n}{\theta^{n+1}},$$

which has no value of θ that makes the derivative equal 0. The derivative does, however, indicate that $L(\theta)$ is a decreasing function of θ. Therefore, we would like to make θ as small as possible. A closer inspection of the likelihood function shows that $L(\theta)$ has a discontinuity. The likelihood function jumps from a value of θ^{-n} to 0 when the value of θ is smaller than any of the observations $x_1, x_2, \ldots$, or x_n. Thus, as θ decreases the value of $L(\theta)$ will increase until $\theta = \max(x_1, x_2, \ldots, x_n)$ when it drops to 0 (see Figure 10.1). Therefore, the maximum value of $L(\theta)$ occurs at

$$\theta = \max(x_1, x_2, \ldots, x_n).$$

Let $Y_n = \max(X_1, X_2, \ldots, X_n)$. Then the maximum likelihood estimator for θ is the nth order statistic that we discussed in Chapter 9. That is,

$$\widehat{\theta} = Y_n.$$

Notice that the maximum of $L(\theta)$ occurred at a critical point where the function was not differentiable. We are sure that readers are gratified to note that their calculus instructors weren't being merely difficult by stressing that nondifferentiable points must also be investigated in the search for extrema. These things actually occur in solving real problems.

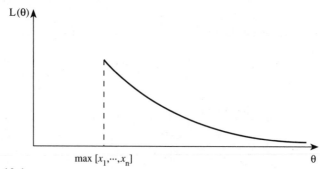

Figure 10.1

It was hinted earlier that there may be no unique value of θ that maximizes the likelihood function. Although somewhat contrived, the following example indicates how this can occur.

Example 10.7

Let $X_1, X_2, \ldots, X_n$ be a random sample from a population having a p.d.f. of

$$f(x \mid \theta) = \begin{cases} \dfrac{1}{2a}, & \theta - a \leq x \leq \theta + a, \quad \text{and} \\ 0, & \text{otherwise.} \end{cases}$$

We assume that the value of a is known and positive. Thus, θ is the parameter we wish to estimate. The likelihood function is

$$L(\theta) = \begin{cases} \dfrac{1}{(2a)^n}, & \theta - a \leq x_i \leq \theta + a, \ i = 1, 2, \ldots, n, \quad \text{and} \\ 0, & \text{otherwise.} \end{cases}$$

Notice that $L(\theta)$ is either $(2a)^{-n}$ or 0 with the maximum value being $(2a)^{-n}$. $L(\theta)$ attains this maximum whenever $\theta - a$ is less than or equal to every observed value and $\theta + a$ is greater than or equal to every observed value. Let Y_1 be the first order statistic and Y_n be the nth order statistic. Then $L(\theta)$ will be a maximum whenever $\theta - a \leq Y_1$ and $\theta + a \geq Y_n$. In other words, the likelihood function is maximized for any value of θ satisfying

$$Y_n - a \leq \theta \leq Y_1 + a.$$

Since the likelihood function attains its maximum value for a range of values of θ, there is no unique maximum likelihood estimator for θ. ∎

There are times when maximum likelihood estimators may be difficult to calculate. In fact, one must sometimes use numerical methods to approximate values of the estimates. The following serves as an example.

Example 10.8

Let $X_1, X_2, \ldots, X_n$ be a random sample from a $G(\alpha, \beta)$ distribution with both α and β unknown. We wish to find maximum likelihood estimators for both α and β. The likelihood function is

$$L(\alpha, \beta) = \frac{1}{[\Gamma(\alpha) \beta^\alpha]^n} \prod_{i=1}^{n} x_i^{\alpha-1} e^{-\sum_{i=1}^{n} x_i/\beta}.$$

Taking logarithms gives

$$\ln L(\alpha, \beta) = -n \ln \Gamma(\alpha) - n\alpha \ln \beta + (\alpha - 1) \sum_{i=1}^{n} \ln x_i - \sum_{i=1}^{n} \frac{x_i}{\beta}.$$

Taking partial derivatives with respect to α and β, respectively, and setting both to 0 gives the following.

$$\frac{\partial}{\partial \alpha} \ln L(\alpha, \beta) = -n \frac{\Gamma'(\alpha)}{\Gamma(\alpha)} - n \ln \beta + \sum_{i=1}^{n} \ln x_i = 0, \quad \text{and}$$

$$\frac{\partial}{\partial \beta} \ln L(\alpha, \beta) = -n \frac{\alpha}{\beta} + \sum_{i=1}^{n} \frac{x_i}{\beta^2} = 0$$

We can solve the second one to get β in terms of α.

$$\beta = \frac{\bar{x}}{\alpha}$$

If we substitute this for β in the first equation we find that we must solve

$$-n \frac{\Gamma'(\alpha)}{\Gamma(\alpha)} - n \ln \frac{\bar{x}}{\alpha} + \sum_{i=1}^{n} \ln x_i = 0$$

for α. There is no closed form solution for both α and β in this case. Once we have an estimate for one we can find the other easily, but that first estimator must be found by the use of some numerical method for finding the root of a function. ∎

Suppose that we have a maximum likelihood estimator, $\hat{p}$, for p in a binomial distribution but we really need an estimator for $q = 1 - p$. It turns out that we do not need to derive a new estimator for q from scratch. We can use the estimator for p to get that $\hat{q} = 1 - \hat{p}$. This useful result is a consequence of the following theorem. We leave the proof as an exercise.

Theorem 10.1

Let $\hat{\theta}$ be a maximum likelihood estimator for θ and let $h(\theta)$ be a function of θ. Then the maximum likelihood estimator for $h(\theta)$ is

$$\widehat{h(\theta)} = h(\hat{\theta}). \tag{10.2}$$

This result is often referred to as the *invariance property* of maximum likelihood estimators.

Example 10.9

We have seen that the maximum likelihood estimator for p in a $B(n, p)$ distribution is $\hat{p} = \frac{p}{n}$. We can use the invariance property to show that the maximum likelihood estimator for $\sqrt{p}$ would be $\widehat{\sqrt{p}} = \sqrt{\frac{\bar{x}}{n}}$. Similarly, the maximum likelihood estimator for $\ln p$ would be $\widehat{\ln p} = \ln\left(\frac{\bar{x}}{n}\right)$. ∎

Exercises 10.2

1. Show that if $g(x)$ is a one-to-one function of x, and $\widehat{\theta}$ is the maximum likelihood estimator for θ, then the maximum likelihood estimator for $g(\theta)$, $\widehat{g(\theta)}$, is $g(\widehat{\theta})$.

2. Let $X_1, X_2, \ldots, X_n$ be a random sample from a Poisson distribution with parameter λ. Find a maximum likelihood estimator for λ.

3. Let $X_1, X_2, \ldots, X_n$ be a random sample from a population with a probability density function of

$$f(x) = \begin{cases} \theta \, x^{\theta-1} , & 0 \le x \le 1, \quad \text{and} \\ 0 , & \text{otherwise.} \end{cases}$$

Find a maximum likelihood estimator for θ.

4. Let $X_1, X_2, \ldots, X_n$ be a random sample from a population with a probability density function of

$$f(x) = \begin{cases} \dfrac{1}{\beta} e^{-x/\beta} , & x > 0, \quad \text{and} \\ 0 , & \text{otherwise.} \end{cases}$$

Find a maximum likelihood estimator for β.

5. Let $X_1, X_2, \ldots, X_n$ be a random sample from a population with a Cauchy distribution whose p.d.f. is

$$f(x) = \frac{1}{\pi[1 + (x - \theta)^2]}, \quad -\infty < x < \infty.$$

Find a maximum likelihood estimator for θ.

6. Let $X_1, X_2, \ldots, X_n$ be a random sample from a population with a geometric distribution with probability of success p. Find a maximum likelihood estimator for p.

8. Let $X_1, X_2, \ldots, X_n$ be a random sample from a $G(\alpha, 1)$ distribution. Find a maximum likelihood estimator for α.

9. Let $X_1, X_2, \ldots, X_n$ be a random sample from a population having a p.d.f. of

$$f(x) = \begin{cases} \dfrac{\theta}{x^2} , & \theta \le x, \quad \text{and} \\ 0 , & \text{otherwise.} \end{cases}$$

Find a maximum likelihood estimator for θ.

10. Let $X_1, X_2, \ldots, X_n$ be a random sample from a population with a $Po(\lambda)$ distribution. Find a maximum likelihood estimator for $P(X = 0)$.

11. Let $X_1, X_2, \ldots, X_n$ be a random sample from a $N(\mu, \sigma^2)$ distribution. Find a maximum likelihood estimator for $E[e^X]$.

12. Let $X_1, X_2, \ldots, X_n$ be a random sample from a $U(a, b)$ distribution. Find maximum likelihood estimators for a and b. In addition, find a maximum likelihood estimator for $b - a$.

13. An ecologist wishes to determine the proportion of cutthroat trout in a lake. She goes out one day and catches and tags M trout, which are returned to the lake. A week later she returns to the lake and catches n cutthroat trout, r of which are tagged. Find the maximum likelihood estimator for the number of cutthroat in the lake. [*Hint:* The number of tagged trout caught has a hypergeometric distribution with $M = Np$. Consider the ratio $\frac{p(x+1)}{p(x)}$.]

14. The ecologist in Exercise 13 decides that, instead of catching n cutthroat trout and counting the number of tagged ones, she will catch and return fish to the lake until she catches r tagged fish. Find a maximum likelihood estimator for the proportion of tagged cutthroat trout in the lake using this sampling method.

15. Let $X, Y,$ and Z be nonnegative discrete random variables where the probability function is

$$f(x, y, z) = \begin{cases} \dfrac{n!}{x!\, y!\, z!}\, p_1^x p_2^y p_3^z, & \begin{array}{l} x, y, z \quad \text{integers}, \\[4pt] x + y + z \le n \end{array} \quad \text{and} \\[12pt] 0 & , \text{otherwise, with } p_1 + p_2 + p_3 = 1. \end{cases}$$

This is a trinomial distribution. Find the maximum likelihood estimators for $p_1, p_2,$ and p_3. In this problem, the likelihood function is the probability function.

16. Let $X_1, X_2, \ldots, X_n$ be a random sample from a double exponential distribution with a p.d.f. of

$$f(x) = \frac{1}{2} e^{-|x - \theta|}, \quad -\infty < x < \infty.$$

Show that the maximum likelihood estimator for θ is the sample median. Keep in mind that $|x_i - \theta|$ is not differentiable whenever some $x_i = \theta$.

17. Use the method of maximum likelihood to find estimators for the 5 parameters, $\mu_X, \mu_Y, \sigma_X^2, \sigma_Y^2,$ and ρ in a bivariate normal distribution based on a sample of n observations.

10.3
Method of Moments

In Example 10.8, we encountered a case where the computation of maximum likelihood estimators can be difficult. In such situations, estimators based on the idea of using sample moments to estimate distribution moments can be a practical alternative. The idea is relatively straightforward. We first define the rth *sample moment*, m_r, to be

$$m_r = \frac{1}{n} \sum_{i=1}^n x_i^r, \quad r = 1, 2, \ldots. \tag{10.3}$$

If the p.d.f. or p.f. for X depends on parameters $\theta_1, \theta_2, \ldots, \theta_k$, then the rth moment, $\mu_r' = E[X^r]$, when it exists, will be a function of those parameters. That is,

$$\mu_r' = g_r(\theta_1, \theta_2, \ldots, \theta_k).$$

The idea behind the *method of moments* is to use the rth sample moment, m_r, to estimate the rth moment of the distribution, μ_r', for $r = 1, 2, \ldots, k$. Since the moments are functions of the parameters, we then solve these for the parameters in terms of the moments. The method of moments then replaces the distribution moments with their estimators to obtain estimates for the parameters. Some examples will illustrate how this works.

Example 10.10   Let $X_1, X_2, \ldots, X_n$ be a random sample from a $B(1, p)$ distribution. We have seen in Chapter 5 that $E[X] = p$. In this case, there is a single parameter so we use m_1 to estimate p. Thus,

$$p = \frac{1}{n} \sum_{i=1}^{n} x_i.$$

If we let $Y = \sum_{i=1}^{n} X_i$ = number of successes observed, then

$$p = \frac{y}{n},$$

or a method of moments estimator for p is

$$\widehat{p} = \frac{Y}{n}.$$

Note that this is also the maximum likelihood estimator for p. ∎

Example 10.11   Let $X_1, X_2, \ldots, X_n$ be a random sample from a $N(\mu, \sigma^2)$ distribution. Here we must estimate both μ and σ^2. Therefore, we use the first 2 moments of the normal distribution. These are $E[X] = \mu$ and $E[X^2] = \sigma^2 + \mu^2$. This means that we estimate these 2 moments as follows.

$$\widehat{\mu} = m_1 = \overline{X}, \quad \text{and}$$

$$\widehat{\sigma^2 + \mu^2} = m_2$$

Solving for μ and σ^2 gives

$$\widehat{\mu} = \overline{X}, \quad \text{and}$$

$$\widehat{\sigma^2} = m_2 - m_1^2 = \frac{1}{n} \sum_{i=1}^{n} X_i^2 - \overline{X}^2.$$

Once again the method of moments estimators are the same as those obtained using maximum likelihood. ∎

In case the reader assumes from these two examples that maximum likelihood and the method of moments always give the same results, we look at some cases where the methods give different estimators.

Example 10.12   Let $X_1, X_2, \ldots, X_n$ be a random sample from a $U(0, \theta)$ distribution. Since there is a single parameter to be estimated we use

$$E[X] = \frac{\theta}{2}.$$

This gives that

$$\theta = 2 E[X],$$

and so a method of moments estimator for θ is

$$\widehat{\theta} = 2\overline{X}.$$

Recall that the maximum likelihood estimator for θ was $\widehat{\theta} = Y_n$, the nth order statistic. ∎

This example also points out one problem that can arise when using the method of moments. In some cases the estimators can give nonsensical results. To illustrate, suppose we have a sample of 4 observations from a $U(0, \theta)$ distribution with values 1, 2, 3, and 14. The maximum likelihood estimate is 14. The method of moments estimate is 10, which is clearly absurd in light of our having observed a value of 14.

The method of moments can sometimes be preferable to maximum likelihood from the standpoint of ease of computation. Recall the difficulty in computing estimates for both parameters of a gamma distribution using maximum likelihood, and compare that with the following.

Example 10.13   Let $X_1, X_2, \ldots, X_n$ be a random sample from a $G(\alpha, \beta)$ distribution. Since we need estimators for 2 parameters we use the first 2 moments. For the gamma distribution the first 2 moments are $E[X] = \alpha\beta$ and $E[X^2] = \alpha\beta^2 + \alpha^2\beta^2 = \beta E[X] + E[X]^2$. Equating sample moments to distribution moments gives

$$\overline{X} = \alpha\beta, \quad \text{and}$$

$$\frac{1}{n}\sum_{i=1}^{n} x_i^2 = \alpha\beta^2 + (\alpha\beta)^2.$$

Solving these for α and β gives that

$$\alpha = \frac{\overline{X}}{\beta}, \quad \text{and}$$

$$\beta = \frac{\dfrac{1}{n}\sum_{i=1}^{n} x_i^2 - \overline{X}}{\overline{X}}.$$

Therefore, the method of moments estimators for α and β are

$$\widehat{\alpha} = \frac{\overline{X}}{\widehat{\beta}}, \quad \text{and}$$

$$\widehat{\beta} = \frac{\dfrac{1}{n}\sum_{i=1}^{n} X_i^2 - \overline{X}}{\overline{X}}.$$

Note that no numerical techniques are required to compute values for these estimates. ∎

It must be pointed out that method of moments estimators do not necessarily exist. In order to apply the method of moments, we must be able to calculate moments for the population distribution. From Chapter 4 we have seen cases where the moments do not exist. In these cases, the method of moments cannot be used since there are no distribution moments to which we can equate the sample moments.

In a certain sense, the method of moments may be used to find *nonparametric* estimates. We would use the sample mean to estimate $E[X]$, m_2 to estimate $E[X^2]$, and so forth. That is, we can find estimators for distribution moments without referring to the specific parameters of the distribution. Hence, the term *nonparametric estimator* is used. In addition to the method of moments estimators, sample percentiles are often used as nonparametric estimators of population quantiles. For example, the sample median is commonly used as an estimator of the distribution median. This estimator has the advantage that the distribution median always exists. For this reason it is common practice to use quantile-based estimators in a nonparametric context. Nonparametric estimators are used when we are unable to determine the family of distributions that describes the population.

Exercises 10.3

1. Let $X_1, X_2, \ldots, X_n$ be a random sample from a Poisson distribution with parameter λ. Find a method of moments estimator for λ.

2. Let $X_1, X_2, \ldots, X_n$ be a random sample from a population with a probability density function of
$$f(x) = \begin{cases} \theta x^{\theta - 1}, & 0 \le x \le 1, \quad \text{and} \\ 0, & \text{otherwise.} \end{cases}$$
Find a method of moments estimator for θ.

3. Let $X_1, X_2, \ldots, X_n$ be a random sample from a population with a geometric distribution with probability of success p. Find a method of moments estimator for p.

4. Let $X_1, X_2, \ldots, X_n$ be a random sample from a population having a p.d.f. of
$$f(x) = \begin{cases} \dfrac{2\theta^2}{x^3}, & \theta \le x, \quad \text{and} \\ 0, & \text{otherwise.} \end{cases}$$
Find a method of moments estimator for θ.

5. Let $X_1, X_2, \ldots, X_n$ be a random sample from a population with a probability density function of
$$f(x) = \begin{cases} \dfrac{1}{\beta} e^{-x/\beta}, & x > 0, \quad \text{and} \\ 0, & \text{otherwise.} \end{cases}$$
Find a method of moments estimator for β.

6. Let $X_1, X_2, \ldots, X_n$ be a random sample from a $U(a, b)$ distribution. Find method of moments estimators for a and b.

7. Let $X, Y,$ and Z be nonnegative discrete random variables where the probability function is

$$f(x,y,z) = \begin{cases} \dfrac{n!}{x!\,y!\,z!}\, p_1^x p_2^y p_3^z\,, & \begin{cases} x, y, z \text{ integers,} \quad \text{and} \\ \quad x + y + z \le n \end{cases} \\ \quad 0 & , \text{ otherwise, with } p_1 + p_2 + p_3 = 1. \end{cases}$$

Find the method of moments estimators for p_1, p_2, and p_3.

8. Let $X_1, X_2, \ldots, X_n$ be a random sample from a double exponential distribution with a p.d.f. of

$$f(x) = \frac{1}{2} e^{-|x-\theta|}, \, -\infty < x < \infty.$$

Find the method of moments estimator for θ.

10.4
Properties of Estimators

We have introduced the reader to two different methods for deriving estimators for parameters. There are other methods that can be found in more advanced texts, such as *The Advanced Theory of Statistics* by M. G. Kendall and A. Stuart, or *Theory of Point Estimation* by E. L. Lehmann. The level of these, however, is well beyond that assumed for this book. Anyway, with a potentially large number of different estimators for the same parameter we need some way to distinguish between them. Therefore, we shall introduce some of the more common ways by which estimators are categorized. Some of these are used simply to classify estimators, while others serve as criteria for selecting a "best" estimator.

10.4.1 Unbiasedness

One of the most common properties by which estimators are classified is that of *unbiasedness*. The idea is that if the average value of a large number of samples is equal to the true value parameter and is not consistently higher or lower, an estimator then is classified as being unbiased. A formal definition of this concept is given below.

Definition 10.1

An estimator $\widehat{\theta}$ for a parameter θ is said to be *unbiased* if

$$E[\widehat{\theta}] = \theta. \tag{10.4}$$

If not, $\widehat{\theta}$ is said to be a *biased* estimator for θ. Furthermore, the quantity

$$B(\widehat{\theta}, \theta) = E[\widehat{\theta}] - \theta \tag{10.5}$$

is called the *bias* of $\widehat{\theta}$.

That is, if an estimator for a parameter has a sampling distribution whose mean is equal to the parameter being estimated, then we have an unbiased estimator and the bias will be zero. If the mean is anything else, then it is a biased estimator and the bias will not be zero. If the bias is negative, then $\widehat{\theta}$ consistently underestimates the actual value of θ. If the bias is positive, then $\widehat{\theta}$ will consistently overestimate the value of θ.

Example 10.14   The maximum likelihood estimator and the method of moments estimator for the parameter p in a binomial distribution was found to be $\widehat{p} = \frac{X}{n}$, where X = number of successes, and n = number of trials. We determine if $\widehat{p}$ is unbiased as follows.

$$E\left[\widehat{p}\right] = E\left[\frac{X}{n}\right]$$

$$= \frac{1}{n}E[X]$$

Since $X \sim B(n,p)$, we note that $E[X] = np$, and

$$E[\widehat{p}] = \frac{np}{n}$$

$$= p.$$

Since $E[\widehat{p}] = p$, Definition 10.1 gives that $\widehat{p}$ is an unbiased estimator for p. ∎

Example 10.15   The maximum likelihood estimator for θ in a $U(0, \theta)$ distribution was found to be $\widehat{\theta}_1 = Y_n$, the largest order statistic. The method of moments estimator was $\widehat{\theta}_2 = 2\bar{X}$. Using Equation 9.8 we find that the sampling distribution for Y_n has a p.d.f. of

$$f_{Y_n}(y) = \begin{cases} n\dfrac{y^{n-1}}{\theta^n}, & 0 \leq y \leq \theta, \quad \text{and} \\ 0, & \text{otherwise.} \end{cases}$$

Then

$$E[\widehat{\theta}_1] = \int_0^\theta y\, n\, \frac{y^{n-1}}{\theta^n}\, dy$$

$$= \frac{n}{n+1}\, \theta.$$

Thus $\widehat{\theta}_1$ is a biased estimator. The bias is

$$B(\widehat{\theta}_1, \theta) = E[\widehat{\theta}_1] - \theta$$

$$= \frac{n}{n+1}\, \theta - \theta$$

$$= \frac{-1}{n+1}\, \theta.$$

On the other hand,

$$E[\widehat{\theta}_2] = E[2\bar{X}]$$

$$= \frac{2}{n}\sum_{i=1}^{n} E[X_i].$$

Since $X_i \sim U(0, \theta)$, then $E[X_i] = \frac{\theta}{2}$. Thus,

$$E[\widehat{\theta}_2] = \frac{2}{n} n \frac{\theta}{2}$$
$$= \theta.$$

This means that, in this case, the method of moments estimator is unbiased.   ■

This example shows that maximum likelihood estimators are not, in general, unbiased. In the exercises it will also be seen that method of moments estimators may also be biased. In many cases it is possible to alter a biased estimator by multiplying by an appropriate constant to get an estimator that is unbiased. An unbiased estimator that is obtained in this manner is said to be *based on* the biased estimator.

Example 10.16   The maximum likelihood estimator for θ in the $U(0, \theta)$ distribution has been shown to be biased. We shall use the fact that $E[aX] = aE[X]$ to obtain an unbiased estimator that is based on the maximum likelihood estimator. In Example 10.15, we found that

$$E[\widehat{\theta}_1] = E[Y_n] = \frac{n}{n+1} \theta.$$

Now if we multiply this expectation by $\frac{n+1}{n}$ we find

$$\frac{n+1}{n} \frac{n}{n+1} \theta = \theta.$$

Thus, if we let

$$\widehat{\theta}_3 = \frac{n+1}{n} Y_n,$$

then

$$E[\widehat{\theta}_3] = \theta.$$

Therefore, $\widehat{\theta}_3 = \frac{n+1}{n} Y_n$ is an unbiased estimator for θ that is based on the maximum likelihood estimator.   ■

An interesting method for reducing and, sometimes, removing bias was suggested by Quenouille in 1956. The process is known as the *jackknife*. It is applicable when it is possible to write the bias, $B(\widehat{\theta}, \theta)$ in the form

$$B(\widehat{\theta}, \theta) = E[\widehat{\theta}] - \theta = \sum_{i=1}^{\infty} \frac{a_i}{n^i}.$$

Here the a_i's may be functions of θ but not of n. Let $\widehat{\theta}_n$ denote the estimator for θ based on a sample of n observations. The idea is to begin with the original sample of n observations, remove a single observation, and compute the estimator. Let $\widehat{\theta}_{(n-1),i}$ denote the estimator $\widehat{\theta}$ computed after removing the ith observation.

There will be n values for $\widehat{\theta}_{(n-1),i}$. Now let $\widetilde{\theta}_{n-1}$ denote the average

$$\widetilde{\theta}_{n-1} = \frac{1}{n} \sum_{i=1}^{n} \widehat{\theta}_{(n-1),i}.$$

The bias of $\widetilde{\theta}_{n-1}$ is

$$E[\widetilde{\theta}_{n-1}] - \theta = \frac{1}{n} \sum_{i=1}^{n} (E[\widetilde{\theta}_{(n-1),i}] - \theta)$$

$$= \sum_{j=1}^{\infty} \frac{a_j}{(n-1)^j}.$$

Consider the estimator

$$\widehat{\theta}_J = n\widehat{\theta}_n - (n-1)\widetilde{\theta}_{n-1}.$$

This is the *jackknife estimator*. The bias of this estimator is

$$E[\widehat{\theta}_J] - \theta = n \sum_{i=1}^{\infty} \frac{a_i}{(n-1)^i} - \sum_{i=1}^{\infty} \frac{a_i}{(n-1)^i}$$

$$= \sum_{i=2}^{\infty} \left(\frac{1}{n^i} - \frac{1}{(n-1)^2} \right)$$

$$= -\frac{a_2}{n^2} - \frac{a_3(2n-1)}{n^2(n-1)^2} - \cdots.$$

Note that, while the bias of the original estimator was of the order $\frac{1}{n}$, the bias of the jackknife estimator has been reduced to being of the order $\frac{1}{n^2}$. In addition, if $a_2 = a_3 = \cdots = 0$, then the jackknife estimator will be unbiased. This is how the jackknife can remove all of the bias in some cases. We shall illustrate the idea with an example.

Example 10.17   Let $X_1, \ldots, X_n$ be a random sample from a $N(\mu, \sigma^2)$ distribution. The maximum likelihood estimator for σ^2 is

$$\widehat{\sigma^2} = \frac{1}{n} \sum_{i=1}^{n} (X_i - \bar{X})^2.$$

The expected value of $\widehat{\sigma^2}$ is

$$E[\widehat{\sigma^2}] = \frac{n-1}{n}\sigma^2 = \sigma^2 - \frac{\sigma^2}{n}.$$

Since the bias is of the order $\frac{1}{n}$, the jackknife estimator will be unbiased. We leave it as an exercise to show that the jackknife estimator is

$$\widehat{\sigma_J^2} = s^2 = \frac{1}{n-1} \sum_{i=1}^{2} (X_i - \bar{X})^2.$$

We have already seen in the exercises in Chapter 4 that $E[s^2] = \sigma^2$. ■

10.4.2 Efficiency

The fact that an estimator is unbiased is no indication that it necessarily does a "good" (whatever that means) job in estimating a parameter. It simply shows that the mean of its sampling distribution is equal to the parameter being estimated. It can turn out that the bulk of the values calculated by an unbiased estimator are further away from the actual value of the parameter than those calculated by some biased estimator. For this reason unbiasedness is not considered to be an *optimal property* of an estimator. It merely is a property that allows us to classify estimators. What we need is some measure of how "close" an estimator is to a parameter. Since estimators are random variables we need to look at how close, on the average, an estimator is to a parameter. This means we want to use the expected value of a measure of distance between a parameter and its estimator. There are a number of ways to measure this distance. The one we shall use is called the *mean square error* or *mse*.

Definition 10.2	Given an estimator $\widehat{\theta}$ for the parameter θ the *mean square error* is $$mse = E[(\widehat{\theta} - \theta)^2]. \qquad (10.6)$$

A useful alternative to Equation 10.6 for computational purposes is based on the definition of the *bias* of $\widehat{\theta}$. We leave it as an exercise to verify the following.

$$mse = \text{Var}(\widehat{\theta}) - \left[B(\widehat{\theta}, \theta)\right]^2 \qquad (10.7)$$

Another possibility for measuring average closeness is the *mean absolute error,* which is $E[|\widehat{\theta} - \theta|]$. It should be clear from the standpoint of using it mathematically that *mse* is more convenient. For instance, there are no differentiability problems to take into account. This is one of the reasons for the popularity of mean square error.

Since mean square error measures how close, on the average, an estimator comes to the actual value of the parameter we can use it as a criterion for determining when one estimator is "better" than another. It would make sense to say that one estimator is better than another if it has a mean square error that is smaller than its competitor for all possible values of the parameter being estimated.

Example 10.18	To illustrate the idea of minimizing the mean square error we shall consider a sample of n observations from a $U(0, \theta)$ distribution. We consider estimators for θ that are of the form $\widehat{\theta}_c = cY_n$, where Y_n is the nth order statistic. The goal is

to determine the value of c that minimizes *mse*. In Example 10.15 we found that

$$E[Y_n] = \frac{n}{n+1}\theta.$$

To compute *mse* we need the second moment of Y_n, which is computed as follows.

$$E[Y_n^2] = \int_0^\theta y^2 n\frac{y^{n-1}}{\theta^n}\, dy = \frac{n}{n+2}\theta^2$$

Therefore,

$$mse(cY_n) = E[(cY_n - \theta)^2] = c^2 E[Y_n^2] - 2c\theta E[Y_n] + \theta^2$$

$$= c^2\frac{n}{n+2}\theta^2 - 2c\frac{n}{n+1}\theta^2 + \theta^2.$$

Thus, to find the value of c that minimizes *mse* we differentiate *mse* with respect to c as follows.

$$\frac{d}{dc}mse = 2c\frac{n}{n+2}\theta^2 - 2\frac{n}{n+1}\theta^2 = 0$$

Solving for c gives

$$c = \frac{n+2}{n+1}.$$

Therefore, the estimator of the form $\widehat{\theta} = cY_n$, which minimizes the mean square error, is $\widehat{\theta} = \frac{n+2}{n+1}Y_n$. Note that this is a biased estimator. ∎

Ideally, we would like to find an estimator that has a mean square error that is at least as small as that of all possible estimators for a given parameter. This turns out to be a very difficult thing to do. Therefore, it is common practice to consider only unbiased estimators and search for an estimator that minimizes *mse* among that group. Note that if we restrict ourselves to unbiased estimators, then mean square error is simplified as follows.

$$mse = E[(\widehat{\theta} - \theta)^2] = E[(\widehat{\theta} - E[\widehat{\theta}])^2] = \text{Var}(\widehat{\theta})$$

This means that the mean square error criterion becomes a search for the estimator whose sampling distribution has the smallest variance. Ignoring biased estimators is not a severe restriction. We stated earlier that in most cases it is possible to find an unbiased estimator that is based on a given biased one.

Unbiased estimators are compared using a quantity known as the *relative efficiency*. The definition follows.

Definition 10.3

> If $\widehat{\theta}_1$ and $\widehat{\theta}_2$ are two unbiased estimators for θ, the efficiency of $\widehat{\theta}_1$ relative to $\widehat{\theta}_2$ is
>
> $$e(\widehat{\theta}_1, \widehat{\theta}_2) = \frac{\text{Var}(\widehat{\theta}_2)}{\text{Var}(\widehat{\theta}_1)}. \qquad (10.8)$$
>
> If $e(\widehat{\theta}_1, \widehat{\theta}_2) > 1$, then $\widehat{\theta}_1$ is said to be *more efficient* than $\widehat{\theta}_2$.

Example 10.19   We have seen that $\widehat{\theta}_3 = \frac{n+1}{n} Y_n$ is an unbiased estimator for θ in a $U(0, \theta)$ distribution. The method of moments estimator, $\widehat{\theta}_2 = 2\bar{X}$, was also unbiased. We wish to determine which of these 2 estimators is the more efficient. To do this we need to compute the variances of each one. For the method of moments estimator recall that the variance of a $U(a, b)$ random variable is $\frac{(b-a)^2}{12}$. Using this and the properties of expected values we proceed as follows.

$$\text{Var}(\widehat{\theta}_2) = \text{Var}(2\bar{X})$$

$$= 4\,\text{Var}(\bar{X})$$

$$= \frac{4}{n^2} \sum_{i=1}^{n} \text{Var}(X_i)$$

$$= \frac{4}{n^2}\, n\, \frac{\theta^2}{12}$$

$$= \frac{\theta^2}{3n}$$

We need to compute the variance of $\widehat{\theta}_3$ directly. Since we have the mean we need to find the second moment. This goes as follows.

$$E[\widehat{\theta}_3^2] = E\left[\left(\frac{n+1}{n} Y_n\right)^2\right]$$

$$= \frac{(n+1)^2}{n^2} \int_0^{\theta} y^2\, n\, \frac{y^{n-1}}{\theta^n}\, dy$$

$$= \frac{(n+1)^2}{n^2} \frac{n}{n+2} \theta^2$$

$$= \frac{(n+1)^2}{n(n+2)} \theta^2$$

Therefore,

$$\text{Var}(\widehat{\theta}_3) = E\left[(\widehat{\theta}_3^2)^2\right] - \theta^2$$

$$= \frac{(n+1)^2}{n(n+2)} \theta^2 - \theta^2$$

$$= \frac{\theta^2}{n(n+2)}.$$

Therefore, the efficiency of $\widehat{\theta}_3$ relative to $\widehat{\theta}_2$ is

$$e(\widehat{\theta}_3, \widehat{\theta}_2) = \frac{\text{Var}(\widehat{\theta}_2)}{\text{Var}(\widehat{\theta}_3)}$$

$$= \frac{\theta^2 / (3n)}{\theta^2 / [n(n + 2)]}$$

$$= \frac{n + 2}{3}.$$

This shows that the unbiased estimator based on the maximum likelihood estimator is more efficient for all samples of size 2 or more. In addition, the 2 are equivalent, in terms of efficiency, for samples consisting of a single observation. This latter observation is not terribly surprising when we note that the 2 estimators are identical if there is only 1 observation. Let X be that single observation. Then $n = 1$, $Y_n = X$, $\bar{X} = X$, and

$$\widehat{\theta}_3 = \frac{n + 1}{n} Y_n = 2X = 2\bar{X} = \widehat{\theta}_2.$$ ∎

We have seen that it is possible for one estimator to be more efficient than another for all values of the parameter being estimated. Therefore, it is logical to ask whether, in a given estimation problem, there is an unbiased estimator that is more efficient than all other unbiased estimators. Sometimes there is, and this motivates the following definition.

Definition 10.4

> An unbiased estimator, $\widehat{\theta}_0$, for θ is said to be a *uniform minimum variance unbiased estimator (umvue)* for θ if, given any other unbiased estimator, $\widehat{\theta}$, for θ it is true that
>
> $$\text{Var}(\widehat{\theta}_0) \le \text{Var}(\widehat{\theta})$$
>
> for all possible values of θ.

Another way of putting this is that $\widehat{\theta}_0$ is a umvue for θ if $e(\widehat{\theta}_0, \widehat{\theta}) \ge 1$ for all possible values of θ.

As might be expected it is a very difficult problem to verify the existence of a umvue and to determine what it is. Most of the theory is beyond the level of this book. The reader with a strong background in analysis is referred to *Theory of Point Estimation* by E. L. Lehmann. We will give a partial answer to this question in this and succeeding sections.

One approach is to look at the existence of a lower limit for the variance of any unbiased estimator. It is possible to find such a lower limit if the population probability density function or probability function is sufficiently well behaved. We shall be somewhat vague as to exactly what is meant by this term. In essence "sufficiently well behaved" will refer to functions that have the property that the derivative of the expectation equals the expectation of the derivative. The result we shall cite is commonly referred to as the *Cramér-Rao Lower Bound* or the *Information Inequality*. We give a proof for the case of a continuous population. The discrete case is identical if the integrals are replaced by summations. In addition, we assume that the distribution depends only on a single unknown parameter. For notational convenience we shall use $L(\theta)$ to denote the joint p.d.f. for the sample. That is,

$$L(\theta) = f(x_1 \mid \theta)f(x_2 \mid \theta) \cdots f(x_n \mid \theta).$$

The proof may be omitted without loss of continuity.

Theorem 10.2

> **Cramér-Rao Lower Bound**   Let $X_1, X_2, \ldots, X_n$ be a random sample from a distribution with probability density function or probability function $f(x \mid \theta)$, with θ being a scalar parameter and $f(x \mid \theta)$ is sufficiently well behaved. If $L(\theta)$ is differentiable with respect to θ and $T = g(X_1, X_2, \ldots, X_n)$ is an unbiased estimator for $\tau(\theta)$ whose variance exists and is finite, then
>
> $$\text{Var}(T) \geq \frac{[\tau'(\theta)]^2}{E\left[\left(\dfrac{\partial \ln L}{\partial \theta}\right)^2\right]}. \tag{10.9}$$

Proof   We first establish a few basic facts that will be useful in the course of the proof. Since $L(\theta)$ is the joint p.d.f. of $X_1, X_2, \ldots, X_n$, we have that

$$\int_{-\infty}^{\infty}\int_{-\infty}^{\infty} \cdots \int_{-\infty}^{\infty} L(\theta)\, dx_1\, dx_2 \cdots dx_n = 1. \tag{i}$$

Also, because t is unbiased for $\tau(\theta)$, we have that

$$\int_{-\infty}^{\infty}\int_{-\infty}^{\infty} \cdots \int_{-\infty}^{\infty} t L(\theta)\, dx_1\, dx_2 \cdots dx_n = E[T] = \tau(\theta). \tag{ii}$$

We now differentiate (i) and (ii) with respect to θ. Then, if we can interchange the order of integration and differentiation (this is what we meant by the term *sufficiently well behaved*), we find that

$$\frac{\partial}{\partial \theta}\int_{-\infty}^{\infty}\int_{-\infty}^{\infty} \cdots \int_{-\infty}^{\infty} L(\theta)\, dx_1\, dx_2 \cdots dx_n = \int_{-\infty}^{\infty}\int_{-\infty}^{\infty} \cdots \int_{-\infty}^{\infty} \frac{\partial L(\theta)}{\partial \theta}\, dx_1\, dx_2 \cdots dx_n$$

$$= \frac{\partial(1)}{\partial \theta} = 0, \tag{iii}$$

and

$$\frac{\partial}{\partial \theta}\int_{-\infty}^{\infty}\int_{-\infty}^{\infty} \cdots \int_{-\infty}^{\infty} t L(\theta)\, dx_1\, dx_2 \cdots dx_n = \int_{-\infty}^{\infty}\int_{-\infty}^{\infty} \cdots \int_{-\infty}^{\infty} t\, \frac{\partial L(\theta)}{\partial \theta}\, dx_1\, dx_2 \cdots dx_n$$

$$= \frac{\partial}{\partial \theta}\tau(\theta) = \tau'(\theta). \tag{iv}$$

Now, note that

$$\frac{\partial L(\theta)}{\partial \theta} = \frac{1}{L(\theta)} \frac{\partial L(\theta)}{\partial \theta} L(\theta) = \frac{\partial \ln L(\theta)}{\partial \theta} L(\theta),$$

and make this substitution in (iv) to get that

$$\tau'(\theta) = \int_{-\infty}^{\infty} \int_{-\infty}^{\infty} \cdots \int_{-\infty}^{\infty} t \frac{\partial \ln L(\theta)}{\partial \theta} L(\theta) \, dx_1 \, dx_2 \cdots dx_n. \qquad \textbf{(v)}$$

We now note that Equation (v) can be written in the form

$$\tau'(\theta) = \int_{-\infty}^{\infty} \int_{-\infty}^{\infty} \cdots \int_{-\infty}^{\infty} [t - \tau(\theta)] \frac{\partial \ln L(\theta)}{\partial \theta} L(\theta) \, dx_1 \, dx_2 \cdots dx_n.$$

We now use the Cauchy-Schwarz Inequality, Theorem 4.15, to finish the proof. To do this we must show that

$$\int_{-\infty}^{\infty} \int_{-\infty}^{\infty} \cdots \int_{-\infty}^{\infty} \tau(\theta) \frac{\partial \ln L(\theta)}{\partial \theta} L(\theta) \, dx_1 \, dx_2 \cdots dx_n = 0.$$

This is shown as follows.

$$\int_{-\infty}^{\infty} \int_{-\infty}^{\infty} \cdots \int_{-\infty}^{\infty} \tau(\theta) \frac{\partial \ln L(\theta)}{\partial \theta} L(\theta) \, dx_1 \, dx_2 \cdots dx_n$$

$$= \tau(\theta) \int_{-\infty}^{\infty} \int_{-\infty}^{\infty} \cdots \int_{-\infty}^{\infty} \frac{\partial L(\theta)}{\partial \theta} \, dx_1 \, dx_2 \cdots dx_n$$

$$= \tau(\theta) \cdot 0 = 0$$

Thus, by the Cauchy-Schwarz Inequality,

$$[\tau'(\theta)]^2 = \left(\int_{-\infty}^{\infty} \int_{-\infty}^{\infty} \cdots \int_{-\infty}^{\infty} [t - \tau(\theta)] \frac{\partial \ln L(\theta)}{\partial \theta} L(\theta) \, dx_1 \, dx_2 \cdots dx_n \right)^2$$

$$\leq \left(\int_{-\infty}^{\infty} \int_{-\infty}^{\infty} \cdots \int_{-\infty}^{\infty} [t - \tau(\theta)]^2 L(\theta) \, dx_1 \, dx_2 \cdots dx_n \right)$$

$$\cdot \left(\int_{-\infty}^{\infty} \int_{-\infty}^{\infty} \cdots \int_{-\infty}^{\infty} \left(\frac{\partial \ln L(\theta)}{\partial \theta} \right)^2 L(\theta) \, dx_1 \, dx_2 \cdots dx_n \right)$$

$$= \mathrm{Var}(T) E \left[\left(\frac{\partial \ln L(\theta)}{\partial \theta} \right)^2 \right].$$

Note: The requirement that we interchange the order of integration and differentiation means that distributions whose ranges depend on the value of the parameter are not covered by this theorem. That is, distributions like the uniform may not be analyzed using the Cramér-Rao Lower Bound.   ■

If $L(\theta)$ is twice differentiable with respect to θ it is often possible to compute the expectation in the Cramér-Rao Lower Bound more easily by making use of the fact

$$E\left[\left(\frac{\partial \ln L(\theta)}{\partial \theta}\right)^2\right] = -E\left[\frac{\partial^2 \ln L(\theta)}{\partial \theta^2}\right]. \qquad \textbf{(10.10)}$$

The proof is left as an exercise.

An unbiased estimator for $\tau(\theta)$ whose variance is equal to the Cramér-Rao Lower Bound is said to be an *efficient* estimator. This is the basis for the following definition of *absolute efficiency* or *efficiency*.

Definition 10.5

> If the Cramér-Rao Lower Bound for an unbiased estimator for $\tau(\theta)$ is defined, then the *efficiency* of an unbiased estimator, $\widehat{\tau(\theta)}$, of $\tau(\theta)$ is
>
> $$e(\widehat{\tau(\theta)}) = \frac{[\tau'(\theta)]^2}{\text{Var}(\widehat{\tau(\theta)})\, E\left[\left(\frac{\partial \ln L(\theta)}{\partial \theta}\right)^2\right]}. \qquad \textbf{(10.11)}$$

Example 10.20

Let X have a $B(n,p)$ distribution. We want to find the lower bound for the variance of an unbiased estimator for p. In the notation of Theorem 10.2

$$\tau(p) = p, \quad \tau'(p) = 1, \quad \text{and} \quad L(p) = \binom{n}{x} p^x (1-p)^{n-x}.$$

Then

$$\ln L(p) = \ln\binom{n}{x} + x \ln p + (n-x) \ln(1-p).$$

Taking the derivative of $\ln L(p)$ with respect to p gives

$$\frac{\partial}{\partial p} \ln L(p) = \frac{x}{p} - \frac{n-x}{1-p}$$

$$= \frac{x - np}{p(1-p)}.$$

If we compute the expectation in the denominator of Equation 10.10 we get

$$E\left[\left(\frac{\partial \ln L(p)}{\partial p}\right)^2\right] = E\left[\left(\frac{x - np}{p(1-p)}\right)^2\right]$$

$$= \frac{E[(x-np)^2]}{[p(1-p)]^2}$$

$$= \frac{\text{Var}(X)}{[p(1-p)]^2}$$

$$= \frac{np(1-p)}{p^2(1-p)^2}$$

$$= \frac{n}{p(1-p)}.$$

Thus, the lower bound for the variance of an unbiased estimator for p is, by Theorem 10.2,

$$\text{Var}(\widehat{p}) \geq \cfrac{1}{\cfrac{n}{[p(1-p)]}} = \frac{p(1-p)}{n}.$$

Note that the variance of the maximum likelihood estimator and the method of moments estimator is

$$\text{Var}(\widehat{p}) = \text{Var}\left(\frac{X}{n}\right)$$

$$= \frac{1}{n^2}\, np(1-p)$$

$$= \frac{p(1-p)}{n}.$$

Since the variance of these estimators equals that of the Cramér-Rao Lower Bound, we see that $\widehat{p} = \frac{X}{n}$ is an efficient estimator for p.

Before leaving this example we wish to illustrate the use of Equation 10.10. The second partial derivative with respect to p is

$$\frac{\partial^2 \ln L(p)}{\partial p^2} = -\frac{x}{p^2} - \frac{n-x}{(1-p)^2}.$$

Then

$$E\left[\frac{\partial^2 \ln L(p)}{\partial p^2}\right] = E\left[\frac{-np^2 + 2px - x}{[p(1-p)]^2}\right]$$

$$= \frac{-np^2 + 2np^2 - np}{p^2(1-p)^2}$$

$$= \frac{-n}{p(1-p)}.$$

In this case, Equation 10.9 seems to be marginally easier in that only a single derivate needed to be taken. The expectations involved were easy to compute in both cases. ∎

Example 10.21

We wish to compute the lower bound for the variance of an unbiased estimator of μ based on a sample of size n from a $N(\mu, \sigma^2)$ distribution. Here

$$L(\mu) = \prod_{i=1}^{n} \frac{1}{\sqrt{2\pi\sigma^2}} e^{-(x_i-\mu)^2/2\sigma^2}$$

$$= (2\pi\sigma^2)^{-n/2} e^{-(1/2\sigma^2)\sum_{i=1}^{n}(x_i-\mu)^2}.$$

The logarithm of $L(\mu)$ is

$$\ln L(\mu) = -\frac{n}{2}\ln(2\pi) - \frac{n}{2}\ln(\sigma^2) - \frac{1}{2\sigma^2}\sum_{i=1}^{n}(x_i - \mu)^2,$$

and

$$\frac{\partial \ln L(\mu)}{\partial \mu} = \frac{1}{2\sigma^2} \sum_{i=1}^{n} (x_i - \mu)$$

$$= \frac{n}{\sigma^2}(\bar{X} - \mu).$$

If we use Equation 10.9 to obtain the denominator we find

$$E\left[\left(\frac{\partial \ln L(\mu)}{\partial \mu}\right)^2\right] = \frac{n^2}{(\sigma^2)^2} E[(\bar{X} - \mu)^2]$$

$$= \frac{n^2}{(\sigma^2)^2} \text{Var}(\bar{X})$$

$$= \frac{n^2}{(\sigma^2)^2} \frac{\sigma^2}{n}$$

$$= \frac{n}{\sigma^2}.$$

Since $\tau(\mu) = \mu$ and $\tau'(\mu) = 1$, Theorem 10.2 gives that, if $\hat{\mu}$ is an unbiased estimator for μ, then

$$\text{Var}(\hat{\mu}) \geq \frac{\sigma^2}{n}.$$

If we use Equation 10.10 we find that

$$\frac{\partial^2 \ln L(\mu)}{\partial \mu^2} = -\frac{n}{\sigma^2}.$$

Since this is a constant we obtain the lower bound with much less effort. Once again note that both the maximum likelihood and the method of moments estimators were $\hat{\mu} = \bar{X}$, which is unbiased with a variance of $\frac{\sigma^2}{n}$. Thus, both methods found the efficient estimator for μ. ■

In the two preceding examples we noted that the umvue for a parameter was efficient and was found by both the method of maximum likelihood and the method of moments. Unfortunately, this is not the case in general. A umvue may simply not exist for a given problem. That is, in some cases, no unbiased estimator exists that has a smallest variance for all possible values of the parameter. Furthermore, even if a umvue does exist there is no guarantee that it will have a variance equal to the Cramér-Rao Lower Bound. For example, in the case of a normal distribution the sample variance s^2 is a umvue for σ^2 when the value of μ is not known. It turns out, however, that its variance is greater than that of the Cramér-Rao Lower Bound. Furthermore, the population distribution may not meet the conditions of Theorem 10.2. Finding umvue estimators in such cases will be considered later.

It turns out that when a umvue does exist whose variance attains the Cramér-Rao Lower Bound there is a simple way of determining this and of finding out what that estimator is. In the proof of Theorem 10.2 we used the Cauchy-Schwarz

Inequality. It turns out that $(E[XY])^2 = E[X^2]E[Y^2]$ if and only if $Y = kX$, where k is some real constant. In our case, this condition means that an unbiased estimator $\widehat{\tau(\theta)}$ for $\tau(\theta)$ will attain the lower bound if and only if

$$\frac{\partial \ln L(\theta)}{\partial \theta} = k[\widehat{\tau(\theta)} - \tau(\theta)], \tag{10.12}$$

where k is a constant whose value may depend on θ but not on $X_1, X_2, \ldots, X_n$. This means that if we can write $\frac{\partial \ln L(\theta)}{\partial \theta}$ in the form of Equation 10.12, then there exists a umvue estimator for $\tau(\theta)$ which attains the lower bound and that $\widehat{\tau(\theta)}$ is that estimator. It further turns out that

$$\text{Var}(\widehat{\tau(\theta)}) = \frac{\tau'(\theta)}{k}.$$

The proof of this is left as an exercise.

Example 10.22

We wish to determine if there is a umvue for σ^2 in a $N(\mu, \sigma^2)$ which attains the Cramér-Rao Lower Bound. Recall from Example 10.5 that

$$\ln L(\mu, \sigma^2) = -\frac{n}{2}\ln(2\pi) - \ln \sigma^2 - \frac{1}{2\sigma^2}\sum_{i=1}^{n}(x_i - \mu)^2.$$

Differentiation with respect to σ^2 gives

$$\frac{\partial \ln L(\mu, \sigma^2)}{\partial \sigma^2} = -\frac{n}{2\sigma^2} + \frac{1}{2(\sigma^2)^2}\sum_{i=1}^{n}(x_i - \mu)^2$$

$$= \frac{n}{2(\sigma^2)^2}\left[\frac{1}{n}\sum_{i=1}^{n}(x_i - \mu)^2 - \sigma^2\right].$$

Since this is in the form of Equation 10.12 we see that the umvue for σ^2 is

$$\widehat{\sigma^2} = \frac{1}{n}\sum_{i=1}^{n}(x_i - \mu)^2.$$

Since $k = \frac{2(\sigma^2)^2}{n}$ and $\frac{\partial}{\partial \sigma^2}\sigma^2 = 1$ we see from the remarks above that

$$\text{Var}(\widehat{\sigma^2}) = \frac{2(\sigma^2)^2}{n}.$$

Note that this estimator requires that μ be known in order to attain the lower bound. It should also be clear that there is no unbiased estimator for σ that attains the lower bound. Keep in mind that this does not imply that there is no umvue for σ. It just means that, if one does exist, it will have a variance that is greater than the Cramér-Rao Lower Bound.   ■

Example 10.23   Let $X_1, X_2, \ldots, X_n$ be a random sample from a $G(1, \beta)$ distribution. In order to determine the existence of an efficient estimator for some $\tau(\beta)$, we compute

$$L(\beta) = \frac{1}{\beta^n} e^{-\sum_{i=1}^{n} x_i/\beta} = \frac{1}{\beta^n} e^{-n\bar{x}/\beta}.$$

The natural logarithm is

$$\ln L(\beta) = -n \ln \beta - n\frac{\bar{x}}{\beta}.$$

Differentiation with respect to β gives

$$\frac{\partial \ln L(\beta)}{\partial \beta} = -\frac{n}{\beta} + \frac{n\bar{x}}{\beta}$$

$$= \frac{n}{\beta^2}(\bar{x} - \beta).$$

Therefore, β has a umvue that is

$$\widehat{\beta} = \bar{X}, \quad \text{and whose variance is}$$

$$\text{Var}(\widehat{\beta}) = \frac{\beta^2}{n}.$$

In these exercises, we showed that this estimator is the maximum likelihood estimator for β.   ∎

It is not coincidental that the minimum variance unbiased estimators found thus far have been unbiased functions of the maximum likelihood estimators. It can be shown that, as long as the probability density function or probability function of the population distribution is sufficiently well behaved, a umvue, if it exists, will be a function of a maximum likelihood estimator. This is one reason why maximum likelihood estimators are popular.

10.4.3 Sufficiency

Consider an experiment where a coin is tossed n times. We have seen that an efficient estimator for the probability p of getting a head is the number of heads observed divided by the number of tosses. This implies that the information about the value of p is most fully used by counting the number of heads. In other words, the number of heads somehow contains all of the information about the value of the parameter that the data could provide. Which observation was a head and which was a tail is superfluous information. Similarly, we have seen that the sample mean is an efficient estimator for the parameter β in a $G(1, \beta)$ distribution. This also implies that the sum of the observations, somehow, contains the maximum amount of information about the true value of β. This is theoretically covered by the notion of *sufficient statistics*. It reflects the idea that some statistic contains all of the information that a sample can provide with regard to the value of a parameter. A formal definition follows.

Definition 10.6

Let $X_1, X_2, \ldots, X_n$ be a random sample from a population whose distribution depends on a parameter θ. A statistic, $T(X_1, X_2, \ldots, X_n)$, is said to be *sufficient* for θ if the conditional distribution of $X_1, X_2, \ldots, X_n$, given that $T = t$, does not depend on θ.

This definition embodies the notion that, once we have a value for a sufficient statistic, there is nothing more that the sample can tell us about the value of a parameter.

Example 10.24

Suppose we have a sample of n observations, $X_1, X_2, \ldots, X_n$, from a $B(1, p)$ population. We wish to determine if $Y = \sum_{i=1}^{n} X_i$ is sufficient for p. Using the definition of the conditional probability function from Chapter 3 we find that

$$p_{X_1, \ldots, X_n | Y}(x_1, \ldots, x_n \mid y) = \frac{p_{X_1, \ldots, X_n, Y}(x, y)}{p_y(y)}$$

$$= \frac{P(X_1 = x_1, X_2 = x_2, \ldots, X_n = x_n, Y = y)}{P(Y = y)}$$

$$= \frac{p^y(1-p)^{n-y}}{\binom{n}{y} p^y(1-p)^{n-y}}$$

$$= \frac{1}{\binom{n}{y}}.$$

This illustrates the idea of sufficient statistics. Since the conditional p.f. of the sample, given the number of heads, does not depend on the probability of obtaining a head, then knowing the number of heads gives us all the information that the sample can divulge about the value of p. ∎

The prospect of computing conditional distributions to determine whether a given statistic is sufficient for a parameter is not very appealing. It turns out that there is a simpler way to ascertain the existence of a sufficient statistic for a parameter. The result is known as the *Fisher Factorization theorem,* which we give below. The proof is beyond the level of this text and is omitted. The interested reader should consult *Theory of Point Estimation* by E. L. Lehmann. We denote the joint p.d.f. or p.f. of the sample by

$$\mathbf{p}_\theta(\mathbf{x}) = \prod_{i=1}^{n} f(x_i \mid \theta).$$

Theorem 10.3

A statistic $T(X_1, X_2, \ldots, X_n)$ is sufficient for the parameter θ if and only if the joint probability density function or joint probability function factors as follows.

$$\mathbf{p}_\theta(\mathbf{x}) = g[T(x_1, x_2, \ldots, x_n) \mid \theta] \, h(x_1, x_2, \ldots, x_n) \qquad \textbf{(10.13)}$$

This theorem states that $T(X_1, X_2, \ldots, X_n)$ is sufficient for the parameter θ if we can factor the joint p.d.f. or p.f. into a product of two quantities. One contains references to θ and $X_1, X_2, \ldots, X_n$ only through $T(X_1, X_2, \ldots, X_n)$. The other contains no references to θ.

Example 10.25

Suppose we have a sample of n observations, $X_1, X_2, \ldots, X_n$ from a $B(1, p)$ distribution. We wish to use the Fisher Factorization theorem to show that $Y = \sum_{I=1}^{n} X_i$ is a sufficient statistic for p. The joint p.f. is

$$\mathbf{p}_p(\mathbf{x}) = \prod_{i=1}^{n} p^{x_i} (1 - p)^{1 - x_i}$$

$$= p^y (1 - p)^{n - y}$$

$$= \left(\frac{p}{1 - p} \right)^y (1 - p)^n.$$

If we let

$$T(X_1, X_2, \ldots, X_n) = Y,$$

$$g[T(X_1, X_2, \ldots, X_n) \mid p] = \left(\frac{p}{1 - p} \right)^y (1 - p)^n, \text{ and}$$

$$h(X_1, X_2, \ldots, X_n) = 1,$$

then we see that

$$\mathbf{p}_p(\mathbf{x}) = g[y \mid p] \, h(x_1, x_2, \ldots, x_n).$$

Therefore, the number of successes is a sufficient statistic for the probability of success in a binomial distribution. ∎

Example 10.26

Let $X_1, X_2, \ldots, X_n$ be a random sample from a $G(1, \beta)$ distribution. We wish to use the Fisher Factorization theorem to show that $Y = \sum_{i=1}^{n} X_i$ is a sufficient statistic for β. We proceed as follows.

$$\mathbf{p}_\beta(\mathbf{x}) = \prod_{i=1}^{n} \frac{1}{\beta} e^{-x_i/\beta}$$

$$= \beta^{-n} e^{-\sum_{i=1}^{n} x_i/\beta}$$

$$= \beta^{-n} e^{-\sum_{i=1}^{n} x_i/\beta} \cdot 1$$

Therefore, with

$$y = T(x_1, x_2, \ldots, x_n) = \sum_{i=1}^{n} x_i,$$

$$g[y \mid \beta] = \beta^{-n} e^{-y/\beta}, \quad \text{and}$$

$$h(x_1, x_2, \ldots, x_n) = 1$$

we see that

$$\mathbf{p}_\beta(\mathbf{x}) = g[y \mid \beta]\, h(x_1, x_2, \ldots, x_n).$$

Therefore, $Y = \sum_{i=1}^{n} X_i$ is a sufficient statistic for β. Notice that the sample mean $\bar{X}$ is also sufficient for β. This shows that there may be many statistics that are sufficient for a given parameter. ∎

When a distribution depends on two parameters, as is the case with the normal, uniform, and gamma distributions, it is desirable to determine a pair of statistics that are *jointly sufficient* for the two parameters. In this case the Factorization theorem can be extended as follows.

Theorem 10.4

> The statistics $T_1(X_1, X_2, \ldots, X_n)$ and $T_2(X_1, X_2, \ldots, X_n)$ are jointly sufficient for the parameters θ_1 and θ_2 if and only if the joint probability density function or joint probability function factors as follows.
>
> $$\mathbf{p}_{\theta_1, \theta_2}(\mathbf{x}) = g[T_1(x_1, x_2, \ldots, x_n), T_2(x_1, x_2, \ldots, x_n) \mid \theta_1, \theta_2]\, h(x_1, x_2, \ldots, x_n)$$
> $$(10.14)$$

Example 10.27

Suppose we have a random sample of n observations from a $N(\mu, \sigma^2)$ distribution. We wish to use Theorem 10.4 to determine the existence of a pair of statistics that are jointly sufficient for μ and σ^2. We proceed as follows.

$$\mathbf{p}_{\mu, \sigma^2}(\mathbf{x}) = \left(2\pi\sigma^2\right)^{-n/2} e^{-(1/2\sigma^2) \sum_{i=1}^{n} (x_i - u)^2}$$

$$= \left(2\pi\sigma^2\right)^{-n/2} e^{-(1/2\sigma^2)\left[\sum_{i=1}^{n} x_i^2 - 2\mu \sum_{i=1}^{n} x_i + n\mu^2\right]}$$

Thus, in the notation of Theorem 10.4 we find that

$$T_1(x_1, x_2, \ldots, x_n) = \sum_{i=1}^{n} x,$$

$$T_2(x_1, x_2, \ldots, x_n) = \sum_{i=1}^{n} x_i^2,$$

$$g[T_1(x_1, \ldots, x_n), T_2(x_1, \ldots, x_n) \mid \mu, \sigma^2] = \left(2\pi\sigma^2\right)^{-n/2}$$

$$\cdot e^{-(1/2\sigma^2)\left[T_2(x_1, \ldots, x_n) - 2\mu T_1(x_1, \ldots, x_n) + n\mu^2\right]},$$

and
$$h(x_1, \ldots, x_n) = 1.$$

Therefore, we see that $\sum_{i=1}^{n} X_i$, and $\sum_{i=1}^{n} X_i^2$ are jointly sufficient for μ and σ^2. ∎

Researchers who studied sufficient statistics found that there was a class of distributions in which all parameters had sufficient statistics. This group is collectively called the *exponential family* of distributions. This family includes distributions such as the normal, binomial, Poisson, and gamma. We begin by defining it for the single parameter case.

Definition 10.7

A random variable X is said to belong to the *single parameter exponential family* if it has a probability density function or probability function that can be written in the form.

$$f(x \mid \theta) = \exp[\tau(\theta)\,T(x) + S(x) + \eta(\theta)], a \le x \le b, \tag{10.15}$$

where

1. neither a nor b depend on θ,
2. $\tau(\theta)$ and $\eta(\theta)$ are real-valued functions of θ, and
3. $T(x)$ and $S(x)$ are real-valued functions of x.

Example 10.28

Let X have a $B(1, p)$ distribution. We wish to show that X is a member of the exponential family. We first note that the p.f. can be written as

$$f(x \mid p) = p^x (1 - p)^{1-x}$$

$$= \left(\frac{p}{1-p}\right)^x (1 - p)$$

$$= \exp\left[\ln\left(\frac{p}{1-p}\right) x + \ln(1 - p)\right].$$

Then, in the notation of Equation 10.15 we see that

$$\tau(p) = \ln\left(\frac{p}{1-p}\right),$$

$$T(x) = x,$$

$$S(x) = 0, \quad \text{and}$$

$$\eta(p) = \ln(1 - p).$$

These functions satisfy the requirements of Definition 10.7. Therefore, the $B(1, p)$ distribution is a member of the one parameter exponential family. ∎

Notice that the $U(0, \theta)$ distribution is not a member of this family. The p.d.f. violates Condition 1 of Definition 10.7.

As we said the importance of the exponential family of distributions is that the existence of sufficient statistics is guaranteed. To see this, assume we have a sample of n observations from a distribution that is a member of the one parameter exponential family. Then

$$\mathbf{p}_\theta(\mathbf{x}) = \prod_{i=1}^{n} \exp[\tau(\theta)T(x_i) + S(x_i) + \eta(\theta)]$$

$$= \exp\left[\tau(\theta)\sum_{i=1}^{n}T(x_i) + \sum_{i=1}^{n}S(x_i) + n\,\eta(\theta)\right]$$

$$= \exp\left[\tau(\theta)\sum_{i=1}^{n}T(x_i) + n\,\eta(\theta)\right] \cdot \exp\left[\sum_{i=1}^{n}S(x_i)\right].$$

Thus, we see by the Factorization theorem that $\sum_{i=1}^{n}T(x_i)$ is a sufficient statistic for θ.

For distributions such as the normal and gamma we need to expand this idea to cover more than one parameter. For this we define the so-called k parameter exponential family.

Definition 10.8

A random variable X is said to belong to the *k parameter exponential family* if it has a probability density function or probability function that can be written in the form

$$f(x \mid \theta_1, \dots, \theta_k) = \exp\left[\sum_{i=1}^{k}\tau_i(\theta_1, \dots, \theta_k)T_i(x) + S(x) + \eta(\theta_1, \dots, \theta_k)\right],$$

$$a \le x \le b,$$

$$(10.16)$$

where

1. neither a nor b depend on $\theta_1, \dots, \theta_k$,
2. $\tau_i(\theta_1, \dots, \theta_k)$ is a real-valued function of $\theta_1, \dots, \theta_k$, for $i = 1, \dots, k$,
3. $\eta(\theta_1, \dots, \theta_k)$ is a real-valued function of $\theta_1, \dots, \theta_k$,
4. $T_i(x)$ is a real-valued function of x, for $i = 1, \dots, k$, and
5. $S(x)$ is a real-valued function of x.

Example 10.29

Let X have a $N(\mu, \sigma^2)$ distribution. We wish to show that it belongs to the two parameter exponential family. To see this we rewrite the p.d.f. as follows.

$$f(x \mid \mu, \sigma^2) = \frac{1}{\sqrt{2\pi\sigma^2}}\exp\left[\frac{-1}{2\sigma^2}(x - \mu)^2\right]$$

$$= \exp\left[\frac{-1}{2\sigma^2}x^2 - \frac{\mu}{\sigma^2}x + \frac{\mu^2}{2\sigma^2} - \frac{1}{2}\ln(2\pi\sigma^2)\right]$$

In the notation of Definition 10.8 we find the following.

$$\tau_1(\mu, \sigma^2) = \frac{-1}{2\sigma^2},$$

$$\tau_2(\mu, \sigma^2) = \frac{-\mu}{\sigma^2},$$

$$T_1(x) = x^2,$$

$$T_2(x) = x,$$

$$S(x) = 0, \quad \text{and}$$

$$\eta(\mu, \sigma^2) = \frac{\mu^2}{2\sigma^2} - \frac{1}{2}\ln(2\pi\sigma^2)$$

Therefore, the normal distribution is a member of the two parameter exponential family. ∎

As was the case with the single parameter exponential family, it turns out that, if a distribution is a member of the k parameter exponential family, there exists a set to k statistics that are jointly sufficient for the k parameters. We leave it as an exercise to show that the set of sufficient statistics for $\theta_1, \ldots, \theta_k$ is

$$\sum_{i=1}^{n} T_1(X_i), \ldots, \sum_{i=1}^{n} T_k(X_i).$$

Thus far, it may not seem very clear why sufficient statistics are of any practical use. We have stated that the distribution of a random sample given the value of a sufficient statistic does not depend on the value of the parameter. This gives credence to the idea that no more information can be obtained from the sample about the value of a parameter. This does not show, however, that more accurate estimates can be obtained through the use of sufficient statistics. It turns out that estimators based on sufficient statistics do have minimum variance under certain conditions. The conditions are based on an additional idea known as *completeness*. This concept is very technical and will be discussed briefly later. For now, we can give some indication as to how the existence of sufficient statistics can improve the performance of estimators. The result is known as the *Rao-Blackwell Theorem*, which we state without proof for the single parameter case.

Theorem 10.5

Rao-Blackwell Theorem   Let $\widehat{\theta}$ be an estimator for θ such that $E[\widehat{\theta}^2]$ exists and is finite for all values of θ. Further, suppose that $T(x_1, \ldots, x_n)$ is sufficient for θ. Now let $\phi(T) = E[\widehat{\theta} \mid T]$. Then, for all values of θ,

$$E[\phi(T)] = E[\widehat{\theta}], \quad \text{and}$$

$$E[(\phi(\theta) - \theta)^2] \le E[(\widehat{\theta} - \theta)^2].$$

(10.17)

This theorem shows that the knowledge of the value of a statistic that is sufficient for a parameter will result in an unbiased estimator that has a smaller mean square error. Therefore, it indicates that sufficient statistics seem to be a good place to begin the search for umvue estimators.

Example 10.30

We wish to give a concrete example of how the Rao-Blackwell theorem guarantees that using sufficient statistics can reduce the variance of estimators. Suppose we have a sample of 2 observations from a $U(0, \theta)$ distribution. We know that $Y_2 = \max(X_1, X_2)$ forms a sufficient statistic for θ. Consider the unbiased estimator $\widehat{\theta} = X_1 + X_2$. This is the method of moments estimator for θ. Since we have seen that the largest order statistic is biased in this case we shall use the Rao-Blackwell theorem to reduce the variance. In Example 6.25 we showed that the joint p.d.f. of $T = X_1 + X_2$ and Y_2 is

$$f_{T,Y_2}(t, y) = \begin{cases} \dfrac{2}{\theta^2} , 0 \le y \le \theta, \ y \le t \le 2y, & \text{and} \\ 0 , \text{otherwise.} \end{cases}$$

Using Equation 9.8 and Equation 3.42 we see that

$$f_{T|Y_2}(t \mid y) = \begin{cases} \dfrac{2}{y} , y \le t \le 2y, & \text{and} \\ 0 , \text{otherwise.} \end{cases}$$

The conditional expectation is then

$$E[T \mid Y_2] = \int_y^{2y} \frac{2t}{y} dt$$
$$= \frac{1}{u} t^2 \Big|_y^{2y}$$
$$= \frac{3}{2} y.$$

Therefore, the Rao-Blackwell theorem shows that $\widehat{\theta} = \frac{3}{2} Y_2 = \frac{n+1}{n} Y_n$ has a smaller variance than does $\widehat{\theta} = 2\bar{X}_n$, at least when $n = 2$. This illustrates the fact that sufficient statistics can produce estimators with smaller variance as claimed by the Rao-Blackwell theorem. ∎

An interesting aspect of the Rao-Blackwell theorem is that it is possible to start with any unbiased estimator, no matter how bad, and use a sufficient statistic to improve it.

Example 10.31

Let $X_1, X_2, \ldots, X_n$ be a random sample from a $Po(\lambda)$ distribution. In the exercises the reader is asked to show that $S = \sum_{i=1}^{n} X_i$ is a sufficient statistic for λ. We wish to estimate $P(X = 0) = e^{-\lambda}$. In the exercises at the end of Section 10.2, we asked you to show that a maximum likelihood estimator for this is $e^{-\bar{X}}$. It turns out that this estimator is not unbiased. An unbiased estimator of $P(X = 0)$ is

$$T = \begin{cases} 1 \text{, if } X_1 = 0, & \text{and} \\ 0 \text{, otherwise.} \end{cases}$$

It is not a very good estimator in that it ignores all but one of the observed values. In order to use the Rao-Blackwell we need to determine the conditional p.f. of T, given $S = s$. We make use of the fact that the sum of k i.i.d. $Po(\lambda)$ random variables has a $Po(k\lambda)$ distribution and proceed as follows.

$$p_{T,S}(1, s) = P(T = 1, S = s)$$

$$= P\left(X_1 = 0, \sum_{i=1}^{n} X_i = s\right)$$

$$= P\left(X_1 = 0, \sum_{i=2}^{n} X_i = s\right)$$

$$= e^{-\lambda} \frac{[(n-1)\lambda]^s \, e^{-(n-1)\lambda}}{s!}$$

$$= \frac{(n-1)^s \lambda^s \, e^{-n\lambda}}{s!}$$

Therefore,

$$p_{T|S}(1 \mid s) = \frac{p_{T,S}(1, s)}{p_S(s)}$$

$$= \frac{(n-1)^s \lambda^s \, e^{-n\lambda} / s!}{n^s \lambda^2 \, e^{-n\lambda} / s!}$$

$$= \left(1 - \frac{1}{n}\right)^s$$

Thus, we find that

$$p_{T|S}(t \mid s) = \begin{cases} \left(1 - \dfrac{1}{n}\right)^s & \text{, if } t = 1, \\ 1 - \left(1 - \dfrac{1}{n}\right)^s & \text{, if } t = 0, \quad \text{and} \\ 0 & \text{, otherwise.} \end{cases}$$

Therefore, by the Rao-Blackwell theorem an unbiased estimator for $P(X = 0)$ having a smaller variance than T is

$$U = E[T \mid S] = \left(1 - \frac{1}{n}\right)^S$$

$$= \left(1 - \frac{1}{n}\right)^{n\bar{X}}.$$

It is not coincidental that the sufficient statistics found thus far have been unbiased functions of the maximum likelihood estimators. We give a result that shows that, in general, sufficient statistics will be functions of maximum likelihood estimators. The same cannot be said for method of moments estimators. This is another reason why maximum likelihood estimators are popular.

Theorem 10.6

> Suppose that $T = f(X_1, X_2, \ldots, X_n)$ is a sufficient statistic for θ and that $\widehat{\theta}$ is a unique maximum likelihood estimator for θ. Then $\widehat{\theta}$ will be a function of T.

Proof   Since $T(X_1, X_2, \ldots, X_n)$ is sufficient for θ we can write the joint p.d.f. or p.f. of $X_1, X_2, \ldots, X_n$ as

$$L(\theta) = \prod_{i=1}^{n} f(x_i \mid \theta)$$

$$= g[T(x_1, x_2, \ldots, x_n) \mid \theta] \, h(x_1, x_2, \ldots, x_n).$$

Therefore, we can see that any value of θ that maximizes $L(\theta)$ will be a function of $g[T(x_1, \ldots, x_n) \mid \theta]$. This shows that the likelihood function will be maximized for a value of the sufficient statistic for θ. Thus, the maximum likelihood estimator for θ will be a function of the sufficient statistic for θ.   ■

10.4.4 Minimal Sufficient Statistics

Suppose that a statistic, T, is sufficient for the parameter θ. Further suppose that there is another statistic, V, such that $T = f(V)$. Then we see from notion of sufficient statistics that V is also sufficient for θ. Now, unless the function $f(\,)$ is one to one, the sufficient statistic T will provide a greater reduction of the original data than will V. To see how this works, suppose we have a sample of n observations from a $G(1, \beta)$ distribution. In Example 10.25, we saw that $T = \sum_{i=1}^{n} X_i$ is sufficient for β. Now, by letting $f(X_1, \ldots, X_n) = \sum_{i=1}^{n} X_i$ we see that $(X_1, \ldots, X_n)$ is also sufficient for β. However, it is clear that T produces a much greater reduction in the data than does simply giving the original sample data. Because of this it is desirable to determine, if possible, the sufficient statistic that produces the greatest reduction of the data. This notion is the motivation for the concept of a *minimal sufficient statistic*. The idea that $T = f(V)$ ensures that T will always result in a reduction of the data that is at least as good as that given by V is the basis for the definition of a minimal sufficient statistic.

Definition 10.9

> Suppose $X_1, \ldots, X_n$ is a random sample from a distribution that has parameters $\theta_1, \ldots, \theta_k$. The set of sufficient statistics $T_1, \ldots, T_k$ is said to be *minimal sufficient* if, given any other set of statistics, $V_1, \ldots, V_k$, which are also sufficient for $\theta_1, \ldots, \theta_k$, there exists a function, f, from R^k to R^k such that $(T_1, \ldots, T_k) = f(V_1, \ldots, V_k)$.

Loosely stated, a sufficient statistic that is a function of every other sufficient statistic is minimal sufficient. While this defines exactly what a minimal sufficient statistic is, it can be a little difficult to use in determining whether a set of statistics is, in fact, minimal sufficient.

To help in this we present a method due to Lehmann and Scheffé. The method assumes that we have two different samples of n observations, $X_1, \ldots, X_n$ and $Y_1, \ldots, Y_n$. Let $L_X(\theta)$ denote the likelihood function

$$L_X(\theta) = \prod_{i=1}^{n} f_X(x_i \mid \theta),$$

and $L_Y(\theta)$ be the likelihood function

$$L_Y(\theta) = \prod_{i=1}^{n} f_X(y_i \mid \theta).$$

Now compute the ratio of the likelihood functions for the two samples

$$\frac{L_X(\theta)}{L_Y(\theta)}.$$

Assume that θ is a vector of k parameters, $\theta_1, \theta_2, \ldots, \theta_k$. If we can find k functions $g_1(t_1, \ldots, t_n)$, $g_2(t_1, \ldots, t_n)$, $\ldots$, $g_K(t_1, \ldots, t_n)$ such that the ratio of likelihood functions does not depend on $\theta_1, \ldots, \theta_k$ if and only if

$$g_i(x_1, \ldots, x_n) = g_i(y_1, \ldots, y_n), \quad \text{for } i = 1, \ldots, k,$$

then $g_i(t_1, \ldots, t_n), i = 1, \ldots, k$ will be jointly minimally sufficient for $\theta_1, \ldots, \theta_k$.

Example 10.32   Suppose we have a sample of size n from a population having a $B(1, p)$ distribution. We wish to determine if there is a minimal sufficient statistic for p. In this case, the likelihood function is

$$L(p) = p^{\sum_{i=1}^{n} t_i}(1 - p)^{n - \sum_{i=1}^{n} t_i}.$$

The ratio of likelihood functions is

$$\frac{L_X(p)}{L_Y(p)} = \frac{p^{\sum_{i=1}^{n} x_i}(1 - p)^{n - \sum_{i=1}^{n} x_i}}{p^{\sum_{i=1}^{n} y_i}(1 - p)^{n - \sum_{i=1}^{n} y_i}}$$

$$= \left(\frac{p}{1 - p}\right)^{\sum_{i=1}^{n} x_i - \sum_{i=1}^{n} y_i}.$$

This ratio will be free of p only when

$$\sum_{i=1}^{n} x_i = \sum_{i=1}^{n} y_i.$$

Therefore, we find that $\sum_{i=1}^{n} X_i$ is a minimal sufficient statistic for p.   ∎

Example 10.33

Suppose we have a sample of n observations from a population having a $N(\mu, \sigma^2)$ distribution. We wish to determine a pair of statistics that will be jointly minimally sufficient for μ and σ^2. Here the likelihood function is

$$L(\mu, \sigma^2) = \left(\frac{1}{2\pi\sigma^2}\right)^{n/2} e^{-\sum_{i=1}^{n}(t_i-\mu)^2/2\sigma^2}.$$

The ratio of likelihood functions is

$$\frac{L_X(\mu, \sigma^2)}{L_Y(\mu, \sigma^2)} = \frac{\left(\frac{1}{2\pi\sigma^2}\right)^{n/2} e^{-\sum_{i=1}^{n}(x_i-\mu)^2/2\sigma^2}}{\left(\frac{1}{2\pi\sigma^2}\right)^{n/2} e^{-\sum_{i=1}^{n}(y_i-\mu)^2/2\sigma^2}}$$

$$= e^{-(1/2\sigma^2)\left[\sum_{i=1}^{n}(x_i-\mu)^2 - \sum_{i=1}^{n}(y_i-\mu)^2\right]}$$

$$= e^{-(1/2\sigma^2)\left[\sum_{i=1}^{n}x_i^2 - \sum_{i=1}^{n}y_i^2\right] + (\mu/\sigma^2)\left[\sum_{i=1}^{n}x_i - \sum_{i=1}^{n}y_i\right]}.$$

Thus, we see that the ratio of likelihood functions will not depend on μ or σ^2 only if

$$\sum_{i=1}^{n} x_i^2 = \sum_{i=1}^{n} y_i^2, \quad \text{and} \quad \sum_{i=1}^{n} x_i = \sum_{i=1}^{n} y_i.$$

Therefore, $\sum_{i=1}^{n} X_i^2$ and $\sum_{i=1}^{n} X_i$ are jointly minimal sufficient statistics for μ and σ^2. ∎

Example 10.34

Suppose we have a sample of n observations from a $U(0, \theta)$ distribution. We wish to determine a minimal sufficient statistic for θ. The likelihood function in this case is

$$L(\theta) = \begin{cases} \dfrac{1}{\theta^n}, & \text{if } \max(x_1, \ldots, x_n) \leq \theta, \quad \text{and} \\ 0, & \text{otherwise.} \end{cases}$$

Let $x_{\max} = \max(x_1, \ldots, x_n)$, and $y_{\max} = \max(y_1, \ldots, y_n)$. The ratio of likelihood functions is

$$\frac{L_X(\theta)}{L_Y(\theta)} = \begin{cases} 1, & \text{if } \max(x_{\max}, y_{\max}) \leq \theta, \\ 0, & \text{if } y_{\max} < x_{\max}, \quad \text{and } y_{\max} < \theta \leq x_{\max}, \\ \text{undefined}, & \text{otherwise.} \end{cases}$$

Thus, the ratio will not depend on θ only if $x_{\max} = y_{\max}$. Therefore, a minimal sufficient statistic for θ is Y_n, the largest order statistic. ∎

10.4.5 Completeness

The preceding sections have shown how we can use the concept of sufficiency to take the information that a sample contains about the values of unknown parameters and summarize it most efficiently. The reader may also have noticed another intriguing coincidence. In those cases, such as the binomial and exponential distributions, where we have already found efficient estimators by using the Cramér-Rao Lower Bound, the efficient estimators were unbiased functions of the minimal sufficient statistics. This would seem to make sense intuitively. If sufficient statistics contain all the relevant information regarding the values of parameters, then efficient estimators should make use of this fact. Also the Rao-Blackwell theorem shows that we can obtain unbiased estimators with smaller variance through the use of sufficient statistics. It turns out that these connections are not coincidental. In a number of cases of practical interest, the minimum variance unbiased estimator is an unbiased function of a minimal sufficient statistic.

The idea is based on the following line of argument. The Rao-Blackwell theorem says, in effect, that we can find an unbiased estimator that is a function of a sufficient statistic that has a smaller variance than an unbiased estimator that is not based on a sufficient statistic. If we can be sure that there is only one function of a sufficient statistic that is unbiased, then the Rao-Blackwell theorem will guarantee that it will have minimum variance. In order for this to occur we appeal to the concept of *completeness*.

Definition 10.10

A family of probability density functions or probability functions $f_X(x \mid \theta)$ is said to be *complete* if for every function $h(x)$ the identity

$$E[h(x)] = 0$$

implies that $h(x) = 0$ at all points for which $f_X(x \mid \theta) > 0$ for some value of θ.

This definition essentially states that a family of distributions is complete if the only unbiased estimator of zero is the zero function. We now tie this idea with that of sufficiency.

Definition 10.11

If a sufficient statistic T has a family of probability density functions or probability functions $f_T(t \mid \theta)$ that is complete, then T is said to be a *complete sufficient statistic*.

Verifying that a given sufficient statistic is complete can, in many cases, be a difficult task. We give some examples of instances where it is not too much work.

Example 10.35

Suppose we have a sample of n observations from a $U(0, \theta)$ distribution. We have seen that the largest order statistic, Y_n, is a minimal sufficient statistic for θ. We would like to determine if it is a complete sufficient statistic. The p.d.f. for Y_n is

$$f_{Y_n}(y) = \begin{cases} \dfrac{n y^{n-1}}{\theta^n}, & 0 \le y \le \theta, \quad \text{and} \\ 0, & \text{otherwise.} \end{cases}$$

To determine completeness we let $h(y)$ be an arbitrary function and assume

$$E[h(y)] = 0.$$

That is,

$$\int_0^\theta h(y)\frac{ny^{n-1}}{\theta^n}\,dy = 0.$$

The idea now is to use the Fundamental Theorem of Calculus. Before doing that we eliminate θ from the integrand by multiplying both sides by θ^n to obtain

$$\int_0^\theta h(y)ny^{n-1}\,dy = 0.$$

Now we differentiate both sides with respect to θ to get

$$h(\theta)n\theta^{n-1} = 0.$$

Since $n\theta^{n-1} > 0$ we divide to find that

$$h(\theta) = 0$$

for all θ. Since $\theta > 0$ we have shown that $h(y) = 0$ for all $y > 0$. Therefore, Y_n is a complete sufficient statistic. $\blacksquare$

Example 10.36   Suppose $X_1, X_2, \ldots, X_n$ are a sample of n observations from a $Po(\lambda)$ distribution. We have seen that $Y = \sum_{i=1}^n X_i$ is a minimal sufficient statistic for λ. We wish to determine if it is also a complete sufficient statistic. The distribution for Y is $Po(n\lambda)$. Therefore, we take $h(y)$ to be an arbitrary function of y and compute

$$\sum_{i=0}^\infty h(i)\frac{(n\lambda)^i e^{-n\lambda}}{i!} = 0.$$

Since $e^{-n\lambda} > 0$ we can divide through to obtain

$$\sum_{i=0}^\infty \frac{h(i)}{i!}(n\lambda)^i = 0.$$

We now use the fact that 2 power series are equal if and only if each of their coefficients is equal to see that $h(i) = 0, i = 0, 1, \ldots$. Since these are precisely the points where the p.f. of Y is not 0 we see that $Y = \sum_{i=1}^n X_i$ is a complete sufficient statistic for λ. $\blacksquare$

Example 10.37   Let $X_1, X_2, \ldots, X_n$ be a random sample from a $N(\mu, \sigma_1^2)$ distribution and $Y_1, Y_2, \ldots, Y_m$ be a random sample from an independent $N(\mu, \sigma_2^2)$ distribution. It is straightforward to determine that a minimal sufficient statistic for $(\mu, \sigma_1^2, \sigma_2^2)$ is

$$\left(\sum_{i=1}^n X_i, \sum_{i=1}^m Y_i, \sum_{i=1}^n X_i^2, \sum_{i=1}^m Y_i^2 \right).$$

Note that $\frac{1}{n}\sum_{i=1}^{n}X_i - \frac{1}{m}\sum_{j=1}^{m}Y_j$ is a nonzero function of the minimal sufficient statistic. Furthermore,

$$E\left[\frac{1}{n}\sum_{i=1}^{n}X_i - \frac{1}{m}\sum_{j=1}^{m}Y_j\right] = 0.$$

Therefore, the minimal sufficient statistic is not complete. This points up a problem in using statistics that are not complete. Suppose we are interested in finding an unbiased estimator for, say, μ which is a function of the minimal sufficient statistic. It turns out that there will be an infinite number of them. If we manage to find one unbiased estimator, we can always obtain another by adding an arbitrary multiple of $\bar{X} - \bar{Y}$ to it. It then becomes an extremely difficult problem to determine exactly which, out of all these unbiased estimators, should be used.   ∎

The two examples in which we proved completeness of a statistic illustrate that it can be a tricky business. In many cases of practical interest, it turns out that we can make use of the properties of the exponential family of distributions. Recall that the Poisson distribution belongs to the exponential family of distributions. It turns out that the exponential family, under certain regularity conditions, can be shown to give rise to complete sufficient statistics. The restrictions necessary yield a subset of the exponential family which is known as the *regular exponential family* of distributions. We state the definition for the k parameter regular exponential family. Simply set k to 1 to obtain the single parameter case.

Definition 10.12

A continuous random variable X is said to belong to the *k parameter regular exponential family of distributions* if it has a probability density function that can be written in the form

$$f(x \mid \theta_1,\ldots,\theta_k) = \exp\left[\sum_{i=1}^{k} \tau_i(\theta_1,\ldots,\theta_k)T_i(x) + S(x) + \eta(\theta_1,\ldots,\theta_k)\right],$$

$$a \leq x \leq b,$$

(10.18)

where

1. Neither a nor b depend on any of $\theta_1,\ldots,\theta_k$;
2. $\tau_i(\theta_1,\ldots,\theta_k), i = 1,\ldots,k$, are nonconstant continuous, independent functions of $\theta_1,\ldots,\theta_k$;
3. $T_i(x)$ is a continuous function for $a < x < b$, for $i = 1,\ldots,k$, and no one is a linear function of the others; and
4. $S(x)$ is a continuous function of x for $a < x < b$.

Definition 10.13

A discrete random variable X is said to belong to the *k parameter regular exponential family of distributions* if it has a probability function that can be written in the form

$$f(x \mid \theta_1, \ldots, \theta_k) = \exp \left[\sum_{i=1}^{k} \tau_i(\theta_1, \ldots, \theta_k) T_i(x) + S(x) + \eta(\theta_1, \ldots, \theta_k) \right],$$

$$x = a_1, a_2, \ldots,$$

(10.19)

where

1. None of $a_1, a_2, \ldots$ depend on any of $\theta_1, \ldots, \theta_k$;
2. $\tau_i(\theta_1, \ldots, \theta_k), i = 1, \ldots, k$, are nonconstant continuous, independent functions of $\theta_1, \ldots, \theta_k$; and
3. $T_i(x)$ is a continuous function for $x = a_1, a_2, \ldots$, for $i = 1, \ldots, k$, and no one is a linear function of the others.

The importance of belonging to the regular exponential family of distributions is pointed out by the following theorem, which we state without proof.

Theorem 10.7

Let $X_1, X_2, \ldots, X_n$ be a random sample from a population whose distribution is in the regular exponential family. Then $T_1(X_1, \ldots, X_n)$, $T_2(X_1, \ldots, X_n)$, $\ldots$, $T_k(X_1, \ldots, X_n)$ are jointly complete sufficient statistics for $\theta_1, \theta_2, \ldots, \theta_k$.

Note here that the joint p.d.f. or p.f. of $X_1, \ldots, X_n$, when written in the form of the exponential family will look like the following.

$$f(x_1, \ldots, x_n \mid \theta_1, \ldots, \theta_k)$$

$$= \exp \left[\sum_{i=1}^{k} \tau_i(\theta_1, \ldots, \theta_k) \sum_{j=1}^{n} T_i(x_j) + \sum_{j=1}^{n} S(x_j) + n\eta(\theta_1, \ldots, \theta_k) \right]$$

(10.20)

Example 10.38

Let $X_1, \ldots, X_n$ be a random sample from a $B(1, p)$ distribution. We wish to use Theorem 10.8 to show that $\sum_{i=1}^{n} X_i$ is a complete sufficient statistic for p. We begin by writing the joint p.f. in the form of Equation 10.20.

$$f(x_1, \ldots, x_n \mid p) = p^{\sum_{i=1}^{n} x_i} (1 - p)^{n - \sum_{i=1}^{n} x_i}$$

$$= \exp \left[\ln \left(\frac{p}{1 - p} \right) \sum_{i=1}^{n} x_i + n \ln(1 - p) \right]$$

In the notation of Equation 10.20, we have

$$\tau(p) = \ln\left(\frac{p}{1-p}\right),$$

$$T(x_i) = x_i$$

$$S(x_i) = 0, \quad \text{and}$$

$$\eta(p) = \ln(1-p).$$

Each of these functions meets the requirements of Definition 10.12. Therefore, by Theorem 10.8 we see that $\sum_{i=1}^{n} X_i$ is a complete sufficient statistic for p. ∎

Example 10.39   Suppose $X_1, \ldots, X_n$ is a random sample from a $N(\mu, \sigma^2)$ distribution. We wish to use Theorem 10.8 to show that $\sum_{i=1}^{n} X_i^2$ and $\sum_{i=1}^{n} X_i$ are joint complete sufficient statistics for μ and σ^2. We begin by writing the joint p.d.f. in the form of Equation 10.20.

$$f(x_1, \ldots, x_n \mid \mu, \sigma^2) = \left(\frac{1}{\sqrt{2\pi\sigma^2}}\right)^{n/2} \exp\left[-\frac{1}{2\sigma^2}\sum_{i=1}^{n}(x_i - \mu)^2\right]$$

$$= \exp\left[-\frac{1}{2\sigma^2}\sum_{i=1}^{n}x_i^2 - \frac{\mu}{\sigma^2}\sum_{i=1}^{n}x_i + n\left(\frac{\mu^2}{2\sigma^2} + \frac{1}{2}\ln(2\pi\sigma^2)\right)\right]$$

In the notation of Equation 10.20 we have

$$\tau_1(\mu, \sigma^2) = \frac{-1}{2\sigma^2},$$

$$\tau_2(\mu, \sigma^2) = \frac{-\mu}{\sigma^2},$$

$$T_1(x_i) = x_i^2,$$

$$T_2(x_i) = x_i,$$

$$S(x_i) = 0, \quad \text{and}$$

$$\eta(\mu, \sigma^2) = \frac{\mu^2}{2\sigma^2} + \frac{1}{2}\ln(2\pi\sigma^2).$$

Each of these meets the requirements of Definition 10.12. Therefore, by Theorem 10.8 we find that $\sum_{i=1}^{n} X_i^2$ and $\sum_{i=1}^{n} X_i$ are joint complete sufficient statistics for μ and σ^2. ∎

As we hinted earlier complete sufficient statistics should be the starting point in our search for minimum variance unbiased estimators for parameters. The following theorem, due to Lehmann and Scheffe, makes this formal.

Theorem 10.8

> **Lehmann-Scheffé Theorem**   Let T be a complete sufficient statistic for θ. If there exists some function $\phi(T)$ such that $E[\phi(T)] = h(\theta)$, then
>
> 1. $\phi(T)$ is unique, and
> 2. $\phi(T)$ has a smaller variance than any other unbiased estimator of $h(\theta)$.

We now have a very powerful method for constructing minimum variance unbiased estimators. Find a complete sufficient statistic for that parameter. Then determine which function of it, if any, is unbiased. That function of the complete sufficient statistic will be the minimum variance unbiased estimator. If such a function cannot be found, then find any unbiased estimator, no matter how bad, condition it on the complete sufficient statistic as we were doing in applying the Rao-Blackwell theorem.

Example 10.40

Suppose we have a sample of n observations from a $N(\mu, \sigma^2)$ distribution. From Example 10.38 we know that the normal p.d.f. belongs to the 2 parameter regular exponential family and, by Theorem 10.7, $\sum_{i=1}^{n} X_i$ and $\sum_{i=1}^{n} X_i^2$ are joint complete sufficient statistics for μ and σ^2. We know that $E\left[\frac{1}{n}\sum_{i=1}^{n} X_i\right] = \mu$. Therefore, by the Lehmann-Scheffé theorem $\bar{X}$ is a minimum variance unbiased estimator for μ. ∎

Example 10.41

Suppose we have a sample of n observations from a $Po(\lambda)$ distribution. We know from Example 10.36 that $\sum_{i=1}^{n} X_i$ is a complete sufficient statistic for λ. Since the mean of a Poisson distribution is λ and the sample mean is an unbiased estimator of $E[X]$ we can see immediately from the Lehmann-Scheffé theorem that $\bar{X} = \frac{1}{n}\sum_{i=1}^{n} X_i$ is a minimum variance unbiased estimator for λ.

Now suppose that we were unable to guess at an unbiased function of $\sum_{i=1}^{n} X_i$. It is then possible to begin with any unbiased estimator and condition on the complete sufficient statistic. Since $E[X] = \lambda$ one possible choice would be to use $\widehat{\lambda} = X_1$. We must now obtain the conditional probability function of X_1 given $\sum_{i=1}^{n} X_i = t$. We proceed as follows.

$$
p_{X_1, \sum_{i=1}^{n} X_i}(s, t) = P\left(X_1 = s, \sum_{i=1}^{n} X_i = t\right)
$$

$$
= P\left(X_1 = s, \sum_{i=2}^{n} X_i = t - s\right)
$$

$$
= P(X_1 = s)P\left(\sum_{i=2}^{n} X_i = t - s\right)
$$

$$
= \frac{\lambda^s e^{-\lambda}}{s!} \frac{[(n-1)\lambda]^{t-s} e^{-(n-1)\lambda}}{(t-s)!}
$$

$$
= \frac{(n-1)^{t-s}\lambda^t e^{-n\lambda}}{s!(t-s)!}
$$

Thus, the conditional distribution is

$$
p_{X_1|\sum_{i=1}^{n} X_i}(s \mid t) = \frac{(n-1)^{t-s}\lambda^t e^{-n\lambda} / s!(t-s)!}{(n\lambda)^t e^{-n\lambda} / t!}
$$

$$
= \frac{t!}{s!(t-s)!}\left(\frac{1}{n}\right)^s \left(1 - \frac{1}{n}\right)^{t-s}.
$$

The conditional distribution is therefore $B(t, \frac{1}{n})$, and the mean of this distribution is $\frac{t}{n}$. Therefore, we find, as before, that the minimum variance unbiased estimator for λ is the sample mean. ∎

Example 10.42

Suppose we have a sample of n observations from a $U(0, \theta)$ distribution. We saw in Example 10.33 that the largest order statistic, Y_n, is a complete sufficient statistic for θ. Furthermore, we found in Example 10.16 that $\frac{n+1}{n}Y_n$ is unbiased. Therefore, we see immediately that $\hat{\theta} = \frac{n+1}{n}Y_n$ is a minimum variance unbiased estimator for θ. ∎

We close our discussion of completeness with an interesting result due to Basu. For this we introduce the notion of an *ancillary statistic*.

Definition 10.14

> A statistic $U(X_1, X_2, \ldots, X_n)$ is said to be an *ancillary statistic* for θ if the distribution of $U(X_1, X_2, \ldots, X_n)$ does not depend on θ.

Basu's result deals with the independence of complete sufficient statistics and ancillary statistics. We state it as a theorem and omit the proof.

Theorem 10.9

> **Basu**   Let $T(X_1, X_2, \ldots, X_n)$ be a complete sufficient statistic for the parameter θ, and let $U(X_1, X_2, \ldots, X_n)$ be an ancillary statistic for θ. Then $T(X_1, X_2, \ldots, X_n)$ and $U(X_1, X_2, \ldots, X_n)$ are independent random variables.

This theorem allows us easily to demonstrate the independence of the sample mean and sample variance in samples from a normal distribution.

Example 10.43

Suppose $X_1, X_2, \ldots, X_n$ is a random sample from a $N(\mu, \sigma^2)$ distribution where σ^2 is known. The joint p.d.f. written in the form of Equation 10.20 is as follows.

$$f(x_1, \ldots, x_n \mid \mu, \sigma^2) = \exp\left[-\frac{\mu}{\sigma^2} \sum_{i=1}^{n} x_i - \frac{1}{2\sigma^2} \sum_{i=1}^{n} x_i^2 + n\left(\frac{\mu^2}{2\sigma^2} + \frac{1}{2}\ln(2\pi\sigma^2) \right) \right]$$

This shows that $\sum_{i=1}^{n} X_i$ is a complete sufficient statistic for μ. In the proof of Theorem 9.6 we found that $\frac{(n-1)s^2}{\sigma^2} = \frac{1}{\sigma^2}\sum_{i=1}^{n}(X_i - \bar{X})^2$ has a $\chi^2(n-1)$ distribution. Since the distribution of s^2 does not depend on μ we note that s^2 is ancillary for μ. Therefore, by Basu's theorem $\frac{(n-1)s^2}{\sigma^2}$ and $\sum_{i=1}^{n} X_i$ are independent, and thus so are s^2 and $\bar{X}$. Note that we still need to determine the distribution of the sample variance in order to demonstrate independence. Also, unlike Theorem 9.6, we assumed that the value of σ^2 is known. We leave it as an exercise to use Basu's theorem to prove independence in the case where σ^2 is not known. ∎

Exercises 10.4

1. Let $X_1, X_2, \ldots, X_n$ be a random sample of size n from a $Po(\lambda)$ distribution. Show that the sample mean is an unbiased estimator for λ and compute its variance.

2. Let $X_1, X_2, \ldots, X_n$ be a random sample from a $N(\mu, \sigma^2)$ distribution. Show that the maximum likelihood estimator

$$\widehat{\sigma^2} = \frac{1}{n} \sum_{i=1}^{n} \left(X_i - \bar{X}\right)^2$$

is not unbiased. Show that

$$B(\widehat{\sigma^2}, \sigma^2) = -\frac{\sigma^2}{n}.$$

3. Verify that the jackknife estimator for σ^2 in a normal distribution is

$$\widehat{\theta}_J = s^2 = \frac{1}{n-1} \sum_{i=1}^{n} \left(X_i - \bar{X}\right)^2.$$

4. Let $X_1, X_2, \ldots, X_n$ be a random sample from a population whose mean is $E[X] = \mu$ and variance is $\text{Var}(X) = 1$. Let an estimator for μ^2 be

$$\widehat{\mu^2} = \bar{X}^2.$$

   Show that $\widehat{\mu^2}$ is not unbiased and compute $B(\widehat{\mu^2}, \mu^2)$.

5. Determine a jackknife estimator for μ^2 in Exercise 4. Is it unbiased?

6. Let $X_1, X_2, \ldots, X_n$ be a random sample from a $B(1, p)$ distribution. Let $Y = \sum_{i=1}^{n} X_i$.
   a. Show that $\widehat{p^2} = \left(\frac{Y}{n}\right)^2$ is a biased estimator.
   b. Show that the jackknife estimator for p^2 is $\frac{Y(Y-1)}{n(n-1)}$. Is it unbiased?

7. Let $Y_1 \le Y_2 \le Y_3 \le Y_4 \le Y_5$ be the order statistics for a sample of 5 observations from a $U(0, \theta)$ distribution. Find a function of Y_3 that is an unbiased estimator for θ and compute its variance. Compute the efficiency of this estimator relative to $\widehat{\theta} = \frac{6}{5} Y_5$.

8. Compute the lower bound for the variance of an unbiased estimator for β based on a sample of size n from a distribution with a probability density function of

$$f(x) = \begin{cases} \beta e^{-\beta x}, & 0 < x, \quad \text{and} \\ 0, & \text{otherwise.} \end{cases}$$

9. Does there exist an unbiased estimator for β in Exercise 1 that achieves the Cramér-Rao Lower Bound?

10. Show that there is no efficient estimator for σ in a $N(0, \sigma^2)$ distribution.

11. Prove

$$E\left[\left(\frac{\partial \ln L(\theta)}{\partial \theta}\right)^2\right] = -E\left[\frac{\partial^2 \ln L(\theta)}{\partial \theta^2}\right].$$

12. Show that, given a sample of size n from a $Po(\lambda)$ distribution, $\sum_{i=1}^{n} X_i$ is sufficient for λ.

13. Show that the largest order statistic is sufficient for θ in a $U(0, \theta)$ distribution.

14. Find a pair of statistics that are jointly sufficient for α and β in a $G(\alpha, \beta)$ distribution.

15. Given a sample of n observations from a population having a probability density function of

$$f(x) = \begin{cases} \theta\, x^{\theta-1}, & 0 < x < 1, \quad \text{and} \\ 0, & \text{otherwise,} \end{cases}$$

a. find the lower bound for an unbiased estimator for θ,

b. find a sufficient statistic for θ, and

c. find the method of moments estimator for θ.

16. Show that

$$\text{Var}(\widehat{\tau(\theta)}) = \frac{\tau'(\theta)}{k},$$

where k is as defined in Equation 10.13.

17. Show that the Poisson distribution is a member of the one parameter exponential family.

18. Show that the gamma distribution is a member of the two parameter exponential family.

19. Let a population have a Cauchy distribution with a probability density function of

$$f(x) = \frac{1}{\pi[1 + (x - \theta)^2]}, \quad -\infty < x < \infty.$$

a. Does this distribution belong to the one parameter exponential family?

b. Given a sample of size n from this distribution, does there exist a sufficient statistic for θ?

20. Let X_1, X_2 be a random sample from a $N(\mu, \sigma^2)$ distribution.

a. Show that the joint distribution of $X = X_1$ and $Y = X_1 + X_2$ is bivariate normal with parameters

$$\mu_X = \mu,$$

$$\mu_Y = 2\mu,$$

$$\sigma_X^2 = \sigma^2,$$

$$\sigma_Y^2 = 2\sigma^2, \quad \text{and}$$

$$\rho = \frac{1}{\sqrt{2}}.$$

Hint: Use moment generating functions.

b. Show that the conditional distribution of X given $Y = y$ is $N\left(\frac{X_1+X_2}{2}, \frac{\sigma^2}{2}\right)$.

c. Show that this verifies the Rao-Blackwell theorem.

21. Find a minimal sufficient statistic for the parameter λ in a $Po(\lambda)$ distribution.

22. Find a minimal sufficient statistic for the parameter β in a $G(1, \beta)$ distribution.

23. Find a pair of minimal sufficient statistics for the parameters a and b in a $U(a, b)$ distribution.

24. Find a pair of minimal sufficient statistics for the parameters α and β in a $G(\alpha, \beta)$ distribution.

25. Suppose we have a sample of n observations from a population having a double exponential distribution with the parameter θ unknown and $\sigma = 1$. Show that the order statistics are minimal sufficient for θ.

26. Suppose that $X_1, X_2, \ldots, X_n$ is a random sample from a $N(\mu, \sigma_1^2)$ distribution. In addition, suppose that $Y_1, Y_2, \ldots, Y_m$ is a random sample from an independent $N(\mu, \sigma_2^2)$ distribution. Show that

$$\left(\sum_{i=1}^{n} X_i, \sum_{j=1}^{m} Y_j, \sum_{i=1}^{n} X_i^2, \sum_{j=1}^{m} Y_j^2 \right)$$

is minimally sufficient for $(\mu, \sigma_1^2, \sigma_2^2)$.

27. Let $X_1, X_2, \ldots, X_n$ be a random sample from a $U(\theta - a, \theta + a)$ distribution where a is known. Let Y_1 and Y_n be the smallest and largest order statistics, respectively.

    a. Show that (Y_1, Y_n) is minimally sufficient for θ.

    b. Show that (Y_1, Y_n) is not complete.

28. Suppose we have a sample of n observations from a $B(1, p)$ distribution. Find a minimum variance unbiased estimator for p in each of the following ways.

    a. Find an unbiased function of a complete sufficient statistic.

    b. Use $\widehat{p} = X_1$ and condition on a complete sufficient statistic.

29. Suppose we have a sample of n observations from a $G(1, \beta)$ distribution. Find a minimum variance unbiased estimator for β in each of the following ways.

    a. Find an unbiased function of a complete sufficient statistic.

    b. Use $\widehat{\beta} = X_1$ and condition on a complete sufficient statistic.

30. Show that, when sampling from a normal population with μ and σ^2 both unknown, the sample variance,

$$s^2 = \frac{1}{n-1} \sum_{i=1}^{n} \left(X_i - \bar{X} \right)^2,$$

is a minimum variance unbiased estimator for σ^2.

31. Suppose $X_1, X_2, \ldots, X_n$ are a random sample from a population having a p.d.f. of

$$f(x \mid \theta) = \begin{cases} e^{x-\theta}, & \theta \leq x, \quad \text{and} \\ 0, & \text{otherwise.} \end{cases}$$

    a. Show that the smallest order statistic, Y_1, is a complete sufficient statistic for θ.

    b. Find a minimum variance unbiased estimator for θ.

    c. Show that Y_1 and $T = \sum_{i=1}^{n}(X_i - Y_1)$ and are independent.

32. Let $X_1, X_2, \ldots, X_n$ be a random sample from a $N(\mu, 1)$ distribution. Use Basu's theorem to show that $\sum_{i=1}^{n} X_i$ and $\sum_{i=1}^{n} a_i X_i$ are independent if $\sum_{i=1}^{n} a_i = 0$.

33. Let $X_1, X_2, \ldots, X_n$ be a random sample from a $N(\mu, \sigma^2)$ distribution with both μ and σ^2 unknown. Use Basu's theorem to demonstrate the independence of $\bar{X}$ and s^2, the sample mean, and the sample variance, respectively.

10.5

Large Sample Properties of Estimators

As we have shown it sometimes happens that a biased estimator can have a mean square error smaller than that of an unbiased estimator. In the optimality criterion used thus far, these estimators would never be considered. To overcome this it is sometimes useful to compare estimators by considering their large sample performance. By the term *large sample* we mean that we are interested in the behavior of an estimator as the sample size n becomes infinitely large.

10.5.1 Consistency

One property that would be desirable is for the values of the estimator to become closer to the parameter being estimated as sample size increases. This, in some sense, means that possession of more information in the form of a larger sample improves our ability to make good estimates of the parameter of interest. Mathematically this idea is embodied in the concept of *consistency*. The formal definition of this property depends on convergence in probability, which was discussed in Chapter 7.

Definition 10.15

A sequence of estimators, $\widehat{\theta}_1, \widehat{\theta}_2, \ldots$, for a parameter θ is said to be *consistent* if, for any $\varepsilon > 0$,

$$\lim_{n \to \infty} P(|\widehat{\theta}_n - \theta| \geq \varepsilon) = 0. \tag{10.21}$$

It should be noted that, although we are actually referring to a sequence of estimators, it is common practice to say that an estimator $\widehat{\theta}$ is consistent. This kind of notational shortcut is fairly common. Anyway, the definition states that a sequence of estimators for a parameter θ is consistent if it converges in probability to θ.

As we saw in our discussion of convergence in probability, using the definition directly to prove convergence can be difficult. As was the case in Chapter 7, our task is made easier when $\text{Var}(\widehat{\theta})$ exists. We can then use Chebyshev's Inequality. Using Theorem 7.2 we get

$$P(|\widehat{\theta} - \theta| \geq \varepsilon) \leq \frac{E[(\widehat{\theta} - \theta)^2]}{\varepsilon^2} = \frac{mse(\widehat{\theta})}{\varepsilon^2}.$$

Thus, our test for consistency becomes that of determining whether the *mse* for the estimator tends to zero as n becomes infinitely large. We summarize this as follows.

Theorem 10.10

Let $\widehat{\theta}_1, \widehat{\theta}_2, \ldots$ be a sequence of estimators for θ. If $\text{Var}(\widehat{\theta}_n)$ exists and is finite and

$$\lim_{n \to \infty} E[(\widehat{\theta}_n - \theta)^2] = 0,$$

then, for any $\varepsilon > 0$,

$$\lim_{n \to \infty} P(|\widehat{\theta}_n - \theta| \geq \varepsilon) = 0.$$

It is sometimes convenient to check that the *mse* converges to zero by using the identity

$$E[(\widehat{\theta} - \theta)^2] = \text{Var}(\widehat{\theta}) + [B(\widehat{\theta}, \theta)]^2.$$

Thus, if the variance and bias of the sequence of estimators both tend to zero in the limit as n becomes infinitely large, then the sequence of estimators is consistent.

Example 10.44

Suppose we have a $N(\mu, \sigma^2)$ population. We have seen that the maximum likelihood estimators for μ and σ^2 are

$$\widehat{\mu} = \bar{X}, \quad \text{and}$$

$$\widehat{\sigma^2} = \frac{1}{n} \sum_{i=1}^{n} (X_i - \bar{X})^2.$$

We already know that $\widehat{\mu}$ is an unbiased estimator for μ. In addition, Theorem 9.6 shows that $\text{Var}(\bar{X}) = \frac{\sigma^2}{n}$. Therefore, we see that

$$B(\bar{X}, \mu) = 0, \quad \text{and} \quad \lim_{n \to \infty} \text{Var}(\bar{X}) = \lim_{n \to \infty} \frac{\sigma^2}{n} = 0.$$

Thus,

$$\lim_{n \to \infty} E[(\bar{X} - \mu)^2] = 0,$$

and $\bar{X}$ is a consistent estimator for μ.

The maximum likelihood estimator for σ^2, however, has been shown to be biased with

$$E[\widehat{\sigma^2}] = \frac{n-1}{n} \sigma^2.$$

Therefore,

$$B(\widehat{\sigma^2}, \sigma^2) = \frac{n-1}{n} \sigma^2 - \sigma^2$$

$$= -\frac{1}{n} \sigma^2.$$

To obtain $\text{Var}(\widehat{\sigma^2})$ we make use of Theorem 9.6. Part 2 of that theorem showed that $\frac{n\widehat{\sigma^2}}{\sigma^2}$ has a chi-squared distribution with $n-1$ degrees of freedom. Thus, by using the fact that the mean and variance of a $G(\alpha, \beta)$ random variable are $\alpha\beta$ and $\alpha\beta^2$, respectively, we proceed as follows.

$$\text{Var}(\widehat{\sigma^2}) = \text{Var}\left(\frac{\sigma^2}{n} \frac{n\widehat{\sigma^2}}{\sigma^2} \right)$$

$$= \frac{(\sigma^2)^2}{n^2} \text{Var}\left(\frac{n\widehat{\sigma^2}}{\sigma^2} \right)$$

$$= \frac{(\sigma^2)^2}{n^2} \frac{n-1}{2} 4$$

$$= \frac{2(n-1)(\sigma^2)^2}{n^2}$$

Thus,

$$\lim_{n \to \infty}[B(\widehat{\sigma^2}, \sigma^2)] = \lim_{n \to \infty}\frac{-\sigma^2}{n} = 0, \quad \text{and}$$

$$\lim_{n \to \infty}\text{Var}(\widehat{\theta^2}) = \lim_{n \to \infty}\frac{2(n-1)(\sigma^2)^2}{n^2} = 0.$$

Thus, the *mse* tends to 0 as n becomes infinitely large, and $\widehat{\sigma^2}$ is consistent for σ^2. ∎

Example 10.45 The maximum likelihood estimator for p in a $B(n, p)$ distribution is $\widehat{p} = \frac{X}{n}$, where X is the number of successes observed. We have already seen that $\widehat{p}$ is an unbiased estimator. From previous discussion we have also seen that it has a variance of

$$\text{Var}(\widehat{p}) = \frac{p(1-p)}{n}.$$

Since

$$B(\widehat{p}, p) = 0, \quad \text{and} \quad \lim_{n \to \infty}\text{Var}(\widehat{p}) = \lim_{n \to \infty}\frac{p(1-p)}{n} = 0,$$

we see that $\widehat{p}$ is a consistent estimator for p. ∎

In the examples given, it has turned out that the maximum likelihood estimators have been consistent. The fact is that, subject to mild conditions on the probability density function or probability function of the population distribution, maximum likelihood estimators are generally consistent. The regularity conditions are given in *Theory of Point Estimation* by E. L. Lehmann, but for our purposes all distributions of practical interest satisfy them. A similar statement holds for the method of moments estimators.

10.5.2 Asymptotic Efficiency

A desirable property for a consistent estimator would be that, at least asymptotically, its variance approaches the Cramér-Rao Lower Bound. This would mean that, while it may make no sense to discuss efficiency of biased estimators for small sample sizes, it would be possible to consider asymptotic efficiency. We formally define this idea as follows. We note here that

$$E\left[\left(\frac{\partial \ln L(\theta)}{\partial \theta}\right)^2\right] = nE\left(\frac{\partial \ln f(x; \theta)}{\partial \theta}\right)^2.$$

Definition 10.16

If $\widehat{\tau(\theta)}$ is a consistent estimator for $\tau(\theta)$, $\text{Var}(\widehat{\tau(\theta)})$ exists and is finite, and the lower bound given by Theorem 10.2 exists, then the *asymptotic efficiency* for $\widehat{\tau(\theta)}$ is

$$\text{a.e.}\ (\widehat{\tau(\theta)}) = \lim_{n \to \infty}\frac{[\tau'(\theta)]^2}{\text{Var}(\widehat{\tau(\theta)})\, nE\left[\left(\frac{\partial \ln f(x; \theta)}{\partial \theta}\right)^2\right]}. \tag{10.22}$$

Furthermore, if a.e. $(\widehat{\tau(\theta)}) = 1$, then $\widehat{\tau(\theta)}$ is said to be *asymptotically efficient*.

Example 10.46

The Cramér-Rao Lower Bound for an unbiased estimator of σ^2 from a $N(\mu, \sigma^2)$ distribution was found in Example 10.21 to be $\frac{2(\sigma^2)^2}{n}$. In Example 10.42 we showed that the variance of the maximum likelihood estimator for σ^2 is $\frac{2(n-1)(\sigma^2)^2}{n^2}$. In the notation of Definition 10.16 $\tau(\sigma^2) = \sigma^2$ which means that $\tau'(\sigma^2) = 1$. Therefore, the asymptotic efficiency of $\widehat{\sigma^2}$ is

$$\text{a.e. } (\widehat{\sigma^2}) = \lim_{n \to \infty} \frac{\dfrac{2(\sigma^2)^2}{n}}{\dfrac{2(n-1)(\sigma^2)^2}{n^2}}$$

$$= \lim_{n \to \infty} \frac{n}{n-1}$$

$$= 1.$$

Therefore, we see that $\widehat{\sigma^2}$ is asymptotically efficient.   ■

It is not, in general, true that consistency implies asymptotic efficiency. We leave it as an exercise to demonstrate that, if a sample of size n is taken from a population having a normal distribution and half of the observations are selected at random and the remainder discarded, then the mean of those observations is consistent but has an asymptotic efficiency of $\frac{1}{2}$.

In addition to consistency it can also be shown that, subject to some mild conditions on the probability density function or probability function, maximum likelihood estimators are asymptotically efficient. In fact, these same regularity conditions guarantee that maximum likelihood estimators belong to the class of *best asymptotically normal (BAN)* estimators. We define this concept as follows.

Definition 10.17

The sequence of estimators $\widehat{\theta}_1, \widehat{\theta}_2, \ldots$ is said to be a *best asymptotically normal (BAN)* estimator for θ if

1. The distribution of

$$\sqrt{n} \frac{\widehat{\theta}_n - \theta}{\sqrt{\text{Var}(\widehat{\theta}_n)}}$$

approaches a $N(0, 1)$ distribution as n becomes infinitely large,
2. $\widehat{\theta}$ is a consistent estimator for all values of θ, and
3. for any other sequence of estimators $\widehat{\theta}'_1, \widehat{\theta}'_2, \ldots$,

$$\lim_{n \to \infty} \text{Var}(\widehat{\theta}_n) \le \lim_{n \to \infty} \text{Var}(\widehat{\theta}'_n)$$

for all values of θ.

Exercises 10.5

1. Show that the largest observation from a $U(0, \theta)$ distribution is a consistent estimator for θ.

2. Show that, given a sample of size n from a $Po(\lambda)$ distribution, $\bar{X}$ is a consistent estimator for λ.

3. Given a sample of n observations from a population having a probability density function of

$$f(x) = \begin{cases} \theta \, x^{\theta-1} \, , & 0 < x < 1, \quad \text{and} \\ 0 \, , & \text{otherwise,} \end{cases}$$

   is the method of moments estimator for θ consistent?

4. Suppose that we begin with a sample of $2n$ observations from a $N(\mu, 1)$ population. From those observations n observations are randomly selected and the $\bar{X}$ computed using the selected observations.

   a. Show that $\bar{X}$ is a consistent estimator for μ.

   b. Show that a.e. $(\bar{X}) = \frac{1}{2}$.

5. Given a sample of n observations from a $N(\mu, \sigma^2)$ distribution, show that the sample variance s^2 is an asymptotically efficient estimator for σ^2.

6. Show that the sample variance s is a best asymptotically normal (BAN) estimator for the variance of a normal distribution where the mean is unknown.

10.6 Bayes Estimators

Until now we have been discussing estimation procedures that have made no prior assumptions regarding the value of a parameter other than those implicit in the population distribution. There are times, however, when the statistician has additional information about likely values for a parameter. In such cases it may make sense to consider the parameter to be a random variable whose distribution reflects this extra information. This means that we should regard the distribution of the population random variable as being a conditional distribution. Before going much further it should be pointed out that it is not always reasonable to treat a parameter as though it were a random variable. For example, if we have no idea about which values are likely to occur, then trying to place a probability distribution on the value of the parameter would be, at a minimum, counterproductive. In such cases, estimation should be done by using the method of maximum likelihood or the method of moments. In this section, we will discuss an estimation procedure based on Bayes rule, which we introduced in Chapter 3. We shall deal only with the case where the distribution has a single unknown parameter.

The general approach to estimation in these cases is best illustrated by means of an example. In order to have something concrete to refer to, we will consider a situation where we are trying to estimate the probability of success in a binomial distribution. The conditional distribution of X, the number of successes in a sample of size n, given a value of p, is

$$f(x \mid p) = \begin{cases} \binom{n}{x} p^x (1-p)^{n-x} \, , & x = 0, 1, \dots, n, \quad \text{and} \\ 0 \, , & \text{otherwise.} \end{cases}$$

Now assume we have information that p is more likely to take on high values, and we can properly describe this by assuming that p follows a beta distribution of the form

$$\pi(p) = \begin{cases} 2p, \, 0 < p < 1, & \text{and} \\ 0, \text{ otherwise.} \end{cases}$$

This distribution is called the *prior distribution of p*. The joint distribution of X and p is then

$$f(x,p) = f(x \mid p)\pi(p)$$

$$= \binom{n}{x} p^x(1-p)^{n-x} 2p$$

$$= \begin{cases} 2\binom{n}{x} p^{x+1}(1-p)^{n-x} 2p, \, x = 0, 1, \ldots, n; \, 0 < p < 1, & \text{and} \\ 0 & \text{, otherwise.} \end{cases}$$

From this we can obtain the distribution of X by integrating over p as follows.

$$g(x) = \int_{-\infty}^{\infty} f(x,p)\,dp$$

$$= \int_0^1 2\binom{n}{x} p^{x+1}(1-p)^{n-x}\,dp$$

$$= \int_0^1 \frac{2\,n!}{x!\,(n-x)!} p^{x+1}(1-p)^{n-x}\,dp.$$

The integrand is similar to the p.d.f. of a beta random variable with $\alpha = x + 2$ and $\beta = n - x + 1$. When the constants are fixed up we obtain

$$g(x) = \frac{2(x+1)}{(n+1)(n+1)} \int_0^1 \frac{(n+2)!}{(x+1)!\,(n-x)!} p^{x+1}(1-p)^{n-x}\,dp$$

$$= \begin{cases} \dfrac{2(x+1)}{(n+2)(n+1)}, \, x = 0, 1, \ldots, n, & \text{and} \\ 0 & \text{, otherwise.} \end{cases}$$

We can now obtain what is called the *posterior distribution of p, given x* by using the definition of conditional probability as follows.

$$h(p \mid x) = \frac{f(x,p)}{g(x)}$$

$$= \frac{2\binom{n}{x} p^{x+1}(1-p)^{n-x}}{\dfrac{2(x+1)}{(n+1)(n+1)}}$$

$$= \begin{cases} \dfrac{(n+2)!}{(x+1)!\,(n-x)!} p^{x+1}(1-p)^{n-x}, \, 0 < p < 1, & \text{and} \\ 0 & \text{, otherwise.} \end{cases}$$

Note that the posterior distribution is a beta distribution with parameters $\alpha = x + 2$ and $\beta = n - x + 1$.

We give a formal definition.

Definition 10.18

The *posterior distribution* of a parameter θ, given the observations $x_1, \ldots, x_n$ is the conditional probability density function or conditional probability function

$$h(\theta \mid x_1, \ldots, x_n) = \frac{f(x_1, \ldots, x_n \mid \theta)\pi(\theta)}{\displaystyle\int_{-\infty}^{\infty} f(x_1, \ldots, x_n \mid \theta)\pi(\theta)\,d\theta}. \qquad (10.23)$$

The posterior distribution gives us information regarding the likelihood of values of p given our sample data. The question remains of how to use this to estimate p. Assume we have an estimator for p. Call it $d(X)$. In Bayesian terminology, $d(X)$ is known as a *decision rule*. To measure how close $d(X)$ is to p we use what is known as a *loss function*. It assigns a numerical value according to the amount by which our decision is incorrect in deciding the true value for p and is denoted by $L(d(X), p)$. Loss functions are determined by the statistician. A common one is *absolute error loss* which is

$$L(d(X), p) = |d(X) - p|\,.$$

Another is *squared error loss* which is

$$L(d(X), p) = (d(X) - p)^2.$$

The goal in Bayes estimation is to find a decision rule that will make the expected loss, conditioned by the values of the observations, a minimum. We make this idea formal in the following definition. For notational convenience let $X = (X_1, \ldots, X_n)$.

Definition 10.19

Given a posterior distribution, $h(\theta \mid x)$ and a loss function $L(d(X), \theta)$ a decision rule $d(X)$ is *Bayes* if it minimizes

$$E[L(d(X), \theta) \mid X = x] = \int_{-\infty}^{\infty} L(d(X), \theta)h(\theta \mid x)\,d\theta.$$

This conditional expectation is called the *posterior risk*. $d(X)$ will then be called the *Bayes estimator for* θ.

For our example we will use squared error loss. From the fact that $E[(X - a)^2]$ is minimized when $a = E[X]$ we see that Bayes estimator for p will be the mean of the posterior distribution. We found that the posterior distribution was beta with $\alpha = x + 2$ and $\beta = n - x + 1$. Since we are using squared error loss the Bayes estimator for p will be the mean of this beta distribution or

$$d(X) = \frac{\alpha}{\alpha + \beta}$$

$$= \frac{X + 2}{n + 3}.$$

The procedure in the above example turns out to give the Bayes estimator whenever we are using squared error loss. In the case of absolute error loss, the median of the posterior distribution will be the Bayes estimator. We state these as a theorem without proof.

Theorem 10.11

Suppose we have a sample of n observations from a population with a probability function of probability distribution function $f(x \mid \theta)$, and suppose that $\pi(\theta)$ is the probability function or probability density function of the prior distribution of the parameter θ.

1. If the loss function is proportional to squared error, then the Bayes estimator for θ is the mean of the posterior distribution of θ, given the observations.

2. If the loss function is proportional to absolute error, then the Bayes estimator for θ is the median of the posterior distribution of θ, given the observations.

Example 10.47

Suppose we have a random sample of size n from a distribution having a p.d.f. of

$$f(x \mid \theta) = \begin{cases} \theta\, e^{-\theta x}, & 0 < x, \quad \text{and} \\ 0, & \text{otherwise.} \end{cases}$$

Suppose that θ has a prior distribution with a p.d.f. of

$$\pi(\theta) = \begin{cases} \eta\, e^{-\eta\theta}, & 0 < \theta, \quad \text{and} \\ 0, & \text{otherwise.} \end{cases}$$

Let the loss function be squared error, or

$$L[d(X), \theta] = [d(X) - \theta]^2.$$

We wish to find the Bayes estimator for θ. First note that

$$f(x_1, \ldots, x_n \mid \theta) = \prod_{i=1}^{n} \theta\, e^{-\theta x_i} = \theta^n\, e^{-n\theta \bar{x}}.$$

Therefore, the joint p.d.f. of $X_1, \ldots, X_n$ and θ is

$$f(x_1, \ldots, x_n, \theta) = f(x_1, \ldots, x_n \mid \theta)\, \pi(\theta)$$

$$= \theta^n\, e^{-n\theta \bar{x}}\, \eta\, e^{-\eta\theta}$$

$$= \theta^n\, \eta\, e^{-\theta[n\bar{x} + \eta]}.$$

The distribution for $X_1, X_2, \ldots, X_n$ is found by integrating with respect to θ as follows.

$$g(x_1, \ldots, x_n) = \int_0^\infty f(x_1, \ldots, x_n, \theta) \, d\theta$$

$$= \int_0^\infty \theta^n \, \eta \, e^{-\theta [n\bar{x} + \eta]} \, d\theta$$

$$= \eta \int_0^\infty \theta^n \, e^{-\theta [n\bar{x} + \eta]} \, d\theta$$

The integrand is almost that of the p.d.f. of a random variable having a gamma distribution with $\alpha = n + 1$ and $\beta = \frac{1}{n\bar{x} + \eta}$. To fix up the constants we multiply and divide by $\Gamma(\alpha) \beta^\alpha = n! \, (n\bar{x} + \eta)^{n+1}$ and proceed as follows.

$$g(x_1, \ldots, x_n) = \frac{\eta \, n!}{(n\bar{x} + \eta)^{n+1}} \int_0^\infty \frac{\theta^n \, (n\bar{x} + \eta)^{n+1}}{n!} \, e^{-\theta [n\bar{x} + \eta]} \, d\theta$$

$$= \frac{\eta \, n!}{(n\bar{x} + \eta)^{n_1}}$$

Thus, the posterior distribution has a p.d.f. of

$$h(\theta \mid x_1, \ldots, x_n) = \frac{f(x_1, \ldots, x_n, \theta)}{g(x_1, \ldots, x_n)}$$

$$= \frac{(n\bar{x} + \eta)^{n+1} \, \theta^n}{n!} \, e^{-\theta [n\bar{x} + \eta]}.$$

This is the p.d.f. of a $G\left(n + 1, \frac{1}{n\bar{x} + \eta}\right)$ random variable. Therefore, under squared error loss, the Bayes estimator for θ is the mean of this distribution or

$$d(X_1, \ldots, X_n) = \alpha \beta$$

$$= \frac{n + 1}{n\bar{X} + \eta}.$$

■

In those instances where the loss function is neither absolute error nor squared error, the posterior risk must be evaluated and minimized directly. We give an example of how this process goes.

Example 10.48   Suppose that we have a binomial distribution with a p.f. of

$$f(x \mid p) = \begin{cases} \binom{n}{x} p^x (1 - p)^{n-x}, & x = 0, \ldots, n, \ 0 < p < 1, \quad \text{and} \\ 0 & , \text{otherwise.} \end{cases}$$

Let p have a prior distribution of

$$\pi(p) = \begin{cases} 1 \, , 0 < p < 1, & \text{and} \\ 0 \, , \text{otherwise.} \end{cases}$$

In addition, we shall use a loss function of

$$L[d(X), p] = \frac{[d(X) - p]^2}{p(1 - p)}.$$

We wish to find a Bayes estimator for p. First the joint distribution of X and p is

$$f(x, p) = f(x \mid p)\pi(p)$$

$$= \begin{cases} \binom{n}{x} p^x (1 - p)^{n-x} \, , x = 0, \dots, n, \ 0 < p < 1, & \text{and} \\ 0 & , \text{otherwise.} \end{cases}$$

We integrate with respect to p to obtain the distribution of x as follows.

$$g(x) = \int_0^1 \binom{n}{x} p^x (1 - p)^{n-x} \, dp$$

$$= \int_0^1 \frac{n!}{x! \, (n - x)!} p^x (1 - p)^{n-x} \, dp$$

The integrand is almost the p.d.f. of a beta random variable with parameters $\alpha = x + 1$ and $\beta = n - x + 1$. To fix up the constants we multiply and divide by $n + 1$ and proceed as follows.

$$g(x) = \frac{1}{n + 1} \int_0^1 \frac{(n + 1)!}{x! \, (n - x)!} p^x (1 - p)^{n-x} \, dp$$

$$= \begin{cases} \frac{1}{n + 1} \, , x = 0, \dots, n, & \text{and} \\ 0 & , \text{otherwise} \end{cases}$$

Thus, the posterior distribution of p given $X = x$ has a p.d.f. of

$$h(p \mid x) = \frac{f(x, p)}{g(x)}$$

$$= \begin{cases} \frac{(n + 1)!}{x! \, (n - x)!} p^x (1 - p)^{n-x} \, , 0 < p < 1, & \text{and} \\ 0 & , \text{otherwise.} \end{cases}$$

Note that $h(p \mid x)$ is the p.d.f. of a beta random variable with $\alpha = x + 1$ and $\beta = n - x + 1$. We now use Equation 10.24 to compute the posterior risk.

$$r_p[d(X)] = \int_0^1 \frac{[d(x) - p]^2}{p(1 - p)} \frac{(n + 1)!}{x! \, (n - x)!} p^x (1 - p)^{n-x} \, dp$$

$$= \int_0^1 [d(x) - p]^2 \frac{(n + 1)!}{x! \, (n - x)!} p^{x-1} (1 - p)^{n-x-1} \, dp$$

It turns out to be slightly less work if we take the derivative with respect to $d(x)$ at this point. Thus,

$$\frac{dr_p[d(x)]}{d[d(x)]} = 2 \int_0^1 [d(x) - p] \frac{(n+1)!}{x!\,(n-x)!} p^{x-1}(1-p)^{n-x-1}\,dp.$$

Setting this equal to 0 and solving for $d(x)$ gives

$$d(x) = \frac{\displaystyle\int_0^1 \frac{(n+1)!}{x!\,(n-x)!} p^x (1-p)^{n-x-1}\,dp}{\displaystyle\int_0^1 \frac{(n+1)!}{x!\,(n-x)!} p^{x-1}(1-p)^{n-x-1}\,dp}.$$

$$= \frac{\dfrac{n+1}{n-x} \displaystyle\int_0^1 \frac{n!}{x!\,(n-x-1)!} p^x (1-p)^{n-x-1}\,dp}{\dfrac{n(n+1)}{x(n-x)} \displaystyle\int_0^1 \frac{(n-1)!}{(x-1)!\,(n-x-1)!} p^{x-1}(1-p)^{n-x-1}\,dp}.$$

Both integrands are now beta p.d.f.'s integrated over their ranges and hence the integrals have a value of 1. Thus,

$$d(x) = \frac{X}{n}$$

is the Bayes estimator. Note that $d(x)$ is not the mean of $h(p \mid x)$ which is $\frac{x+1}{n+2}$. ∎

Exercises 10.6

1. Let X be a single observation from a Poisson distribution with a p.f. of

$$f(x \mid \lambda) = \begin{cases} \dfrac{\lambda^x e^{-\lambda}}{x!}, & x = 0, 1, 2, \ldots, \quad \text{and} \\ 0, & \text{otherwise.} \end{cases}$$

Let λ have a prior distribution with a p.d.f. of

$$\pi(\lambda) = \begin{cases} \theta e^{-\theta\lambda}, & \lambda > 0, \quad \text{and} \\ 0, & \text{otherwise.} \end{cases}$$

Let the loss function be squared error, or

$$L[d(x), \lambda] = [d(x) - \lambda]^2.$$

Find the posterior distribution $h(\lambda \mid x)$ and determine the Bayes estimator for λ.

2. Let X be a continuous random variable. Show that $E[|X - a|]$ is minimized when a is the median of the distribution of X. Recall that the median of a distribution is that point $\tilde{x}$ such that

$$\frac{1}{2} = \int_{-\infty}^{\tilde{x}} f(t)\,dt,$$

where $f(x)$ is the probability density function for X.

3. Suppose we have a single observation from a $N(\mu, \sigma^2)$ distribution, and let μ have a $N(0, 1)$ prior distribution.

   a. Find the posterior distribution of μ, given x.

   b. Using squared error loss find the estimator for μ that is Bayes with respect to $\pi(\mu)$.

4. Suppose we have a single observation from a $U(0, \theta)$ distribution, and let θ have a prior distribution whose p.d.f. is

$$\pi(\theta) = \begin{cases} \dfrac{1}{\theta^2}, & 1 < \theta, \quad \text{and} \\ 0, & \text{otherwise.} \end{cases}$$

   Using squared error loss find the Bayes estimator for θ with respect to $\pi(\theta)$.

5. Solve Exercise 4 using absolute error loss.

6. Suppose we have a single observation from a $U(0, \theta)$ distribution, and let θ have a prior distribution with a p.d.f. of

$$\pi(\theta) = \begin{cases} \theta\, e^{-\theta}, & \theta > 0, \quad \text{and} \\ 0, & \text{otherwise.} \end{cases}$$

   Let the loss function be

$$L[d(x), \theta] = 2[d(x) - \theta]^2.$$

   Find the Bayes estimator for θ.

10.7
Robust
Estimation

The estimation techniques we have discussed thus far have been aimed primarily at deriving estimators for particular parameters of a presumed underlying family of distributions. The only exception to this has been the method of moments. There, it was mentioned that the sample moments can be used to estimate the distribution moments without referring to how these moments are related to the parameter values. This is why they are sometimes referred to as *nonparametric estimators* of the population moments.

Using sample moments in this fashion is not without some risk, however. In the case of distributions with heavy tail weight such as the double exponential distribution, the nonparametric estimator for the population mean from the method of moments would be the sample mean. In this case it turns out that the sample median, which is not derivable from the method of moments, has a smaller variance. The sample median is the maximum likelihood estimator for the mean of a double exponential distribution. This is one reason why maximum likelihood estimators are very popular. If the choice of the underlying family of distributions is based on past experience or the application of the methods discussed in Chapter 8, there is a distinct possibility that the true population will be slightly different from the model used to derive the estimators. This means that, even though a given estimator may have minimum variance for the distribution used, it may not be very good for the actual distribution. In those cases where the actual population distribution is different from the one assumed, we would like for our estimators to have small variance over a range of distributions. Estimators that exhibit this property are said to be *robust estimators*. To be a little more specific we give the following definition.

Definition 10.20

> A statistical procedure is said to be *robust* if its behavior is relatively insensitive to deviations from the assumptions on which it is based.

In our case, the behavior of an estimator will be taken to be its variance, and the assumptions on which it is based will be the family of distributions that is used.

To illustrate how the variance of an estimator can be affected by deviations from the presumed underlying population model consider estimating the mean of a standard normal distribution. Specifically, we shall assume that $X_1, X_2, \ldots, X_n$ are a random sample from a $N(0, 1)$ distribution. Suppose, however, that the population actually follows a *contaminated* normal distribution. This means that, for $0 \leq \varepsilon \leq 1$, $100(1 - \varepsilon)\%$ of the observations come from a $N(0, 1)$ distribution, and the remaining $100\varepsilon\%$ of the observations come from a $N(0, 16)$ distribution. We have seen that the minimum variance unbiased estimator for μ in an uncontaminated normal population is the sample mean. A less efficient alternative would be the sample median. In a computer simulation study, 10,000 random samples of 100 observations each were taken. The mean and median for each sample was computed along with the sample variance of each. For various values of ε the results are summarized below.

ε	0.0	0.1	0.2	0.3	0.4	0.5
$\mathbf{Var(\bar{X})}$	0.010	0.025	0.039	0.055	0.070	0.084
$\mathbf{Var(\widetilde{X})}$	0.015	0.018	0.021	0.026	0.032	0.041

Note that as ε increases the heavily tailed $N(0, 16)$ distribution introduces more extreme observations that cause the variance of the sample mean to increase rather quickly compared to the sample median.

The goal of robust estimation in this problem is to derive estimators with variance near that of the sample mean when the distribution is standard normal while having the variance remain relatively stable as ε increases. From the above discussion one possibility would be the sample median. Another simple method that has been suggested to achieve this is to use what is called an *α-trimmed mean*. Let $0 \leq \alpha \leq 0.5$, and define $k = [n\alpha]$, where $[x]$ is the greatest integer that is less than or equal to x. The reader may have been introduced to this function in a calculus course. The sample is then ordered, the k highest and lowest observations are discarded, and the mean of the remaining $n - 2k$ observations is computed. That is, if $X_1, X_2, \ldots, X_n$ is the original random sample and $Y_1 \leq Y_2 \leq \cdots \leq Y_n$ are the ordered observations, then the α-trimmed mean would be

$$\bar{X}_\alpha = \frac{Y_{k+1} + Y_{k+2} + \cdots + Y_{n-k}}{n - 2k}. \tag{10.24}$$

The 0-trimmed mean is the sample mean, and the 0.5-trimmed mean is the sample median. The 0.25-trimmed mean is commonly referred to as the *midmean*. The α-trimmed mean is a member of a larger class of estimators called *L-estimators*, where the L refers to the fact that these estimators are based on a *linear* combination of the order statistics. Trimming has been found to contribute to keeping the variance relatively stable in the presence of extreme observations.

A monumental study was conducted by D. F. Andrews, et al. (1972) where a large number of robust estimators for location were compared. In that study, the contaminating normal distribution was $N(0,9)$ with various values of ε. The results for the estimators discussed here are given in Table 10.1.

Table 10.1   Comparison of Robust Estimators for Location Distribution:
$(1 - \varepsilon)100\% N(0,1)$ and $\varepsilon\,100\% N(0,9)$
$(n = 20)$

			ε		
20 × Variance	**0.00**	**0.05**	**0.10**	**0.15**	**0.25**
Mean	1.000	1.420	1.883	2.259	3.007
0.05–trimmed mean	1.022	1.156	1.389	1.637	2.270
0.10–trimmed mean	1.056	1.166	1.308	1.471	1.926
Midmean	1.199	1.271	1.406	1.495	1.790
Median	1.498	1.516	1.704	1.747	2.156

Source: Reprinted from D. F. Andrews, et al., *Robust Estimates of Location, Survey, and Advances* (New Jersey: Princeton University Press, 1972).

From this table it can be seen that those estimators that restrict the effects of extreme observations (sometimes referred to as *outliers)* show a substantial improvement in the stability of their variances over a range of contamination. We must point out that the robust estimators discussed here are only a small subset of those that have been proposed for estimating location. We have not attempted to discuss robust estimators for other problems, such as scale.

Exercises 10.7

1. Let a sample of size 9 be

$$10, \ 7, \ 9, \ 12, \ 21, \ 13, \ 8, \ 11, \quad \text{and } 9.$$

Let $\alpha = 0.25$ and compute $\bar{X}_{0.25}$, the midmean.

2. Let a sample of size 40 be

27	81	61	40	82	28	57	90	41	13
53	11	43	31	36	11	47	39	49	24
71	35	49	75	99	74	84	61	51	33
73	82	76	48	91	77	81	48	33	69

Let $\alpha = 0.10$ and compute $\bar{X}_{0.10}$.

Chapter Summary

Maximum Likelihood Estimator: Given a set of observations $x_1, \ldots, x_n$ from a population having a p.f. or p.d.f. of $f(x \mid \theta)$, a maximum likelihood estimator for θ is the value of θ that maximized the value of the likelihood function,

$$L(\theta) = f(x_1 \mid \theta)f(x_2 \mid \theta) \cdots f(x_n \mid \theta).$$ **(Section 10.2)**

Invariance Property of Maximum Likelihood Estimators: If $\widehat{\theta}$ is a maximum likelihood estimator for θ, then $h(\widehat{\theta})$ is a maximum likelihood estimator for $h(\theta)$.   **(Section 10.2)**

Method of Moments: Given a sample from a distribution with parameters $\theta_1, \ldots, \theta_k$, whose first k moments exist and are functions of $\theta_1, \ldots, \theta_k$, find the method of moments estimators as follows:

1. Estimate $\mu_r = E[X^r]$ with $m_r = \frac{1}{n}\sum_{i=1}^{n} x_i^r$ for $r = 1, \ldots, k$;
2. Solve the resulting system of equations in $\theta_1, \ldots, \theta_k$ for each parameter in terms of $m_1, \ldots, m_k$.   **(Section 10.3)**

Unbiasedness: An estimator $\widehat{\theta}$ for θ is unbiased if

$$E[\widehat{\theta}] = \theta.$$ **(Section 10.4.1)**

Bias: The bias of an estimator, $B(\widehat{\theta}, \theta)$ is

$$B(\widehat{\theta}, \theta) = E[\widehat{\theta}] - \theta.$$ **(Section 10.4.1)**

Mean Square Error: The mean square error, *mse*, of an estimator is

$$mse = E[(\widehat{\theta} - \theta)^2]$$
$$= \text{Var}(\widehat{\theta}) - \left[B(\widehat{\theta}, \theta)\right]^2.$$

For an unbiased estimator $mse = \text{Var}(\widehat{\theta})$.   **(Section 10.4.2)**

Efficiency: Given two unbiased estimators, $\widehat{\theta}_1$ and $\widehat{\theta}_1$, the efficiency of $\widehat{\theta}_2$ relative to $\widehat{\theta}_2$ is

$$e\left(\widehat{\theta}_1, \widehat{\theta}_2\right) = \frac{\text{Var}(\widehat{\theta}_2)}{\text{Var}(\widehat{\theta}_1)}.$$

If $e\left(\widehat{\theta}_1, \widehat{\theta}_2\right) > 1$, then $\widehat{\theta}_1$ is more efficient than $\widehat{\theta}_2$.   **(Section 10.4.2)**

Uniform Minimum Variance Unbiased Estimator: An unbiased estimator, $\widehat{\theta}_0$, is a uniform minimum variance unbiased estimator if, given any other unbiased estimator, $\widehat{\theta}$,

$$\text{Var}(\widehat{\theta}_0) \leq \text{Var}(\widehat{\theta})$$

for all values of θ.   **(Section 10.4.2)**

Cramér-Rao Lower Bound: Let $T = g(X_1, \ldots, X_n)$ be an unbiased estimator for a scalar parameter $\tau(\theta)$ from a population having a p.f. or p.d.f. of $f(x \mid \theta)$. Then, subject to certain regularity conditions,

$$\text{Var}(T) \geq \frac{[\tau'(\theta)]^2}{E\left[\left(\dfrac{\partial \ln L(\theta)}{\partial \theta}\right)^2\right]}.$$

If $L(\theta)$ is twice differentiable with respect to θ

$$E\left[\left(\frac{\partial \ln L(\theta)}{\partial \theta}\right)^2\right] = -E\left[\frac{\partial^2 \ln L(\theta)}{\partial \theta^2}\right].$$

An unbiased estimator whose variance equals the Cramér-Rao Lower Bound is said to be efficient.

The efficiency of an unbiased estimator of $\tau\theta$ is

$$e\left(\widehat{\tau\theta}\right) = \frac{[\tau'(\theta)]^2}{\text{Var}(\widehat{\tau(\theta)})\, E\left[\left(\frac{\partial \ln L(\theta)}{\partial \theta}\right)^2\right]}. \qquad \textbf{(Section 10.4.2)}$$

Sufficiency: Let $X_1, X_2, \ldots, X_n$ be a random sample from a population whose distribution depends on a parameter θ. A statistic, $T(X_1, X_2, \ldots, X_n)$ is said to be sufficient for θ if the conditional distribution of $X_1, X_2, \ldots, X_n$, given that $T = t$, does not depend on θ. **(Section 10.4.3)**

Fisher Factorization Theorem: A statistic $T(X_1, X_2, \ldots, X_n)$ is sufficient for the parameter θ if and only if the joint probability density function or joint probability function factors as follows.

$$\mathbf{p}_\theta(\mathbf{x}) = g[T(x_1, x_2, \ldots, x_n) \mid \theta]\, h(x_1, x_2, \ldots, x_n).$$

The statistics $T_1(X_1, X_2, \ldots, X_n)$ and $T_2(X_1, X_2, \ldots, X_n)$ are jointly sufficient for the parameters θ_1 and θ_2 if and only if the joint probability density function or joint probability function factors as follows.

$$\mathbf{p}_{\theta_1, \theta_2}(\mathbf{x}) = g[T_1(x_1, x_2, \ldots, x_n), T_2(x_1, x_2, \ldots, x_n) \mid \theta_1, \theta_2]\, h(x_1, x_2, \ldots, x_n).$$
$$\textbf{(Section 10.4.3)}$$

Exponential Family: A random variable X is said to belong to the k parameter exponential family if it has a probability density function or probability function that can be written in the form

$$f(x \mid \theta_1, \ldots, \theta_k) = \exp\left[\sum_{i=1}^{k} \tau_i(\theta_1, \ldots, \theta_k)\, T_i(x) + S(x) + \eta(\theta_1, \ldots, \theta_k)\right],$$

$$a \leq x \leq b,$$

where

1. Neither a nor b depend on $\theta_1, \ldots, \theta_k$,

2. $\tau_i(\theta_1, \ldots, \theta_k)$ is a real-valued function of $\theta_1, \ldots, \theta_k$, for $i = 1, \ldots, k$,

3. $\eta(\theta_1, \ldots, \theta_k)$ is a real-valued function of $\theta_1, \ldots, \theta_k$,

4. $T_i(x)$ is a real-valued function of x, for $i = 1, \ldots, k$, and

5. $S(x)$ is a real-valued function of x.   **(Section 10.4.3)**

Rao-Blackwell Theorem: Let $\widehat{\theta}$ be an estimator for θ such that $E[\widehat{\theta}^2]$ exists and is finite for all values of θ. Further, suppose that $T(x_1, \ldots, x_n)$ is sufficient for θ. Now let $\phi(T) = E[\widehat{\theta} \mid T]$. Then, for all values of θ,

$$E[\phi(T)] = E[\widehat{\theta}], \quad \text{and}$$

$$E[(\phi(\theta) - \theta)^2] \le E[(\widehat{\theta} - \theta)^2].   \quad \textbf{(Section 10.4.3)}$$

Suppose that $T = f(X_1, X_2, \ldots, X_n)$ is a sufficient statistic for θ and that $\widehat{\theta}$ is a unique maximum likelihood estimator for θ. Then $\widehat{\theta}$ will be a function of T.

Minimal Sufficiency: Suppose $X_1, \ldots, X_n$ is a random sample from a distribution that has parameters $\theta_1, \ldots, \theta_k$. The set of sufficient statistics $T_1, \ldots, T_k$ is said to be minimal sufficient if, given any other set of statistics, $V_1, \ldots, V_k$, which are also sufficient for $\theta_1, \ldots, \theta_k$, there exists a function, f, from R^k to R^k such that $(T_1, \ldots, T_k) = f(V_1, \ldots, V_k)$.   **(Section 10.4.4)**

Completeness: A family of p.d.f.'s or p.f.'s $f_X(x \mid \theta)$ is said to be complete if, for every function $h(x)$, the identity

$$E[h(x)] = 0$$

implies that $h(x) = 0$ at all points for which $f_X(x \mid \theta) > 0$ for some value of θ.

If a sufficient statistic T has a family of probability density functions or probability functions $f_T(t \mid \theta)$ that is complete, then T is said to be a complete sufficient statistic.   **(Section 10.4.5)**

Continuous Regular Exponential Family: A continuous random variable X is said to be a member of the k parameter regular exponential family if it has a probability density function that can be written in the form

$$f(x \mid \theta_1, \ldots, \theta_k) = \exp\left[\sum_{i=1}^{k} \tau_i(\theta_1, \ldots, \theta_k)T_i(x) + S(x) + \eta(\theta_1, \ldots, \theta_k)\right],$$

$$a \le x \le b,$$

where

1. Neither a nor b depend on any of $\theta_1, \ldots, \theta_k$,

2. $\tau_i(\theta_1, \ldots, \theta_k), i = 1, \ldots, k$, are nonconstant continuous, independent functions of $\theta_1, \ldots, \theta_k$,

3. $T_i(x)$ is a continuous function for $a < x < b$, for $i = 1, \ldots, k$, and no one is a linear function of the others,

4. $S(x)$ is a continuous function of x for $a < x < b$.   **(Section 10.4.5)**

Discrete Regular Exponential Family: A discrete random variable X is said to be a member of the k parameter regular exponential family if it has a probability function that can be written in the form

$$f(x|\theta_1,\ldots,\theta_k)$$

$$= \exp\left[\sum_{i=1}^{k} \tau_i(\theta_1,\ldots,\theta_k)T_i(x) + S(x) + \eta(\theta_1,\ldots,\theta_k)\right], x = a_1, a_2,\ldots,$$

where

1. none of $a_1, a_2,\ldots$ depend on any of $\theta_1,\ldots,\theta_k$;
2. $\tau_i(\theta_1,\ldots,\theta_k), i = 1,\ldots,k$, are nonconstant continuous, independent functions of $\theta_1,\ldots,\theta_k$;
3. $T_i(x)$ is a continuous function for $x = a_1, a_2,\ldots$, for $i = 1,\ldots,k$, and no one is a linear function of the others. **(Section 10.4.5)**

Let $X_1, X_2,\ldots,X_n$ be a random sample from a distribution in the regular exponential family. Then $T_1(X_1,\ldots,X_n)$, $T_2(X_1,\ldots,X_n)$, $\ldots$, $T_k(X_1,\ldots,X_n)$ are jointly complete sufficient statistics for $\theta_1, \theta_2,\ldots,\theta_k$. **(Section 10.4.5)**

Lehmann-Scheffé Theorem: Let T be a complete sufficient statistic for θ. If there exists some function $\phi(T)$ such that $E[\phi(T)] = h(\theta)$, then

1. $\phi(T)$ is unique, and
2. $\phi(T)$ has a smaller variance than any other unbiased estimator of $h(\theta)$. **(Section 10.4.5)**

Ancillary Statistic: A statistic $U(X_1, X_2,\ldots,X_n)$ is said to be ancillary for θ if the distribution of $U(X_1, X_2,\ldots,X_n)$ does not depend on θ. **(Section 10.4.5)**

Basu's Theorem: Let $T(X_1,\ldots,X_n)$ be a complete sufficient statistic for the parameter θ, and let $U(X_1,\ldots,X_n)$ be ancillary for θ. Then $T(X_1,\ldots,X_n)$ and $U(X_1,\ldots,X_n)$ are independent random variables. **(Section 10.4.5)**

Consistency: A sequence of estimators, $\widehat{\theta}_1, \widehat{\theta}_2,\ldots$, for a parameter θ is said to be *consistent* if, for any $\varepsilon > 0$,

$$\lim_{n\to\infty} P(|\widehat{\theta}_n - \theta| \geq \varepsilon) = 0.$$

Let $\widehat{\theta}_1, \widehat{\theta}_2,\ldots$ be a sequence of estimators for θ. If $\mathrm{Var}(\widehat{\theta}_n)$ exists and is finite and

$$\lim_{n\to\infty} E[(\widehat{\theta}_n - \theta)^2] = 0,$$

then, for any $\varepsilon > 0$,

$$\lim_{n\to\infty} P(|\widehat{\theta}_n - \theta| \geq \varepsilon) = 0. \qquad \text{(Section 10.5.1)}$$

Asymptotic Efficiency: If $\widehat{\tau(\theta)}$ is a consistent estimator for $\tau(\theta)$, $\mathrm{Var}(\widehat{\tau(\theta)})$ exists and is finite and the lower bound given by Theorem 10.2 exists then the *asymptotic efficiency* for $\widehat{\tau(\theta)}$ is

$$\mathrm{a.e.}\ (\widehat{\tau(\theta)}) = \lim_{n\to\infty} \frac{[\tau'(\theta)]^2}{\mathrm{Var}(\widehat{\tau(\theta)})\, n\, E\left[\left(\dfrac{\partial\, \ln f(x;\theta)}{\partial\theta}\right)^2\right]}.$$

Furthermore, if a.e. $(\widehat{\tau(\theta)}) = 1$, then $\widehat{\tau(\theta)}$ is said to be asymptotically efficient. **(Section 10.5.2)**

Best Asymptotical Normal (BAN): The sequence of estimators $\widehat{\theta}_1, \widehat{\theta}_2, \ldots$ is said to be a best asymptotically normal (BAN) estimator for θ if

1. The distribution of

$$\sqrt{n}\,\frac{\widehat{\theta}_n - \theta}{\sqrt{\mathrm{Var}(\widehat{\theta}_n)}}$$

approaches a $N(0,1)$ distribution as n becomes infinitely large,

2. $\widehat{\theta}$ is a consistent estimator for all values of θ, and

3. for any other sequence of estimators $\widehat{\theta}'_1, \widehat{\theta}'_2, \ldots,$

$$\lim_{n\to\infty} \mathrm{Var}(\widehat{\theta}_n) \leq \lim_{n\to\infty} \mathrm{Var}(\widehat{\theta}'_n)$$

for all values of θ. **(Section 10.5.2)**

Bayes Estimation:

Posterior Distribution: The posterior distribution of a parameter θ, given the observations $x_1, \ldots, x_n$ is the conditional probability density function or conditional probability function

$$h(\theta \mid x_1, \ldots, x_n) = \frac{f(x_1, \ldots, x_n \mid \theta)\pi(\theta)}{\int_{-\infty}^{\infty} f(x_1, \ldots, x_n \mid \theta)\pi(\theta)\,d\theta}.$$ **(Section 10.6)**

Bayes Estimator: Suppose we have a posterior distribution, $h(\theta \mid x)$ and a loss function $L(d(X), \theta)$. A decision rule $d(X)$ is Bayes if it minimizes

$$E[L(d(X), \theta) \mid X = x] = \int_{-\infty}^{\infty} L(d(X), \theta)h(\theta \mid x)\,d\theta.$$

Suppose we have a sample of n observations from a population with a probability function of probability distribution function $f(x \mid \theta)$, and suppose that $\pi(\theta)$ is the probability function or probability density function of the prior distribution of the parameter θ.

1. If the loss function is proportional to squared error, then the Bayes estimator for θ is the mean of the posterior distribution of θ, given the observations.

2. If the loss function is proportional to absolute error, then the Bayes estimator for θ is the median of the posterior distribution of θ, given the observations.   **(Section 10.6)**

Robustness: A statistical procedure is said to be robust if its behavior is relatively insensitive to deviations from the assumptions on which it is based.   **(Section 10.7)**

α-Trimmed Mean: Let $X_1,\ldots,X_n$ be a random sample and $Y_1 \leq \cdots \leq Y_n$ be the ordered observations. Further, let for $0 \leq \alpha \leq 0.5$, $k = [n\alpha]$. Then α-trimmed mean is

$$\bar{X}_\alpha = \frac{Y_{k+1} + Y_{k+2} + \cdots + Y_{n-k}}{n - 2k}.$$ **(Section 10.7)**

Review Exercises

1. Let $X_1, X_2, \ldots, X_n$ be a random sample of size n from a population with a probability density function of

$$f(x) = \begin{cases} \beta e^{-\beta(x-\theta)}, & \theta < x, \quad \text{and} \\ 0, & \text{otherwise.} \end{cases}$$

   a. Find the maximum likelihood estimators for β and θ.
   b. Find the method of moments estimators for β and θ.
   c. Show that the maximum likelihood estimator for θ is biased.

2. Given a sample of size n from a distribution with a probability density function of

$$f(x) = \begin{cases} \dfrac{2x}{\theta^2}, & 0 < x < \theta, \quad \text{and} \\ 0, & \text{otherwise.} \end{cases}$$

   a. compute an unbiased estimator for θ based on the maximum likelihood estimator for θ,
   b. compute a method of moments estimator for θ and show that it is unbiased.
   c. In terms of relative efficiency, which estimator do you recommend?
   d. Are your two estimators consistent?

3. Let $X_1, X_2, \ldots, X_n$ be a random sample of size n from a $N(\mu, \sigma^2)$ population, and $Y_1, Y_2, \ldots, Y_m$ be a random sample of size m from an independent $N(\tau, \sigma^2)$ population. Find maximum likelihood estimators for μ, τ, and σ^2.

4. Let $X_1, X_2, \ldots, X_n$ be a random sample from a $N(\theta, \theta^2)$ distribution. Compare the problem of finding a maximum likelihood estimator for θ with that of finding a method of moments estimator for θ.

5. Let $X_1, X_2, \ldots, X_n$ be a random sample from a distribution having a p.d.f. of

$$f(x) = \begin{cases} \dfrac{\theta}{x^2}, & \theta \leq x, \quad \text{and} \\ 0, & \text{otherwise.} \end{cases}$$

   a. Show that $Y_1 = \min(X_1, \ldots, X_n)$ is a maximum likelihood estimator for θ.

    b. Find a function of Y_1 that is unbiased for θ.

    c. Show that Y_1 is a minimal sufficient statistic for θ.

6. Show that the maximum likelihood estimator for the variance of a normal distribution,

$$\widehat{\sigma^2} = \frac{1}{n} \sum_{i=1}^{n} \left(X_i - \bar{X} \right)^2,$$

has a smaller mean square error than does the sample variance,

$$s^2 = \frac{1}{n-1} \sum_{i=1}^{n} \left(X_i - \bar{X} \right)^2.$$

7. Let $X_1, X_2, \ldots, X_n$ be a random sample from a $G(\alpha, \beta)$ distribution.

    a. Show that

$$\sum_{i=1}^{n} X_i, \quad \text{and} \quad \prod_{i=1}^{n} X_i$$

    are jointly complete and sufficient for α and β.

    b. Find a uniform minimum variance unbiased estimator for $E[X] = \alpha\beta$.

8. Let X amd Y be i.i.d. $U(0, \theta)$ random variables. We have already shown that $U = \max(X, Y)$ is a complete sufficient statistic for θ. Use Basu's theorem to show that $T = \frac{(X+Y)}{U}$ and U are independent random variables.

9. Suppose you have a sample of size 1 from a population having a Weibull distribution with a probability density function of

$$f(x \mid \theta) = \begin{cases} e^{-(x-\theta)} & \theta < x, \quad \text{and} \\ 0 & , \text{otherwise.} \end{cases}$$

Let θ have a prior distribution that is $G(1, 1)$. Find a Bayes estimator for θ using squared error loss.

    The next two problems ask you to think about how you might apply the method of maximum likelihood and the method of moments to estimation problems that are somewhat different than those discussed in the chapter.

10. It is known that the diameter of apples from a particular orchard can be modeled using a $N(\mu, \sigma^2)$ distribution. Harvested apples are sorted into three groups. Those with a diameter less than d inches are sold to applesauce manufacturers, those with a diameter between d inches and D inches are sold to frozen pie manufacturers, and those with diameters above D inches are sold as eating apples. On a given day n_d small apples, n medium apples, and n_D large apples are harvested. How would you use this information to estimate μ and σ^2? That is, estimate μ and σ^2 using only $d, D, n_d, n,$ and n_D.

11. A chemist wishes to determine the spacing between certain atoms in a crystal structure. The particular atoms, $A, B, C,$ and $D,$ are known to be in a straight line as shown below.

$$A - - - - - - B - - - - - - C - - - - - - D$$

Observations are available for the distances $AB, AC, AD, BC, BD,$ and $CD.$ The apparatus is known to make observations that have errors that follow a $N(0, \sigma^2)$ distribution. Discuss how you would use these six observations to estimate $AB, AC,$ and $AD.$

11

Confidence Intervals

11.1
Introduction

In Chapter 10 we discussed methods for deriving estimators for the values of distribution parameters. We referred to these as *point estimators* for the parameters. Values of estimates computed using these estimators generally differ from the actual parameter value by varying amounts. In the case of unbiased estimation, we used the variance of the sampling distribution of the estimator to measure the "average" error.

In this chapter, we wish to discuss another type of estimation, called an *interval estimate,* for estimating parameter values. We motivate the approach in the following manner. A point estimate is an approximation to the true value of a parameter. The reader has encountered approximation techniques before. In a calculus course, for example, Taylor polynomials are introduced to approximate the values of complicated functions. Once the polynomial was derived additional analysis was performed to place an upper bound on the possible error. If the resulting Taylor polynomial returned a value of b when $x = a$, and the maximum error was found to be E, then we stated with complete certainty that $f(a) = b \pm E$. We would like to do the same thing when estimating parameter values. Unfortunately, a similar type of error analysis for random variables does not give us any useful results.

Example 11.1

We have seen that the efficient estimator for μ in a $N(\mu, \sigma^2)$ distribution is $\bar{X}$, the sample mean. It has a $N(\mu, \frac{\sigma^2}{n})$ distribution. We would like to find a and b so that $P(a < \bar{X} - \mu < b) = 1$. We can achieve this only by including all possible values of $\bar{X} - \mu$. This means that we obtain that $P(-\infty < \bar{X} - \mu < \infty) = 1$.

Thus, all we can state with complete certainty is that $-\infty < \mu < \infty$. By requiring complete certainty we have learned nothing new about the true value of μ by using the sample mean.   ■

For this reason we choose a probability that is *near 1*. Typical values are 0.9, 0.95, and 0.99. If we let $0 \le \alpha \le 1$, this probability is commonly denoted by $1 - \alpha$. The reason for this somewhat odd notation will be explained in the next chapter. Suppose we wish to find an interval estimate for the parameter θ. Our goal is to derive statistics l and u for the endpoints of an interval such that

$$P[l(X_1,\ldots,X_n) < \theta < u(X_1,\ldots,X_n)] = 1 - \alpha. \tag{11.1}$$

After sampling we compute values for these estimators, say

$$l(x_1,\ldots,x_n) = a, \quad \text{and} \quad u(x_1,\ldots,x_n) = b.$$

At this point, however, we cannot state that $P(a < \theta < b) = 1 - \alpha$ since a, b, and θ are all constants. This means that the probability is 1 if $a < \theta < b$, and 0 if not. We can, however, say the following. Using the long run proportion interpretation for probability $l(X_1,\ldots,X_n)$ and $u(X_1,\ldots,X_n)$ will result in intervals, $(1 - \alpha)100\%$ of which will contain θ. Thus, we can say that we are $(1 - \alpha)100\%$ *confident* that $a < \theta < b$. For this reason $a < \theta < b$ is called a $(1 - \alpha)100\%$ *confidence interval for* θ. a and b are called the *lower* and *upper confidence limits,* respectively. Remember that the term $(1 - \alpha)100\%$ *confidence level* means that, if a large number of samples are drawn and the statistics $l(X_1,\ldots,X_n)$ and $u(X_1,\ldots,X_n)$ are computed to obtain the end points, then in the long run $(1 - \alpha)100\%$ of all those intervals will contain θ and $\alpha 100\%$ will not. For a particular sample the interval $a < \theta < b$ is one such interval. It may or may not contain θ. To illustrate what we mean consider Figure 11.1.

Figure 11.1 shows the results of constructing one hundred 95% confidence intervals for the mean of a normal distribution with $\mu = 10$. Note that 95 of the intervals contain the true mean of 10 and 5 do not. In this case, we have been fortunate to get exactly 95 out of 100 intervals to contain the true mean. The 95% figure is only true in a limiting sense. Not every set of 100 randomly chosen intervals will contain exactly 95 intervals that cover μ.

In this chapter, we shall discuss two methods that can be used to derive confidence intervals for parameters. One is based on our ability to find quantities with known sampling distributions such as the t and χ^2 distributions from Chapter 9. The other is based on using the distribution of a point estimator for a parameter directly. Following that we shall discuss a method based on the asymptotic properties of maximum likelihood estimators for deriving confidence intervals for large samples. Finally, we shall show how to construct a nonparametric confidence interval for the median of a distribution.

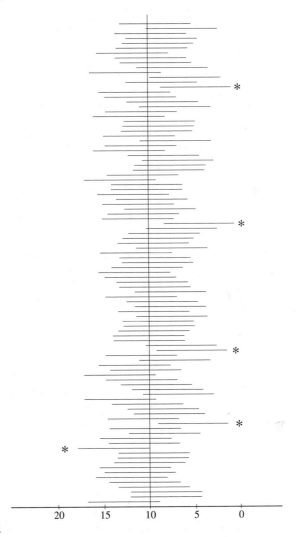

Figure 11.1

11.2 ━━━━━
**Pivotal
Quantity
Method**

The method we will discuss in this section is fairly easy to apply when it works. It relies heavily on our knowledge of sampling distributions. The idea is best introduced by means of an example.

Example 11.2

Let $X_1, X_2, \ldots, X_n$ be a random sample from a $N(\mu, \sigma^2)$ with σ^2 known. We wish to derive a $(1 - \alpha)100\%$ confidence interval for μ. The maximum likelihood estimator for μ is $\bar{X}$, which we know has a $N(\mu, \frac{\sigma^2}{n})$ distribution. This distribution depends on the value of μ, which is unknown. However, the distribution of

$$Z = \frac{\bar{X} - \mu}{\dfrac{\sigma}{\sqrt{n}}}$$

is $N(0, 1)$, which does not depend on μ. Therefore, we can use Table 3 in Appendix A to find values c and d so that

$$P(c < Z < d) = 1 - \alpha.$$

Our interval will be shortest if we let c and d be symmetric about 0. This means that $c = -d$. Then the fact that $P(Z \geq d) + P(Z \leq -d) = \alpha$ implies that $P(Z \geq d) = \frac{\alpha}{2}$. If we recall the α subscripting notation from Chapter 9 this means that $d = -c = z_{\alpha/2}$. We then proceed as follows to derive a confidence interval.

$$1 - \alpha = P(-z_{\alpha/2} < Z < z_{\alpha/2})$$

$$= P\left(-z_{\alpha/2} < \frac{\bar{X} - \mu}{\dfrac{\sigma}{\sqrt{n}}} < z_{\alpha/2}\right)$$

$$= P\left(\bar{X} - z_{\alpha/2}\frac{\sigma}{n} < \mu < \bar{X} + z_{\alpha/2}\frac{\sigma}{n}\right)$$

This is of the form

$$P(l < \theta < u) = 1 - \alpha, \quad \text{where}$$

$$l = \bar{X} - z_{\alpha/2}\frac{\sigma}{\sqrt{n}}, \quad \text{and}$$

$$u = \bar{X} + z_{\alpha/2}\frac{\sigma}{\sqrt{n}}.$$

Now, l and u are statistics since they depend on no unknown quantities. This means that a $(1 - \alpha)100\%$ confidence interval for μ in a normal distribution with σ^2 known is

$$\bar{X} - z_{\alpha/2}\frac{\sigma}{\sqrt{n}} < \mu < \bar{X} + z_{\alpha/2}\frac{\sigma}{\sqrt{n}}. \tag{11.2}$$

■

Looking at this example we note the following. We began with an estimator, $\hat{\theta}$, for the parameter θ. We then found a random variable, call it $C(\hat{\theta}, \theta)$, which satisfied the following conditions.

1. $C(\hat{\theta}, \theta)$ contained $\hat{\theta}$, θ and no other unknown parameters.
2. The distribution of $C(\hat{\theta}, \theta)$ did not depend on θ or any other unknown parameters.

A random variable that satisfies these conditions is called a *pivotal quantity.* We then found values c and d such that

$$P[C(\widehat{\theta}, \theta) \le c] = P[C(\widehat{\theta}, \theta) \ge d] = \frac{\alpha}{2}.$$

Then the expression

$$P[c < C(\widehat{\theta}, \theta) < d] = 1 - \alpha$$

was manipulated to obtain a probability statement of the form of Equation 11.1. The difficult part of this method is coming up with pivotal quantity $C(\widehat{\theta}, \theta)$. In fact, it is not always possible to obtain one. This is what we meant by the qualifier "when it works." In addition, Example 11.2 illustrates another common practice in confidence interval estimation. For a $(1 - \alpha)100\%$ confidence the probability α is split in half with $\frac{\alpha}{2}$ being allocated to each tail of the distribution of $C(\widehat{\theta}, \theta)$. It should be noted that one-sided confidence intervals do exist, but we shall not discuss them here.

Example 11.3   Let $X_1, X_2, \ldots, X_n$ be a random sample from a $N(\mu, \sigma^2)$ where this time the value of σ^2 is unknown. As in Example 11.2

$$Z = \frac{\bar{X} - \mu}{\sqrt{\sigma^2 / n}}$$

has a distribution that is $N(0, 1)$ and, hence, does not depend on μ. However, Z contains the unknown parameter σ^2. If we replace σ^2 with its unbiased estimator, s^2, then we obtain a quantity that does not contain any unknown parameters except for μ. From Exercise 2 of Section 9.3 we know that

$$\frac{\frac{\bar{X} - \mu}{\sigma} / \sqrt{n}}{\sqrt{(n-1)s^2 / [\sigma^2(n-1)]}} = \frac{\bar{X} - \mu}{s / \sqrt{n}}$$

has a t distribution with $n - 1$ degrees of freedom. This distribution does not depend on μ or any other unknown parameters. Therefore,

$$C(\widehat{\mu}, \mu) = \frac{\bar{X} - \mu}{s / \sqrt{n}}$$

is a pivotal quantity and can serve as the basis for a confidence interval for μ. Since the t distribution is symmetric we split α and obtain a confidence interval as follows.

$$1 - \alpha = P\left(-t(n-1)_{\alpha/2} < \frac{\bar{X} - \mu}{s / \sqrt{n}} < t(n-1)_{\alpha/2}\right)$$

$$= P\left(\bar{X} - t(n-1)_{\alpha/2}\frac{s}{\sqrt{n}} < \mu < \bar{X} + t(n-1)_{\alpha/2}\frac{s}{\sqrt{n}}\right)$$

Therefore, a $(1 - \alpha)100\%$ confidence interval for μ in a normal distribution with σ^2 unknown is

$$\bar{X} - t(n - 1)_{\alpha/2} s / \sqrt{n} < \mu < \bar{X} + t(n - 1)_{\alpha/2} s / \sqrt{n}. \qquad \textbf{(11.3)}$$

∎

Example 11.4   Recruits in the U. S. Navy are given a physical fitness test to assess their overall condition. As part of that examination they are required to run 300 yards. Times for this distance are known to be approximately normally distributed. A sample of 10 recruits gives the following times, in seconds.

$$
\begin{array}{ccccc}
47.1 & 45.5 & 42.5 & 47.5 & 38.1 \\
47.3 & 44.1 & 54.5 & 45.4 & 41.1
\end{array}
$$

We wish to construct a 95% confidence interval for the mean time for the 300-yard run. Since this is a case where we have a normal distribution with an unknown variance the method derived in Example 11.3 is appropriate. Since $n = 10$, the t statistic will have 9 degrees of freedom. A confidence level of 95% gives that $\alpha = 0.05$. Thus, Table 5 in Appendix A gives

$$t(n - 1)_{\alpha/2} = t(9)_{0.025} = 2.262.$$

From the 10 observations we compute

$$\bar{X} = 45.3, \quad \text{and} \quad s = 4.416.$$

Therefore, Equation 11.3 gives

$$45.3 - (2.262)\frac{4.416}{\sqrt{10}} < \mu < 45.3 + (2.262)\frac{4.416}{\sqrt{10}},$$

or

$$42.14 < \mu < 48.46.$$

∎

Example 11.5   Let $X_1, X_2, \ldots, X_n$ be a random sample from a $N(\mu, \sigma^2)$ distribution. We wish to derive a $(1 - \alpha)100\%$ confidence interval for σ^2. We know that s^2 is an unbiased estimator for σ^2, which is based on the maximum likelihood estimator. In addition, from Theorem 9.6 we know that

$$\frac{(n - 1)s^2}{\sigma^2}$$

has a $\chi^2(n - 1)$ distribution. Note that this quantity contains s^2, σ^2, and no other unknown parameters. Furthermore, its distribution does not depend on σ^2. Therefore, we can use

$$C(s^2, \sigma^2) = \frac{(n - 1)s^2}{\sigma^2}$$

as a pivotal quantity to construct a confidence interval for σ^2. Choose c and d so that $P[C(s^2, \sigma^2) \leq c] = P[C(s^2, \sigma^2) \geq d] = \frac{\alpha}{2}$. We then find that $c = \chi^2(n-1)_{1-\alpha/2}$ and $d = \chi^2(n-1)_{\alpha/2}$. We then obtain a confidence interval as follows.

$$1 - \alpha = P\left[\chi^2(n-1)_{1-\alpha/2} < \frac{(n-1)s^2}{\sigma^2} < \chi^2(n-1)_{\alpha/2}\right]$$

$$= P\left[\frac{(n-1)s^2}{\chi^2(n-1)_{\alpha/2}} < \sigma^2 < \frac{(n-1)s^2}{\chi^2(n-1)_{1-\alpha/2}}\right]$$

Therefore, a $(1 - \alpha)100\%$ confidence interval for σ^2 in a normal distribution is

$$\frac{(n-1)s^2}{\chi^2(n-1)_{\alpha/2}} < \sigma^2 < \frac{(n-1)s^2}{\chi^2(n-1)_{1-\alpha/2}}. \tag{11.4}$$

■

Example 11.6   We wish to use the data from Example 11.4 to construct a 90% confidence for the variance of the times, in seconds, to run 300 yards. Since the population is normal, Equation 11.4 can be used. The χ^2 statistic will have 9 degrees of freedom. Ninety percent confidence means that $\alpha = 0.10$. Therefore, we use Table 4 in Appendix A to find that

$$\chi^2(n-1)_{\alpha/2} = \chi^2(9)_{0.05} = 16.919, \quad \text{and}$$

$$\chi^2(n-1)_{1-\alpha/2} = \chi^2(9)_{0.95} = 3.325.$$

Using the data in Example 11.4 we find that $s^2 = 19.503$. Thus, Equation 11.4 gives that a 90% confidence interval for the variance in the times for the 300 yard run is

$$\frac{(9)(19.503)}{16.919} < \sigma^2 < \frac{(9)(19.503)}{3.325},$$

or

$$10.375 < \sigma^2 < 52.791.$$

■

These ideas can be applied to derive confidence intervals for comparing parameters from two independent populations.

Example 11.7   Let $X_1, X_2, \ldots, X_n$ be a random from a population having a $N(\mu_X, \sigma^2)$ distribution and $Y_1, Y_2, \ldots, Y_m$ be a random sample from an independent population having a $N(\mu_Y, \sigma^2)$ distribution. We wish to derive a $(1 - \alpha)100\%$ confidence interval for $\mu_X - \mu_Y$. We assume that the common variance σ^2 is unknown. From the exercises in Section 10.8 we find that unbiased estimators for μ_X, μ_Y, and σ^2 are

$$\widehat{\mu_X} = \overline{X}, \quad \widehat{\mu_Y} = \overline{Y}, \quad \text{and} \quad \widehat{\sigma^2} = \frac{(n-1)s_X^2 + (m-1)s_Y^2}{n+m-2}.$$

Now we can build a quantity with a t distribution by referring to Exercise 3 of Section 9.3. There it was shown that

$$\frac{\bar{X} - \bar{Y} - (\mu_X - \mu_Y)}{\sqrt{\dfrac{(n-1)s_X^2 + (m-1)s_Y^2}{n+m-2} \left[\dfrac{1}{n} + \dfrac{1}{m}\right]}}$$

has a t distribution with $n + m - 2$ degrees of freedom. Note that this quantity contains both $\mu_X - \mu_Y$ and an estimator for it. In addition, it depends on no other unknown parameters, and it has a distribution that does not depend on $\mu_X - \mu_Y$. Let

$$s_{\bar{X}-\bar{Y}}^2 = \frac{(n-1)s_X^2 + (m-1)s_Y^2}{n+m-2}\left[\frac{1}{n} + \frac{1}{m}\right]. \tag{11.5}$$

We see that a confidence interval can be constructed by using

$$\frac{\bar{X} - \bar{Y} - (\mu_X - \mu_Y)}{s_{\bar{X}-\bar{Y}}}.$$

Following the derivation for the case of the mean for a single population we find that a $(1 - \alpha)100\%$ confidence interval for $\mu_X - \mu_Y$ is

$$\bar{X} - \bar{Y} - t(n+m-2)_{\alpha/2}\, s_{\bar{X}-\bar{Y}} < \mu_X - \mu_Y < \bar{X} - \bar{Y} + t(n+m-2)_{\alpha/2}\, s_{\bar{X}-\bar{Y}}. \tag{11.6}$$

The details are left as an exercise.   ■

Example 11.8

A study was conducted to compare 2 different procedures for assembling components. Both procedures were implemented and run for a month to allow employees to learn each procedure. Then each was observed for 8 days with the following results. Values are number of components assembled per day.

Procedure A	106	107	76	113	104	113	95	99
Procedure B	100	99	78	111	97	102	94	86

After some investigation it appeared appropriate to assume that the data from each procedure were approximately normally distributed with a common variance. We shall use Equation 11.6 to construct a 99% confidence interval for the differences in the mean number of components assembled by the 2 methods. Since $n = m = 8$ the t statistic will have 14 degrees of freedom. A confidence level of 99% gives that $\alpha = 0.01$. Thus, Table 5 in Appendix A gives that

$$t(n + m - 2)_{\alpha/2} = t(14)_{0.005} = 2.977.$$

We shall let the output from Method A be $X_1, \ldots, X_8$, and the output from Method B be $Y_1, \ldots, Y_8$. Then we compute that

$$n = 8, \quad \bar{X} = 101.6, \quad s_X^2 = 145.696,$$
$$m = 9, \quad \bar{Y} = 95.9, \quad \text{and} \quad s_Y^2 = 104.125.$$

An application of Equation 11.6 gives that a 99% confidence interval for the difference in the mean number of components assembled using the 2 procedures is

$$-10.9 < \mu_X - \mu_Y < 22.3.$$

Note that the confidence interval contains 0. This indicates that we cannot reasonably conclude that there is any difference between the 2 methods in terms of number of components produced. ∎

Example 11.9

Let $X_1, X_2 \ldots, X_n$ be a random sample from a $N(\mu_X, \sigma_X^2)$ distribution and $Y_1, Y_2, \ldots, Y_m$ be a random sample from an independent $N(\mu_Y, \sigma_Y^2)$ distribution. We wish to derive a $(1 - \alpha)100\%$ confidence interval for $\frac{\sigma_X^2}{\sigma_Y^2}$. Unbiased estimators for σ_X^2 and σ_Y^2 based on the maximum likelihood estimators are s_X^2 and s_Y^2, respectively. Using Theorem 9.6 we see that $\frac{(n-1)s_X^2}{\sigma_X^2}$ has a $\chi^2(n - 1)$ distribution, and $\frac{(m-1)s_Y^2}{\sigma_Y^2}$ has a $\chi^2(m - 1)$ distribution. In addition, these 2 random variables are independent. This means that

$$\frac{\dfrac{(n-1)s_X^2 / \sigma_X^2}{(n-1)}}{\dfrac{(m-1)s_Y^2 / \sigma_Y^2}{(m-1)}} = \frac{\sigma_Y^2 \, s_X^2}{\sigma_X^2 \, s_Y^2}$$

has an F distribution with $(n - 1, m - 1)$ degrees of freedom. Note that this quantity contains both $\frac{\sigma_X^2}{\sigma_Y^2}$ and an estimator for it. In addition, it depends on no other unknown parameters, and its distribution does not depend on $\frac{\sigma_X^2}{\sigma_Y^2}$. Therefore, we can construct a confidence interval as follows.

$$1 - \alpha = P\left[F(n-1, m-1)_{1-\alpha/2} < \frac{\sigma_Y^2 \, s_X^2}{\sigma_X^2 \, s_Y^2} < F(n-1, m-1)_{\alpha/2} \right]$$

$$= P\left[\frac{s_X^2 / s_Y^2}{F(n-1, m-1)_{\alpha/2}} < \frac{\sigma_X^2}{\sigma_Y^2} < \frac{s_X^2 / s_Y^2}{F(n-1, m-1)_{1-\alpha/2}} \right]$$

Therefore, a $(1 - \alpha)100\%$ confidence interval for σ_X^2 / σ_Y^2 is

$$\frac{s_X^2 / s_Y^2}{F(n-1, m-1)_{\alpha/2}} < \frac{\sigma_X^2}{\sigma_Y^2} < \frac{s_X^2 / s_Y^2}{F(n-1, m-1)_{1-\alpha/2}}. \tag{11.7}$$

∎

Example 11.10

We wish to use the data in Example 11.9 to construct a 95% confidence interval for $\frac{\sigma_X^2}{\sigma_Y^2}$. This is sometimes useful when checking the assumption of equality of variances that is made in the confidence interval for $\mu_X - \mu_Y$. Since the distributions are approximately normal we shall use Equation 11.7. The F statistic will have 7 degrees of freedom for both the numerator and denominator. A 95%

confidence level gives that $\alpha = 0.05$. Thus, using Table 6 in Appendix A we find that

$$F(n - 1, m - 1)_{\alpha/2} = F(7, 7)_{0.025} = 4.99, \quad \text{and}$$

$$F(n - 1, m - 1)_{1-\alpha/2} = F(7, 7)_{0.975} = \frac{1}{F(7, 7)_{0.025}} = \frac{1}{4.99}.$$

Equation 11.7 then gives that a 95% confidence interval for the ratio of variances of the number of units assembled by both procedures is

$$0.29 < \frac{\sigma_X^2}{\sigma_Y^2} < 7.16.$$

Since the confidence interval contains 1, it is reasonable to conclude that the variances are equal.   ∎

All of the examples thus far have dealt with samples from normal populations. This is because the sampling distributions for statistics from such populations are well known. This method can be used for parameters of other distributions as the following example shows.

Example 11.11   Let $X_1, X_2, \ldots, X_n$ be a random sample from a $G(1, \beta)$ distribution. We wish to construct a $(1 - \alpha)100\%$ confidence interval for β. The maximum likelihood estimator for β is $\bar{X}$. The distribution for $\bar{X}$ can be found by using moment generating functions.

$$M_{\bar{X}}(t) = E[e^{t\bar{X}}]$$

$$= \left(E[e^{(t/n)X}] \right)^n$$

$$= \left(\frac{1}{1 - \dfrac{\beta t}{n}} \right)^n$$

This is the moment generating function for a $G(n, \frac{\beta}{n})$ random variable. This distribution depends on the value of β. We note that if we can replace t in the moment generating function by $\frac{t}{\beta}$ the distribution would then only depend on n. For this reason we consider the moment generating function for $\frac{\bar{X}}{\beta}$ which is

$$M_{\bar{X}/\beta}(t) = M_{\bar{X}}\left(\frac{t}{\beta} \right)$$

$$= \left(\frac{1}{1 - \dfrac{t}{n}} \right)^n.$$

This shows that $\bar{X}/\beta$ has a $G(n, \frac{1}{n})$ distribution. Thus, we have a random variable that contains β, its estimator, and the distribution depends only on the sample size. The only problem is a practical one of finding c and d such that

$$P\left(\frac{\bar{X}}{\beta} \leq c\right) = P\left(\frac{\bar{X}}{\beta} \geq d\right) = \frac{\alpha}{2}.$$

To overcome this we leave it as an exercise to show that $\frac{2n\bar{X}}{\beta}$ has a $\chi^2(2n)$ distribution. Therefore, we can obtain a confidence interval as follows.

$$1 - \alpha = P\left(\chi^2(2n)_{1-\alpha/2} < \frac{2n\bar{X}}{\beta} < \chi^2(2n)_{\alpha/2}\right)$$

$$= P\left(\frac{2n\bar{X}}{\chi^2(2n)_{\alpha/2}} < \beta < \frac{2n\bar{X}}{\chi^2(2n)_{1-\alpha/2}}\right)$$

Thus, a $(1 - \alpha)100\%$ confidence interval for β is

$$\frac{2n\bar{X}}{\chi^2(2n)_{\alpha/2}} < \beta < \frac{2n\bar{X}}{\chi^2(2n)_{1-\alpha/2}}.$$ ∎

Exercises 11.2

1. Let $X_1, X_2, \ldots, X_n$ be a random sample from a $U(0, \theta)$ distribution, and let Y_n be the largest observation. Show that the quantity Y_n / θ satisfies the requirements of this section, and use it to construct a $(1 - \alpha)100\%$ confidence interval for θ.

2. Let $X_1, X_2, \ldots, X_n$ be a random sample from a $N(\mu, 1)$ distribution. Show that the confidence interval for μ given in Example 11.2, where the probability α was allocated equally in the upper and lower tails, gives a confidence interval that is shorter than would be obtained by any other method for allocating α.

3. Let $X_1, X_2, \ldots, X_n$ be a random sample from a population with a $N(\mu_X, \sigma^2)$ distribution, and let $Y_1, Y_2, \ldots, Y_m$ be a random sample from an independent population having a $N(\mu_Y, \sigma^2)$ distribution. Use Examples 11.5 and 11.7 to suggest a procedure for constructing a $(1 - \alpha)100\%$ confidence interval for σ^2.

4. Find a 98% confidence interval for the mean of a normal distribution with $\sigma^2 = 4$ from the following sample.

$$1.2 \quad 3.7 \quad 2.1 \quad 5.7 \quad 4.1 \quad 7.2$$

5. Find a 99% confidence interval for the mean of a normal distribution whose variance is unknown for the data given in Exercise 4.

6. Find a 90% confidence interval for σ^2 using the data given in Exercise 4. Assume that the data are normally distributed.

7. A study was conducted to measure the time for a particular fabric used in baby clothing to catch fire. The data are given below.

$$9.3 \quad 8.4 \quad 7.9 \quad 12.4 \quad 9.1 \quad 11.2 \quad 10.4 \quad 9.7 \quad 8.9$$

Assume a normal distribution. Find a 90% confidence interval for the mean time for the fabric to catch fire.

8. Using the data in Exercise 7 construct a 95% interval for the variance of the time for the fabric to catch fire.

9. A consumer group conducted a study to determine the reliability of a certain brand of compact disk player. A random sample of 10 owners of the player gave the following results. Data are in years of reliable service.

$$4.4 \quad 5.3 \quad 6.2 \quad 8.0 \quad 4.0$$
$$3.9 \quad 7.7 \quad 5.2 \quad 6.8 \quad 6.6$$

Find a 99% confidence interval for the mean lifetime of this brand of compact disk player.

10. Using the data in Exercise 9 find a 95% confidence interval for the variance in the lifetime of this brand of compact disk player.

11. Two different arrangements for typewriter keys are being investigated for the effect on typing speed. High school freshmen who have never typed before are divided into two groups. Group A used keyboard style A, and group B was trained on keyboard style B with the following results. All data are in words per minute.

Group A	44	39	52	41	46	60	48
Group B	40	47	54	49	46	59	

Construct a 95% confidence interval for the difference in mean typing speed for the 2 types of keyboards. Assume that the distributions are normal with equal variances. Interpret your results.

12. Use the data from Exercise 11 to construct a 95% confidence interval for the ratio of the variances of typing speed. Does this interval support the assumption of equality of variances?

13. Fill in the details of the derivation of the confidence interval for the difference between the means of two independent normal populations discussed in Example 11.7.

14. It has been suggested that recent college graduates spend more time watching television to obtain news information than reading newspapers. In an effort to determine if this is true, a college journalism department surveyed a number of college graduates. The participants were divided into 2 groups. Group 1 consisted of graduates who had been out of school less than 5 years. Group 2 consisted of graduates who had been out of school more than 10 years. The results are summarized below. Data are in hours per week.

Group	Number	Mean	s
Less than 5 years	12	15.6	2.3
More than 10 years	11	10.1	2.2

Assume normal distributions and construct a 90% confidence interval for the differences in hours per week for the two groups. Interpret your results.

15. Use the data in Exercise 14 to construct a 90% confidence interval for the ratio of the variances of the 2 groups. Interpret your results.

16. Let $X_1, X_2, \ldots, X_n$ be a random sample from a distribution having a p.d.f. of

$$f(x) = \begin{cases} e^{-(x-\theta)}, & \theta \le x, \quad \text{and} \\ 0, & \text{otherwise.} \end{cases}$$

a. Let $Y_1 = \min(X_1, \ldots, X_n)$ and show that $Y_1 - \theta$ is a pivotal quantity.
b. Use $Y_1 - \theta$ to construct a $(1 - \alpha)100\%$ confidence interval for θ.
c. Use the confidence interval from Part b to find a 90% confidence interval for θ for the following data.

$$11 \quad 8 \quad 9.1 \quad 6.8 \quad 7.2$$

11.3
Method Based on Sampling Distributions

The method we used in the previous section required that we find a quantity that contained both the parameter and its estimator with the additional restriction that the distribution of that quantity did not depend on any unknown parameters. In addition, there was the practical necessity that the distribution of the quantity make it easy to compute probabilities. As we have seen this can be a difficult task. In fact, in many situations, such a quantity does not exist. Because of this we shall discuss a method that does not require that we find such a quantity. All we need for this method to succeed is the distribution of the estimator for the parameter.

Example 11.12

Suppose we have a random sample of 10 observations from a $N(\mu, 1)$ distribution.

$$\begin{array}{ccccc} 10.464 & 10.060 & 11.486 & 10.022 & 13.394 \\ 10.137 & 7.474 & 9.366 & 9.528 & 9.445 \end{array}$$

If we use the confidence interval from Example 11.2 with a confidence level of 95% we obtain that $\bar{x} = 10.138$, and

$$9.517 < \mu < 10.758.$$

Let $p = \frac{\alpha}{2} = 0.025$. Now we ask the following question. What value of μ will we need so that

$$P\left(Z \le \frac{\bar{x} - \mu}{\frac{\sigma}{\sqrt{n}}}\right) = P\left(Z \le \frac{10.138 - \mu}{0.316}\right)$$

$$= 0.025?$$

Table 3 of Appendix A gives that

$$\frac{10.138 - \mu}{0.316} = -1.96.$$

Solving for μ gives

$$\mu = 10.138 + 1.96\frac{1}{\sqrt{10}} = 10.758.$$

Now let $p = 0.025$ and determine the value of μ that we need so that

$$P\left(Z \geq \frac{\bar{x} - \mu}{\frac{\sigma}{\sqrt{n}}}\right) = P\left(Z \geq \frac{10.138 - \mu}{0.316}\right)$$

$$= 0.025.$$

Using Table 3 of Appendix A again gives that

$$\frac{10.138 - \mu}{0.316} = 1.96.$$

Solving for μ this time gives that

$$\mu = 10.138 - 1.96\frac{1}{\sqrt{10}} = 9.517.$$

Now if we ask for all values of μ such that

$$P\left(Z \leq \frac{10.138 - \mu}{0.316}\right) \geq 0.025, \quad \text{and} \quad P\left(Z \geq \frac{10.138 - \mu}{0.316}\right) \geq 0.025$$

we find that

$$9.517 < \mu < 10.758.$$

This is precisely the range of values for μ in our 95% confidence interval.   ■

This example points to what our method will be. We shall begin with an estimator for θ, $\hat{\theta}$, and determine its sampling distribution. Then select two probability levels, p_1 and p_2, so that $p_1 + p_2 = \alpha$. Normally we let $p_1 = p_2$. We then take a sample and compute the value of $\hat{\theta}$. Suppose $\hat{\theta} = t$. Next determine the value of θ, call it θ_l, so that

$$P(\hat{\theta} \geq t) = p_1, \tag{11.8}$$

and another value of θ, call it θ_u, so that

$$P(\hat{\theta} \leq t) = p_2. \tag{11.9}$$

Then a $(1 - \alpha)100\%$ confidence interval for θ will be

$$\theta_l < \theta < \theta_u.$$

We show how this works by means of an example.

Example 11.13   Suppose we wish to construct a $(1 - \alpha)100\%$ confidence interval for θ in a $U(0, \theta)$ distribution based on a sample of n observations. The maximum likelihood estimator for θ is $\hat{\theta} = Y_n$, the largest order statistic. We know from Chapter 9 that the p.d.f. of Y_n is

$$f_{Y_n}(y) = \begin{cases} n\dfrac{y^{n-1}}{\theta^n}, & 0 \le y \le \theta, \quad \text{and} \\ \\ 0, & \text{otherwise.} \end{cases}$$

Let $p_1 = p_2 = \frac{\alpha}{2}$. We now use Equation 11.8 to determine the value of θ_l. We assume that the observed value for Y_n is y_n, and let $\theta = \theta_l$

$$\frac{\alpha}{2} = \int_{y_n}^{\theta_l} n\frac{y^{n-1}}{\theta_l^n}\, dy$$

$$= 1 - \left(\frac{y_n}{\theta_l}\right)^n.$$

Solving for θ_l gives

$$\theta_l = \frac{y_n}{\left[1 - \frac{\alpha}{2}\right]^{1/n}}.$$

Next use Equation 11.9 to determine the value of θ_u. We still assume that the observed value for Y_n is y_n, and this time let $\theta = \theta_u$

$$\frac{\alpha}{2} = \int_0^{y_n} n\frac{y^{n-1}}{\theta_u^n}\, dy$$

$$= \left(\frac{y_n}{\theta_u}\right)^n.$$

Solving for θ_u gives

$$\theta_u = \frac{y_n}{\left[\frac{\alpha}{2}\right]^{1/n}}.$$

Therefore, a $(1 - \alpha)100\%$ confidence interval for θ will be

$$\frac{y_n}{\left[1 - \frac{\alpha}{2}\right]^{1/n}} < \theta < \frac{y_n}{\left[\frac{\alpha}{2}\right]^{1/n}}.$$

It should be pointed out that the quantity $\frac{Y_n}{\theta}$ was shown in the exercises for Section 11.2 to meet the requirements for using the method based on sampling distributions. Thus, it was not necessary to obtain the confidence interval in this manner. On the other hand, making such an observation is not easy, especially if one is new to statistics. In the preceding example we have been able to solve the probability statements for the parameter. This will not always be the case.   ∎

We now give an example of a confidence interval where the method of Section 11.2 cannot be used.

Example 11.14   Let X have a binomial distribution with parameters n and p. We wish to derive a $(1 - \alpha)100\%$ confidence interval for p. There is no quantity that satisfies the requirements of Section 11.2. Therefore, we must use the more general method.

The maximum likelihood estimator for p was seen in Section 10.2 to be $\hat{p} = \frac{X}{n}$, where X is the number of successes in the n trials. In Section 6.2 we showed that the probability function for $\hat{p}$ is

$$f_{\hat{p}}(x \mid p) = \begin{cases} \binom{n}{nx} p^{nx}(1-p)^{n-nx}, & x = 0, \frac{1}{n}, \frac{2}{n}, \ldots, 1, \quad \text{and} \\ 0 & , \text{ otherwise.} \end{cases}$$

Assume that $\hat{p} = t$. Then the method requires that we find 2 values, p_l and p_u, such that

$$\frac{\alpha}{2} = \sum_{x=0}^{t} \binom{n}{nx} p_u^{nx}(1-p_u)^{n-nx}, \quad \text{and} \tag{11.10}$$

$$\frac{\alpha}{2} = \sum_{x=t}^{1} \binom{n}{nx} p_l^{nx}(1-p_l)^{n-nx}. \tag{11.11}$$

These 2 equations can be solved for p_l and p_u by trial and error, but this is quite tedious for even moderate values of n. We can, however, simplify the task by making use of Equation 5.28. Recall that it related the binomial and beta distributions as follows.

$$\int_0^x \frac{n!}{(\alpha-1)!(n-\alpha+1)!} t^{\alpha-1}(1-t)^{n-\alpha}\,dt = \sum_{i=\alpha}^{n} \binom{n}{i} x^i(1-x)^{n-i}$$

The left-hand side is commonly referred to as the *incomplete beta function with parameters α and $\beta = n - \alpha$*. It is tabulated in a number of places such as the *CRC Handbook of Tables for Probability and Statistics*. In addition, numerical methods may be used to determine the value of x for which the integral equals a specified value. Thus, by using either tables or numerical computation the problem becomes one of determining values p_l and p_u that satisfy

$$\frac{\alpha}{2} = \int_0^{p_l} \frac{n!}{(na-1)!(n-na+1)!} t^{na-1}(1-t)^{n-na}\,dt, \quad \text{and} \tag{11.12}$$

$$\frac{\alpha}{2} = 1 - \int_0^{p_u} \frac{n!}{(na-1)!(n-na+1)!} t^{na-1}(1-t)^{n-na}\,dt. \tag{11.13}$$

A $(1 - \alpha)100\%$ confidence interval for p would be

$$p_l < p < p_u.$$

Since this process would be quite difficult for investigators with limited mathematical backgrounds the results of this method are frequently summarized in charts such as the one shown in Figure 11.2. There is a separate chart for each confidence level. The one shown is for 95% confidence intervals. With these a confidence interval is obtained by simply selecting the chart for the desired confidence level and then using the values for n and $\hat{p}$ to find the values for p_l and p_u. ∎

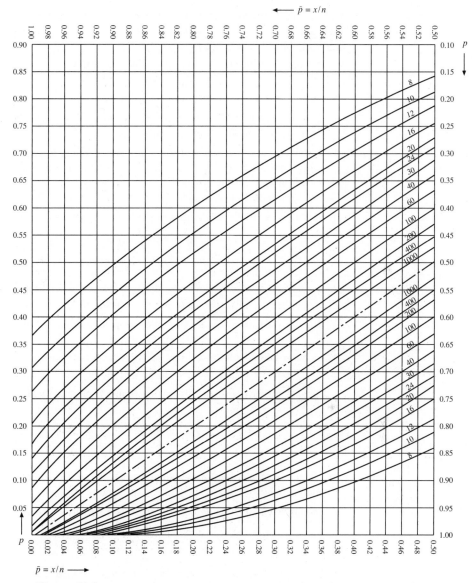

$\bar{p} = x/n$

$\bar{p} = x/n \longrightarrow$

Figure 11.2   Source: *Biometrika Tables for Statisticians,* Vol. 1, Third Edition (1966) with permission of the Biometrika Trustees.

Example 11.15   One hundred registered voters were polled as to which candidate they favored in an upcoming election. Sixty voters indicated a preference for the incumbent. We wish to find a 95% confidence interval for p, the proportion favoring the incumbent. We enter Figure 11.2 with $n = 100$ and $\widehat{p} = 0.60$ and find

$$0.50 < p < 0.70.$$

Exercises 11.3

1. Let $X_1, X_2, \ldots, X_n$ be a random sample from a $Po(\lambda)$ distribution. Describe how to construct a $(1 - \alpha)100\%$ confidence interval for λ if you are given a computer program that can compute $P(X \leq a)$ for any value of a and λ.

2. Use the method based on sampling distributions to derive a $(1 - \alpha)100\%$ confidence interval for the parameter β of a $G(1, \beta)$ distribution.

3. A coin is tossed 100 times with 45 heads occurring. Construct a 95% confidence interval for the probability of getting a head.

4. A political candidate claims that the majority of voters favor his stand on a particular issue. A poll of 60 voters shows that 35 agree with his position. Construct a 95% confidence interval for the proportion favoring the candidate's position. Does this support the claim made by the candidate?

5. Suppose you have a single observation from a population whose p.d.f. is

$$f(x) = \begin{cases} \theta\, x^{\theta-1} & , 0 \leq x \leq 1, \quad \text{and} \\ 0 & , \text{otherwise.} \end{cases}$$

   a. Use Part a to derive a $(1 - \alpha)100\%$ confidence interval for θ.
   b. Construct a 95% confidence interval when $X = 0.3$.

6. Suppose we have a single observation from a geometric distribution with probability of success of p.

   a. Show how to construct a $(1 - \alpha)100\%$ confidence interval for p.
   b. Find a 95% confidence interval for p if the first success occurred on trial 5.

7. Suppose we have a single observation from a distribution whose p.d.f. is

$$f(x) = \begin{cases} \dfrac{2\theta^2}{x^3} & , \theta \leq x, \quad \text{and} \\ 0 & , \text{otherwise.} \end{cases}$$

Use the method based on sampling distributions to derive a $(1 - \alpha)100\%$ confidence interval for θ.

8. Suppose we have a single observation from a Cauchy distribution whose p.d.f. is

$$f(x) = \frac{1}{\pi[1 + (x - \theta)^2]}, -\infty < x < \infty.$$

Use the method based on sampling distributions to derive a $(1 - \alpha)100\%$ confidence interval for θ.

11.4
Large Sample Confidence Intervals

The methods described thus far have some practical shortcomings. The method of Section 11.2 depends on our ability to find a quantity that contains the parameter and its estimator such that the distribution does not depend on the parameter. Such a quantity may not exist. The method of Section 11.3 can involve some tedious calculations. For these reasons we shall discuss a method that can be used if we have a large sample. It is based on the asymptotic distribution of maximum likelihood estimators. Recall in Chapter 10 that we stated that maximum likelihood estimators, under fairly general conditions, are consistent and have a limiting distribution, which is normal. Maximum likelihood estimators are also asymptotically efficient. This means that in large samples the maximum likelihood estimator for θ, $\widehat{\theta}$, will have a distribution that is approximately normal with a mean of θ. In addition, if the Cramér-Rao Lower Bound exists, the limiting variance, $\sigma^2_{\widehat{\theta}}$, will be

$$\sigma^2_{\widehat{\theta}} = \frac{1}{E\left[\left(\frac{\partial \ln L}{\partial \theta}\right)^2\right]}.$$

This means that the quantity

$$Z = \frac{\widehat{\theta} - \theta}{\sigma_{\widehat{\theta}}}$$

will have a distribution that is approximately standard normal. Thus, in an approximate sense, we have a quantity that meets the requirements of Section 11.2. A large sample $(1 - \alpha)100\%$ confidence interval can then be derived by starting with the probability statement

$$P\left[-z_{\alpha/2} < \frac{\widehat{\theta} - \theta}{\sigma_{\widehat{\theta}}} < z_{\alpha/2}\right] \approx 1 - \alpha.$$

How we proceed from here will vary depending on whether or not $\sigma_{\widehat{\theta}}$ depends on θ.

Example 11.16   We wish to obtain a large sample confidence interval for the parameter p in a binomial distribution based on n trials. In Chapter 10 we showed that the maximum likelihood estimator for p is $\widehat{p} = \frac{X}{n}$, X is the number of successes in the n trials. We also showed that $\widehat{p}$ is an efficient estimator whose variance attains the Cramér-Rao Lower Bound of

$$\sigma^2_{\widehat{p}} = \frac{p(1 - p)}{n}.$$

Therefore, we begin with

$$P\left[-z_{\alpha/2} < \frac{\widehat{p} - p}{\sqrt{\dfrac{p(1 - p)}{n}}} < z_{\alpha/2}\right] \approx 1 - \alpha.$$

Following the method of Section 11.2 we could now solve these inequalities for p. The result will be a little messy in that it entails finding the roots of a quadratic equation in p. A simpler approach is as follows.

Since $\widehat{p}$ is a maximum likelihood estimator for p we know that it converges in probability to p. Then, by Theorem 7.9 we know that the sequence of ratios of the form

$$\frac{\widehat{p}(1 - \widehat{p})}{p(1 - p)}$$

converges in probability to one. Thus, we may use Theorems 7.8 and 7.12 to assert that the sequence of quantities of the form

$$\frac{(\widehat{p} - p) \Big/ \sqrt{\dfrac{p(1-p)}{n}}}{\widehat{p}(1-\widehat{p}) \Big/ \sqrt{p(1-p)}} = \frac{\widehat{p} - p}{\sqrt{\dfrac{\widehat{p}(1-\widehat{p})}{n}}}$$

will have an approximate standard normal distribution. We can use this to construct a confidence interval. Notice that the net effect of this has been that we replaced p by its estimator each time it occurred in $\sigma_{\widehat{p}}^2$. It turns out that this is a general result as long as the population distribution is sufficiently well behaved for the maximum likelihood estimators to have an asymptotic distribution that is normal.

Continuing with our example, to obtain the final confidence interval we write

$$P\left[-z_{\alpha/2} < \frac{\widehat{p} - p}{\sqrt{\widehat{p}(1-\widehat{p})/n}} < z_{\alpha/2}\right] \approx 1 - \alpha.$$

Solving for p gives

$$P\left[\widehat{p} - z_{\alpha/2}\sqrt{\frac{\widehat{p}(1-\widehat{p})}{n}} < p < \widehat{p} + z_{\alpha/2}\sqrt{\frac{\widehat{p}(1-\widehat{p})}{n}}\right] \approx 1 - \alpha.$$

Thus, an approximate large sample $(1 - \alpha)100\%$ confidence interval for p in a binomial distribution is

$$\widehat{p} - z_{\alpha/2}\sqrt{\frac{\widehat{p}(1-\widehat{p})}{n}} < p < \widehat{p} + z_{\alpha/2}\sqrt{\frac{\widehat{p}(1-\widehat{p})}{n}}. \tag{11.14}$$

This is the form of the confidence interval for p that appears in most elementary statistics texts. Investigators have found that the large sample confidence interval is valid for all but the most extreme values of p when $n \geq 60$.   ■

Example 11.17   We wish to compute a large sample 95% confidence interval for p, the proportion of registered voters favoring the incumbent candidate in Example 11.15. Recall that $n = 100$ and $X = 60$. Therefore, $\widehat{p} = 0.60$. For 95% confidence we find that $z_{\alpha/2} = z_{0.025} = 1.96$. Then Equation 11.14 gives that

$$0.60 - (1.96)\sqrt{\frac{(0.60)(0.40)}{100}} < p < 0.60 + (1.96)\sqrt{\frac{(0.60)(0.40)}{100}},$$

or

$$0.50 < p < 0.70.$$

In this case, the approximate method agrees with the exact one to at least 2 decimal places.   ■

We now state the general procedure for constructing large sample confidence intervals.

1. Determine a maximum likelihood estimator, $\widehat{\theta}$, for θ. Find maximum likelihood estimators for any other unknown parameters.
2. Obtain $\sigma_{\widehat{\theta}}$ either directly or by using the Cramér-Rao Lower Bound.
3. At each occurrence of θ in $\sigma_{\widehat{\theta}}$ substitute $\widehat{\theta}$. Replace any other unknown parameters in $\sigma_{\widehat{\theta}}$ with their maximum likelihood estimators. Call the resulting quantity $s_{\widehat{\theta}}$.
4. Construct a $(1 - \alpha)100\%$ confidence interval for θ by using

$$\widehat{\theta} - z_{\alpha/2}\, s_{\widehat{\theta}} < \theta < \widehat{\theta} + z_{\alpha/2}\, s_{\widehat{\theta}}. \qquad \textbf{(11.15)}$$

Example 11.18

We wish to derive a large sample confidence interval for the parameter β based on a sample of size n from a $G(1, \beta)$ distribution. We shall illustrate each step of the method.

1. The maximum likelihood estimator for β is

$$\widehat{\beta} = \bar{X}.$$

2. The variance of $\widehat{\beta}$ is

$$\sigma_{\widehat{\beta}}^2 = \text{Var}(\bar{X}) = \frac{\text{Var}(X)}{n} = \frac{\beta^2}{n}.$$

3. Since $\sigma_{\widehat{\beta}}^2$ depends on the value of β we replace it with $\bar{X}$ to obtain

$$s_{\widehat{\beta}}^2 = \frac{\bar{X}^2}{n}.$$

4. An application of Equation 11.15 gives that an approximate $(1 - \alpha)100\%$ confidence interval for β is

$$\bar{X} - z_{\alpha/2}\, \frac{\bar{X}}{\sqrt{n}} < \beta < \bar{X} + z_{\alpha/2}\, \frac{\bar{X}}{\sqrt{n}}. \qquad \textbf{(11.16)}$$

∎

Exercises 11.4

1. Derive a large sample confidence interval for the parameter λ in a Poisson distribution.
2. Derive a large sample confidence interval for the parameter θ in a double exponential distribution with $\sigma = 1$. The maximum likelihood estimator for θ is the sample median.
3. Let $X_1, X_2, \ldots, X_n$ be a random sample from a distribution with a p.d.f. of

$$f(x) = \begin{cases} \theta\, x^{\theta - 1} , & 0 \leq x \leq 1, \quad \text{and} \\ 0 , & \text{otherwise.} \end{cases}$$

Derive a large sample confidence interval for θ.

4. Let $X_1, X_2, \ldots, X_n$ be a random sample from a distribution having a p.d.f. of

$$f(x) = \begin{cases} e^{-(x-\theta)}, & \theta \leq x, \quad \text{and} \\ 0, & \text{otherwise.} \end{cases}$$

The method of moments estimator for θ is $\widehat{\theta} = \bar{x} - 1$. Derive a large sample confidence interval for θ.

5. Let $X_1, X_2, \ldots, X_n$ be a random sample from a $N(\mu, \sigma^2)$ distribution. Derive a large sample confidence interval for σ^2.

6. A political candidate claims that the majority of voters favor his stand on a particular issue. A poll of 60 voters shows that 35 agree with his position. Construct a 95% confidence interval for the proportion favoring the candidate's position. Does this support the claim made by the candidate? Compare this with the result obtained in Exercise 4 of Section 11.3.

7. A disk drive for a personal computer is tested for time to failure. One hundred fifty drives are started and stopped repeatedly until each fails to operate properly. The time to failure is known to follow a gamma distribution with parameters $\alpha = 1$ and β. The 150 drives last a mean time of 250 hours. Find a 90% confidence interval for β, the mean time to failure.

8. Solve the following for p directly.

$$P\left[-z_{\alpha/2} < \frac{\widehat{p} - p}{\sqrt{\dfrac{p(1-p)}{n}}} < z_{\alpha/2} \right] \approx 1 - \alpha$$

That is, use the quadratic formula rather than appeal to Theorem 7.12. Compare your results with Example 11.16.

11.5
A Nonparametric Confidence Interval

In some situations, it may be impractical or impossible to collect a large sample for analysis. With small samples it is extremely difficult to determine the population distribution from the data. For this reason it is often advisable to construct confidence intervals for population quantities that are not parameters of a particular family of distributions. The median is commonly used in such cases as a measure of location. The methods we have discussed thus far have depended on having knowledge of the population distribution or the sampling distribution of an estimator. In the nonparametric setting, we need procedures where the quantities used have distributions that do not depend on the population distribution. The population median is a particularly convenient quantity for constructing confidence intervals because we can make use of the definition of the population median. Throughout this discussion we assume that we are dealing with a continuous population distribution.

Recall that the median is the point where the probability of lying below it is equal to the probability of lying above it. That is, if $\widetilde{\mu}$ is the median of the distribution and X is any observation from that distribution, then $P(X \leq \widetilde{\mu}) = \frac{1}{2}$. This means that, given a sample of n observations from a population whose median is $\tilde{\mu}_0$, the distribution of the number of observations falling below $\tilde{\mu}_0$ will follow a $B(n, \frac{1}{2})$ distribution. We can then construct a confidence interval for the median by using the binomial distribution.

We begin by determining the number of observations such that

$$\frac{\alpha}{2} = \sum_{i=0}^{a} \binom{n}{i} \left(\frac{1}{2}\right)^n, \text{ and}$$

$$\frac{\alpha}{2} = \sum_{b}^{n} \binom{n}{i} \left(\frac{1}{2}\right)^n.$$

This shows that the population median will be below the order statistic, Y_a, $\frac{\alpha}{2}(100)\%$ of the time and above the order statistic, Y_b, $\frac{\alpha}{2}(100)\%$ of the time. Thus, in terms of our discussion of Section 11.3, a $(1 - \alpha)100\%$ confidence interval for the median of a distribution would be

$$Y_a < \tilde{\mu} < Y_b.$$

If we divide the upper and lower tail probabilities equally we find that $b = n + 1 - a$. Therefore, our confidence interval becomes

$$Y_a < \tilde{\mu} < Y_{n+1-a}. \tag{11.17}$$

In summary we find by using the method of Section 11.3 that a $(1 - \alpha)100\%$ confidence interval for the population median is given by Equation 11.17, where a is chosen so that

$$\frac{\alpha}{2} = \sum_{i=0}^{a} \binom{n}{i} \left(\frac{1}{2}\right)^n.$$

In practice, a is usually chosen to come as close to achieving $\frac{\alpha}{2}$ as possible. This will ensure that the actual confidence level is *near* $(1 - \alpha)100\%$.

Example 11.19

A drug is suspected of causing an elevated heart rate in a certain group of high-risk patients. Fifteen patients from this group were given the drug. The changes in heart rates were found to be as follows.

$$
\begin{array}{ccccc}
10 & 8 & 12 & 6 & 11 \\
-1 & 5 & 0 & 7 & 2 \\
1 & 4 & 6 & 3 & 9
\end{array}
$$

We wish to construct a 90% confidence interval for the median change in heart rate. For 90% confidence we use $\alpha = 0.10$. From Table 1 in Appendix A we find that the $B(15, \frac{1}{2})$ distribution gives the following.

$$P(X \leq 3) = 0.0176.$$

$$P(X \leq 4) = 0.0592.$$

We note that $a = 4$ comes closest to achieving $\frac{\alpha}{2} = 0.05$. Therefore, we sort the data and use the fourth observation from the bottom and the fourth observation from the top for our confidence limits. The sorted data are given below.

$$
\begin{array}{rrrrr}
-1 & 0 & 1 & 2 & 3 \\
4 & 5 & 6 & 6 & 7 \\
8 & 9 & 10 & 11 & 12
\end{array}
$$

The values chosen are 2 and 9. Therefore, our confidence interval is

$$2 < \tilde{\mu} < 9$$

with an actual confidence level of 88.16%.   ■

Exercises 11.5

1. Use the data in Example 11.4 to construct a 90% confidence interval for the median time to run 300 yards.

2. A certain chemical in powder form must be mixed with water before use. The manufacturer wishes to determine what the directions should say regarding the time spent stirring the mixture before the powder dissolves. Twenty standard doses are each mixed in a gallon of water. The times, in seconds, for each to dissolve completely are given below.

$$
\begin{array}{lllll}
15.1 & 11.9 & 13.2 & 12.5 & 14.2 \\
13.3 & 15.0 & 15.3 & 12.4 & 12.9 \\
14.1 & 13.5 & 12.4 & 15.4 & 12.0 \\
13.7 & 12.9 & 13.0 & 14.6 & 13.1
\end{array}
$$

Construct a 95% confidence interval for the median time for the chemical to dissolve. Based on this, what do you recommend the manufacturer state in his directions to users of the chemical?

3. Construct a 99% confidence interval for the median weight gain for the cattle data given in Table 8.2. What is the exact confidence level?

4. Construct a 95% confidence interval for the metal wire data given in Exercise 3 at the end of Section 8.3.

Chapter Summary

Pivotal Quantities: A pivotal quantity satisfies the following conditions.

1. $C(\widehat{\theta}, \theta)$ contained $\widehat{\theta}$, θ and no other unknown parameters.
2. The distribution of $C(\widehat{\theta}, \theta)$ did not depend on θ or any other unknown parameters.

Find values c and d such that

$$P[C(\widehat{\theta}, \theta) \le c] = P[C(\widehat{\theta}, \theta) \ge d] = \frac{\alpha}{2}.$$

Then manipulate the expression

$$P[c < C(\widehat{\theta}, \theta) < d] = 1 - \alpha$$

to obtain a probability statement of the form

$$P[l(X_1,\ldots,X_n) < \theta < u(X_1,\ldots,X_n)] = 1 - \alpha.$$

Given a sample of n observations from a $N(\mu,\sigma^2)$ distribution, we have the following.

1. A $(1 - \alpha)100\%$ confidence interval for μ with σ^2 known is

$$\bar{X} - z_{\alpha/2} < \mu < \bar{X} + z_{\alpha/2}.$$

2. A $(1 - \alpha)100\%$ confidence interval for μ with σ^2 unknown is

$$\bar{X} - t(n - 1)_{\alpha/2} < \mu < \bar{X} + t(n - 1)_{\alpha/2}.$$

3. A $(1 - \alpha)100\%$ confidence interval for σ^2

$$\frac{(n - 1)s^2}{\chi^2(n - 1)_{\alpha/2}} < \sigma^2 < \frac{(n - 1)s^2}{\chi^2(n - 1)_{1-\alpha/2}}.$$

Given a sample of size n from a $N(\mu_X,\sigma_X^2)$ distribution and a sample of size m from an independent $N(\mu_Y,\sigma_Y^2)$ distribution, we have the following.

1. If $\sigma_X^2 = \sigma_Y^2$, a $(1 - \alpha)100\%$ confidence interval for $\mu_X - \mu_Y$ is

$$\bar{X} - \bar{Y} - t(n + m - 2)_{\alpha/2}\, s_{\bar{X}-\bar{Y}} < \mu_X - \mu_Y < \bar{X} - \bar{Y} + t(n + m - 2)_{\alpha/2}\, s_{\bar{X}-\bar{Y}}$$

where

$$s_{\bar{X}-\bar{Y}}^2 = \frac{(n - 1)s_X^2 + (m - 1)s_Y^2}{n + m - 2}\left[\frac{1}{n} + \frac{1}{m}\right],$$

2. a $(1 - \alpha)100\%$ confidence interval for $\dfrac{\sigma_X^2}{\sigma_Y^2}$ is

$$\frac{s_X^2 / s_Y^2}{F(n - 1, m - 1)_{\alpha/2}} < \frac{\sigma_X^2}{\sigma_Y^2} < \frac{s_X^2 / s_Y^2}{F(n - 1, m - 1)_{1-\alpha/2}}. \qquad \textbf{(Section 11.2)}$$

Method Based on Sampling Distributions:

1. Find an estimator for θ, $\hat{\theta}$, and determine its sampling distribution.
2. Select two probability levels, p_1 and p_2, so that $p_1 + p_2 = \alpha$. Usually $p_1 = p_2$.
3. Take a sample and compute $\hat{\theta}$. Suppose $\hat{\theta} = t$.
4. Determine the value of θ, call it θ_l, so that

$$P(\hat{\theta} \geq t) = p_1.$$

5. Determine the value of θ, call it θ_u, so that

$$P(\hat{\theta} \leq t) = p_2.$$

6. A $(1 - \alpha)100\%$ confidence interval for θ will be

$$\theta_l < \theta < \theta_u. \qquad \text{(Section 11.3)}$$

Large Sample Confidence Intervals:

1. Determine a maximum likelihood estimator, $\widehat{\theta}$, for θ. Find maximum likelihood estimators for any other unknown parameters.
2. Obtain $\sigma_{\widehat{\theta}}$ either directly or by using the Cramér-Rao Lower Bound.
3. At each occurrence of θ in $\sigma_{\widehat{\theta}}$ substitute $\widehat{\theta}$. Replace any other unknown parameters in $\sigma_{\widehat{\theta}}$ with their maximum likelihood estimators. Call the resulting quantity $s_{\widehat{\theta}}$.
4. Construct a $(1 - \alpha)100\%$ confidence interval for θ by using

$$\widehat{\theta} - z_{\alpha/2}\, s_{\widehat{\theta}} < \theta < \widehat{\theta} + z_{\alpha/2}\, s_{\widehat{\theta}}. \qquad \text{(Section 11.4)}$$

Review Exercises

1. State what happens to the length of a confidence interval for the mean of a normal distribution under each of the following.
   a. The sample size is increased.
   b. The variance increases.
   c. Confidence level is increased.

2. A 95% confidence interval for the mean of a normal distribution based on a sample of 100 observations is found to be

$$205 < \mu < 239.$$

   One hundred additional samples of size 100 each were then taken from the same population and their sample means computed. What can be said about the proportion of these 100 sample means that would be expected to fall in the interval $(205, 239)$?

3. How large of a sample must be taken from a normal population whose variance is 64 so that a 99% confidence interval for μ will be no more than 4 units wide?

4. How might you answer Exercise 3 if the variance were unknown?

5. Use the large sample confidence interval for the parameter p in a binomial distribution to determine the sample size necessary to ensure that a 95% confidence interval for p will be no wider than 0.06. *(Hint: For what value of p is $\sigma_{\widehat{p}}^2$ the largest?)*

6. A sample of 50 voters revealed that only 8 were registered as Independents. Construct a 95% confidence interval for the proportion of Independents in the voting population.

7. Suppose you have a single observation from a population whose distribution has a p.d.f. of

$$f(x) = \begin{cases} \dfrac{2x}{\theta^2} & , \ 0 \le x \le \theta, \quad \text{and} \\[2mm] 0 & , \ \text{otherwise.} \end{cases}$$

   a. Use the method of Section 11.3 to derive a confidence interval for θ.
   b. Use this to construct a 90% confidence interval when the observed value is 0.75.

8. Let $X \sim B(n, p_X)$ and $Y \sim B(m, p_Y)$, where X and Y are independent. Show that a large sample confidence interval for $p_X - p_Y$ is given by

$$\widehat{p}_X - \widehat{p}_Y - z_{\alpha/2}\sqrt{\frac{\widehat{p}_X(1 - \widehat{p}_X)}{n} + \frac{\widehat{p}_Y(1 - \widehat{p}_Y)}{m}}$$

$$< p_X - p_Y < \widehat{p}_X - \widehat{p}_Y + z_{\alpha/2}\sqrt{\frac{\widehat{p}_X(1 - \widehat{p}_X)}{n} + \frac{\widehat{p}_Y(1 - \widehat{p}_Y)}{m}}.$$

9. Construct a 95% confidence interval for the median of a population using the following data.

$$12 \quad 9 \quad 6 \quad 11 \quad 8 \quad 7 \quad 15 \quad 18 \quad 14$$

10. Suppose the data in Exercise 5 of Section 8.3 come from a population with a normal distribution.
    a. Construct a 99% confidence interval for μ.
    b. Construct a 90% confidence interval for σ^2.

11. Construct a 90% confidence interval for the population median using the data from Exercise 6 of Section 8.3.

12. A vocational school is investigating 2 different methods of training automotive students to perform a certain repair. The school has established that a competent mechanic should complete this job in an average time of 60 minutes. The following data from the 2 training methods represent the difference, in minutes, between a student's actual time to perform the repair and 60 minutes. Assume that the data are from independent normal populations.

Method A	2.4	0.7	1.9	1.0	−0.1	1.4	1.0	
Method B	−0.2	−2.5	0.1	0.9	0.4	−0.6	0.8	−1.2

Construct a 90% confidence interval for the difference in the mean times to complete the repair. What, if anything, can you conclude about the 2 training methods?

13. Use the data in Exercise 12 to construct a 95% confidence interval for the ratio of the variances for the 2 training methods. Does your interval make the assumption of equality of variances seem justified?

14. Let $X_1, \ldots, X_n$ be a random sample from a $N(\mu_X, \sigma_X^2)$ distribution and $Y_1, \ldots, Y_m$ be a random sample from an independent $N(\mu_Y, \sigma_Y^2)$ distribution. Derive a large sample $(1 - \alpha)100\%$ confidence interval for $\mu_X - \mu_Y$.

15. Let $X_1, \ldots, X_n$ be a random sample from a $G(1, \beta_X)$ distribution and $Y_1, \ldots, Y_n$ be a random sample from an independent $G(1, \beta_Y)$ distribution.
    a. Show that $\frac{2n\bar{X}}{\beta_X}$ has a $\chi^2(2n)$ distribution.
    b. Use the result of Part a to show that

    $$\frac{\bar{X}\,\beta_Y}{\bar{Y}\,\beta_X}$$

    is a pivotal quantity for $\frac{\beta_X}{\beta_Y}$.
    c. Use the result of Part b to construct a $(1 - \alpha)100\%$ confidence interval for $\frac{\beta_X}{\beta_Y}$.

12

Hypothesis Testing

12.1
Introduction

The two major problems treated by statistical theory are estimation and hypothesis testing. Point and interval estimation have been discussed in the previous two chapters. In this chapter we introduce the concepts of hypothesis testing. Estimation uses information from random samples to determine, as well as possible, the value of a population parameter. In hypothesis testing, the problem is to use sample data to determine which of two statements regarding a distribution is correct. These two problems are connected.

Recall in Example 11.4 we constructed a 95% confidence interval for the mean of a normal distribution. The confidence limits were found to be

$$42.14 < \mu < 48.46.$$

Now suppose that we wished to determine whether or not it was reasonable, based on this sample, to conclude that the true population mean could be 45. Since this confidence interval contains $\mu = 45$ we would be inclined to accept the proposition that the mean could actually be this value. This latter problem is the sort of question that hypothesis testing seeks to answer. If the investigator wishes to know what values for a population parameter are reasonable, then a confidence interval is needed. However, if the only issue is whether a particular value for the population parameter is in a confidence interval, then a test of hypotheses would be appropriate. The point here is, however, that the two approaches to using the sample data are closely related.

We illustrate some of the basic ideas in hypothesis testing with an example.

Example 12.1

Suppose we have a die that is known to be either fair or loaded so that the numbers 4, 5, or 6 occur on 70% of the rolls. We decide to roll the die a number of times and use the outcomes of the rolls to determine whether it seems more reasonable to conclude that the die is fair than to conclude it is loaded. Suppose the die is rolled 20 times, and the number of times a 4, 5, or 6 occurs is noted. Call this number X. X will have a binomial distribution with $n = 20$. We shall consider success to be observing a 4, 5, or 6. If the die is fair, then $p = 0.5$. If the die is loaded, then $p = 0.7$. Therefore, we wish to determine whether $p = 0.5$ or $p = 0.7$. These two statements about the value of p are called *hypotheses*. There are always exactly 2 hypotheses. One of them is referred to as the *null hypothesis* and is denoted by H_0. The other is known as the *alternative hypothesis* and is denoted by H_1. Suppose we choose $p = 0.5$ as the null hypothesis. The hypothesis testing problem is then stated that we wish to test

$$H_0 : p = 0.5 \quad \text{vs.} \quad H_1 : p = 0.7.$$

In this problem we will use the observed value of X to determine which is the more reasonable value for p. Since X is a number computed from sample data that will be used to choose between the 2 hypotheses it is called the *test statistic*. Since it is possible to observe any integer value of X between 0 and 20 under either hypothesis we can never be completely sure we shall decide on the correct value of p by using X. We do know, however, that large values for X have a higher probability of occurrence when the alternative hypothesis is true than under the null hypothesis. Thus, it is reasonable to conclude $p = 0.7$ if we observe a value of X that is too large. Therefore, we must decide what value of X is large enough to indicate that the alternative hypothesis is more plausible than the null hypothesis. Suppose, for the sake of having a specific example, we decide that any value of X equal to or exceeding 15 will indicate that the null hypothesis is unreasonable. In the language of hypothesis testing, we would say that the *null hypothesis is rejected if $X \geq 15$* and the *null hypothesis is accepted if $X < 15$*. The set of values of the test statistic that leads us to reject the null hypothesis in favor of the alternative is called the *critical region* or the *rejection region* for the test. To conduct the test we would roll the die 20 times and compute X. If X is less than 15 the null hypothesis is accepted, and if X is 15 or more the null hypothesis is rejected. ∎

All tests of hypotheses have the components mentioned in the preceding example. They include

1. a null and alternative hypotheses and
2. a critical region.

A test statistic commonly exists for a test, but it is not absolutely necessary. All we really need is a partition of the sample space into a critical region and a non-critical region. This partition is often determined by values of a test statistic.

It should be pointed out that statistical hypotheses are not necessarily statements about distribution parameters. For example, it is perfectly legitimate to test

$$H_0 : X \text{ and } Y \text{ are independent random variables,} \quad \text{vs.}$$

$$H_1 : X \text{ and } Y \text{ are not independent random variables.}$$

It is logical at this point to inquire as to how one determines which of the two hypotheses should be the null and which one the alternative. In all cases, at least one hypothesis will contain a statement involving equality. In our discussion, the null hypothesis will serve as the point of reference. For this reason the null hypothesis will always contain equality. Whenever we refer to the null hypothesis, the point where the equality occurs will be used. Another rule of thumb is that if there is a choice, the hypothesis one wishes to conclude is false is usually made the null hypothesis. Combining these two ideas means that the hypothesis that one wishes to reject is usually formulated to contain an equality statement and is made the null hypothesis. Again this is for technical reasons that will be discussed in Section 12.2. In some cases, it may not be obvious which hypothesis contains an equality statement. For example, the hypotheses regarding the independence of X and Y given above are equivalent to testing

$$H_0 : P(X \in A, Y \in B) = P(X \in A)P(Y \in B), \text{ for all events } A \text{ and } B, \quad \text{vs.}$$

$$H_1 : P(X \in A, Y \in B) \neq P(X \in A)P(Y \in B), \text{ for some events } A \text{ and } B.$$

When written in this fashion we see that H_0 contains an equality statement and H_1 does not. Therefore, according to our rule of thumb H_0 and H_1 were chosen correctly.

There is a classification scheme for statistical hypotheses. If the population distribution is completely specified when the hypothesis is assumed to be true, then that hypothesis is said to be *simple*. If not, the hypothesis is called *composite*. Both hypotheses in Example 12.1 are simple. Both hypotheses concerning the independence of X and Y given above are composite. As another example, suppose we have a normally distributed population and we wish to test

$$H_0 : \mu = 100 \quad \text{vs.} \quad H_1 : \mu > 100.$$

If the value of σ^2 is known, then the null hypothesis is simple and the alternative hypothesis is composite. If the value of σ^2 is not known, then both hypotheses are composite.

The goal of this chapter is to develop some general techniques for deriving tests of hypotheses. The method will depend on whether the hypotheses are simple or composite. By deriving a test it is meant that, given the null and alternative hypotheses, we wish to determine an appropriate test statistic and critical region. To choose between competing test statistics and critical regions we need to discuss how tests are compared in terms of quality.

12.2 ▬▬▬▬
Types of Errors

Whenever a test of hypotheses is conducted an error can occur. In Example 12.1, a fair die can give 15 or more successes in 20 rolls and be judged loaded. Likewise, a loaded die can give fewer than 15 successes in 20 rolls and be judged fair. In either case, the rule for deciding which hypothesis is more reasonable would lead to a wrong conclusion. Since the errors we just mentioned occur under different circumstances we are led to the following definition.

Definition 12.1

> The action of rejecting the null hypothesis when it is true is called a *Type I error*. The action of accepting the null hypothesis when it is false is called a *Type II error*.

Since any testing procedure can result in making an error we are interested in how often a given test will commit each type of error. Thus, it is of interest to determine the probability that a test will result in each type of error. In fact, these probabilities are the basis for determining whether one test statistic and critical region is better than another one for the same hypotheses. The values of these probabilities are called α and β. They are defined as follows.

$$\alpha = P(\text{Type I error}) = P(\text{Reject } H_0 \mid H_0 \text{ true})$$
$$\beta = P(\text{Type II error}) = P(\text{Accept } H_0 \mid H_0 \text{ false}) \tag{12.1}$$

Example 12.2

We wish to compute α and β for the test that was discussed in Example 12.1. Recall that it was a test of

$$H_0 : p = 0.5 \quad \text{vs.} \quad H_1 : p = 0.7.$$

The test statistic is X = number of outcomes of 4, 5, or 6 in 20 rolls. The null hypothesis was rejected if $X \geq 15$. Thus, the critical region is $\{X : X \geq 15\}$. The probability of a Type I error is found as follows.

$$\alpha = P(\text{Type I error})$$
$$= P(\text{Reject } H_0 \mid H_0 \text{ true})$$
$$= P(X \geq 15 \mid p = 0.5)$$

Under the assumption that H_0 is true X has a $B(20, 0.5)$ distribution. Therefore, using Table 1 in Appendix A we find that

$$\alpha = P(X \geq 15 \mid p = 0.5)$$
$$= 1 - P(X < 15 \mid p = 0.5)$$
$$= 1 - P(X \leq 14 \mid p = 0.5)$$
$$= 0.0207.$$

The probability of a Type II error is found as follows.

$$\beta = P(\text{Type II error})$$
$$= P(\text{Accept } H_0 \mid H_0 \text{ false})$$
$$= P(X < 15 \mid p = 0.7)$$

When H_1 is true X has a $B(20, 0.7)$ distribution. Thus, the binomial tables give

$$\beta = P(X < 15 \mid p = 0.7)$$
$$= P(X \le 14 \mid p = 0.7)$$
$$= 0.5836.$$

The interpretation of these is as follows. If this test is conducted a large number of times with a fair die, we will erroneously conclude it is loaded about 2% of the time. On the other hand, if the die is loaded, then the test will incorrectly indicate that it is fair almost 60% of the time.

Since β seems awfully high we would like to alter the critical region in order to reduce it. Suppose we decide to reject H_0 if $X \ge 12$. Proceeding as above the reader should verify that

$$\alpha = 0.2517 \quad \text{and} \quad \beta = 0.1133. \qquad \blacksquare$$

This example illustrates a problem in hypothesis testing. For a given test statistic and sample size any attempt to reduce β will lead to an increase in α, and vice versa. β was made smaller by making the critical region larger. This means that H_0 is rejected more often, which leads to an increase in α. This also means that for a test with a fixed sample size it is impossible to simultaneously reduce both α and β. We shall discuss later how this can be dealt with by allowing the sample size to be variable, but for now we shall discuss only tests where the sample size is fixed beforehand by the experimenter. Under this condition tests are constructed by fixing α at some small value and searching for a test statistic and critical region that minimizes β. Typical values for α are 0.1, 0.05, and 0.01. Because α is fixed it is commonly referred to as the *size* of the test, and the critical region is called a *size α critical region*. Other standard terminology is that α is called the *significance level* of the test.

The general procedure is for an investigator to select a value for α and a test statistic T. Once the general shape of the critical region, R, is determined, then a critical value is determined so that $\alpha = P(T \in R \mid H_0 \text{ true})$. From this discussion we can see why the null hypothesis is chosen to be the one containing the equality statement. To compute α the null hypothesis is assumed to be true. If the null hypothesis is simple there is no question about how to compute α. When the null hypothesis is composite we must assure that the probability of committing a Type I error never exceeds the predetermined value for any point in the null hypothesis. For this reason the point in the null hypothesis that is used to compute α for a composite null hypothesis is where equality occurs. To see how this works we wish to adjust Example 12.1 slightly to test

$$H_0 : p \le 0.5 \quad \text{vs.} \quad H_1 : p > 0.5.$$

Let the critical region still be to reject H_0 if $X \ge 15$. For various values of p in H_0 we compute the probability of rejecting H_0 as shown Table 12.1.

Table 12.1

p	$P(X \geq 15)$
0.5	0.0207
0.4	0.0016
0.3	0.0000
0.2	0.0000
0.1	0.0000

Thus, we see that the probability of incorrectly rejecting H_0 is highest if we choose the value of p in the null hypothesis that is closest to the values of p in the alternative hypothesis.

Example 12.3   In the test of

$$H_0 : p = 0.5 \quad \text{vs.} \quad H_1 : p = 0.7$$

with a test statistic of X = number of outcomes of 4, 5, or 6 in 20 rolls, the general form for the critical region is to reject H_0 if $X \geq a$. We wish to find a critical region whose size is as close to 0.05 as possible. Using Table 1 in Appendix A we find that

$$P(X \geq 13 \mid p = 0.5) = 0.1316,$$

$$P(X \geq 14 \mid p = 0.5) = 0.0577, \quad \text{and}$$

$$P(X \geq 15 \mid p = 0.5) = 0.0207.$$

Therefore, the critical region that rejects H_0 when $X \geq 14$ comes closest to 0.05 and has an actual size of 0.0577.   ■

There is a method whereby a test such as the one above can have a size of *exactly* 0.05. It results in what is known as a *randomized test*. In the preceding example, it is clear that we want always to reject H_0 if $X \geq 15$. However, if we could reject H_0 when $X = 14$ only some of the time, then any value of α between 0.0207 and 0.0577 can be achieved depending on what proportion of the time we reject H_0 when $X = 14$. The question then becomes one of determining the correct proportion of observations of 14 for which to reject H_0. In this example we are searching for a value of k so that

$$0.05 = P(X \geq 15 \mid p = 0.5) + k P(X = 14 \mid p = 0.5)$$

$$= 0.0207 + k (0.0370).$$

Solving this for k gives

$$k = 0.792.$$

Thus, a randomized test would be conducted as follows. Reject H_0 if $X \geq 15$. Accept H_0 if $X \leq 13$. If $X = 14$ observe a binomial random variable, Y, with parameters $n = 1$ and $p = 0.792$. Reject H_0 if a success occurs, otherwise accept H_0. There are many computer programs that can generate values of random variables that have a binomial distribution. A common way is to use a random number generator that produces observations of a $U(0, 1)$ random variable. If the output of the random number generator is less than or equal to k, then a success occurs.

The need to conduct randomized tests to achieve exact significance levels arises only when the test statistic being used has a discrete distribution. In fact, in practice randomized tests are rarely used. In most cases, the significance level is an approximate measure of the desire not to commit a Type I error. As a result most investigators will use a nonrandomized test with a significance level as close as possible to the specified value. A more conservative approach is to use a nonrandomized with an exact significance level that is as close as possible to the specified amount without exceeding it. This latter approach can result in rather large values for β. The primary use for randomized test is in theoretical considerations. They are of use there because randomization ensures that we can achieve a critical region of size α in any situation.

Another quantity related to α and β is commonly used as a measure of quality for testing procedures. It is known as *power* and is defined as follows.

Definition 12.2

> Let x be a fixed point in the alternative hypothesis. The *power* of a test, denoted by η is
> $$\eta = P(\text{Reject } H_0 \mid x \in H_1). \tag{12.2}$$

From the definition it is clear that $\eta = 1 - \beta$, and it represents the probability of correctly rejecting the null hypothesis. When both hypotheses are simple there is only one value for η for a given test procedure. However, if the alternative hypothesis is composite, then η depends on the particular point in the alternative hypothesis that is being assumed. In the die example, if the alternative hypothesis were $H_1 : p > 0.5$ (the die is loaded in favor of 4, 5, or 6) the power would depend on the particular value of p. This gives rise to the notion of a *power function*. The definition uses Ω to denote the set of possible values for a parameter θ, Ω_0 to denote the values of θ under H_0, and Ω_1 to denote the values of θ under H_1.

Definition 12.3

> In testing $H_0 : \theta \in \Omega_0$ vs. $H_1 : \theta \in \Omega_1$, the *power function*, denoted $\eta(\theta)$, is
> $$\eta(\theta) = P(\text{Reject } H_0 \mid \theta), \quad \text{for } \theta \in \Omega_0 \cup \Omega_1. \tag{12.3}$$
>
> In addition, the *operating characteristic (OC)* of the test, denoted $\beta(\theta)$, is
> $$\beta(\theta) = P(\text{Accept } H_0 \mid \theta), \quad \text{for } \theta \in \Omega_0 \cup \Omega_1. \tag{12.4}$$

From this definition we see that

$$\eta(\theta) = 1 - \beta(\theta).$$

Example 12.4   We wish to compute the power function and operating characteristic for the randomized test of

$$H_0 : p = 0.5 \quad \text{vs.} \quad H_1 : p > 0.5,$$

where the size of the critical region is 0.05. As we have seen, the critical region rejects H_0 if $X \geq 15$ and if $X = 14$ with probability 0.792. Thus, the power function is

$$\eta(p) = P(X \geq 15 \mid p) + (0.792) P(X = 14 \mid p), \quad \text{for } p \in [0.5, 1],$$

and the operating characteristic is

$$\beta(p) = 1 - \eta(p).$$

Using Table 1 in Appendix A for selected values of p we find the values shown in Table 12.2.

Table 12.2

p	0.5	0.6	0.7	0.8	0.9	1.0
$\eta(p)$	0.050	0.224	0.568	0.891	0.996	1.000
$\beta(p)$	0.950	0.776	0.432	0.109	0.004	0.000

These results are typical for all power functions and operating characteristics. The power function begins at a value of α under H_0 and increases toward 1 as the value of p moves further away from the null hypothesis. The graphs of the power function and operating characteristic are shown in Figures 12.1 and 12.2, respectively.

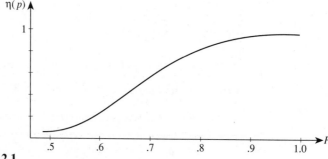

Figure 12.1

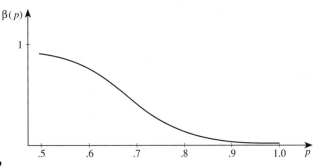

Figure 12.2

There is a quantity closely related to α whose use is popular among practitioners. It is known as the *p-value* for the test. What it does is give a numerical value to indicate how unusual it would be to have a given outcome occur if the null hypothesis is true. It is computed as follows.

In a given test, the general form of the rejection region is known by using methods that we will discuss later. Suppose we use T as a test statistic. Many tests have rejections regions that are of the form $T \geq a$, $T \leq a$, or $|T| \geq a$. For example, the test on the die in Example 12.3 rejected the null hypothesis if $X \geq a$. Suppose we take a sample and compute a value for T, call it t. The *p*-value is the size of the critical region if $a = t$. Suppose in the die example, we observe $x = 17$ outcomes of 4, 5, or 6. Thus, $a = 17$, and the *p*-value for this run of the test would be

$$p\text{-value} = P(X \geq 17 \,|\, p = 0.5) = 0.0015.$$

Since this is a very small value indicating that this result is highly unlikely if H_0 were true, the common practice is to say that the outcome of the test is *highly significant*. That is, there is strong evidence that H_0 is not true.

In terms of our discussion of size α tests, *p*-values have the following interpretation. If, on a given run of a test, the computed *p*-value is less than α, then t would lie in a size α critical region, and we would reject H_0. If not, we would accept H_0. In the previous example, if $x = 17$ we would reject $H_0 : p = 0.5$ in favor of $H_1 : p > 0.5$ for any critical region whose size is $\alpha = 0.0015$ or higher.

There seem to be a couple of reasons for the rise in use of *p*-values. First, there is the widespread availability of computer statistical packages. Rather than ask the user to provide a value for α on each test, these packages simply report the *p*-value. Second, many times investigators wish to make a statement regarding the strength of the evidence leading to a rejection of the null hypothesis. The *p*-value approach permits this. A *p*-value of 0.001 is stronger support for rejecting H_0 than a *p*-value of 0.01. It is common practice in journals for professions that use statistics to cite *p*-values for any tests that are conducted.

While *p*-values are useful in practice, they are of little use in statistical theory. The statistician is interested in determining the general shape for a "good" critical region in a given test. That is, theoretical interest is in determining that $X \geq a$ gives a test that has higher power than one whose critical region might be of the form $a \leq X \leq b$. *p*-values can be computed only if the shape of the critical region is determined. Since our goal here is to use theoretical considerations in deriving good tests we will not consider p-values any further.

Exercises 12.2

1. A bowl contains 20 chips, r of which are red and the remainder black. To test

$$H_0 : r = 10 \quad \text{vs.} \quad H_1 : r = 15,$$

3 chips are selected at random without replacement from the bowl, and H_0 is rejected if all 3 chips are red. Calculate α and β for this test.

2. An investigator tested for attitudinal differences toward drug use before and after a presentation by a well-known expert. He found a difference that was significant at a level between 0.05 and 0.03.

a. State the null and alternative hypotheses in words.

b. Which level indicates a higher level of significance—0.05 or 0.03?

c. If his α is 0.05, will he reject H_0? What if his α level is 0.03? Which α level has the higher value for β?

3. To test

$$H_0 : \lambda = 0.1 \quad \text{vs.} \quad H_1 : \lambda > 0.1,$$

where λ is the parameter of a Poisson distribution, a sample of 10 observations is taken and H_0 is rejected if $\sum_{i=1}^{10} X_i \geq 5$.

a. Calculate α.

b. Compute the power function of the test using λ values of 0.1, 0.2, 0.3, 0.4, and 0.5.

c. Graph the power function.

4. For the test in Exercise 3 determine a critical value in order to obtain a significance level closest to 0.01. Assume the critical region is of the form that rejects H_0 if $\sum_{i=1}^{10} X_i \geq a$.

5. For the test in Exercise 3 give a critical region using randomization that achieves an exact size of 0.01. Assume the critical region is of the form that rejects H_0 if $\sum_{i=1}^{10} X_i \geq a$.

6. For the test in Exercise 3 calculate the p-value if a sample of 10 observations results in $\sum_{i=1}^{10} x_i = 4$.

7. Suppose the test in Exercise 1 was altered to reject the null hypothesis if the number of red chips observed, X, fell in the critical region $X \geq a$.

a. Find the value of a that gives a significance level closest to 0.1.

b. Find a randomized critical region that achieves an exact size of 0.1.

8. Suppose in Example 1 we wished to test

$$H_0 : p = 0.5 \quad \text{vs.} \quad H_1 : p \neq 0.5$$

by using a critical region that rejects the null hypothesis if the number of outcomes of 4, 5, or 6 is either too large or too small. That is, we would reject H_0 if

$$X \geq a \quad \text{or} \quad X \leq b.$$

Find values of a and b that will give significance levels as close to 0.05 as possible and such that $P(X \leq b) = P(X \geq a)$.

12.3 ══════

Testing Simple Hypotheses

When testing a given set of hypotheses at a specified significance level we would like to be able to derive a test statistic and a critical region that has the smallest possible value for β or, equivalently, the highest power. This is possible only in certain cases. In those situations where this goal can be achieved, we say that we wish to derive what is known as a *best critical region*. It turns out that if both hypotheses are simple, then we can always find such a critical region. To help in the discussion we make the idea of a best critical region more precise with the following definition.

Definition 12.4

> When testing a simple hypothesis against a simple alternative hypothesis at a significance level of α, we say that the critical region C, associated with a test statistic T, is a *best size α critical region* if, given any other test statistic, U with a critical region C^* of size α,
>
> $$P(T \in C \mid H_1 \text{ true}) \geq P(U \in C^* \mid H_1 \text{ true}).$$
>
> A size α test procedure that has a best critical region is said to be a *best test* of size α.

This definition says that a best critical region of size α is the one whose power is at least as great as that for all size α critical regions.

═══════════════════════════════════

Example 12.5

We wish to make the notion of a best critical region clear with an elementary example. Suppose that a population has 1 of 2 possible discrete distributions. They are shown in Table 12.3.

Table 12.3

x	1	2	3	4	5	6	7	8	9	10
H_0	0.02	0.05	0.23	0.01	0.02	0.36	0.04	0.03	0.13	0.11
H_1	0.08	0.21	0.02	0.13	0.06	0.05	0.15	0.10	0.09	0.11

We assume that a single observation is taken with the value of X being our test statistic. We wish to find the best size 0.05 critical region. When H_0 is true we see that there are 5 sets of values of X whose probability sums to 0.05. They are

$$C_1 = \{2\}, \quad C_2 = \{1,4,5\}, \quad C_3 = \{1,8\}, \quad C_4 = \{5,8\}, \quad \text{and } C_5 = \{4,7\}.$$

If we select 1 of these regions to be our rejection region, then the probability of incorrectly rejecting H_0 is 0.05. Thus, these are the size 0.05 critical regions for this test. We now compute the probability of each region when H_1 is true. This is the power of each one. The results are as follows.

$$P(C_1 \mid H_1) = 0.21,$$
$$P(C_2 \mid H_1) = 0.27,$$

$$P(C_3 \mid H_1) = 0.18,$$

$$P(C_4 \mid H_1) = 0.16, \quad \text{and}$$

$$P(C_5 \mid H_1) = 0.28.$$

Since C_5 has a power higher than any of the other critical regions it is the best size 0.05 critical region for this test. If there is a tie among critical regions for the highest power, then each would be a best critical region. Nothing in Definition 12.4 claimed that a best critical region had to be unique. ∎

As we mentioned earlier when testing two simple hypotheses it is always possible to determine a best critical region. The existence of a best critical region is guaranteed by the following theorem. In addition, the theorem provides a means for determining what the best critical region is. We give a simplified version of the theorem. The interested reader with a good background in analysis can find a rigorous statement and proof in *Testing Statistical Hypotheses* by E. L. Lehmann. To make the statement of the theorem less complicated we introduce the following notation. We assume that a sample of size n is being taken.

1. Let $f(x_i \mid H_0)$ be the p.f. or p.d.f. under H_0.
2. Let $f(x_i \mid H_1)$ be the p.f. or p.d.f. under H_1.
3. Let $L_0 = f(x_1 \mid H_0)f(x_2 \mid H_0)\cdots f(x_n \mid H_0)$ be the likelihood function under H_0.
4. Let $L_1 = f(x_1 \mid H_1)f(x_2 \mid H_1)\cdots f(x_n \mid H_1)$ be the likelihood function under H_1.

Theorem 12.1

Neyman-Pearson Lemma   In testing the simple hypothesis $H_0 : \theta = \theta_0$ against the simple alternative hypothesis $H_1 : \theta = \theta_1$, let k be a fixed positive number and C be a subset of the sample space such that

1. $\dfrac{L_0}{L_1} \le k$, if $(x_1, x_2, \ldots, x_n) \in C$,

2. $\dfrac{L_0}{L_1} \ge k$, if $(x_1, x_2, \ldots, x_n) \notin C$, and

3. $P[(X_1, X_2, \ldots, X_n) \in C] = \alpha$. Then C is the best size α critical region for testing H_0 vs. H_1.

There appears to be a certain ambiguity when the ratio of likelihood functions equals k. All this means is that when $\frac{L_0}{L_1} = k$ we can use randomization, if necessary, to obtain a size α critical region. The statement of the theorem in Lehmann's book makes this clearer.

Proof   We shall give a proof for the case where the random variables are continuous. The proof for the discrete case is identical with sums replacing integrals. We first note that if C is the only size α, the theorem is automatically proven. So suppose that there is another size α critical region, call it C^*. To make things more readable let

$$\int \cdots \int_R L\, dx_1 \cdots dx_n, \quad \text{be denoted by} \quad \int_R L,$$

where R is some region of R^n. We wish to show that

$$\int_C L_1 \geq \int_{C^*} L_1.$$

Since both C and C^* are size α critical regions we know that

$$\alpha = \int_C L_0 = \int_{C^*} L_0.$$

We want to use this along with the fact that for points in C we have that $\frac{L_0}{L_1} \leq k$ to carry out the proof. First consider Figure 12.3.

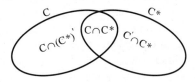

Figure 12.3

We see that C is the union of $C \cap C^*$ and $C \cap (C^*)'$. Also C^* is the union of $C \cap C^*$ and $C' \cap C^*$. Thus,

$$\alpha = \int_{C \cap C^*} L_0 + \int_{C \cap (C^*)'} L_0$$

$$= \int_{C \cap C^*} L_0 + \int_{C' \cap C^*} L_0.$$

Thus, we find that

$$\int_{C \cap (C^*)'} L_0 = \int_{C' \cap C^*} L_0.$$

Now, item 1 in the statement of the theorem gives that $L_1 \geq \frac{1}{k} L_0$ for each point in C. Therefore,

$$\int_{C \cap (C^*)'} L_1 \geq \frac{1}{k} \int_{C \cap (C^*)'} L_0.$$

Similarly, item (2) in the statement of the theorem gives that $L_1 \leq \frac{1}{k} L_0$ for each point not in C. Thus,

$$\int_{C' \cap C^*} L_1 \leq \frac{1}{k} \int_{C' \cap C^*} L_0.$$

We now combine these as follows.

$$\int_{C \cap (C^*)'} L_1 \geq \frac{1}{k} \int C \cap (C^*) L_0$$

$$= \int_{C' \cap C^*} L_0$$

$$\geq \frac{1}{k} \int_{C' \cap C^*} L_1$$

We then finish the proof as follows.

$$\int_C L_1 = \int_{C \cap C*} L_1 + \int_{C \cap (C*)'} L_1$$

$$\geq \int_{C \cap C*} L_1 + \int_{C' \cap C*} L_1$$

$$= \int_{C*} L_1$$

∎

We now give some examples of how the Neyman-Pearson Lemma can be used to derive best tests. The general idea is to begin with the ratio of likelihood functions that defines the critical region in the Lemma. Then this is manipulated until an equivalent critical region is obtained that is in terms of a statistic whose distribution we know. This is crucial since we need to be able to provide a critical value to anyone using the test on actual data. This general approach for deriving critical regions will be used quite often in the rest of the chapter.

Example 12.6   Suppose $X_1, X_2, \ldots, X_n$ constitute a random sample from a $B(1, p)$ distribution, and we wish to test

$$H_0 : p = p_0 \quad \text{vs.} \quad H_1 : p = p_1, \quad \text{where } p_1 > p_0.$$

Note that this is the general form of the test we considered in Example 12.1. Since this is a test of 2 simple hypotheses we can use Theorem 12.1. If we let $Y = \sum_{i=1}^{n} X_i$ we obtain

$$L_0 = \prod_{i=1}^{n} p_0^{x_i}(1 - p_0)^{1 - x_i} = p_0^y(1 - p_0)^{n-y}, \quad \text{and}$$

$$L_1 = \prod_{i=1}^{n} p_1^{x_i}(1 - p_1)^{1 - x_i} = p_1^y(1 - p_1)^{n-y}.$$

Then, by Theorem 12.1, there exists a positive number k such that a best critical region of size α will reject H_0 whenever

$$\frac{L_0}{L_1} = \frac{p_0^y(1 - p_0)^{n-y}}{p_1^y(1 - p_1)^{n-y}} \leq k. \tag{*}$$

We now need to determine the value of k that will make this critical region have a size of α. This is generally quite hard to do because we need to know the distribution of this ratio of likelihood functions. On the other hand, we do know that the distribution of Y is $B(n, p_0)$ when H_0 is true. If we can manipulate (*) to get an equivalent statement regarding values of Y, then the binomial tables can be used to obtain a best size α critical region. Thus, we proceed as follows.

Equation (*) can be rewritten

$$\left[\frac{p_0}{p_1}\right]^y \left[\frac{1-p_0}{1-p_1}\right]^{n-y} = \left[\frac{p_0(1-p_1)}{p_1(1-p_0)}\right]^y \left[\frac{1-p_0}{1-p_1}\right]^n \le k.$$

Then

$$\left[\frac{p_0(1-p_1)}{p_1(1-p_0)}\right]^y \le k \left[\frac{1-p_1}{1-p_0}\right]^n.$$

Taking logarithms gives

$$y \ln\left[\frac{p_0(1-p_1)}{p_1(1-p_0)}\right] \le \ln k + n \ln\left(\frac{1-p_1}{1-p_0}\right).$$

Since $p_1 > p_0$ we note that $p_0(1-p_1) < p_1(1-p_0)$. This means that the logarithm on the left-hand side is negative. Therefore, solving for y gives

$$y \ge \frac{\ln k + n \ln\left(\frac{1-p_1}{1-p_0}\right)}{\ln\left[\frac{p_0(1-p_1)}{p_1(1-p_0)}\right]}.$$

The right-hand side is a constant. Let's call it a. Thus, we find that a best size α critical region for testing $H_0 : p = p_0$ vs. $H_1 : p = p_1$ with $p_1 > p_0$ rejects H_0 whenever $y \ge a$.   ∎

Example 12.7   Let $X_1, X_2, \ldots, X_n$ be a random sample from a $N(\mu, \sigma^2)$ distribution where the value of σ^2 is known. We wish to test

$$H_0 : \mu = \mu_0 \quad \text{vs.} \quad H_1 : \mu = \mu_1,$$

where $\mu_1 < \mu_0$. Since σ^2 is known both hypotheses are simple, and we can use the Neyman-Pearson Lemma to derive a best critical region. The likelihood functions are

$$L_0 = (2\pi\sigma^2)^{-n/2} \exp\left[-\frac{1}{2\sigma^2}\sum_{i=1}^{n}(x_i - \mu_0)^2\right], \quad \text{and}$$

$$L_1 = (2\pi\sigma^2)^{-n/2} \exp\left[-\frac{1}{2\sigma^2}\sum_{i=1}^{n}(x_i - \mu_1)^2\right].$$

Then, according to Theorem 12.1, the best critical region rejects H_0 if

$$\frac{L_0}{L_1} = \frac{(2\pi\sigma^2)^{-n/2} \exp\left[-\frac{1}{2\sigma^2}\sum_{i=1}^{n}(x_i - \mu_0)^2\right]}{(2\pi\sigma^2)^{-n/2} \exp\left[-\frac{1}{2\sigma^2}\sum_{i=1}^{n}(x_i - \mu_1)^2\right]} \le k.$$

Once again the task of determining the value of k such that it gives a size α critical region is difficult. Therefore, we wish to find an equivalent critical region in terms of the sample mean $\bar{X}$. From Theorem 9.6, we know that $\bar{X} \sim N(\mu_0, \frac{\sigma^2}{n})$ when H_0 is true. We proceed as follows by making use of the fact that $n\bar{x} = \sum_{i=1}^{n} x_i$.

$$\frac{L_0}{L_1} = \frac{\exp\left[-\dfrac{1}{2\sigma^2} \sum_{i=1}^{n} (x_i - \mu_0)^2\right]}{\exp\left[-\dfrac{1}{2\sigma^2} \sum_{i=1}^{n} (x_i - \mu_1)^2\right]}$$

$$= \frac{\exp\left[-\dfrac{1}{2\sigma^2}\left(\sum_{i=1}^{n} x_i^2 - 2n\mu_0\bar{x} + n\mu_0^2\right)\right]}{\exp\left[-\dfrac{1}{2\sigma^2}\left(\sum_{i=1}^{n} x_i^2 - 2n\mu_1\bar{x} + n\mu_1^2\right)\right]}$$

$$= \exp\left\{\dfrac{1}{2\sigma^2}[2n\bar{x}(\mu_0 - \mu_1) - n(\mu_0^2 - \mu_1^2)]\right\}$$

$$\leq k$$

We now take logarithms to get

$$\frac{1}{2\sigma^2}[2n\bar{x}(\mu_0 - \mu_1) - n(\mu_0^2 - \mu_1^2)] \leq \ln k.$$

Solving for $\bar{x}$ and using the fact that $\mu_0 > \mu_1$ implies that $(\mu_0 - \mu_1) > 0$ gives

$$\bar{x} \leq \frac{2\sigma^2 \ln k + n(\mu_0^2 - \mu_1^2)}{n(\mu_0 - \mu_1)}.$$

The right-hand side is a constant. Therefore, a best critical region for testing $H_0 : \mu = \mu_0$ vs. $H_1 : \mu = \mu_1$, with $\mu_1 < \mu_0$ rejects the null hypothesis whenever

$$\bar{x} \leq a.$$

Since $\bar{X} \sim N(\mu_0, \frac{\sigma^2}{n})$ when H_0 is true we note that

$$\sqrt{n}\,\frac{\bar{X} - \mu_0}{\sigma} \sim N(0, 1).$$

Therefore, we can use Table 3 of Appendix A to obtain a size α critical region. We find the value of c such that

$$\Phi(c) = \alpha,$$

where $\Phi(x)$ is the distribution function of a standard normal random variable. Then we reject H_0 whenever

$$\sqrt{n}\,\frac{\bar{x} - \mu_0}{\sigma} \leq c.$$

Example 12.8

Let $X_1, X_2, \ldots, X_n$ be a random sample from a $U(0, \theta)$ distribution. We wish to test

$$H_0 : \theta = \theta_0 \quad \text{vs.} \quad H_1 : \theta = \theta_1,$$

where $\theta_1 > \theta_0$. Since the hypotheses are simple we can use Theorem 12.1 to derive a best critical region. The likelihood functions are as follows.

$$L_0 = \begin{cases} \dfrac{1}{\theta_0^n}, & 0 \le x_i \le \theta_0, \ i = 1, \ldots, n, \quad \text{and} \\ 0, & \text{otherwise} \end{cases}$$

$$L_1 = \begin{cases} \dfrac{1}{\theta_1^n}, & 0 \le x_i \le \theta_1, \ i = 1, \ldots, n, \quad \text{and} \\ 0, & \text{otherwise} \end{cases}$$

By letting Y_n be the largest observation these become

$$L_0 = \begin{cases} \dfrac{1}{\theta_0^n}, & 0 \le y_n \le \theta_0, \quad \text{and} \\ 0, & \text{otherwise.} \end{cases}$$

$$L_1 = \begin{cases} \dfrac{1}{\theta_1^n}, & 0 \le y_n \le \theta_1, \quad \text{and} \\ 0, & \text{otherwise.} \end{cases}$$

Now the ratio of the likelihood functions is

$$\frac{L_0}{L_1} = \begin{cases} 0, & \theta_0 < y_n \le \theta_1, \\ \left(\dfrac{\theta_1}{\theta_0}\right)^n, & 0 \le y_n \le \theta_0, \quad \text{and} \\ \text{undefined}, & \text{otherwise.} \end{cases}$$

Theorem 12.1 states that a best critical region rejects H_0 if

$$\frac{L_0}{L_1} \le k.$$

Since k must be positive, any value of Y_n above θ_0 will cause us to reject H_0. However

$$P(Y_n > \theta_0 \mid \theta = \theta_0) = 0.$$

Thus, for any $0 < k < \left(\frac{\theta_1}{\theta_0}\right)^n$, the critical region will have size 0. On the other hand

$$P(0 \le Y_n \mid \theta = \theta_0) = 1.$$

This means that a critical region with $k > \left(\frac{\theta_1}{\theta_0}\right)^n$ will have size 1. Therefore,

$$k = \left(\frac{\theta_1}{\theta_0}\right)^n.$$

This implies that a critical region that rejects H_0 whenever

$$y_n \geq a$$

is a best critical region. The value of a is determined as follows. When H_0 is true Y_n has a p.d.f. of

$$f_{Y_n}(y) = \begin{cases} n \dfrac{y^{n-1}}{\theta_0^n}, & 0 \leq y \leq \theta_0, \quad \text{and} \\ 0, & \text{otherwise.} \end{cases}$$

Thus, we compute α as follows.

$$\alpha = P(Y_n \geq a \mid \theta = \theta_0)$$

$$= \int_a^{\theta_0} n \frac{y^{n-1}}{\theta_0^n} \, dy$$

$$= 1 - \left(\frac{a}{\theta_0} \right)^n$$

Solving for a gives

$$a = \theta_0 (1 - \alpha)^{1/n}.$$

Therefore, a best size α critical region for testing $H_0 : \theta = \theta_0$ vs. $H_1 : \theta = \theta_1$ rejects the null hypothesis whenever

$$y_n \geq \theta_0 (1 - \alpha)^{1/n}. \qquad \blacksquare$$

Exercises 12.3

1. A bag contains 4 balls, r of which are red. Two balls are drawn with replacement to test the hypotheses

$$H_0 : r = 0 \quad \text{vs.} \quad H_1 : r = 2.$$

   a. List all the critical regions for which $\alpha \leq \frac{1}{2}$.
   b. Which of the critical regions found in Part a minimizes the value of β?

2. Show that the best critical region for testing

$$H_0 : \lambda = \lambda_0 \quad \text{vs.} \quad H_1 : \lambda = \lambda_1,$$

   where $\lambda_1 > \lambda_0$, based on a sample of n observations from a Poisson distribution with parameter λ rejects H_0 whenever

$$\sum_{i=1}^{n} x_i \geq a.$$

3. Let $X_1, X_2, \ldots, X_n$ be a random sample from a $N(0, \sigma^2)$ distribution. We wish to test

$$H_0 : \sigma^2 = \sigma_0^2 \quad \text{vs.} \quad H_1 : \sigma^2 = \sigma_1^2,$$

   where $\sigma_1^2 < \sigma_0^2$.

a. Show that a best critical region rejects H_0 whenever

$$\frac{1}{\sigma_0^2} \sum_{i=1}^{n} x_i^2 \geq a.$$

b. If $n = 10$, find a so that $\alpha = 0.1$.

4. Let $X_1, X_2, \ldots, X_n$ be a random sample from a distribution with a p.d.f. of

$$f(x) = \begin{cases} \beta e^{-\beta x}, & 0 < x, \quad \text{and} \\ 0 & , \text{ otherwise.} \end{cases}$$

We wish to test

$$H_0 : \beta = \beta_0 \quad \text{vs.} \quad H_1 : \beta = \beta_1,$$

where $\beta_1 < \beta_0$. Show that a best critical region rejects H_0 whenever $\bar{x} \geq a$.

5. Let $X_1, X_2, \ldots, X_n$ be a random sample from a $G(\alpha, 1)$ distribution. We wish to test

$$H_0 : \alpha = \alpha_0 \quad \text{vs.} \quad H_1 : \alpha = \alpha_1,$$

where $\alpha_0 < \alpha_1$. Find a best critical region in terms of $\prod_{i=1}^{n} X_i$.

6. Let $X_1, X_2, \ldots, X_n$ be a random sample from a distribution having a p.d.f. of

$$f(x) = \begin{cases} e^{-(x-\theta)}, & \theta \leq x, \quad \text{and} \\ 0 & , \text{ otherwise.} \end{cases}$$

We wish to test

$$H_0 : \theta = \theta_0 \quad \text{vs.} \quad H_1 : \theta = \theta_1,$$

where $\theta_0 > \theta_1$. Find a best critical region in terms of $Y_1 = \min(X_1, \ldots, X_n)$, the smallest order statistic.

12.4
Uniformly Most Powerful Tests

We now consider the problem of testing with composite hypotheses. The goal is to derive a critical region that will be a best critical region for testing H_0 against all points in H_1. As we shall see this is possible in some so-called *one-sided tests*. These are tests of the form

$$H_0 : \theta = \theta_0 \quad \text{vs.} \quad H_1 : \theta > \theta_0,$$

or

$$H_0 : \theta = \theta_0 \quad \text{vs.} \quad H_1 : \theta < \theta_0.$$

Recall Example 12.6 where $X_1, X_2, \ldots, X_n$ was a random sample from a $B(1, p)$ distribution, and we wished to test

$$H_0 : p = p_0 \quad \text{vs.} \quad H_1 : p = p_1,$$

where $p_1 > p_0$. The Neyman-Pearson Lemma showed that a best critical region is of a form that rejects H_0 whenever $y \geq a$, where y is the number of successes in the sample. Since the critical value a depends only on p_0 this single critical region will be a best critical region for every fixed value of p_1 that is larger than p_0. In addition, we have seen that the size of the critical is at most α for every value of $p \leq p_0$. Thus, if we are testing $H_0 : p \leq p_0$ vs. $H_1 : p > p_0$ the critical region that rejects the null hypothesis whenever $y \geq a$ where

$$P(Y \geq a \mid p = p_0) = \alpha$$

will have a size of at most α for every value of p in H_0. In addition, it is a best critical region for every value of $p > p_0$. A test with these properties is said to be *uniformly most powerful*. We make this a little more precise in the following definition.

Definition 12.5

> In testing the simple hypothesis H_0 against the composite alternative H_1, the critical region C is a *uniformly most powerful critical region of size* α if it is a best critical region of size at most α for testing H_0 against every simple hypothesis in Ω_1. A test that is based on the uniformly most powerful critical region is said to be a *uniformly most powerful test of size* α for testing H_0 against H_1.

One method for deriving uniformly most powerful tests, when they exist, is based on the Neyman-Pearson Lemma. This is because it gives best critical regions when testing two simple hypotheses. Suppose, for example, we are testing

$$H_0 : \theta = \theta_0 \quad \text{vs.} \quad H_1 : \theta > \theta_0.$$

If there are no other unknown parameters, then the null hypothesis is simple. We then select an arbitrary point in the alternative hypothesis. Call it θ_1. Replace the alternative hypothesis with $H_1 : \theta = \theta_1$. If this new hypothesis is simple, then we can use the Neyman-Pearson Lemma to derive a best critical region. If this best critical region does not depend on the value of θ_1 that was chosen, then the critical region is uniformly most powerful for testing the original hypotheses. If it is not possible to replace the original hypotheses with two simple ones in this manner, or if the best critical region from the Neyman-Pearson Lemma does depend on the value of θ_1, then there is no uniformly most powerful test for the original hypotheses. In that case, we refer the reader to the next section for the likelihood ratio method for deriving tests.

Example 12.9

Suppose $X_1, X_2, \ldots, X_n$ is a random sample from a $G(1, \beta)$ distribution. We wish to test

$$H_0 : \beta = \beta_0 \quad \text{vs.} \quad H_1 : \beta < \beta_0.$$

Fix β at $\beta_1 < \beta_0$ in the alternative hypothesis. The test of

$$H_0 : \beta = \beta_0 \quad \text{vs.} \quad H_1 : \beta = \beta_1$$

now involves two simple hypotheses. Thus, we can use Theorem 12.1 to obtain a best critical region. The likelihood functions are

$$L_0 = \beta_0^{-n} e^{-n\bar{x}/\beta_0}, \quad \text{and}$$

$$L_1 = \beta_1^{-n} e^{-n\bar{x}/\beta_1}.$$

Therefore, a best critical region rejects $H_0 : \beta = \beta_0$ whenever

$$\frac{L_0}{L_1} = \frac{\beta_0^{-n} e^{-n\bar{x}/\beta_0}}{\beta_1^{-n} e^{-n\bar{x}/\beta_1}} \leq k,$$

or, after some simplification,

$$\left(\frac{\beta_1}{\beta_0}\right)^n \exp\left[n\bar{x}\frac{\beta_0 - \beta_1}{\beta_0\beta_1}\right] \leq k.$$

After taking logarithms and solving for $\bar{x}$ we find that an equivalent critical region is

$$\bar{x} \leq \left[\ln k + n \ln\left(\frac{\beta_0}{\beta_1}\right)\right] \frac{\beta_0\beta_1}{\beta_0 - \beta_1}.$$

Since this is of the form $\bar{x} \leq a$ for any $\beta_1 < \beta_0$, and the actual value of a depends only on the distribution for $\bar{x}$ when $\beta = \beta_0$, the critical region is independent of the actual value for β_1. Thus, a uniformly most powerful test for $H_0 : \beta \geq \beta_0$ vs. $H_1 : \beta < \beta_0$ will reject H_0 whenever $\bar{x} \leq a$. To find the value of a for a given significance level α we note that under H_0 we assume that $\beta = \beta_0$. In Example 11.11 we indicated that

$$\frac{2n\bar{X}}{\beta} \sim \chi^2(2n).$$

Therefore, when H_0 is true

$$\frac{2n\bar{X}}{\beta_0} \sim \chi^2(2n).$$

Thus, rejecting the null hypothesis whenever $\bar{x} \leq a$ is equivalent to rejecting H_0 whenever $2n\frac{\bar{x}}{\beta_0} \leq b = \frac{2na}{\beta_0}$. We can use Table 4 of Appendix A to determine the critical value b.   ∎

Example 12.10   Let $X_1, X_2, \ldots, X_n$ be a random sample from a geometric distribution with parameter p. We wish to test

$$H_0 : p = p_0 \quad \text{vs.} \quad H_1 : p > p_0.$$

To begin we fix $p = p_1 > p_0$ in the alternative hypothesis. Then

$$H_0 : p = p_0 \quad \text{and} \quad H_1 : p = p_1$$

are both simple hypotheses, and we can use Theorem 12.1 to derive a best critical region. If we let $Y = \sum_{i=1}^{n} X_i$ the likelihood functions are

$$L_0 = p_0^n (1 - p_0)^{y-n}, \quad \text{and}$$

$$L_1 = p_1^n (1 - p_1)^{y-n}.$$

Then, according to the Neyman-Pearson Lemma, the best critical region rejects $H_0 : p = p_0$ whenever

$$\frac{L_0}{L_1} = \frac{p_0^n (1 - p_0)^{y-n}}{p_1^n (1 - p_1)^{y-n}} \leq k,$$

or

$$\left[\frac{p_0(1 - p_1)}{p_1(1 - p_0)} \right]^n \left[\frac{1 - p_0}{1 - p_1} \right]^y \leq k.$$

Taking logarithms gives that the best critical region is equivalent to

$$n \ln \left[\frac{p_0(1 - p_1)}{p_1(1 - p_0)} \right] + y \ln \left[\frac{1 - p_0}{1 - p_1} \right] \leq \ln k.$$

For any value of $p_1 > p_0$ we note that $1 - p_1 < 1 - p_0$. Therefore, solving for y gives that

$$y \leq \frac{\ln k + n \ln[p_1(1 - p_0)] - n \ln[p_0(1 - p_1)]}{\ln(1 - p_0) - \ln(1 - p_1)}.$$

Since the best critical region of the form $y \leq a$ for every value of $p_1 > p_0$, a uniformly most powerful test of the original hypotheses, exists and has a critical region that rejects the original null hypothesis whenever

$$y \leq a.$$

For a specified significance level a can be computed using a negative binomial distribution with parameters $r = n$ and $p = p_0$. ∎

The next example illustrates that, even though fixing the value of the parameter in each hypothesis creates a pair of simple hypotheses, it can happen that there is no uniformly most powerful test.

Example 12.11   Let $X_1, X_2, \ldots, X_n$ be a random sample from a $N(\mu, 1)$ distribution. We wish to test

$$H_0 : \mu = \mu_0 \quad \text{vs.} \quad H_1 : \mu \neq \mu_0.$$

The null hypothesis is already simple. If we fix μ at $\mu_1 \neq \mu_0$, the hypothesis $H_1 : \mu = \mu_1$ is also simple. Thus, Theorem 12.1 will give a best critical region for testing $H_0 : \mu = \mu_0$ vs. $H_1 : \mu = \mu_1$. The likelihood functions are

$$L_0 = (2\pi)^{-n/2} \exp \left[-\frac{1}{2} \sum_{i=1}^{n} (x_i - \mu_0)^2 \right], \quad \text{and}$$

$$L_1 = (2\pi)^{-n/2} \exp \left[-\frac{1}{2} \sum_{i=1}^{n} (x_i - \mu_1)^2 \right].$$

Therefore, the best critical region rejects H_0 whenever

$$\frac{L_0}{L_1} = \frac{(2\pi)^{-n/2} \exp\left[-\frac{1}{2}\sum_{i=1}^{n}(x_i - \mu_0)^2\right]}{(2\pi)^{-n/2} \exp\left[-\frac{1}{2}\sum_{i=1}^{n}(x_i - \mu_1)^2\right]} \le k,$$

or, after some simplification,

$$\exp[n\bar{x}(\mu_0 - \mu_1) + n(\mu_1^2 - \mu_0^2)] \le k.$$

Solving for $\bar{x}$ shows that we should reject H_0 if

$$\bar{x} \le \frac{\ln k + n(\mu_0^2 - \mu_1^2)}{n(\mu_0 - \mu_1)}, \quad \text{if } \mu_0 > \mu_1, \quad \text{and}$$

$$\bar{x} \ge \frac{\ln k + n(\mu_0^2 - \mu_1^2)}{n(\mu_0 - \mu_1)}, \quad \text{if } \mu_0 < \mu_1.$$

Since the critical region depends on the particular value that is chosen for μ_1, there is no uniformly most powerful test for the original hypotheses. Tests of the form $H_0 : \theta = \theta_0$ vs. $H_1 : \theta \ne \theta_0$ are known as *two-sided tests*. In general, for any two-sided test there is a one-sided test that will have higher power for any given value in the alternative hypothesis. To illustrate, suppose we are testing $H_0 : \mu = 0$ vs. $H_1 : \mu \ne 0$ at a significance level of $\alpha = 0.05$ based on a sample of 100 observations from a $N(\mu, 10)$ distribution. We leave it as an exercise to show that a uniformly most powerful test for H_0 vs. $H_1 : \mu > 0$ rejects the null hypothesis if

$$\bar{x} \ge 16.4,$$

and a uniformly most powerful test for H_0 vs. $H_1 : \mu < 0$ rejects H_0 if

$$\bar{x} \le -16.4.$$

A commonly used test for the original hypotheses rejects H_0 if

$$|\bar{x}| \ge 19.6.$$

The next section will provide a method for deriving such tests. Figure 12.4 shows the power functions for these three tests. Note that the power for one of the one-sided tests always exceeds that of the two-sided test. This shows the idea that, for any value of μ in the alternative, there is a one-sided critical region with higher power than the two-sided critical region. Since there is another test with higher power, the two-sided test fails to meet the definition of a uniformly most powerful test.

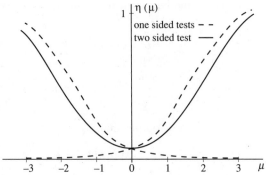

Figure 12.4

The important thing to keep in mind when determining if a test is uniformly most powerful is not a comparison of tests for the same two composite hypotheses. We are interested in the behavior when we fix at a single point in the null hypothesis and a single point in the alternative hypothesis. If our critical region has power at least as high as any other critical region for testing those two fixed hypotheses, and if this is true no matter which pair of points we should choose, then we have a uniformly most powerful test. If for any pair of points there is a test of those two particular hypotheses with higher power, then our critical region is not uniformly most powerful.

The preceding examples illustrate the general procedure for deriving uniformly most powerful tests. We would now like to consider some cases where this process can be made easier. The key idea in this method is a property of a family of distributions known as *monotone likelihood ratio*.

Definition 12.6

> Let X be a random variable with probability density function of probability function $f(x \mid \theta)$, where θ is a scalar parameter. Let $\theta_2 > \theta_1$ be a function of X. The family of distributions is said to have *monotone likelihood ratio in $T(x)$* if there exists a function $T(x)$ such that the ratio
>
> $$\frac{f(x \mid \theta_2)}{f(x \mid \theta_1)}$$
>
> is a nondecreasing function of $T(x)$.

Note the following:

1. The function $T(x)$ must be the same for every pair of parameter values, θ_1 and θ_2.
2. An equivalent statement is that

$$\frac{f(x \mid \theta_1)}{f(x \mid \theta_2)}$$

is a nonincreasing function of $T(x)$.

Example 12.12

Let $X_1, \ldots, X_n$ be a random sample from a $Po(\lambda)$ distribution. We wish to determine if the family of distributions of $\mathbf{X} = (X_1, \ldots, X_n)$ has monotone likelihood ratio. The joint probability function of $\mathbf{X}$ is

$$f(x_1, \ldots, x_n \mid \lambda) = \frac{\lambda^{\sum_{i=1}^n x_i} e^{-n\lambda}}{\prod_{i=1}^n x_i!}.$$

This is the $f(x \mid \theta)$ of Definition 12.6. Now select two arbitrary values of λ so that $\lambda_2 > \lambda_1$ and form the likelihood ratio as follows.

$$\frac{f(x_1, \ldots, x_n \mid \lambda_2)}{f(x_1, \ldots, x_n \mid \lambda_1)} = \frac{\dfrac{\lambda_2^{\sum_{i=1}^n x_i} e^{-n\lambda_2}}{\prod_{i=1}^n x_i!}}{\dfrac{\lambda_1^{\sum_{i=1}^n x_i} e^{-n\lambda_1}}{\prod_{i=1}^n x_i!}}$$

$$= \left(\frac{\lambda_2}{\lambda_1}\right)^{\sum_{i=1}^n x_i} e^{-n(\lambda_2 - \lambda_1)}$$

Let $T(x_1, \ldots, x_n) = \sum_{i=1}^n x_i$. Since $\frac{\lambda_2}{\lambda_1} > 1$ we see that the likelihood ratio is of the form

$$k\, a^{T(x_1, \ldots, x_n)},$$

which is an exponential function having a positive derivative. Therefore, the family of distributions of $\mathbf{X}$ has monotone likelihood ratio in $\sum_{i=1}^n x_i$.

It turns out that uniformly most powerful tests can be constructed using the function $T(x)$ when the family of distributions has monotone likelihood ratio. We state the result as a theorem. For notational convenience X is taken to be either a vector $\mathbf{X}$ or a scalar.

Theorem 12.2

Let the family of probability density function or probability functions, $f(x \mid \theta)$ of the random variable X depend on a single unknown parameter θ, and let $f(x \mid \theta)$ have monotone likelihood ratio in $T(x)$. Then for testing

$$H_0 : \theta = \theta_0 \quad \text{vs.} \quad H_1 : \theta > \theta_0$$

there exists a uniformly most powerful test which will

Reject H_0 if $T(x) \geq C$,   and

Accept H_0 if $T(x) \leq C$.

The constant C is chosen so that

$$\alpha = P[T(x) \geq C \mid \theta = \theta_0].$$

Note the following:

1. As was the case in the statement of the Neyman-Pearson Lemma the ambiguity about whether to accept or reject when $T(x) = C$ allows us to use randomization to achieve an exact size α critical region.
2. If each of the inequalities in the statement of the theorem is reversed we obtain a uniformly most powerful test for $H_0 : \theta = \theta_0$ vs. $H_1 : \theta < \theta_0$.

Proof We shall use the method for deriving uniformly most powerful tests that was illustrated in the preceding examples. That is, we shall select θ_1 from the alternative hypothesis and apply the Neyman-Pearson Lemma to the test of

$$H_0 : \theta = \theta_0 \quad \text{vs.} \quad H_1 : \theta = \theta_1.$$

Let $f(x_1, \ldots, x_n \mid \theta)$ be the joint p.d.f. of p.f. of the sample. Since θ is the only unknown parameter this is a test of a simple null hypothesis versus a simple alternative hypothesis.

Therefore, according to the Neyman-Pearson Lemma

$$\text{If } \frac{f(x_1, \ldots, x_n \mid \theta_0)}{f(x_1, \ldots, x_n \mid \theta_1)} \leq k, \text{ reject } H_0, \quad \text{and}$$

$$\text{If } \frac{f(x_1, \ldots, x_n \mid \theta_0)}{f(x_1, \ldots, x_n \mid \theta_1)} \geq k, \text{ accept } H_0.$$

Now we use the fact the joint distribution of the sample has monotone likelihood ratio in $T(x_1, \ldots, x_n)$ to note that, since $\theta_1 > \theta_0$, the ratio of joint p.d.f.'s or p.f.'s is a decreasing function of $T(x_1, \ldots, x_n)$. Thus, the test based on the Neyman-Pearson Lemma is equivalent to

$$\text{if } T(x_1, \ldots, x_n) \geq C, \text{ reject } H_0, \quad \text{and}$$

$$\text{if } T(x_1, \ldots, x_n) \leq C, \text{ accept } H_0.$$

Since this critical region does not depend on the particular value of θ_1 we see that this test is uniformly most powerful.   ■

To develop a uniformly most powerful test using this theorem all we need to do is determine if the family of distributions of our random sample has monotone likelihood ratio in some function $T(x)$. Then we simply use $T(x)$ for our test statistic.

Example 12.13   Let $X_1, \ldots, X_n$ be a random sample from a $Po(\lambda)$ distribution. We wish to test

$$H_0 : \lambda = \lambda_0 \quad \text{vs.} \quad H_1 : \lambda > \lambda_1.$$

In Example 12.12 we found that the family of distributions of $(X_1, \ldots, X_n)$ has monotone likelihood ratio in $T(x_1, \ldots, x_n) = \sum_{i=1}^{n} x_i$. Therefore, by Theorem 12.2 we see that a uniformly most powerful test will reject H_0 whenever

$$\sum_{i=1}^{n} x_i \geq C.$$

C is chosen to obtain the desired significance level. For example, suppose that $\theta_0 = 0.5$, $n = 10$, and we choose $\alpha = 0.05$. We wish to determine C so that the actual significance level comes as close to 0.05 as possible without resorting to randomization. We know that, if $X_1, \ldots, X_n$ and i.i.d. $Po(\lambda)$ random variables, then $\sum_{i=1}^{n} X_i$ will have a $Po(n\lambda)$ distribution. Therefore, we would compute α as follows.

$$\alpha = P\left(\sum_{i=1}^{n} X_i \geq C \mid \lambda = 0.5\right)$$

Since $\sum_{i=1}^{n} X_i \sim Po(5)$ when H_0 is true we use Table 2 of Appendix A to see that $C = 9$ comes closest to 0.05 with an actual significance level of 0.0681.   ∎

Example 12.14   Let $X_1, \ldots, X_n$ be a random sample from a $G(1, \beta)$ distribution. We wish to derive a uniformly most powerful test of

$$H_0 : \beta = \beta_0 \quad \text{vs.} \quad H_1 : \beta < \beta_0$$

using monotone likelihood ratio. The joint p.d.f. of the sample is

$$f(x_1, \ldots, x_n \mid \beta) = \beta^{-n} e^{-\sum_{i=1}^{n} x_i/\beta}.$$

To determine if this family has monotone likelihood ratio we let $\beta_2 > \beta_1$ and form the likelihood ratio as follows.

$$\frac{f(x_1, \ldots, x_n \mid \beta_2)}{f(x_1, \ldots, x_n \mid \beta_1)} = \frac{\beta_2^{-n} e^{-\sum_{i=1}^{n} x_i/\beta_2}}{\beta_1^{-n} e^{-\sum_{i=1}^{n} x_i/\beta_1}}$$

$$= \left(\frac{\beta_1}{\beta_2}\right)^n e^{(1/\beta_1 - 1/\beta_2)\sum_{i=1}^{n} x_i}$$

Let $T(x) = T(x_1, \ldots, x_n) = \sum_{i=1}^{n} x_i$. Since $\beta_2 > \beta_1$ we see that the likelihood ratio is of the form

$$ke^{aT(x)}, \quad \text{where } a > 0.$$

Therefore, the family of distributions of the sample has monotone likelihood ratio in $T(x_1, \ldots, x_n) = \sum_{i=1}^{n} x_i$. Therefore, Theorem 12.2 states that a uniformly most powerful test will reject H_0 whenever

$$\sum_{i=1}^{n} x_i \geq C.$$

We would choose C so that

$$P\left(\sum_{i=1}^{n} X_i \geq C \middle| \beta = \beta_0\right) = \alpha.$$

To determine C in practice we can use the result given in Example 11.11 that $2\sum_{i=1}^{n} X_i / \beta \sim \chi^2(2n)$. For instance, if $\beta_0 = 3$, $n = 5$, and $\alpha = 0.05$ Table 4 in Appendix A would show that we reject H_0 whenever

$$\frac{2\sum_{i=1}^{n} X_i}{3} \geq \chi^2(10)_{0.05} = 18.3,$$

or

$$\sum_{i=1}^{n} X_i \geq 27.45.$$

The reader may have noticed that the population distributions in both of the preceding examples came from the single parameter exponential family. It turns out that monotone likelihood ratio is a property for many members of this family. We state this as a theorem.

Theorem 12.3

> Suppose that $(X_1, \ldots, X_n)$ has a distribution that is a member of the single parameter exponential family. That is, it has a probability density function or probability function of the form
>
> $$f(x_1, \ldots, x_n \mid \theta) = e^{\tau(\theta)T(x_1, \ldots, x_n) + S(x_1, \ldots, x_n) + \eta(\theta)}.$$
>
> If $\tau(\theta)$ is a nondecreasing function of θ, then the family of distributions of $(X_1, \ldots, X_n)$ has monotone likelihood ratio in $T(X_1, \ldots, X_n)$.

Proof   Let $\theta_2 > \theta_1$. Since $\tau(\theta)$ is a nondecreasing function of θ we see that $\tau(\theta_2) \geq \tau(\theta_1)$. Forming the likelihood ratio gives

$$\frac{f(x_1, \ldots, x_n \mid \theta_2)}{f(x_1, \ldots, x_n \mid \theta_1)} = \frac{e^{\tau(\theta_2)T(x_1, \ldots, x_n) + S(x_1, \ldots, x_n) + \eta(\theta_2)}}{e^{\tau(\theta_1)T(x_1, \ldots, x_n) + S(x_1, \ldots, x_n) + \eta(\theta_1)}}$$

$$= e^{(\tau(\theta_2) - \tau(\theta_1))T(x_1, \ldots, x_n) + \eta(\theta_2) - \eta(\theta_1)}.$$

Since this is of the form

$$e^{aT(x) + b}, \quad \text{where } a > 0$$

we see that the likelihood ratio is a nondecreasing function of $T(x_1, \ldots, x_n)$, and we are done.

We can combine this result with Theorem 12.2 to show immediately that there will always be a uniformly most powerful test for testing one-sided alternatives in the single parameter exponential family. We state it as corollary and leave the proof as an exercise.

Corollary 12.1

Let $X_1, \ldots, X_n$ be a random sample from a distribution that is a member of the single parameter exponential family of the form

$$f(x_1, \ldots, x_n \mid \theta) = e^{\tau(\theta)T(x_1, \ldots, x_n) + S(x_1, \ldots, x_n) + \eta(\theta)}.$$

1. For testing

$$H_0 : \tau(\theta) = \tau(\theta_0) \quad \text{vs.} \quad H_1 : \tau(\theta) > \tau(\theta_0)$$

there exists a uniformly most powerful test that will reject H_0 if

$$T(X_1, \ldots, X_n) \geq C.$$

2. For testing

$$H_0 : \tau(\theta) = \tau(\theta_0) \quad \text{vs.} \quad H_1 : \tau(\theta) < \tau(\theta_0)$$

there exists a uniformly most powerful test that will reject H_0 if

$$T(X_1, \ldots, X_n) \leq C.$$

Example 12.15

We wish to show how to apply this corollary to the testing problem given in Example 12.14. Here

$$f(x_1, \ldots, x_n \mid \beta) = e^{-(1/\beta) \sum_{i=1}^{n} x_i - n \ln \beta}.$$

From this we see that this is a member of the single parameter exponential family with

$$\tau(\beta) = \frac{1}{\beta}, \quad \text{and} \quad T(x_1, \ldots, x_n) = \sum_{i=1}^{n} x_i.$$

In Example 12.14 the desired hypotheses were

$$H_0 : \beta = \beta_0 \quad \text{vs.} \quad H_1 : \beta < \beta_0.$$

To apply the corollary we must rephrase these in terms of $\tau(\beta)$. Thus, we see that an equivalent pair of hypotheses is

$$H_0 : \frac{1}{\beta} = \frac{1}{\beta_0} \quad \text{vs.} \quad H_1 : \frac{1}{\beta} > \frac{1}{\beta_0}.$$

Therefore, Part 1 of Corollary 12.1 gives that a uniformly most powerful test will reject the null hypothesis if

$$\sum_{i=1}^{n} X_i \geq C.$$

∎

Exercises 12.4

1. Let $X_1, X_2, \ldots, X_n$ be a random sample from a $N(\mu, 1)$ distribution. Show that a uniformly most powerful test of

$$H_0 : \mu = 0 \quad \text{vs.} \quad H_1 : \mu > 0$$

rejects the null hypothesis whenever

$$\bar{x} \geq a.$$

  a. Use the Neyman-Pearson Lemma.
  b. Use monotone likelihood ratio.
  c. Let $n = 100$ and find the value of a for a significance level of $\alpha = 0.1$.

2. Show that a uniformly most powerful test of

$$H_0 : \theta = 10 \quad \text{vs.} \quad H_1 : \theta < 10$$

based on a sample of size n from a $U(0, \theta)$ distribution rejects the null hypothesis if $y_n \leq a$, where y_n is the value of the largest order statistic. Find the value of a when $n = 5$ and $\alpha = 0.01$.

3. Suppose we have a sample of n observations from a $N(0, \sigma^2)$ distribution. Show that a uniformly most powerful test of

$$H_0 : \sigma = 1 \quad \text{vs.} \quad H_1 : \sigma > 1$$

rejects the null hypothesis whenever

$$\sum_{i=1}^{n} x_i^2 \geq a.$$

  a. Use the Neyman-Pearson Lemma.
  b. Use monotone likelihood ratio.
  c. Determine a critical region so that $\alpha = 0.1$ when $n = 20$.

4. Suppose we have a single observation from a population with a probability density function of

$$f(x) = \begin{cases} \theta x^{\theta - 1} & , \ 0 \leq x \leq 1, \quad \text{and} \\ 0 & , \ \text{otherwise.} \end{cases}$$

  a. Show that a uniformly most powerful test of

$$H_0 : \theta = \theta_0 \quad \text{vs.} \quad H_1 : \theta < \theta_0$$

  rejects H_0 if $x \leq a$.
  b. Find a such that $\alpha = 0.05$ if $\theta_0 = 2$.

5. Let $X_1, \ldots, X_n$ be a random sample from a Cauchy distribution. That is, it has a p.d.f. of

$$f(x \mid \theta) = \frac{1}{\pi[1 + (x - \theta)^2]}, \quad -\infty < x < \infty.$$

Show that the family of joint p.d.f.'s of the sample does not have monotone likelihood ratio.

6. Let $X_1, \ldots, X_n$ be a random sample from a geometric distribution with probability of success p. Use Corollary 3.1 to derive a uniformly most powerful test of

$$H_0 : p = p_0 \quad \text{vs.} \quad H_1 : p > p_0.$$

7. Let $X_1, \ldots, X_n$ be a random sample from a $B(1, p)$ distribution. Use Corollary 3.1 to derive a uniformly most powerful test of

$$H_0 : p = p_0 \quad \text{vs.} \quad H_1 : p < p_0.$$

8. Let $X_1, \ldots, X_n$ be a random sample from a distribution having a p.d.f. of

$$f(x) = \begin{cases} \dfrac{2x}{\theta^2} e^{-x^2/\theta^2} & ,0 < x, \quad \text{and} \\ 0 & , \text{otherwise.} \end{cases}$$

Find a uniformly most powerful test of

$$H_0 : \theta = \theta_0 \quad \text{vs.} \quad H_1 : \theta < \theta_0.$$

12.5
Likelihood
Ratio Tests

In practice there are relatively few testing situations that give rise to uniformly most powerful tests and even fewer cases where best tests can be derived. To cover the other testing situations we introduce a general method based on maximum likelihood estimation and the ratio of likelihood functions used in the Neyman-Pearson Lemma. This method derives what are called *likelihood ratio tests*. This method was first proposed in 1928 by Neyman and Pearson. The popularity of the method is due to the fact that it can be applied to an extremely wide variety of problems, and in a number of cases it derives tests that are in some sense optimal. We shall not be concerned with these optimality issues. The interested reader should consult *Testing Statistical Hypotheses* by E. L. Lehmann.

An outline of the method is as follows.

1. Let $L(\theta)$ be the likelihood function. Let Ω_0 be the set of parameter values under the null hypothesis and Ω_1 be the set of parameter values under the alternative hypothesis. Further, let $\Omega = \Omega_0 \cup \Omega_1$. Ω is called the *parameter space*. Let L_0 be the likelihood function where the null hypothesis is assumed to be true, and let L be the likelihood function for the entire parameter space.

2. Find the maximum values for L_0 and L. Call them L_0^* and L^*, respectively. This entails replacing all parameters not fixed by the parameter space being assumed for that likelihood function with their maximum likelihood estimators.

3. If the value of L_0^* is much smaller than L^*, then the likelihood function is maximized at a point in Ω_1. This would indicate that the alternative hypothesis is more reasonable for the observed sample data.

4. Let the *likelihood ratio statistic* be defined by

$$\lambda = \frac{L_0^*}{L^*}. \tag{12.5}$$

Since small values of λ indicate that the alternative hypothesis is more reasonable we reject H_0 if

$$\lambda \leq \lambda_0, \tag{12.6}$$

where λ_0 is chosen to obtain a size α critical region.

The general method should be clearer if we can see an example.

Example 12.16   Let $X_1, X_2, \ldots, X_n$ be a random sample from a $N(\mu, 1)$ distribution. We have seen that there is no uniformly most powerful test for

$$H_0 : \mu = \mu_0 \quad \text{vs.} \quad H_1 : \mu \neq \mu_0.$$

Therefore, we shall derive a likelihood ratio test for this problem. In the notation of the method we have that

$$\Omega_0 = \{\mu_0\}, \quad \Omega_1 = \mathbb{R} \setminus \{\mu_0\}, \quad \Omega = \mathbb{R}, \text{ where } \mathbb{R} = (-\infty, \infty), \quad \text{and}$$

$$L(\mu) = (2\pi)^{-n/2} \exp\left[-\frac{1}{2}\sum_{i=1}^{n}(x_i - \mu)^2\right].$$

Since Ω_0 contains only a single point and there are no other unknown parameters in the function the maximum occurs at $\mu = \mu_0$. Therefore, in the notation of the method,

$$L_0^* = (2\pi)^{-n/2} \exp\left[-\frac{1}{2}\sum_{i=1}^{n}(x_i - \mu_0)^2\right].$$

In Ω, the only unspecified parameter is μ, which is free to be any real number. Thus, $L = L(\mu)$, and it will be a maximum when μ equals the value of its maximum likelihood estimator. From Example 10.5 we know that the maximum likelihood estimator for μ is

$$\widehat{\mu} = \bar{X},$$

the sample mean. Thus, we find that

$$L^* = (2\pi)^{-n/2} \exp\left[-\frac{1}{2}\sum_{i=1}^{n}(x_i - \bar{x})^2\right].$$

Therefore, the likelihood ratio statistic is

$$\lambda = \frac{L_0^*}{L^*} = \frac{(2\pi)^{-n/2} \exp\left[-\frac{1}{2}\sum_{i=1}^{n}(x_i - \mu_0)^2\right]}{(2\pi)^{-n/2} \exp\left[-\frac{1}{2}\sum_{i=1}^{n}(x_i - \bar{x})^2\right]}.$$

After some simplification we obtain

$$\lambda = \exp\left[-\frac{1}{2}n(\bar{x}-\mu_0)^2\right].$$

Then the likelihood ratio test procedure gives that H_0 is rejected whenever

$$\lambda = \exp\left[-\frac{1}{2}n(\bar{x}-\mu_0)^2\right] \leq \lambda_0.$$

We must now determine the value of λ_0 in order to obtain a critical region of size α. This requires that we know the distribution of λ. This is a difficult problem, but we do know that the distribution of $\bar{X}$, when $\mu = \mu_0$, is $N(\mu_0, \frac{1}{n})$. Therefore, it is desirable to find a set of values for the sample mean that correspond to the points where $\lambda \leq \lambda_0$. This will give us the general shape for a rejection region in terms of the value of $\bar{X}$. We can then use the known distribution for $\bar{X}$ to obtain critical values. Since the equation for λ is of the form

$$e^{-(x-a)^2}$$

we know the following about the graph of $\lambda = f(\bar{x})$.

1. The graph will pass through the point $(\bar{x}, \lambda) = (\mu_0, 1)$.
2. The graph is symmetric about the point $\bar{x} = \mu_0$.
3. The graph will decrease as we move away from that point in either direction.

This is shown in Figure 12.5.

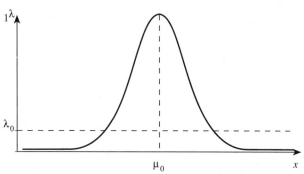

Figure 12.5

Notice that the points where $\lambda \leq \lambda_0$ correspond to values of $\bar{x}$ that are both too far above μ_0 or too far below μ_0. Because of the symmetry about μ_0 we can say that there is some number a for which $\lambda \leq \lambda_0$ whenever $|\bar{x} - \mu_0| \geq a$. By letting $Z = \sqrt{n}(\bar{X} - \mu_0)$ and $b = \sqrt{n}a$ we see that this rejection region is equivalent to rejecting the null hypothesis whenever $|z| \geq \sqrt{b}$, where Z has a $N(0,1)$ distribution. If we recall the α subscripting notation from Chapter 9 it turns out that we get a size α critical region when $b = z_{\alpha/2}$. Thus, a test of

$H_0 : \mu = \mu_0$ vs. $H_1 : \mu \neq \mu_0$ with a significance level of α rejects H_0 whenever

$$|z| = |\sqrt{n}\,(\bar{x} - \mu_0)| \geq z_{\alpha/2}.$$

This is where the α subscripting notation came from.   ■

Example 12.17   We wish to repeat the previous example except that this time the population has an unknown variance σ^2. Thus, both

$$H_0 : \mu = \mu_0 \quad \text{and} \quad H_1 : \mu \neq \mu_0$$

are composite hypotheses. The likelihood function is

$$L(\mu, \sigma^2) = (2\pi\sigma^2)^{-n/2} \exp\left[-\frac{1}{2\sigma^2} \sum_{i=1}^{n} (x_i - \mu)^2\right].$$

In Ω_0:   Now $\Omega_0 = \{(\mu_0, \sigma^2) : \sigma^2 > 0\}$, and

$$L_0 = (2\pi\sigma^2)^{-n/2} \exp\left[-\frac{1}{2\sigma^2} \sum_{i=1}^{n} (x_i - \mu_0)^2\right].$$

Therefore, we need a maximum likelihood estimator for σ^2. We leave it to the reader to verify that it is

$$\widehat{\sigma^2} = \frac{1}{n} \sum_{i=1}^{n} (x_i - \mu_0)^2.$$

Substituting this in $L(\Omega_0)$ and simplifying gives

$$L_0^* = \left[2\pi \frac{1}{n} \sum_{i=1}^{n} (x_i - \mu_0)^2\right]^{-n/2} \exp\left(-\frac{n}{2}\right).$$

In Ω: Here $\Omega = \{(\mu, \sigma^2) : -\infty < \mu < \infty, 0 < \sigma^2\}$. Since Ω places no restrictions on the values of μ or σ^2, $L = L(\mu, \sigma^2)$ and the maximum likelihood estimators are those we derived in Example 10.5. Therefore,

$$\hat{\mu} = \bar{X}, \quad \text{and} \quad \widehat{\sigma^2} = \frac{1}{n} \sum_{i=1}^{n} (x_i - \bar{x})^2.$$

Substituting these in L and simplifying gives

$$L^* = \left[2\pi \frac{1}{n} \sum_{i=1}^{n} (x_i - \bar{x})^2\right]^{-n/2} \exp\left(-\frac{n}{2}\right).$$

Thus, the likelihood ratio statistic is, after some simplification,

$$\lambda = \frac{L_0^*}{L^*}$$

$$= \left[\frac{\displaystyle\sum_{i=1}^{n}(x_i - \bar{x})^2}{\displaystyle\sum_{i=1}^{n}(x_i - \mu_0)^2} \right]^{n/2},$$

and we reject H_0 whenever $\lambda \leq \lambda_0$. It turns out that we can get an equivalent critical region in terms of a statistic that has a t distribution. First we note that

$$\sum_{i=1}^{n}(x_i - \mu_0)^2 = \sum_{i=1}^{n}(x_i - \bar{x} + \bar{x} - \mu_0)^2$$

$$= \sum_{i=1}^{n}(x_i - \bar{x})^2 + \sum_{i=1}^{n}(\bar{x} - \mu_0)^2 - 2(\bar{x} - \mu_0)\sum_{i=1}^{n}(x_i - \bar{x})$$

$$= \sum_{i=1}^{n}(x_i - \bar{x})^2 + n(\bar{x} - \mu_0)^2.$$

This, coupled with the fact that

$$(n - 1)s^2 = \sum_{i=1}^{n}(x_i - \bar{x})^2,$$

allows us to rewrite λ as

$$\lambda = \left[1 + \frac{n(\bar{x} - \mu_0)^2}{\displaystyle\sum_{i=1}^{n}(x_i - \bar{x})^2} \right]^{-n/2}$$

$$= \left[\frac{1}{1 + \dfrac{n}{n-1}\dfrac{(\bar{x} - \mu_0)^2}{s^2}} \right]^{n/2}.$$

If we let

$$t = \sqrt{n}\,\frac{\bar{x} - \mu_0}{s},$$

then we obtain that

$$\lambda = \left[\frac{1}{1 + \dfrac{t^2}{n-1}} \right]^{n/2}.$$

In Section 9.3 it is shown that t has a Student's T distribution with $n-1$ degrees of freedom. Since λ becomes small whenever $|t|$ becomes large we see that there is a value a such that $\lambda \leq \lambda_0$ whenever $|t| \geq a$. Again using the α subscripting notation we find that $a = t(n-1)_{\alpha/2}$. Therefore, an equivalent critical region rejects H_0 if

$$|t| = \left| \sqrt{n} \frac{\bar{x} - \mu_0}{s} \right| \geq t(n-1)_{\alpha/2}.$$

This is the critical region that most elementary texts recommend for this test. It is popularly called the t test. ∎

The likelihood ratio procedure can also be applied to one-sided tests. If we proceed as above for the two possible one-sided alternatives we find that, in each case, the critical region can be based on the test statistic

$$t = \sqrt{n} \frac{\bar{x} - \mu_0}{s},$$

where, when $H_0 : \mu = \mu_0$ is true, t will have a t distribution with $n-1$ degrees of freedom. A summary of the critical regions in each case are given in Table 12.4.

Table 12.4   Null hypothesis: $H_0 : \mu = \mu_0$

Alternative Hypothesis	Critical Region
$H_1 : \mu > \mu_0$	$t \geq t(n-1)_\alpha$
$H_1 : \mu < \mu_0$	$t \leq -t(n-1)_\alpha$
$H_1 : \mu \neq \mu_0$	$|t| \geq t(n-1)_{\alpha/2}$

Example 12.18   A consumer testing agency is investigating the lifetime of size AA batteries. Battery lifetimes have been found to be approximated by a normal distribution. The manufacturer claims that they have a mean lifetime of 10 hours under normal conditions. To assess this claim a sample of 10 batteries is tested with the following times to failure.

$$9.5 \quad 9.3 \quad 9.2 \quad 9.9 \quad 9.9$$
$$8.6 \quad 9.6 \quad 10.2 \quad 9.8 \quad 10.1$$

In cases like this, we are usually interested in whether the actual mean lifetime is less than that claimed by the manufacturer. Therefore, we are testing

$$H_0 : \mu = 10 \quad \text{vs.} \quad H_1 : \mu < 10.$$

We shall use a significance level of 0.05. The test statistic will be

$$t = \sqrt{n} \frac{\bar{x} - \mu_0}{s},$$

where $\mu_0 = 10$ and $n = 10$. From Table 12.4 and using Table 5 of Appendix A we see that the critical region will reject H_0 if

$$t < -t(n-1)_\alpha = -t(9)_{0.05} = -1.833.$$

From the data we calculate

$$\bar{x} = 9.61,$$
$$s = 0.4818, \quad \text{and}$$
$$t = -2.560.$$

Since the value of t is less than -1.833 we reject the null hypothesis and conclude, based on this sample, that the mean lifetime of this company's AA batteries is less than 10 hours.   ■

Example 12.19   Suppose that $X_1, X_2, \ldots, X_n$ is a random sample from a $N(\mu_X, \sigma^2)$ and that $Y_1, Y_2, \ldots, Y_m$ is a random sample from an independent $N(\mu_Y, \sigma^2)$ population, where σ^2 is not known. We wish to test

$$H_0 : \mu_X = \mu_Y \quad \text{vs.} \quad H_1 : \mu_X \neq \mu_Y.$$

Both hypotheses are composite and cannot be reduced to simple hypotheses. Therefore, the likelihood ratio test procedure must be used. The reader should verify that the likelihood function for this problem is

$$L(\mu_X, \mu_Y, \sigma^2) = \prod_{i=1}^{n} \frac{1}{\sqrt{2\pi\sigma^2}} e^{-\frac{1}{2\sigma^2}(x_i - \mu_X)^2} \prod_{j=1}^{m} \frac{1}{\sqrt{2\pi\sigma^2}} e^{-\frac{1}{2\sigma^2}(y_j - \mu_Y)^2}$$

$$= (2\pi\sigma^2)^{-(n+m)/2} \exp\left\{ -\frac{1}{2\sigma^2} \left[\sum_{i=1}^{n} (x_i - \mu_X)^2 + \sum_{j=1}^{m} (y_j - \mu_Y)^2 \right] \right\}.$$

In Ω_0:   Let θ be the common value of μ_X and μ_Y. Then

$$\Omega_0 = \{(\theta, \sigma^2) : -\infty < \theta < \infty, \sigma^2 > 0\}.$$

Thus,

$$L_0 = (2\pi\sigma^2)^{-(n+m)/2} \exp\left\{ -\frac{1}{2\sigma^2} \left[\sum_{i=1}^{n} (x_i - \theta)^2 + \sum_{j=1}^{m} (y_j - \theta)^2 \right] \right\}.$$

Since both θ and σ^2 are unknown we need maximum likelihood estimators for each. We leave it as an exercise to verify that we obtain

$$\widehat{\theta} = \frac{n\bar{x} + m\bar{y}}{n + m}, \quad \text{and}$$

$$\widehat{\sigma^2} = \frac{\displaystyle\sum_{i=1}^{n}(x_i - \widehat{\theta})^2 + \sum_{j=1}^{m}(y_j - \widehat{\theta})^2}{n + m}.$$

Substituting these in the likelihood function and simplifying gives

$$
L_0^* = \left[2\pi \, \frac{\displaystyle\sum_{i=1}^{n}(x_i - \widehat{\theta})^2 + \sum_{j=1}^{m}(y_j - \widehat{\theta})^2}{n + m} \right]^{-(n+m)/2} \exp\left[-\frac{1}{2}(n + m) \right].
$$

In Ω: Here

$$
\Omega = \{(\mu_X, \mu_Y, \sigma^2) : -\infty < \mu_X < \infty, \, -\infty < \mu_Y < \infty, \, \sigma^2 > 0\},
$$

and $L = L(\mu_X, \mu_Y, \sigma^2)$. It was an exercise in Chapter 10 to show that the maximum likelihood estimators for μ_X, μ_Y, and σ^2 are

$$
\widehat{\mu_X} = \bar{x},
$$
$$
\widehat{\mu_Y} = \bar{y}, \quad \text{and}
$$
$$
\widehat{\sigma^2} = \frac{\displaystyle\sum_{i=1}^{n}(x_i - \bar{x})^2 + \sum_{j=1}^{m}(y_j - \bar{y})^2}{n + m}.
$$

After substituting in L and simplifying we find that

$$
L^* = \left[2\pi \, \frac{\displaystyle\sum_{i=1}^{n}(x_i - \bar{x})^2 + \sum_{j=1}^{m}(y_j - \bar{y})^2}{n + m} \right]^{-(n+m)/2} \exp\left[-\frac{1}{2}(n + m) \right].
$$

Forming the likelihood ratio statistic and simplifying results in

$$
\lambda = \left[\frac{\displaystyle\sum_{i=1}^{n}(x_i - \bar{x})^2 + \sum_{j=1}^{m}(y_j - \bar{y})^2}{\displaystyle\sum_{i=1}^{n}(x_i - \widehat{\theta})^2 + \sum_{j=1}^{m}(y_j - \widehat{\theta})^2} \right]^{(n+m)/2},
$$

where $\widehat{\theta} = \frac{n\bar{x} + m\bar{y}}{n+m}$. In an exercise in Chapter 9 it is shown that

$$
t = \frac{(\bar{x} - \bar{y}) - (\mu_X - \mu_Y)}{\sqrt{\dfrac{(n-1)s_x^2 + (m-1)s_y^2}{n + m - 2}\left[\dfrac{1}{n} + \dfrac{1}{m}\right]}}
$$

has a t distribution with $n + n - 2$ degrees of freedom. The quantity under the square root is called the *pooled estimate* of the variance of the differences in means and is denoted by

$$s^2_{\bar{x}-\bar{y}} = \frac{(n-1)s^2_x + (m-1)s^2_y}{n+m-2} \left[\frac{1}{n} + \frac{1}{m} \right].$$

Thus, when H_0 is true we see that

$$t = \frac{\bar{x} - \bar{y}}{s_{\bar{x}-\bar{y}}}$$

will have a t distribution with $n + m - 2$ degrees of freedom. If we proceed in a fashion similar to Example 12.16 the likelihood ratio becomes

$$\lambda = \left[1 + \frac{t^2}{n+m-2} \right]^{-(n+m)/2}.$$

The details are left as an exercise. From this we can see that $\lambda \leq \lambda_0$ whenever $|t| \geq a$. Using the α subscripting notation we see that $a = t(n + m - 2)_{\alpha/2}$. Thus, an equivalent critical region can be based on a statistic that has a t distribution. ■

   As before, the likelihood ratio procedure can also be applied to one-sided tests. If we proceed as in Example 12.19 for the two possible one-sided alternatives we find that, in each case, the critical region can be based on the test statistic

$$t = \frac{\bar{x} - \bar{y}}{s_{\bar{x}-\bar{y}}},$$

where, when $H_0 : \mu_X = \mu_Y$ is true, t will have a t distribution with $n + m - 2$ degrees of freedom. To summarize, the critical regions in each case are given in Table 12.5.

Table 12.5   Null hypothesis: $H_0: \mu_X = \mu_Y$

Alternative Hypothesis	Critical Region
$H_1 : \mu_X > \mu_Y$	$t \geq t(n + m - 2)_\alpha$
$H_1 : \mu_X < \mu_Y$	$t \leq -t(n + m - 2)_\alpha$
$H_1 : \mu_X \neq \mu_Y$	$|t| \geq t(n + m - 2)_{\alpha/2}$

Example 12.20

It has been claimed by numerous sources that standardized tests administered to school children are biased in favor of white students at the expense of black and minority students. To assess this claim, an educator randomly selected 10 white students and 12 black students from an inner city school and gave each one a standard test of reading comprehension. Letting X denote a score for a black student and Y denote a score for a white student the scores showed the following results. Reading comprehension scores have been found to follow a normal distribution.

Black (X)	White (Y)
72	79
74	80
75	78
76	76
73	78
74	77
77	79
75	76
73	78
78	75
77	
75	

We wish to test

$$H_0 : \mu_X = \mu_Y \quad \text{vs.} \quad H_1 : \mu_X < \mu_Y.$$

Let $\alpha = 0.01$. If we assume that the population variances are equal, then we can use the results of Example 12.18 and Table 12.5. Therefore, our test statistic will be

$$t = \frac{\bar{x} - \bar{y}}{s_{\bar{x}-\bar{y}}},$$

and, using Table 5 of Appendix A, we will reject H_0 if

$$t \le -t(n + m - 2)_\alpha = -t(20)_{0.01} = -2.528.$$

From the data we calculate the following.

Black (X)	White (Y)
$n = 12$	$m = 10$
$\bar{x} = 74.917$	$\bar{y} = 77.600$
$s_x^2 = 3.356$	$s_y^2 = 2.489$

Therefore, we find that

$$t = -3.639.$$

Since $t < -t(20)_{0.01}$ we reject the null hypothesis and conclude that black students do not perform as well as white students on this standard reading comprehension test.   ∎

Example 12.21   Suppose we have an observation from a $B(n, p)$ distribution. We wish to test

$$H_0 : p = p_0 \quad \text{vs.} \quad H_1 : p \ne p_0.$$

Here the null hypothesis is simple, and the alternative hypothesis is composite. As it is a two-sided test there is no uniformly most powerful test. Therefore, we shall derive a likelihood ratio test. The likelihood function for this problem is

$$L(p) = \binom{n}{x} p^x (1 - p)^{n-x}.$$

In Ω_0:   Since $\Omega_0 = \{p_0\}$ is a single point we find that

$$L_0^* = \binom{n}{x} p_0^x (1 - p_0)^{n-x}.$$

In Ω: Here $\Omega = \{p : 0 \le p \le 1\}$ places no restriction on the value of p. Therefore, we need a maximum likelihood estimator for p. From Example 10.4 we know that this is

$$\bar{p} = \frac{x}{n}.$$

This gives that

$$L^* = \binom{n}{x} \left(\frac{x}{n}\right)^x \left(1 - \frac{x}{n}\right)^{n-x}.$$

Therefore, after some simplification, the likelihood ratio is

$$\lambda = \left(\frac{np_0}{x}\right)^x \left(\frac{n - np_0}{n - x}\right)^{n-x},$$

and we reject H_0 whenever $\lambda \le \lambda_0$. We wish to find an equivalent rejection region in terms of x, the number of successes observed. To do this we shall graph λ versus x. It is clear that $\lambda = 1$ when $x = np_0$. To obtain the rest of the graph consider the logarithm of λ.

$$\ln(\lambda) = x \ln(np_0) - x \ln(x) + (n - x)\ln(n - np_0) - (n - x)\ln(n - x).$$

Now, if we treat λ as if it were a differentiable function of x we find that

$$\frac{d}{dx}\ln(\lambda) = \ln\left(\frac{p_0}{1 - p_0}\right) - \ln\left(\frac{\frac{x}{n}}{1 - \frac{x}{n}}\right).$$

Thus,

$$\frac{d}{dx}\ln(\lambda) = \begin{cases} < 0, \ x > np_0, & \text{and} \\ > 0, \ x < np_0. \end{cases}$$

The graph then is increasing for $x < np_0$, $\lambda = 1$ for $x = np_0$, and the graph is decreasing for $x > np_0$. We must now determine the behavior when $x = 0$ and $x = n$. At each point we note that $L^* = 1$, and find that

$$\lambda = \begin{cases} (1 - p_0)^n, \ x = 0, & \text{and} \\ p_0^n \qquad\ , \ x = n. \end{cases}$$

Thus, if λ is a continuous function of x, the graph is as shown in Figure 12.6. The actual graph will be a series of steps, but the general shape is the same as shown. Note that $\lambda \le \lambda_0$ occurs for values of x that are smaller than some value a or larger than another value b. As we have seen earlier, in order to achieve an exact size α test we may have to use randomization. If randomization is not used it may turn out in some cases that the critical region is empty. That is, λ_0 may be smaller than both p_0^n and $(1 - p_0)^n$. If this is not the case, then we will determine *a* and *b* so that

$$\frac{\alpha}{2} = \sum_{x=0}^{a} \binom{n}{x} p_0^x (1 - p_0)^{n-x}, \quad \text{and}$$

$$\frac{\alpha}{2} = \sum_{x=b}^{n} \binom{n}{x} p_0^x (1 - p_0)^{n-x}.$$

If these probabilities cannot be reached exactly, there are 2 possible approaches if one does not wish to randomize. One is to find the *a* and *b* values that give a probability as close to $\frac{\alpha}{2}$ as possible. This gives a critical region with size *near* α. On the other hand, if it is necessary for the size to be *no larger than* α, then choose the values of *a* and *b* that come closest to achieving $\frac{\alpha}{2}$ *without exceeding* it.

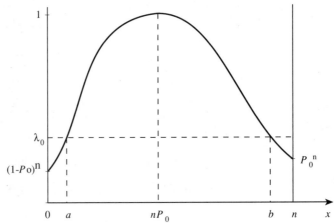

Figure 12.6

Recall from Section 11.4 that each of the one-sided tests has a uniformly most powerful critical region. The results are summarized in Table 12.6.

Table 12.6 $H_0 : p = p_0$

Alternative Hypothesis	Critical Region
$H_1 : p > p_0$	$x \geq a$, where $\sum_{i=a}^{n} \binom{n}{i} p_0^i (1 - p_0)^{n-i} = \alpha$
$H_1 : p < p_0$	$x \leq a$, where $\sum_{i=0}^{a} \binom{n}{i} p_0^i (1 - p_0)^{n-i} = \alpha$
$H_1 : p \neq p_0$	$x \leq a$, where $\sum_{i=0}^{a} \binom{n}{i} p_0^i (1 - p_0)^{n-i} = \frac{\alpha}{2}$, or $x \geq b$, where $\sum_{i=b}^{n} \binom{n}{i} p_0^i (1 - p_0)^{n-i} = \frac{\alpha}{2}$

Example 12.22   An opinion survey in the town of Cranby showed that 70% of the registered voters favored some form of gun control. In the town of Wapner a random sample of

20 registered voters showed that 11 favored some form of gun control. We wish to test at the 0.05 level whether the proportion of registered voters in Wapner in favor of gun control is significantly different from that of Cranby. If we let p be the proportion in favor of gun control, then we are testing

$$H_0 : p = 0.70 \quad \text{vs.} \quad H_1 : p \neq 0.70.$$

Let X = number of people in favor of gun control in our sample. Then, according to Example 12.20 we would reject the null hypothesis if

$$x \geq a \quad \text{or} \quad x \leq b,$$

where

$$0.025 = \sum_{i=0}^{a} \binom{20}{i} (0.70)^i (0.3)^{20-i}, \text{ and}$$

$$0.025 = \sum_{i=b}^{20} \binom{20}{i} (-0.70)^i (0.30)^{20-i}.$$

If we use the values of a and b that come closest to achieving a value of $\frac{\alpha}{2}$, then Table 1 of Appendix A gives that we reject H_0 whenever

$$x \leq 10 \quad \text{or} \quad x \geq 18,$$

with an actual significance level of 0.0526. Our sample showed that $x = 11$. Therefore, we cannot reject the null hypothesis.   ■

The likelihood ratio tests have some interesting properties. First, if a test has a best critical region then the likelihood ratio test will be identical to the test based on the Neyman-Pearson Lemma. The proof of this is straightforward and is left as an exercise. Secondly, there are a number of situations where an equivalent critical region in terms of a random variable with a known distribution cannot be found. In small sample cases we would have to use computer simulation to approximate the exact distribution of λ. We would then use this simulated distribution to obtain a value for λ_0. In large sample situations, it can be shown that $-2 \ln \lambda$ has an approximate chi-squared distribution. The proof is beyond the level of this book and is omitted. The interested reader with a good background in real analysis can consult a book such as *Mathematical Statistics* by S. S. Wilks.

Theorem 12.4

> Under the same regularity conditions necessary for maximum likelihood estimators to have an asymptotic normal distribution, $-2 \ln \lambda$ has an asymptotic chi-squared distribution with degrees of freedom equal to the difference between the dimension of Ω and the dimension of Ω_0.

This theorem says, in effect, that when the sample size is large we can use Theorem 12.4 to reject H_0 if

$$-2 \ln \lambda \geq \chi^2(n)_\alpha, \quad \text{where } n = \dim(\Omega) - \dim(\Omega_0). \tag{12.7}$$

To see that this result is at least reasonable we note that $0 \leq \lambda \leq 1$. This means that $-2 \ln \lambda \geq 0$ and $\lim_{\lambda \to 0} -2 \ln \lambda = \infty$. So, at least the range of values is that of a chi-squared random variable. Now, as λ gets small $-2 \ln \lambda$ gets large. Therefore, when $\lambda \leq \lambda_0$ we see that $-2 \ln \lambda \geq a$.

Example 12.23   Consider the test of

$$H_0 : \mu = \mu_0 \quad \text{vs.} \quad H_1 : \mu \neq \mu_0$$

for a normal population with an unknown variance that was discussed in Example 12.16. We wish to derive a large sample critical region based on Theorem 12.4. First, we note that

$$\Omega = \{(\mu, \sigma^2) : -\infty < \mu < \infty, \sigma^2 > 0\}.$$

Therefore, $\dim(\Omega) = 2$. Similarly,

$$\Omega_0 = \{(\mu_0, \sigma^2) : \sigma^2 > 0\}.$$

This gives that $\dim(\Omega_0) = 1$. Then according to Theorem 12.4 the asymptotic distribution of $-2 \ln \lambda$ will be chi-squared with 1 degree of freedom. Thus, we would reject the null hypothesis if

$$-2 \ln \lambda \geq \chi^2(1)_\alpha. \qquad \blacksquare$$

We close our discussion of likelihood ratio tests with an example where Theorem 12.4 must be used to obtain a critical region.

Example 12.24   Consider the joint distribution of two discrete random variables, X and Y, depicted by Table 12.7.

Table 12.7

		Y				
		1	2	$\cdots$	c	
	1	p_{11}	p_{12}	$\cdots$	p_{1c}	$p_{1\cdot}$
	2	p_{21}	p_{22}	$\cdots$	p_{2c}	$p_{2\cdot}$
X	$\vdots$	$\vdots$	$\vdots$	$\ddots$	$\vdots$	$\vdots$
	r	p_{r1}	p_{r2}	$\cdots$	p_{rc}	$p_{r\cdot}$
		$p_{\cdot 1}$	$p_{\cdot 2}$	$\cdots$	$p_{\cdot c}$	

In Table 12.7 we have used the notation

$$p_{i\cdot} = \sum_{j=1}^{c} p_{ij}, \quad \text{and} \quad p_{\cdot j} = \sum_{i=1}^{r} p_{ij}.$$

Two-way tables such as this are often referred to as *contingency tables*. They find wide application in the social sciences. Our goal here is to test the null hypothesis that X and Y are independent random variables against the alternative hypothesis that they are not. Stated formally we have

$$H_0 : p_{ij} = p_{i\cdot}\, p_{\cdot j}, \quad \text{for} \quad \begin{cases} i = 1, \ldots, r, & \text{and} \\ j = 1, \ldots, c, & \text{vs.} \end{cases}$$

$$H_1 : p_{ij} \neq p_{i\cdot}\, p_{\cdot j}, \quad \text{for some } i \text{ and } j.$$

Since both hypotheses are composite we shall derive a likelihood ratio test. Let

$$n_{ij} = \text{number of observations in cell } (i,j),$$

$$n_{i\cdot} = \sum_{j=1}^{c} n_{ij}, \quad \text{the number of observations in row } i,$$

$$n_{\cdot j} = \sum_{i=1}^{r} n_{ij}, \quad \text{the number of observations in column } j, \quad \text{and}$$

$$n = \sum_{i=1}^{r} \sum_{j=1}^{c} n_{ij}, \quad \text{the total number of observations.}$$

The likelihood function is

$$L(p_{11}, \ldots, p_{rc}) = p_{11}^{n_{11}} p_{12}^{n_{12}} \cdots p_{1c}^{n_{1c}} p_{21}^{n_{21}} \cdots p_{rc}^{n_{rc}}.$$

In Ω_0:   Under the null hypothesis $p_{ij} = p_{i\cdot}p_{\cdot j}$. Thus, the likelihood function becomes

$$L_0 = (p_{1\cdot}p_{\cdot 1})^{n_{11}}(p_{1\cdot}p_{\cdot 2})^{n_{12}} \cdots (p_{r\cdot}p_{\cdot c})^{n_{rc}}.$$

We need maximum likelihood estimators for $p_{1\cdot}, p_{2\cdot}, \ldots, p_{r\cdot}$ and $p_{\cdot 1}, p_{\cdot 2}, \ldots, p_{\cdot c}$. We leave it as an exercise to show that

$$\widehat{p_{i\cdot}} = \frac{n_{i\cdot}}{n}, i = 1, 2, \ldots, r, \quad \text{and}$$

$$\widehat{p_{\cdot j}} = \frac{n_{\cdot j}}{n}, j = 1, 2, \ldots, c.$$

Thus,

$$L_0^* = \left(\frac{n_{1\cdot}n_{\cdot 1}}{n^2}\right)^{n_{11}} \left(\frac{n_{1\cdot}n_{\cdot 2}}{n^2}\right)^{n_{12}} \cdots \left(\frac{n_{r\cdot}n_{\cdot c}}{n^2}\right)^{n_{rc}}.$$

In Ω: Here $L = L(p_{11}, \ldots, p_{rc})$. We need maximum likelihood estimators for p_{ij}, for $i = 1, 2, \ldots, r$, and $j = 1, 2, \ldots, c$. Again it is left as an exercise to verify that

$$\widehat{p}_{ij} = \frac{n_{ij}}{n}, \quad \text{for} \begin{cases} i = 1, 2, \ldots, r, \quad \text{and} \\ j = 1, 2, \ldots, c. \end{cases}$$

Substituting these in the likelihood function gives

$$L^* = \left(\frac{n_{11}}{n}\right)^{n_{11}} \left(\frac{n_{n12}}{n}\right)^{n_{12}} \cdots \left(\frac{n_{rc}}{n}\right)^{n_{rc}}.$$

The likelihood ratio is, after some simplification,

$$\lambda = \prod_{i=1}^{r} \prod_{j=1}^{c} \left(\frac{n_{i}.n_{.j}}{nn_{ij}}\right)^{n_{ij}}.$$

The null hypothesis is rejected when $\lambda \le \lambda_0$. The distribution of λ when the null hypothesis is true is extremely difficult to obtain. Therefore, we shall use Theorem 12.4 to obtain a large sample distribution. In Ω_0 we estimated $r + c$ quantities. However, it turns out that the restrictions

$$\sum_{i=1}^{r} p_{i.} = 1 \quad \text{and} \quad \sum_{j=1}^{c} p_{.j} = 1$$

mean that we only needed to estimate $(r - 1) + (c - 1)$ probabilities. Therefore, Ω_0 has dimension $(r - 1) + (c - 1)$. By a similar argument we find that the dimension of Ω is $rc - 1$. Then, by Theorem 12.4 we see that $-2 \ln \lambda$ will have an asymptotic distribution that is chi-squared with degrees of freedom of $rc - 1 - [(r - 1) + (c - 1)] = (r - 1)(c - 1)$. ■

Example 12.25   A plant manager was concerned that there was a relationship between the shift on which an item was produced and the number of defective items produced. He already knew that the graveyard shift (midnight to 8 A.M.) tended to produce more defective items overall. What he was interested in was whether there was a difference in pattern on an item by item basis. The numbers of defective items per shift were observed for a week with results as shown in Table 12.8.

Table 12.8

	Component			
	A	B	C	Total
Days	8	9	14	31
Swing	10	10	14	34
Graveyard	13	12	17	42
Total	31	31	45	107

We are testing

$$H_0 : p_{ij} = p_{i\cdot}\, p_{\cdot j}, \quad \begin{cases} i = 1, 2, 3, \\ j = 1, 2, 3, \end{cases} \quad \text{vs.}$$

$$H_1 : p_{ij} \neq p_{i\cdot}\, p_{\cdot j}, \quad \text{for some } i, j.$$

A significance level of 0.05 will be used. We shall compute the likelihood ratio statistic and use the large sample distribution of $-2 \ln \lambda$. Since $r = c = 3$ we use Table 4 of Appendix A to find that the null hypothesis will be rejected if

$$-2 \ln \lambda \geq \chi^2(4)_{0.05} = 9.488.$$

Referring to Example 12.23 we compute

$$-2 \ln \lambda = -2 \left(8 \ln \frac{(31)(31)}{(8)(107)} + 9 \ln \frac{(31)(31)}{(9)(107)} + \cdots + 17 \ln \frac{(42)(45)}{(17)(107)} \right)$$

$$= 7.304.$$

Since this value is less than 9.488 we do not reject the null hypothesis and conclude that there is no difference in the pattern of defective items produced according to shift. ∎

Exercises 12.5

1. Show that when testing two simple hypotheses the likelihood ratio procedure and the Neyman-Pearson Lemma give the same test when both exist.

2. Fill in the missing steps in the derivation of the two-sample test based on a t statistic in Example 12.19.

3. Show that the maximum likelihood estimators given in Example 12.24 are correct.

4. Let $X_1, X_2, \ldots, X_n$ be a random sample from a $Po(\lambda)$ population.
   a. Show that a likelihood ratio test of

   $$H_0 : \lambda = \lambda_0 \quad \text{vs.} \quad H_1 : \lambda \neq \lambda_0$$

   rejects H_0 if

   $$\bar{x} \leq a \quad \text{or} \quad \bar{x} \geq b.$$

   b. Find a and b that achieve an approximate significance level $\alpha = 0.1$ if $\lambda_0 = 0.1$ and $n = 10$.

5. Let $X_1, X_2, \ldots, X_n$ be a random sample from a $N(\mu, \sigma^2)$ population, where σ^2 is not known. Show that a likelihood ratio test of

   $$H_0 : \mu = \mu_0 \quad \text{vs.} \quad H_1 : \mu > \mu_0$$

   has a critical region that is equivalent to rejecting H_0 if

   $$\frac{\bar{x} - \mu_0}{s / \sqrt{n}} > a$$

6. Let $X_1, X_2, \ldots, X_n$ be a random sample from a geometric distribution with probability of success p. Find a likelihood ratio test for

$$H_0 : p = p_0 \quad \text{vs.} \quad H_1 : p \neq p_0.$$

7. Let $X_1, X_2, \ldots, X_n$ be a random sample from a distribution having a p.d.f. of

$$f(x) = \begin{cases} \dfrac{2x}{\theta^2} e^{-x^2/\theta^2}, & 0 \leq x, \quad \text{and} \\ 0, & \text{otherwise.} \end{cases}$$

Derive a likelihood ratio test of

$$H_0 : \theta = \theta_0 \quad \text{vs.} \quad H_1 : \theta \neq \theta_0.$$

8. Let $X_1, X_2, \ldots, X_n$ be a random sample from a $N(\mu, \sigma^2)$ population.

a. Derive a likelihood ratio test of

$$H_0 : \mu = \sigma \quad \text{vs.} \quad H_1 : \mu \neq \sigma.$$

b. Use the large sample approximation for $-2 \ln \lambda$ to test these hypotheses at the 0.05 level when

$$n = 100, \quad \bar{x} = 15 \quad \text{and} \quad \sum_{i=1}^{100} (x_i - \bar{x})^2 = 1350.$$

9. Let $X_1, X_2, \ldots, X_n$ be a random sample from a $N(\mu, \sigma^2)$ population. Show that a likelihood ratio test of

$$H_0 : \sigma^2 = \sigma_0^2 \quad \text{vs.} \quad H_1 : \sigma^2 \neq \sigma_0^2$$

rejects the null hypothesis if

$$\frac{(n-1)s^2}{\sigma_0^2} \geq \chi^2(n-1)_{\alpha/2} \quad \text{or} \quad \frac{(n-1)s^2}{\sigma_0^2} \leq \chi^2(n-1)_{1-\alpha/2}.$$

10. A manufacturer of automatic drip coffee makers guarantees that his product will last, on the average, 2 years with a standard deviation of 6 months. Five of his coffee makers lasted the following number of years.

$$1.3 \quad 1.6 \quad 2.0 \quad 2.3 \quad 2.8$$

Do these data support the claim that the standard deviation is 6 months? Use $\alpha = 0.05$ and the results of Exercise 9.

11. Let $X_1, \ldots, X_n$ be a random sample from a $N(\mu, \sigma^2)$ population and $Y_1, \ldots, Y_m$ be a random sample from a $N(\tau, \eta^2)$ population.

a. Show that a likelihood ratio test of

$$H_0 : \sigma_2 = \eta^2 \quad \text{vs.} \quad H_1 : \sigma^2 \neq \eta^2$$

rejects H_0 if

$$\frac{s_X^2}{s_Y^2} \geq F(n-1, m-1)_{\alpha/2} \quad \text{or} \quad \frac{s_X^2}{s_Y^2} \leq F(n-1, m-1)_{1-\alpha/2}.$$

    b. Find a size 0.05 critical region when $n = 11$ and $m = 5$.

12. Do the data in Example 12.20 support the assumption that the 2 population variances are equal? Use $\alpha = 0.1$.

13. When testing for a difference between the means of two groups it is often useful to use an experimental design technique known as *paired comparisons*. For example, if we wish to determine if the average length of the dominant arm of people is longer than that of the other arm the general difference in arm lengths would tend to cover up any differences between the length of a person's left and right arms. Therefore, taking a sample of dominant arm lengths and an independent sample of nondominant arm lengths would not permit the detection of a difference, if one exists. Paired comparisons remove the effects of such variability by looking at closely related measurements. That is, if population 1 is dominant arm lengths and population 2 is nondominant arm lengths we would devise a scheme that would result in a natural pairing between members of each population. In this case, we would pair the length of the dominant arm of a person with that of the same person's nondominant arm. Another example would be to use the score on a test of a person before a training program with the same person's score after completing the training program. In such cases, we would compute the difference between the score of an observation from group 1 and its pair from group 2. This has the effect of reducing the problem of comparing the difference between the means of 2 groups to that of testing whether the mean *difference* is equal to 0 or not. Thus, paired comparison tests become a special case of testing the mean of a single group of difference scores. Use this idea to test whether there is a difference between the length of the dominant arm and nondominant arm based on the following sample of college students. All lengths are in inches. Assume that arm lengths follow a normal distribution, and use a significance level of 0.05.

Student	Dominant Arm	Nondominant Arm
1	32.00	31.75
2	29.00	29.00
3	33.75	33.50
4	34.25	34.00
5	30.50	29.75
6	29.25	29.50
7	33.75	33.75
8	31.75	31.50
9	32.50	32.00
10	31.50	31.25

14. A study was conducted to determine whether men and women who were convicted of petty larceny received different treatment when it came to sentencing. Fifty men and 50 women convicted of petty larceny were selected at random, and their sentences were compared to the average sentence for this crime. The results are tabulated in the following table.

Sex	More Severe	Average	More Lenient
Men	20	7	23
Women	37	9	14

Use the large sample approximation for the likelihood ratio statistic to test for a difference in sentencing patterns according to sex of the offender. Use $\alpha = 0.05$.

12.6
Chi-Square Tests

There is another method for testing independence in contingency tables that is in widespread use. It is a special case of a class of tests that were developed by Karl Pearson around 1900. The tests are based on the limiting distribution of a multinomial random vector. The multinomial distribution was discussed in Section 5.4 of Chapter 5. Let $X_1, X_2, \ldots, X_k$ have a multinomial distribution with parameters $n, p_1, p_2, \ldots$ and p_k. In more advanced courses, it is shown that

$$Q = \sum_{i=1}^{k} \frac{(X_i - np_i)^2}{np_i} \tag{12.8}$$

has a limiting distribution that is $\chi^2(k-1)$. We can motivate this by outlining a particularly nice proof due to Fisher. It involves very little analysis. The details can be found in *The Advanced Theory of Statistics*, vol. 2, by M. G. Kendall and A. Stuart. Suppose that $X_1, X_2, \ldots, X_k$ are independent $Po(np_i)$ random variables. It turns out that the conditional probability function of $X_1, \ldots, X_n$, given that $\sum_{i=1}^{k} X_i = n$ is

$$p_{X_1, \ldots, X_n | \sum_{i=1}^{k} X_i}(x_1, \ldots, x_n \mid n) = \frac{n!}{x_1! \cdots x_n!} p_1^{x_1} \cdots p_k^{x_k},$$

which is our original multinomial distribution. Now, the Central Limit theorem gives that each

$$Y_i = \frac{X_i - np_i}{\sqrt{np_i}}$$

has a limiting standard normal distribution as n becomes infinitely large. Therefore, each

$$\frac{(X_i - np_i)^2}{np_i}$$

has an asymptotic $\chi(1)$ distribution. In this case, there is a constraint that $\sum_{i=1}^{k} X_i = n$. This is equivalent to $\sum_{i=1}^{k} \sqrt{np_i} Y_i = 0$. We now need a fact that we have not discussed in this text. It can be shown that if $Z_1, \ldots, Z_k$ are i.i.d. $N(0, 1)$ random variables, then the sum $\sum_{i=1}^{n} Z_i^2$ subject to the constraint $\sum_{i=1}^{k} a_i Z_i = 0$ has a $\chi^2(k-1)$ distribution. Therefore, we see that Q has an asymptotic $\chi^2(k-1)$ distribution.

The quantity Q is the basis for all of the tests to be discussed in this section. The tests are called *chi-square tests* because of the limiting distribution of Q. We shall consider two of these tests. The first is known as a *goodness-of-fit* test.

Goodness-of-fit tests are useful for determining if sample data appear to follow a particular type of probability distribution. Recall in Chapter 8 we were interested in using sample data to arrive at a reasonable choice for a population distribution. The goodness-of-fit test is a formal way to verify that the chosen distribution is consistent with the sample data. To illustrate how this test is conducted suppose that we wish to test the null hypothesis

$$H_0 : p_1 = p_{1_0}, p_2 = p_{2_0}, \ldots, p_k = p_{k_0}, \quad \text{vs.}$$

$$H_1 : p_i \neq p_{i_0}, \quad \text{for some } i.$$

If the null hypothesis is true we would anticipate that the observed value of $\frac{X_i}{n}$, the proportion of observations found in category i, would be fairly close to p_i, the true proportion of the population that is in category i. Another way of saying this is that X_i should be fairly close to np_{i_0}. np_{i_0} is commonly referred to as the *expected number of observations* in category i and denoted by e_i. The actual value of X_i after sampling is known as the *observed value* for category i and is denoted by o_i. If the null hypothesis is not true, then at least one o_i should be very different from e_i. Thus, if H_0 is true we would anticipate that the value of Q will be near zero; if H_0 is false, then Q should be much larger than zero. This gives us a basis for conducting a test of H_0 vs. H_1. A critical region would reject the null hypothesis if

$$Q \geq a.$$

The exact distribution of Q is quite complicated and is not known in general. For this reason tests based on Q use the limiting chi-square distribution. This means that if n is large a critical region would reject H_0 if

$$Q \geq \chi^2(k - 1)_\alpha.$$

There are a number of rules of thumb for determining if n is large enough for the limiting distribution to be fairly accurate. It is generally felt, however, that the chi-square approximation can be safely used as long as each $np_{i_0} \geq 5$. If this is not the case the test can be used by combining categories until each np_{i_0} is 5 or more.

Example 12.26   A recent poll indicated that 70% of the population favored some sort of control of gun sales, 15% were against any controls, 11% were undecided, and the remaining 4% had no opinion. A national lobbying group opposed to any gun control conducted a vigorous "guns don't kill, people do" campaign in an attempt to alter public opinion. A poll of 100 people interviewed after the campaign was completed found that 65 still favored controls, 25 were opposed to controls, 7 were undecided, and 3 had no opinion. We wish to test the hypothesis that the campaign was ineffectual. If the campaign did not have any effect on public opinion, then the actual proportions would be the same as they were before the campaign. This means that our null hypothesis will be

$$H_0 : p_1 = 0.7, \quad p_2 = 0.15, \quad p_3 = 0.11, \quad p_4 = 0.04.$$

The alternative hypothesis would be that at least one of these proportions is different.

We first check to see if the expected numbers of observations are large enough to use the chi-square approximation

$$np_{1_0} = 100(0.7) = 70,$$

$$np_{2_0} = 100(0.15) = 15,$$

$$np_{3_0} = 100(0.11) = 11, \quad \text{and}$$

$$np_{4_0} = 100(0.04) = 4.$$

The expected number of observations in the "no opinion" category is too low. Therefore, we must combine this category with another one. The most logical choice seems to be the "undecided" group. This way the population would be divided into a "for" group, an "against" group, and an "other" group. The expected number of observations in the new "other" group is now 15, which is large enough to make the chi-square approximation valid. Thus, the new null hypothesis is

$$H_0 : p_1 = 0.7, \quad p_2 = 0.15, \quad p_3 = 0.15.$$

We shall use a significance level of 0.05. Since $k = 3$, we use Table 4 of Appendix A to see that the null hypothesis will be rejected if

$$Q \geq \chi^2(k - 1)_\alpha = \chi^2(2)_{0.05} = 5.991.$$

Equation 12.8 gives

$$Q = \frac{[65 - (100)(0.7)]^2}{(100)(0.7)} + \frac{[25 - (100)(0.15)]^2}{(100)(0.15)} + \frac{[10 - (100)(0.15)]^2}{(100)(0.15)}$$

$$= 0.357 + 6.667 + 1.667$$

$$= 8.691.$$

Since this is larger than 5.991 we reject the null hypothesis and conclude that the campaign did alter the distribution of opinions regarding gun control. ∎

We indicated that the goodness-of-fit test is a formal way to determine if a given type of distribution is supported by sample data. For example, the question arises as to whether a given random sample could have come from a normally distributed population. An example is the cattle data from Chapter 8. There, it was postulated that the distribution appeared to be normal. We would like to use the goodness-of-fit idea to test this assumption. To do this we must use the presumed distribution to create a multinomial random vector. Assume we believe that the random variable X has a distribution with probability density function or probability function $f_0(x)$. To create a multinomial random vector the interval, I, of possible values of the random variable X is partitioned into k disjoint subintervals, $I_1, I_2 \ldots, I_k$. If we let

$$p_i = P(X \in I_i),$$

then $p_1, p_2, \ldots, p_k$ and n are the parameters of a multinomial distribution. We can then use the chi-square goodness-of-fit test to test

$$H_0 : f(x) = f_0(x) \quad \text{vs.} \quad H_1 : f(x) \neq f_0(x)$$

as long as the expected number of observations in each interval is at least 5. The procedure is best explained by using an example.

Example 12.27

We wish to test the claim that the following 50 observations came from a chi-square distribution with 4 degrees of freedom.

4.40	4.62	1.25	2.35	4.59	4.88	4.27	6.54	8.46	3.00
0.74	1.50	3.56	1.59	5.22	9.49	0.95	2.80	3.41	6.19
4.82	3.19	3.69	1.91	10.45	6.68	17.30	8.82	4.66	1.46
3.09	1.72	2.28	1.73	0.95	0.28	1.78	0.66	2.86	5.25
5.41	0.47	3.14	3.13	2.94	1.60	0.96	0.48	11.55	0.50

Therefore, we are testing

$$H_0 : X \sim \chi^2(4) \quad \text{vs.} \quad H_1 : X \not\sim \chi^2(4).$$

We shall use $\alpha = 0.05$. We shall partition the range of values for the chi-square distribution into intervals that have a probability of 0.1 of occurrence of each. Since $p_{1_0} = p_{2_0} = \cdots = p_{10_0} = 0.1$ each np_{i_0} will have a value of 5. A statistical computer package was used to obtain the deciles of the $\chi^2(4)$ distribution. We then set up the following 10 categories and counted the number of observations in each one. The results are shown in Table 12.9.

Table 12.9

Category	Interval	Observed	Expected
1	[0, 1.06)	9	5
2	[1.06, 1.65)	5	5
3	[1.65, 2.19)	4	5
4	[2.19, 2.75)	2	5
5	[2.75, 3.36)	8	5
6	[3.36, 4.04)	3	5
7	[4.04, 4.88)	6	5
8	[4.88, 5.99)	4	5
9	[5.99, 7.78)	3	5
10	[7.78, ∞)	6	5

Since each expected value is at least 5 the chi-square approximation gives that we will reject the null hypothesis if

$$Q \geq \chi^2(9)_{0.05} = 16.919.$$

Using the values in Table 12.9 we compute the value of Q as follows.

$$Q = \frac{(9-5)^2}{5} + \frac{(5-5)^2}{5} + \frac{(4-5)^2}{5} + \frac{(2-5)^2}{5} + \frac{(8-5)^2}{5}$$

$$+ \frac{(3-5)^2}{5} + \frac{(6-5)^2}{5} + \frac{(4-5)^2}{5} + \frac{(3-5)^2}{5} + \frac{6-5)^2}{5}$$

$$= 3.1 + 0 + 0.2 + 1.8 + 1.8 + 1.8 + 0.2 + 0.2 + 0.8 + 0.2$$

$$= 10.1$$

Since this value is less than 16.919 we do not reject H_0 and conclude that it is reasonable to assume that the 50 observations came from a chi-squared distribution with 4 degrees of freedom. ∎

This test can be generalized to cover situations where the null hypothesis is that a sample comes from some member of a hypothesized family of distribution. For these cases as well as other tests of goodness-of-fit, the reader is directed to texts such as *Statistical Methods* by G. Snedecor and W. Cochran.

We now show how to apply the chi-square test procedure to test the independence of the row and column variables in a contingency table. Recall that the test is

$$H_0 : p_{ij} = p_{i\cdot} \, p_{\cdot j}, \quad \begin{cases} i = 1, 2, \ldots, r, & \text{and} \\ j = 1, 2, \ldots, c, & \text{vs.} \end{cases}$$

$$H_1 : p_{ij} \neq p_{i\cdot} \, p_{\cdot j}, \quad \text{for some } i, j.$$

In this situation we can view $p_{i\cdot_0}$ as being the population proportions in a multinomial distribution with $k = r$ and $p_{\cdot j_0}$ the population proportions in a multinomial distribution with $k = c$. When H_0 is true $p_{ij_0} = p_{i\cdot_0} p_{\cdot j_0}$. Thus, the chi-square statistic is

$$Q = \sum_{i=1}^{r} \sum_{j=1}^{c} \frac{(X_{ij} - np_{i\cdot_0} p_{\cdot j_0})^2}{np_{i\cdot_0} p_{\cdot j_0}}.$$

At this point we note that the null hypothesis does not specify the values for $p_{i\cdot_0}$ and $p_{\cdot j_0}$. Thus, in order to proceed, we must estimate them. Recall from Example 12.24 that the maximum likelihood estimators are

$$\widehat{p_{i\cdot_0}} = \frac{n_{i\cdot}}{n}, i = 1, 2, \ldots, r, \quad \text{and}$$

$$\widehat{p_{\cdot j_0}} = \frac{n_{\cdot j}}{n}, j = 1, 2, \ldots, c.$$

Substituting into Q and simplifying gives

$$Q = \sum_{i=1}^{r} \sum_{j=1}^{c} \frac{\left(X_{ij} - \frac{n_{i\cdot} n_{\cdot j}}{n} \right)^2}{\frac{n_{i\cdot} n_{\cdot j}}{n}}. \tag{12.9}$$

If all of the $\frac{n_i \cdot n_j}{n}$ are 5 or more, then Q will have an approximate chi-squared distribution with $(r-1)(c-1)$ degrees of freedom. This means that we would reject the null hypothesis if

$$Q \geq \chi^2([r-1][c-1])_\alpha.$$

Example 12.28

We wish to perform the chi-square test on the same contingency table used in Example 12.24. Recall that we were interested in whether there was a relationship between workshift and the number of defective components produced for each of 3 different items. We shall use $\alpha = 0.05$. The data are repeated below.

| | \multicolumn{4}{c}{Component} |
	A	B	C	Total
Days	8	9	14	31
Swing	10	10	14	34
Graveyard	13	12	17	42
Total	31	31	45	107

Thus, $r = c = 3$. Also

$$n_1. = 31, \quad n_2. = 34, \quad n_3. = 42,$$
$$n._1 = 31, \quad n._2 = 31, \quad n._3 = 45, \quad \text{and}$$
$$n = 107.$$

The smallest $\frac{n_i \cdot n_j}{n}$ is $\frac{(31)(31)}{107} = 8.98$, which is greater than 5. Therefore, we can use the chi-squared approximation for Q. We use Table 4 of Appendix A to find that the null hypothesis should be rejected if

$$Q \geq \chi^2([r-1][c-1])_\alpha = \chi^2(4)_{0.05} = 9.488.$$

Using Equation 12.9 we compute

$$Q = \frac{\left(8 - \frac{(31)(31)}{107}\right)^2}{\frac{(31)(31)}{107}} + \frac{\left(9 - \frac{(31)(31)}{107}\right)^2}{\frac{(31)(31)}{107}} + \frac{\left(14 - \frac{(31)(45)}{107}\right)^2}{\frac{(31)(45)}{107}}$$

$$+ \frac{\left(10 - \frac{(34)(31)}{107}\right)^2}{\frac{(34)(31)}{107}} + \frac{\left(10 - \frac{(34)(31)}{107}\right)^2}{\frac{(34)(31)}{107}} + \frac{\left(14 - \frac{(34)(45)}{107}\right)^2}{\frac{(34)(45)}{107}}$$

$$+ \frac{\left(13 - \frac{(42)(31)}{107}\right)^2}{\frac{(42)(31)}{107}} + \frac{\left(12 - \frac{(42)(31)}{107}\right)^2}{\frac{(42)(31)}{107}} + \frac{\left(17 - \frac{(42)(45)}{107}\right)^2}{\frac{(42)(45)}{107}}$$

$$= 0.285.$$

Since this is less than 9.488 we do not reject H_0, and we conclude that the pattern of defects produced on a shift does not depend on the item produced. Note that the value of this chi-square statistic is different from $-2 \ln \lambda$ found in Example 12.24.   ∎

Exercises 12.6

1. The goodness-of-fit test is one of many methods used to ascertain if a sequence of digits is random. On the average a random sequence should have the same proportion of each digit 0 through 9. Use the random number table given in Table 8.1 to select 100 digits in the range 0 through 9. Let p_i be the proportion of the digit i in the population and test

$$H_0 : p_0 = p_1 = \cdots = p_9 = 0.1 \quad \text{vs.}$$

$$H_1 : p_i \neq 0.1, \quad \text{for some } i$$

at a significance level of $\alpha = 0.1$.

2. A historically famous example of the Poisson distribution is the number of fatalities caused by the kick of a horse in the Prussian Army Corps during the period 1875–1984 as reported by L. v. Bortkiewicz. The data are summarized below.

Number of Deaths	0	1	2	3	4	≥ 5
Frequency	109	65	22	3	1	0

Use the goodness-of-fit test to assess the claim that these data follow a Poisson distribution with parameter $\lambda = 0.6$. Note here that $n = 200$.

3. A researcher feels that, due to genetic theory, the proportion of offspring from a particular mating should occur in the ratios

$$\frac{1}{6}, \quad \frac{1}{3}, \quad \frac{1}{3}, \quad \frac{1}{6}.$$

An experiment was conducted with the following results.

Theoretical Proportion	$\frac{1}{6}$	$\frac{1}{3}$	$\frac{1}{3}$	$\frac{1}{6}$
Observed Frequency	42	62	64	31

Do these data support the theoretical distribution at the 0.05 level?

4. Did the following data come from a $G(1, 50)$ distribution?

27	38	46	4	38	62	25	9	5	40
53	21	44	23	13	21	64	53	84	22
7	33	24	76	49	43	58	37	25	43
27	62	47	87	9	71	28	14	23	56

5. Did the following data come from a population with a $U(0, 100)$ distribution?

72	18	16	4	28	82	75	9	4	30
35	11	34	13	63	11	74	93	94	42
7	53	94	57	99	47	48	16	15	33
37	82	67	84	9	77	18	84	33	96

6. Use the chi-square test to determine if there is a difference in the treatment of men and women who are convicted of petty larceny using the data in Exercise 4 of Section 12.5. Let $\alpha = 0.05$.

7. A study was conducted to determine the relationship, if any, between the status of disease (advanced or beginning) and the response to a new treatment with the following results.

	Much Improvement	Moderate Improvement	Slight Improvement	No Improvement
Beginning	12	37	44	53
Advanced	4	19	17	14

Use the chi-square test of Equation 12.9 to determine whether the effectiveness of the treatment depends on the severity of the disease. If it does, comment on the form of that dependency. Use $\alpha = 0.05$.

8. Three different sites are being considered for a regional landfill. A landfill is a mixed blessing in that the ecological impact is offset by the ecomonic benefit to the local residents. To get information regarding public opinion a random sample of residents in each region is taken. The data are given below.

	Region A	Region B	Region C
For	21	27	34
Against	14	39	27
No Opinion	7	13	11

Do these data indicate that the pattern of attitudes is the same in each region? Use $\alpha = 0.1$.

12.7 Large Sample Tests

There are a number of instances where tests can be derived based on the large sample properties of the estimators for parameters. In Chapter 7 we have seen that, because of the Central Limit theorem, the distribution of sample means can be approximated by a normal distribution. In addition, in Chapter 10 we mentioned that maximum likelihood estimators have a large sample distribution that is approximately normal. A large number of commonly used tests exploit these ideas. In this section we illustrate them with a few examples.

Example 12.29

Suppose that we wish to test the hypothesis that a candidate's position on a certain issue is preferred by the majority of the electorate. That is, we wish to test that

$$H_0 : p = 0.5 \quad \text{vs.} \quad H_1 : p > 0.5,$$

where p is proportion favoring the candidate's position on that issue. Let X be the number of voters in favor of the candidate's position. If $(0.5)n \geq 5$, then the distribution of X, when H_0 is true, according to the De Moivre-Laplace theorem, is

$N(0.5, \frac{1}{(4n)})$. This means that, if the null hypothesis is true, then the distribution of

$$z = \frac{\frac{x}{n} - 0.5}{\sqrt{\frac{1}{4n}}}$$

is approximately standard normal. If $\frac{x}{n}$ is significantly larger than $p_0 = 0.5$ we would be inclined to reject the null hypothesis in favor of the alternative. Therefore, a critical region would reject H_0 if

$$z = \frac{\frac{x}{n} - 0.5}{\frac{0.5}{\sqrt{n}}} \geq a.$$

Thus, a large sample critical region will reject the null hypothesis when

$$z \geq z_\alpha.$$

■

Example 12.30   A sample of 400 voters showed that 210 were in favor of a candidate's position of abortion. Does this indicate, at the 0.05 level, that the population at large is in favor of her position on this issue? If we let p be the proportion in favor of her position on abortion, then we are testing

$$H_0 : p = 0.5 \quad \text{vs.} \quad H_1 : p > 0.5.$$

If X is the number of members of the sample that are in favor of the candidate's position, then, if H_0 is true, X will have a distribution that is approximately normally distributed with a mean of 0.5 and a variance of $\frac{1}{1600}$. Thus, according to Example 12.29 we will reject H_0 if

$$z = \frac{\frac{x}{n} - 0.5}{\frac{0.5}{20}} \geq z_{0.05} = 1.645.$$

From the data we compute

$$z = \frac{\frac{210}{400} - 0.5}{\frac{0.5}{20}} = 1.00.$$

Since this value is not greater than 1.645 we do not reject the null hypothesis and cannot conclude that more than half of the electorate favor the candidate's position. ■

The preceding example illustrates many of the ideas behind large sample tests. To begin, if the estimator for a parameter is based on the sample mean or on maximum likelihood principles, then they have distributions that are asymptotically normal. The Central Limit theorem guarantees this for the sample means, and the theory of maximum likelihood estimation does the same for the others. This means that, for large values of n, the estimator for θ, $\widehat{\theta}$, will be approximately normally distributed with a mean of θ and a variance of $Var(\widehat{\theta})$. This means that a large sample test of $H_0 : \theta = \theta_0$ will reject the null hypothesis when

Alternative	Critical Region
$\theta > \theta_0$	$z \geq z_\alpha$
$\theta < \theta_0$	$z \leq -z_\alpha$
$\theta \neq \theta_0$	$|z| \geq z_{\alpha/2}$

where

$$z = \frac{\widehat{\theta} - \theta_0}{\sqrt{Var(\widehat{\theta}_0)}}.$$

Example 12.31   Suppose we have a sample of 50 observations from a $Po(\lambda)$ distribution with a sample mean of 4.7, and we wish to test

$$H_0 : \lambda = 5 \quad \text{vs.} \quad H_1 : \lambda < 5$$

at the 0.05 level. From Chapter 10 we know that an unbiased estimator for λ is $\overline{X}$. Therefore, the Central Limit theorem guarantees that, for large n, $\overline{X}$ will have a distribution that is approximately normal with mean λ and variance $\frac{\lambda}{n}$. Thus, when the null hypothesis is true,

$$z = \frac{\overline{X} - \lambda_0}{\sqrt{\dfrac{\lambda_0}{n}}}$$

will have an approximate $N(0, 1)$ distribution. This means that, from the preceding discussion, we will reject the null hypothesis if

$$z \leq -z_\alpha.$$

Therefore, we will reject H_0 if $z \leq z_{0.05} = -1.64$. From the sample we compute that

$$z = \frac{4.7 - 5}{\sqrt{\dfrac{5}{50}}} = -0.949.$$

Therefore, a large sample test will not reject H_0. ∎

In other cases, we may know that the estimator of a parameter, under certain circumstances, has a distribution that is asymptotically normal. The problem often is that the variance may depend on an unknown parameter. If, however, the estimator for that unknown parameter can be shown to be consistent, then we can use Theorem 7.12 to derive a statistic whose limiting distribution is standard normal.

Example 12.32   Suppose that $X_1, X_2, \ldots, X_n$ are a random sample from a $B(1, p_X)$ distribution, and that $Y_1, Y_2, \ldots, Y_m$ are a random sample from an independent $B(1, p_Y)$ distribution. We wish to test

$$H_0 : p_X = p_Y \quad \text{vs.} \quad H_1 : p_X \neq p_Y.$$

From Chapter 10 we know that the best unbiased estimator for p is $\frac{X}{n}$, where X is the number of successes in the sample. Therefore,

$$\widehat{p}_X = \bar{x} = \frac{\sum\limits_{i=1}^{n} x_i}{n}, \quad \text{and} \quad \widehat{p}_Y = \bar{y} = \frac{\sum\limits_{j=1}^{m} y_j}{m}$$

are the best unbiased estimators for p_X and p_Y, respectively. Similarly, it can be shown that when H_0 is true

$$\widehat{p} = \frac{n\bar{x} + m\bar{y}}{n + m}$$

is the best unbiased estimator for the common proportion of successes. In addition, when the null hypothesis is true we can show that $\widehat{p}$ is a consistent estimator for the common value of p. This means that, if we let p be the common value when H_0 is true, then

$$\frac{\widehat{p}(1 - \widehat{p})}{p(1 - p)}$$

will converge in probability to one. Now, when the null hypothesis is true, the variance of $\bar{x} - \bar{y}$ will be

$$p(1 - p)\left(\frac{1}{n} + \frac{1}{m}\right).$$

This means that

$$\frac{\widehat{p}_X - \widehat{p}_Y}{\sqrt{p(1 - p)\left(\frac{1}{n} + \frac{1}{m}\right)}}$$

will have an asymptotic standard normal distribution. Then, since

$$\frac{\widehat{p}(1 - \widehat{p})}{p(1 - p)}$$

converges in probability to one, Theorem 7.12 guarantees that

$$z = \frac{\widehat{p}_X - \widehat{p}_Y}{\sqrt{\widehat{p}(1 - \widehat{p})\left(\frac{1}{n} + \frac{1}{m}\right)}}$$

will have an asymptotic standard normal distribution when the null hypothesis is true. Therefore, a large sample test can be based on

$$z = \frac{\widehat{p}_X - \widehat{p}_Y}{\sqrt{\widehat{p}(1 - \widehat{p})\left(\dfrac{1}{n} + \dfrac{1}{m}\right)}}.$$

Furthermore, it will reject the null hypothesis if

$$|z| \leq z_{\alpha/2}.$$ ∎

Example 12.33   A large state university has 2 campuses. On one campus a sample of 200 students showed that 54 of them smoked. On the other campus 30 out of 100 students sampled smoked. We wish to determine if this indicates that the same proportion of students on each campus smokes. If we let p_1 and p_2 be the proportion of students that smoke on campus 1 and 2, respectively, then we are testing

$$H_0 : p_1 = p_2 \quad \text{vs.} \quad H_1 : p_1 \neq p_2.$$

If we use a significance level of 0.05 and a test statistic of

$$z = \frac{\widehat{p}_1 - \widehat{p}_2}{\sqrt{\widehat{p}(1 - \widehat{p})\left(\dfrac{1}{n} + \dfrac{1}{m}\right)}},$$

we would reject the null hypothesis if

$$|z| \geq z_{\alpha/2} = z_{0.025} = 1.96.$$

From the samples we find that

$$\widehat{p}_1 = \frac{54}{200} = 0.27, \quad \widehat{p}_2 = \frac{30}{100} = 0.30, \quad \text{and } \widehat{p} = \frac{54 + 30}{300} = 0.28.$$

Then we compute

$$z = \frac{0.27 - 0.30}{\sqrt{(0.28)(0.72)\left(\dfrac{1}{200} + \dfrac{1}{100}\right)}} = -0.173.$$

Since the absolute value of this is well below 1.96 we do not reject the null hypothesis and conclude that the same proportion of students smoke on both campuses. ∎

Exercises 12.7   1. Suppose we have a sample of n observations from a $G(1, \beta)$ distribution. Show that a large sample test of

$$H_0 : \beta = \beta_0 \quad \text{vs.} \quad H_1 : \beta > \beta_0$$

will reject the null hypothesis if

$$z = \frac{\bar{X} - \beta_0}{\sqrt{\dfrac{\beta_0^2}{n}}} \geq z_\alpha.$$

2. An assembly line in a factory must be shut down and repaired if the proportion of defective components exceeds 5%. A random sample of one day's output shows that 8 out of 100 components are defective. Should the assembly line be shut down? Let $\alpha = 0.05$.

3. A political scientist believes that the proportion of Republicans in favor of gun control is less than that of the Democrats. In a sample, 92 out of 200 Republicans and 102 out of 200 Democrats were in favor of gun control. Do these data support the political scientist's belief at the 0.05 level?

4. Suppose that we have a sample of n observations from a population with a mean of μ_1 and a variance of σ_1^2 and a sample of m observations from a population with a mean of μ_2 and a variance of σ_2^2.

   a. Show that, if s_1^2 and s_2^2 converge in probability to σ_1^2 and σ_2^2, respectively, then

   $$Z = \frac{\bar{X}_1 - \bar{X}_2}{\sqrt{\dfrac{s_1^2}{n} + \dfrac{s_2^2}{m}}}$$

   will have an asymptotic standard normal distribution if $\mu_1 = \mu_2$.

   b. Give a large sample critical region for testing

   $$H_0 : \mu_1 = \mu_2 \quad \text{vs.} \quad H_1 : \mu_1 \neq \mu_2.$$

5. Suppose we have a sample of n observations from a distribution having a p.d.f. of

   $$f(x) = \begin{cases} \theta x^{\theta - 1}, & 0 \leq x \leq 1, \quad \text{and} \\ 0, & \text{otherwise.} \end{cases}$$

   Derive a large sample test of

   $$H_0 : \theta = \theta_0 \quad \text{vs.} \quad H_1 : \theta \neq \theta_0.$$

12.8 Sequential Probability Ratio Test

Until now we have been considering the sample size as being a quantity fixed in advance by the investigator. This means that, while α is fixed and known, we have relatively little control over the value for β. Thus, in some situations even the use of a test that is uniformly most powerful does not mean that β will not be so large that we have a high chance of not detecting deviations from the null hypothesis. Of course, this problem can be overcome by increasing the sample size or increasing the value of α. In this section we shall discuss an alternative approach.

Suppose we agree not to fix the sample size in advance, but instead do something like the following.

1. We sample an item.

2. Based on the data accumulated so far we either

   a. accept the null hypothesis and stop,

b. reject the null hypothesis and stop, or

c. go back to Step 1 and repeat.

The idea of conducting a testing procedure in this *sequential* fashion was originated by Abraham Wald in the mid-1940's. As we shall see, this type of procedure permits us to fix α and β in advance. However, it does require that the sampling situation be such that we can return at will to the population to get another unit. This procedure has been applied successfully in the area of statistical quality control where the output from a production run is available, and additional items can be examined easily.

Consider the case where we are testing a simple hypothesis $H_0 : \theta = \theta_0$ against a simple alternative hypothesis $H_1 : \theta = \theta_1$. As we have seen, the Neyman-Pearson Lemma states that a best test rejects the null hypothesis whenever

$$\frac{L_0}{L_1} \leq k,$$

where k is a positive constant. Let $L_0(n)$ and $L_1(n)$ denote the likelihood functions for a sample of size n under the null and alternative hypotheses, respectively. Now consider the sequence of likelihood ratios

$$\frac{L_0(1)}{L_1(1)}, \frac{L_0(2)}{L_1(2)}, \ldots$$

Let a and b be two positive constants with $a < b$. We can then define the following testing procedure.

1. If $\frac{L_0(n)}{L_1(n)} \leq a$, stop and reject H_0.

2. If $\frac{L_0(n)}{L_1(n)} \geq b$, stop and accept H_0.

3. If $a < \frac{L_0(n)}{L_1(n)} < b$, continue sampling.

We shall show shortly that, if we let

$$a = \frac{\alpha}{1 - \beta}, \quad \text{and} \quad b = \frac{1 - \alpha}{\beta},$$

then $P(\text{Type I error}) = \alpha$ and $P(\text{Type II error}) = \beta$. This procedure is known as the *sequential probability ratio test*. We shall illustrate its use with an example.

Example 12.34

We wish to test whether the population of viewers of a college basketball game contains as many men as women against the alternative that 70% of the viewers are men. This is a test of

$$H_0 : p = 0.5 \quad \text{vs.} \quad H_1 : p = 0.7,$$

where p is the probability of a randomly selected viewer being a man. Let $\alpha = 0.05$ and $\beta = 0.10$. Thus, we have

$$a = \frac{\alpha}{1 - \beta} = \frac{0.05}{.90} = 0.0556,$$

$$b = \frac{1-\alpha}{\beta} = \frac{0.95}{0.1} = 9.5,$$

$$L_0(n) = (0.5)^n, \quad \text{and}$$

$$L_1(n) = (0.7)^x (0.3)^{n-x},$$

where $x =$ the number of men observed. Thus, the sequential probability ratio test becomes the following.

1. Reject H_0 if

$$\frac{(0.5)^n}{(0.7)^x (0.3)^{n-x}} \le 0.0556.$$

2. Accept H_0 if

$$\frac{(0.5)^x}{(0.7)^x (0.3)^{n-x}} \ge 9.5.$$

3. Continue sampling if

$$0.0556 < \frac{(0.5)^n}{(0.7)^x (0.3)^{n-x}} < 9.5.$$

To show how the test might proceed suppose the sampling was as follows.

n	Sex	$\dfrac{L_0(n)}{L_1(n)}$
1	M	0.714
2	W	1.190
3	W	1.984
4	W	3.307
5	W	5.511
6	W	9.186
7	M	6.561
8	W	10.935

At this point we stop and accept H_0. The results of this test are shown graphically in Figure 12.7.

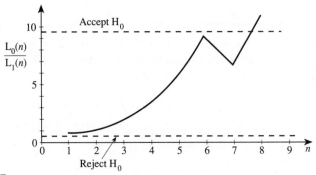

Figure 12.7

At this point it is natural to ask what prevents the probability ratio from always remaining between a and b. The answer is nothing. It is possible that the sequential test will never terminate. However, in more advanced texts it is shown that the sequential probability ratio test will terminate in a finite number of steps with probability one. In addition, it can also be shown that the expected value of the sample size using the sequential probability ratio test is less than the sample size of a fixed sample test that achieve the same values of α and β. Thus, on the average we can achieve at least the same results by using a sequential probability ratio test and do it with a smaller sample.

We shall now give a justification of the relationship between α, β, a, and b. We first consider the calculation of α and β. Let C_n be the critical region after n items have been sampled, and let X_n be the result of the first n items (i.e., $X_n = (x_1, x_2, \ldots, x_n)$). Then

$$\alpha = P(\text{Reject } H_0 \mid H_0 \text{ true})$$

$$= \sum_{n=1}^{\infty} P(X_n \in C_n \mid H_0 \text{ true}).$$

The second equality is justified because the test terminates with probability one. Now, if we assume continuous distributions, then we get the following. The argument in the discrete case is identical with summations replacing integrals.

$$\alpha = \sum_{n=1}^{\infty} P(X_n \in C_n \mid H_0 \text{ true})$$

$$= \sum_{n=1}^{\infty} \int \cdots \int_{C_n} L_0(n)\, dx_1 \cdots dx_n$$

Now X_n is in C_n only if $\dfrac{L_0(n)}{L_1(n)} \le a$, or, equivalently if $L_0(n) \le a L_1(n)$. Thus, we get the following.

$$\alpha \le \sum_{n=1}^{\infty} \int \cdots \int_{C_n} a L_1(n)\, dx_1 \cdots dx_n$$

$$= a \sum_{n=1}^{\infty} \int \cdots \int_{C_n} L_1(n)\, dx_1 \cdots dx_n$$

$$= a \sum_{n=1}^{\infty} P(X_n \in C_n \mid H_1 \text{ true})$$

$$= a P(\text{reject } H_0 \mid H_1 \text{ true})$$

$$= a(1 - \beta)$$

Thus, we find that $a \ge \frac{\alpha}{1-\beta}$. Proceeding in a similar fashion we find that $b \le \frac{1-\alpha}{\beta}$. The details are left as an exercise. These inequalities are shown in Figure 12.8.

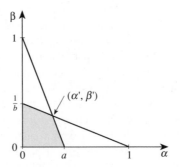

Figure 12.8

This means that for specified values of a and b, the possible values for α and β are those in the shaded region. Thus, α and β cannot be larger than the values in that region. If we pick α and β to be the intersection of the two lines (denoted α' and β' in Figure 12.8) and determine a and b accordingly, the actual values of α and β cannot be very much larger than α' and β', respectively. In fact, experiments have shown that for typical values of α' and β' (i.e., 0.1, 0.05, 0.01) the differences between α and α' and β and β' are small enough to be considered negligible.

Example 12.35   Suppose a population has a $N(\mu, 1)$ distribution. We wish to derive a sequential probability ratio test for

$$H_0 : \mu = \mu_0 \quad \text{vs.} \quad H_1 : \mu = \mu_1,$$

where $\mu_0 > \mu_1$. Assume that $\alpha = \beta = 0.1$. The likelihood functions are

$$L_0(n) = (2\pi)^{-n/2} \exp\left[-\frac{1}{2} \sum_{i=1}^{n} (x_i - \mu_0)^2\right], \quad \text{and}$$

$$L_1(n) = (2\pi)^{-n/2} \exp\left[-\frac{1}{2} \sum_{i=1}^{n} (x_i - \mu_1)^2\right].$$

Forming the sequential probability ratio and simplifying gives that

$$\frac{L_0(n)}{L_1(n)} = \exp\left[n\bar{x}_n(\mu_1 - \mu_0)] - \frac{1}{2}n(\mu_0^2 - \mu_1^2)\right].$$

Here $\bar{x}_n$ represents the sample mean for n items. With $\alpha = \beta = 0.1$ we find that $a = \frac{1}{9}$ and $b = 9$. Thus, the sequential probability ratio test keeps sampling as long as

$$\frac{1}{9} < \exp\left[n\bar{x}_n(\mu_0 - \mu_1) - \frac{1}{2}n(\mu_0^2 - \mu_1^2)\right] < 9.$$

Since the decision on whether or not to continue sampling depends on the value of the sample mean, we wish to derive cutoff values based on it. We begin by taking logarithms and solving for $\bar{x}_n$ to obtain the following.

$$\frac{1}{2}(\mu_0^2 - \mu_1^2) - \frac{\ln 9}{n(\mu_0 - \mu_1)} < \bar{x}_n < \frac{1}{2}(\mu_0^2 - \mu_1^2) + \frac{\ln 9}{n(\mu_0 - \mu_1)}$$

If we let

$$a_n = \frac{1}{2}(\mu_0^2 - \mu_1^2) - \frac{\ln 9}{n(\mu_0 - \mu_1)}, \quad \text{and}$$

$$b_n = \frac{1}{2}(\mu_0^2 - \mu_1^2) + \frac{\ln 9}{n(\mu_0 - \mu_1)}$$

then the sequential probability ratio test will

1. accept H_0 if $\bar{x}_n \geq b_n$,
2. reject H_0 if $\bar{x}_n \leq a_n$, and
3. continue sampling if $a_n < \bar{x}_n < b_n$.

$\blacksquare$

Exercises 12.8

1. In the derivation of the sequential probability ratio test show that $b \leq \frac{1-\alpha}{\beta}$.

2. Refer to Exercise 1 of Section 11.6. Use Table 8.1 to conduct a sequential test of

$$H_0 : p_0 = 0.1 \quad \text{vs.} \quad H_1 : p_0 = 0.2,$$

where p_0 is the proportion of zeros. Let $\alpha = \beta = 0.1$.

3. Suppose we have a $G(1, \lambda)$ population.
   a. Show that a sequential test for

   $$H_0 : \lambda = 1 \quad \text{vs.} \quad H_1 : \lambda = 0.1$$

   can be based on $\sum_{i=1}^{n} x_i$.
   b. Find a_n and b_n when $\alpha = 0.1$ and $\beta = 0.2$.

4. Suppose we have a $N(0, \sigma^2)$ population.
   a. Show that a sequential test for

   $$H_0 : \sigma^2 = 1 \quad \text{vs.} \quad H_1 : \sigma^2 = 2$$

   can be based on $\sum_{i=1}^{n} x_i^2$.
   b. Find a_n and b_n when $\alpha = 0.01$ and $\beta = 0.05$.

5. Suppose we have a distribution whose p.d.f. if

   $$f(x) = \begin{cases} e^{-(x-\theta)}, & \theta \leq x, \quad \text{and} \\ 0, & \text{otherwise.} \end{cases}$$

   Show that a sequential test for

   $$H_0 : \theta = 4 \quad \text{vs.} \quad H_1 : \theta = 2$$

   can be based on $Y_1 = \min(X_1, \ldots, X_n)$.

6. Suppose we have a geometric distribution.

   a. Show that a sequential test for

$$H_0 : p = 0.6 \quad \text{vs.} \quad H_1 : p = 0.3$$

   can be based on $\sum_{i=1}^{n} x_i$.

   b. Find a_n and b_n for $\alpha = \beta = 0.01$.

12.9
Nonparametric Tests

Until now we have been concerned with testing hypotheses regarding the value of a parameter of a population distribution. The Neyman-Pearson and likelihood ratio procedures are all aimed at this type of problem. There does exist, however, a class of tests that are directed at testing hypotheses regarding population properties without making any connection between that property and any parameter of the population distribution. These are known as *nonparametric* or *distribution-free* tests. They are based on the idea that no assumption is made that the population distribution must be a member of any specified family of distributions. These procedures work regardless of what the actual distribution is. This gives rise to the term *distribution-free*. For many parametric tests there are nonparametric counterparts.

If a testing question can be answered by using a nonparametric test, and if such a test does not require that we make any, possibly erroneous, assumptions about the population distribution, then why would we ever want to use a parametric test? The reason is power. If a parametric test is applied to a sample from the assumed distribution, its power will always exceed that of its nonparametric counterpart. On the other hand certain parametric tests are extremely dependent on the underlying population distribution. That is, the distribution of the test statistic is very sensitive to small deviations from the assumed distribution. This means that a parametric test with an assumed significance level when, say, the population is normal may have an actual significance level that is very different if the population happens to be something else. Tests that are relatively insensitive to departures from the assumed distribution are said to be *robust*. The test on the mean of a normal distribution based on the t-statistic, for example, has been found to be quite robust against departures from normality. This, however, is not always the case. In addition, even the distribution of the t-statistic is not totally insensitive to departures from normality. For these reasons tests that do not require any specific family of distributions are of practical interest.

In this section, we shall briefly discuss three tests on the population median: the *sign test,* the *Wilcoxon Signed Rank test,* and the *Wilcoxon Rank Sum test.* These illustrate many of the ideas that underlie nonparametric testing in general. Discussions of tests for other situations can be found in books such as *Nonparametrics: Statistical Methods Based on Ranks* by E. L. Lehmann and *Nonparametric Statistical Methods* by M. Hollander and D. A. Wolfe.

12.9.1 Sign Test

Recall that the median of a distribution is that point, $\widetilde{\mu}$, where

$$P(X < \widetilde{\mu}) \leq 0.5 \leq P(X \leq \widetilde{\mu}).$$

We consider the problem of testing the population median $\widetilde{\mu}$.

$$H_0 : \widetilde{\mu} = \widetilde{\mu}_0 \quad \text{vs.} \quad H_1 : \widetilde{\mu} \neq \widetilde{\mu}_0$$

For simplicity we shall assume that the underlying population distribution is continuous. Then, we can say with certainty that the median satisfies

$$P(X \leq \widetilde{\mu}) = \frac{1}{2}.$$

The *sign test* is based on the definition of the median. It proceeds as follows.

1. If an observation is less than $\widetilde{\mu}_0$, give it a minus sign.
2. If an observation is greater than $\widetilde{\mu}_0$, give it a plus sign.
3. If an observation is equal to $\widetilde{\mu}_0$, disregard it.
4. Let n be the number of observations that were given a sign.
5. Let R^+ be the number of observations with plus signs.

R^+ will be our test statistic. If $\widetilde{\mu}_0$ is the actual median, then R^+ should have a value near $\frac{n}{2}$. Thus, we would tend to reject H_0 if R^+ is much larger or much smaller than $\frac{n}{2}$. Therefore, the critical region has a form that will reject H_0 if

$$R^+ \leq a \quad \text{or} \quad R^+ \geq b, \quad \text{where } P(R^+ \leq a) + P(R^+ \geq b) = \alpha.$$

To determine values for a and b to achieve a specified significance level, we need to determine the distribution of R^+. This turns out to be relatively easy. Each of the n observations used to compute R^+ was given either a plus sign or a minus sign. If H_0 is true, then the probability that an observation selected at random will receive a plus sign is $\frac{1}{2}$. This is because the probability that an observation will be greater than the median is $\frac{1}{2}$. Thus, the sample can be viewed as n independent trials, the outcome of each being either a plus sign or a minus sign, and the probability of a plus sign is $\frac{1}{2}$. This means that, when H_0 is true, R^+ will have a binomial distribution with parameters n and $p = \frac{1}{2}$. Therefore, we can determine a critical region such that

$$P(R^+ \leq a) = \sum_{i=0}^{a} \binom{n}{i} \left(\frac{1}{2}\right)^n = \frac{\alpha}{2}, \quad \text{and}$$

$$P(R^+ \geq b) = \sum_{i=b}^{n} \binom{n}{i} \left(\frac{1}{2}\right)^n = \frac{\alpha}{2}.$$

If $\frac{\alpha}{2}$ cannot be achieved exactly, common practice is to choose a and b so that the probabilities come as close to $\frac{\alpha}{2}$ as possible. In those cases where it must be assured that the actual significance level does not exceed α, then choose a and b such that the probabilities are as close as possible to $\frac{\alpha}{2}$ without exceeding it. We illustrate this test with an example.

Example 12.36   An archeologist has uncovered a site containing 9 thigh bones, and she wishes to determine if they might belong to a certain ancient tribe that is known to have a median thigh bone length of 33 cm. The 9 thigh bones have the following lenghts in cm.

$$31 \quad 30 \quad 33 \quad 34 \quad 32 \quad 31 \quad 30 \quad 32 \quad 29$$

We shall use the sign test with $\alpha = 0.05$ to test

$$H_0 : \tilde{\mu} = 33 \text{ cm} \quad \text{vs.} \quad H_1 : \tilde{\mu} \neq 33 \text{ cm.}$$

We allocate signs as follows. An asterisk means that the observation is equal to $\tilde{\mu}_0$ and is disregarded.

Observation	Sign
31	−
30	−
33	*
34	+
32	−
31	−
30	−
32	−
29	−

Therefore, $n = 8$ and $R^+ = 1$. Using Table 1 in Appendix A with these values we find

$$P(R^+ = 0) = P(R^+ = 8) = 0.0039, \quad \text{and}$$

$$P(R^+ \leq 1) = P(R^+ \geq 7) = 0.0351.$$

Since 0.0351 is closest to $\frac{\alpha}{2} = 0.025$ a critical region rejects the null hypothesis if

$$R^+ \leq 1 \quad \text{or} \quad R^+ \geq 7$$

with an actual significance level of 0.0702. Since $R^+ = 1$ we reject H_0 and conclude that the thigh bones do not provide supporting evidence that the site belongs to the ancient tribe.   ■

The critical regions for the one-sided alternatives are obtained in the same manner as we have just described. We present them in Table 12.10.

Table 12.10 $H_0: \tilde{\mu} = \tilde{\mu}_0$

Alternative Hypothesis	Critical Region
$H_1 : \tilde{\mu} > \tilde{\mu}_0$	$R^+ \geq a$, where $\sum_{i=a}^{n} \binom{n}{i} \left(\frac{1}{2}\right)^n = \alpha$
$H_1 : \tilde{\mu} < \tilde{\mu}_0$	$R^+ \leq a$, where $\sum_{i=0}^{a} \binom{n}{i} \left(\frac{1}{2}\right)^n = \alpha$
$H_1 : \tilde{\mu} \neq \tilde{\mu}_0$	$R^+ \leq a$, where $\sum_{i=0}^{a} \binom{n}{i} \left(\frac{1}{2}\right)^n = \frac{\alpha}{2}$, or $R^+ \geq b$, where $\sum_{i=b}^{n} \binom{n}{i} \left(\frac{1}{2}\right)^n = \frac{\alpha}{2}$

When n is large we can use the De Moivre-Laplace theorem to get a test statistic that has a normal distribution. According to the theorem, when H_0 is true and n is large R^+ will have an approximate normal distribution with a mean of $np = \frac{n}{2}$ and a variance of $np(1 - p) = \frac{n}{4}$. This means that

$$Z = \frac{R^+ - \dfrac{n}{2}}{\sqrt{n/4}} = \frac{(2R^+ - n)}{\sqrt{n}}$$

will have an approximate standard normal distribution. Since this is a two-tailed test a large sample critical region would reject H_0 if

$$|z| \geq z_{\alpha/2}.$$

The De Moivre-Laplace theorem gives a good approximation whenever $np \geq 5$ and $n(1 - p) \geq 5$. This means that we can use the large sample test whenever $n \geq 10$. The critical regions for the large sample sign test are summarized in Table 12.11.

Table 12.11 $H_0: \tilde{\mu} = \tilde{\mu}_0$ $Z = \dfrac{2R^+ - n}{\sqrt{n}}$

Alternative Hypothesis	Critical Region
$H_1 : \tilde{\mu} > \tilde{\mu}_0$	$z \geq z_\alpha$
$H_1 : \tilde{\mu} < \tilde{\mu}_0$	$z \leq -z_\alpha$
$H_1 : \tilde{\mu} \neq \tilde{\mu}_0$	$|z| \geq z_{\alpha/2}$

Example 12.37   A company claims that providing background music will increase worker satisfaction. Fifteen workers are selected at random and administered a questionnaire. They are then placed in an environment with background music for 1 month and then readministered the questionnaire. Both sets of responses are rated according

to level of satisfaction. A score of 0 indicates low satisfaction, and a score of 10 indicates high satisfaction. The results are shown below.

Worker	1	2	3	4	5	6	7	8	9	10	11	12	13	14	15
Before	5	5	6	5	5	6	7	8	5	6	7	8	6	6	6
After	6	7	6	6	5	7	8	8	6	5	6	8	7	5	7

This type of experimental procedure is known as *paired comparisons*. We introduced it in the exercises at the end of Section 12.5. Since the same units are measured in each case the "before" and "after" readings are certainly dependent. The way to test such a situation is to consider the difference between the 2 scores for a unit to be a single observation on that unit. We are thus treating the sample as being 15 observations on the population of differences. In this case, if the music had no effect on satisfaction level the median difference would be 0. Therefore, if we subtract the before score from the after score, then we are testing

$$H_0 : \tilde{\mu} = 0 \quad \text{vs.} \quad H_1 : \tilde{\mu} > 0.$$

We shall use the sign test. The differences, "after" − "before," and their corresponding signs are given below. An asterisk indicates that the difference is zero, and we disregard that observation.

Difference	Sign
1	+
2	+
0	*
1	+
0	*
1	+
1	+
0	*
1	+
−1	−
−1	−
0	*
1	+
−1	−
1	+

In this case, $n = 11$, and we can use the large sample statistic. If we let $\alpha = 0.5$ the null hypothesis will be rejected if

$$Z = \frac{(2R^+ - 11)}{\sqrt{11}} \geq 1.64.$$

From the sample we compute that $R^+ = 8$. This gives a value of $z = 1.508$. Since this is not greater than 1.64 we do not reject the null hypothesis. Therefore, the background music does not appear to increase worker satisfaction.   ∎

12.9.2 Wilcoxon Signed Rank Test

The sign test is easy to apply, but it does not use much of the information about the median contained by the data. It reduces the data to indicators that merely say whether or not each observation was above or below the hypothesized median. One way to improve on this is to order the data in some fashion and replace each observation by its rank (1 for the smallest observation, 2 for the next smallest, and so on). Tests based on ranks are quite popular in practice. Most statistical computer packages implement them for a variety of problems.

The *Wilcoxon Signed Rank test* for testing

$$H_0 : \tilde{\mu} = \tilde{\mu}_0 \quad \text{vs.} \quad H_1 : \tilde{\mu} \neq \tilde{\mu}_0$$

uses ranks in the following manner.

1. Compute the absolute differences $y = |x - \tilde{\mu}_0|$ for each observation. If $y = 0$ disregard that observation. With each y associate a plus sign if $x > \tilde{\mu}_0$, and a minus sign if $x < \tilde{\mu}_0$.
2. Let n be the number of observations associated with a sign.
3. Assign each y a value equal to its rank.
4. Let W^+ be the sum of the ranks associated with plus signs.

   Note: If, in Step 3, two y's are equal, then we have what is known as a *tie*. In such cases, we assign each y a rank equal to the average of the ranks each should receive if they had not been tied.

Example 12.38

We shall illustrate these steps for the data in Example 12.35. Recall that, in this example, $\tilde{\mu} = 33$. The results of Steps 1 to 3 are given below.

X	$y = \mid x - 33 \mid$	Sign	Rank
31	2	−	4.5
30	3	−	6.5
33	0	*	*
34	1	+	2
32	1	−	2
31	2	−	4.5
30	3	−	6.5
32	1	−	2
29	4	−	8

Using this we find that $n = 8$ and calculate that $W^+ = 2$.

If $\tilde{\mu}_0$ is the actual median, then the observations should be fairly evenly distributed about it. This means that the sum of the ranks with plus signs should be about the same as the sum of the ranks with minus signs. Thus, if the largest sum is too small or too large we would tend to reject the null hypothesis. This means that a size α critical region would reject H_0 if

$$W^+ \leq a, \quad \text{where } P(W^+ \leq a) = \frac{\alpha}{2}, \quad \text{or}$$

$$W^+ \geq b, \quad \text{where } P(W^+ \geq b) = \frac{\alpha}{2}.$$

The exact distribution of W^+ is extremely complicated and can be evaluated only by enumerating all possible values and counting. For this reason the distribution has been computed for n between 1 and 20 and is given in Table 7 in Appendix A.   ■

Example 12.39   This is a continuation of Example 12.37. For $n = 8$ and $\alpha = 0.05$ Table 7 in Appendix A indicates that we should reject H_0 if $W^+ \leq 4$ or if $W^+ \geq 31$ with an actual size of $\alpha = 0.054$. Since $W^+ = 2$ we reject H_0 and conclude that the data do not support the hypothesis that the thigh bones did not come from a population with a median thigh bone length of 33 cm.   ■

Table 7 can also be used to give critical values for one-sided tests on the median. The analysis is similar to that for the two-sided test. The results are summarized in Table 12.12.

Table 12.12 $H_0: \tilde{\mu} = \tilde{\mu}_0$ (use Table 7 in Appendix A)

Alternative Hypothesis	Critical Region
$H_1 : \tilde{\mu} > \tilde{\mu}_0$	$W^+ \geq a, \quad$ where $P(W^+ \geq a) = \alpha$
$H_1 : \tilde{\mu} < \tilde{\mu}_0$	$W^+ \leq a, \quad$ where $P(W^+ \leq a) = \alpha$
$H_1 : \tilde{\mu} \neq \tilde{\mu}_0$	$W^+ \leq a, \quad$ where $P(W^+ \leq a) = \frac{\alpha}{2}$ or $W^+ \geq b, \quad$ where $P(W^+ \geq b) = \frac{\alpha}{2}$

Example 12.40   In 1988, the federal government reported that the median price for a home in the United States was \$85,300. A sample of 10 houses for sale in the Syracuse, New York area gave the following results.

$$83,900 \quad 69,900 \quad 75,900 \quad 87,900 \quad 114,900$$
$$104,900 \quad 49,900 \quad 92,500 \quad 50,000 \quad 60,000$$

It has been commonly thought that Syracuse housing prices are below the national norm. We wish to see if the data support this idea. That is, if house prices are in thousands of dollars, we wish to test

$$H_0 : \tilde{\mu} = 85.3 \quad \text{vs.} \quad H_1 : \tilde{\mu} < 85.3$$

at the 0.05 level. The ranks are computed as shown in the following table.

X	$y = \lvert x - 85.3 \rvert$	Sign	Rank
83.9	1.4	−	1
69.9	15.4	−	5
75.9	9.4	−	4
87.9	2.6	+	2
114.9	29.6	+	8
104.9	19.6	+	6
49.9	35.4	−	10
92.5	7.2	+	3
50.0	35.3	−	9
60.0	25.3	−	7

In this instance, $n = 10$, and Table 7 indicates that we should reject H_0 if $W^+ \leq 11$ with an actual size of $\alpha = 0.053$. In this case, we compute that $W^+ = 2 + 8 + 6 + 3 = 19$. Since this value is greater than 11 we do not reject the null hypothesis. Thus, we conclude that housing prices in Syracuse are not less than the national median. ∎

Although it is beyond the level of this text it can be shown that, when n is sufficiently large and H_0 is true, the distribution of W^+ is approximately normal with

$$E[W^+] = \frac{1}{4} n(n + 1) \quad \text{and} \quad \text{Var}(W^+) = \frac{n(n + 1)(2n + 1)}{24}.$$

It turns out that this approximation is valid when $n > 20$. This means that, when $n > 20$,

$$Z = \frac{W^+ - \frac{1}{4}n(n + 1)}{\sqrt{n(n + 1)(2n + 1)/24}}$$

will have an approximate standard normal distribution. Large sample critical regions will be as shown in Table 12.13.

Table 12.13 $H_0: \tilde{\mu} = \tilde{\mu}_0$

$$Z = \frac{W^+ - \frac{1}{4}n(n + 1)}{\sqrt{n(n + 1)(2n + 1)/24}}$$

Alternative Hypothesis	Critical Region
$H_1 : \tilde{\mu} > \tilde{\mu}_0$	$z \leq z_\alpha$
$H_1 : \tilde{\mu} < \tilde{\mu}_0$	$z \geq z_\alpha$
$H_1 : \tilde{\mu} \neq \tilde{\mu}_0$	$\lvert z \rvert \geq z_{\alpha/2}$

12.9.3 Wilcoxon Rank Sum Test

When testing the equality of the medians of two independent populations a rank sum test very similar to the Wilcoxon Signed Rank test is quite popular. It is called the *Wilcoxon Rank Sum test*. The idea behind this test is quite straightforward. We shall discuss the problem of testing

$$H_0 : \tilde{\mu}_1 = \tilde{\mu}_2 \quad \text{vs.} \quad H_1 : \tilde{\mu}_1 \neq \tilde{\mu}_2$$

in some detail. Assume that we have n observations from population 1 and m observations from population 2 with $n \leq m$. The procedure is as follows.

1. Combine the two samples into a single sample of $n + m$ observations keeping track of each observation's parent population.
2. Place the $m + n$ observations in ascending order and assign ranks.
3. Sum the ranks for the observations from population 2. Call it R.
4. Let $W = R - \frac{1}{2} m(m + 1)$.

Our test statistic will be W. When H_0 is false we would expect that the value of W would be very large or very small. That is, the observations from population 2 would be generally larger or smaller than those of population 1 and would get the bulk of the large or small ranks, respectively. Thus, we would be inclined to reject the null hypothesis when

$$W \leq a \quad \text{or} \quad W \geq b,$$

where a and b are chosen so that

$$P(W \leq a) = P(W \geq b) = \frac{\alpha}{2}.$$

For small values of n and m the exact distribution of W, when H_0 is true, is tabulated in Table 8 in Appendix A. The reason why the distribution of W rather than that of R is tabulated is space. The smallest value that R can assume is $\frac{1}{2} m(m + 1)$. Therefore, it is convenient to use W since its values always begin at 0. Also it is easily seen that the maximum value for W is nm. An additional measure to keep the size of the table under control is to set up the test such that $m \geq n$. These practical considerations are the reason for our seemingly arbitrary decision to sum the ranks from population 2 and to use W as our test statistic.

The critical regions for the one-sided tests are arrived at in a similar manner. The small sample critical regions for the Wilcoxon Rank Sum test are given in Table 12.14.

Table 12.14 $H_0 : \tilde{\mu}_1 = \tilde{\mu}_2$ (use Table 8 in Appendix A)

Alternative Hypothesis	Critical Region
$H_1 : \tilde{\mu}_1 > \tilde{\mu}_2$	$W \leq a$, where $P(W \leq a) = \alpha$
$H_1 : \tilde{\mu}_1 < \tilde{\mu}_2$	$W \geq a$, where $P(W \geq a) = \alpha$
$H_1 : \tilde{\mu}_1 \neq \tilde{\mu}_2$	$W \leq a$, where $P(W \leq a) = \frac{\alpha}{2}$, or $W \geq b$, where $P(W \geq b) = \frac{\alpha}{2}$

Example 12.41   It is felt that prejudice against women in executive positions causes them to be promoted later than men. Thus, it is conjectured that the median number of years until promotion to vice-president for female junior executives is greater than that of male junior executives. A sample of 6 female vice-presidents and 8 male vice-presidents gives the following results.

Years to Promotion	
Male	**Female**
18	16
20	20
19	22
17	19
14	25
21	21
15	
19	

Since the number of males is larger than the number of females we assign the female executives to population 1 and the male executives to population 2. We wish to test

$$H_0 : \tilde{\mu}_1 = \tilde{\mu}_2 \quad \text{vs.} \quad H_1 : \tilde{\mu}_1 > \tilde{\mu}_2.$$

We shall let $\alpha = 0.05$. Since $n = 6$ and $m = 8$ we find from Table 8 in Appendix A that we should reject the null hypothesis if

$$W \leq 11$$

with an actual significance level of $\alpha = 0.0539$. Combining the 2 samples and assigning ranks gives the following. The male observations have a line drawn over them.

Value	$\overline{14}$	$\overline{15}$	16	$\overline{17}$	$\overline{18}$	$\overline{19}$	$\overline{19}$	19	$\overline{20}$	20	$\overline{21}$	21	22	25
Rank	1	2	3	4	5	7	7	7	9.5	9.5	11.5	11.5	13	14

From this we find that $R = 1 + 2 + 4 + 5 + 7 + 7 + 9.5 + 11.5 = 47$ and

$$W = R - \frac{1}{2} m(m + 1) = 47 - \frac{1}{2}(8)(9) = 11.$$

Since W is in the critical region we reject the null hypothesis and conclude that it does take longer for women than for men to get promoted to higher executive positions. ∎

When the sample sizes are large the distribution of the Wilcoxon Rank Sum test, when H_0 is true, can be approximated by a normal distribution. It can be shown that, when the null hypothesis is true,

$$E[W] = \frac{nm}{2} \quad \text{and} \quad \text{Var}(W) = \frac{nm(n + m + 1)}{12}.$$

Therefore, when the sample sizes are large

$$Z = \frac{W - \dfrac{nm}{2}}{\sqrt{nm(n + m + 1)/12}}$$

will have an approximate standard normal distribution. This approximation has been found to be valid when both n and m are greater than 10. The large sample critical regions are summarized in Table 12.15.

Table 12.15 $H_0: \tilde{\mu}_1 = \tilde{\mu}_2$

$$Z = \frac{W - \dfrac{nm}{2}}{\sqrt{nm(n + m + 1)/12}}$$

Alternative Hypothesis	Critical Region
$H_1 : \tilde{\mu}_1 > \tilde{\mu}_2$	$z \leq z_\alpha$
$H_1 : \tilde{\mu}_1 < \tilde{\mu}_2$	$z \geq z_\alpha$
$H_1 : \tilde{\mu}_1 \neq \tilde{\mu}_2$	$|z| \geq z_{\alpha/2}$

Exercises 12.9

1. Girl Scout cookies are either sold door to door or at tables in shopping centers. In an effort to determine if the median sales depend on where the cookies are sold, 8 Girl Scouts were selected at random. Each spent 1 day selling cookies door to door and 1 day at a table in a local shopping center. The daily sales, in dollars, were as follows.

Girl Scout	Door to Door	Shopping Mall
1	82	114
2	96	102
3	122	100
4	88	96
5	94	92
6	90	86
7	94	116
8	98	104

Using paired comparisons determine whether or not there is a difference in the sales method at the 0.05 level by using

a. the sign test, and

b. the Wilcoxon Signed Rank test.

2. Two methods for determining the serum cholesterol level in human blood were applied to each of 16 blood specimens with the following results.

Sample	Method A	Method B	Sample	Method A	Method B
1	202	204	9	188	189
2	241	241	10	194	196
3	224	224	11	212	215
4	255	256	12	199	199
5	198	200	13	205	204
6	206	208	14	178	179
7	288	291	15	249	247
8	231	228	16	233	234

Use paired comparisons to determine whether there is a difference between the results of the two methods. Let $\alpha = 0.1$.

a. Use the sign test.

b. Use the Wilcoxon Signed Rank test.

c. Assume the differences have a normal distribution, and use the t test from Section 11.5. Does the assumption of normality seem justified?

3. A coffee dispensing machine is supposed to dispense 6 oz servings of coffee. The machine was used at 6 randomly selected times and the output measured. The results, in ounces, are as follows.

$$5.7 \quad 6.0 \quad 5.6 \quad 6.1 \quad 5.8 \quad 5.9$$

Determine whether or not the machine is dispensing less than 6 oz per serving by using the Wilcoxon Signed Rank test at the 0.05 level.

4. A small pilot study was conducted to assess differences between low and high income families. One item of interest was the number of children per family. The data are given below.

Family	Income Level	Number of Children	Family	Income Level	Number of Children
1	Low	3	9	Low	4
2	Low	6	10	High	3
3	High	2	11	High	4
4	Low	1	12	Low	5
5	High	2	13	High	2
6	High	0	14	High	1
7	Low	4	15	Low	4
8	Low	2			

Use the Wilcoxon Rank Sum test at a significance level of 0.05 to test whether the median number of children is the same for low and high income families.

5. Show that, when H_0 is true, the mean and variance of the Wilcoxon Signed Rank test are

$$E[W^+] = \frac{1}{4} n(n + 1), \quad \text{and} \quad \text{Var}(W^+) = \frac{n(n + 1)(2n + 1)}{24}.$$

6. Show that, when H_0 is true, the mean and variance of the Wilcoxon Rank Sum test are

$$E[W] = \frac{nm}{2}, \quad \text{and} \quad \text{Var}(W) = \frac{nm(n + m + 1)}{12}.$$

7. A college makes the claim that the median score for incoming freshmen on a standardized test is 55. A random sample of incoming freshmen for one particular class gave
the following results for this test. The scores have been placed in descending order.

$$72 \quad 70 \quad 68 \quad 68 \quad 65 \quad 62 \quad 61 \quad 59 \quad 58 \quad 55 \quad 53$$
$$53 \quad 49 \quad 46 \quad 44 \quad 41 \quad 41 \quad 36 \quad 35 \quad 32 \quad 30 \quad 29$$

a. Use the sign test with $\alpha = 0.1$ to determine if these data suppport the college's
claim.

b. Use the Wilcoxon Signed Rank test with $\alpha = 0.1$ to determine if these data support
the college's claim.

8. A political scientist conducted a study of the effectiveness of the use of campaign
advertising where the candidate stressed negative aspects of his opponent (a negative
campaign) as compared to advertising where the candidate stressed what he would
do to handle issues of concern to the electorate (a positive campaign). Subjects were
asked to view videotapes of campaign ads and then to complete a questionnaire. The
political scientist then computed an overall score to measure the degree to which the
ad had influenced the subject to vote for the candidate. Higher scores indicate a higher
inclination to vote for the candidate. The data are as follows.

Negative Campaign	65	118	122	72	130	194	77
Positive Campaign	133	163	89	204	65	182	79

Use the Wilcoxon Rank Sum test to determine if the data support a claim that advertising style makes no difference in affecting voter preferences. Use a significance level
of 0.05.

**Chapter
Summary**

All tests of hypotheses have

1. null and alternative hypotheses, and
2. a critical region.

If the population distribution is completely specified when the hypothesis is assumed to be true, then that hypothesis is said to be simple. If not, the
hypothesis is called composite.   **(Section 12.1)**

Types of Errors: The action of rejecting the null hypothesis when it is true is
called a Type I error. The action of accepting the null hypothesis when it is
false is called a Type II error.   **(Section 12.2)**

Power: Let x be a fixed point in the alternative hypothesis. The power of a test,
denoted by η, is

$$\eta = P(\text{Reject } H_0 \mid x \in H_1).$$ **(Section 12.2)**

Power Function: In testing $H_0 : \theta \in \Omega_0$ vs. $H_1 : \theta \in \Omega_1$, the power function,
denoted $\eta(\theta)$, is

$$\eta(\theta) = P(\text{Reject } H_0 \mid \theta), \quad \text{for } \theta \in \Omega_0 \cup \Omega_1.$$ **(Section 12.2)**

Operating Characteristic: The operating characteristic (OC) of a test, denoted $\beta(\theta)$, is

$$\beta(\theta) = P(\text{Accept } H_0 \mid \theta), \quad \text{for } \theta \in \Omega_0 \cup \Omega_1. \qquad \textbf{(Section 12.2)}$$

Best Critical Region: When testing a simple hypothesis against a simple alternative hypothesis at a significance level of α, we say that the critical region C associated with a test statistic T is a best size α critical region if, given any other test statistic U, with a critical region C^*, of size α,

$$P(T \in C \mid H_1 \text{ true}) \geq P(U \in C^* \mid H_1 \text{ true}).$$

A size α test procedure that has a best critical region is said to be a best test of size α.

Given a sample of size n,

1. let $f(x_i \mid H_0)$ be the p.f. or p.d.f. under H_0,
2. let $f(x_i \mid H_1)$ be the p.f. or p.d.f. under H_1,
3. let $L_0 = f(x_1 \mid H_0)f(x_2 \mid H_0)\cdots f(x_n \mid H_0)$ be the likelihood function under H_0,
4. let $L_1 = f(x_1 \mid H_1)f(x_2 \mid H_1)\cdots f(x_n \mid H_1)$ be the likelihood function under H_1. **(Section 12.3)**

Neyman-Pearson Lemma: In testing the simple hypothesis $H_0 : \theta = \theta_0$ against the simple alternative hypothesis $H_1 : \theta = \theta_1$, let k be a fixed number and C be a subset of the sample space such that

1. $\dfrac{L_0}{L_1} \leq k$, if $(x_1, x_2, \ldots, x_n) \in C$,

2. $\dfrac{L_0}{L_1} \geq k$, if $(x_1, x_2, \ldots, x_n) \notin C$, and

3. $P[(X_1, X_2, \ldots, X_n) \in C] = \alpha$. Then C is the best size α critical region for testing H_0 vs. H_1. **(Section 12.3)**

Uniformly Most Powerful Test: In testing the simple hypothesis H_0 against the composite alternative H_1, the critical region C is a *uniformly most powerful critical region of size* α if it is a best critical region of size at most α for testing H_0 against every simple hypothesis in Ω_1. A test that is based on the uniformly most powerful critical region is said to be a uniformly most powerful test of size α for testing H_0 against H_1. **(Section 12.4)**

Monotone Likelihood Ratio: Let X be a random variable with probability density function of probability function $f(x \mid \theta)$, where θ is a scalar parameter. Let $\theta_2 > \theta_1$ be a function of X. The family of distributions is said to have monotone likelihood ratio in $T(x)$ if there exists a function $T(x)$ such that the ratio

$$\frac{f(x \mid \theta_2)}{f(x \mid \theta_1)}$$

is a nondecreasing function of $T(x)$.

Let the family of probability density function or probability functions, $f(x \mid \theta)$ of the random variable X depend on a single unknown parameter θ, and let $f(x \mid \theta)$ have monotone likelihood ratio in $T(x)$. Then for testing

$$H_0 : \theta = \theta_0, \quad \text{vs.} \quad H_1 : \theta > \theta_0$$

there exists a uniformly most powerful test that will

$$\text{Reject } H_0 \text{ if } T(x) \geq C, \quad \text{and}$$
$$\text{Accept } H_0 \text{ if } T(x) \leq C.$$

The constant C is chosen so that

$$\alpha = P[T(x) \geq C \mid \theta = \theta_0].$$

Suppose that $(X_1, \ldots, X_n)$ has a distribution that is a member of the single parameter exponential family. That is, it has a probability density function or probability function of the form

$$f(x_1, \ldots, x_n \mid \theta) = e^{\tau(\theta)T(x_1, \ldots, x_n) + S(x_1, \ldots, x_n) + \eta(\theta)}.$$

If $\tau(\theta)$ is a nondecreasing function of θ, then the family of distributions of $(X_1, \ldots, X_n)$ has monotone likelihood ratio in $T(X_1, \ldots, X_n)$.

Let $X_1, \ldots, X_n$ be a random sample from a distribution that is a member of the single parameter exponential family of the form

$$f(x_1, \ldots, x_n \mid \theta) = e^{\tau(\theta)T(x_1, \ldots, x_n) + S(x_1, \ldots, x_n) + \eta(\theta)}.$$

1. For testing

$$H_0 : \tau(\theta) = \tau(\theta_0) \quad \text{vs.} \quad H_1 : \tau(\theta) > \tau(\theta_0)$$

there exists a uniformly most powerful test that will reject H_0 if

$$T(X_1, \ldots, X_n) \geq C.$$

2. For testing

$$H_0 : \tau(\theta) = \tau(\theta_0) \quad \text{vs.} \quad H_1 : \tau(\theta) < \tau(\theta_0)$$

there exists a uniformly most powerful test that will reject H_0 if

$$T(X_1, \ldots, X_n) \leq C. \qquad \text{(Section 12.4)}$$

Likelihood Ratio Tests:

1. Let $L(\theta)$ be the likelihood function. Let Ω_0 be the set of parameter values under the null hypothesis and Ω_1 be the set of parameter values under the

alternative hypothesis. Further, let $\Omega = \Omega_0 \cup \Omega_1$. Ω is called the *parameter space*. Let L_0 be the likelihood function where the null hypothesis is assumed to be true, and let L be the likelihood function for the entire parameter space.

2. Find the maximum values for L_0 and L. Call them L_0^* and L^*, respectively. This entails replacing all parameters not fixed by the parameter space being assumed for that likelihood function with their maximum likelihood estimators.

3. If the value of L_0^* is much smaller than L^*, then the likelihood function is maximized at a point in Ω_1. This would indicate that the alternative hypothesis is more reasonable for the observed sample data.

4. Let the *likelihood ratio statistic* be defined by

$$\lambda = \frac{L_0^*}{L^*}.$$

Since small values of λ indicate that the alternative hypothesis is more reasonable we reject H_0 if

$$\lambda \leq \lambda_0,$$

where λ_0 is chosen to obtain a size α critical region.

Under the same regularity conditions necessary for maximum likelihood estimators to have an asymptotic normal distribution, $-2 \ln \lambda$ has an asymptotic chi-squared distribution with degrees of freedom equal to the difference between the dimension of Ω and the dimension of Ω_0.

Let $X_1, X_2, \ldots, X_k$ have a multinomial distribution with parameters $n, p_1, p_2, \ldots$ and p_k. Then

$$Q = \sum_{i=1}^{k} \frac{(X_i - np_i)^2}{np_i}$$

has a limiting distribution that is $\chi^2(k-1)$.   (Section 12.5)

Goodness-of-Fit Tests:

1. $H_0 : p_1 = p_{1_0}, p_2 = p_{2_0}, \ldots, p_k = p_{k_0}$,   vs.

$H_1 : p_i \neq p_{i_0}$, for some i.

2. Partition observations into k intervals. Let X_i be the number of observations in interval i.

3. Test statistic is

$$Q = \sum_{i=1}^{k} \frac{(X_i - np_{i_0})^2}{np_{i_0}}.$$

4. If each $np_{i_0} \geq 5$ then reject H_0 if

$$Q \geq \chi^2(k-1)_\alpha.$$ (Section 12.6)

Tests of Independence:

1.
$$H_0 : p_{ij} = p_{i\cdot}\,p_{\cdot j}, \begin{cases} i = 1, 2, \ldots, r, & \text{and} \\ j = 1, 2, \ldots, c, & \text{vs.} \end{cases}$$

$$H_1 : p_{ij} \neq p_{i\cdot}\,p_{\cdot j}, \quad \text{for some } i, j.$$

2. Test statistic:

$$Q = \sum_{i=1}^{r} \sum_{j=1}^{c} \frac{\left(X_{ij} - \dfrac{n_{i\cdot}n_{\cdot j}}{n}\right)^2}{\dfrac{n_{i\cdot}n_{\cdot j}}{n}}.$$

3. If all of the $\dfrac{n_{i\cdot}n_{\cdot j}}{n} \geq 5$ then reject H_0 when

$$Q \geq \chi^2([r-1][c-1])_{\alpha}. \qquad \text{(Section 12.6)}$$

Large Sample Tests:

1. Let $\widehat{\theta}$ be an estimator that has an asymptotic distribution that is normal.
2. Test statistic

$$z = \frac{\widehat{\theta} - \theta_0}{\sqrt{\text{Var}(\widehat{\theta_0})}}.$$

3. Reject the null hypothesis of

$$H_0 : \theta = \theta_0$$

when we have the following.

Alternative	Critical Region
$\theta > \theta_0$	$z \geq z_{\alpha}$
$\theta < \theta_0$	$z \leq -z_{\alpha}$
$\theta \neq \theta_0$	$|z| \geq z_{\alpha/2}$

(Section 12.7)

Sequential Probability Ratio Test: When testing two simple hypotheses

$$H_0 : \theta = \theta_0 \quad \text{vs.} \quad H_1 : \theta = \theta_1,$$

1. Let $L_0(n)$ and $L_1(n)$ denote the likelihood functions for a sample of size n under the null and alternative hypotheses, respectively.
2. Let a and b be 2 positive constants with $a < b$ defined by

$$a = \frac{\alpha}{1 - \beta}, \quad \text{and}$$

$$b = \frac{1 - \alpha}{\beta}.$$

3. Test procedure

$$\text{If } \frac{L_0(n)}{L_1(n)} \leq a, \quad \text{stop and reject } H_0.$$

$$\text{If } \frac{L_0(n)}{L_1(n)} \geq b, \quad \text{stop and accept } H_0.$$

$$\text{If } a < \frac{L_0(n)}{L_1(n)} < b, \quad \text{continue sampling.}$$

(Section 12.8)

Sign Test:

$$H_0 : \tilde{\mu} = \tilde{\mu}_0 \quad \text{vs.} \quad H_1 : \tilde{\mu} \neq \tilde{\mu}_0.$$

Given a sample of m observations,

1. if an observation is less than $\tilde{\mu}_0$, give it a minus sign;
2. if an observation is greater than $\tilde{\mu}_0$, give it a plus sign;
3. if an observation is equal to $\tilde{\mu}_0$, disregard it;
4. let n be the number of observations that were given a sign;
5. let R^+, the number of observations with plus signs, be the test statistic;
6. reject H_0 if

$$R^+ \leq a \quad \text{or} \quad R^+ \geq b, \text{ where } P(R^+ \leq a) + P(R^+ \geq b) = \alpha,$$

where R^+ has a $B(n, 0.5)$ distribution.

When n is large $\sqrt{n}(2R^+ - n)$ has an approximate standard normal distribution.

(Section 12.9)

Wilcoxon Signed Rank Test: For testing

$$H_0 : \tilde{\mu} = \tilde{\mu}_0 \quad \text{vs.} \quad H_1 : \tilde{\mu} \neq \tilde{\mu}_0,$$

do the following.

1. Compute the absolute differences $y = |x - \tilde{\mu}_0|$ for each observation. If $y = 0$ disregard that observation. With each y associate a plus sign if $x > \tilde{\mu}_0$, and a minus sign if $x < \tilde{\mu}_0$.
2. Let n be the number of observations associated with a sign.
3. Assign each y a value equal to its rank.
4. Let W^+ be the sum of the ranks associated with plus signs.
5. Reject H_0 if

$$W^+ \leq a, \text{ where } P(W^+ \leq a) = \frac{\alpha}{2}, \quad \text{or}$$

$$W^+ \geq b, \text{ where } P(W^+ \geq b) = \frac{\alpha}{2}.$$

a and b are obtained from Table 7 in Appendix A.

6. When n is large

$$Z = \frac{W^+ - \dfrac{n(n+1)}{4}}{\sqrt{n(n+1)(2n+1)/24}}$$

has an approximate standard normal distribution.   **(Section 12.9)**

Wilcoxon Rank Sum Test: For testing the equality of the medians of two independent populations

$$H_0 : \tilde{\mu}_1 = \tilde{\mu}_2 \quad \text{vs.} \quad H_1 : \tilde{\mu}_1 \neq \tilde{\mu}_2,$$

assume there are n observations from population 1 and m observations from population 2 with $n \leq m$.

1. Combine the two samples into a single sample of $n + m$ observations, keeping track of each observation's parent population.
2. Place the $m + n$ observations in ascending order and assign ranks.
3. Sum the ranks for the observations from population 2. Call it R.
4. Let $W = R - \frac{1}{2} m(m + 1)$ be the test statistic.
5. Reject H_0 when
$$W \leq a \quad \text{or} \quad W \geq b,$$

where a and b are chosen so that

$$P(W \leq a) = P(W \geq b) = \frac{\alpha}{2}.$$

Obtain a and b from Table 8 in Appendix A.
6. When n is large

$$Z = \frac{W - \dfrac{nm}{2}}{\sqrt{nm(n+m+1)/12}}$$

will have an approximate standard normal distribution.   **(Section 12.9)**

Review Exercises

1. There are 2 boxes. Box I contains 5 red balls and 5 black balls. Box II contains 2 red balls and 8 black balls. A box is selected at random and 2 balls removed without replacement. It is agreed to conclude that Box II was selected if both balls are black. The null hypothesis is that Box II was selected, and the alternative hypothesis is that Box I was selected.

   a. State the null and alternative hypotheses mathematically.
   b. What is the test statistic?
   c. What is the critical region in terms of values of the test statistic?
   d. Calculate α and β.

2. Suppose we have a sample of n observations from a population with a p.d.f. of

$$f(x) = \begin{cases} \theta\, x^{\theta-1} , & 0 \le x \le 1,\ 1 \le \theta, \quad \text{and} \\ 0 , & \text{otherwise.} \end{cases}$$

Find a best critical region for testing

$$H_0 : \theta = \theta_0 \quad \text{vs.} \quad H_1 : \theta = \theta_1,$$

where $\theta_1 > \theta_0$ in terms of
$$T = \prod_{i=1}^{n} x_i.$$

3. Suppose we have a sample of n observations from a population with a p.d.f. of

$$f(x) = \begin{cases} \dfrac{2x}{\theta^2} , & 0 \le x \le \theta,\ 0 < \theta, \quad \text{and} \\ 0 , & \text{otherwise.} \end{cases}$$

Find a uniformly most powerful test of

$$H_0 : \theta = \theta_0 \quad \text{vs.} \quad H_1 : \theta > \theta_0.$$

In terms of $Y_n = \max[X_1, X_2, \dots, X_n]$.

4. Suppose we have a sample of n observations from a population with a p.d.f. of

$$f(x) \begin{cases} \sqrt{\dfrac{2}{\pi\theta}}\, \exp\left(-\dfrac{x^2}{2\theta}\right) , & 0 \le x,\ \theta > 0, \quad \text{and} \\ 0 , & \text{otherwise.} \end{cases}$$

This is the upper half of a normal distribution. Show that a likelihood ratio test of

$$H_0 : \theta = \theta_0 \quad \text{vs.} \quad H_1 : \theta \ne \theta_0$$

rejects H_0 if

$$\sum_{i=1}^{n} x_i^2 \le a \quad \text{or} \quad \sum_{i=1}^{n} x_i^2 \ge b.$$

5. Consider the *goodness-of-fit* test that was discussed in Section 11.6. That is,

$$H_0 : p_1 = p_{1_0},\ p_2 = p_{2_0}, \dots, p_k = p_{k_0}, \quad \text{vs.}$$
$$H_1 : p_i \ne p_{i_0}, \quad \text{for some } i.$$

Show that the likelihood ratio test rejects H_0 whenever

$$\lambda = \prod_{i=1}^{k} \left(\frac{np_{i_0}}{n_i}\right)^{n_i} \le \lambda_0.$$

Note: $L = p_1^{n_1} p_2^{n_2} \dots p_k^{n_k}$ and $n_1 + n_2 + \cdots + n_k = n$.

6. In Exercise 5 show that when n is large a critical region rejects H_0 if $-2 \ln \lambda \ge \chi^2(k-1)_\alpha$.

7. Did the following data come from a population with a Weibull distribution having parameters $\alpha = 1$, $\beta = 1$, and $\theta = 0$?

$$
\begin{array}{cccccccccc}
42 & 15 & 26 & 4 & 32 & 82 & 67 & 9 & 8 & 40 \\
55 & 17 & 43 & 33 & 41 & 44 & 31 & 53 & 84 & 52 \\
33 & 54 & 74 & 87 & 89 & 57 & 49 & 26 & 45 & 23 \\
67 & 32 & 57 & 74 & 9 & 37 & 18 & 34 & 27 & 62
\end{array}
$$

8. Suppose a population has a $Po(\lambda)$ distribution.

   a. Show that a sequential probability ratio test for

   $$H_0 : \lambda = 0.01, \quad \text{vs.} \quad H_1 : \lambda = 0.05$$

   can be based on the statistic $\sum_{i=1}^{n} x_i$.

   b. Find a_n and b_n for $\alpha = 0.1$ and $\beta = 0.05$.

9. Let $X_1, X_2, \ldots, X_n$ be a random sample from a population having a $N(\mu_X, \sigma_X^2)$ distribution and let $Y_1, Y_2, \ldots, Y_m$ be a random sample from an independent $N(\mu_Y, \sigma_Y^2)$ distributed population. Show that a large sample test of

   $$H_0 : \mu_X = \mu_Y \quad \text{vs.} \quad H_1 : \mu_X \neq \mu_Y$$

   can be based on

   $$Z = \frac{\bar{X} - \bar{Y}}{\sqrt{\dfrac{s_X^2}{n} + \dfrac{s_Y^2}{m}}},$$

   where Z has a standard normal distribution and

   $$\bar{X} = \frac{1}{n} \sum_{i=1}^{n} X_i,$$

   $$\bar{Y} = \frac{1}{m} \sum_{i=1}^{m} Y_i,$$

   $$s_X^2 = \frac{1}{n-1} \sum_{i=1}^{n} (X_i - \bar{X})^2 \quad \text{and}$$

   $$s_Y^2 = \frac{1}{m-1} \sum_{i=1}^{m} (Y_i - \bar{Y})^2.$$

10. Students in the sixth grade at the Bear Road School challenged the sixth grade students at the Pebble Hill School to a spelling contest. Six students were randomly selected from Bear Road and 8 were chosen from Pebble Hill. Each student was given a spelling test with the following results. Higher scores are better.

Bear Road	Pebble Hill
89	89
82	79
96	91
87	94
87	84
77	75
	88
	89

Is there a difference in spelling ability between the 2 schools at the 0.05 level? Use the Wilcoxon Rank Sum test.

The following problems may require that you make some assumptions regarding the population distribution and then apply one of the testing procedures developed in either the text or the exercises. In each case

a. specify precisely what assumptions must be made to conduct a parametric test,

b. comment briefly on how you might verify your assumptions, and

c. conduct the parametric test.

11. An investor is trying to determine which of 2 stocks to buy. He decides to watch both stocks for a 20-day period and buy the one with the higher long-term average closing price. If the average closing prices are equal he will invest half his money in each one. He found that the mean closing price of stock A was 104.33 with a standard deviation of 5.56, and the mean closing price of stock B was 145.24 with a standard deviation of 9.33. Should the investor place all of his money in a single stock?

12. The highway department is planning to extend a freeway through a suburban area into a city in an effort to ease rush hour traffic congestion. Residents of the affected suburb and residents of the city were polled as to their opinion of this project with the following results.

	For Project	Against Project	Undecided
City	67	41	12
Suburb	47	55	23

Is there a difference between the preferences in the two regions?

13. A new method for processing federal income tax forms has been proposed that might reduce the time per form and result in reduced costs. To test the system one office processed 100 forms in the current way showing a mean processing time of 23.1 days with a standard deviation of 2.3 days. Another office processed 120 forms in the new manner with a mean time of 21.6 days with a standard deviation of 1.9 days. Is the new system better?

14. An experiment was conducted to study the effects of different types of fertilization (self-fertilization vs. cross-fertilization by bees) on the growth of a particular type of plant. Fifteen plants were subjected to self-fertilization, and 15 plants were cross-fertilized. The results, in inches, are given below.

Cross-Fertilization	Self-Fertilization	Cross-Fertilization	Self-Fertilization
23.50	17.38	18.25	16.50
12.00	20.38	21.63	18.00
21.00	20.00	23.25	16.25
22.00	20.00	21.00	18.00
19.13	18.38	22.13	12.75
21.50	18.63	23.00	15.50
22.13	18.63	12.00	18.00
20.38	15.25		

Is there a difference between the fertilization methods?

15. Some people feel there is a tendency for gas station operators in a certain area to overcharge for repairs that are performed on out-of-state automobiles that stop for gasoline or directions. To check this out the state consumer affairs agency took a car with a mechanical problem whose average repair cost was known to be $75, put on out-of-state license plates, and visited 64 randomly selected gasoline stations. Of those visited, 37 charged more than $75 and 27 charged $75 or less. Does this indicate that the majority of the gasoline stations overcharge?

13 | Regression and Correlation

In this chapter we investigate linear relationships in sample data. Until now we have considered estimating and testing hypotheses about population quantities that have a single value. An example is the mean, μ, of a normal population. Now we consider situations where the population quantity to be estimated or tested depends on the value of some other variable. As a simple example consider a situation where the stopping distance of an automobile is being measured. This distance will depend on a number of uncontrollable factors such as road surface, tire tread, and ambient conditions. This means that the distance it takes a car to go from a speed of 50 miles per hour to a complete stop will be a random variable centered about some mean value. However, it is common knowledge that it takes a greater distance to stop a car going 60 miles per hour than it does to stop a car going 25 miles per hour. Statistically, we would interpret this situation to be that the mean stopping distance depends on the speed at which the automobile was originally traveling. More specifically, if Y is the stopping distance for an automobile and X is the original speed of the automobile, we would expect that $E[Y] = f(X)$, where $f(\ )$ is some function. Since the mean value of Y depends in some way on X it is customary to refer to Y as being the *dependent variable* and X as being the *independent variable*. Here the term "dependent" refers to functional dependence not stochastic dependence. The particular functional relationship between $E[Y]$ and X that will interest us is a linear one.

This linear relationship will be considered from two distinct viewpoints. If we assume that the values of X are not random variables, but are fixed, then we have what is known as a *linear regression* problem. In this problem, the values of Y for a given value of x are randomly distributed about $E[Y] = \mu_{Y|x} = \alpha + \beta x$. If we let e be a random variable such that $E[e] = 0$ and $\text{Var}(e) = \sigma^2$ then this problem can be modeled by

$$Y = \alpha + \beta x + e.$$

Note that $\mu_{Y|x} = E[\alpha + \beta x + e] = E[\alpha] + E[\beta x] + E[e] = \alpha + \beta x$. This model is called a *linear model*. Such a model is appropriate in our stopping distance example. In a stopping distance experiment it is common practice for automobiles to be driven at predetermined speeds such as multiples of 10 miles per hour. This means that X will not be a random variable. The determination of the coefficients in the model and statistical inferences concerning them is called *regression analysis*. The term "regression" was coined by Francis Galton in his work *Natural Inheritance,* which was published in 1889. In that work he noted that although tall men tended to have tall sons, the average height of the sons was less than that of their fathers. That is, there appeared to be a *regression,* or going back, of the son's heights toward the average height for all men.

The second situation of interest is called a *correlation* problem. Here *both X and Y* are random variables. In this case we have a model of the form

$$E[Y] = \alpha + \beta E[X].$$

From an experimental viewpoint this means that we are observing random vectors (X, Y) drawn from some bivariate population. A simple example of a correlation experiment would be one where we wish to determine if there is a relation between the height and weight of people. An investigator would typically collect a random sample of people and measure both quantities rather than select a set of heights for people and then take sample observations from only people with those heights.

In linear regression, investigators are first interested in determining if x and Y are linearly related. If so, we wish to estimate the values for α and β or test hypotheses about α and β. The quantity of interest in correlation studies is the correlation between X and Y, $\rho_{XY} = E[(Y - \mu_Y)(X - \mu_X)]$, which we discussed in Chapter 4. Typically, we seek estimators for ρ_{XY} or wish to test hypotheses about the value of ρ_{XY}.

In this chapter we begin by investigating the ideas behind the regression problem in the case where there is a single independent variable. This problem is commonly referred to as *simple regression*. This will permit us to introduce most of the important ideas in a setting that will not obscure what is going on. We shall then go on to consider the case where there are many independent variables. This more general case is usually called *multiple regression*. This can be further generalized to the so-called *multivariate regression* problem where there are many independent and many dependent variables, but we shall leave discussion of this to courses in multivariate statistical analysis. Following our discussion of regression problems we shall consider the correlation problem.

13.2
Simple Linear Regression

The situation we are considering begins with a sample of n observations of the form $(x_1, y_1), (x_2, y_2), \ldots, (x_n, y_n)$, where x is the independent variable and Y is the dependent variable. The first thing to be determined is whether or not a straight line seems to describe the general relationship between the two variables. The primary descriptive technique for determining the form of relationship between X and Y is the *scatter diagram*. A scatter diagram is constructed by simply plotting the sample observations in Cartesian coordinates. The general pattern of the points gives an indication of the type of relation, if any, that might be appropriate. Some examples of scatter diagrams are given in Figure 13.1.

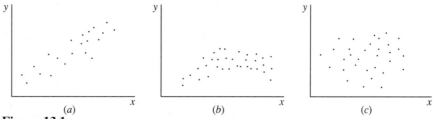

Figure 13.1

Part *a* shows a scatter diagram where the relation between *X* and *Y* is linear with a positive slope. Part *b* indicates that *X* and *Y* are related, but the form is more like a higher degree polynomial than a straight line. Part *c* is typical for a scatter diagram when there is no relation between *X* and *Y*.

Example 13.1

An employer is considering adopting an aptitude test for screening applicants for management trainee positions. To evaluate the usefulness of the examination, 10 randomly selected new management trainees are given the test. The new employees are then evaluated by their supervisors after 1 year. The results are given below.

Test (X)	20	25	30	35	40	44	48	52	55	67
Eval (Y)	53	66	52	63	66	72	78	69	75	73

The scatter diagram of these points is shown in Figure 13.2. The points appear generally to follow a linear trend with a slightly positive slope as is shown by the straight line that is superimposed on the plot.

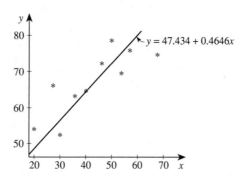

Figure 13.2

Once we have decided that a straight line adequately describes the relation between the independent and dependent variables, the problem becomes one of finding estimators for α and β. We shall denote $\widehat{\alpha}$ by a and $\widehat{\beta}$ by b. This is a departure from the custom set forth in Chapter 10, but it is standard practice in the regression setting. There are a number of criteria that can be used to determine the slope and intercept of a linear equation that "best" fits the observed scatter diagram. The one that we shall use is by far the most widely used. The criterion is known as *least squares*. It has the advantage of being easy to apply. In addition, when the regression model assumes that the error term e has a normal distribution, the estimators derived using least squares turn out also to be the maximum likelihood estimators for the problem. The reader may have encountered the least squares concept in a linear algebra course where it is often discussed as a practical application of matrices.

We begin by assuming that we have a sample of n observations,

$$(x_1, y_1), (x_2, y_2), \ldots, (x_n, y_n).$$

Now we consider a linear equation of the form $y = a + bx$, and let $\widehat{y}_i = a + bx_i$, for $i = 1, 2, \ldots, n$. The least squares criterion states that we wish to find the slope, b, and intercept, a, that minimizes

$$
\begin{aligned}
S &= \sum_{i=1}^{n} (y_i - \widehat{y}_i)^2 \\
&= \sum_{i=1}^{n} (y_i - a - bx_i)^2.
\end{aligned}
\tag{13.1}
$$

These will be the *least squares estimators* for α and β. Since this is a minimization problem in 2 variables, we take partial derivatives with respect to a and b and equate each to 0 as follows.

$$\frac{\partial S}{\partial a} = 2na - 2n\bar{y} + 2nb\bar{x} = 0$$

$$\frac{\partial S}{\partial b} = 2b\sum_{i=1}^{n} x_i^2 - 2\sum_{i=1}^{n} x_i y_i + 2na\bar{x} = 0$$

Solving these equations simultaneously for a and b gives

$$a = \bar{y} - b\bar{x}, \quad \text{and}$$

$$b = \frac{\displaystyle\sum_{i=1}^{n} x_i y_i - n\bar{x}\bar{y}}{\displaystyle\sum_{i=1}^{n} x_i^2 - n\bar{x}^2}.
\tag{13.2}$$

We leave it to the reader to verify that these values of a and b do indeed minimize S. It is also interesting to note that the regression equation $\widehat{y} = a + bx$ always passes through the point $(\bar{x}, \bar{y})$. Later we shall see that estimates for Y when $x = \bar{x}$ have the highest accuracy.

Example 13.2

We wish to determine the least squares estimates for a and b for the data in Example 13.1. For those data we compute

$$n = 10,$$

$$\bar{x} = 41.6,$$

$$\bar{y} = 66.7,$$

$$\sum_{i=1}^{n} x_i^2 = 19208, \quad \text{and}$$

$$\sum_{i=1}^{n} x_i y_i = 28631.$$

From these we use Equation 13.2 to compute

$$b = \frac{28631 - 10(41.6)(66.7)}{19208 - 10(41.6)^2}$$

$$= \frac{883.8}{1902.4} = 0.4646, \quad \text{and}$$

$$a = 66.7 - 0.4646(41.6) = 47.434.$$

Thus our least squares regression line is

$$\hat{y} = 47.434 + 0.4646x.$$

This line is plotted in the scatter diagram of Figure 13.2. We can use this equation to estimate the mean value of Y for various values of x. For example, if $x = 50$ we would estimate that $\hat{y} = 47.434 + 0.4646(50) = 70.664.$ ∎

We now show the equivalence of least squares estimators and maximum likelihood estimators when the error term in the regression model given in Section 13.1 has a normal distribution. To be specific we assume that the observations follow

$$Y_i = \alpha + \beta x_i + e_i, \quad \text{where} \quad \begin{cases} e_i \sim N(0, \sigma^2), \\ \text{for } i = 1, 2, \ldots, n, \quad \text{and} \\ e_i \text{ and } e_j \text{ are uncorrelated if } i \neq j. \end{cases} \quad \textbf{(13.3)}$$

These assumptions have the following effects.

1. Since uncorrelated normal random variables are independent, $Y_1, Y_2, \ldots, Y_n$ are independent random variables.

2. Since each e_i has a common variance of σ^2, the dispersion about the line $E[Y_i] = \alpha + \beta x_i$ is the same regardless of the value of x. This property is referred to as *homoscedasticity*.

3. Each $Y_i \sim N(\alpha + \beta x_i, \sigma^2)$, for $i = 1, 2, \ldots, n$.

From the facts just listed we can see that the likelihood function for this problem will be

$$L(\alpha, \beta) = \prod_{i=1}^{n} \frac{1}{\sqrt{2\pi\sigma^2}} e^{-[y_i - (\alpha + \beta x_i)]^2 / 2\sigma^2}$$

$$= (2\pi\sigma^2)^{-n/2} e^{-\sum_{i=1}^{n}(y_i - \alpha - \beta x_i)^2 / 2\sigma^2}.$$

Note that the likelihood function will be a maximum when the exponent is a minimum. The sum in the exponent is precisely the quantity that we were minimizing in the least squares procedure. We shall proceed to compute the estimators for α, β, and σ^2. Taking the natural logarithm gives

$$\ln(L) = -\frac{n}{2}\ln(2\pi) - \frac{n}{2}\ln(\sigma^2) - \frac{1}{2\sigma^2}\sum_{i=1}^{n}(y_i - \alpha - \beta x_i)^2.$$

By comparison with Equation 13.1 we can see that the solutions for $\widehat{\alpha}$ and $\widehat{\beta}$ will be identical to those of a and b from the least squares problem. Thus, we only need to compute $\widehat{\sigma^2}$. We proceed as follows.

$$\frac{\partial}{\partial \sigma^2} \ln(L) = -\frac{n}{2\sigma^2} + \frac{1}{2(\sigma^2)^2}\sum_{i=1}^{n}(y_i - \widehat{\alpha} - \widehat{\beta}x_i)^2 = 0$$

Solving for σ^2 gives that

$$\widehat{\sigma^2} = \frac{1}{n}\sum_{i=1}^{n}(y_i - \widehat{\alpha} - \widehat{\beta}x_i)^2,$$

where

$$\widehat{\alpha} = \bar{y} - \widehat{\beta}\bar{x}, \quad \text{and} \quad \widehat{\beta} = \frac{\displaystyle\sum_{i=1}^{n} x_i y_i - n\bar{x}\bar{y}}{\displaystyle\sum_{i=1}^{n} x_i^2 - n\bar{x}^2}.$$

It can be shown that an unbiased estimator for σ^2 is

$$s^2 = \frac{n}{n-2}\widehat{\sigma^2}. \tag{13.4}$$

To use these estimators to construct confidence intervals or to conduct tests of hypotheses we must determine their properties. We have just demonstrated that the least squares estimators are the maximum likelihood estimators when the regression model assumes a normal distribution. Since maximum likelihood estimators are not always unbiased, we must determine $E[a]$ and $E[b]$. We begin with $E[b]$. Recall that in the regression model, the x's are fixed and $E[Y] = \alpha + \beta x$. This implies that

$$E[\bar{Y}] = \alpha + \beta \bar{x}.$$

We proceed as follows.

$$E[b] = E\left[\frac{\displaystyle\sum_{i=1}^{n} x_i Y_i - n\bar{x}\,\bar{Y}}{\displaystyle\sum_{i=1}^{n} x_i^2 - \bar{x}^2}\right]$$

$$= \frac{\displaystyle\sum_{i=1}^{n} x_i(\alpha + \beta x_i) - n\bar{x}(\alpha + \beta\bar{x})}{\displaystyle\sum_{i=1}^{n} x_i^2 - n\bar{x}^2}$$

$$= \frac{\beta \displaystyle\sum_{i=1}^{n} x_i^2 - n\beta\bar{x}^2}{\displaystyle\sum_{i=1}^{n} x_i^2 - n\bar{x}^2}$$

$$= \beta$$

Thus, b is an unbiased estimator for β. Once we have this result the expected value for a follows easily.

$$E[a] = E[\bar{Y}] - E[b]\bar{x}$$
$$= \alpha + \beta\bar{x} - \beta\bar{x}$$
$$= \alpha$$

Therefore, the least squares estimators for both α and β are unbiased.

We now turn to the variances of a and b. Since the x's are fixed we know that $\text{Var}(x) = 0$. Using the fact that $\text{Var}(\alpha + \beta x) = \beta^2 \text{Var}(x)$ we find that $\text{Var}(Y) = \text{Var}(e) = \sigma^2$. In addition, recall that the Y's are independent. This implies that $\text{Var}(\bar{Y}) = \frac{\sigma^2}{n}$. Thus, we obtain $\text{Var}(b)$ as follows.

$$\text{Var}(b) = \frac{\text{Var}\left(\displaystyle\sum_{i=1}^{n} x_i Y_i - n\bar{x}\,\bar{Y}\right)}{\left(\displaystyle\sum_{i=1}^{n} x_i^2 - n\bar{x}^2\right)^2}$$

$$= \frac{\displaystyle\sum_{i=1}^{n} x_i^2 \,\text{Var}(Y_i) - 2n\bar{x}\,\text{Cov}\left(\displaystyle\sum_{i=1}^{n} x_i Y_i, \bar{Y}\right) + n^2\bar{x}^2 \,\text{Var}(\bar{Y})}{\left(\displaystyle\sum_{i=1}^{n} x_i^2 - n\bar{x}^2\right)^2}$$

For the covariance term in the numerator we find the following.

$$\text{Cov}\left(\sum_{i=1}^{n} x_i Y_i, \bar{Y}\right) = \sum_{i=1}^{n} x_i \, \text{Cov}(Y_i, \bar{Y})$$

$$= \sum_{i=1}^{n} x_i \frac{1}{n} \sum_{j=1}^{n} \text{Cov}(Y_i, Y_j)$$

$$= \sum_{i=1}^{n} \frac{1}{n} x_i \sigma^2$$

$$= \bar{x} \sigma^2$$

Thus, we find that

$$\text{Var}(b) = \frac{\displaystyle\sum_{i=1}^{n} x_i^2 \sigma^2 - 2n\bar{x}^2 \sigma^2 + n\bar{x}^2 \sigma^2}{\left(\displaystyle\sum_{i=1}^{n} x_i^2 - n\bar{x}^2\right)^2}$$

$$= \frac{\sigma^2}{\displaystyle\sum_{i=1}^{n} x_i^2 - n\bar{x}^2}.$$

Using this result we obtain Var(a) as follows.

$$\text{Var}(a) = \text{Var}(\bar{Y} - b\bar{x})$$

$$= \text{Var}(\bar{Y}) + \bar{x}^2 \, \text{Var}(b) - 2\bar{x} \, \text{Cov}(\bar{Y}, b)$$

$$= \frac{\sigma^2}{n} + \bar{x}^2 \left[\frac{\sigma^2}{\displaystyle\sum_{i=1}^{n} x_i - n\bar{x}^2}\right] - 2 \, \text{Cov}(\bar{Y}, b)$$

We need to find $\text{Cov}(\bar{Y}, b)$. We begin by rewriting b in a more convenient manner.

$$b = \frac{\displaystyle\sum_{i=1}^{n} x_i Y_i - n\bar{x}\bar{Y}}{\displaystyle\sum_{i=1}^{n} x_i^2 - n\bar{x}^2}$$

$$= \frac{\displaystyle\sum_{i=1}^{n} Y_i(x_i - \bar{x})}{\displaystyle\sum_{i=1}^{n} x_i^2 - n\bar{x}^2}$$

$$= \sum_{i=1}^{n} k_i Y_i$$

Using this form we proceed as follows.

$$\text{Cov}(\bar{Y}, b) = \text{Cov}\left(\sum_{i=1}^{n} \frac{1}{n} Y_i, \sum_{i=1}^{n} k_i Y_i\right)$$

$$= \sum_{i=1}^{n} \frac{k_i}{n} \text{Var}(Y_i) + 2 \sum_{i=1}^{n-1} \sum_{j=i+1}^{n} \frac{k_j}{n} \text{Cov}(Y_i, Y_j)$$

Now $\text{Cov}(Y_i, Y_j) = 0$, when $i \neq j$, and

$$\sum_{j=1}^{n} \frac{k_j}{n} \text{Var}(Y_j) = \frac{\sigma^2}{n} \sum_{j=1}^{n} \frac{x_j - \bar{x}}{\sum_{i=1}^{n} x_i^2 - n\bar{x}^2}$$

$$= \frac{\sigma^2}{n} \frac{n\bar{x} - n\bar{x}}{\sum_{i=1}^{n} x_i^2 - n\bar{x}^2}$$

$$= 0.$$

Thus, $\text{Cov}(\bar{Y}, b) = 0$, and

$$\text{Var}(a) = \frac{\sigma^2}{n} + \frac{\bar{x}^2 \sigma^2}{\sum_{i=1}^{n} x_i^2 - n\bar{x}^2}$$

$$= \sigma^2 \left[\frac{1}{n} + \frac{\bar{x}^2}{\sum_{i=1}^{n} x_i^2 - n\bar{x}^2}\right]$$

$$= \frac{\sigma^2 \sum_{i=1}^{n} x_i^2}{n\left(\sum_{i=1}^{n} x_i^2 - n\bar{x}^2\right)}.$$

We leave it as an exercise to show that

$$\text{Cov}(a, b) = \frac{-\sigma^2 \bar{x}}{\sum_{i=1}^{n} x_i^2 - n\bar{x}^2}.$$

Using the results given thus far the problem of determining the distributions of a and b is not too difficult. First recall that the distribution of Y_i is $N(\alpha + \beta x_i, \sigma^2)$. Using this fact, Theorem 9.1 and the fact that we can write b as a linear combination of the Y's shows that b has a normal distribution. In addition, since $a = \bar{Y} - b\bar{x}$, we see that a is also normally distributed. However, the construction of confidence intervals and testing of hypotheses regarding α and β requires a more detailed investigation of the joint distribution of a, b, and $\widehat{\sigma^2}$. We summarize the results in the following theorem. The proof is omitted. The interested reader can find a proof using linear algebra in *An Introduction to Linear Statistical Models,* vol. 1, by Franklin A. Graybill.

Theorem 13.1

For the linear model defined by Equation 13.3 the maximum likelihood estimators for α, β, and σ^2 are as given in Equation 13.2 and Equation 13.4. In addition,

1. a and b have a joint bivariate normal distribution with parameters

$$\mu_a = \alpha,$$

$$\mu_b = \beta,$$

$$\sigma_a^2 = \frac{\sigma^2 \sum_{i=1}^{n} x_i^2}{n \left(\sum_{i=1}^{n} x_i^2 - n\bar{x}^2 \right)},$$

$$\sigma_b^2 = \frac{\sigma^2}{\sum_{i=1}^{n} x_i^2 - n\bar{x}^2}, \quad \text{and}$$

$$\rho = \frac{-\bar{x}}{\sqrt{\frac{1}{n} \sum_{i=1}^{n} x_i^2}}.$$

2. $\frac{n\widehat{\sigma^2}}{\sigma^2}$ has a chi-square distribution with $n - 2$ degrees of freedom.

3. a and b are independent of $\widehat{\sigma^2}$.

Note: It also turns out that these maximum likelihood estimators are uniform minimum variance unbiased, consistent, and asymptotically efficient. The demonstration of these facts is beyond the level of this text.

Theorem 13.1 gives us a basis for testing hypotheses concerning α and β. We shall consider β. In either case the arguments are similar. Let

$$(n - 1)s_X^2 = \sum_{i=1}^{n} x_i^2 - n\bar{x}^2.$$

Theorem 13.1 says that b has a normal distribution with the mean and variance as given. Therefore,

$$z = \frac{b - \beta}{\sqrt{\dfrac{\sigma^2}{(n - 1)s_X^2}}}.$$

will have a $N(0, 1)$ distribution. If the value of σ^2 is unknown we may then use $\widehat{\sigma^2}$ to obtain a test statistic that will have a t distribution as follows.

$$t = \frac{(b - \beta)\Big/\sqrt{\dfrac{\sigma^2}{(n-1)s_X^2}}}{\sqrt{\dfrac{\widehat{\sigma^2}}{\sigma^2}}} \tag{13.5}$$

$$= (b - \beta)\sqrt{\frac{(n-1)s_X^2}{s^2}}$$

Since this is the ratio of a $N(0, 1)$ random variable to the square root of an independent $\chi^2(n - 2)$ that is divided by its degrees of freedom, it follows from Definition 9.1 that t will have a Student's t distribution with $n - 2$ degrees of freedom.

Example 13.3

We wish to use the data in Example 12.1 to test

$$H_0 : \beta = 0 \quad \text{vs.} \quad H_1 : \beta \neq 0$$

with a significance level of $\alpha = 0.05$. This is the test that is commonly run to determine if the independent variable is of any use in predicting values for the dependent variable. When the null hypothesis is true the preceding discussion indicates that

$$t = b\sqrt{\frac{(n-1)s_X^2}{s^2}}$$

will have a t distribution with $n - 2$ degrees of freedom. Since this is a two-sided test we would reject the null hypothesis if we observe $|t| \geq t(n - 2)_{\alpha/2}$. For this particular case $n = 10$, and using Table 5 in Appendix A we would reject H_0 if $|t| \geq t(8)_{0.025} = 2.306$. Recall from Example 13.2 that

$$\bar{x} = 41.6$$

$$\sum_{i=1}^{n} x_i^2 = 19208$$

$$a = 47.434, \quad \text{and}$$

$$b = 0.4640.$$

From the data we calculate

$$(n - 2)s^2 = \sum_{i=1}^{n}(y_i - a - bx_i)^2 = 277.5497.$$

Thus Equation 13.5 gives that

$$t = 0.4646\sqrt{\frac{8[19208 - (10)(41.6)^2]}{277.5497}}$$

$$= 0.4646\sqrt{54.834}$$

$$= 3.44.$$

Since the observed value for t exceeds $t(8)_{0.025}$ we reject the null hypotheses and conclude that the sample indicates that there appears to be a relation between aptitude test scores and employee ratings.

Note that we have a test statistic that contains β, its estimator, and whose distribution does not depend on the value of β or any other unknown quantities. This means that we have a pivotal quantity for β. Therefore, we can construct a confidence interval for the value of β by proceeding as explained in Section 11.2.

$$P\left(-t(n-2)_{\alpha/2} \le (b-\beta)\sqrt{\frac{s_X^2}{s^2}} \le t(n-2)_{\alpha/2}\right) = 1 - \alpha$$

Solving for β gives a $(1 - \alpha)100\%$ confidence interval of

$$b - t(n-2)_{\alpha/2}\sqrt{\frac{s^2}{(n-1)s_X^2}} \le \beta \le b + t(n-2)_{\alpha/2}\sqrt{\frac{s^2}{(n-1)s_X^2}}. \tag{13.6}$$

■

Example 13.4   We wish to construct a 99% interval for β for the data in Example 13.1 by using Equation 13.1. Here $\alpha = 0.01$, which gives $t(8)_{0.005} = 3.355$. From the previous example we know that the value beneath the square root is $\frac{1}{54.834} = 0.018$. Thus, a 99% confidence interval for β will be

$$0.4646 - 3.355\sqrt{0.018} \le \beta \le 0.4646 + 3.355\sqrt{0.018}$$

or

$$0.01115 \le \beta \le 0.9177.$$

■

If we use arguments similar to the ones used to derive Equation 13.5 it is left as an exercise to show that

$$t = (a - \alpha)\sqrt{\frac{n(n-1)s_X^2}{\sum_{i=1}^{n} x_i^2 s^2}} \tag{13.7}$$

has a t distribution with $n - 2$ degrees of freedom. This quantity can then be used to construct confidence intervals for α or test hypotheses regarding α.

It is worthwhile to note that Theorem 13.1 indicates that the accuracy of the estimates for α and β can be affected by the experimenter's choices for the values of the independent variable x. First note that

$$\sigma_a^2 = \frac{\sigma^2 \sum_{i=1}^{n} x_i^2}{n(n-1)s_X^2} \geq \frac{\sigma^2}{n}.$$

Therefore, by choosing values of x so that $\bar{x} = 0$ we can produce estimates for α that have the smallest possible variance for a given sample size. As far as estimates for β are concerned recall that

$$\sigma_b^2 = \frac{\sigma^2}{(n-1)s_X^2}.$$

This variance can be made small by making the denominator large. Thus, the variance for b can be reduced by choosing values for x that are as far apart as possible.

We now consider the predictions made by the simple linear regression equation. Recall that the linear equation, $\hat{y} = a + bx$, derived by the least squares method produces estimates for $\mu_{Y|x} = \alpha + \beta x$. Each time we compute a regression line from a random sample we are, in effect, observing one possible linear equation in a population consisting of all possible linear equations. In addition, we remind ourselves that the actual value of Y that will be observed for a given value of x is a $N(\alpha + \beta x, \sigma^2)$ random variable. So the value actually observed will be different from $\mu_{Y|x}$. Therefore, the predicted value for y by using the regression equation will be in error from two different sources:

1. a and b are randomly distributed about α and β, and
2. y is randomly distributed about $\mu_{Y|x}$.

Let y_0 denote the actual value of y that will be observed for the value of x_0 and consider the random variable

$$d = y_0 - a - bx_0.$$

Since d is a linear combination of normally distributed random variables it follows that it, too, has a normal distribution. Now

$$E[d] = E[y_0] - E[a] - E[b]x_0 = \alpha + \beta x_0 - \alpha - \beta x_0 = 0.$$

As seems to be usual the variance of d requires more effort.

$$\begin{aligned}
\text{Var}(d) &= E[(d - \mu_d)^2] = E[d^2] \\
&= E[(y_0 - a - bx_0)^2] = E[(y_0 - \alpha - \beta x_0 - (a - \alpha) - (b - \beta)x_0)^2] \\
&= E[(y_0 - \alpha - \beta x_0)^2] + E[(a - \alpha)^2] + x_0^2 E[(b - \beta)^2] \\
&\quad - 2E[(y_0 - \alpha + \beta x_0)(a - \alpha)] - 2x_0 E[(y_0 - \alpha - \beta x_0)(b - \beta)] \\
&\quad + 2x_0 E[(a - \alpha)(b - \beta)] \\
&= \text{Var}(y_0) + \text{Var}(a) + x_0^2 \text{Var}(b) + 2x_0 \text{Cov}(a, b)
\end{aligned}$$

$$= \sigma^2 + \frac{\sigma^2 \sum\limits_{i=1}^{n} x_i^2}{n(n-1)s_X^2} + \frac{\sigma^2 x_0^2}{(n-1)s_X^2} + -2 \frac{\sigma^2 \bar{x} x_0}{(n-1)s_X^2}$$

$$= \sigma^2 \left[1 + \frac{1}{n} + \frac{(x_0 - \bar{x})^2}{(n-1)s_X^2} \right]$$

We can use these results to construct what is called a *prediction interval* for y_0. We first note that, under the assumptions of the regression model, the quantity

$$t = \frac{\dfrac{(y_0 - a - bx_0)}{\sigma_{y_0 - a - bx_0}}}{\sqrt{\dfrac{s^2}{\sigma^2}}}$$

$$= \frac{y_0 - a - bx_0}{s \sqrt{\dfrac{n+1}{n} + \dfrac{(x_0 - \bar{x})^2}{(n-1)s_X^2}}}$$

will have a t distribution with $n-2$ degrees of freedom. Therefore, we have a pivotal quantity for y_0 and can use the method of Section 11.2 to construct a $(1 - \alpha)100\%$ prediction interval for y_0. We leave it to the reader to verify that this method results in a prediction interval of the following form.

$$a + bx_0 - t(n-2)_{\alpha/2} s \sqrt{\frac{n+1}{n} + \frac{(x_0 - \bar{x})^2}{(n-1)s_X^2}}$$

$$\leq y_0 \leq a + bx_0 + t(n-2)_{\alpha/2} s \sqrt{\frac{n+1}{n} + \frac{(x_0 - \bar{x})^2}{(n-1)s_X^2}}$$

(13.8)

Example 13.5   We wish to construct a 95% prediction interval for the performance rating of an employee who scores 50 on the screening test described in Example 13.1. Under the assumptions of our regression model we can use Equation 13.8. From previous examples we recall that

$$n = 10,$$

$$\bar{x} = 41.6,$$

$$a = 47.434,$$

$$b = 0.4646,$$

$$\widehat{\sigma^2} = 27.75497, \text{ and}$$

$$\sum_{i=1}^{n} (x_i - \bar{x})^2 = 1902.4.$$

Therefore,

$$s\sqrt{\frac{n+1}{n} + \frac{(x_0 - \bar{x})^2}{(n-1)s_X^2}} = 6.28.$$

From Table 5 in Appendix A we find that $t(8)_{0.025} = 2.306$. Using these we find that a 95% prediction interval for the evaluation rating of an employee who scores 50 on the screening test is

$$47.434 + (0.4646)(50) - (2.306)(6.28) \leq y_0 \leq 47.434 + (0.4646)(50) + (2.306)(6.28),$$

or

$$56.18 \leq y_0 \leq 85.15. \qquad \blacksquare$$

By looking at Equation 13.8 a little more closely we note the following properties of prediction intervals.

1. The quantity under the square root always is greater than 1. This means that a prediction interval will, on average, have a width that is greater than $2t(n-2)_{\alpha/2}\sigma$. This reflects the fact that we always will be less sure about the value of any particular member of a population than about the average value.

2. As the value of x_0 moves away from $\bar{x}$, the average length of the prediction intervals increases. This highlights the fact that we are more confident in using regression equations for values near the center of the data than for points near either extreme.

3. It should be pointed out that for prediction intervals to have any real meaning—that is, that they cover the "true" value $(1-\alpha)100\%$ of the time—each prediction interval that we construct should be based on separate estimates for α, β, and σ^2.

It is also possible to construct prediction intervals for the *mean* value, $\mu_{Y|x_0}$, of an observation with $x = x_0$. By considering

$$d' = \mu_{Y|x_0} - a - bx_0$$

we can proceed in a manner similar to that used previously to derive a $(1-\alpha)100\%$ interval for $\mu_{Y|x_0}$. It is left as an exercise to show that we obtain

$$a + bx_0 - t(n-2)_{\alpha/2}s\sqrt{\frac{1}{n} + \frac{(x_0 - \bar{x})^2}{(n-1)s_X^2}} \tag{13.9}$$

$$\leq \mu_{Y|x_0} \leq a + bx_0 + t(n-2)_{\alpha/2}s\sqrt{\frac{1}{n} + \frac{(x_0 - \bar{x})^2}{(n-1)s_X^2}}$$

Example 13.6

We wish to construct a 95% prediction interval for $\mu_Y \mid x_0$ when $x_0 = 50$. The components of Equation 13.9 are the same as in Example 13.5 except that

$$\sqrt{\frac{1}{n} + \frac{(x_0 - \bar{x})^2}{(n-1)s_X^2}} = 0.370.$$

Therefore, a 95% prediction interval for $\mu_{Y|x_0}$ when $x_0 = 50$ is

$$47.434 + (0.4646)(50) - (2.306)(5.894)(0.414)$$

$$\leq \mu_{Y|x_0} \leq 47.434 + (0.4646)(50) + (2.306)(5.894)(0.414),$$

or

$$65.635 \leq \mu_{Y|x_0} \leq 75.693.$$

Note that this interval is narrower than the one for a single observation that we obtained in Example 13.5. ∎

Recall in our discussion of the properties of estimators in Chapter 10 one criterion for optimality was that an unbiased estimator for a parameter have the smallest possible variance. This was the notion of a *uniform minimum variance unbiased estimator*. It turns out that a and b are unbiased but that they do not always have the smallest possible variance. However, if we restrict ourselves to only those estimators that are linear combinations of $y_1, y_2, \ldots, y_n$ it is possible to define a slightly restricted optimality criterion for such estimators.

Definition 13.1

> Consider the problem of estimating a parameter θ based on a random sample consisting of $y_1, y_2, \ldots, y_n$. If an estimator $\widehat{\theta}$, which is a linear combination of the sample observations, has a distribution whose variance is less than or equal to that of any other estimator that is also a linear combination of the sample observations, then $\widehat{\theta}$ is said to be a *best linear unbiased estimator (BLUE)* for θ.

This definition is important because it turns out that the least squares estimators for the regression problem are best linear unbiased estimators. This result is established by the so-called *Gauss-Markov theorem*, which we state without proof. The interested reader should consult *An Introduction to Linear Statistical Models*, vol. 1, by Franklin A. Graybill.

Theorem 13.2

> **Gauss-Markov** Let the simple regression model, $Y = \alpha + \beta x_i + e_i$, for $i = 1, 2, \ldots, n$, be such that each x_i is fixed, each Y_i is an observable random variable and each e_i is an unobservable random variable. In addition, suppose that the random variables e_i each have $E[e_i] = 0$, $\text{Var}(e_i) = \sigma^2$ and $\text{Cov}(e_i, e_j) = 0$, if $i \neq j$. Then the least squares estimators for α and β are best linear unbiased estimators.

Before concluding our discussion of simple linear regression we wish to mention a commonly used measure of the quality of a least squares curve fit. It is known as the *coefficient of determination* and denoted by r^2. It is based on the following idea. If we had no knowledge of a relation between the X and Y and were required to use a sample of values of Y to predict the value of Y for a particular value of X, we would probably be inclined to guess $\bar{Y}$. Now the sum of squared deviations of our predicted value of Y from the observed values of Y would be

$$\sum_{i=1}^{n}(Y_i - \bar{Y})^2.$$

This quantity is often referred to as the *total sum of squares* and denoted by SST. Now using our regression line $\hat{Y} = a + bX$ the differences $Y_i - \hat{Y}_i$, for $i = 1, 2, \ldots, n$, are known as *residuals*. The sum of squares minimized by linear regression is

$$\sum_{i=1}^{n}(Y_i - \hat{Y}_i)^2$$

and is referred to as the *residual sum of squares* or the *sum of squares for error* and denoted by SSE. The term *error* arises from the fact that each residual is the difference between the actual value of Y_i and the value estimated by the regression line—that is, the error involved in using the regression line. The difference between these two sums of squares is known as the *sum of squares due to regression*, which is denoted by SSR and is

$$\text{SSR} = \text{SST} - \text{SSE}$$

$$= \sum_{i=1}^{n}(Y_i - \bar{Y})^2 - \sum_{i=1}^{n}(Y_i - \hat{Y}_i)^2 \tag{13.10}$$

$$= \sum_{i=1}^{n}(\hat{Y}_i - \bar{Y})^2.$$

Clearly, if SSE $= 0$ then the regression line passes through each observed value of Y. It is common practice in such a case to say that the regression line has *explained* all of the observed variation in Y. This terminology has a certain element of danger in that it is easy to misinterpret this as implying that there is a cause-and-effect relation between X and Y. Least squares curve fitting simply observes a relationship without trying to make any interpretation of the causes for the existence of that relationship. It is more often the case that the regression line does not pass through each of the observed points. In such cases it is of interest to determine what proportion of the observed variation about $\bar{Y}$ is "explained" by the regression line. In other words, we wish to determine what proportion SSR is of SST. This proportion is what is called the *coefficient of determination*, r^2. It is

$$r^2 = \frac{\text{SSR}}{\text{SST}} = 1 - \frac{\text{SSE}}{\text{SST}}. \tag{13.11}$$

This can be written as

$$r^2 = \frac{\displaystyle\sum_{i=1}^{n}(\hat{Y}_i - \bar{Y})^2}{(Y_i - \bar{Y})^2}$$

$$= \frac{\left[\displaystyle\sum_{i=1}^{n}(X_i - \bar{X})(Y_i - \bar{Y})\right]^2}{\displaystyle\sum_{i=1}^{n}(X_i - \bar{X})^2 \sum_{i=1}^{n}(Y_i - \bar{Y})^2}$$

(13.12)

$$= \frac{\left[\displaystyle\sum_{i=1}^{n}X_i Y_i - n\bar{X}\bar{Y}\right]^2}{\left[\displaystyle\sum_{i=1}^{n}X_i^2 - n\bar{X}^2\right]\left[\displaystyle\sum_{i=1}^{n}Y_i^2 - n\bar{Y}^2\right]}.$$

The details of this derivation are left as an exercise. It should be noted that r^2 is the square of the correlation between Y_i and $\hat{Y}_i$.

Example 13.7   We wish to compute the coefficient of determination for the data in Example 13.1. From Example 13.2 we recall that

$$n = 10,$$

$$\bar{x} = 41.6,$$

$$\bar{y} = 66.7,$$

$$\sum_{i=1}^{n} x_i^2 = 19208, \quad \text{and}$$

$$\sum_{i=1}^{n} x_i y_i = 28631.$$

From the data we obtain

$$\sum_{i=1}^{n} y_i^2 = 45177.$$

Therefore, Equation 13.11 gives

$$r^2 = \frac{[28631 - (10)(41.6)(66.7)]^2}{[19208 - (10)(41.6)^2][45177 - (10)(66.7)^2]}$$

$$= 0.596.$$

Therefore, the amount of variation observed in evaluation scores that is explained by our regression line is almost 60%.   ■

Exercises 13.2   1. Construct scatter diagrams for each of the following sets of data and indicate the general type of equation that best describes the relation, if any, between the independent variable X and the dependent variable Y.

a.

(X)	5	10	15	20	25	30	35	40	45	50
(Y)	12	15	21	25	26	32	38	39	45	49

b.

(X)	1	2	3	4	5	6	7	8	9	10
(Y)	4	6	8	7	9	9	12	11	12	13

c.

(X)	5	10	15	20	25	30	35	40	45	50
(Y)	17	14	12	18	15	11	13	19	14	12

d.

(X)	10	20	30	40	50	60	70	80	90	100
(Y)	133	135	131	128	119	116	107	106	98	93

2. For each set of data in Exercise 1 that exhibits a linear relation between the independent variable X and the dependent variable Y, do the following.

   a. Compute a, b, $\widehat{\sigma^2}$ and give the regression equation.

   b. Draw your regression equation in the scatter diagram.

   c. Use your regression equation to predict the mean value for Y when $X = 17$.

3. Sometimes the nature of the problem indicates that the regression line should pass through the origin. This would be the case in trying to fit a model relating stopping distance and automobile speed. This means that our regression model becomes

$$Y = \beta X + e.$$

Show that the least squares estimator for β in this model is

$$b = \frac{\displaystyle\sum_{i=1}^{n} x_i y_i}{\displaystyle\sum_{i=1}^{n} x_i^2}.$$

Assume that the error terms are i.i.d. $N(0, \sigma^2)$ random variables.

4. Verify that the least squares estimators given in Equation 13.2 do indeed minimize S as given in Equation 13.1.

5. Verify that the estimators given in Equation 13.2 always pass through the point $(\bar{X}, \bar{Y})$.

6. Show that

$$\text{Cov}(a, b) = \frac{-\sigma^2 \bar{X}}{\displaystyle\sum_{i=1}^{n} x_i^2 - n\bar{X}^2}.$$

7. Use the fact proven in Exercise 6 to determine conditions for the independence of a and b.

8. Show that

$$t = (a - \alpha) \sqrt{\frac{n(n-2)\left(\sum_{i=1}^{n} x_i^2 - n\bar{x}^2\right)}{\left(\sum_{i=1}^{n} x_i^2\right)\sum_{i=1}^{n}(y_i - a - bx_i)^2}}$$

has a t distribution with $n - 2$ degrees of freedom.

9. Let us denote the intercept in the linear regression model as β_0. Use the results of Exercise 8 to show that a $(1 - \alpha)\%$ confidence interval for β_0 is given by

$$\hat{\beta}_0 \pm t(n-2)_{\alpha/2} \sqrt{\frac{\widehat{\sigma^2}\sum_{i=1}^{n} x_i^2}{n\left[\sum_{i=1}^{n} x_i^2 - n\bar{x}^2\right]}}.$$

10. Derive the prediction interval given in Equation 13.9.

11. Verify that SST = SSR + SSE.

12. Fill in the details of the derivation of r^2 given in Equation 13.12.

The following table gives data from an experiment where rats were given various doses of alcohol and then required to navigate a maze. It shows the alcohol level, X, on a scale from 0 to 100 and time to navigate the maze, Y, in seconds.

(X)	0	5	10	15	20	25	30	35	40	45	50	55	60	65	70	75	80
(Y)	102	97	90	82	77	72	63	77	72	65	57	50	43	38	31	21	16

Exercises 13 to 20 refer to these data. Assume that the model given in Equation 13.3 is appropriate.

13. Determine the least squares estimates, a and b. Use these to predict the time through the maze for an alcohol level of 58.

14. Test the hypothesis that the slope is 0 against the two-sided alternative at the 0.05 level.

15. Test the hypothesis that the intercept is 100 against the alternative that it is greater than than 100 at the 0.01 level.

16. Construct a 95% confidence interval for the intercept.

17. Construct a 90% confidence interval for the slope.

18. Construct a 95% prediction interval for a single observation of the time through the maze when the actual alcohol level is 72.

19. Construct a 95% prediction interval for the mean time through the maze when the actual alcohol level is 72. Compare your results with Exercise 18.

20. Compute the coefficient of determination for the regression equation.

The following data are part of a study conducted by a sociologist. Among other things respondents were asked to provide their annual income and the amount that was placed in savings in the past year. The data, in thousands of dollars, are given below.

Income	29.3	30.5	31.0	31.9	32.9	33.5	34.4	35.0	36.7	38.2	38.7	39.6	41.8
Savings	0.27	0.26	0.14	0.18	0.55	0.65	0.58	0.72	1.09	1.27	1.10	1.83	2.23

The remaining exercises all refer to these data. Assume that the model given in Equation 13.3 is appropriate.

21. Determine the least squares estimates, a and b, to use income to predict savings. Use these to predict savings when income is $32,000.

22. Test the hypothesis that the slope is 0 against the 2-sided alternative at the 0.01 level.

23. Test the hypothesis that the intercept is 0 against the alternative that it is less than 0 at the 0.05 level.

24. Construct a 99% confidence interval for the intercept.

25. Construct a 95% confidence interval for the slope.

26. Construct a 90% prediction interval for a single observation when the income is $37,500.

27. Construct a 90% prediction interval for the mean savings when the actual income is $37,500. Compare your results with Exercise 26.

28. Compute the coefficient of determination for the regression equation.

13.3
Matrix Calculus

In this section we shall discuss some topics from matrix analysis that are useful in regression analysis. Some are not covered in standard linear algebra courses. They shall make our subsequent discussion of regression models with more than one independent variable easier to understand. We shall use the standard notational conventions for denoting scalars, vectors, and matrices. For example, scalars shall be given as ordinary lower-case letters such as x and y. Vectors shall be column vectors and be denoted by boldface lower-case letters such as $\mathbf{x}$ and $\mathbf{y}$. Matrices will be denoted by upper-case boldface letters such as $\mathbf{X}$ and $\mathbf{Y}$. In most cases the dimensions of vectors and matrices should be clear from the context. If not, we shall denote a vector $\mathbf{x}$ of length p by $\mathbf{x}_p$ and a matrix $\mathbf{X}$ having r rows and c columns by $\mathbf{X}_{r \times c}$. A row vector $\mathbf{x}$ will be denoted by $\mathbf{x}'$, and the transpose of the matrix $\mathbf{X}$ will be denoted by $\mathbf{X}'$.

We begin by defining a function that is basic in the applications of matrix algebra to probability and statistics.

Definition 13.2

The function $y = f(x_1, x_2, \ldots, x_p)$ such that

$$y = \sum_{i=1}^{p} \sum_{j=1}^{p} a_{ij} x_i x_j, \tag{13.13}$$

where $a_{ij}, i = 1, 2, \ldots, p, j = 1, 2, \ldots, p$ is a set of constants and $-\infty < x_i < \infty, i = 1, 2, \ldots, p$ is said to be a *quadratic form* in the variables x_i.

Note that a quadratic form is a scalar quantity. The formal definition is cumbersome to work with. It turns out that there is a convenient way to write quadratic forms using vectors and matrices. Let $\mathbf{x}_p$ be a vector and $\mathbf{A}_{p \times p}$ be a matrix such that

$$\mathbf{x} = \begin{pmatrix} x_1 \\ x_2 \\ \vdots \\ x_p \end{pmatrix} \quad \text{and} \quad \mathbf{A} = \begin{pmatrix} a_{11} & a_{12} & \cdots & a_{1p} \\ a_{21} & a_{22} & \cdots & a_{2p} \\ \vdots & \vdots & & \vdots \\ a_{p1} & a_{p2} & \cdots & a_{pp} \end{pmatrix}.$$

A term-by-term comparison will show that we can write the quadratic form, y, as

$$y = \mathbf{x}'\mathbf{A}\mathbf{x}. \tag{13.14}$$

It is common practice simply to refer to $\mathbf{x}'\mathbf{A}\mathbf{x}$ as a quadratic form and to $\mathbf{A}$ as the matrix of the quadratic form. The term quadratic form arises from consideration of quadratic equations of the form $ax^2 + bx + c$. If we write ax^2 as xax we see that the quadratic form $\mathbf{x}'\mathbf{A}\mathbf{x}$ is the matrix equivalent of the second degree term in a quadratic equation.

To illustrate how matrix notation can be of use in probability theory we introduce the idea of a *covariance matrix*. Consider the random vector

$$\mathbf{x} = \begin{pmatrix} x_1 \\ x_2 \\ \vdots \\ x_p \end{pmatrix}.$$

Now, if we denote $\text{Cov}(x_i, x_j)$ by σ_{ij} and note that $\text{Var}(x_i) = \text{Cov}(x_i, x_i)$, we can define the *covariance matrix*, $\boldsymbol{\Sigma}$, for $\mathbf{x}$ by

$$\boldsymbol{\Sigma} = \begin{pmatrix} \sigma_{11} & \sigma_{12} & \cdots & \sigma_{1p} \\ \sigma_{21} & \sigma_{22} & \cdots & \sigma_{2p} \\ \vdots & \vdots & & \vdots \\ \sigma_{p1} & \sigma_{p2} & \cdots & \sigma_{pp} \end{pmatrix}.$$

In addition, by recalling the definition of *correlation* from Chapter 4 we can write $\sigma_{ij} = \rho_{ij} \sigma_i \sigma_j$, where $\sigma_i = \sqrt{\text{Var}(x_i)}$.

Example 13.8   We discussed the *bivariate normal distribution* in Section 5.10. We had a random vector of size 2 and the probability density function had parameters

$$\mu_X = E[X], \ \mu_y = E[Y], \ \sigma_X^2 = \text{Var}(X), \ \sigma_Y^2 = \text{Var}(Y) \quad \text{and} \quad \rho = \rho_{XY}.$$

The joint probability density function was given in Equation 5.32 to be

$$f_{X,Y}(x,y) = \frac{1}{2\pi\sqrt{\sigma_X^2 \sigma_Y^2 (1 - \rho^2)}} e^{-(1/2)Q^2}, \quad \text{for} \ -\infty < x < \infty \quad \text{and} \quad -\infty < y < \infty,$$

where

$$Q = \frac{1}{1 - \rho^2} \left[\frac{(x - \mu_X)^2}{\sigma_X^2} - 2\rho \frac{(x - \mu_X)}{\sigma_X} \frac{(y - \mu_Y)}{\sigma_Y} + \frac{(y - \mu_Y)^2}{\sigma_Y^2} \right].$$

Our goal is to rewrite the joint p.d.f. using matrix notation. The covariance matrix for (X, Y) would be

$$\Sigma = \begin{pmatrix} \sigma_X^2 & \rho \sigma_X \sigma_Y \\ \rho \sigma_X \sigma_Y & \sigma_Y^2 \end{pmatrix}.$$

Note that the determinant of Σ is

$$|\Sigma| = \sigma_X^2 \sigma_Y^2 (1 - \rho^2),$$

and the inverse of Σ is

$$\Sigma^{-1} = \frac{1}{\sigma_X^2 \sigma_Y^2 (1 - \rho^2)} \begin{pmatrix} \sigma_Y^2 & -\rho \sigma_X \sigma_Y \\ -\rho \sigma_X \sigma_Y & \sigma_X^2 \end{pmatrix}.$$

If we let the vector $\mathbf{v}$ be such that

$$\begin{pmatrix} v_1 \\ v_2 \end{pmatrix} = \begin{pmatrix} X - \mu_X \\ Y - \mu_Y \end{pmatrix},$$

then Equation 5.32 can be written in the form

$$f_{X,Y}(x, y) = \frac{1}{2\pi \sqrt{|\Sigma|}} e^{-(1/2)\mathbf{v}' \Sigma^{-1} \mathbf{v}}.$$

■

In Section 13.2 we were interested in finding the values a and b that minimized $S = \sum_{i=1}^{n} (y_i - a - bx_i)^2$. At the time we computed $\frac{\partial S}{\partial a}$ and $\frac{\partial S}{\partial b}$. The reader will recall from multivariate calculus that we were actually computing the components of the *gradient* of S with respect to the vector $\mathbf{x}$. In our discussion, we shall define this formally to be the *derivative of S with respect to* $\mathbf{x}$.

Definition 13.3

Let f be a function of the p real variables $x_1, x_2, \ldots, x_p$. The derivative of f with respect to the vector $\mathbf{x}$ is denoted by $\frac{\partial f}{\partial \mathbf{x}}$ and is defined to be

$$\frac{\partial f}{\partial \mathbf{x}} = \begin{pmatrix} \frac{\partial f}{\partial x_1} & \frac{\partial f}{\partial x_2} & \cdots & \frac{\partial f}{\partial x_p} \end{pmatrix}'. \tag{13.15}$$

Example 13.9

Suppose that $f(x_1, x_2, x_3)$ is defined to be

$$f(x_1, x_2, x_3) = 1 - e^{-x_1 - 2x_2 - 3x_3}, \quad \text{for } 0 < x_i, \quad i = 1, 2, 3.$$

Equation 13.15 gives that

$$\frac{\partial f}{\partial \mathbf{x}} = \begin{pmatrix} e^{-x_1 - 2x_2 - 3x_3} \\ 2\,e^{-x_1 - 2x_2 - 3x_3} \\ 3\,e^{-x_1 - 2x_2 - 3x_3} \end{pmatrix}.$$

■

Definition 13.3 permits us to give some theorems concerning differentiation with respect to a vector that will be useful in our discussion of multiple regression.

Theorem 13.3

Let $f(\mathbf{x})$ be a linear combination of the elements of the vector $\mathbf{x}_p$ defined by

$$f(\mathbf{x}) = \sum_{i=1}^{p} a_i x_i = \mathbf{a}'\mathbf{x} = \mathbf{x}'\mathbf{a},$$

where

$$\mathbf{a} = \begin{pmatrix} a_1 & a_2 & \cdots & a_p \end{pmatrix}',$$

and the a_i are real constants. Then

$$\frac{\partial f}{\partial \mathbf{x}} = \mathbf{a}. \tag{13.16}$$

Proof   From Definition 13.3 we see that the ith element of $\frac{\partial f}{\partial \mathbf{x}}$ is

$$\left(\frac{\partial f}{\partial \mathbf{x}} \right)_i = \frac{\partial}{\partial x_i} \sum_{j=1}^{p} a_j x_j = a_i,$$

and we are done.   ■

This result can be extended to cover quadratic forms. We leave the proof as an exercise.

Theorem 13.4

Let $y(\mathbf{x})$ be a quadratic form in the p real variables $x_1, x_2, \ldots, x_p$, be defined by

$$y(\mathbf{x}) = \mathbf{x}'\mathbf{A}\mathbf{x},$$

where $\mathbf{A}$ is a real, symmetric $p \times p$ matrix. Then

$$\frac{\partial y}{\partial \mathbf{x}} = 2\mathbf{A}\mathbf{x}. \tag{13.17}$$

We give two theorems whose proofs can be found in most linear algebra texts. They are useful in deriving least squares estimators.

Theorem 13.5

> Let $\mathbf{A}$ be an $m \times n$ matrix of rank r. Then the rank of $\mathbf{A}'\mathbf{A}$ is r, and the rank of $\mathbf{AA}'$ is r.

Theorem 13.6

> Let $\mathbf{A}$ be an $m \times n$ matrix of rank r and $\mathbf{B}$ be an $n \times p$ matrix of rank s. Then the rank of $\mathbf{AB}$ is less than or equal to $\min(r, s)$.

Example 13.10

These 3 theorems allow us to give a matrix version of the least squares estimators that were derived in Section 13.2. Notation will be simplified if we let $b_0 = a$ and $b_1 = b$. Then we have $y_i = b_0 + b_1 x_i$, for $i = 1, 2, \ldots, n$. Now let

$$\mathbf{y} = \begin{pmatrix} y_1 \\ y_2 \\ \vdots \\ y_n \end{pmatrix}, \quad \mathbf{X} = \begin{pmatrix} 1 & x_1 \\ 1 & x_2 \\ \vdots & \vdots \\ 1 & x_n \end{pmatrix}, \quad \text{and} \quad \mathbf{b} = \begin{pmatrix} b_0 \\ b_1 \end{pmatrix}.$$

Using these we can write $\mathbf{y} = \mathbf{Xb}$. Then the sum of squares, S, we wished to minimize can be written in matrix notation as

$$\begin{aligned} S &= (\mathbf{y} - \mathbf{Xb})'(\mathbf{y} - \mathbf{Xb}) \\ &= \mathbf{y}'\mathbf{y} - \mathbf{y}'\mathbf{Xb} - \mathbf{b}'\mathbf{X}'\mathbf{y} + \mathbf{b}'\mathbf{X}'\mathbf{Xb}. \end{aligned} \tag{13.18}$$

We wish to evaluate $\frac{\partial S}{\partial \mathbf{b}}$. The first term in the expansion of S is a constant with respect to $\mathbf{b}$ and will have a 0 derivative. We can apply Theorem 13.3 to the second and third terms. The last term is a quadratic form, and we can use Theorem 13.4 if the matrix $\mathbf{X}'\mathbf{X}$ is symmetric. This is seen to be the case by recalling that a matrix, $\mathbf{A}$, is symmetric if and only if $\mathbf{A}' = \mathbf{A}$ and noting that $(\mathbf{X}'\mathbf{X})' = \mathbf{X}'(\mathbf{X}')'$ $= \mathbf{X}'\mathbf{X}$. An application of Theorems 13.3 and 13.4 gives

$$\frac{\partial S}{\partial \mathbf{b}} = -2\mathbf{X}'\mathbf{y} + 2\mathbf{X}'\mathbf{Xb} = 0,$$

or

$$\mathbf{X}'\mathbf{Xb} = \mathbf{X}'\mathbf{y}. \tag{13.19}$$

This is a set of 2 equations that are known as the *normal equations*. A unique solution for $\mathbf{b}$ exists if and only if $\mathbf{X}'\mathbf{X}$ is a nonsingular matrix. As long as $x_i \neq x_j$, for some i and j the matrix $\mathbf{X}$ will have a rank of 2. Therefore, according to Theorem 13.5, $\mathbf{X}'\mathbf{X}$ will have rank 2. Since $\mathbf{X}'\mathbf{X}$ is a 2×2 matrix it is of full rank and, hence, nonsingular. Therefore, we obtain

$$\mathbf{b} = (\mathbf{X}'\mathbf{X})^{-1}\mathbf{X}'\mathbf{y}. \tag{13.20}$$

We leave it as an exercise to verify that b_0 and b_1 are identical to the estimators for α and β that we obtained in Section 13.2. ∎

There is an interesting and useful relation between the elements of $\mathbf{X}'\mathbf{X}$ and the variance of a and b. Note that

$$(\mathbf{X}'\mathbf{X})^{-1} = \frac{1}{\displaystyle\sum_{i=1}^{n} x_i^2 - n\bar{x}^2} \begin{pmatrix} \displaystyle\sum_{i=1}^{n} x_i^2 & -n\bar{x} \\ -n\bar{x} & n \end{pmatrix}.$$

If we recall the variances of a and b from Section 13.2 and let a_{ij} denote the ijth element of $(\mathbf{X}'\mathbf{X})^{-1}$ we see that

$$\mathrm{Var}(a) = \frac{\sigma^2 \displaystyle\sum_{i=1}^{n} x_i^2}{n\left(\displaystyle\sum_{i=1}^{n} x_i^2 - n\bar{x}^2 \right)} = a_{11}\sigma^2, \quad \text{and}$$

$$\mathrm{Var}(b) = \frac{\sigma^2}{\displaystyle\sum_{i=1}^{n} x_i^2 - n\bar{x}^2} = a_{22}\sigma^2.$$

Similarly, we find that

$$\mathrm{Cov}(a, b) = \frac{-\sigma^2 \bar{x}}{\displaystyle\sum_{i=1}^{n} x_i^2 - n\bar{x}^2} = a_{12}\sigma^2.$$

These relations will be helpful in the next section where we consider the situation with more than one independent variable.

Exercises 13.3

1. Let $\mathbf{x}' = [1\ 2\ 3\ 4\ 5]$, and find a square matrix, $\mathbf{A}$, such that
   a. $\mathbf{x}'\mathbf{A}\mathbf{x} = 1 + 4 + 9 + 16 + 25$,
   b. $\mathbf{x}'\mathbf{A}\mathbf{x} = \frac{(1+2+3+4+5)^2}{5}$, and
   c. $\mathbf{x}'\mathbf{A}\mathbf{x} = (1-3)^2 + (2-3)^2 + (3-3)^2 + (4-3)^2 + (5-3)^2$.

2. In each case, find a square matrix $\mathbf{A}$ with integer elements such that
   a. $\mathbf{x}'\mathbf{A}\mathbf{x} = 3x_1^2 + 3x_1 x_2 + 4x_2^2$,
   b. $\mathbf{x}'\mathbf{A}\mathbf{x} = 2x_1^2 + 2x_2^2 + 2x_3^2 - x_1 x_2 - x_1 x_3 - x_2 x_3$, and
   c. $\mathbf{x}'\mathbf{A}\mathbf{x} = x_1^2 + x_2^2 + x_3^2 + 6x_1 x_2 + 2x_1 x_3 + 4x_2 x_3$.

3. Find the correlation matrix for the normal random variables that have the quadratic form $2x_1^2 + 3x_2^2 - 2x_1 x_2$ in the distribution. What is the correlation between x_1 and x_2?

4. For each of the following functions determine $\frac{\partial f}{\partial \mathbf{x}}$.
   a. $f(x_1, x_2, x_3) = 4x_1^2 - 3x_2^2 + 5x_2 x_3$.
   b. $f(x_1, x_2, x_3) = \frac{x_1}{x_3} \cos(x_2) - x_1 \sin(x_2)$.

5. It is possible to define the *derivative of a function with respect to a matrix* as follows. Assume that the $m \times n$ matrix $\mathbf{X}$ is

$$\mathbf{X} = \begin{pmatrix} x_{11} & x_{12} & \cdots & x_{1n} \\ x_{21} & x_{22} & \cdots & x_{2n} \\ \vdots & \vdots & \ddots & \vdots \\ x_{m1} & x_{m2} & \cdots & x_{mn} \end{pmatrix}.$$

We then define

$$\frac{\partial f}{\partial \mathbf{X}} = \begin{pmatrix} \dfrac{\partial f}{\partial x_{11}} & \dfrac{\partial f}{\partial x_{12}} & \cdots & \dfrac{\partial f}{\partial x_{1n}} \\ \dfrac{\partial f}{\partial x_{21}} & \dfrac{\partial f}{\partial x_{22}} & \cdots & \dfrac{\partial f}{\partial x_{2n}} \\ \vdots & \vdots & \ddots & \vdots \\ \dfrac{\partial f}{\partial x_{m1}} & \dfrac{\partial f}{\partial x_{m2}} & \cdots & \dfrac{\partial f}{\partial x_{mn}} \end{pmatrix}.$$

With this definition show that, if $f(\mathbf{X}) = \mathbf{a}'\mathbf{X}\mathbf{b}$, then

$$\frac{\partial f}{\partial \mathbf{X}} = \mathbf{a}\mathbf{b}'.$$

6. Verify that the matrix forms for the least squares estimators given in Equation 13.20 are equivalent to those of Equation 13.2. That is, show that $a = b_0$ and $b = b_1$.

7. Use Equation 13.20 to compute the least squares estimators for the data in Example 13.1. Determine $(\mathbf{X}'\mathbf{X})^{-1}$ by inspection of $(\mathbf{X}'\mathbf{X})$.

13.4
Multiple Regression

We now turn to a situation where k independent variables are to be used to predict the dependent variable. This means that an experimenter predetermines n sets of values for the independent variables $x_1, x_2, \ldots, x_k$. These values are then used to observe values for $Y_1, Y_2, \ldots, Y_n$. During our discussion it will be assumed that the observations follow

$$Y_i = \beta_0 + \beta_1 x_{1i} + \beta_2 x_{2i} + \cdots + \beta_k x_{ki} + e_i, \tag{13.21}$$

where

$$\begin{cases} e_i \sim N(0, \sigma^2) \\ \text{for } i = 1, 2, \ldots, n, \quad \text{and} \\ e_i \text{ and } e_j \text{ are uncorrelated if } i \neq j. \end{cases}$$

Similar to the simple regression situation—that is, when $k = 1$—these assumptions have the following effects.

1. $Y_1, Y_2, \ldots, Y_n$ are independent random variables.
2. The Y's are homoscedastic about the line $E[Y] = \beta_0 + \beta_1 x_1 + \beta_2 x_2 + \cdots + \beta_k x_k$.
3. Each $Y_i \sim N(\beta_0 + \beta_1 x_{1i} + \beta_2 x_{2i} + \cdots + \beta_k x_{ki}, \sigma^2)$, for $i = 1, 2, \ldots, n$.

The notation for this problem is more complicated and harder to follow than was the case in simple regression. For this reason we wish to use matrix notation to try to simplify matters. If we define

$$\boldsymbol{\beta} = \begin{pmatrix} \beta_0 \\ \beta_1 \\ \beta_2 \\ \vdots \\ \beta_k \end{pmatrix}, \quad \mathbf{y} = \begin{pmatrix} y_1 \\ y_2 \\ \vdots \\ y_n \end{pmatrix}, \quad \mathbf{X} = \begin{pmatrix} 1 & x_{11} & x_{12} & \cdots & x_{1k} \\ 1 & x_{21} & x_{22} & \cdots & x_{2k} \\ \vdots & \vdots & \vdots & \ddots & \vdots \\ 1 & x_{n1} & x_{n2} & \cdots & x_{nk} \end{pmatrix}, \quad \text{and} \quad \mathbf{e} = \begin{pmatrix} e_1 \\ e_2 \\ \vdots \\ e_n \end{pmatrix},$$

then

$$\mathbf{y} = \mathbf{X}\boldsymbol{\beta} + \mathbf{e}.$$

The least squares estimator for β_i is denoted as b_i, for $i = 0, 1, 2, \ldots, k$. These estimators will minimize the sum of squares

$$
\begin{aligned}
S &= \sum_{i=1}^{n} (y_i - b_0 - b_1 x_{i1} - b_2 x_{2i} - \cdots - b_k x_{ik})^2 \\
&= (\mathbf{y} - \mathbf{Xb})'(\mathbf{y} - \mathbf{Xb}) \\
&= \mathbf{y}'\mathbf{y} - \mathbf{y}'\mathbf{Xb} - (\mathbf{Xb})'\mathbf{y} + \mathbf{b}'\mathbf{X}'\mathbf{Xb}.
\end{aligned}
\tag{13.22}
$$

The matrix $\mathbf{X}$ is of full rank, which means that $\mathbf{X}'\mathbf{X}$ is nonsingular. By comparing Equation 13.21 with Example 13.10 we see immediately that the *normal equations* are

$$\mathbf{X}'\mathbf{Xb} = \mathbf{X}'\mathbf{y}. \tag{13.23}$$

Thus, the least squares estimators are

$$\mathbf{b} = (\mathbf{X}'\mathbf{X})^{-1}\mathbf{X}'\mathbf{y}. \tag{13.24}$$

Notice how the use of matrix notation has removed any need to be concerned with exactly how many independent variables are being used to predict the value of the dependent variable. Any least squares estimation problem has a solution of the same form.

In Section 13.2 we showed how, under the regression model with normally distributed error terms, the least squares estimators were identical to those obtained by the method of maximum likelihood. The same thing is true in the multiple regression situation. Therefore, $b_0, b_1, \ldots, b_k$ are the maximum likelihood estimators for $\beta_0, \beta_1, \ldots, \beta_k$, respectively. In addition, we leave it as an exercise to show that an unbiased estimator based on the maximum likelihood estimator for σ^2 is

$$\widehat{\sigma^2} = \frac{1}{n - (k + 1)} \sum_{i=1}^{n} (y_i - b_0 - b_1 x_1 - \cdots - b_k x_k)^2. \tag{13.25}$$

It should be noted that the summation can be written in matrix notation as

$$\sum_{i=1}^{n} (y_i - b_0 - b_1 x_{1i} - \cdots - b_k x_{ki})^2 = \mathbf{y}'\mathbf{y} - \mathbf{b}'\mathbf{X}'\mathbf{y}. \tag{13.26}$$

Example 13.11   In the situation of Example 13.1, interviewers were also asked to rate each job applicant on a scale from 0 to 100 as to how they felt the interviewee would perform if hired. We wish to use this interview score as an additional independent variable to try to obtain better predictions for performance. The data are given below.

Test (X_1)	20	25	30	35	40	44	48	52	55	67
Interview (X_2)	70	88	65	72	64	80	79	68	76	74
Evaluation (Y)	53	66	52	63	66	72	78	69	75	73

Writing these data in matrix form gives that

$$\mathbf{X} = \begin{pmatrix} 1 & 20 & 70 \\ 1 & 25 & 88 \\ 1 & 30 & 65 \\ 1 & 35 & 72 \\ 1 & 40 & 64 \\ 1 & 44 & 80 \\ 1 & 48 & 79 \\ 1 & 52 & 68 \\ 1 & 55 & 76 \\ 1 & 67 & 74 \end{pmatrix}, \quad \text{and} \quad \mathbf{y} = \begin{pmatrix} 53 \\ 66 \\ 52 \\ 63 \\ 67 \\ 72 \\ 78 \\ 69 \\ 75 \\ 73 \end{pmatrix}.$$

The normal equations are

$$\begin{pmatrix} 10 & 416 & 736 \\ 416 & 19208 & 30616 \\ 736 & 30616 & 54666 \end{pmatrix} \begin{pmatrix} b_0 \\ b_1 \\ b_3 \end{pmatrix} = \begin{pmatrix} 668 \\ 28671 \\ 49478 \end{pmatrix}.$$

Inverting and multiplying gives

$$\begin{pmatrix} b_0 \\ b_1 \\ b_2 \end{pmatrix} = \begin{pmatrix} 6.873 \\ 0.464 \\ 0.552 \end{pmatrix}.$$

Thus, the regression equation is

$$\hat{y} = 6.873 + 0.464x_1 + 0.552x_2.$$

Using this we would predict that the mean evaluation score for an applicant scoring 50 on the screening examination and receiving an interviewer score of 70 would be

$$\hat{y} = 6.873 + 0.464(50) + 0.552(70) = 68.71.$$

The estimate for σ^2 is found by computing

$$\mathbf{y}'\mathbf{y} = \sum_{i=1}^{n} y_i^2 = 45310.$$

Thus, Equation 13.25 gives

$$\widehat{\sigma^2} = \frac{1}{7} \left[45310 - (6.873 \quad 0.464 \quad 0.552) \begin{pmatrix} 668 \\ 28671 \\ 49478 \end{pmatrix} \right] = 18.19.$$

■

If we follow the pattern of development used in Section 13.2, the next step is to consider the properties of the estimators $b_0, b_1, \ldots, b_k$. These results are the natural generalizations of the results of Section 13.2, and we omit the details of the development. The following theorem summarizes these properties.

Theorem 13.7

If the assumptions of the linear regression model, Equation 13.21, are true, then the maximum likelihood estimators for $\beta_0, \beta_1, \ldots, \beta_k$ and σ^2 have the following properties.

1. $E[b_i] = \beta_i$, for $i = 0, 1, \ldots, k$.
2. $\text{Var}(b_i) = a_{ii}\sigma^2$, where a_{ij} is the (i,j)th entry of $\mathbf{X'X}^{-1}$, for $i = 0, 1, \ldots, k$.
3. $\text{Cov}(b_i, b_j) = a_{ij}\sigma^2$.
4. $b_0, b_1, \ldots, b_k$ each has a normal distribution.
5. An unbiased estimator for σ^2 is

$$\widehat{\sigma^2} = \frac{\displaystyle\sum_{i=1}^{n}(y_i - b_0 - b_1x_1 - b_2x_2 - \cdots - b_kx_k)^2}{n - (k+1)} = \frac{\mathbf{y'y} - \mathbf{b'X'y}}{n - (k+1)}.$$

6. $\frac{[n-(k+1)]\widehat{\sigma^2}}{\sigma^2}$ is a random variable that has a χ^2 distribution with $n - (k+1)$ degrees of freedom. In addition, $\widehat{\sigma^2}$ is independent from each $b_i, i = 0, 1, \ldots, k$.

We can use these properties to test hypotheses concerning the β_i and to construct confidence intervals. Properties 1, 2, and 4 show that the random variables

$$Z_i = \frac{b_i - \beta_i}{\sqrt{a_{ii}}\sigma}, \quad i = 0, 1, \ldots, k, \qquad (13.27)$$

each has a standard normal distribution. If the true value for σ^2 is not known, we can substitute $\widehat{\sigma^2}$ and obtain

$$t_i = \frac{b_i - \beta_i}{\sqrt{a_{ii}}\widehat{\sigma}}, \quad i = 0, 1, \ldots, k, \qquad (13.28)$$

each of which has a Student's t distribution with $n - (k+1)$ degrees of freedom. Therefore, to test

$$H_0 : \beta_i = \beta_{i0} \quad \text{vs.} \quad H_1 : \beta_i \neq \beta_{i0}$$

we would compute

$$t = \frac{b_i - \beta_{i0}}{\sqrt{a_{ii}}\,\widehat{\sigma}}$$

and reject H_0 if $|t| > t[n - (k + 1)]_{\alpha/2}$. The most commonly run test of this type is for the case $\beta_{i0} = 0$.

Example 13.12

We wish to conduct two tests. The first is

$$H_0 : \beta_1 = 0 \quad \text{vs.} \quad H_1 : \beta_1 \neq 0.$$

Note that

$$(\mathbf{X'X})^{-1} = \begin{pmatrix} 11.933 & -2.199 \times 10^{-2} & -0.1483 \\ -2.199 \times 10^{-2} & 5.25 \times 10^{-4} & 1.0 \times 10^{-6} \\ -0.1483 & 1.0 \times 10^{-6} & 2.014 \times 10^{-3} \end{pmatrix}.$$

Therefore, $a_{11} = 5.25 \times 10^{-4}$. Under the assumptions of the regression model the test statistic is

$$t = \frac{b_1}{\sqrt{a_{11}}\,\widehat{\sigma}}.$$

This has a Student's t distribution with 7 degrees of freedom. Using a significance level of 0.05 we would reject H_0 whenever

$$|t| > t(7)_{0.025} = 2.365.$$

The computed value for t is

$$t = \frac{0.464}{\sqrt{5.25 \times 10^{-4}}(4.265)} = 4.748.$$

Since this is well above 2.365 we reject H_0 and conclude that the data indicate that x_1 is useful in making predictions regarding y.

To test

$$H_0 : \beta_2 = 0 \quad \text{vs.} \quad H_1 : \beta_2 \neq 0$$

we note that $a_{22} = 2.014 \times 10^{-3}$. Again we shall use $\alpha = 0.05$. The test statistic in this case is

$$t = \frac{b_2}{\sqrt{a_{22}}\,\widehat{\sigma}},$$

which has a Student's t distribution with 7 degrees of freedom. Therefore, we will reject H_0 whenever $|t| > 2.365$. The computed value for t is

$$t = \frac{0.552}{\sqrt{2.014 \times 10^{-3}}(4.265)} = 2.798.$$

Therefore, we again reject H_0 and conclude that x_2 is also useful in making predictions of the mean value for y. It should be noted that, in this second test, we would not have rejected H_0 if a value of $\alpha = 0.01$ was chosen. ∎

To construct confidence intervals for the $\beta's$ we note that, under the assumptions of the regression model

$$t_i = \frac{b_i - \beta_i}{\sqrt{a_{ii}}\hat{\sigma}}, \quad i = 0, 1, \ldots, k,$$

has a Student's t distribution with $n - (k + 1)$ degrees of freedom. This is a pivotal quantity and meets the requirements of Section 11.2. Thus, for a $(1 - \alpha)100\%$ confidence interval for β_i we would write

$$P(-t[n - (k + 1)]_{\alpha/2} \leq t \leq t[n - (k + 1)]_{\alpha/2}) = 1 - \alpha,$$

and solve for β_i to obtain

$$b_i - t[n - (k + 1)]_{\alpha/2}\sqrt{a_{ii}}\hat{\sigma} \leq \beta_i \leq b_i + t[n - (k + 1)]_{\alpha/2}\sqrt{a_{ii}}\hat{\sigma}. \qquad \textbf{(13.29)}$$

Example 13.13   We wish to construct a 99% confidence interval for β_1. We note that $\alpha = 0.01$. Thus,

$$t(7)_{0.005} = 3.499,$$
$$b_1 = 0.464,$$
$$a_{11} = 5.25 \times 10^{-4}, \quad \text{and}$$
$$\widehat{\sigma^2} = 18.19.$$

Equation 13.29 gives

$$0.464 - (3.499)\sqrt{5.25 \times 10^{-4}}(4.265) \leq \beta_1 \leq 0.464 + (3.499)\sqrt{5.25 \times 10^{-4}}(4.265),$$

and a 99% confidence interval for β_1 is

$$0.122 \leq \beta_1 \leq 0.806.$$

∎

It is often desirable to test the hypothesis that *some* of the regression coefficients are zero against the alternative that at least one of them is not. That is, if we assume that the regression model is of the form

$$E[Y] = \beta_0 + \beta_1 x_1 + \cdots + \beta_k x_k,$$

we are interested in testing

$$H_0 : \beta_{l+1} = \beta_{l+2} = \cdots = \beta_k = 0 \quad \text{vs.}$$
$$H_1 : \beta_j \neq 0, \text{ for some } j \text{ in } \{l + 1, l + 2, \ldots, k\}.$$

This test is very useful in determining whether or not we can replace the full model with

$$E[Y] = \beta_0 + \beta_1 x_1 + \beta_2 x_2 + \cdots + \beta_l x_x$$

without impairing our ability to make predictions of $E[Y]$. This test is perfectly general because we can always rearrange and relabel the β's to place a problem in this form. The test statistic is motivated as follows. Suppose we begin by fitting the linear model

$$E[Y] = \beta_0 + \beta_1 x_1 + \cdots + \beta_l x_l$$

to the data using least squares. In this context, the model is usually called the *reduced model*. The linear model using all $k + 1$ coefficients is called the *complete model*. The amount of variation left about the regression line will be

$$\mathbf{y}'\mathbf{y} - \mathbf{b}'_{l+1} \mathbf{X}'_{n \times (l+1)} \mathbf{y}.$$

This quantity is commonly called the *sum of squares for error* (SSE). Since the above sum of squares for error is for fitting the reduced model we shall denote it by SSE_r.

Now fit the complete model to the data and compute the amount of variation left about the new regression line. We shall call this quantity SSE_c and compute it by

$$\text{SSE}_c = \mathbf{y}'\mathbf{y} - \mathbf{b}'_{k+1} \mathbf{X}'_{n \times (k+1)} \mathbf{y}.$$

If $\beta_{l+1}, \beta_{l+2}, \ldots, \beta_k$ are of value in predicting Y, we would expect that SSE_c will be significantly smaller that SSE_r. Another way of stating this, which will be useful for our purposes, is that if H_0 is not true, then $\text{SSE}_r - \text{SSE}_c$ should be significantly larger than 0.

The next task is to convert this idea into a usable test statistic. We shall derive a likelihood ratio test for the problem. Since we are assuming that the regression model is given by Equation 13.21 the likelihood function is

$$L = \left(2\pi\sigma^2\right)^{-n/2} \exp\left[\frac{1}{2\sigma^2} \left(\mathbf{y}'\mathbf{y} - \boldsymbol{\beta}'\mathbf{X}'\mathbf{y}\right)\right].$$

In Ω_0:   Using the notation just introduced and Theorem 13.6 we see that

$$\widehat{\boldsymbol{\beta}}_{l+1} = \mathbf{b}_{l+1} = \left(\mathbf{X}'\mathbf{X}\right)^{-1}_{(l+1) \times (l+1)} \mathbf{X}'_{n \times (l+1)} \mathbf{y}, \quad \text{and}$$

$$\widehat{\sigma^2} = \frac{\text{SSE}_r}{n - (l+1)}.$$

This gives that

$$L_0^* = \left(\frac{n - (l+1)}{2\pi \text{SSE}_r}\right)^{-n/2} \exp\left[\frac{n - (l+1)}{2}\right].$$

In Ω: Since we are using the complete model the maximum likelihood estimators are

$$\widehat{\boldsymbol{\beta}} = \mathbf{b}_{k+1} = \left(\mathbf{X}'\mathbf{X}\right)^{-1}_{(k+1) \times (k+1)} \mathbf{X}'_{n \times (k+1)} \mathbf{y}, \quad \text{and}$$

$$\widehat{\sigma^2} = \frac{\text{SSE}_c}{n - (k+1)}.$$

Evaluating the likelihood function gives

$$L^* = \left(\frac{n - (k + 1)}{2\pi\text{SSE}_c}\right)^{-n/2} \exp\left[\frac{n - (k + 1)}{2}\right].$$

Therefore, the likelihood ratio test will reject H_0 when

$$\lambda = \frac{L_0^*}{L^*} \leq \lambda_0.$$

We leave it as an exercise to show that this is equivalent to rejecting H_0 whenever

$$\frac{\text{SSE}_r - \text{SSE}_c}{\text{SSE}_c} \geq k.$$

Theorem 13.7 shows that when H_0 is true

$$\frac{\text{SSE}_r}{\sigma^2}$$

has a χ^2 distribution with $n - (l + 1)$ degrees of freedom, and

$$\frac{\text{SSE}_c}{\sigma^2}$$

has a χ^2 distribution with $n - (k + 1)$ degrees of freedom. Although it is beyond the level of this book it can be shown that

$$\frac{\text{SSE}_r - \text{SSE}_c}{\sigma^2}$$

has a χ^2 distribution with $k - l$ degrees of freedom, and it is independent of $\dfrac{\text{SSE}_c}{\sigma^2}$. Thus, by recalling the definition of the F distribution from Chapter 9, we see that

$$F = \frac{(\text{SSE}_r - \text{SSE}_c)/(k - l)}{\text{SSE}_c/[n - (k + 1)]} \tag{13.30}$$

has an F distribution with $(k - l)$ and $(n - [k + 1])$ degrees of freedom.

Example 13.14   Continuing with the example we have been using in this section we wish to test

$$H_0 : \beta_1 = \beta_2 = 0, \quad \text{vs.} \quad H_1 : \beta_1 \neq 0 \text{ or } \beta_2 \neq 0$$

using $\alpha = 0.05$. Here the reduced model is

$$E[Y] = \beta_0.$$

This means that $l = 0$. In addition, it is easily verified that $\widehat{\beta_0} = b_0 = \bar{Y}$, and $\mathbf{X'y} = n\bar{Y}$. Therefore,

$$\text{SSE}_r = \mathbf{y'y} - n\bar{Y}^2$$

$$= 45310 - 10(66.8) = 687.6.$$

In the complete model, $k = 2$, and SSE_c is simply the numerator in the estimator for σ^2 that we computed in Example 13.11. Therefore,

$$\text{SSE}_c = 127.32.$$

Now

$$F = \frac{(\text{SSE}_r - \text{SSE}_c) / (k - l)}{\text{SSE}_c / [n - (k + 1)]}$$

will have an F distribution with 2 and 7 degrees of freedom, and we use Table 6 in Appendix A to find that H_0 would be rejected if $F \geq F(2, 7)_{0.05} = 4.74$. We now compute F and find that

$$F = \frac{(687.6 - 127.32) / 2}{127.32 / 7} = 15.40.$$

Therefore, we reject H_0 and conclude that the coefficients for x_1 and x_2 are not both 0.

It should be noted that, given a regression model of the form

$$E[Y] = \beta_0 + \beta_1 x_1 + \beta_2 x_2 + \cdots + \beta_k x_k,$$

the form of the test just discussed where the null hypothesis is

$$H_0 : \beta_1 = \beta_2 = \cdots = \beta_k = 0$$

is part of the standard output for the regression modules of the most widely used computer packages. ∎

Recall in Section 13.2 we introduced the *coefficient of determination, r^2,* as the proportion of the variation in Y that was "explained" by the regression model. It was computed as $r^2 = \frac{\text{SSR}}{\text{SST}}$. If we make the same computation in a multiple regression model, we obtain what is known as the *coefficient of multiple determination,* which is denoted as R^2. It is still computed as

$$R^2 = \frac{\text{SSR}}{\text{SST}}.$$

As before

$$\text{SST} = \mathbf{y'y} - n\bar{Y}^2.$$

From the foregoing discussion it should be easy to see that

$$\text{SSR} = \text{SST} - \text{SSE}_c$$

$$= \mathbf{y'y} - n\bar{Y}^2 - \mathbf{y'y} + \mathbf{b'X'y}$$

$$= \mathbf{b'X'Xb} - n\bar{Y}^2.$$

Thus, we find that

$$R^2 = \frac{\mathbf{b'X'Xb} - n\bar{Y}^2}{\mathbf{y'y} - n\bar{Y}^2}. \tag{13.31}$$

Example 13.15   We wish to compute the value for R^2. First we compute

$$\mathbf{b'X'Xb} = \begin{pmatrix} 6.8727 & 0.4642 & 0.5519 \end{pmatrix} \begin{pmatrix} 10 & 416 & 736 \\ 416 & 19208 & 30616 \\ 736 & 30616 & 54666 \end{pmatrix} \begin{pmatrix} 6.8727 \\ 0.4642 \\ 0.5519 \end{pmatrix}$$

$$= 45183.$$

Then by using Equation 13.30 we find that

$$R^2 = \frac{45183 - 10(66.8)^2}{45310 - 10(66.8)^2} = \frac{560.28}{687.6} = 0.815.$$

∎

This discussion of simple and multiple regression has barely scratched the surface of what is a very big and complicated subject. We have avoided the practical question of how to determine which independent variables are the "best" predictors. There are a number of procedures that have been developed to help with this problem. Among the more popular are the so-called *stepwise regression* methods. They permit the investigator to systematically include or remove independent variables in the model. There are a number of criteria for determining which variable to add or remove. For example, one is assessment of each variable's contribution to the F ratio for the model.

In addition, since the testing and confidence interval procedures we have discussed are based on the assumptions of homoscedasticity and normal distributions for the error terms, we would be interested in being able to verify that such assumptions are reasonable. A common way for dealing with this is to examine the *residuals*. The residual for observation i is commonly denoted as R_i and is defined to be

$$R_i = Y_i - \widehat{Y}_i, i = 1, 2, \ldots, n. \tag{13.32}$$

A number of procedures have been developed to examine the distribution of the residuals. For example, the techniques discussed in Chapter 8 can be used to assess normality.

A third issue that comes to mind is that of fitting models when the relationship between the variables appears to be nonlinear in nature. There are a number of "tricks of the trade" that can be used to "linearize" data so that linear regression methods can be used. For example, it is often possible to transform the data using logarithms or other functions so as to make the relationship between the transformed variables appear linear. In addition, there exist nonlinear curve fitting techniques. These issues, as well as many others, are covered at length in books on regression analysis such as *Applied Regression Analysis* by N. R. Draper and H. Smith.

Exercises 13.4

For Exercises 1–8 refer to the following set of observations.

X_1	1	2	3	4	5	6	7	8	9	10
X_2	4	8	6	2	5	0	9	4	7	3
Y	10.7	7.3	13.1	18.8	17.6	27.8	14.1	25.9	22.0	29.8

1. Fit the model $E[Y] = \beta_0 + \beta_1 x_1 + \beta_2 x_2$ to the data. Estimate the mean value of Y when $x_1 = 3.7$ and $x_2 = 4.4$.

2. Estimate σ^2.

3. Find a 95% confidence interval for β_2.

4. Use the F statistic of Equation 13.30 to test $H_0 : \beta_1 = 0$ vs. $H_1 : \beta_1 \neq 0$ using $\alpha = 0.01$.

5. Test the same hypotheses as Exercise 4 using the t-statistic.

6. Test $H_0 : \beta_1 = \beta_2 = 0$ against the alternative that they are not both 0 using $\alpha = 0.1$.

7. Compute R^2.

8. Process the data using the regression module of a computer statistical package such as MINITAB, SAS, BMDP, or SPSS and compare the results with those of the preceding questions.

9. Fill in the details of the derivation of Equation 13.30. (*Hint:* It will be helpful to resort to the trick you learned in Calculus of using 1 in the right form at the appropriate time.)

   A researcher wanted to use the results of 3 standard achievement tests to predict the intelligence, as measured by IQ scores, of students. Data for a number of students show the following results.

Test 1(X_1)	46	31	52	25	85	95
Test 2(X_2)	85	120	101	121	91	84
Test 3(X_3)	39	35	38	41	42	49
IQ(Y)	133	128	93	91	103	91

For exercises 10–17 refer to the following set of observations.

10. Fit the model $E[Y] = \beta_0 + \beta_1 x_1 + \beta_2 x_2 + \beta_3 x_3$ to the data. Estimate the mean value of Y when $x_1 = 75$, $x_2 = 100$, and $x_3 = 40$.

11. Estimate σ^2.

12. Find a 99% confidence interval for β_1.

13. Test $H_0 : \beta_3 = 0$ vs. $H_1 : \beta_3 \neq 0$ using $\alpha = 0.05$.

14. Test the same hypotheses as Exercise 12.3 using the F statistic of Equation 13.30.

15. Compute R^2.

16. Test $H_0 : \beta_1 = \beta_2 = \beta_3 = 0$ against the alternative that they are not all 0 using $\alpha = 0.05$.

17. Process the data using the regression module of a computer statistical package such as MINITAB, SAS, BMDP, or SPSS and compare the results with those of the preceding questions.

18. Show that

$$\sum_{i=1}^{n} (y_i - b_0 - b_1 x_{1i} - \cdots - b_k x_{ki})^2 = \mathbf{y}'\mathbf{y} - \mathbf{b}'\mathbf{X}'\mathbf{y}.$$

19. In Example 13.14 verify that, in the reduced model, $b_0 = \bar{Y}$.

20. From a computational standpoint there are a number of drawbacks to solving the normal equations directly to obtain the least squares estimates. For example, while the matrix $\mathbf{X}$ has full rank, it can happen that $\mathbf{X}'\mathbf{X}$ will be *almost* singular. In such a case, small changes in the dependent variables can produce wildly different estimates for the regression coefficients. In numerical analysis, this phenomenon is called *instability*. An alternative computational approach has been found to be much more stable. It is based on the use of orthogonal transformations and the fact that it can be shown that, for any $n \times k + 1$ matrix that has full rank, there exists an $n \times n$ orthogonal matrix $\mathbf{Q}$ (i.e., $\mathbf{Q}'\mathbf{Q} = \mathbf{I}$) such that

$$\mathbf{Q}'\mathbf{X} = \mathbf{R}.$$

$\mathbf{R}$ is an $n \times k + 1$ of the form

$$\mathbf{R} = \begin{pmatrix} \mathbf{R}_{k+1} \\ \mathbf{0} \end{pmatrix}$$

with $\mathbf{R}_{k+1}$ being a $k + 1 \times k + 1$ upper triangular matrix of rank $k + 1$ (that is, there are no 0 entries along the main diagonal) and $\mathbf{0}$ being an $n - (k + 1) \times k + 1$ matrix of 0's. The details of determining $\mathbf{Q}$ and $\mathbf{R}$ can be found in *Matrix Computations* by G. H. Golub and C. F. Van Loan. Let

$$\mathbf{y}^* = \mathbf{Q}'\mathbf{y} = \begin{pmatrix} \mathbf{y}^*_{k+1} \\ \mathbf{y}^*_{n-(k+1)} \end{pmatrix}, \quad \text{and}$$

$$\mathbf{e}^* = \mathbf{Q}'\mathbf{e} = \begin{pmatrix} \mathbf{e}^*_{k+1} \\ \mathbf{e}^*_{n-(k+1)} \end{pmatrix}.$$

a. Show that the regression model $\mathbf{Y} = \mathbf{X}\boldsymbol{\beta} + \mathbf{e}$ becomes

$$\begin{pmatrix} \mathbf{y}^*_{k+1} \\ \mathbf{y}^*_{n-(k+1)} \end{pmatrix} = \begin{pmatrix} \mathbf{R}_{k+1} \\ \mathbf{0} \end{pmatrix} \boldsymbol{\beta} + \begin{pmatrix} \mathbf{e}^*_{k+1} \\ \mathbf{e}^*_{n-(k+1)} \end{pmatrix}.$$

b. Use the fact that, if $\mathbf{Q}$ is an orthogonal matrix, then $\|\mathbf{x}\|^2 = \|\mathbf{Q}\mathbf{x}\|^2$, where $\|\mathbf{y}\|^2 = \mathbf{y}'\mathbf{y}$, to show that

$$\|\mathbf{Y} - \mathbf{X}\boldsymbol{\beta}\|^2 = \|\mathbf{Y}^*_{k+1} - \mathbf{X}^*_{k+1}\boldsymbol{\beta}\|^2 + \|\mathbf{Y}^*_{n-(k+1)}\|^2.$$

c. From Part b deduce that the least squares estimators are the solutions to the system

$$\mathbf{X}^*_{k+1}\boldsymbol{\beta} = \mathbf{Y}^*_{k+1}.$$

d. Show that

$$\|\mathbf{Y}^*_{n-(k+1)}\|^2 = \mathbf{y}'\mathbf{y} - \mathbf{b}'\mathbf{X}'\mathbf{y}.$$

Use the following data for Exercises 21–23.

X	1	1.2	1.4	1.6	1.8	2	2.2	2.4	2.6	2.8	3
Y		5.4	3.3	5.1	5.7	6.6	7.2	8.5	9.7	11.7	14.2	16.1

21. When a scatter diagram indicates that a polynomial function might adequately describe the relation between the independent variable x and the dependent variable y it is possible to use multiple linear regression to determine the coefficients. For example, a cubic model would be

$$Y = \beta_0 + \beta_1 x + \beta_2 x^2 + \beta_3 x^3 + e.$$

Note that, if we consider the independent variables to be x, x^2, and x^3, this is a multiple regression model. Use this idea to fit a quadratic polynomial to the following data.

22. Test the adequateness of the quadratic polynomial using the F statistic of Equation 13.30 at the 0.05 level.

23. Use a computer package such as MINITAB, SAS, BMDP, or SPSS to fit a linear, a quadratic, and a cubic polynomial to the data. Carefully review the results and determine which model is the "best" one. State the reasons for your decision.

13.5
Correlation

In regression analysis, we were assuming that the independent variables were fixed by the experimenter and, hence, were not random variables. The error term used in the models meant that the dependent variable was random. It is possible to extend the regression model to include randomness in the independent variables. Models based on this idea are called *generalized regression* models. A discussion of this is well beyond the scope of this text. The purpose of this section is to investigate the case where two random variables, X and Y, are jointly distributed and assess the degree to which they are related. In Section 4.4 we discussed covariance and correlation as measures of the degree to which X and Y were linearly related. Recall that

$$\rho_{XY} = \frac{\text{Cov}(X, Y)}{\sqrt{\text{Var}(X) \ \text{Var}(Y)}}$$

has the following properties:

1. $-1 \le \rho_{XY} \le 1$.
2. If $|\rho_{XY}| = 1$ then X and Y are perfectly linearly related.
3. If X and Y are independent then $\rho_{XY} = 0$, but the converse is not necessarily true.
4. An important case where X and Y are independent if and only if $\rho_{XY} = 0$ is when X and Y have a joint bivariate normal distribution. See Section 5.10.

In probability we were given the joint distribution of X and Y, and the problem was to compute the correlation. From a statistical standpoint the joint distribution of X and Y is assumed to be a member of a family of distributions. This means that, in an estimation setting, the problem is to find estimators and construct confidence intervals for the correlation between X and Y. In testing, we wish to determine if the correlation has a specified value. As was the case in regression model, the most commonly used model is to assume that the two variables are jointly normally distributed. That is, X and Y have a joint distribution that is bivariate normal. The bivariate normal distribution was discussed in Section 5.10. Recall that the joint probability density function is

$$f_{X,Y}(x, y) = \frac{1}{2\pi\sqrt{\sigma_X^2\sigma_Y^2(1 - \rho^2)}}\exp\left\{-\frac{1}{2(1 - \rho^2)}\left[\frac{(x - \mu_X)^2}{\sigma_X^2}\right.\right.$$

$$\left.\left. -2\rho\left(\frac{x - \mu_X}{\sigma_X}\right)\left(\frac{y - \mu_Y}{\sigma_Y}\right) + \frac{(y - \mu_Y)^2}{\sigma_Y^2}\right]\right\},$$

where

$$E[X] = \mu_X,$$

$$E[Y] = \mu_Y,$$

$$\text{Var}(X) = \sigma_X^2,$$

$$\text{Var}(Y) = \sigma_Y^2, \quad \text{and}$$

$$\rho_{XY} = \rho.$$

Our first task is to obtain a maximum likelihood estimator for ρ. The likelihood function for this problem is

$$L(\mu_x, \mu_y, \sigma_X^2, \sigma_Y^2, \rho) = \left[2\pi\sqrt{\sigma_X^2\sigma_Y^2(1 - \rho^2)}\right]^{-n}\exp\left\{-\frac{1}{2(1 - \rho^2)}\sum_{i=1}^{n}\left[\frac{(x_i - \mu_X)^2}{\sigma_X^2}\right.\right.$$

$$\left.\left. -2\rho\left(\frac{x_i - \mu_X}{\sigma_X}\right)\left(\frac{y_i - \mu_Y}{\sigma_Y^2}\right) + \frac{(y_i - \mu_Y)^2}{\sigma_Y^2}\right]\right\}.$$

By taking the logarithm and computing the 5 partial derivatives it is a straightforward, but tedious, job to show that the maximum likelihood estimators are

$$\widehat{\mu_X} = \bar{x},$$

$$\widehat{\mu_Y} = \bar{y},$$

$$\widehat{\sigma_X^2} = \frac{1}{n}\sum_{i=1}^{n}(x_i - \bar{x})^2,$$

$$\widehat{\sigma_Y^2} = \frac{1}{n}\sum_{i=1}^{n}(y_i - \bar{y})^2, \quad \text{and} \tag{13.33}$$

$$\widehat{\rho} = \frac{\displaystyle\sum_{i=1}^{n}(x_i - \bar{x})(y_i - \bar{y})}{\sqrt{\displaystyle\sum_{i=1}^{n}(x_i - \bar{x})^2\sum_{i=1}^{n}(y_i - \bar{y})^2}}.$$

It is common practice to denote $\hat{\rho}$ as r and refer to it as the *sample correlation*. For computational purposes a more convenient form for r is

$$r = \frac{\sum\limits_{i=1}^{n} x_i y_i - n\bar{x}\bar{y}}{\sqrt{\left(\sum\limits_{i=1}^{n} x_i^2 - n\bar{x}^2\right)\left(\sum\limits_{i=1}^{n} y_i^2 - n\bar{y}^2\right)}}. \tag{13.34}$$

When written this way the reader will notice the similarity with the coefficient of determination given in Equation 13.12. It is the square of Equation 13.34. This similarity sometimes leads to a tendency on the part of students to think of correlation and the coefficient of determination as being essentially the same thing. Because of this we wish to repeat what was said earlier. The crucial difference lies not in the computational formula but in the underlying models for the 2 cases. The coefficient of determination measures the degree to which a straight line with fixed values for x describes the variation observed in y. Correlation measures the degree to which 2 *random* variables are linearly related.

Example 13.16   Students in a statistics class were measured in various ways. Two items measured were height and left arm length, to the nearest quarter inch. The data follows.

Height X	Left Arm Y	Height X	Left Arm Y
65.00	27.50	72.00	29.00
73.00	29.00	65.00	28.00
66.00	26.50	62.00	27.00
63.00	24.50	62.25	25.50
61.00	24.00	74.00	28.00
64.00	25.25	71.00	30.00
70.00	29.50	65.00	27.00
65.00	27.00	66.00	25.50
61.00	26.00	71.00	28.00
64.00	26.00	65.00	27.25
74.00	29.00	75.00	30.00
62.50	25.00	70.00	29.00
59.00	24.00	67.00	27.75

From these we compute the following.

$$n = 26 \qquad \bar{x} = 27.125 \qquad \bar{y} = 66.644$$

$$\sum_{i=1}^{n} x_i^2 = 19212 \qquad \sum_{i=1}^{n} y_i^2 = 116007 \qquad \sum_{i=1}^{n} x_i y_i = 47179$$

Using Equation 13.34 gives

$$r = \frac{47179 - (26)(27.125)(66.644)}{\sqrt{[19212 - (26)(27.125)^2][116007 - (26)(66.644)^2]}}$$

$$= 0.858.$$

This shows a strong linear relationship between height and left arm length with a positive slope. That is, as height increases so does the length of one's left arm. This doesn't seem to be a very surprising finding. ∎

In order to construct confidence intervals and to conduct tests of hypotheses concerning ρ_{XY} we need to know something about the distribution of r. Unfortunately, this is an extremely difficult problem. Not much is known about the exact distribution of r except in the case when X and Y are independent normal random variables. This distribution is useful when testing $H_0 : \rho = 0$ against, say, $H_1 : \rho \neq 0$. This test, which seeks to determine if there is a relation between X and Y, is one of the most common uses for correlation analysis. We give the distribution as a theorem without proof.

Theorem 13.8

Let $(X_1, Y_1), (X_2, Y_2), \ldots, (X_n, Y_n)$ be a random sample from a population with a bivariate normal distribution, and let r be the sample correlation. If $\rho = 0$ then

$$t = \sqrt{n - 2}\,\frac{r}{\sqrt{1 - r^2}} \tag{13.35}$$

has a t distribution with $n - 2$ degrees of freedom.

Example 13.17

We wish to use Theorem 13.8 to test

$$H_0 : \rho = 0 \quad \text{vs.} \quad H_1 : \rho \neq 0$$

at the 0.01 level using the sample data from Example 13.16. Recall that $n = 26$ and $r = 0.858$. If we assume that X and Y are jointly normally distributed then

$$t = \sqrt{n - 2}\,\frac{r}{\sqrt{1 - r^2}}$$

will have a t distribution with 24 degrees of freedom. Because this is a two-sided test we would reject H_0 if $|t| \geq t(24)_{0.005}$. From Table 5 of Appendix A we find that $t(24)_{0.005} = 2.797$. Equation 13.34 gives

$$t = \sqrt{24}\,\frac{0.858}{\sqrt{1 - (0.858)^2}} = 8.183.$$

Therefore, we would reject H_0 and conclude that there appears to be a linear relationship between height and left arm length. ∎

Theorem 13.8 is not useful for testing any null hypothesis concerning ρ_{XY} or for constructing confidence intervals. In addition, Theorem 13.8 cannot be used if the joint distribution of X and Y is not bivariate normal. For these other cases about all that can be done is to appeal to the asymptotic distribution of r. The most widely used large sample approximation is known as the *Fisher Z-transformation.* Its development is beyond the level of this text. Therefore, we state it as a theorem and omit the proof.

Theorem 13.9

Let $(X_1, Y_2), (X_2, Y_2), \ldots, (X_n, Y_n)$ be a random sample from a bivariate population where X and Y have correlation ρ. Let r be the sample correlation. Define

$$Z_r = \frac{1}{2} \ln \left(\frac{1+r}{1-r} \right), \quad \text{and}$$

$$Z_\rho = \frac{1}{2} \ln \left(\frac{1+\rho}{1-\rho} \right).$$

(13.36)

Then

$$\lim_{n \to \infty} P \left(\frac{Z_r - Z_\rho}{\frac{1}{\sqrt{n-3}}} \le a \right) = \Phi(a),$$

(13.37)

where $\Phi(a)$ is the distribution function of a $N(0,1)$ random variable.

This theorem states, in effect, that for large samples the distribution of Z_r can be approximated using a normal distribution with a mean of Z_ρ and a variance of $\frac{1}{n-3}$. As is always the case with large sample results it is natural to ask how large a sample is needed before the theorem can be used. Practical experience indicates that the Fisher Z-transformation works well as long as $n \ge 50$.

Example 13.18

A political scientist feels that candidates for the House of Representatives who are running on the issues will make relatively few promises during a campaign but are committed enough to carry through on those promises. On the other hand, a candidate who is merely trying to gather as many votes as possible by appealing to everyone will promise many things and carry through on few of them. In particular, she felt that if the theory were correct there should be a correlation of less than -0.75 between promises made and promises kept after election. To test this idea she sampled 60 successful congressional candidates and computed a value of $r = -0.78$. We wish to test

$$H_0 : \rho = -0.75 \quad \text{vs.} \quad H_1 : \rho < -0.75$$

with $\alpha = 0.01$. Since n is large enough we can use Theorem 13.9 to conduct the test using a normal approximation. Our test statistic will be

$$z = \frac{Z_r - Z_{\rho 0}}{\dfrac{1}{\sqrt{n-3}}},$$

and we will reject H_0 if $z \leq -z_{0.01} = -2.33$. The critical value was found in Table 3 of Appendix A. Now Equation 13.36 gives

$$Z_r = \frac{1}{2} \ln \left[\frac{1 + (-0.78)}{1 - (-0.78)} \right] = -1.045,$$

and

$$Z_{\rho 0} = \frac{1}{2} \ln \left[\frac{1 + (-0.75)}{1 - (-0.75)} \right] = -0.973.$$

Then we use Equation 13.37 to compute

$$z = \frac{-1.045 - (-0.973)}{\dfrac{1}{\sqrt{57}}} = -0.543.$$

Therefore, we cannot reject the null hypothesis. Note that we would have rejected $H_0 : \rho = 0$ in favor of $H_1 : \rho < 0$. This indicates that, while there does appear to be a relation between promises made and promises kept, it is not strong enough to support a claim that it is less than -0.75.   ■

The preceding example invites the question of exactly how one is supposed to interpret a correlation. For example, what does it really mean for 2 variables to have a correlation of -0.75? As was the case for variance in Chapter 4, it turns out that there is no precise meaning that can be attached to particular values of correlation. We can say that a correlation of ± 1 indicates a perfect linear relation between the variables with a slope having the same sign as the correlation. Also, a correlation of 0 indicates that there is no linear relation. However, the interpretation of correlation breaks down when we try to compare, say a correlation of -0.70 with a correlation of -0.80. About all we can say is that the correlation of -0.80 shows a stronger linear relation than does -0.70. It is impossible to attach anything meaningful to that one-tenth difference between them. An experimenter may feel that a relation is "strong" if $|\rho| > 0.80$ and "weak" otherwise, but that is merely a subjective judgment on his or her part.

We now turn to the construction of confidence intervals for ρ. When the sample size is large Theorem 13.9 gives us a quantity that meets the requirements of Section 11.2. That is, we can construct a confidence interval for ρ based on

$$z = \frac{Z_r - Z_\rho}{\dfrac{1}{\sqrt{n-3}}}.$$

We can use the normal distribution to set up the probability statement

$$P\left(-z_{\alpha/2} \le \frac{Z_r - Z_\rho}{1/\sqrt{n-3}} \le z_{\alpha/2}\right) = 1 - \alpha.$$

Solving the inequalities inside the probability statement for Z_ρ gives

$$Z_r - \frac{z_{\alpha/2}}{\sqrt{n-3}} \le Z_\rho \le Z_r + \frac{z_{\alpha/2}}{\sqrt{n-3}}.$$

We can now utilize the fact that $x = Z_\rho$ is a one-to-one transformation to solve for ρ. We leave it as an exercise to show that, if $x = Z_\rho$, then

$$\rho = \frac{e^{2x} - 1}{e^{2x} + 1}.$$

Combining these we find that a large sample $(1 - \alpha)100\%$ confidence interval for ρ is given by

$$\frac{\exp\left[2\left(Z_r - \dfrac{z_{\alpha/2}}{\sqrt{n-3}}\right)\right] - 1}{\exp\left[2\left(Z_r - \dfrac{z_{\alpha/2}}{\sqrt{n-3}}\right)\right] + 1} \le \rho \le \frac{\exp\left[2\left(Z_r + \dfrac{z_{\alpha/2}}{\sqrt{n-3}}\right)\right] - 1}{\exp\left[2\left(Z_r + \dfrac{z_{\alpha/2}}{\sqrt{n-3}}\right)\right] + 1}. \quad (13.38)$$

Example 13.19

We wish to construct a 95% confidence interval for ρ for the correlation of Example 13.18. Recall that $n = 60$ and $Z_r = -1.045$. For 95% confidence we find in Table 3 of Appendix A that $z_{.025} = 1.96$. Therefore, an application of Equation 13.38 gives

$$\frac{\exp\left[2\left(-1.045 - \dfrac{1.96}{\sqrt{57}}\right)\right] - 1}{\exp\left[2\left(-1.045 - \dfrac{1.96}{\sqrt{57}}\right)\right] + 1} \le \rho \le \frac{\exp\left[2\left(-1.045 + \dfrac{1.96}{\sqrt{57}}\right)\right] - 1}{\exp\left[2\left(-1.045 + \dfrac{1.96}{\sqrt{57}}\right)\right] + 1}, \text{ or}$$

$$-0.863 \le \rho \le -0.656$$

As we mentioned earlier it is difficult to attach a firm meaning to the range of values covered by a confidence interval such as this. It covers values that indicate a fairly strong linear relationship with a negative slope. Beyond that little else can be said. ∎

Before leaving this discussion of correlation we need to address the issue of correlation and causation. It must be stressed that the existence of a strong correlation between two random variables in no way implies, from a statistical standpoint, that there is a cause-and-effect relation between them. For example, researchers have noted a strong correlation between smoking and the incidence of lung cancer. While there are probably valid reasons for stating that smoking contributes to the risk of contracting lung cancer, the reasons must be biological and not statistical in nature. Correlations can be used as supporting evidence but cannot be cited as "proof" in such cases. The hazard in using correlation incorrectly is highlighted by the phenomenon known as *spurious correlation*. One way this manifests itself is when a correlation is observed between two variables but there exists a third variable that causes the original two to vary in the manner observed.

Suppose that we observe a positive correlation between coffee consumption and grade point average in college students. One possible interpretation of the results is that coffee is "brain food" and, hence, makes people smarter. Another possibility is that receiving good grades stimulates one's desire for coffee. Both interpretations are, of course, absurd. A more likely interpretation is that students who study more tend to consume more coffee to stay awake and also students who study more tend to receive better grades. Therefore, a third variable—hours spent studying—would be a plausible underlying cause for the observed correlation between coffee consumption and grades. Social scientists have developed a variety of techniques for testing situations like this. An excellent reference is *Social Statistics* by H. M. Blalock.

Exercises 13.5

Use the following data for Exercises 1 and 2. The data are partial results from a heart transplant program. The data given are survival time, in days, for patients who had died before the cutoff date for the study and the age of the recipient, in years.

Survival Time (days) X	Age (years) Y	Survival Time (days) X	Age (years) Y
15	54	1961	33
3	40	136	52
46	42	1	54
623	51	836	44
126	48	60	64
64	54	1996	49
1350	54	0	41
23	56	47	62
279	49	54	49
1024	43	44	36
10	56	994	48
39	42	51	47
730	58		

Source: From Miller, R. G., and Halpern, J. W., *Regression with Censored Data*, *Biometrika* 69 (1982), 521–531.

1. Compute the sample correlation, r. Interpret your results.

2. Assuming that X and Y have a joint bivariate normal distribution, test for a nonzero correlation at the 0.05 level.

   Use the following data for Exercises 3 and 4. A criminologist feels that there is a relationship between sunspot activity and violent crime. The following data give the average number of sunspots per month and the number of violent crimes during that month for a sample of 9 months.

Sunspots X	Violent Crimes Y
54.0	54
73.1	67
69.9	51
36.2	43
46.4	49
2.1	32
19.1	34
17.2	39
36.7	48

3. Compute the sample correlation, r. Interpret your results.

4. Test for a nonzero correlation at the 0.1 level under the assumption that X and Y have a joint bivariate normal distribution.

5. Data for 67 randomly selected automobiles showed a correlation between automobile weight, in pounds, and gasoline consumption, in miles per gallon, of -0.879. Construct a 99% confidence interval for the actual correlation between weight and gasoline consumption.

6. Do the automobile data from Exercise 3 support a claim that the correlation between automobile weight and gasoline consumption is less than -0.85? Let $\alpha = 0.05$.

7. Show that, if the automobile data from Exercise 3 were based on automobile weight being measured in kilograms and mileage in kilometers per liter, the correlation would still be -0.879. That is, show that if $T = aX + b$ and $U = cY + d$ then $r_{XY} = r_{TU}$.

8. Show that if $Y = aX + b$, where $a > 0$, then $r_{XY} = 1$.

9. Show that $Z_\rho = \frac{1}{2} \ln \left(\frac{1+\rho}{1-\rho} \right)$ is a one-to-one transformation.

10. Show that $\rho = \frac{e^{2Z_\rho} - 1}{e^{2Z_\rho} + 1}$.

11. Give another example involving three variables that might give rise to a spurious correlation between two of them.

13.6 Nonparametric Methods

The discussion of regression and correlation analysis in this chapter has relied heavily on the assumption of normal distributions for the random variables. There are, however, situations involving small samples where it is not possible to verify that such assumptions are justified. When sample sizes are very small it is virtually impossible to tell anything about the true distribution of the population. Researchers have determined that many of the estimators and test statistics that we have discussed—particularly those having a t distribution—are relatively insensitive to departures from normality. There are times, however, when we wish to use a procedure that is free from depending on any such assumptions. For example, investigation of regression residuals might show that the error term appears to be highly skewed making the use of normality assumptions questionable. Also, there are times that many investigators like to compare the results of parametric and nonparametric procedures to make sure the results are consistent. If they do not, then the parametric model is probably not correct for the population. For instance, if a test using a t statistic when the population is normal indicates that the null hypothesis should be accepted and a nonparametric test for the same hypotheses indicates that the null hypothesis should be rejected, then it would be wise to go back and carefully look at the data. In this section, we wish to give a brief introduction to how nonparametric ideas have been applied to regression and correlation problems. We shall consider a nonparametric estimator for the slope, β, in simple regression and a rank-based measure of relationship.

We begin with estimation of the slope, β, in simple regression. The model of Equation 13.3 is altered slightly for this situation.

$$Y_i = \alpha + \beta x_i + e_i, \quad \text{where} \quad \begin{cases} E[e_i] = 0, \\ \text{Var}(e_i) = \sigma^2, \quad \text{and} \\ e_i \text{ and } e_j \text{ are uncorrelated if } i \neq j, \end{cases} \quad \textbf{(13.39)}$$

for $i = 1, 2, \ldots, n$. Note that the model has just removed the assumption of normally distributed error terms. Theil has proposed a procedure for estimating β that is based on the following idea. Suppose we take all possible pairs of points in the scatter diagram, connect each pair with a straight line, and compute the slopes of those lines. It seems logical that β should lie somewhere near the "center" of all of these slopes. Since we wish to have a nonparametric estimator use the median of the slopes. To be a little more precise, suppose we have a random sample $(x_1, y_1), (x_2, y_2), \ldots, (x_n, y_n)$ from a population that obeys the model of Equation 13.39. There are $N = \binom{n}{2}$ pairs of observations. Let

$$m_{ij} = \frac{y_j - y_i}{x_j - x_i}, 1 \leq i < j \leq n,$$

be the slopes of straight lines connecting each (x_j, y_j) and (x_i, y_i). Then the Theil estimator for β is

$$\widehat{\beta} = \text{median}(m_{ij}). \quad \textbf{(13.40)}$$

It is important to note that, since we are computing a slope for all possible pairs of data points, the Theil estimator is useful only in samples where each of the x's is distinct. If not, some slopes will have a 0 denominator and will not be defined. This difficulty can be overcome if the procedure is altered to consider only those pairs of observations for which a slope can be computed. In Chapter 8 we observed that the sample median is less sensitive to extreme observations than the sample mean. We can use this idea to compare the least squares estimator for β with the Theil estimator. It is possible to show that the least squares estimator,

$$b = \frac{\displaystyle\sum_{i=1}^{n} (y_i - \bar{y})(x_i - \bar{x})}{\displaystyle\sum_{i=1}^{n} (x_i - \bar{x})^2},$$

is a linear combination of the m_{ij}'s. It is also a weighted average of the slopes

$$\frac{y_i - \bar{y}}{x_i - \bar{x}}.$$

That is, there are weights, w_{ij}, such that

$$b = \sum_{i=1}^{n-1} \sum_{j=i+1}^{n} w_{ij} m_{ij}.$$

Because of this the least squares estimator will be affected by extreme observations in much the same way as the sample mean was in Chapter 8. Similarly, the fact that the Theil estimator uses a median for its computation means that it will be less sensitive to extreme observations than the least squares estimators. This means that large differences between the two estimators can be indication of extremes in the data. The influence of extreme observations, sometimes called *outliers,* on least squares estimators is referred to as *leverage.*

Example 13.20   We wish to compute the Theil estimator for β for the management trainee data of Example 13.1. The observations are repeated below.

Test (X)	20	25	30	35	40	44	48	52	55	67
Evaluation (Y)	53	66	52	63	66	72	78	69	75	73

There are $\binom{10}{2} = 45$ pairs of observations. For example,

$$m_{12} = \frac{66 - 53}{25 - 20} = 2.6.$$

The 45 values for m_{ij} are given below.

i	1	1	1	1	1	1	1	1	1	2	2	2	2	2	2
j	2	3	4	5	6	7	8	9	10	3	4	5	6	7	8
m_{ij}	2.6	−0.1	0.67	0.65	0.79	0.89	0.5	0.63	0.43	−2.8	−0.3	0.0	0.32	0.52	0.11

i	2	2	3	3	3	3	3	3	3	4	4	4	4	4	4
j	9	10	4	5	6	7	8	9	10	5	6	7	8	9	10
m_{ij}	0.3	0.17	2.2	1.4	1.43	1.44	0.77	0.92	0.57	0.6	1.0	1.15	0.35	0.6	0.31

i	5	5	5	5	5	6	6	6	6	7	7	7	8	8	9
j	6	7	8	9	10	7	8	9	10	8	9	10	9	10	10
m_{ij}	1.5	1.5	0.25	0.6	0.26	1.5	−0.38	0.27	0.04	−2.25	−0.43	−0.26	2.0	0.27	−0.17

After placing the slopes in ascending order we find the median is the 23rd order statistic, or

$$\widehat{\beta} = 0.52.$$

This value compares favorably with the least squares estimate of $b = 0.4646$. ∎

We illustrate the leverage effect on least squares estimators with the following example.

Example 13.21   Consider the following data. We wish to compute the least squares estimator and the Theil estimator for β.

Y	5.8	8.3	10.1	12.0	13.8	12.4
X	1	2	3	4	5	8

There are 15 pairs of points. The slopes of the lines connecting them are as shown below.

i	1	1	1	1	1	2	2	2	2	3	3	3	4	4	5
j	2	3	4	5	6	3	4	5	6	4	5	6	5	6	6
m_{ij}	2.5	2.15	2.07	2.0	0.94	1.8	1.85	1.83	0.68	1.9	1.85	0.46	1.80	0.1	−0.47

From these we find that the Theil estimator for β gives

$$\widehat{\beta} = 1.83.$$

The least squares estimator is computed from Equation 13.2 and is

$$b = 0.904.$$

The reason for the large difference between them is due to the outlier at $(8, 12.4)$. If this point is removed, the least squares estimator changes to $b = 1.97$, while the Theil estimator only changes to $\widehat{\beta} = 1.875$. The true equation was

$$Y = 4 + 2x + e.$$

The data along with the straight lines from the least squares and Theil estimators are shown in Figure 13.3. The intercept for the Theil estimator was calculated by taking the median of the y-intercepts of the 15 straight line equations connecting each pair of points. Clearly the nonparametric estimator for slope gives a truer picture of the trend displayed by the bulk of the data. ∎

In order to test hypotheses or construct confidence intervals for β based on the Theil estimator, we need to know the distribution for $\widehat{\beta}$. For small samples this is quite complicated. When sample sizes increase to, say, n larger than 40 we can appeal to an approximation based on the normal distribution. The application of these ideas is somewhat complicated and we shall not discuss them further.

The interested reader is directed to a text on nonparametric statistics such as *Nonparametric Statistical Methods* by M. Hollander and D. A. Wolfe.

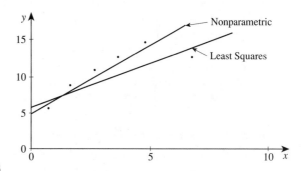

Figure 13.3

In the case of correlation analysis, a popular alternative to the use of the sample correlation r for continuous data is *Spearman's rho*. It is based on ranks in the following way. Suppose we begin with the sample correlation r. For each x value we assign it a rank using the procedure that was given in Section 12.8. Denote the rank of x_i by r_i. Similarly, for each y value we assign it a rank. Denote the rank of y_i by q_i. Now, use Equation 13.33 with r_i replacing x_i and q_i replacing y_i. This gives us Spearman's rho which we denote by r_S. Therefore, Spearman's rho is the sample correlation of the ranks and is computed by

$$r_S = \frac{\sum_{i=1}^{n}(r_i - \bar{r})(q_i - \bar{q})}{\sqrt{\sum_{i=1}^{n}(r_i - \bar{r})^2 \sum_{i=1}^{n}(q_i - \bar{q})^2}}. \tag{13.41}$$

If we define $d_i = r_i - q_i$, we leave it as an exercise to show that an alternative way to write r_S is

$$r_S = 1 - \frac{6\sum_{i=1}^{n}d_i^2}{n^3 - n}. \tag{13.42}$$

This form shows that Spearman's rho measures how the two variables, when converted to ranks, differ from each other. If the two sets of ranks agree completely, then this would be an indication that there is a strong relation between them. If their orders are completely reversed, then this would also indicate a strong inverse relation between the variables. Equation 13.42 is the form that appears in most applied statistics literature. It can be shown that, like r,

$$-1 \leq r_S \leq 1.$$

Thus, it takes on values that look like those of a correlation. Many researchers prefer to refer to Spearman's rho as a *measure of association* rather than a correlation.

Presumably this terminology is intended to emphasize the fact that this measure is not an alternative estimator for ρ in a bivariate normal population. It is important to note that the assumption of continuity in the data implies that there will be no ties in the values of x or y. However, given the measurement process, Spearman's rho is commonly used in cases where rounding of data introduces a moderate number of ties. If, however, the data are truly discrete, then the number of ties becomes too large to make Spearman's rho useful. In such cases, other statistics must be used to measure the degree of relationship between x and y. A discussion of these is beyond the scope of this text. The interested reader is referred to a text such as *Social Statistics* by H. M. Blalock, Jr.

Example 13.22

We wish to compute Spearman's rho for the student data of Example 13.16. The original data along with their ranks are given in Table 13.1.

Table 13.1

Rank r	Height X	Left Arm Y	Rank q	Rank r	Height X	Left Arm Y	Rank q
12	65.00	27.50	12	22	72.00	29.00	21.5
23	73.00	29.00	21.5	12	65.00	28.00	18
15.5	66.00	26.50	10	4	62.00	27.00	12
7	63.00	24.50	3	5	62.25	25.50	6.5
2.5	61.00	24.00	1.5	24.5	74.00	28.00	18
8.5	64.00	25.25	6.5	20.5	71.00	30.00	25.5
18.5	70.00	29.50	24	12	65.00	27.00	12
15.5	65.00	27.00	12	15.5	66.00	25.50	6.5
2.5	61.00	26.00	8.5	20.5	71.00	28.00	18
8.5	64.00	26.00	8.5	12	65.00	27.25	14
24.5	74.00	29.00	21.5	26	75.00	30.00	25.5
6	62.50	25.00	5	18.5	70.00	29.00	21.5
1	59.00	24.00	1.5	17	67.00	27.75	16

Using these results we compute that

$$\sum_{i=1}^{n} d_i^2 = \sum_{i=1}^{n} (r_i - q_i)^2 = 421.5.$$

Therefore, Equation 13.42 gives that

$$r_S = 1 - \frac{6(421.5)}{(26)^3 - 25} = 0.856.$$

Note that this result corresponds well with $r^2 = 0.858$ from Example 13.16. ∎

It must be pointed out that, although Spearman's rho is a measure of the strength of relationship between two variables, it is difficult to interpret exactly what type of relationship is being estimated. Many texts simply state that there is a population quantity, ρ_S, for which r_S is an estimator without addressing the question of what exactly ρ_S is. The distribution of Spearman's rho has been evaluated for small samples for the case when $\rho_S = 0$. Since such a distribution is only useful for testing $H_0 : \rho_S = 0$ against $H_1 : \rho_S \neq 0$ Table 9 of Appendix A gives one-tail critical values for selected values of α. As the number of pairs of observations increases there is a large sample approximation to the distribution of r_S when $\rho_S = 0$ that is based on the normal distribution. It turns out that when $n \geq 10$ the sampling distribution of r_S is well approximated by a normal distribution with mean 0 and variance $\frac{1}{n-1}$.

Example 13.23   We wish to test

$$H_0 : \rho_S = 0 \quad \text{vs.} \quad H_1 : \rho_S \neq 0$$

at the $\alpha = 0.05$ level for the height versus left arm length data. We shall do this using both the exact distribution and the normal approximation. From Example 13.22 we recall that $r_S = 0.856$. Since $n = 26$ and $\frac{\alpha}{2} = 0.025$ we find from Table 9 of Appendix A that the critical value for r_s is 0.392. This means that we would reject H_0 at the 0.05 level if $|r_S| \geq 0.392$. Since $r_S = 0.856$ the exact distribution indicates that we should reject H_0.

A test using the large sample approximation will use the standard normal statistic

$$z = \frac{r_S - 0}{\dfrac{1}{\sqrt{n-1}}}.$$

Table 3 of Appendix A shows that we should reject H_0 if $|z| \geq z_{\alpha/2} = 1.96$. We compute z to be

$$z = \frac{0.856}{\dfrac{1}{\sqrt{26-1}}} = 4.28.$$

Therefore, the large sample approximation also indicates that we should reject H_0.   ■

Exercises 13.6   A professor at Occult University has conjectured that there is a relation between sunspot activity and violent crime. As supporting evidence he gives the following data.

Observed Sunspots (X)	2	3	4	5	6	7	8	9	10	11
Violent Crimes (Y)	5	8	9	12	10	13	15	15	18	17

Use these data for Exercises 1–5.

1. Use Theil's estimator to compute the number of violent crimes that occur per sunspot.

2. Compare your result in Exercise 1 with the least squares estimate for the same quantity. Comment on your results.

3. Use Spearman's rho to measure the relation between sunspots and violent crime.

4. Use the exact distribution to test $H_0 : \rho_S = 0$ against $H_1 : \rho_S > 0$ at the 0.01 level.

5. Repeat the test of Exercise 4 using the normal approximation.

6. Show that, if

$$w_{ij} = \frac{\dfrac{1}{n}(x_j - x_i)}{\displaystyle\sum_{i=1}^{n}(x_i - \bar{x})^2},$$

then the least squares estimator for β is

$$b = \sum_{i=1}^{n-1}\sum_{i+1}^{n} w_{ij}m_{ij}.$$

7. Assume that x and y are continuous random variables and, hence, will have no ties.
   a. Show that
   $$\sum_{i=1}^{n}(r_i - \bar{r})^2 = \sum_{i=1}^{n}(q_i - \bar{q})^2 = \frac{n^3 - n}{12}.$$

   b. Use Part a to show that Equation 13.41 follows from Equation 13.40.

8. State conditions that are necessary for Spearman's rho to achieve a value of 1 and a value of -1. Give examples to illustrate your arguments.

9. Assuming there are no ties in the data $(x_1, y_1), (x_2, y_2), \ldots, (x_n, y_n)$ we can order the ranks of the x values. Therefore, the ranked data can be viewed as looking like $(1, t_1), (2, t_2), \ldots, (n, t_n)$, where t_i is the rank of the y value associated with the x value having a rank of i. When $\rho_S = 0$ the particular permutation of the t's should be a random choice from the $n!$ different permutations of the integers 1 to n.
   a. Use this idea to show that, in Spearman's rho,
   $$E\left[\sum_{i=1}^{n} d_i^2\right] = \frac{n^3 - n}{6}.$$

   b. In addition, show that
   $$\text{Var}\left(\sum_{i=1}^{n} d_i^2\right) = \frac{n^2(n+1)^2(n-1)}{36}.$$

   c. Use Parts a and b to show that, if $\rho_S = 0$, then
   $$E[r_s] = 0 \quad \text{and} \quad \text{Var}(r_S) = \frac{1}{n-1}.$$

Chapter Summary

Scatter Diagram: A plot of (X, Y) pairs to determine type of relationship that might exist between X and Y.   **(Section 13.2)**

Least Squares Estimators: Given

$$\widehat{Y} = aX + b,$$

find slope b, and intercept, a, that minimizes

$$S = \sum_{i=1}^{n} (y_i - a - bx_i)^2$$

$$a = \bar{y} - b\bar{x}, \quad \text{and}$$

$$b = \frac{\displaystyle\sum_{i=1}^{n} x_i y_i - n\bar{x}\,\bar{y}}{\displaystyle\sum_{i=1}^{n} x_i^2 - n\bar{x}^2}.$$

Using the following model

$$Y_i = \alpha + \beta x_i + e_i, \quad \text{where} \quad \begin{cases} e_i \sim N(0, \sigma^2), \\ \text{for } i = 1, 2, \ldots, n, \quad \text{and} \\ e_i \text{ and } e_j \text{ are uncorrelated if } i \neq j. \end{cases}$$

a and b are also the maximum likelihood estimators. An unbiased estimator for σ^2 is

$$s^2 = \frac{n}{n-2} \hat{\sigma}^2.$$

For the linear model defined by the maximum likelihood estimators for α, β are as given above. In addition, we have the following.

1. a and b have a joint bivariate normal distribution with parameters

$$\mu_a = \alpha,$$

$$\mu_b = \beta,$$

$$\sigma_a^2 = \frac{\sigma^2 \displaystyle\sum_{i=1}^{n} x_i^2}{n \left(\displaystyle\sum_{i=1}^{n} x_i^2 - n\bar{x}^2 \right)},$$

$$\sigma_b^2 = \frac{\sigma^2}{\displaystyle\sum_{i=1}^{n} x_i^2 - n\bar{x}^2}, \quad \text{and}$$

$$\rho = \frac{-\bar{x}}{\sqrt{\dfrac{1}{n} \displaystyle\sum_{i=1}^{n} x_i^2}}.$$

2. $\dfrac{n\hat{\sigma}^2}{\sigma^2}$ has a chi-square distribution with $n - 2$ degrees of freedom.

3. a and b are independent of $\widehat{\sigma^2}$.

$$t = (b - \beta)\sqrt{\frac{(n-1)s_X^2}{s^2}} \sim t(n-2).$$

Test of

$$H_0 : \beta = 0 \quad \text{vs.} \quad H_1 : \beta \neq 0$$

will reject H_0 whenever

$$|t| \geq t(b-2)_{\alpha/2}.$$

A $(1 - \alpha)100\%$ confidence interval for β is

$$b - t(n-2)_{\alpha/2}\sqrt{\frac{s^2}{(n-1)s_X^2}} \leq \beta \leq b + t(n-2)_{\alpha/2}\sqrt{\frac{s^2}{(n-1)s_X^2}}.$$

$t = (a - \alpha)\sqrt{\frac{n(n-1)s_X^2}{\sum_{i=1}^{n}x_i^2 s^2}}$ has a t distribution with $n-2$ degrees of freedom.

A $(1 - \alpha)100\%$ prediction interval for a particular value of Y at x_0 is

$$a + bx_0 - t(n-2)_{\alpha/2}s\sqrt{\frac{n+1}{n} + \frac{(x_0 - \bar{x})^2}{(n-1)s_X^2}}$$

$$\leq y_0 \leq a + bx_0 + t(n-2)_{\alpha/2}s\sqrt{\frac{n+1}{n} + \frac{(x_0 - \bar{x})^2}{(n-1)s_X^2}}.$$

A $(1 - \alpha)100\%$ interval for $\mu_{Y|x_0}$ is

$$a + bx_0 - t(n-2)_{\alpha/2}s\sqrt{\frac{1}{n} + \frac{(x_0 - \bar{x})^2}{s_X^2}}$$

$$\leq \mu_{Y|x_0} \leq a + bx_0 + t(n-2)_{\alpha/2}s\sqrt{\frac{1}{n} + \frac{(x_0 - \bar{x})^2}{s_X^2}}.$$

(Section 13.2)

BLUE: Consider the problem of estimating a parameter θ based on a random sample consisting of $y_1, y_2, \ldots, y_n$. If an estimator $\widehat{\theta}$, which is a linear combination of the sample observations, has a distribution whose variance is less than or equal to that of any other estimator that is also a linear combination of the sample observations, then $\widehat{\theta}$ is said to be a *best linear unbiased estimator (BLUE)* for θ.   **(Section 13.2)**

Gauss-Markov Theorem: Let the simple regression model, $Y = \alpha + \beta x_i + e_i$, for $i = 1, 2, \ldots, n$, be such that each x_i is fixed, each Y_i is an observable random variable and each e_i is an unobservable random variable. In addition, suppose that the random variables e_i each have $E[e_i] = 0$, $\text{Var}(e_i) = \sigma^2$ and $\text{Cov}(e_i, e_j) = 0$, if $i \neq j$. Then the least squares estimators for α and β are best linear unbiased estimators.

Coefficient of determination, r^2 is

$$r^2 = \frac{\left[\sum_{i=1}^{n} X_i Y_i - n\bar{X}\bar{Y}\right]^2}{\left[\sum_{i=1}^{n} X_i^2 - n\bar{X}^2\right]\left[\sum_{i=1}^{n} Y_i^2 - n\bar{Y}^2\right]}. \qquad \textbf{(Section 13.2)}$$

Matrix Calculus: The function $y = f(x_1, x_2, \ldots, x_p)$ such that

$$y = \sum_{i=1}^{p} \sum_{j=1}^{p} a_{ij} x_i x_j$$

$$= y = \mathbf{x}'\mathbf{A}\mathbf{x},$$

where $a_{ij}, i = 1, 2, \ldots, p, j = 1, 2, \ldots, p$ is a set of constants and $-\infty < x_i < \infty, i = 1, 2, \ldots, p$ is said to be a *quadratic form* in the variables x_i.

The covariance matrix, $\boldsymbol{\Sigma}$, for the random vector $\mathbf{x}$ is

$$\boldsymbol{\Sigma} = \begin{pmatrix} \sigma_{11} & \sigma_{12} & \cdots & \sigma_{1p} \\ \sigma_{21} & \sigma_{22} & \cdots & \sigma_{2p} \\ \vdots & \vdots & & \vdots \\ \sigma_{p1} & \sigma_{p2} & \cdots & \sigma_{pp} \end{pmatrix}.$$

Let f be a function of the p real variables $x_1, x_2, \ldots, x_p$. The derivative of f with respect to the vector $\mathbf{x}$ is denoted by $\frac{\partial f}{\partial \mathbf{x}}$ and is defined to be

$$\frac{\partial f}{\partial \mathbf{x}} = \begin{pmatrix} \dfrac{\partial f}{\partial x_1} \\ \dfrac{\partial f}{\partial x_2} \\ \vdots \\ \dfrac{\partial f}{\partial x_p} \end{pmatrix}.$$

Let $f(\mathbf{x})$ be a linear combination of the elements of the vector $\mathbf{x}_p$ defined by

$$f(\mathbf{x}) = \sum_{i=1}^{p} a_i x_i = \mathbf{a}'\mathbf{x} = \mathbf{x}'\mathbf{a},$$

where

$$\mathbf{a} = \begin{pmatrix} a_1 \\ a_2 \\ \vdots \\ a_p \end{pmatrix},$$

and the a_i are real constants. Then

$$\frac{\partial f}{\partial \mathbf{x}} = \mathbf{a}.$$

Let $y(\mathbf{x})$ be a quadratic form in the p real variables $x_1, x_2, \ldots, x_p$, be defined by

$$y(\mathbf{x}) = \mathbf{x}'\mathbf{A}\mathbf{x},$$

where $\mathbf{A}$ is a real, symmetric $p \times p$ matrix. Then

$$\frac{\partial y}{\partial \mathbf{x}} = 2\mathbf{A}\mathbf{x}.$$

Let $\mathbf{A}$ be an $m \times n$ matrix of rank r. Then the rank of $\mathbf{A}'\mathbf{A}$ is r, and the rank of $\mathbf{A}\mathbf{A}'$ is r.

Let $\mathbf{A}$ be an $m \times n$ matrix of rank r and $\mathbf{B}$ be an $n \times p$ matrix of rank s. Then the rank of $\mathbf{A}\mathbf{B}$ is less than or equal to $\min(r, s)$.

The normal equations for simple regression can be written in matrix form as

$$\mathbf{X}'\mathbf{X}\mathbf{b} = \mathbf{X}'\mathbf{y},$$

and the least squares estimators become

$$\mathbf{b} = (\mathbf{X}'\mathbf{X})^{-1}\mathbf{X}'\mathbf{y}. \qquad \text{(Section 13.3)}$$

Multiple Regression:

Model

$$Y_i = \beta_0 + \beta_1 x_{1i} + \beta_2 x_{2i} + \cdots + \beta_k x_{ki} + e_i,$$

where

$$\begin{cases} e_i \sim N(0, \sigma^2) \\ \text{for } i = 1, 2, \ldots, n, \quad \text{and} \\ e_i \text{ and } e_j \text{ are uncorrelated if } i \neq j. \end{cases}$$

Define

$$\boldsymbol{\beta} = \begin{pmatrix} \beta_0 \\ \beta_1 \\ \beta_2 \\ \vdots \\ \beta_k \end{pmatrix}, \; \mathbf{y} = \begin{pmatrix} y_1 \\ y_2 \\ \vdots \\ y_n \end{pmatrix}, \mathbf{X} = \begin{pmatrix} 1 & x_{11} & x_{12} & \cdots & x_{1k} \\ 1 & x_{21} & x_{22} & \cdots & x_{2k} \\ \vdots & \vdots & \vdots & & \vdots \\ 1 & x_{n1} & x_{n2} & \cdots & x_{nk} \end{pmatrix}, \text{ and } \mathbf{e} = \begin{pmatrix} e_1 \\ e_2 \\ \vdots \\ e_n \end{pmatrix}.$$

Then the model becomes

$$\mathbf{y} = \mathbf{X}\boldsymbol{\beta} + \mathbf{e}.$$

The normal equations are

$$\mathbf{X}'\mathbf{X}\mathbf{b} = \mathbf{X}'\mathbf{y}.$$

The least squares estimators are

$$\mathbf{b} = (\mathbf{X'X})^{-1}\mathbf{X'y}.$$

$b_0, b_1, \ldots, b_k$ are the maximum likelihood estimators for $\beta_0, \beta_1, \ldots, \beta_k$, respectively.

An unbiased estimator based on the maximum likelihood estimator for σ^2 is

$$\widehat{\sigma^2} = \frac{1}{n-(k+1)} \sum_{i=1}^{n} (y_i - b_0 - b_1 x_1 - \cdots - b_k x_k)^2$$

$$= \frac{1}{n-(k+1)} \mathbf{y'y} - \mathbf{b'X'y}.$$

If the assumptions of the linear regression model are true, then the maximum likelihood estimators for $\beta_0, \beta_1, \ldots, \beta_k$ and σ^2 have the following properties.

1. $E[b_i] = \beta_i$, for $i = 0, 1, \ldots, k$.
2. $\text{Var}(b_i) = a_{ii}\sigma^2$, where a_{ij} is the (i,j)th entry of $\mathbf{X'X}^{-1}$, for $i = 0, 1, \ldots, k$.
3. $\text{Cov}(b_i, b_j) = a_{ij}\sigma^2$.
4. $b_0, b_1, \ldots, b_k$ each has a normal distribution.
5. An unbiased estimator for σ^2 is

$$\widehat{\sigma^2} = \frac{\displaystyle\sum_{i=1}^{n}(y_i - b_0 - b_1 x_1 - b_2 x_2 - \cdots - b_k x_k)^2}{n-(k+1)} = \frac{\mathbf{y'y} - \mathbf{b'X'y}}{n-(k+1)}.$$

6. $\frac{[n-(k+1)]\widehat{\sigma^2}}{\sigma^2}$ is a random variable that has a χ^2 distribution with $n - (k + 1)$ degrees of freedom. In addition, $\widehat{\sigma^2}$ is independent from each $b_i, i = 0, 1, \ldots, k$.

To test

$$H_0 : \beta_i = \beta_{i0} \quad \text{vs.} \quad H_1 : \beta_i \neq \beta_{i0}.$$

Test statistic

$$t = \frac{b_i - \beta_{i0}}{\sqrt{a_{ii}}\,\widehat{\sigma}},$$

has a t distribution with $n - (k + 1)$ degrees of freedom. Reject H_0 if $|t| \geq t[n - (k + 1)]_{\alpha/2}$.

A $(1 - \alpha)100\%$ confidence interval for β_i is

$$b_i - t[n - (k + 1)]_{\alpha/2}\sqrt{a_{ii}}\,\widehat{\sigma} \leq \beta_i \leq b_i + t[n - (k + 1)]_{\alpha/2}\sqrt{a_{ii}}\,\widehat{\sigma}.$$

To test

$$H_0 : \beta_{l+1} = \ldots = \beta_k = 0 \text{ vs. } \beta_j \neq 0, \text{ for some } j \text{ in } \{l + 1, \ldots, k\}.$$

Let

$$SSE_c = \mathbf{y'y} - \mathbf{b}'_{k+1}\mathbf{X}'_{n\times(k+1)}\mathbf{y}, \quad \text{and}$$

$$SSE_r = \mathbf{y'y} - \mathbf{b}'_{l+1}\mathbf{X}'_{n\times(l+1)}\mathbf{y}.$$

Test statistic

$$F = \frac{(SSE_r - SSE_c)/(k-l)}{SSE_c/[n-(k+1)]}$$

has an F distribution with $(k-l)$ and $(n-[k+1])$ degrees of freedom. Reject H_0 whenever

$$F(k-1, n-[k+1])_\alpha.$$

The coefficient of multiple determination is

$$R^2 = \frac{\mathbf{b'X'Xb} - n\bar{Y}^2}{\mathbf{y'y} - n\bar{Y}^2}. \qquad \textbf{(Section 13.4)}$$

Correlation: Let (X, Y) have a bivariate normal distribution. The maximum likelihood estimator for the correlation ρ is the sample correlation

$$r = \frac{\displaystyle\sum_{i=1}^{n} x_i y_i - n\bar{x}\bar{y}}{\sqrt{\left(\displaystyle\sum_{i=1}^{n} x_i^2 - n\bar{x}^2\right)\left(\displaystyle\sum_{i=1}^{n} y_i^2 - n\bar{y}^2\right)}}.$$

Let $(X_1, Y_1), (X_2, Y_2), \ldots, (X_n, Y_n)$ be a random sample from a population with a bivariate normal distribution, and let r be the sample correlation. If $\rho = 0$ then

$$t = \sqrt{n-2}\frac{r}{\sqrt{1-r^2}}$$

has a t distribution with $n-2$ degrees of freedom.

To test

$$H_0 : \rho = 0, \quad \text{vs.} \quad H_1 : \rho \neq 0.$$

Test statistic

$$t = \sqrt{n-2}\frac{r}{\sqrt{1-r^2}}$$

has a t distribution with $(n-2)$ degrees of freedom. Reject H_0 whenever $|t| \geq t(n-2)_{\alpha/2}$.

Let $(X_1, Y_1), (X_2, Y_2), \ldots, (X_n, Y_n)$ be a random sample from a bivariate population where X and Y have correlation ρ. Let r be the sample correlation.

Define

$$Z_r = \frac{1}{2} \ln \left(\frac{1+r}{1-r} \right), \quad \text{and}$$

$$Z_\rho = \frac{1}{2} \ln \left(\frac{1+\rho}{1-\rho} \right).$$

Then

$$\lim_{n \to \infty} P \left(\frac{Z_r - Z_\rho}{\frac{1}{\sqrt{n-3}}} \le a \right) = \Phi(a),$$

where $\Phi(a)$ is the distribution function of a $N(0,1)$ random variable.

To test

$$H_0 : \rho = \rho_0 \quad \text{vs.} \quad H_1 : \rho \neq \rho_0.$$

Test statistic

$$z = \frac{Z_r - Z_{\rho_0}}{\frac{1}{\sqrt{n-3}}},$$

and reject H_0 if $|z| \neq z_{\alpha/2}$.

A large sample $(1 - \alpha)100\%$ confidence interval for ρ is

$$\frac{\exp \left[2 \left(Z_r - \frac{z_{\alpha/2}}{\sqrt{n-3}} \right) \right] - 1}{\exp \left[2 \left(Z_r - \frac{z_{\alpha/2}}{\sqrt{n-3}} \right) \right] + 1} \le \rho \le \frac{\exp \left[2 \left(Z_r + \frac{z_{\alpha/2}}{\sqrt{n-3}} \right) \right] - 1}{\exp \left[2 \left(Z_r + \frac{z_{\alpha/2}}{\sqrt{n-3}} \right) \right] + 1}.$$

(Section 13.5)

Nonparametric Methods:

Simple regression

Model

$$Y_i = \alpha + \beta x_i + e_i, \quad \text{where} \quad \begin{cases} E[e_i] = 0 \\ \text{Var}(e_i) = \sigma^2, \quad \text{and} \\ e_i \text{ and } e_j \text{ are uncorrelated if } i \neq j, \end{cases}$$

for $i = 1, 2, \ldots, n$.

Let

$$m_{ij} = \frac{y_j - y_i}{x_j - x_i}, 1 \le i < j \le n,$$

be the slopes of straight lines connecting each (x_j, y_j) and (x_i, y_i).

The Theil estimator for β is

$$\hat{\beta} = \text{median}(m_{ij}).$$

Correlation: Let r_i be the rank of x_i and q_i be the rank of y_i.

Spearman's rho is

$$r_S = \frac{\displaystyle\sum_{i=1}^{n}(r_i - \bar{r})(q_i - \bar{q})}{\sqrt{\displaystyle\sum_{i=1}^{n}(r_i - \bar{r})^2 \sum_{i=1}^{n}(q_i - \bar{q})^2}}.$$

Let $d_i = r_i - q_i$, then

$$r_S = 1 - \frac{6\displaystyle\sum_{i=1}^{n} d_i^2}{n^3 - n}.$$

For large n and $\rho_S = 0$, r_S has an approximate $N\left(0, \frac{1}{(n-1)}\right)$ distribution. **(Section 13.6)**

Review Exercises

A college admissions office is attempting to develop a mathematical model to predict success among incoming Freshmen. The data for 8 randomly selected students in the current Freshman class are given below.

Freshman GPA (Y)	2.2	2.5	3.0	3.5	3.2	2.4	3.1	3.3
High School GPA (X_1)	3.3	3.5	3.7	3.4	3.8	3.0	3.9	2.9
SAT Score (X_2)	1154	1317	1422	1247	1078	1247	1390	1423

Use these data for Exercises 1–6.

1. Use least squares to fit a model

$$y = \beta_0 + \beta_1 x_1 + \beta_2 x_2 + e$$

to the data. Use the model to predict the Freshman grade point average for an incoming student with a high school grade point average of 2.8 and combined SAT scores of 1250.

2. Assuming that the error terms are independent and normally distributed test the hypothesis that a combined SAT score can be used to help predict a student's first year grade point average.

3. Assuming that the error terms are normally distributed, test, at the 0.05 level, the hypothesis that $\beta_1 = \beta_2 = 0$ against the alternative that at least one differs from zero.

4. Is there a correlation between high school grade point average and Freshman grade point average? Assume normally distributed data and use a significance level of 0.01.

5. Compute a Theil estimate of β_1.

6. Use Spearman's rho to measure the association between high school grade point average and Freshman grade point average.

7. Do the data indicate an association between the two variables at the 0.01 level?

8. Suppose we wish to use least squares to estimate a linear model of the form

$$y = \alpha + \beta x + e.$$

In addition, we wish to use n values for x in the interval $[-1, 1]$.

a. What values of x should be used to minimize $Var(a)$?

b. What values of x should be used to minimize $Var(b)$?

9. Describe how you might transform data from each of the following models by using a function of the form $V_i = g(Y_i), i = 1, 2, \ldots, n$, so that multiple linear regression could be used to estimate the parameters.

a. $y = \beta_0 x^{\beta_1} e$

b. $y = \beta_0 \frac{\beta_1 x_1}{\beta_2 x_2} e$

c. $y = (\beta_0 + \beta_1 x_1 + \beta_2 x_2 + e)^{1/2}$

14

Analysis of Variance

14.1
Introduction

In regression analysis we assumed that a linear model was appropriate to describe the observed data. The goal from a statistical standpoint was to determine the values of the coefficients in that model and to test hypotheses concerning them. Most investigators use regression models as predictive tools. In fact, it frequently is the first method discussed when one is introduced to economic forecasting. The tests performed are primarily aimed at assessing the validity of the model. Analysis of variance, on the other hand, is primarily interested in the testing of hypotheses concerning linear models. Before going any further we must emphasize that the term "analysis of variance" does not imply that we will be analyzing the variance that we introduced in Chapter 4. It is perhaps more correct to say that we are analyzing means by identifying sources of variability of data. Analysis of variance can be looked at in a number of ways. In its simplest form it can be considered an extension of the two group test we derived in Section 12.5. There we were presented with two normally distributed populations and wished to test the hypothesis that their means were equal. The so-called *one-way analysis of variance* is the natural generalization of this to a test of equality of the means of more than two independent, normally distributed populations. We shall discuss this problem in Section 14.3.

If, however, experimental subjects can be classified according to two different types of populations, we have a more complicated situation. For example, suppose we are interested in whether or not children who view television shows containing violent behavior tend to behave more aggressively. This would suggest we have two independent populations—those children who view violent television programs and those children who do not. However, it might be logical to suspect that young boys might, because of the way society treats male and female children, tend to be more aggressive in their behavior than young girls. In addition, it might be reasonable to ask if young boys watching violent television programming are more or less affected than young girls watching the same shows. Thus, it seems that we have three possible contributing factors to aggressive behavior in children:

1. observing violent television shows,
2. sex of the child, and
3. a difference in the way male and female children react to violence on television.

We have two primary contributors to aggressive behavior and what appears to be an interaction between these two. The statistical procedure that is applied to such problems is known as *two-way analysis of variance with interaction*. We discuss these types of problems in Section 14.4. The problem just outlined leads one to ask how a researcher should go about collecting data efficiently to assess the contribution of each of the three possible factors to aggressiveness in behavior. The key word here is *efficiently*. It is certainly possible to conduct three separate experiments, but practitioners have devised methods that can use a single experiment. Because of this we begin the chapter with a brief discussion of the factors one should consider in designing an experiment that will address the questions of interest to an experimenter.

Suppose that we conduct an experiment that compares the mean response of members of three independent populations to a particular stimulus. For example, we might be interested in the difference in growth of corn, if any, when it is treated with three different kinds of pesticide. If it is determined that there appears to be a difference (that is, the null hypothesis of equality is rejected), the next step is to determine, as well as possible, the exact nature of the difference. This is an estimation problem. Unfortunately, in the type of linear models we will be discussing in this chapter, it will not be possible simply to compute estimators like we did in the regression case. The reason is that the data matrix, $\mathbf{X}$, is not of full rank, which means that $\mathbf{X}'\mathbf{X}$ will be singular. All is not lost, however, as we shall see in Section 14.5.

Finally, in keeping with the tradition of the preceding chapters, we shall take a brief look at some of the nonparametric alternatives to the standard procedures. The classical procedures are, as was the case in regression, based on the assumption of normally distributed data. Testing procedures based on ranks can be used when such assumptions are not warranted.

14.2 Experimental Design

Experiments that make use of analysis of variance procedures are inherently different than those discussed thus far. For the most part the preceding chapters have been interested in collecting samples from existing populations. For example, a pollster might sample members of the voting population to ascertain which of two mayoral candidates is currently ahead. A sociologist interested in racial discrimination could sample members of the work force in an attempt to determine if blacks earn less than whites in the same occupations. The point is that these populations are existing physical entities, and we simply go out and select a sample from them. In experimental design, however, the populations do not exist until the researcher creates them. In the Salk polio vaccine trial, for instance, the population of children inoculated with the vaccine did not exist until those conducting the trial administered the vaccine. In other words, in this situation the experimenter does something to create a difference between populations. In our example in the introduction, the researcher creates a population of children who view violent television programs by actually setting up a situation where a selected group of children is exposed to such shows. We emphasize that the experimenter does not actually "create" a population but, rather, attempts to generate typical sample units from such a population. It is common practice to refer to this process of creating a difference as a *treatment*.

Treatment can be accomplished in a number of different ways. In the polio vaccine experiment, all patients were actually given injections (treatments). For some the injection contained the vaccine, and for others the injection contained a harmless solution—a placebo. In the children and television example we have two different methods of treatment: one in which the children as a group watched violent or nonviolent television programming and another in which the sex of the child was viewed as a treatment. The experiment was not aimed at showing that there is a biological reason for boys to be more or less aggressive than girls. Instead, it was hypothesized that, by being a particular sex, society (that is, parents, peers, etc.) would administer a treatment that might tend to make children of one sex more aggressive than those of the other. The variables that the experimenter is able to control completely in the design of an experiment are called *independent variables*. Using this terminology, vaccination, type of television program seen, and the sex of a child are things that are under the control of the researcher and, hence, are independent variables. Another term for an independent variable is *factor*. Therefore, as we have described them, the Salk polio vaccine trial would be called a *single-factor experiment,* and the children and television investigation would be called a *two-factor experiment*. It is also common practice to refer to a factor as a *main effect*.

One last piece of terminology should be mentioned before getting on with our discussion of experimental design. In the polio vaccine trial, patients were given the treatment in one of two different ways. They either received the vaccine or the placebo. These different treatment methods are referred to as the treatment or factor *levels*. In the children's experiment, the treatment based on the child's sex had two levels—male and female. The television program treatment also had two levels—violent and nonviolent. This treatment could, however, have had three levels, say, no violence, moderate violence, and high violence. It is the task of the designer of the experiment to determine the factors to be used, the levels of each factor, and how to administer the treatments to obtain the desired information.

Let us begin by considering a single-factor experiment. A number of methods have been proposed in both professional and popular literature to help people cope with stress. Among these are transcendental meditation (TM), regular exercise (E), and diet (D). We wish to design an experiment to determine if there is a difference in the abilities of these three methods to reduce stress. Before reading further you might stop for a minute and consider how to design an experiment to accomplish this. We assume that we have the ability to measure the stress level in an individual and to obtain patients who have been identified as having high stress levels. The type of experiment we shall describe for this problem is known as a *completely randomized design*. The idea is simple and quite intuitive. Since there are three treatment levels we randomly select a group of people suffering from high stress. Then we randomly assign each of these people to one of the three stress reduction programs. We can, if we wish, make the selections in such a way that equal numbers are assigned to each treatment level. The important thing is to ensure that each subject has an equal chance of being placed in each program. Such assignments can be made by using random number tables such as shown in Table 8.1 (see page 330) or by using computer random number generators.

Example 14.1   To illustrate the use of random number tables to assign subjects to treatment levels suppose that we have access to a group of 1000 people suffering from high stress. We wish to conduct a completely randomized experiment where 3 people are assigned to each group. We would begin by numbering the members of the population from 000 to 999. Then beginning, say, at the top of the second group of 5-digit numbers in Table 8.1 and using the last 3 digits and proceeding down the column, we would select the first 9 distinct 3-digit numbers and assign them to treatments in some predetermined order such as TM, E, D, TM, E, D, TM, E, D. Following this scheme our experimental units would be assigned as follows.

Transcendental Meditation	Exercise	Diet
899,735,390	629,035,300	496,716,674

■

As a second example suppose we wish to examine the effect of various types of fertilizers on the yield of wheat crops. For simplicity's sake, assume that there are four different fertilizers called A, B, C, and D. A completely randomized design might set aside four plots of land and then randomly assign the four fertilizers to the plots, one per plot. While this is a perfectly valid way to design an experiment it has some drawbacks. There are many variables that affect wheat crops besides type of fertilizer. Such things as soil type, drainage, and wind also contribute to crop yields. Since we are interested in being able to detect, as well as possible, differences caused by fertilizers, it is desirable to reduce the effects of any other influences. In engineering parlance these other factors constitute *noise*, which might tend to drown out the fertilizer differences. It is much like trying to listen to a long distance phone call over lines with a lot of static. One way to reduce this noise in our wheat experiment is to break the plots up into smaller subplots, called *blocks*. Thus, each plot could be divided into four blocks. Within each plot we would randomly assign one of the fertilizers to each block. This would eliminate whatever differences there might be as a result of the location of the plots. This type of design is known as *randomized complete blocks*. The term "complete" simply means that each plot contains each treatment level. As an example, a randomized complete block experiment with four plots might look like Figure 14.1.

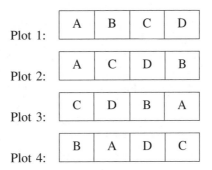

Plot 1: | A | B | C | D |

Plot 2: | A | C | D | B |

Plot 3: | C | D | B | A |

Plot 4: | B | A | D | C |

Figure 14.1

A randomized complete block design can also be viewed as a 2-factor experiment. Sometimes it is interesting to know if blocking the data has a significant effect on the reduction of noise in the data. In such cases, blocking is viewed as a second factor in the experiment. In many experiments, the researcher wishes to investigate more than 1 factor on a given response. For example, in the wheat study we might also want to study the effects of various methods of pest control on crop yield. Since data are typically expensive to obtain it is desirable to design an experiment where information regarding both factors can be obtained. Suppose that it is desired to grow wheat with each type of fertilizer in 4 plots and each type of pest control method in 4 plots. In the terminology of experimental design, the experiment is said to have 4 *replications*. If we were to conduct 2 single-factor experiments we would need 16 plots to study fertilizers and 16 plots to investigate pest control for a total of 32 plots. Now suppose that we form all possible fertilizer-pesticide combinations. If we denote the fertilizers as F_1, F_2, F_3, and F_4, and the pesticides as P_1, P_2, P_3, and P_4 these will be $F_1P_1, F_1P_2, F_1P_3, \ldots, F_4P_4$. There are a total of 16 such combinations. Further note that in these 16 combinations each fertilizer and each pesticide appear 4 times. Thus, if we plant 16 plots with one fertilizer-pesticide combination in each we have achieved our goal at half the cost of conducting 2 single-factor experiments. An experiment such as we have just described is called a *factorial experiment*. An additional benefit of a factorial experiment over multiple single-factor experiments is that we can also ascertain if there appears to be what is called an *interaction* between the factors. In other words, the combination of a particular fertilizer and pesticide may affect crop yield in a manner that differs from that of each taken separately. For instance, there may be a chemical reaction between a fertilizer and a pesticide that does not occur with other pesticides and that kills wheat plants.

In some cases, it is desirable to have additional structure imposed when designing blocked experiments. Suppose that the plots in Figure 14.1 constitute a large square area that has been divided into 16 square subplots. It is common that soil types vary over regions in bands. For example, there may be a vein of particularly rich soil passing down the left edge of the plot. To control for the systematic effect of this soil difference we would like to make sure that our design assigns each type of fertilizer to this rich soil. In other cases, the soil patterns in the plot may be unknown and we would like to make sure that our design does not inadvertently assign one type of fertilizer to all of the rich soil deposits. A popular way of doing this is to use a *Latin square design.* Some readers may have already been exposed to Latin squares in abstract algebra or combinatorics courses. A Latin square is a design where an area is divided into an equal number of rows and columns. In the wheat example, since there are 4 different types of fertilizer, we would divide our area into 4 rows and 4 columns. Then each type of fertilizer is assigned to 4 subplots such that each type of fertilizer appears only once in any row or column. This would then constitute a 4×4 *Latin square.* An example of one such Latin square is given in Figure 14.2.

A	B	C	D
B	C	D	A
C	D	A	B
D	A	B	C

Figure 14.2

As we have seen in previous chapters randomization is necessary in order to carry out the basic statistical tasks of estimation and hypothesis testing. In addition, replication is also necessary to make sure that we have samples of size larger than one unit. If we only have one observation of a treatment level it will be impossible to perform the types of analysis we have discussed in earlier chapters. For instance, suppose we have a single factor with two levels and we have a single observation at each level. The likelihood ratio test developed in Section 12.5 cannot be performed because the *t*-statistic will have zero degrees of freedom. Thus, we need at least two observations on at least one of the levels. In addition, it is also necessary to have an untreated group as a *control* to assess whether any treatment effect can be attributed to the factor being tested or not.

The particular experimental designs discussed here are only a few of the many that are used by researchers. They are used by applied statisticians in all fields. Some excellent references include the following:

1. *The Design and Analysis of Experiments* by O. Kempthorne.
2. *Design and Analysis of Industrial Experiments* edited by O. L. Davies.
3. *Statistics in Research* by B. Ostle.

Before closing we wish to mention some general guidelines for designing an experiment. The application of these can make the difference between a study that arrives at meaningful results and one that is misleading. In any well-designed experiment, almost all of the mental work should take place before a single piece of data is collected.

1. Make sure that the objectives of the experiment are clearly set forth and understood. The number of factors should be decided upon. For example, in the children and aggression situation do we wish to test for a difference in the level of aggressive behavior or do we wish to estimate the amount of difference? What is our definition of aggressive behavior and how do we measure it? What constitutes violence in a television program? Are cartoon actions or actions along the lines of "The Three Stooges" violence for the purposes of the experiment? Do we wish to investigate a possible difference in the level of aggressive behavior due to sex? What about an interaction effect?

2. The details of conducting the experiment should be formulated before collecting any data. Such things as whether or not a control group is needed, whether blocking is desirable, and how many replications can be afforded should be discussed and settled. All physical aspects of the experiment should be finalized. In the children and television experiment, if aggression is to be measured by a group of observers it must be decided how many observers to use and how to collate their measurements. The exact details of the environment in which the children will be observed should be determined.

3. The statistical analyses that will be performed on the data must be decided upon beforehand. This is important for a couple of reasons. The analysis used can determine what type of data to collect. For example, if we wish to estimate the proportion of a population above a given level for a specified attribute, then data can be collected as counts. However, if we wish to estimate the mean level of an attribute in a population, then we must collect data on attribute values. Nothing is more irritating to a statistician than to have a researcher collect some data and then ask what can be done with them. More than likely the answer will be that the data are of little value because they were not collected with a specific goal or set of procedures in mind.

Exercises 14.2

1. Suppose we wish to create a completely randomized design to assess the differences, if any, in three competing methods of selling a consumer product. They are direct mail advertising, television advertising, and telephone advertising. Discuss how we might set up such an experiment. Specifically address what hypotheses are to be tested and how to collect data.

2. It is desired to determine if there is a difference between three competing approaches to health insurance. One method reimburses patients only for hospital expenses. Another believes that preventive medicine is cheaper in the long run and reimburses for all office visits. The third is somewhere in the middle and reimburses only those visits that are demonstrated to be directly connected to some medical condition for which doctor's care is necessary. Discuss how you might measure such differences and how to set up and conduct an experiment to determine if a difference exists.

3. Use the random number tables or a computer random number generator to select sample units for a single-factor completely randomized design. The population has 10,000 members, there are 5 different levels of the factor involved, and we wish to assign 4 subjects to each treatment level.

4. How many subjects are necessary to conduct a factorial experiment where there are two factors with three levels to each factor? Give an example of how to assign subjects to factor combinations in such an experiment.

5. Give 2 different examples of a 3×3 Latin square design. How many different 3×3 Latin squares are possible?

6. Suppose we wished to measure the difference, if any, in the strength of grip between the dominant hand of a person and the nondominant hand. Two procedures are proposed.

   a. Select 10 people and measure the grip strength in the dominant hand of each. Select 10 other people and measure the grip strength in the nondominant hand of each.

   b. Select 10 people and measure the grip strength in both hands.

   Which procedure would be preferable in this situation? Explain.

7. How many distinct 4×4 Latin square designs are there?

8. Two $n \times n$ Latin squares, A and B, are said to be *orthogonal* if all n^2 ordered pairs of the form (a_{ij}, b_{ij}) are distinct. For example, the following two 3×3 Latin squares are orthogonal.

A	B	C
B	C	A
C	A	B

A	B	C
C	A	B
B	C	A

The ordered pairs are as follows.

$$(A,A) \quad (B,B) \quad (C,C)$$
$$(B,C) \quad (C,A) \quad (A,B)$$
$$(C,B) \quad (A,C) \quad (B,A)$$

Note that no two ordered pairs are the same. Find 2 orthogonal 4×4 Latin squares.

9. It was desired to determine the effect of spraying local marshes for mosquitoes on the general health levels of people living in an isolated rural area. State health officials proposed that a study be conducted wherein the general health of the population in the area would be measured before the study. Then the marshes in the area would be sprayed for mosquitoes over a number of years. The general health of the populace would be measured once a year during the course of the experiment by medical teams from the state university using specially equipped mobile medical treatment vehicles. Comment on the shortcomings in this design. Specifically, if an improvement in the general health of the population is observed would it be statistically justifiable to attribute the improvement to the spraying program? How can the experiment be redesigned to give more meaningful results?

10. Discuss how to design an experiment to test whether students in quantitatively oriented majors such as mathematics and the sciences are better than students in nonquantitatively oriented majors in the ability to estimate the volume of irregularly shaped objects.

11. Discuss how to design an experiment to test whether practice improves students' abilities to estimate the volume of irregularly shaped objects.

12. Suppose that an experimenter is interested in determining if the method of packaging D cell batteries affects battery lifetime. There are 3 different types of packaging. Describe how to conduct such an experiment if you are given 45 batteries.

13. Answer Exercise 12 if you are provided with the additional information that the batteries came from 3 different production runs. You have 15 batteries from each run.

14.3
One-Way
Analysis of
Variance

We now consider the problem of testing the equality of the means of more than 2 independent populations. In Section 12.5, we derived a likelihood ratio test for $H_0 : \mu_1 = \mu_2$ vs. $H_1 : \mu_1 \neq \mu_2$ when both populations were normally distributed with common variance, σ^2. We now wish to investigate testing

$$H_0 : \mu_1 = \mu_2 = \cdots = \mu_k \quad \text{vs.} \quad H_1 : \text{ not all the } \mu_i\text{'s are equal.}$$

The test will be based on the following assumptions.

1. We have n_i observations from population $i, i = 1, 2, \ldots, k$.
2. Let Y_{ij} be jth observation from population i, $i = 1, 2, \ldots, k$, $j = 1, 2, \ldots, n_i$. Each Y_{ij} has a $N(\mu_i, \sigma^2)$ distribution.

This is the natural extension of the two-sample problem considered in Example 12.18. It can be written as a linear model in the following way.

$$Y_{ij} = \mu_i + e_{ij}, \quad \text{where} \quad \begin{cases} & i = 1, 2, \ldots, k, \\ e_i \sim N(0, \sigma^2), \text{ for} & j = 1, 2, \ldots, n_i, \\ e_{ij} \text{ and } e_{kl} \text{ are uncorrelated if } (i, j) \neq (k, l) \end{cases} \quad \textbf{(14.1)}$$

This would be a model for a single-factor completely randomized experimental design. It turns out that an alternative formulation of this model is more commonly used for analysis of variance. This other model begins with an overall mean response for the populations taken as a whole. That is, there is a parameter, μ, that is

$$\mu = \frac{1}{n} \sum_{i=1}^{k} n_i \mu_i.$$

This is commonly called the *grand mean*. The differences in the treatment levels are measured as differences from this grand mean. Thus, we define these differences as follows

$$\mu_i = \mu + \tau_i, \text{ for } i = 1, 2, \ldots, k.$$

The effect of treatment level i is now measured by τ_i. This formulation implies that

$$\sum_{i=1}^{k} n_i \tau_i = 0.$$

Using this notation, our model for a single-factor completely randomized design becomes the following.

$$Y_{ij} = \mu + \tau_i + e_{ij}, \text{ where } \begin{cases} \displaystyle\sum_{i=1}^{k} n_i \tau_i = 0, \\ \\ e_{ij} \sim N(0, \sigma^2), \text{ for } \begin{array}{l} i = 1, 2, \ldots, k, \\ j = 1, 2, \ldots, n_i, \end{array} \\ \\ e_{ij} \text{ and } e_{kl} \text{ are uncorrelated if } (i, j) \neq (k, l) \end{cases} \quad (14.2)$$

With this model our test of the equality of group means becomes

$$H_0 : \tau_1 = \tau_2 = \cdots = \tau_k = 0 \quad \text{vs.} \quad H_1 : \text{ not all } \tau\text{'s equal to } 0. \quad (14.3)$$

Compare Equation 14.2 with the multiple regression model of Equation 13.20. There is a strong similarity between the 2 models. To make it more apparent consider the following. If we let $\beta_0 = \mu$, and $\beta_i x_i = \tau_i$, for $i = 1, 2, \ldots, k$, then we can write

$$Y_{ij} = \mu + \tau_i + e_{ij},$$

as

$$Y_{ij} = \beta_0 + \sum_{i=1}^{k} \beta_i x_i + e_{ij},$$

where $x_l = 1$, if $l = i$, and 0, otherwise. This is the reason why regression and analysis of variance together form what is known as the *general linear model*. This is also why the test of $\beta_1 = \beta_2 = \cdots = \beta_k = 0$ in multiple regression is an example of one-way ANOVA. The fact that, in the one-way analysis of variance model, $\sum_{i=1}^{k} n_i \tau_i = 0$ means, however, that we cannot use least squares estimators from the multiple regression model to estimate μ and $\tau_i, i = 1, \ldots, k$. The matrix $\mathbf{X'X}$ in the normal equations will not be of full rank.

Example 14.2

In Section 14.2 we used the example of a comparison of 3 competing methods for reduction of stress in individuals. They were transcendental meditation (TM), exercise (E), and diet (D). For illustrative purposes we shall consider implementing a completely randomized design where we assign 5 patients diagnosed as having high stress levels to each treatment for a total sample of 15 subjects. Each will have their stress level measured before and after treatment. Let Y_{ij} be the difference in stress measurements (after-before) for subject j in treatment i, for $i = 1, 2, 3$ and $j = 1, 2, 3, 4, 5$. The observations are as follows.

Transcendental Meditation	Exercise	Diet
−5	−10	−6
−11	−7	−4
−4	−14	−3
−6	−9	−1
−7	−11	−5

Before getting into the development of the test statistic we wish to introduce some notation that is standard in analysis of variance of literature. It is referred to as *dot notation*. We used it without giving it a name in Example 12.23. If we have a doubly subscripted set of variables such as the Y_{ij} in our case we would introduce the following notation.

$$n = \sum_{i=1}^{k} n_i,$$

$$Y_{i.} = \sum_{j=1}^{n_i} Y_{ij},$$

$$Y_{.j} = \sum_{i=1}^{k} Y_{ij},$$

$$Y_{..} = \sum_{i=1}^{k} \sum_{j=1}^{n_i} Y_{ij},$$

$$\bar{Y}_{i.} = \frac{Y_{i.}}{n_i}, \quad \text{and}$$

$$\bar{Y}_{..} = \frac{Y_{..}}{n}$$

This notation generalizes in the natural way to variables with more than two subscripts. For example,

$$X_{i.k} = \sum_{j} X_{ijk}, \ X_{..k} = \sum_{i} \sum_{j} X_{ijk}, \quad \text{and} \quad X_{...} = \sum_{i} \sum_{j} \sum_{k} X_{ijk}.$$

To derive a test for the hypotheses in Equation 14.3 we can use the likelihood ratio procedure. The development of a test statistic follows very closely that of Example 12.18. For this reason most of the details are omitted. It is more convenient to use Equation 14.1 for our model. Using this the likelihood function for this problem is

$$L(\mu, \tau_1, \tau_2, \ldots, \tau_k) = (2\pi\sigma^2)^{-n/2} \exp\left[\frac{-1}{2\sigma^2} \sum_{i=1}^{k} \sum_{j=1}^{n_i} (y_{ij} - \mu_i)^2\right].$$

In Ω_0 we find that, since $\mu_1 = \mu_2 = \cdots = \mu_k = \mu$,

$$\hat{\mu} = \bar{y}_{..}, \quad \text{and}$$

$$\widehat{\sigma^2} = \frac{1}{n} \sum_{i=1}^{k} \sum_{j=1}^{n_i} (y_{ij} - \bar{y}_{..})^2.$$

In Ω we find the following:

$$\widehat{\mu}_i = \bar{y}_{i.}, \text{ for } i = 1, 2, \ldots, k, \quad \text{and}$$

$$\widehat{\sigma^2} = \frac{1}{n} \sum_{i=1}^{k} \sum_{j=1}^{n_i} (y_{ij} - \bar{y}_{i.})^2$$

After substituting these into the likelihood functions and simplifying, we get that the likelihood ratio is

$$\lambda = \left[\frac{\displaystyle\sum_{i=1}^{k} \sum_{j=1}^{n_i} (y_{ij} - \bar{y}_{i.})^2}{\displaystyle\sum_{i=1}^{k} \sum_{j=1}^{n_i} (y_{ij} - \bar{y}_{..})^2} \right]^{n/2}.$$

Now, if we add and subtract $\bar{y}_{i.}$ in the interior sums in the denominator we get that

$$\sum_{i=1}^{k} \sum_{j=1}^{n_i} (y_{ij} - \bar{y}_{..})^2 = \sum_{i=1}^{k} n_i (\bar{y}_{i.} - \bar{y}_{..})^2 + \sum_{i=1}^{k} \sum_{j=1}^{n_i} (y_{ij} - \bar{y}_{i.})^2. \qquad \textbf{(14.4)}$$

It is common practice to write Equation 14.4 as

$$\text{SSTotal} = \text{SST} + \text{SSE},$$

where *SSTotal* is called the *total sum of squares, SST* is called the *treatment sum of squares,* and *SSE* is the *error sum of squares.* This *partitioning* of the total sum of squares is the way that analysis of variance tests are constructed. If we then let $F = \frac{\text{SST}/(k-1)}{\text{SSE}/(n-k)}$, the likelihood ratio becomes

$$\lambda = \left[\frac{1}{1 + \dfrac{k-1}{n-k} F} \right]^{n/2}, \qquad \textbf{(14.5)}$$

and we would reject H_0 if

$$\lambda \leq \lambda_0.$$

From Equation 14.5 we can see that this is equivalent to rejecting H_0 whenever

$$F \geq a, \qquad \textbf{(14.6)}$$

where a is chosen to achieve the desired value for α.

Now that we have an equivalent rejection region in terms of F the next task is to determine the distribution for F. Given our choice of letters and divisors it is, hopefully, not too surprising that it will have an F distribution with $k-1$ and $n-k$ degrees of freedom. We could, at this point, proceed to prove that, when H_0 is true, $\frac{\text{SST}}{\sigma^2}$ has a $\chi^2(k-1)$ distribution, $\frac{\text{SSE}}{\sigma^2}$ has a $\chi^2(n-k)$ distribution, and that they are independent to establish the distribution for F. This, however, would not give us a general method for establishing the distribution of other ratios that will be developed later. For this reason we wish to present, without proof, some theorems regarding the distribution of quadratic forms. Central to this discussion is the notion of an *idempotent* matrix, which we now define.

Definition 14.1

An $n \times n$ matrix $\mathbf{A}$ is said to be *idempotent* if

$$\mathbf{AA} = \mathbf{A}.$$

From this definition we can see that an example of an idempotent matrix is the identity matrix. A less trivial example is

$$\mathbf{A} = \begin{pmatrix} \dfrac{1}{2} & \dfrac{1}{2} \\ \dfrac{1}{2} & \dfrac{1}{2} \end{pmatrix}.$$

With this definition we are now ready to state an important theorem regarding the distribution of quadratic forms. The reader interested in a proof should consult *An Introduction to Linear Statistical Models*, vol. 1, by F. A. Graybill.

Theorem 14.1

Let the random vector

$$\mathbf{x} = (X_1, X_2, \ldots, X_n)'$$

consist of independent and identically distributed $N(\mu, \sigma^2)$ random variables. Furthermore, let

$$Y = \mathbf{x}'\mathbf{Ax}$$

where rank($\mathbf{A}$) = r. Then

$$\frac{Y}{\sigma^2} \sim \chi^2(r)$$

if and only if $\mathbf{A}$ is an idempotent matrix and

$$\sum_{i=1}^{n} \sum_{j=1}^{n} a_{ij} = 0.$$

Example 14.3

Suppose that $X_1, X_2, \ldots, X_n$ are i.i.d. $N(\mu, \sigma^2)$ random variables and that we wish to establish the result of Theorem 9.6 that

$$\frac{(n-1)s^2}{\sigma^2} = \frac{\displaystyle\sum_{i=1}^{n}(x_i - \bar{x})^2}{\sigma^2} \sim \chi^2(n-1).$$

We need to find $\mathbf{A}$ so that

$$\mathbf{x}'\mathbf{Ax} = \sum_{i=1}^{n}(x_i - \bar{x})^2.$$

Consider the case where $n = 3$. We begin by expanding the sum of squares as follows.

$$\sum_{i=1}^{3} (x_i - \bar{x})^2 = \left(\frac{2}{3}x_1 - \frac{1}{3}x_2 - \frac{1}{3}x_3 \right)^2$$

$$+ \left(\frac{2}{3}x_2 - \frac{1}{3}x_1 - \frac{1}{3}x_3 \right)^2$$

$$+ \left(\frac{2}{3}x_3 - \frac{1}{3}x_1 - \frac{1}{3}x_2 \right)^2$$

$$= \frac{2}{3}(x_1^2 + x_2^2 + x_3^2 - x_1 x_2 - x_1 x_3 - x_2 x_3)$$

Now consider the 3×3 matrix

$$\mathbf{A} = \begin{pmatrix} \frac{2}{3} & -\frac{1}{3} & -\frac{1}{3} \\ -\frac{1}{3} & \frac{2}{3} & -\frac{1}{3} \\ -\frac{1}{3} & -\frac{1}{3} & \frac{2}{3} \end{pmatrix},$$

and, by comparing terms we see that

$$\mathbf{x}'\mathbf{A}\mathbf{x} = \sum_{i=1}^{3} (x_i - \bar{x})^2.$$

It is also easily verified that $\mathbf{A}$ is idempotent and rank($\mathbf{A}$) = 2. Recall that one method learned in linear algebra to determine the rank of a matrix is to obtain a basis for the row or column space of a matrix by using elementary row operations to reduce the matrix to upper triangular form. The rank is the dimension of the row space or the column space, where the dimension of a vector space is the number of vectors in a basis for that space. Therefore, Theorem 14.1 gives that

$$\sum_{i=1}^{3} \frac{(x_i - \bar{x})^2}{\sigma^2} \sim \chi^2(2).$$

This 3×3 example generalizes in a straightforward manner to the $n \times n$ case to obtain an idempotent matrix, $\mathbf{A}$, whose rank is $n - 1$.

$$\mathbf{A}_1 = \begin{pmatrix} \frac{n-1}{n} & -\frac{1}{n} & \cdots & -\frac{1}{n} \\ -\frac{1}{n} & \frac{n-1}{n} & \cdots & -\frac{1}{n} \\ \vdots & \vdots & \ddots & \vdots \\ -\frac{1}{n} & -\frac{1}{n} & \cdots & \frac{n-1}{n} \end{pmatrix}$$

Therefore, by Theorem 14.1 we see that

$$\frac{(n-1)s^2}{\sigma^2} \sim \chi^2(n-1).$$

∎

The next theorem provides a method for determining when two quadratic forms are independent random variables.

Theorem 14.2

Let the random vector

$$\mathbf{x} = (X_1, X_2, \ldots, X_n)'$$

consist of independent and identically distributed $N(\mu, \sigma^2)$ random variables. Furthermore, let

$$Y_1 = \mathbf{x}'\mathbf{A}\mathbf{x}, \quad \text{and} \quad Y_2 = \mathbf{x}'\mathbf{B}\mathbf{x}.$$

Then Y_1 and Y_2 are independent random variables if and only $\mathbf{AB} = \mathbf{0}$.

Before proceeding we present without proof an additional theorem that will be useful in later sections. It is a variation of a theorem due to Cochran.

Theorem 14.3

Cochran   Let

$$\mathbf{X} = (X_1, X_2, \ldots, X_n)'$$

consist of independent and identically distributed $N(\mu, \sigma^2)$ random variables. Suppose there exist quadratic forms

$$Y_1 = \mathbf{x}'\mathbf{A}_1\mathbf{x}, Y_2 = \mathbf{x}'\mathbf{A}_2\mathbf{x}, \ldots, Y_k = \mathbf{x}'\mathbf{A}_k\mathbf{x},$$

such that

$$\sum_{i=1}^{n}(X_i - \bar{X})^2 = Y_1 + Y_2 + \cdots + Y_k.$$

Then $Y_1, Y_2, \ldots, Y_n$ are independent random variables if and only if

$$\sum_{i=1}^{k} \text{rank}(A_i) = n - 1.$$

Returning to the one-way analysis of variance problem we now wish to show that, when H_0 is true, $\frac{\text{SST}}{\sigma^2} \sim \chi^2(k-1)$. We leave it as an exercise to show that, when H_0 is true, $\frac{\text{SSE}}{\sigma^2} \sim \chi^2(n-k)$.

Example 14.4   We need to determine an $n \times n$ idempotent matrix $\mathbf{A}$ so that

$$\mathbf{y}'\mathbf{A}\mathbf{y} = \sum_{i=1}^{k} n_i (\bar{y}_{i.} - \bar{y}_{..})^2.$$

We leave it to the reader to verify that a matrix of the form

$$\mathbf{A} = \begin{pmatrix} \mathbf{A}_1 & \mathbf{B} & \cdots & \mathbf{B} \\ \mathbf{B} & \mathbf{A}_2 & \cdots & \mathbf{B} \\ \vdots & \vdots & \ddots & \vdots \\ \mathbf{B} & \mathbf{B} & \cdots & \mathbf{A}_k \end{pmatrix},$$

where $\mathbf{A}_i$ is an $n_i \times n_i$ matrix each of whose elements is $\frac{1}{n_i} - \frac{1}{n}$ and the $\mathbf{B}$'s represent elements of the form $-\frac{1}{n}$ will work. For example, in a 3-level experiment where $n_1 = 1, n_2 = 2$, and $n_3 = 3$, we would have

$$\mathbf{A} = \begin{pmatrix} \frac{5}{6} & -\frac{1}{6} & -\frac{1}{6} & -\frac{1}{6} & -\frac{1}{6} & -\frac{1}{6} \\ -\frac{1}{6} & \frac{1}{3} & \frac{1}{3} & -\frac{1}{6} & -\frac{1}{6} & -\frac{1}{6} \\ -\frac{1}{6} & \frac{1}{3} & \frac{1}{3} & \frac{1}{3} & -\frac{1}{6} & \frac{1}{6} \\ -\frac{1}{6} & -\frac{1}{6} & -\frac{1}{6} & \frac{1}{6} & \frac{1}{6} & \frac{1}{6} \\ -\frac{1}{6} & -\frac{1}{6} & -\frac{1}{6} & \frac{1}{6} & \frac{1}{6} & \frac{1}{6} \\ -\frac{1}{6} & -\frac{1}{6} & -\frac{1}{6} & \frac{1}{6} & \frac{1}{6} & \frac{1}{6} \end{pmatrix}.$$

It is a little tedious, but straightforward, to check that $\mathbf{A}$ is idempotent. The rank is shown to be $k - 1$ by noting that, out of the n rows, only k are different. Therefore, row operations would leave us with, at most, k nonzero rows. However, by adding $k - 1$ of them together you get the negative of the unused row. From there it is easily verified that the remaining $k - 1$ rows are linearly independent. In addition, when H_0 is true we verify that $\lambda = 0$. Therefore, by Theorem 14.1 we find that, when H_0 is true, $\frac{\text{SST}}{\sigma^2}$ has a $\chi^2(k - 1)$ distribution.

Since $\frac{\text{SST}}{\sigma^2} \sim \chi^2(k - 1)$ and $\frac{\text{SSE}}{\sigma^2} \sim \chi^2(n - k)$, we leave it as an exercise to show that Theorem 14.2 gives that $\frac{\text{SST}}{\sigma^2}$ and $\frac{\text{SSE}}{\sigma^2}$ are independent random variables. Therefore, when H_0 is true,

$$F = \frac{\dfrac{\text{SST}}{(k - 1)}}{\dfrac{\text{SSE}}{(n - k)}}$$

has an F distribution with $k - 1$ and $n - k$ degrees of freedom. Therefore, we can now say that the likelihood ratio test for the hypotheses in Equation 14.3 is equivalent to one that rejects H_0 whenever

$$F = \frac{\dfrac{\text{SST}}{(k - 1)}}{\dfrac{\text{SSE}}{(n - k)}} \geq F(k - 1, n - k)_\alpha. \qquad (14.7)$$

■

It is common practice to summarize the results of an experiment in what is called an *analysis of variance or ANOVA table*. Before showing what such a table looks like we mention that $\frac{\text{SST}}{(k-1)}$ is referred to as the *mean square for treatment* and denoted by MST, and $\frac{\text{SSE}}{(n-k)}$ is called the *mean square for error* and denoted by MSE. Then we have that $F = \frac{\text{MST}}{\text{MSE}}$. The general form for an ANOVA table is given in Figure 14.3. The column headings should be self-explanatory.

Source of Variation	d.f.	SS	MS	F
Treatment	$k - 1$	SST	MST	$\dfrac{\text{MST}}{\text{MSE}}$
Error	$n - k$	SSE	MSE	
Total	$n - 1$	SSTotal		

Figure 14.3

As we have seen in previous chapters there are ways to compute statistics that are more efficient than the formal definitions. At this point we wish to give the so-called *machine formulas* for the sums of squares in one-way analysis of variance. We leave their verification as exercises.

$$\text{SSTotal} = \sum_{i=1}^{k} \sum_{j=1}^{n_i} y_{ij}^2 - \frac{y_{..}^2}{n},$$

$$\text{SST} = \sum_{i=1}^{k} \frac{y_{i.}^2}{n_i} - \frac{y_{..}^2}{n}, \qquad (14.8)$$

$$\text{SSE} = \text{SSTotal} - \text{SST}$$

Example 14.5   Assuming that the stress level measurements are normally distributed with constant variance, we wish to test

$$H_0 : \tau_1 = \tau_2 = \tau_3 = 0 \quad \text{vs.} \quad H_1 : \text{not all of the } \tau\text{'s are } 0$$

at the 0.05 level. To begin we compute

$$n_1 = n_2 = n_3 = 5,$$
$$y_{1.} = -33,$$
$$y_{2.} = -51,$$
$$y_{3.} = -19,$$
$$y_{..} = -103, \quad \text{and}$$

$$\sum_{i=1}^{3} \sum_{j=1}^{n_i} y_{ij}^2 = 881.$$

From these we get

$$\text{SSTotal} = 881 - \frac{(-103)^2}{15} = 173.73,$$

$$\text{SST} = \frac{(-33)^3}{5} + \frac{(-51)^2}{5} + \frac{(-19)^2}{5} - \frac{(-103)^2}{15} = 102.93, \quad \text{and}$$

$$\text{SSE} = 173.73 - 102.93 = 70.8.$$

Therefore, the ANOVA table is as follows.

Source of Variation	d.f.	SS	MS	F
Treatment	2	102.93	51.47	8.72
Error	12	70.80	5.90	
Total	14	173.73		

Now, for 2 and 12 degrees of freedom, Table 6 of Appendix A gives that $F_{.05} = 3.89$. Therefore, we reject the hypothesis that there is no difference in the effects of the 3 regimes on stress level at the 0.05 level.   ■

Exercises 14.3

1. Verify Equation 14.4.

2. Given $x_{ij}, i = 1, 2, \ldots, k, j = 1, 2, \ldots, n$, find idempotent matrices, $\mathbf{A}$, for each of the following quadratic forms.
   a. $\mathbf{x}'\mathbf{x} = \mathbf{x}'\mathbf{A}\mathbf{x}$
   b. $\dfrac{x_{i.}^2}{n} = \mathbf{x}'\mathbf{A}\mathbf{x}$

3. Find an idempotent matrix, $\mathbf{A}$, such that

$$\mathbf{y}'\mathbf{A}\mathbf{y} = \sum_{i=1}^{k} \sum_{j=1}^{n_i} (y_{ij} - \bar{y}_{i.})^2.$$

4. Use Exercise 3 and Theorem 14.1 to show that $\dfrac{\text{SSE}}{\sigma^2} \sim \chi^2(n - k)$.

5. Verify directly that the product of the matrices in the quadratic forms for SST and SSE have elements that are all zero.

6. In psychology there is a phenomenon known as the *Stroop effect*. Basically, it involves the confusion in response to visual stimuli. For example, it takes a person less time to read the word "red" when it is written in black ink than when it is written in blue ink. An experiment was conducted where students were separated into 3 groups. Each student was given a list containing 30 entries to read. Group 1 was asked to read a list of numbers. Group 2 was asked to count the number of dots in a number of rows. Group 3 was asked to count the number of digits in a row where, for example, a row containing 4 digits might look like "1111." The results are given below. Times are in seconds.

Group

1	2	3
11	14	19
10	17	21
14	16	23
9	12	21
7	11	15

Use one-way analysis of variance to test for a difference in the mean time to complete the list for the 3 groups at the 0.01 level.

7. An experiment was carried out to determine if there is a difference in the yield of corn crops when exposed to different concentrations in various fertilizer components. For simplicity, we designate the treatments as 1, 2, 3, and 4. The results, in pounds of corn per plant, are as follows.

Treatment

1	2	3	4
4.96	3.94	6.35	5.56
5.34	3.87	5.99	4.40
5.20	3.86	4.69	4.43
2.90	1.63	4.00	
2.92	2.92		
2.56			

Is there a difference in the treatments at the 0.05 level?

8. An insurance company wishes to determine if there is a difference in the average time to process claim forms among its 3 central processing facilities. Data are gathered over a period of 3 weeks as shown below. The data represent the average number of days to process a form.

Facility 1	Facility 2	Facility 3
1.25	2.30	1.45
0.87	1.45	0.65
1.11	1.45	2.99
1.94	1.22	2.13

Is there a difference in processing time at the 0.1 level?

9. It is felt that ambient humidity has an effect on the lifetime of flashlight batteries. Five flashlight batteries were assigned to each of 3 different humidity levels at a constant temperature of 75° F. The lifetimes, in minutes, under steady load are as follows.

Relative Humidity

10%	30%	50%
105	130	72
155	83	90
121	153	99
154	127	113
147	117	101

Test whether humidity has an effect on battery life at the 0.01 level.

10. A manufacturer wishes to determine if there is a difference in the tensile strength of wool yarn due to the method in which the yarn is dyed. There are 4 popular methods for applying dye to yarn. To assess the effect of dye method a skein of yarn was selected and 16 pieces of yarn were cut from it. The pieces were then randomly assigned such that each method was used to dye 4 pieces. The results are as follows. The numbers represent the breaking strength in pounds.

Method 1	Method 2	Method 3	Method 4
52	61	42	58
47	53	60	49
60	43	39	51
53	57	48	55

Does the dye method appear to make a difference at the 0.05 level?

14.4
Two-Way Analysis of Variance

Our discussion of two-factor experiments will consider a number of different models. We shall begin with the *randomized complete block* design. As we mentioned in Section 14.2 it is sometimes desirable to subdivide experiments into blocks to eliminate as much noise as possible. While the real goal of such an experiment is to test the equality of levels for the treatment effect, it is sometimes of interest to also test for a difference among blocks. The description we gave of randomized complete block designs in Section 14.2 was heavily agricultural in nature, but blocking can be of use in many other settings. For example, it is a fairly common practice among psychological researchers to test subjects on all levels of a treatment to reduce underlying variability that might mask actual differences in responses to the treatments. In such a case, each subject could be considered a block. It may be argued that such a practice violates the underlying assumption that the populations at each treatment level are independent. To overcome this, experimenters ensure that a long enough time elapses between tests at each level so that any effects from other tests have faded. In addition, the treatment levels are administered in different orders for each subject. We begin with an example of an experiment using blocks.

Example 14.6

An experiment is conducted to investigate the effect of 4 different chemicals used in growing alfalfa on the population of beneficial insects. A field was subdivided into 40 square plots in a pattern of 10 rows and 4 columns. The 4 treatments, A, B, C, and D, were assigned as shown below. The numbers in parentheses are the plot numbers.

(1)A	(5)B	(9)C	(13)D	(17)D	(21)C	(25)B	(29)A	(33)B	(37)D
(2)B	(6)C	(10)D	(14)A	(18)C	(22)B	(26)A	(30)C	(34)D	(38)A
(3)C	(7)D	(11)A	(15)B	(19)B	(23)A	(27)D	(31)D	(35)A	(39)C
(4)D	(8)A	(12)B	(16)C	(20)A	(24)D	(28)C	(32)B	(36)C	(40)B

After the treatments were applied for a number of weeks each plot was sampled for insects with the following numbers recorded.

Plot	1	2	3	4	5	6	7	8	9	10
Count	19	29	17	15	10	15	13	5	13	7

Plot	11	12	13	14	15	16	17	18	19	20
Count	15	11	7	24	11	29	8	16	15	8

Plot	21	22	23	24	25	26	27	28	29	30
Count	12	12	16	31	7	10	17	16	8	16

Plot	31	32	33	34	35	36	37	38	39	40
Count	13	25	17	12	15	21	13	40	24	29

This experiment was designed with 10 blocks, each containing the 4 treatment levels. ∎

To develop tests for a difference in treatment levels and a difference in blocks we shall make use of Theorems 14.1 and 14.2 to partition the total sum of squares into sums of squares having independent chi-square distributions. We begin with a statement of the model for this experiment. Let k be the number of treatment levels, b be the number of blocks, and Y_{ij} be the observed response for treatment level i in block j. Then the model we shall assume is

$$Y_{ij} = \mu + \tau_i + \beta_j + e_{ij}, \quad \text{where} \quad \begin{cases} \sum_{i=1}^{k} n_i \tau_i = 0, \\[2mm] \sum_{j=1}^{b} n_j \beta_j = 0, \\[2mm] e_{ij} \sim N(0, \sigma^2), \quad \text{and} \\[1mm] e_{ij} \text{ and } e_{kl} \text{ are uncorrelated if } (i,j) \neq (k,l). \end{cases} \quad (14.9)$$

Note that this is a logical extension from the single-factor model used in the previous section. μ is still the grand mean, and the τ's are still the contributions due to the treatment levels. The new component, β_j, is the contribution to the observed response from being in block j. In this model, we wish to be able to test the following two sets of hypotheses.

1. $H_0 : \tau_1 = \tau_2 = \cdots = \tau_k = 0,$ vs. H_1 : not all τ's equal 0.
2. $H_0 : \beta_1 = \beta_2 = \cdots = \beta_b = 0,$ vs. H_1 : not all β's equal 0.

In the single-factor case, we began our partitioning by adding and subtracting $\bar{y}_{i.}$ in each term of the total sum of squares. We shall proceed in an analogous manner here by first adding and subtracting $\bar{y}_{i.}$ and $\bar{y}_{.j}$ in each term of the total sum of squares as follows.

$$y_{ij} - \bar{y}_{..} = (\bar{y}_{i.} - \bar{y}_{..}) + (\bar{y}_{.j} - \bar{y}_{..}) + (y_{ij} - \bar{y}_{i.} - \bar{y}_{.j} + \bar{y}_{..})$$

This is the way we shall partition the total sum of squares. We leave it as an exercise to show that the following partition is valid.

$$\sum_{i=1}^{k} \sum_{j=1}^{b} (y_{ij} - \bar{y}_{..})^2 = b \sum_{i=1}^{k} (\bar{y}_{i.} - \bar{y}_{..})^2 + k \sum_{j=1}^{b} (\bar{y}_{.j} - \bar{y}_{..})^2 + \sum_{i=1}^{k} \sum_{j=1}^{b} (y_{ij} - \bar{y}_{i.} - \bar{y}_{.j} + \bar{y}_{..})^2$$

$$(14.10)$$

It is common practice to write this partition as

$$\text{SSTotal} = \text{SST} + \text{SSB} + \text{SSE}, \qquad \text{where}$$

$$\text{SSTotal} = \sum_{i=1}^{k} \sum_{j=1}^{b} (y_{ij} - \bar{y}_{..})^2,$$

$$\text{SST} = b \sum_{i=1}^{k} (\bar{y}_{i.} - \bar{y}_{..})^2,$$

$$\text{SSB} = k \sum_{j=1}^{b} (\bar{y}_{.j} - \bar{y}_{..})^2, \quad \text{and}$$

$$\text{SSE} = \sum_{i=1}^{k} \sum_{j=1}^{b} (y_{ij} - \bar{y}_{i.} - \bar{y}_{.j} + \bar{y}_{..})^2.$$

SSTotal is the total sum of squares, SST is the treatment sum of squares, SSB is the block sum of squares, and SSE is the error sum of squares. To determine the distributions of SST, SSB, and SSE from Theorem 14.1 we shall use the results of Example 14.3. SST and SSB have the same general form as SST in the single-factor model. Therefore, Example 14.3 gives us an idempotent matrix, A_1, with rank $k-1$ that will generate the first sum on the right-hand side of Equation 14.10. It also gives the other idempotent matrix, A_2, with rank $b-1$ that will generate the second sum. The elements of both matrices sum to 0. It can also be shown that $A_1 A_2 = 0$. This verification is computationally intensive and not very enlightening. We leave it as an exercise to show that this is true in a small special case. The third component of the right-hand side of Equation 14.10 can be shown to be the quadratic form with an idempotent matrix, $\mathbf{A}_3$, whose rank is $(k-1)(b-1)$. We begin by noticing that

$$\sum_{i=1}^{k}\sum_{j=1}^{b} y_{ij}^2 = \text{SST} + \text{SSB} + \text{SSE} + n\bar{y}_{..}^2.$$

An idempotent matrix, $\mathbf{A}_4$, with rank 1, that gives $\mathbf{y}'\mathbf{A}_4\mathbf{y} = n\bar{y}_{..}^2$ is

$$\mathbf{A}_4 = \begin{pmatrix} \dfrac{1}{n} & \dfrac{1}{n} & \cdots & \dfrac{1}{n} \\ \dfrac{1}{n} & \dfrac{1}{n} & \cdots & \dfrac{1}{n} \\ \vdots & \vdots & \ddots & \vdots \\ \dfrac{1}{n} & \dfrac{1}{n} & \cdots & \dfrac{1}{n} \end{pmatrix}.$$

It can also be shown that $\mathbf{A}_1\mathbf{A}_4 = \mathbf{A}_2\mathbf{A}_4 = \mathbf{0}$. In addition, we see that

$$\mathbf{A}_3 = \mathbf{I} - \mathbf{A}_1 - \mathbf{A}_2 - \mathbf{A}_4.$$

We can also note that

$$\begin{aligned} \mathbf{A}_1\mathbf{A}_3 &= \mathbf{A}_1(\mathbf{I} - \mathbf{A}_1 - \mathbf{A}_2 - \mathbf{A}_4) \\ &= \mathbf{A}_1 - \mathbf{A}_1\mathbf{A}_1 - \mathbf{A}_1\mathbf{A}_2 - \mathbf{A}_1\mathbf{A}_4 \\ &= \mathbf{A}_1 - \mathbf{A}_1 = \mathbf{0}, \\ \mathbf{A}_2\mathbf{A}_3 &= \mathbf{A}_2(\mathbf{I} - \mathbf{A}_1 - \mathbf{A}_2 - \mathbf{A}_4) \\ &= \mathbf{A}_2 - \mathbf{A}_2\mathbf{A}_1 - \mathbf{A}_2\mathbf{A}_2 - \mathbf{A}_2\mathbf{A}_4 \\ &= \mathbf{A}_2 - \mathbf{A}_2 = \mathbf{0}, \quad \text{and} \\ \mathbf{A}_4\mathbf{A}_3 &= \mathbf{A}_4(\mathbf{I} - \mathbf{A}_1 - \mathbf{A}_2 - \mathbf{A}_4) \\ &= \mathbf{A}_4 - \mathbf{A}_1\mathbf{A}_4 - \mathbf{A}_1\mathbf{A}_4 - \mathbf{A}_4\mathbf{A}_4 \\ &= \mathbf{A}_4 - \mathbf{A}_4 = \mathbf{0}. \end{aligned}$$

We can also see that $\mathbf{A}_3$ is idempotent as follows.

$$\begin{aligned} \mathbf{A}_3\mathbf{A}_3 &= (\mathbf{I} - \mathbf{A}_1 - \mathbf{A}_2 - \mathbf{A}_4)(\mathbf{I} - \mathbf{A}_1 - \mathbf{A}_2 - \mathbf{A}_4) \\ &= \mathbf{II} - \mathbf{A}_1 - \mathbf{A}_2 - \mathbf{A}_4 - \mathbf{A}_1 + \mathbf{A}_1 + \mathbf{A}_1\mathbf{A}_2 + \mathbf{A}_1\mathbf{A}_4 \\ &\quad - \mathbf{A}_2 + \mathbf{A}_2\mathbf{A}_1 + \mathbf{A}_2\mathbf{A}_2 + \mathbf{A}_2\mathbf{A}_4 - \mathbf{A}_4 + \mathbf{A}_4\mathbf{A}_1 + \mathbf{A}_4\mathbf{A}_2 + \mathbf{A}_4\mathbf{A}_4 \\ &= \mathbf{I} - \mathbf{A}_1 - \mathbf{A}_2 - \mathbf{A}_4 \\ &= \mathbf{A}_3 \end{aligned}$$

Then Theorem 14.2 gives that SST, SSB, and SSE are independent random variables, and Theorem 14.3 gives that the rank of $\mathbf{A}_3$ is

$$(kb - 1) - (k - 1) - (b - 1) = (k - 1)(b - 1)$$

Therefore, Theorem 14.1 gives that $\frac{SSE}{\sigma^2} \sim \chi^2[(k - 1)(b - 1)]$.
In summary, this argument has shown that

$$\frac{SST}{\sigma^2} \sim \chi^2(k - 1),$$

$$\frac{SSB}{\sigma^2} \sim \chi^2(b - 1), \quad \text{and}$$

$$\frac{SSE}{\sigma^2} \sim \chi^2([k - 1][b - 1]).$$

In addition, all three random variables are independent.

Since we have independent chi-square random variables we can again form F ratios for test statistics. The only question remaining is which ratio to use to test the equality of the τ's and which one to use to test the equality of the β's. For this we must consider the expected values of the mean squares associated with each of these sums of squares. We obtain these as follows. Note that $n = kb$.

$$MST = \frac{SST}{k - 1} = \frac{b}{k - 1}\left[\sum_{i=1}^{k}\frac{y_{i.}^2}{b} - \frac{y_{..}^2}{n}\right],$$

$$MSB = \frac{SSB}{b - 1} = \frac{k}{b - 1}\left[\sum_{j=1}^{b}\frac{y_{.j}^2}{k} - \frac{y_{..}^2}{n}\right], \quad \text{and}$$

$$MSE = \frac{SSE}{(k - 1)(b - 1)} = \frac{1}{(k - 1)(b - 1)}\left[\sum_{i=1}^{k}\sum_{j=1}^{b}y_{ij}^2 - \sum_{i=1}^{k}\frac{y_{i.}^2}{b} - \sum_{j=1}^{b}\frac{y_{.j}^2}{k} + \frac{y_{..}^2}{n}\right]$$

Now,

$$E[y_{ij}^2] = \sigma^2 + (\mu + \tau_i + \beta_j)^2,$$

$$E\left[\frac{y_{i.}^2}{b}\right] = \sigma^2 + b(\mu + \tau_i)^2,$$

$$E\left[\frac{y_{.j}^2}{k}\right] = \sigma^2 + k(\mu + \beta_j)^2, \quad \text{and}$$

$$E\left[\frac{y_{..}^2}{n}\right] = \sigma^2 + n\mu^2.$$

Combining these gives

$$E[MST] = \frac{1}{k - 1}\left\{\sum_{i=1}^{k}[\sigma^2 + b\mu^2 + b\tau_i^2 + 2\mu\tau_i] - \sigma^2 - n\mu^2\right\}$$

$$= \sigma^2 + \frac{b}{k - 1}\sum_{i=1}^{k}\tau_i^2,$$

$$E[\text{MSB}] = \frac{1}{b-1} \left\{ \sum_{j=1}^{b} [\sigma^2 + k\mu^2 + k\beta_j^2 + 2\mu\beta_j] - \sigma^2 - n\mu^2 \right\}$$

$$= \sigma^2 + \frac{k}{b-1} \sum_{j=1}^{b} \beta_j^2, \quad \text{and}$$

$$E[\text{MSE}] = \frac{1}{(k-1)(b-1)}$$

$$\times \left\{ \sum_{i=1}^{k} \sum_{j=1}^{b} \left[\sigma^2 + \mu^2 + \tau_i^2 + \beta_j^2 + 2\mu\tau_i + 2\mu\beta_j + 2\tau_i\beta_j \right] \right.$$

$$- \left[k\sigma^2 + n\mu^2 + b\sum_{i=1}^{k} \tau_i^2 \right] - \left[b\sigma^2 + n\mu^2 + k\sum_{j=1}^{b} \beta_j^2 \right]$$

$$\left. + [\sigma^2 + n\mu^2] \right\}$$

$$= \sigma^2.$$

From these expected mean squares we can see that the F ratio

$$F = \frac{\text{MST}}{\text{MSE}}$$

should have a value near 1 when all of the τ's are 0 and it should be greater than 1 when they differ. This indicates that we should use this F ratio to test $H_0 : \tau_1 = \tau_2 = \cdots = \tau_k = 0$. In addition, when this null hypothesis is true, F will have an F distribution with $k - 1$ and $(k - 1)(b - 1)$ degrees of freedom. By a similar argument we should use the F ratio

$$F = \frac{\text{MSB}}{\text{MSE}}$$

to test the null hypothesis $H_0 : \beta_1 = \beta_2 = \cdots = \beta_b = 0$. Also, when this null hypothesis is true F will have an F distribution with $b - 1$ and $(k - 1)(b - 1)$ degrees of freedom. As we did in the single factor case, the results of these tests are usually summarized in an ANOVA table like the one shown in Figure 14.4.

Source of Variation	d.f.	SS	MS	F
Treatment	$k - 1$	SST	$\dfrac{\text{SST}}{k-1}$	$\dfrac{\text{MST}}{\text{MSE}}$
Blocks	$b - 1$	SSB	$\dfrac{\text{SSB}}{b-1}$	$\dfrac{\text{MSB}}{\text{MSE}}$
Error	$(k-1)(b-1)$	SSE	$\dfrac{\text{SSE}}{(k-1)(b-1)}$	
Total	$kb - 1$	SSTotal		

Figure 14.4

As we saw in the single-factor case there are ways to compute statistics that are more efficient than the formal definitions. At this point we give the so-called machine formulas for the sums of squares in randomized complete block design. We leave their verification as exercises.

$$\text{SSTotal} = \sum_{i=1}^{k} \sum_{j=1}^{n_i} y_{ij}^2 - \frac{y_{..}^2}{kb},$$

$$\text{SST} = \sum_{i=1}^{k} \frac{y_{i.}^2}{b} - \frac{y_{..}^2}{kb}, \qquad (14.11)$$

$$\text{SSB} = \sum_{j=1}^{b} \frac{y_{.j}^2}{b} - \frac{y_{..}^2}{kb}, \quad \text{and}$$

$$\text{SSE} = \text{SSTotal} - \text{SST} - \text{SSB}$$

Example 14.7

We would like to analyze the insect data from Example 14.6. From the plot numbering scheme we see that $y_{11} = 19, y_{21} = 29, y_{31} = 17, y_{41} = 15, y_{12} = 5, y_{22} = 10, y_{23} = 15, y_{24} = 13$, and so on. Thus, the necessary quantities for Equation 14.11 are as follows.

$$k = 4 \qquad b = 10 \qquad y_{..} = 641 \qquad \sum_{i=1}^{k}\sum_{j=1}^{b} y_{ij}^2 = 12593$$

$$y_{1.} = 160 \quad y_{2.} = 166 \quad y_{3.} = 179 \qquad\qquad y_{4.} = 136$$

$$y_{.1} = 80 \quad y_{.2} = 43 \quad y_{.3} = 46 \qquad\qquad y_{.4} = 71 \qquad y_{.5} = 47$$

$$y_{.6} = 71 \quad y_{.7} = 50 \quad y_{.8} = 62 \qquad\qquad y_{.9} = 65 \qquad y_{.10} = 106$$

We then obtain the following.

$$\text{SSTotal} = 12593 - \frac{(641)^2}{40} = 2320.98,$$

$$\text{SST} = \left[\frac{(160)^2}{10} + \frac{(166)^2}{10} + \frac{(179)^2}{10} + \frac{(136)^2}{10}\right] - \frac{(641)^2}{40}$$

$$= 97.28,$$

$$\text{SSB} = \left[\frac{(80)^2}{4} + \frac{(43)^2}{4} + \frac{(46)^2}{4} + \frac{(71)^2}{4} + \frac{(47)^2}{4}\right.$$

$$\left. + \frac{(71)^2}{4} + \frac{(50)^2}{4} + \frac{(62)^2}{4} + \frac{(65)^2}{4} + \frac{(106)^2}{4}\right] - \frac{(641)^2}{40}$$

$$= 843.23, \quad \text{and}$$

$$\text{SSE} = 2320.98 - 97.28 - 843.23 = 1380.48$$

The ANOVA table is given on the next page.

Source of Variation	d.f.	SS	MS	F
Treatment	3	97.28	32.43	0.63
Blocks	9	843.23	93.69	1.83
Error	27	1380.48	51.29	
Total	39	2320.988		

If we use $\alpha = 0.05$ and approximate 39 degrees of freedom by 40 degrees of freedom we find in Table 6 of Appendix A that

$$F(3, 40)_{.05} = 2.84 \quad \text{and} \quad F(9, 40)_{.05} = 2.12.$$

Since $\frac{\text{MST}}{\text{MSE}} < 2.84$ we do not reject $H_0 : \tau_1 = \tau_2 = \tau_3 = \tau_4 = 0$ at the 0.05 level. Therefore, we conclude that the 4 chemical treatments do not differ in terms of effect on beneficial insect populations. Also, since $\frac{\text{MSB}}{\text{MSE}} < 2.12$ we do not reject $H_0 : \beta_1 = \beta_2 = \cdots = \beta_{10} = 0$ at the 0.05 level. Therefore, the insect populations did not seem to vary from plot to plot.   ■

The other two-factor model we wish to discuss is called the *two-factor factorial design*. In this model, two different treatments are applied to the same experimental units. It is desired to test for a difference in the effects due to each treatment and for a possible interaction between the two treatments. This was the situation in our children and television violence example in Section 14.2. For the sake of discussion we shall call the two treatments A and B and assume that there are a levels for treatment A and b levels for treatment B. This means that there will be ab possible different levels for an interaction between them. We shall refer to the interaction as AB. In order to conduct all three tests for main effects and interactions it is necessary that we have replication. Remember that replication is the assignment of more than one experimental unit to a given treatment combination. For the sake of simplicity, we shall consider the case where each treatment combination has the same number, r, of replications.

If we let μ be the grand mean, α_i be the ith level of treatment A, β_j be the jth level of treatment B, and $(\alpha\beta)_{ij}$ be the interaction between the ith level of treatment A and the jth level of treatment B, the model for this design is

$$Y_{ijk} = \mu + \alpha_i + \beta_j + (\alpha\beta)_{ij} + e_{ijk}, \quad \text{where} \quad \begin{cases} \displaystyle\sum_{i=1}^{a} br\alpha_i = 0, \\[2mm] \displaystyle\sum_{j=1}^{b} ar\beta_j = 0, \\[2mm] \displaystyle\sum_{i=1}^{a}\sum_{j=1}^{b} r(\alpha\beta)_{ij} = 0, \\[2mm] k = 1, 2, \ldots, r, \quad \text{and} \\[1mm] e_{ijk} \text{ are uncorrelated } N(0, \sigma^2) \\ \quad \text{random variables.} \end{cases} \tag{14.12}$$

This model requires that we use abr experimental units.

Example 14.8   A market research firm conducted a study to determine the difference, if any, between 2 types of advertising campaigns for selling 3 different types of snack foods. The snack foods tested were peanuts, popcorn, and potato chips. Advertising campaign A stressed the health and nutritional aspects of the foods such as no preservatives, no added salt, and no cholesterol. Campaign B emphasized the connection between having fun and the consumption of the snack foods. Each campaign/snack food combination was replicated 3 times with the following results. The data represent the number sold in a week.

Snack Food	Campaign A	Campaign B
Peanuts	622, 596, 604	798, 813, 807
Popcorn	643, 651, 638	801, 816, 800
Potato chips	567, 589, 553	879, 858, 862

Following the approach we used in the randomized complete blocks design the partitioning of the total sum of squares is motivated by the following.

$$y_{ijk} - \bar{y}_{...} = (\bar{y}_{i..} - \bar{y}_{...}) + (\bar{y}_{.j.} - \bar{y}_{...}) + (\bar{y}_{ij.} - \bar{y}_{i..} - \bar{y}_{.j.} + \bar{y}_{...}) + (y_{ijk} - \bar{y}_{ij.})$$

We leave it as an exercise to verify that this scheme results in the following partition.

$$\sum_{i=1}^{a}\sum_{j=1}^{b}\sum_{k=1}^{r}(y_{ijk} - \bar{y}_{...})^2 = br\sum_{i=1}^{a}(\bar{y}_{i..} - \bar{y}_{...})^2 + ar\sum_{j=1}^{b}(\bar{y}_{.j.} - \bar{y}_{...})^2$$

$$+ r\sum_{i=1}^{a}\sum_{j=1}^{b}(\bar{y}_{ij.} - \bar{y}_{i..} - \bar{y}_{.j.} + \bar{y}_{...})^2 \qquad \text{(14.13)}$$

$$+ \sum_{i=1}^{a}\sum_{j=1}^{b}\sum_{k=1}^{r}(y_{ijk} - \bar{y}_{ij.})^2$$

This is commonly written as

$$\text{SSTotal} = \text{SSA} + \text{SSB} + \text{SS(AB)} + \text{SSE},$$

where

$$\text{SSTotal} = \sum_{i=1}^{a}\sum_{j=1}^{b}\sum_{k=1}^{r}(y_{ijk} - \bar{y}...)^2,$$

$$\text{SSA} = br\sum_{i=1}^{a}(\bar{y}_{i..} - \bar{y}_{...})^2,$$

$$\text{SSB} = ar\sum_{j=1}^{b}(\bar{y}_{.j.} - \bar{y}_{...})^2,$$

$$SS(AB) = r \sum_{i=1}^{a} \sum_{j=1}^{b} (\bar{y}_{ij.} - \bar{y}_{i..} - \bar{y}_{.j.} + \bar{y}_{...})^2, \quad \text{and}$$

$$SSE = \sum_{i=1}^{a} \sum_{j=1}^{b} \sum_{k=1}^{r} (y_{ijk} - \bar{y}_{ij.})^2.$$

SSTotal is the total sum of squares, SSA is the sum of squares for treatment A, SSB is the sum of squares for treatment B, SS(AB) is the interaction sum of squares, and SSE is the error sum of squares.

The distributions of the sums of squares are determined in the same way we proceeded in the randomized complete blocks design. We leave it as an exercise to verify that there exist idempotent matrices $\mathbf{A}_1$, $\mathbf{A}_2$, $\mathbf{A}_3$, and $\mathbf{A}_4$, such that

$$\frac{SSA}{\sigma^2} \sim \chi^2(a - 1),$$

$$\frac{SSB}{\sigma^2} \sim \chi^2(b - 1),$$

$$\frac{SS(AB)}{\sigma^2} \sim \chi^2([a - 1][b - 1]), \quad \text{and} \tag{14.14}$$

$$\frac{SSE}{\sigma^2} \sim \chi^2(ab[r - 1]).$$

The derivation of the expected mean squares is also identical to what we did for the randomized complete block design. Therefore, we leave it as an exercise to verify that

$$E[MSA] = E\left[\frac{SSA}{a - 1}\right] = \sigma^2 + \frac{rb}{a - 1} \sum_{i=1}^{a} \alpha_i^2,$$

$$E[MSB] = E\left[\frac{SSB}{b - 1}\right] = \sigma^2 + \frac{ra}{b - 1} \sum_{j=1}^{b} \beta_j^2,$$

$$E[MS(AB)] = E\left[\frac{SS(AB)}{(a - 1)(b - 1)}\right] = \sigma^2 + \frac{r}{(a - 1)(b - 1)} \sum_{i=1}^{a} \sum_{j=1}^{b} (\alpha_i \beta_j)^2, \tag{14.15}$$

$$E[MSE] = E\left[\frac{SSE}{ab(r - 1)}\right] = \sigma^2.$$

These results give us the proper F ratios to test the different null hypotheses in this design. They are as follows.

Null Hypothesis	F Ratio
$H_0 : \alpha_i = 0, i = 1, \ldots, a$	$\dfrac{MSA}{MSE}$
$H_0 : \beta_j = 0, j = 1, \ldots, b$	$\dfrac{MSB}{MSE}$
$H_0 : (\alpha\beta)_{ij} = 0, \begin{array}{l} i = 1, \ldots, a \\ j = 1, \ldots, b \end{array}$	$\dfrac{MS(AB)}{MSE}$

The ANOVA table for this design is as shown in Figure 14.5.

Source of Variation	d.f.	SS	MS	F
Treatment A	$a - 1$	SSA	$\dfrac{\text{SSA}}{a - 1}$	$\dfrac{\text{MST}}{\text{MSE}}$
Treatment B	$b - 1$	SSB	$\dfrac{\text{SSB}}{b - 1}$	$\dfrac{\text{MSB}}{\text{MSE}}$
Interaction	$(a - 1)(b - 1)$	SS(AB)	$\dfrac{\text{SS(AB)}}{(a - 1)(b - 1)}$	$\dfrac{\text{MS(AB)}}{\text{MSE}}$
Error	$ab(r - 1)$	SSE	$\dfrac{\text{SSE}}{ab(r - 1)}$	
Total	$rab - 1$	SSTotal		

Figure 14.5

The machine formulas for the sums of squares in the factorial design are as follows.

$$\text{SSTotal} = \sum_{i=1}^{a} \sum_{j=1}^{b} \sum_{k=1}^{r} y_{ijk}^2 - \frac{y_{...}^2}{rab},$$

$$\text{SSA} = \frac{1}{rb} \sum_{i=1}^{a} y_{i..}^2 - \frac{y_{...}^2}{rab},$$

$$\text{SSB} = \frac{1}{ra} \sum_{j=1}^{b} y_{.j.}^2 - \frac{y_{...}^2}{rab}, \tag{14.16}$$

$$\text{SS(AB)} = \frac{1}{r} \sum_{i=1}^{a} \sum_{j=1}^{b} y_{ij.}^2 - \frac{1}{rb} \sum_{i=1}^{a} y_{i..}^2 - \frac{1}{ra} \sum_{j=1}^{b} y_{.j.}^2 + \frac{y_{...}^2}{rab}, \quad \text{and}$$

$$\text{SSE} = \text{SSTotal} - \text{SSA} - \text{SSB} - \text{SS(AB)}$$

Example 14.9   We wish to carry out the analysis of the advertising data from Example 14.8. The quantities needed for Equation 14.16 are as follows.

$$a = 2 \qquad b = 3 \qquad r = 3$$

$$y_{1..} = 5463 \qquad y_{2..} = 7434$$

$$y_{.1.} = 4240 \qquad y_{.2.} = 4349 \qquad y_{.3.} = 4308$$

$$y_{11.} = 1822 \qquad y_{12.} = 1932 \qquad y_{13.} = 1709$$

$$y_{21.} = 2418 \qquad y_{22.} = 2417 \qquad y_{23.} = 2599$$

$$y_{...} = 12897$$

$$\sum_{i=1}^{a} \sum_{j=1}^{b} \sum_{k=1}^{r} y_{ijk}^2 = 9473757$$

The sums of squares from Equation 14.16 are as follows.

$$\text{SSTotal} = 9473757 - \frac{(12897)^2}{18} = 233056.5,$$

$$\text{SSA} = \frac{1}{9}[(5463)^2 + (7434)^2] - \frac{(12897)^2}{18} = 215824.5,$$

$$\text{SSB} = \frac{1}{6}[(4240)^2 + (4349)^2 + (4308)^2] - \frac{(12897)^2}{18} = 1010.33,$$

$$\text{SS(AB)} = \frac{1}{3}[(1822)^2 + (1932)^2 + (1709)^2 + (2418)^2 + (2417)^2 + (2599)^2]$$

$$- \frac{1}{9}[(5463)^2 + (7434)^2] - \frac{1}{6}[(4240)^2 + (4349)^2 + 4308)^2]$$

$$+ \frac{(12897)^2}{18} = 14599, \quad \text{and}$$

$$\text{SSE} = 233056.5 - 215824.5 - 1010.33 - 14599 = 1604.67$$

Using these, the ANOVA table for these data is given below.

Source of Variation	d.f.	SS	MS	F
Treatment A	1	215824.5	215824.5	1613.97
Treatment B	2	1010.33	505.17	3.78
Interaction	2	14599	7299.5	54.59
Error	12	1604.67	133.72	
Total	17	233056.5		

If we use a significance level of 0.05, Table 6 of Appendix A gives the following critical regions for the F ratios.

$$F(1, 12)_{.05} = 4.75 \quad \text{and} \quad F(2, 12)_{.05} = 3.89$$

Therefore, the data indicate that we should reject the null hypothesis of no difference between the two advertising campaigns, we cannot reject the null hypothesis of equality of sales level for the different snack foods, and we should reject the hypothesis that there is no interaction between type of advertising and snack food. That is, the sales of snack foods can be enhanced or depressed depending on the type of advertising of the particular snack food.   ■

It must be pointed out that the existence of an interaction effect in the above example means that we must be very careful about how to interpret the main effects tests. Differences in the level of one main effect can make it appear that there is a difference in the level of the other main effect when there really is none present. Also an interaction effect can tend to cover up a difference due to the level of a main effect. Figure 14.6 gives some examples of how interactions can manifest themselves. The plot shows cell means and illustrates some of the possibilities for a two-factor design with two treatment levels in one main effect and three levels in the other.

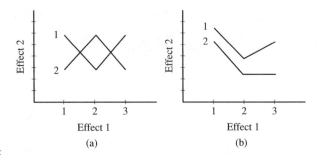

Figure 14.6

One way to get around the effects of interactions is to test for a difference in level of Treatment A at each separate level of Treatment B. This will remove the possible spurious differences due to an interaction between the two treatments.

If the test for interaction had indicated no interaction effect many practitioners recommend that the analysis be rerun as a two-way analysis of variance with no interaction effect. They feel that more often than not a more sensitive test for the main effects is the result. There is no theoretical justification for this, but it seems to work well in most cases.

The experimental designs that we have analyzed in this and the preceding section are only a few of a huge number of possibilities. The Latin square design that was discussed in Section 14.2 is a special case of the two-factor factorial design where both treatments have the same number of levels, and there is known to be no interaction between them. In addition, there are other ways to block experiments. One such way is known as a randomized incomplete block design. As the name implies, there are designs where the blocks are not complete. That is, not every treatment level is in each block.

There are also different ways to look at the treatment levels themselves. The way we have been using them results in what are called *fixed effects models*. In essence, this means that the number of treatment levels are finite and we are looking at them all. It is possible, however, to view the treatment levels being used as constituting a random sampling from a larger population of levels. Such a treatment is known as a *random effect*. Suppose, for example, a company wishes to compare three different machines and also determine if machine output is operator dependent. We can certainly view the three machines as constituting the different levels of a fixed effects treatment. However, the operators used are probably a sample from all possible operators. Therefore, we must view the operator treatment as being a random effect. Models that contain only fixed effect treatments are called *fixed effects models*. Models that contain only random effect treatments are called *random effects models*. In the example we just gave, there was one fixed effect and one random effect. Logically enough such models are called *mixed models*. These distinctions are important because the F ratios that are used with fixed effects differ from those used for random effects. Excellent theoretical coverage of these topics can be found in *An Introduction to Linear Statistical Models*, vol. 1, by F. A. Graybill. A more applied treatment can be found in *Statistics in Research* by B. Ostle and R. W. Mensing.

Exercises 14.4

1. Verify the partition of Equation 14.10.

2. In the randomized complete block design suppose that $k = 3$ and $b = 2$. Determine the two idempotent matrices that will generate SST and SSB. Furthermore, show that the product of these 2 matrices is the 0 matrix.

3. Verify the machine formulas in Equation 14.11.

4. A manufacturer of clothing for infants was interested in comparing 4 different flame retarding treatments for cloth. Three different pieces of cloth were selected from stock. Each piece was divided into 4 equal parts which then had one of the flame retarding chemicals applied. Each piece of cloth was then exposed to an open flame. The times, in minutes, for each to catch fire are recorded in the following table.

		Treatment		
Cloth	**1**	**2**	**3**	**4**
Flannel	6.5	8.0	10.0	10.0
Wool	11.0	12.5	14.5	29
Cotton	12.5	15.0	17.0	16.0

Complete the analysis of variance for this randomized complete block design. Interpret your results at the 0.05 level.

5. An automobile manufacturer wishes to analyze 4 different exterior paints. Four cars are selected and one-quarter of the exterior surface is covered with each type of paint in a Latin square design where the columns of the square correspond to the left front, right front, left rear, and right rear, respectively. Each row is an automobile. After 6 months of exposure the paints are judged on quality. The quality ratings are as follows.

		Paint		
Automobile	**1**	**2**	**3**	**4**
1	37	49	33	41
2	42	38	34	48
3	45	40	40	40
4	49	39	38	42

Carry out the analysis of variance for this Latin squares design at the 0.1 level. Interpret your results.

6. Verify the validity of the partition given in Equation 14.13.

7. Use Theorem 14.1 and derivations in the text to show that the sums of squares in the two-factor factorial design are distributed as given in Equation 14.13 if $a = b = r = 2$.

8. Verify that the expected mean squares for the two-factor factorial design are as shown in Equation 14.15.

9. A psychologist investigated the effect of distracting conditions on the ability to find particular nonsense letter combinations among other nonsense letter combinations. She was also interested in the effects, if any, of IQ on this ability. The distracting conditions of interest were lights, noises, and smells. To carry this out she divided the population into three IQ groups—low, medium, and high. Two people from each group were then

assigned to each distraction. In addition, two more from each group were placed in a control group having no distractions. The number of correct nonsense syllables located in a fixed period of time are given below.

IQ	Noise	Light	Smell	Control
High	25, 29	30, 27	22, 25	24, 21
Medium	24, 23	21, 23	23, 22	20, 21
Low	20, 18	22, 24	20, 21	24, 22

Complete the analysis of variance for this factorial design at the 0.05 level, and interpret your results.

10. Recall the experiment described in Section 14.2 to study the effects of television violence on children. A group of 6 boys and 6 girls were randomly divided in half. Three of each watched an hour of television containing no violence. The remaining watched an hour of television that contained violent behavior. They were then sent out to play where they were rated as to their aggressiveness. The data are as follows.

Sex	Violence	No Violence
Female	62, 57, 64	44, 50, 43
Male	70, 73, 71	51, 53, 49

Complete the analysis of variance for this factorial design at the 0.01 level, and interpret your results.

11. Fill in the missing entries in the following ANOVA table. Interpret the results at the 0.1 level.

Source of Variation	d.f.	SS	MS	F
Treatment A	3	120.45		
Treatment B		2428.66	1214.33	
Interaction				3.65
Error			190.95	
Total	24	9217.50		

12. A personal computer magazine was interested in the differences between brands of floppy disks. In particular the staff was interested in the effects of brand on the lifetime of the read/write heads in various brands of disk drives. Four brands of floppy disks were used along with 3 different brands of disk drives. Two disks of one brand were assigned to each brand of drive. Head lifetimes, in hundreds of hours, are given in the following table.

	Disk A	Disk B	Disk C	Disk D
Drive I	12.4, 14.3	11.7, 12.6	16.3, 14.25	11.4, 13.1
Drive II	14.1, 12.9	13.0, 12.3	14.3, 15.2	14.6, 13.9
Drive III	12.0, 11.8	15.2, 14.4	12.8, 12.1	13.7, 14.2

Complete the analysis of variance for this factorial design at the 0.1 level. Interpret your results.

14.5
Estimation

We now turn to a brief discussion of estimation of the parameters of the analysis of variance model. Aside from the general interest in being able to estimate values for model parameters, estimation is a natural follow-up to the tests we have been discussing. If we happen to reject the null hypothesis that all treatment levels are equal to 0, it becomes natural to try to determine as well as possible how the difference manifests itself. In many experiments, the investigator has some idea of how the treatment levels should differ and he or she wishes to confirm this through estimation. We will give an elementary overview of some of the issues involved in this problem. In particular, we shall confine our attention to the single-factor model of Equation 14.1. We shall also ignore the constraint in the single-factor model that $\sum_{i=1}^{k} n_i \tau_i = 0$. These ideas generalize in a straightforward manner to the more complicated models. If we follow the pattern from Chapter 13 our initial goal is to try to determine least squares estimators for $\mu, \tau_1, \tau_2, \ldots$ and τ_k from the normal equations. In this case, the matrix formulation of the model is

$$
\begin{pmatrix} y_{11} \\ y_{12} \\ \vdots \\ y_{kn_k} \end{pmatrix} =
\begin{pmatrix}
\left.\begin{matrix} 1 & 1 & 0 & \cdots & 0 \\ 1 & 1 & 0 & \cdots & 0 \\ \vdots & \vdots & \vdots & & \vdots \\ 1 & 1 & 0 & \cdots & 0 \end{matrix}\right\} n_1 \text{ times} \\
\left.\begin{matrix} 1 & 0 & 1 & \cdots & 0 \\ 1 & 0 & 1 & \cdots & 0 \\ \vdots & \vdots & \vdots & & \vdots \\ 1 & 0 & 1 & \cdots & 0 \end{matrix}\right\} n_2 \text{ times} \\
\vdots \\
\left.\begin{matrix} 1 & 0 & 0 & \cdots & 1 \\ 1 & 0 & 0 & \cdots & 1 \\ \vdots & \vdots & \vdots & & \vdots \\ 1 & 0 & 0 & \cdots & 1 \end{matrix}\right\} n_k \text{ times}
\end{pmatrix}
\begin{pmatrix} \mu \\ \tau_1 \\ \tau_2 \\ \vdots \\ \tau_k \end{pmatrix} +
\begin{pmatrix} e_{11} \\ e_{12} \\ \vdots \\ e_{kn_k} \end{pmatrix}. \quad \textbf{(14.17)}
$$

By letting $\boldsymbol{\beta}' = (\mu, \tau_1, \ldots, \tau_k)$ we see that Equation 14.17 can be written in the form

$$\mathbf{y} = \mathbf{X}\boldsymbol{\beta} + \mathbf{e}.$$

Therefore, the *normal equations* would be

$$\mathbf{X}'\mathbf{X}\boldsymbol{\beta} = \mathbf{X}'\mathbf{y}$$

as before.

Example 14.10   Assume that $k = 3$, $n_1 = 1$, $n_2 = 2$, and $n_3 = 3$. Then Equation 14.17 becomes

$$
\begin{pmatrix} y_{11} \\ y_{21} \\ y_{22} \\ y_{31} \\ y_{32} \\ y_{33} \end{pmatrix} =
\begin{pmatrix} 1 & 1 & 0 & 0 \\ 1 & 0 & 1 & 0 \\ 1 & 0 & 1 & 0 \\ 1 & 0 & 0 & 1 \\ 1 & 0 & 0 & 1 \\ 1 & 0 & 0 & 1 \end{pmatrix}
\begin{pmatrix} \mu \\ \tau_1 \\ \tau_2 \\ \tau_3 \end{pmatrix} +
\begin{pmatrix} e_{11} \\ e_{21} \\ e_{22} \\ e_{31} \\ e_{32} \\ e_{33} \end{pmatrix}.
$$

The normal equations would be

$$\begin{pmatrix} 6 & 1 & 2 & 3 \\ 1 & 1 & 0 & 0 \\ 2 & 0 & 2 & 0 \\ 3 & 0 & 0 & 3 \end{pmatrix} \begin{pmatrix} \mu \\ \tau_1 \\ \tau_2 \\ \tau_3 \end{pmatrix} = \begin{pmatrix} y_{..} \\ y_{1.} \\ y_{2.} \\ y_{3.} \end{pmatrix}.$$

∎

From Example 14.10 we notice that the matrix $\mathbf{X'X}$ is 4×4 but that its rank is 3. Therefore, it is not of full rank, which means that it is singular and does not possess an inverse. Thus, the normal equations will have either no solution or an infinite number of solutions. The reader will recall from a linear algebra course that the way to tell which case exists is from the augmented matrix $(\mathbf{X'X}|\mathbf{X'y})$ and then perform elementary row operations to place this matrix in reduced row-echelon form. This is the *Gaussian elimination* procedure. If augmented matrix in reduced row-echelon form had a row whose entries are all 0 except for the rightmost column, then the system has no solutions. Otherwise, there is at least 1 solution. In our case this would mean that there would be an infinite number of solutions. If we perform Gaussian elimination of the normal equations of Example 14.10 we obtain the following.

$$\begin{pmatrix} 1 & 1 & 0 & 0 & y_{1.} \\ 0 & 1 & -1 & 0 & y_{1.} - \frac{1}{2}y_{2.} \\ 0 & 0 & 1 & -1 & \frac{1}{2}y_{2.} - \frac{1}{3}y_{3.} \\ 0 & 0 & 0 & 0 & 0 \end{pmatrix}$$

This indicates that for this case there are an infinite number of solutions for μ, τ_1, τ_2, and τ_3. This situation is certainly undesirable. What it means is that 2 different investigators could use the same data and, in all likelihood, come up with entirely different least squares estimates for the same model parameters.

The analysis of this example suggests that a logical question would be whether or not, in every single factor model, there will always be an infinite number of solutions for the normal equations. It turns out that this question can be answered affirmatively. The matrix $\mathbf{X}$ from Equation 14.17 will always have dimension $n \times (k+1)$ and rank k. This is seen from the fact that $\mathbf{X}$ has exactly k different rows, one for each different treatment level. Furthermore, none of these k rows can be obtained by taking a linear combination of the remaining $k-1$ rows. There is no way to place a 1 in column 2, for example, as each of the other rows has a 0 there. Therefore, $\mathbf{X}$ is always less than full rank. Now in standard linear algebra courses you learned that the rank of a matrix is equal to the number of nonzero rows that exist after the matrix has been placed in row-echelon form. The system of linear equations

$$\mathbf{y} = \mathbf{Ax}$$

will have at least 1 solution if the augmented matrix $(\mathbf{A}|\mathbf{y})$ in row-echelon form has exactly as many nonzero rows as the matrix $\mathbf{A}$ in row-echelon form. In terms of rank this means that the system will have at least 1 solution if

$$\text{Rank}(\mathbf{A}|\mathbf{y}) = \text{rank}(\mathbf{A}).$$

To show that the normal equations will always have an infinite number of solutions we wish to verify that

$$\text{Rank}(\mathbf{X}'\mathbf{X}|\mathbf{X}'\mathbf{y}) = \text{rank}(\mathbf{X}'\mathbf{X}).$$

We shall do this by showing that the 2 ranks are both greater than or equal to each other and less than or equal each other. First, note that the augmented matrix has 1 more column than does $\mathbf{X}'\mathbf{X}$ with all of the other columns being identical. This implies that

$$\text{Rank}(\mathbf{X}'\mathbf{X}|\mathbf{X}'\mathbf{y}) \geq \text{rank}(\mathbf{X}'\mathbf{X}) = k.$$

Also

$$\text{Rank}(\mathbf{X}|\mathbf{y}) \geq \text{rank}(\mathbf{X}).$$

Now we see that the augmented matrix can be written as

$$(\mathbf{X}'\mathbf{X}|\mathbf{X}'\mathbf{y}) = \mathbf{X}'(\mathbf{X}|\mathbf{y}).$$

Since $\text{rank}(\mathbf{X}') \leq \text{rank}(\mathbf{X}|\mathbf{y})$, Theorem 13.5 gives that

$$\text{Rank}(\mathbf{X}'\mathbf{X}|\mathbf{X}'\mathbf{y}) = \text{rank}[\mathbf{X}'(\mathbf{X}|\mathbf{y})] \leq \text{rank}(\mathbf{X}') = k.$$

Therefore, the normal equations will always have an infinite number of solutions.

Since the single-factor model will always have multiple estimators for the model parameters, theoretical statisticians then turned to address the issue of the existence of unbiased estimators for linear combinations of the model parameters. It turns out that something can be done here.

Suppose we consider the single-factor model as defined in Equation 14.1. That is,

$$\mu_i = \mu + \tau_i, i = 1, 2, \ldots, k.$$

The matrix formulation for this model is as follows.

$$
\begin{pmatrix} y_{11} \\ y_{12} \\ \vdots \\ y_{kn_k} \end{pmatrix}
=
\left(
\begin{array}{ccccc}
\left. \begin{array}{ccccc} 1 & 0 & 0 & \cdots & 0 \\ 1 & 0 & 0 & \cdots & 0 \\ \vdots & \vdots & \vdots & & \vdots \\ 1 & 0 & 0 & \cdots & 0 \end{array} \right\} & n_1 \text{ times} \\
\left. \begin{array}{ccccc} 0 & 1 & 0 & \cdots & 0 \\ 0 & 1 & 0 & \cdots & 0 \\ \vdots & \vdots & \vdots & & \vdots \\ 0 & 1 & 0 & \cdots & 0 \end{array} \right\} & n_2 \text{ times} \\
\vdots \\
\left. \begin{array}{ccccc} 0 & 0 & 0 & \cdots & 1 \\ 0 & 0 & 0 & \cdots & 1 \\ \vdots & \vdots & \vdots & & \vdots \\ 0 & 0 & 0 & \cdots & 1 \end{array} \right\} & n_k \text{ times}
\end{array}
\right)
\begin{pmatrix} \mu + \tau_1 \\ \mu + \tau_2 \\ \vdots \\ \mu + \tau_k \end{pmatrix}
+
\begin{pmatrix} e_{11} \\ e_{12} \\ \vdots \\ e_{kn_k} \end{pmatrix}.
$$

$$(14.18)$$

Let $\boldsymbol{\beta}' = (\mu + \tau_1, \mu + \tau_2, \ldots, \mu + \tau_k)$. Then the model can be written as

$$\mathbf{y} = \mathbf{X}\boldsymbol{\beta} + \mathbf{e}.$$

In this case, $\mathbf{X}$ is an $n_1 + n_2 + \cdots + n_k$ by k matrix whose rank is k. This means that if we write the normal equations for this model,

$$\mathbf{X}'\mathbf{X}\boldsymbol{\beta} = \mathbf{X}'\mathbf{y},$$

$\mathbf{X}'\mathbf{X}$ will be a $k \times k$ matrix whose rank is k. Therefore, there will be a unique least squares solution for $\mu + \tau_i, i = 1, 3, \ldots, k$. If we let

$$Y_{i\cdot} = \sum_{j=1}^{n_i} Y_{ij},$$

the sum of the observations from group i for $i = 1, 2, \ldots, k$, then the normal equations will be

$$\begin{pmatrix} n_1 & 0 & 0 & \cdots & 0 \\ 0 & n_2 & 0 & \cdots & 0 \\ \vdots & \vdots & \vdots & \ddots & \vdots \\ 0 & 0 & 0 & \cdots & n_k \end{pmatrix} \begin{pmatrix} \mu + \tau_1 \\ \mu + \tau_2 \\ \vdots \\ \mu + \tau_k \end{pmatrix} = \begin{pmatrix} Y_{1\cdot} \\ Y_{2\cdot} \\ \vdots \\ y_{k\cdot} \end{pmatrix}.$$

Solving these gives that

$$\widehat{\mu + \tau_i} = \bar{Y}_i, i = 1, 2, \ldots, k,$$

where $\bar{Y}_i = \dfrac{1}{n_i} Y_{i\cdot}$. The single factor model assumes that the observations are normally distributed. This implies that

$$\widehat{\mu + \tau_i} = \bar{Y}_i \sim N\left(\mu + \tau_i, \frac{\sigma^2}{n_i}\right).$$

In addition, we know from Section 14.3 that

$$\widehat{\sigma^2} = \frac{\text{SSE}}{n - k}$$

is an unbiased estimator for σ^2, and it has a $\chi^2(n - k)$ distribution. Also we can use Theorem 9.6 and the fact that the groups were assumed to be independent to see that $\widehat{\sigma^2}$ is independent from $\bar{Y}_i$.

The main interest in one-way analysis of variance is in the values of the individual τ_i. Since the model is not of full rank we cannot obtain unique estimates for these quantities. We can, however, use the unique estimators for the $\mu + \tau_i$ to compare τ_i to τ_j. Consider the problem of estimating $\tau_i - \tau_j$. A unique estimator, based on the least squares estimators will be

$$\widehat{\tau_i - \tau_j} = \bar{Y}_i - \bar{Y}_j. \tag{14.19}$$

We can use these results to construct confidence intervals for $\tau_i - \tau_j$. It is easy to see that

$$\widehat{\tau_i - \tau_j} \sim N\left(\tau_i - \tau_j, \sigma^2\left[\frac{1}{n_i} + \frac{1}{n_j}\right]\right).$$

Therefore,

$$T = \frac{\bar{Y}_i - \bar{Y}_j - (\tau_i - \tau_j)}{\sqrt{\widehat{\sigma^2}\left[\frac{1}{n_i} + \frac{1}{n_j}\right]}} \tag{14.20}$$

will have a t distribution with $n - k$ degrees of freedom. Recall that $n = n_1 + n_2 + \cdots + n_k$. Therefore, we have a pivotal quantity. This means that if we follow the procedure of Section 11.2 we find that a $(1 - \alpha)100\%$ confidence interval for $\tau_i - \tau_j$ is

$$\bar{Y}_i - \bar{Y}_j - t(n - k)_{1-\alpha/2}\sqrt{\widehat{\sigma^2}\left(\frac{1}{n_i} + \frac{1}{n_j}\right)}$$

$$\leq \tau_i - \tau_j \leq \bar{Y}_i - \bar{Y}_j + t(n - k)_{1-\alpha/2}\sqrt{\widehat{\sigma^2}\left(\frac{1}{n_i} + \frac{1}{n_j}\right)}. \tag{14.21}$$

Example 14.11

In Section 14.3 we tested for a difference in patient stress levels after subjects were placed in one of three programs—transcendental meditation, exercise, or diet. A difference in stress levels was indicated. Advocates for transcendental meditation claim that their method is at least as effective in lowering stress as vigorous exercise. Let τ_1 represent the effect due to transcendental meditation, τ_2 the effect due to exercise, and τ_3 the effect due to diet and consider $\tau_1 - \tau_2$. From the data given in Example 14.2 we find from Equation 14.18 that

$$\mathbf{X'X} = \begin{pmatrix} 5 & 0 & 0 \\ 0 & 5 & 0 \\ 0 & 0 & 5 \end{pmatrix} \quad \text{and} \quad \mathbf{X'y} = \begin{pmatrix} -33 \\ -51 \\ -19 \end{pmatrix}.$$

Therefore, an estimate for $\tau_1 - \tau_2$ is

$$\widehat{\tau_1 - \tau_2} = \bar{Y}_1 - \bar{Y}_2$$

$$= \frac{1}{5}(-33) - \frac{1}{5}(-51)$$

$$= 3.6.$$

In addition a 95% confidence interval for $\tau_1 - \tau_2$ is found as follows. From Example 14.5 we know that SSE = 70.8, $n = 15$, and $k = 3$. Therefore,

$$\widehat{\sigma^2} = \frac{70.8}{12} = 5.90.$$

In addition, we find from Table 5 of Appendix A that $t(12)_{0.025} = 2.179$. Thus,

Equation 14.22 gives that

$$3.6 - 2.179\sqrt{5.90 \left(\frac{1}{5} + \frac{1}{5}\right)} \leq \tau_1 - \tau_2 \leq 3.6 - 2.179\sqrt{5.90 \left(\frac{1}{5} + \frac{1}{5}\right)},$$

or

$$0.35 \leq \tau_1 - \tau_2 \leq 6.95.$$

Thus, with 95% confidence it appears that exercise is more effective than transcendental meditation for stress reduction.   ■

Exercises 14.5

1. Use the data in Exercise 7 of Section 14.3 to estimate the difference between group 1 and group 3.

2. In the model of Exercise 7 of Section 14.3 to construct a 99% confidence interval for $\tau_3 - \tau_2$. Interpret your results.

3. Use the data in Exercise 8 of Section 14.3 to estimate $\mu + \tau_2$.

4. Use the data in Exercise 8 of Section 14.3 to construct a 90% confidence interval for $\tau_1 - \tau_3$. Interpret your results.

5. Use the data in Exercise 9 of Section 14.3 to estimate the difference between the mean number of days of facility 2 and facility 3. That is, estimate $\tau_2 - \tau_3$.

6. Use the data in Exercise 9 of Section 14.3 to construct a 95% confidence interval for $\tau_2 - \tau_3$. Interpret your results.

7. Show how you might use the ideas discussed in this section to estimate $2\tau_1 - \tau_2 - \tau_3$. Use the data in Exercise 9 of Section 14.3 to estimate $2\tau_1 - \tau_2 - \tau_3$. Interpret your results.

8. Show how you might use the ideas discussed in this section to construct a $(1 - \alpha)100\%$ confidence interval for $2\tau_1 - \tau_2 - \tau_3$. Use the data in Exercise 9 of Section 14.3 to construct a 95% confidence interval for $2\tau_1 - \tau_2 - \tau_3$.

9. Consider a single-factor analysis of variance model with 4 treatment levels. Suppose we suspect that the effect of treatment levels 1, 3, and 4 are all equal but differ from that of treatment level 2. Suggest a linear combination of τ_1, τ_2, τ_3, and τ_4 that will permit us to determine if this is true. Be sure that the coefficients sum to 0. In addition, give the t-statistic that would be used to construct a confidence interval.

10. In single-factor models, statisticians often use what are known as *contrasts*. Assume that there are k treatment levels and let $S_i = \sum_{j=1}^{n_j} Y_{ij}$ be the sum of the n_i observations of treatment level i, $i = 1, 2, \ldots, k$. Then a contrast is defined to be

$$c = c_1 S_1 + c_2 + S_2 + \cdots + c_k S_k$$

subject to the condition that

$$\sum_{i=1}^{k} n_i c_i = 0.$$

Constrasts are useful in that they can be used to account for differences in sample sizes for treatment levels. Assume that a single-factor model has 4 treatment levels and that the sample sizes for those levels are

$$n_1 = 4, \quad n_2 = 6, \quad n_3 = 9, \quad \text{and} \quad n_4 = 5.$$

Give contrasts that will compare the following.

a. τ_1 and τ_3.

b. τ_2 against τ_1 and τ_4.

c. τ_3 against τ_1, τ_2, and τ_4.

14.6
Nonparametric Methods

The techniques used for estimating parameters and testing hypotheses in the analysis of variance models that we have discussed in the preceding sections have been based on the assumption that the error terms are uncorrelated normal random variables. The F- and t-statistics developed there have been found by investigators to be fairly robust to departures from normality, but they are not totally insensitive to them. Therefore, in those cases where it is known that the error terms are decidedly nonnormal we need tests that do not rely on such assumptions. In some settings, conservative investigators like to corroborate the results of the classical F tests by also performing nonparametric tests. In this section we shall present two popular tests for the single-factor case. In a completely randomized design the nonparametric test of the equality of treatment levels is the *Kruskal-Wallis test*. In a randomized block design with one observation per block we shall present the *Friedman test*. Both of these tests actually test the equality of treatment level *medians* rather than means as the F tests do. In the case of symmetric distributions for the error terms, these are the same. It should also be noted that both of these tests do assume that the error terms are independent and identically distributed. They just do not require the additional constraint that the distributions be normal.

We shall begin with a discussion of the *Kruskal-Wallis test*. Here we assume a model of the form

$$Y_{ij} = \tilde{\mu} + \tilde{\tau}_i + e_{ij}, \text{ where } \begin{cases} \displaystyle\sum_{i=1}^{k} \tilde{\tau}_i = 0, \quad \text{and} \\[2mm] e_{ij} \text{ are independent and identically distributed} \\ \text{with median 0, for} \\ i = 1, 2, \ldots, k, \\ j = 1, 2, \ldots, n_i. \end{cases}$$

(14.22)

We are testing

$$H_0 : \tilde{\tau}_1 = \tilde{\tau}_2 = \cdots = \tilde{\tau}_k = 0 \quad \text{vs.} \quad H_1 : \text{not all } \tilde{\tau}\text{'s equal } 0.$$

A convenient way to motivate the form of the test statistic is to consider the F-statistic from Section 14.3 and replace the y's with their ranks. That is, we begin by combining all of the observations and assigning to each y_{ij}, its rank, call it R_{ij}.

Ties are handled in the usual manner. We shall use the following notation.

$$n = \sum_{i=1}^{k} n_i,$$

$$R_{i.} = \sum_{j=1}^{n_i} R_{ij}, i = 1, 2, \ldots, k,$$

$$\bar{R}_{i.} = \frac{R_{i.}}{n_i}, i = 1, 2, \ldots, k, \qquad (14.23)$$

$$R_{..} = \sum_{i=1}^{k} R_{i.} = \frac{n(n+1)}{2}, \quad \text{and}$$

$$\bar{R}_{..} = \frac{R_{..}}{n} = \frac{n+1}{2}$$

Now the F-statistic for testing the completely randomized design was based on the ratio

$$\frac{\text{SST}}{\text{SSE}} = \frac{\sum_{i=1}^{k} n_i (\bar{y}_{i.} - \bar{y}_{..})^2}{\sum_{i=1}^{k} \sum_{j=1}^{n_i} (y_{ij} - \bar{y}_{i.})^2}.$$

If the treatment medians differ significantly from the grand median this fact will also be reflected in the average treatment rank differing from the overall average rank of $\frac{n+1}{2}$. Then a logical quantity for detecting a difference in treatment levels would be to use SST computed on the ranks. In other words use

$$\sum_{i=1}^{k} n_i (\bar{R}_{i.} - \bar{R}_{..})^2.$$

This quantity is multiplied by a constant that will give it a large sample distribution that is approximately chi-squared. Thus, the Kruskal-Wallis statistic is

$$H = \frac{12}{n(n+1)} \sum_{i=1}^{k} n_i (\bar{R}_{i.} - \bar{R}_{..})^2. \qquad (14.24)$$

A convenient computational form for H is

$$H = \frac{12}{n(n+1)} \sum_{i=1}^{k} \frac{R_{i.}^2}{n_i} - 3(n+1). \qquad (14.25)$$

An inspection of this statistic shows that if the average ranks are all equal then H will equal 0. As the difference between the average ranks increases so will the value of H. Therefore, we would be inclined to reject H_0 when the value of

H becomes large enough. That is, we would reject H_0 if

$$H \geq c, \tag{14.26}$$

where the constant c is chosen to achieve a specified value for α. We leave it as an exercise to show that when $k = 2$, the Kruskal-Wallis is equivalent to the Wilcoxon Rank Sum test. The exact distribution for H is complicated. Essentially what must be done is to enumerate all possible values for H. The distribution is tabulated in Table 10 of Appendix A for $k = 3$ and values of n_i less than or equal to 5.

Example 14.12   We wish to use the Kruskal-Wallis statistic to test for a difference in stress levels for the data given in Example 14.2. We shall let those using transcendental meditation be group 1, those using exercise be group 2, and the dieters be group 3. Then we are testing

$$H_0 : \tilde{\tau}_1 = \tilde{\tau}_2 = \tilde{\tau}_3 \quad \text{vs.} \quad H_1 : \tilde{\tau}_1, \tilde{\tau}_2, \tilde{\tau}_3 \text{ not all equal.}$$

We shall use $\alpha = 0.05$. Since $n_1 = n_2 = n_3 = 5$ and $k = 3$ we find from Table 10 of Appendix A that we should reject H_0 when $H \geq 5.66$ with an exact significance level of 0.051. To compute the value for H, the first task is to combine the 15 observations and place them in ascending order while keeping track of the group to which each belongs. Then we assign ranks. Thus, we have the following.

Y	-14	-11	-11	-10	-9	-7	-7	-6	-6	-5	-5	-4	-4	-3	-1
Group	2	2	1	2	2	2	1	1	3	1	3	1	3	3	3
Rank	1	2.5	2.5	4	5	6.5	6.5	8.5	8.5	10.5	10.5	12.5	12.5	14	15

From this we find that

$$R_{1.} = 40.5, \quad R_{2.} = 19, \quad \text{and} \quad R_{3.} = 60.5.$$

As a check that these are correct we note that $R_{1.} + R_{2.} + R_{3.} = 120 = \frac{n(n+1)}{2}$. Then from Equation 14.26 we compute

$$H = \frac{12}{15(16)} \left[\frac{(40.5)^2}{5} + \frac{(19)^2}{5} + \frac{(60.5)^2}{5} \right] - 3(16)$$

$$= 8.615.$$

Since $H > 5.66$ we reject H_0 and conclude that there is a difference between the 3 methods in terms of stress reduction.   ■

As can be seen from Table 10 of Appendix A it would be computationally prohibitive to determine the exact distribution for H whenever k or n gets large. In such cases we would like to approximate the exact distribution by something simpler. As we mentioned earlier the unintuitive constant used in Equation 14.25

is included so that we can use chi-squared tables for larger samples. We state the result as a theorem and omit the proof.

Theorem 14.4

When $H_0 : \tilde{\tau}_1 = \tilde{\tau}_2 = \cdots = \tilde{\tau}_k$ is true then, as n becomes large,

$$H = \frac{12}{n(n+1)} \sum_{i=1}^{k} n_i (\bar{R}_{i.} - \bar{R}_{..})^2$$

has a limiting distribution that is chi-squared with $k - 1$ degrees of freedom.

This means that for large samples we can reject H_0 when

$$H \geq \chi^2(k-1)_\alpha. \tag{14.27}$$

Researchers have found through simulation studies that the chi-square approximation is acceptable when group sample sizes exceed 5 with $k \geq 3$.

Example 14.13

The sample sizes from Example 14.12 are just on the borderline for use of the chi-square approximation. Therefore, to see how good the approximation is we wish to compare the critical value from Table 10 of Appendix A with those obtained using Table 4. Since $k = 3$ in this case we use 2 degrees of freedom for the chi-square table.

α	0.10	0.05	0.025	0.01
Exact	4.560	5.720	6.740	7.990
Approx.	4.605	5.991	7.378	9.210

As we can see the approximation is pretty good for the lower values of α. As might be expected, as we get further out in the tail of the distribution the approximation becomes less accurate. ∎

In the randomized complete blocks design, we shall assume that we have k different treatment levels and b blocks. In each block, we assign one experimental unit to each treatment level. Our model is

$$Y_{ij} = \tilde{\mu} + \tilde{\tau}_i + \tilde{\beta}_j + e_{ij}, \text{ where } \begin{cases} \sum_{i=1}^{k} \tilde{\tau}_i = 0, \\[2mm] \sum_{j=1}^{b} \tilde{\beta}_j = 0, \quad \text{and} \\[2mm] e_{ij} \text{ are independent and identically} \\ \text{distributed with median 0, for} \\ i = 1, 2, \ldots, k, \\ j = 1, 2, \ldots, b. \end{cases} \tag{14.28}$$

We are interested in testing

$$H_0 : \tilde{\tau}_1 = \tilde{\tau}_2 = \cdots = \tilde{\tau}_k = 0 \quad \text{vs.} \quad H_1 : \text{not all } \tilde{\tau}\text{'s equal 0.}$$

A test statistic for this problem is developed using an approach similar to what we used to arrive at the Kruskal-Wallis statistic. However, rather than combine the entire sample, we order the y's within each block and then assign each its rank. To eliminate the differences due to blocks we then sum the ranks for each treatment level. We shall use the following quantities. Let R_{ij} denote the rank of the observation for treatment level i in block j.

$$R_{i.} = \sum_{j=1}^{b} R_{ij},$$

$$\bar{R}_{i.} = \frac{R_{i.}}{b},$$

$$R_{..} = \sum_{i=1}^{k} R_{i.} = \frac{bk(k+1)}{2}, \quad \text{and} \qquad (14.29)$$

$$\bar{R}_{..} = \frac{k+1}{2} = \frac{R_{..}}{bk}.$$

If the treatment medians differ significantly from the grand median this fact should be reflected in a difference in the average of the ranks for the treatment levels from the overall average rank of $\frac{k+1}{2}$. This implies that a logical quantity for detecting a difference in treatment levels would be

$$\sum_{i=1}^{j} (\bar{R}_{i.} - \bar{R}_{..})^2.$$

When this quantity is multiplied by a constant that will give it a large sample distribution that is approximately chi-squared we obtain what is known as the *Friedman statistic*. It is given by

$$S = \frac{12b}{k(k+1)} \sum_{i=1}^{k} (\bar{R}_{i.} - \bar{R}_{..})^2. \qquad (14.30)$$

A convenient computational form for S is

$$S = \frac{12}{bk(k+1)} \sum_{i=1}^{k} R_{i.}^2 - 3b(k+1). \qquad (14.31)$$

An inspection of this statistic shows that if the average ranks for each of the treatment levels are all equal then S will equal 0. As the difference between the average ranks increases so will the value of S. Therefore, we would be inclined to reject H_0 when the value of S becomes large enough. That is, we would reject H_0 if

$$S \geq c, \qquad (14.32)$$

where the constant c is chosen to achieve a specified value for α. As was the case for the Kruskal-Wallis test the exact distribution for S is complicated. Essentially we must enumerate all possible values for S. Therefore, the distribution is tabulated in Table 11 of Appendix A for $k = 3, 4, 5$, and various values of b. It is interesting to note that when $k = 2$ the Friedman test reduces to a form of the sign test that was discussed in Section 12.9. This can be seen from the fact that when $k = 2$ the ranks are all 1 and 2. Suppose that d of the blocks in treatment level 1 receive a rank of 1. Then $R_{1.} = d + 2(b - d) = 2b - d$, and $R_{2.} = b - d + 2d = b + d$. Equation 14.31 becomes

$$S = 4b \left(\frac{d}{b} - \frac{1}{2} \right)^2 .$$

To see that this is like the sign test all we need do is note that d represents the number of blocks where the observation at treatment level 2 had the highest value, or alternately the observation at treatment level 2 would have been given a plus sign.

Example 14.14

Researchers wish to investigate the effect of 4 different preventive maintenance programs on the amount of down time for machines in a production line. A factory runs 4 parallel production lines, and each line has 6 different types of machines. To eliminate the possible side effects due to the differing types of machines it is decided to assign the different maintenance programs randomly to each of the 4 production lines and to treat the various machines as blocks. Therefore, in the model of Equation 14.29 we have that $k = 4$ and $b = 6$. The down times, in hours, for a 1-month period for the machines are as follows.

Machine	Method 1	Method 2	Method 3	Method 4
A	6.2	4.4	5.2	6.0
B	7.0	5.3	4.9	5.1
C	1.0	0.9	1.9	1.1
D	3.4	2.9	3.3	3.5
E	4.1	3.3	3.9	3.6
F	2.4	1.9	2.3	2.5

We wish to test

$$H_0 : \tilde{\tau}_1 = \tilde{\tau}_2 = \tilde{\tau}_3 = \tilde{\tau}_4 = 0 \quad \text{vs.} \quad H_1 : \tilde{\tau}_1, \tilde{\tau}_2, \tilde{\tau}_3, \tilde{\tau}_4 \text{ not all } 0$$

at a level of $\alpha = 0.05$. Since $k = 3$ and $b = 6$ we find in Table 11 of Appendix A that we should reject H_0 if $S \geq 7.4$ to achieve an exact significance level of 0.056. To compute the value for S we first assign ranks in each machine type as follows.

Machine	Method 1	Method 2	Method 3	Method 4
A	4	1	2	3
B	4	3	1	2
C	2	1	4	3
D	3	1	2	4
E	4	1	3	2
F	3	1	2	4

From these we compute

$$R_{1.} = 20, \quad R_{2.} = 8, \quad R_{3.} = 14, \quad \text{and} \quad R_{4.} = 18.$$

Therefore, Equation 14.32 gives

$$S = \frac{12}{6(4)(5)}[(20)^2 + (8)^2 + (14)^2 + (18)^2] - 3(6)(5)$$

$$= 11.3.$$

Since the computed value for S is greater than the critical value of 7.4 we reject H_0 and conclude that there is a difference in down time among the 4 different preventive maintenance programs.   ■

As the number of blocks becomes large, the Friedman statistic has a distribution that can be approximated by a chi-square distribution. We state the result as a theorem without proof.

Theorem 14.5

When $H_0 : \tilde{\tau}_1 = \tilde{\tau}_2 = \cdots = \tilde{\tau}_k$ is true then, as b becomes large,

$$S = \frac{12b}{k(k+1)} \sum_{i=1}^{k} (\bar{R}_{i.} - \bar{R}_{..})^2$$

has a limiting distribution that is chi-squared with $k - 1$ degrees of freedom.

Therefore, for large values of b an approximate size α rejection region for H_0 will be

$$S \geq \chi^2(k - 1)_\alpha. \tag{14.33}$$

As the number of treatment levels increases it has been found that the number of blocks required to use the limiting distribution decreases. In fact, it turns out that the chi-square approximation gives adequate results when k and b exceed those for which the exact distribution is given in Table 11 of Appendix A.

Example 14.15

We wish to use the chi-square approximation to test the data given in Example 14.6. These data were measuring the effects of different chemicals on beneficial insect populations in alfalfa crops. We note that, since $k = 4$ and $b = 10$, the rejection region of Equation 14.34 is appropriate. Therefore, using $\alpha = 0.05$ Table 4 of Appendix A gives that we should reject

$$H_0 : \tilde{\tau}_1 = \tilde{\tau}_2 = \tilde{\tau}_3 = \tilde{\tau}_4 = 0$$

if $S \geq \chi^2(3)_{0.05} = 7.815$. Ranking the observations in each block gives the following.

Block	Chemical A	Chemical B	Chemical C	Chemical D
1	3	4	2	1
2	1	2	4	3
3	4	2	3	1
4	3	2	4	1
5	1.5	3	4	1.5
6	3	1.5	1.5	4
7	2	1	3	4
8	1	4	3	2
9	2	3	4	1
10	4	3	2	1

Then the rank sums are

$$R_{1.} = 24.5, \quad R_{2.} = 25.5, \quad R_{3.} = 30.5, \quad \text{and} \quad R_{4.} = 19.5.$$

Therefore, we compute

$$S = \frac{12}{10(4)(5)}[(24.5)^2 + (25.5)^2 + (30.5)^2 + (19.5)^2] - 3(10)(5)$$

$$= 3.66.$$

Since $S < 7.815$ we do not reject H_0 and conclude that there is no difference in the effect of the various chemicals on the beneficial insect population in alfalfa crops.   ∎

Exercises 14.6

1. Show that, when $k = 2$, Equation 14.25 becomes equivalent to the Wilcoxon Rank Sum test described in Section 12.8.

2. Perform the Kruskal-Wallis test at the 0.01 level on the data in Exercise 7 of Section 14.3. Interpret your results.

3. Perform the Kruskal-Wallis test at the 0.05 level on the data in Exercise 8 of Section 14.3. Use the chi-square approximation. Interpret your results.

4. Perform the Kruskal-Wallis test at the 0.1 level on the data in Exercise 9 of Section 14.3. Interpret your results.

5. Derive the exact distribution for the Kruskal-Wallis statistic when $n_1 = n_2 = n_3 = 2$. Compare your results with Table 10 of Appendix A.

6. Perform the Friedman test at the 0.05 level on the data in Exercise 4 of Section 14.4. Interpret your results.

7. Perform the Friedman test at the 0.1 level on the data in Exercise 5 of Section 14.4. Interpret your results.

8. Show that Equations 14.31 and 14.32 are equivalent.

9. Three groups of students are taught elementary algebra by using 3 different methods. At the end of the semester each student is given a standardized test. Their results are as follows.

Method 1	Method 2	Method 3
92	87	88
89	84	70
90	77	73
73	82	
88		

Test if there is a difference between the 3 methods at the 0.05 level.

10. A testing agency is interested in determining if there is a difference in the mileage in 3 brands of gasoline. To do this they use 3 different cars, each of which is driven with each brand of gasoline. The results are given below.

	Gasoline		
Car	A	B	C
1	20	28	18
2	32	19	17
3	23	27	20

Chapter Summary

Complete-Randomized Design: Given k treatment levels,

1. Determine how many experimental units to assign to each treatment level.
2. Randomly assign each experimental unit to the treatment levels.

(Section 14.2)

Randomized Complete Block Design: Given k treatment levels and b blocks, note that

1. Each block gets one observation per treatment level,
2. In each block the k observations are randomly assigned to each of the treatment levels.   **(Section 14.2)**

Two-Factor Factorial Design: Given one treatment having r levels and one treatment having c levels, assign $n(> 1)$ observations to each possible treatment combination. This permits testing both main effects and interactions.

An $n \times n$ matrix $\mathbf{A}$ is said to be idempotent if

$$\mathbf{AA} = \mathbf{A}. \qquad \text{(Section 14.3)}$$

Let the random vector

$$\mathbf{x} = (X_1, X_2, \ldots, X_n)'$$

consist of independent and identically distributed $N(\mu, \sigma^2)$ random variables. Furthermore, let

$$Y = \mathbf{x}'\mathbf{Ax}$$

where $\text{rank}(\mathbf{A}) = r$. Then

$$\frac{Y}{\sigma^2} \sim \chi^2(r)$$

if and only if $\mathbf{A}$ is an idempotent matrix and

$$\sum_{i=1}^{n} \sum_{j=1}^{n} a_{ij} = 0. \qquad \textbf{(Section 14.3)}$$

Let the random vector

$$\mathbf{x} = (X_1, X_2, \ldots, X_n)'$$

consist of independent and identically distributed $N(\mu, \sigma^2)$ random variables. Furthermore, let

$$Y_1 = \mathbf{x}'\mathbf{A}\mathbf{x}, \quad \text{and} \quad Y_2 = \mathbf{x}'\mathbf{B}\mathbf{x}.$$

Then Y_1 and Y_2 are independent random variables if and only $\mathbf{AB} = 0$.
$\qquad \textbf{(Section 14.3)}$

Let

$$\mathbf{X} = (X_1, X_2, \ldots, X_n)'$$

consist of independent and identically distributed $N(\mu, \sigma^2)$ random variables. Suppose there exist quadratic forms

$$Y_1 = \mathbf{x}'\mathbf{A}_1\mathbf{x}, Y_2 = \mathbf{x}'\mathbf{A}_2\mathbf{x}, \ldots Y_k = \mathbf{x}'\mathbf{A}_k\mathbf{x},$$

such that

$$\sum_{i=1}^{n} (X_i - \bar{X})^2 = Y_1 + Y_2 + \cdots + Y_k.$$

Then $Y_1, Y_2, \ldots, Y_n$ are independent random variables if and only if

$$\sum_{i=1}^{k} \text{rank}(A_i) = n - 1. \qquad \textbf{(Section 14.3)}$$

One-Way Analysis of Variance:
  Model

$$Y_{ij} = \mu + \tau_i + e_{ij}, \quad \text{where} \quad \begin{cases} \displaystyle\sum_{i=1}^{k} n_i \tau_i = 0, \\[2mm] e_{ij} \sim N(0, \sigma^2), \quad \text{for} \quad \begin{aligned} & i = 1, 2, \ldots, k, \\ & j = 1, 2, \ldots, n_i, \end{aligned} \\[2mm] e_{ij} \text{ and } e_{kl} \text{ are uncorrelated if } (i,j) \neq (k.l). \end{cases}$$

Test

$$H_0 : \tau_1 = \tau_2 = \cdots = \tau_k = 0 \quad \text{vs.} \quad H_1 : \text{not all } \tau\text{'s equal to 0.}$$

ANOVA Table

Source of Variation	d.f.	SS	MS	F
Treatment	$k-1$	SST	MST	$\dfrac{\text{MST}}{\text{MSE}}$
Error	$n-k$	SSE	MSE	
Total	$n-1$	SSTotal		

where

$$\text{SSTotal} = \sum_{i=1}^{k} \sum_{j=1}^{n_i} y_{ij}^2 - \frac{y_{..}^2}{n},$$

$$\text{SST} = \sum_{i=1}^{k} \frac{y_{i.}^2}{n_i} - \frac{y_{..}^2}{n},$$

$$\text{SSE} = \text{SSTotal} - \text{SST}.$$

Reject H_0 whenever $F \geq F(k-1, n-k)_\alpha$.   (Section 14.3)

Two-Way Analysis of Variance:

Model of Randomized Complete Block Design:

$$Y_{ij} = \mu + \tau_i + \beta_j + e_{ij}, \quad \text{where} \quad \begin{cases} \displaystyle\sum_{i=1}^{k} n_i \tau_i = 0, \\[2ex] \displaystyle\sum_{j=1}^{b} n_j \beta_j = 0, \\[2ex] e_{ij} \sim N(0, \sigma^2), \quad \text{and} \\[1ex] e_{ij} \text{ and } e_{kl} \text{ are uncorrelated if } (i,j) \neq (k,l). \end{cases}$$

Tests

$$H_0 : \tau_1 = \tau_2 = \cdots = \tau_k = 0 \quad \text{vs.} \quad H_1 : \text{not all } \tau\text{'s equal } 0.$$

$$H_0 : \beta_1 = \beta_2 = \cdots = \beta_b = 0 \quad \text{vs.} \quad H_1 : \text{not all } \beta\text{'s equal } 0.$$

ANOVA Table

Source of Variation	d.f.	SS	MS	F
Treatment	$k-1$	SST	$\dfrac{\text{SST}}{k-1}$	$\dfrac{\text{MST}}{\text{MSE}}$
Blocks	$b-1$	SSB	$\dfrac{\text{SSB}}{b-1}$	$\dfrac{\text{MSB}}{\text{MSE}}$
Error	$(k-1)(b-1)$	SSE	$\dfrac{\text{SSE}}{(k-1)(b-1)}$	
Total	$kb-1$	SSTotal		

where

$$SSTotal = \sum_{i=1}^{k} \sum_{j=1}^{n_i} y_{ij}^2 - \frac{y_{..}^2}{kb},$$

$$SST = \sum_{i=1}^{k} \frac{y_{i.}^2}{b} - \frac{y_{..}^2}{kb},$$

$$SSB = \sum_{j=1}^{b} \frac{y_{.j}^2}{k}, \quad \text{and}$$

$$SSE = SSTotal - SST - SSB. \qquad \textbf{(Section 14.4)}$$

Factorial Design Model:

$$Y_{ijk} = \mu + \alpha_i + \beta_j + (\alpha\beta)_{ij} + e_{ijk}, \quad \text{where} \quad \begin{cases} \sum_{i=1}^{a} br\alpha_i = 0, \\[2mm] \sum_{j=1}^{b} ar\beta_j = 0, \\[2mm] \sum_{i=1}^{a} \sum_{j=1}^{b} r(\alpha\beta)_{ij} = 0, \\[2mm] k = 1, 2, \ldots, r, \quad \text{and} \\[1mm] e_{ijk} \text{ are uncorrelated } N(0, \sigma^2) \\ \quad \text{random variables.} \end{cases}$$

Tests

Null Hypothesis	F Ratio
$H_0 : \alpha_i = 0, i = 1, \ldots, a$	$\dfrac{MSA}{MSE}$
$H_0 : \beta_j = 0, j = 1, \ldots, b$	$\dfrac{MSB}{MSE}$
$H_0 : (\alpha\beta)_{ij} = 0, \quad \begin{aligned} i &= 1, \ldots, a \\ j &= 1, \ldots, b \end{aligned}$	$\dfrac{MS(AB)}{MSE}$

ANOVA Table

Source of Variation	d.f.	SS	MS	F
Treatment A	$a - 1$	SSA	$\dfrac{SSA}{a-1}$	$\dfrac{MST}{MSE}$
Treatment B	$b - 1$	SSB	$\dfrac{SSB}{b-1}$	$\dfrac{MSB}{MSE}$
Interaction	$(a-1)(b-1)$	SS(AB)	$\dfrac{SS(AB)}{(a-1)(b-1)}$	$\dfrac{MS(AB)}{MSE}$
Error	$ab(r-1)$	SSE	$\dfrac{SSE}{ab(r-1)}$	
Total	$rab - 1$	SSTotal		

where

$$\text{SSTotal} = \sum_{i=1}^{a} \sum_{j=1}^{b} \sum_{k=1}^{r} y_{ijk}^2 - \frac{y_{...}^2}{rab},$$

$$\text{SSA} = \frac{1}{rb} \sum_{i=1}^{a} y_{i..}^2 - \frac{y_{...}^2}{rab},$$

$$\text{SSB} = \frac{1}{ra} \sum_{j=1}^{b} y_{.j.}^2 - \frac{y_{...}^2}{rab},$$

$$\text{SS(AB)} = \frac{1}{r} \sum_{i=1}^{a} \sum_{j=1}^{b} y_{ij.}^2 - \frac{1}{rb} \sum_{i=1}^{a} y_{i..}^2 - \frac{1}{ra} \sum_{j=1}^{b} y_{.j.}^2 + \frac{y_{...}^2}{rab}, \quad \text{and}$$

$$\text{SSE} = \text{SSTotal} - \text{SSA} - \text{SSB} - \text{SS(AB)}.$$

Unique Estimator for $\tau_i - \tau_j$ in one way ANOVA is

$$\widehat{\tau_i - \tau_j} = \bar{Y}_i - \bar{Y}_j.$$

$$\widehat{\tau_i - \tau_j} \sim N\left(\tau_i - \tau_j, \sigma^2 \left[\frac{1}{n_i} + \frac{1}{n_j}\right]\right).$$

A $(1 - \alpha)100\%$ confidence interval for $\tau_i - \tau_j$ is

$$\bar{Y}_i - \bar{Y}_j - t(n-k)_{1-\alpha/2} \sqrt{\widehat{\sigma^2}\left(\frac{1}{n_i} + \frac{1}{n_j}\right)}$$

$$\leq \tau_i - \tau_j \leq \bar{Y}_i - \bar{Y}_j + t(n-k)_{1-\alpha/2} \sqrt{\widehat{\sigma^2}\left(\frac{1}{n_i} + \frac{1}{n_j}\right)}.$$

(Section 14.4)

Kruskal-Wallis Test:

Model

$$Y_{ij} = \tilde{\mu} + \tilde{\tau}_i + e_{ij}, \text{ where } \begin{cases} \sum_{i=1}^{k} \tilde{\tau}_i = 0, \quad \text{and} \\ \\ e_{ij} \quad \text{are independent and identically distributed} \\ \quad \text{with median 0, for} \\ \quad i = 1, 2, \ldots, k, \\ \quad j = 1, 2, \ldots, n_i. \end{cases}$$

Test

$$H_0 : \tilde{\tau}_1 = \tilde{\tau}_2 = \cdots = \tilde{\tau}_k = 0 \quad \text{vs.} \quad H_1 : \text{not all } \tilde{\tau}\text{'s equal 0}.$$

Test Statistic

$$H = \frac{12}{n(n+1)} \sum_{i=1}^{k} n_i (\bar{R}_{i.} - \bar{R}_{..})^2$$

Reject H_0 whenever

$$H \geq c,$$

where the constant, c, is found in Table 10 of Appendix A.

When $H_0 : \tilde{\tau}_1 = \tilde{\tau}_2 = \cdots = \tilde{\tau}_k$ is true then, as n becomes large,

$$H = \frac{12}{n(n+1)} \sum_{i=1}^{k} n_i (\bar{R}_{i.} - \bar{R}_{..})^2$$

has a limiting distribution that is chi-squared with $k - 1$ degrees of freedom.   **(Section 14.6)**

Friedman Test:
   Model

$$Y_{ij} = \tilde{\mu} + \tilde{\tau}_i + \tilde{\beta}_j + e_{ij}, \quad \text{where} \quad \begin{cases} \displaystyle\sum_{i=1}^{k} \tilde{\tau}_i = 0, \\[2mm] \displaystyle\sum_{j=1}^{b} \tilde{\beta}_j = 0, \quad \text{and} \\[2mm] e_{ij} \quad \text{are independent and identically} \\ \quad \text{distributed with median 0, for} \\ \quad i = 1, 2, \ldots, k, \\ \quad j = 1, 2, \ldots, b. \end{cases}$$

Test

$$H_0 : \tilde{\tau}_1 = \tilde{\tau}_2 = \cdots = \tilde{\tau}_k = 1 \quad \text{vs.} \quad H_1 : \text{ not all } \tilde{\tau}\text{'s equal 0.}$$

Test Statistic

$$S = \frac{12b}{k(k+1)} \sum_{i=1}^{k} (\bar{R}_{i.} - \bar{R}_{..})^2$$

Reject H_0 whenever

$$S \geq c,$$

where the constant, c, is found in Table 11 of Appendix A.

When $H_0 : \tilde{\tau}_1 = \tilde{\tau}_2 = \cdots = \tilde{\tau}_k$ is true then, as b becomes large,

$$S = \frac{12b}{k(k+1)} \sum_{i=1}^{k} (\bar{R}_{i.} - \bar{R}_{..})^2$$

has a limiting distribution that is chi-squared with $k - 1$ degrees of freedom.   **(Section 14.6)**

Review Exercises

1. A consumer advocate has asserted that health insurance companies are being short-sighted in their approach to health care. He feels that by paying for the cost of preventive health care, patients will run less risk of becoming seriously and expensively ill. This, he feels, would result in less total outlay of insurance funds over the long term than the present policy of paying only for those procedures and medications that arise from the treatment of an acute medical problem. A sociologist is interested in assessing the validity of this claim. He wishes to study the long-term costs of various schemes for reimbursing patients for medical expenses. The payment methods are:

   1. Patients are reimbursed only for corrective procedures and medications of a non-cosmetic nature performed in a hospital.

   2. Patients are reimbursed for corrective procedures and medications of a noncosmetic nature performed by a physician.

   3. Patients are reimbursed for corrective and preventive procedures and medications that are prescribed by a physician.

   To investigate the possible differences between these plans a group of 15 people from the same age group is assigned randomly in groups of 5 to each payment method. The total amount of insurance payments over a period of 10 years for each is observed. The results, to the nearest dollar, for each are as follows.

Payment Method 1	Payment Method 2	Payment Method 3
4721	3548	2433
3729	4318	2890
3914	3468	3016
4017	3722	3153
4435	3515	2219

   a. Test for a difference in the total outlays for each method at the 0.05 level using the F test.

   b. Test for a difference in the payment methods at the 0.05 level using the Kruskal-Wallis test.

   c. Construct a 95% confidence interval for $\tau_3 - \tau_1$. Interpret your results.

2. A study was conducted to determine the difference in the various water softening methods. There are 4 different methods of interest. To obtain data 5 cities throughout the country were selected at random and each of the different softening methods were used on tap water in that city. This experiment is a randomized block design where the cities serve as blocks. Water softness is measured on a scale of 0–100 with a score of 100 indicating that the water is free of chemical salts. The data are as follows.

City (Block)	Treatment Method			
	A	B	C	D
1	90	88	86	87
2	88	88	86	86
3	92	86	85	87
4	91	86	86	84
5	89	87	86	86

   a. Complete the appropriate analysis of variance at the 0.05 level and interpret your results.

   b. Test for a difference in water softening methods using the Friedman test at the 0.05 level. Interpret your results.

3. A government agency is interested in the rate of drug-related crime in U.S. cities. The agency wishes to determine if such crimes vary from region to region of the country and also if there is a difference according to the type of city. The country is divided into 4 regions, and cities are categorized as to whether the major activity in the city was governmental, industrial, or service related. In each region 3 cities of each type were chosen at random. Crime data for a 10-year period was obtained for each city and the average number of drug-related crimes per year computed. The data are as follows.

	City Type		
Region	**Governmental**	**Industrial**	**Service**
Northeast	3.7, 4.5, 5.8	4.9, 5.1, 3.6	3.1, 2.2, 1.8
Southeast	8.2, 6.7, 9.1	4.8, 6.1, 5.5	4.8, 5.1, 3.9
Northwest	4.5, 4.1, 4.9	4.0, 3.8, 4.2	3.9, 3.3, 2.8
Southwest	10.9, 9.8, 7.4	5.9, 6.4, 6.0	3.8, 4.7, 5.2

Carry out a two-factor analysis of variance testing for both main effects and an interaction effect at the 0.01 level. Interpret your results.

4. A consumer organization is interested in determining the difference, if any, in the ability of 4 different brands of household sponges to absorb water. Four pieces of each brand of sponge were randomly chosen and their water absorbency, in cc, measured. The data are as follows.

Sponge 1	**Sponge 2**	**Sponge 3**	**Sponge 4**
10	8	2	14
7	9	6	11
9	6	5	9
8	7	7	10

Test for a difference in sponge abosrbency at the 0.05 level.

5. A fast food chain is interested in determining the effect on sales, if any, of the name given to a new entree. Three different names are being considered. The item was marketed and sold under each of the names. The sales, in units, were recorded for 4 randomly selected days. The data are as follows.

Name 1	**Name 2**	**Name 3**
72	69	59
68	80	77
53	70	64
73	63	60

Does name appear to make a difference in sales? Use the 0.1 level.

6. An experiment was conducted to determine the effect of the spacing between apple trees and yield, in bu/tree. Three spacings were chosen and the yield of 5 trees at each spacing obtained. The results are as follows.

10 ft.	15 ft.	20 ft.
4.0	6.0	5.0
4.5	5.7	4.9
4.4	5.1	5.2
5.0	5.2	4.8
4.9	5.0	4.7

Test whether spacing seems to make a difference at the 0.05 level.

7. A telephone marketing company is trying to decide between 3 different styles of order forms for people manning the telephones. Five employees were chosen at random and each used each form for 1 week. The average time to complete a form was computed. The results are as follows.

Employee (Block)	Form A	Form B	Form C
1	47	40	51
2	50	49	46
3	49	47	47
4	46	45	52
5	48	43	47

Does the difference in forms appear to result in a difference in the time needed to fill out an order? Use the 0.1 level.

8. In certain regions, radon can be a hazard in household basements. A local government intends to institute a program to help homeowners determine if they have a radon problem. There are 3 different devices in common use that have widely differing costs. In an effort to determine if there is a significant difference in the readings by each of the devices, 4 test homes are selected at random and equipped with each device. The readings are as follows.

House (Block)	Device 1	Device 2	Device 3
1	17	15	18
2	22	19	20
3	20	21	23
4	14	12	15

Test for a difference between the 3 radon measurement devices at the 0.05 level.

9. An orchid grower is interested in the possible effects of different type of fertilizer and growth medium on seedling growth. In particular, she is trying to decide between 3 types of fertilizer and 2 different types of growth medium. To do this she randomly selected 12 cymbidium seedlings and grew 2 seedlings in each possible combination of fertilizer and growth medium and measured the time, in days, for each to grow 1 inch. The results are as follows.

	Fertilizer Type		
Medium	**1**	**2**	**3**
A	5, 7	4.2, 5.1	6.1, 5.9
B	6.2, 5.8	3.8, 4.2	7.1, 6.5

Conduct a 2-factor analysis of variance on these data and interpret your results. Run all tests at the 0.1 level.

10. Analyze the data in Exercise 4 using the Kruskal-Wallis test.

11. Analyze the data in Exercise 5 using the Kruskal-Wallis test.

12. Analyze the data in Exercise 6 using the Kruskal-Wallis test.

13. Analyze the data in Exercise 7 using the Friedman test.

14. Analyze the data in Exercise 8 using the Friedman test.

15. Five litters of four animals each are to be used to study four treatments. Treatments are assigned at random within each litter such that each animal is given a different treatment.

    a. Give an appropriate model for this experiment.

    b. Write out the ANOVA table giving the sources of variation, degrees of freedom, and expected mean squares.

16. A litter of eight animals is to be used to study two treatments in each of two different environments. Two animals will be assigned to each combination of treatment and environment.

    a. Give an appropriate model for this experiment.

    b. Write out the ANOVA table giving the sources of variation, degrees of freedom, and expected mean squares.

A

Statistical Tables

Table 1   **Binomial Distribution**

$$P(X \le r) = \sum_{i=0}^{r} \binom{n}{i} p^i (1 - p)^{n-i}$$

						p					
n	r	0.05	0.10	0.15	0.20	0.25	0.30	0.35	0.40	0.45	0.50
2	0	0.9025	0.8100	0.7225	0.6400	0.5625	0.4900	0.4225	0.3600	0.3025	0.2500
	1	0.9975	0.9900	0.9775	0.9600	0.9375	0.9100	0.8775	0.8400	0.7975	0.7500
	2	1.0000	1.0000	1.0000	1.0000	1.0000	1.0000	1.0000	1.0000	1.0000	1.0000
3	0	0.8574	0.7290	0.6141	0.5120	0.4219	0.3430	0.2746	0.2160	0.1664	0.1250
	1	0.9928	0.9720	0.9393	0.8960	0.8438	0.7840	0.7182	0.6480	0.5748	0.5000
	2	0.9999	0.9990	0.9966	0.9920	0.9844	0.9730	0.9571	0.9360	0.9089	0.8750
	3	1.0000	1.0000	1.0000	1.0000	1.0000	1.0000	1.0000	1.0000	1.0000	1.0000
4	0	0.8145	0.6561	0.5220	0.4096	0.3164	0.2401	0.1785	0.1296	0.0915	0.0625
	1	0.9860	0.9477	0.8905	0.8192	0.7383	0.6517	0.5630	0.4752	0.3910	0.3125
	2	0.9995	0.9963	0.9880	0.9728	0.9492	0.9163	0.8735	0.8208	0.7585	0.6875
	3	1.0000	0.9999	0.9995	0.9984	0.9961	0.9919	0.9850	0.9744	0.9590	0.9375
	4	1.0000	1.0000	1.0000	1.0000	1.0000	1.0000	1.0000	1.0000	1.0000	1.0000
5	0	0.7738	0.5905	0.4437	0.3277	0.2373	0.1681	0.1160	0.0778	0.0503	0.0313
	1	0.9774	0.9185	0.8352	0.7373	0.6328	0.5282	0.4284	0.3370	0.2562	0.1875
	2	0.9988	0.9914	0.9734	0.9421	0.8965	0.8369	0.7648	0.6826	0.5931	0.5000
	3	1.0000	0.9995	0.9978	0.9933	0.9844	0.9692	0.9460	0.9130	0.8688	0.8125
	4	1.0000	1.0000	0.9999	0.9997	0.9990	0.9976	0.9947	0.9898	0.9815	0.9688
	5	1.0000	1.0000	1.0000	1.0000	1.0000	1.0000	1.0000	1.0000	1.0000	1.0000

Table 1   Binomial Distribution (continued)

$$P(X \leq r) = \sum_{i=0}^{r} \binom{n}{i} p^i (1 - p)^{n-i}$$

n	r	0.05	0.10	0.15	0.20	0.25	0.30	0.35	0.40	0.45	0.50
6	0	0.7351	0.5314	0.3771	0.2621	0.1780	0.1176	0.0754	0.0467	0.0277	0.0156
	1	0.9672	0.8857	0.7765	0.6554	0.5339	0.4202	0.3191	0.2333	0.1636	0.1094
	2	0.9978	0.9842	0.9527	0.9011	0.8306	0.7443	0.6471	0.5443	0.4415	0.3438
	3	0.9999	0.9987	0.9941	0.9830	0.9624	0.9295	0.8826	0.8208	0.7447	0.6563
	4	1.0000	0.9999	0.9996	0.9984	0.9954	0.9891	0.9777	0.9590	0.9308	0.8906
	5	1.0000	1.0000	1.0000	0.9999	0.9998	0.9993	0.9982	0.9959	0.9917	0.9844
	6	1.0000	1.0000	1.0000	1.0000	1.0000	1.0000	1.0000	1.0000	1.0000	1.0000
7	0	0.6983	0.4783	0.3206	0.2097	0.1335	0.0824	0.0490	0.0280	0.0152	0.0078
	1	0.9556	0.8503	0.7166	0.5767	0.4449	0.3294	0.2338	0.1586	0.1024	0.0625
	2	0.9962	0.9743	0.9262	0.8520	0.7564	0.6471	0.5323	0.4199	0.3164	0.2266
	3	0.9998	0.9973	0.9879	0.9667	0.9294	0.8740	0.8002	0.7102	0.6083	0.5000
	4	1.0000	0.9998	0.9988	0.9953	0.9871	0.9712	0.9444	0.9037	0.8471	0.7734
	5	1.0000	1.0000	0.9999	0.9996	0.9987	0.9962	0.9910	0.9812	0.9643	0.9375
	6	1.0000	1.0000	1.0000	1.0000	0.9999	0.9998	0.9994	0.9984	0.9963	0.9922
	7	1.0000	1.0000	1.0000	1.0000	1.0000	1.0000	1.0000	1.0000	1.0000	1.0000
8	0	0.6634	0.4305	0.2725	0.1678	0.1001	0.0576	0.0319	0.0168	0.0084	0.0039
	1	0.9428	0.8131	0.6572	0.5033	0.3671	0.2553	0.1691	0.1064	0.0632	0.0352
	2	0.9942	0.9619	0.8948	0.7969	0.6785	0.5518	0.4278	0.3154	0.2201	0.1445
	3	0.9996	0.9950	0.9786	0.9437	0.8862	0.8059	0.7064	0.5941	0.4770	0.3633
	4	1.0000	0.9996	0.9971	0.9896	0.9727	0.9420	0.8939	0.8263	0.7396	0.6367
	5	1.0000	1.0000	0.9998	0.9988	0.9958	0.9887	0.9747	0.9502	0.9115	0.8555
	6	1.0000	1.0000	1.0000	0.9999	0.9996	0.9987	0.9964	0.9915	0.9819	0.9648
	7	1.0000	1.0000	1.0000	1.0000	1.0000	0.9999	0.9998	0.9993	0.9983	0.9961
	8	1.0000	1.0000	1.0000	1.0000	1.0000	1.0000	1.0000	1.0000	1.0000	1.0000
9	0	0.6302	0.3874	0.2316	0.1342	0.0751	0.0404	0.0207	0.0101	0.0046	0.0020
	1	0.9288	0.7748	0.5995	0.4362	0.3003	0.1960	0.1211	0.0705	0.0385	0.0195
	2	0.9916	0.9470	0.8591	0.7382	0.6007	0.4628	0.3373	0.2318	0.1495	0.0898
	3	0.9994	0.9917	0.9661	0.9144	0.8343	0.7297	0.6089	0.4826	0.3614	0.2539
	4	1.0000	0.9991	0.9944	0.9804	0.9511	0.9012	0.8283	0.7334	0.6214	0.5000
	5	1.0000	0.9999	0.9994	0.9969	0.9900	0.9747	0.9464	0.9006	0.8342	0.7461
	6	1.0000	1.0000	1.0000	0.9997	0.9987	0.9957	0.9888	0.9750	0.9502	0.9102
	7	1.0000	1.0000	1.0000	1.0000	0.9999	0.9996	0.9986	0.9962	0.9909	0.9805
	8	1.0000	1.0000	1.0000	1.0000	1.0000	1.0000	0.9999	0.9997	0.9992	0.9980
	9	1.0000	1.0000	1.0000	1.0000	1.0000	1.0000	1.0000	1.0000	1.0000	1.0000
10	0	0.5987	0.3487	0.1969	0.1074	0.0563	0.0282	0.0135	0.0060	0.0025	0.0010
	1	0.9139	0.7361	0.5443	0.3758	0.2440	0.1493	0.0860	0.0464	0.0233	0.0107
	2	0.9885	0.9298	0.8202	0.6778	0.5256	0.3828	0.2616	0.1673	0.0996	0.0547
	3	0.9990	0.9872	0.9500	0.8791	0.7759	0.6496	0.5138	0.3823	0.2660	0.1719
	4	0.9999	0.9984	0.9901	0.9672	0.9219	0.8497	0.7515	0.6331	0.5044	0.3770
	5	1.0000	0.9999	0.9986	0.9936	0.9803	0.9527	0.9051	0.8338	0.7384	0.6230
	6	1.0000	1.0000	0.9999	0.9991	0.9965	0.9894	0.9740	0.9452	0.8980	0.8281
	7	1.0000	1.0000	1.0000	0.9999	0.9996	0.9984	0.9952	0.9877	0.9726	0.9453

Table 1   Binomial Distribution (continued)

$$P(X \le r) = \sum_{i=0}^{r} \binom{n}{i} p^i (1-p)^{n-i}$$

n	r	0.05	0.10	0.15	0.20	0.25	0.30	0.35	0.40	0.45	0.50
	8	1.0000	1.0000	1.0000	1.0000	1.0000	0.9999	0.9995	0.9983	0.9955	0.9893
	9	1.0000	1.0000	1.0000	1.0000	1.0000	1.0000	1.0000	0.9999	0.9997	0.9990
	10	1.0000	1.0000	1.0000	1.0000	1.0000	1.0000	1.0000	1.0000	1.0000	1.0000
11	0	0.5688	0.3138	0.1673	0.0859	0.0422	0.0198	0.0088	0.0036	0.0014	0.0005
	1	0.8981	0.6974	0.4922	0.3221	0.1971	0.1130	0.0606	0.0302	0.0139	0.0059
	2	0.9848	0.9104	0.7788	0.6174	0.4552	0.3127	0.2001	0.1189	0.0652	0.0327
	3	0.9984	0.9815	0.9306	0.8389	0.7133	0.5696	0.4256	0.2963	0.1911	0.1133
	4	0.9999	0.9972	0.9841	0.9496	0.8854	0.7897	0.6683	0.5328	0.3971	0.2744
	5	1.0000	0.9997	0.9973	0.9883	0.9657	0.9218	0.8513	0.7535	0.6331	0.5000
	6	1.0000	1.0000	0.9997	0.9980	0.9924	0.9784	0.9499	0.9006	0.8262	0.7256
	7	1.0000	1.0000	1.0000	0.9998	0.9988	0.9957	0.9878	0.9707	0.9390	0.8867
	8	1.0000	1.0000	1.0000	1.0000	0.9999	0.9994	0.9980	0.9941	0.9852	0.9673
	9	1.0000	1.0000	1.0000	1.0000	1.0000	1.0000	0.9998	0.9993	0.9978	0.9941
	10	1.0000	1.0000	1.0000	1.0000	1.0000	1.0000	1.0000	1.0000	0.9998	0.9995
	11	1.0000	1.0000	1.0000	1.0000	1.0000	1.0000	1.0000	1.0000	1.0000	1.0000
12	0	0.5404	0.2824	0.1422	0.0687	0.0317	0.0138	0.0057	0.0022	0.0008	0.0002
	1	0.8816	0.6590	0.4435	0.2749	0.1584	0.0850	0.0424	0.0196	0.0083	0.0032
	2	0.9804	0.8891	0.7358	0.5583	0.3907	0.2528	0.1513	0.0834	0.0421	0.0193
	3	0.9978	0.9744	0.9078	0.7946	0.6488	0.4925	0.3467	0.2253	0.1345	0.0730
	4	0.9998	0.9957	0.9761	0.9274	0.8424	0.7237	0.5833	0.4382	0.3044	0.1938
	5	1.0000	0.9995	0.9954	0.9806	0.9456	0.8822	0.7873	0.6652	0.5269	0.3872
	6	1.0000	0.9999	0.9993	0.9961	0.9857	0.9614	0.9154	0.8418	0.7393	0.6128
	7	1.0000	1.0000	0.9999	0.9994	0.9972	0.9905	0.9745	0.9427	0.8883	0.8062
	8	1.0000	1.0000	1.0000	0.9999	0.9996	0.9983	0.9944	0.9847	0.9644	0.9270
	9	1.0000	1.0000	1.0000	1.0000	1.0000	0.9998	0.9992	0.9972	0.9921	0.9807
	10	1.0000	1.0000	1.0000	1.0000	1.0000	1.0000	0.9999	0.9997	0.9989	0.9968
	11	1.0000	1.0000	1.0000	1.0000	1.0000	1.0000	1.0000	1.0000	0.9999	0.9998
	12	1.0000	1.0000	1.0000	1.0000	1.0000	1.0000	1.0000	1.0000	1.0000	1.0000
13	0	0.5133	0.2542	0.1209	0.0550	0.0238	0.0097	0.0037	0.0013	0.0004	0.0001
	1	0.8646	0.6213	0.3983	0.2336	0.1267	0.0637	0.0296	0.0126	0.0049	0.0017
	2	0.9755	0.8661	0.6920	0.5017	0.3326	0.2025	0.1132	0.0579	0.0269	0.0112
	3	0.9969	0.9658	0.8820	0.7473	0.5843	0.4206	0.2783	0.1686	0.0929	0.0461
	4	0.9997	0.9935	0.9658	0.9009	0.7940	0.6543	0.5005	0.3530	0.2279	0.1334
	5	1.0000	0.9991	0.9925	0.9700	0.9198	0.8346	0.7159	0.5744	0.4268	0.2905
	6	1.0000	0.9999	0.9987	0.9930	0.9757	0.9376	0.8705	0.7712	0.6437	0.5000
	7	1.0000	1.0000	0.9998	0.9988	0.9944	0.9818	0.9538	0.9023	0.8212	0.7095
	8	1.0000	1.0000	1.0000	0.9998	0.9990	0.9960	0.9874	0.9679	0.9302	0.8666
	9	1.0000	1.0000	1.0000	1.0000	0.9999	0.9993	0.9975	0.9922	0.9797	0.9539
	10	1.0000	1.0000	1.0000	1.0000	1.0000	0.9999	0.9997	0.9987	0.9959	0.9888
	11	1.0000	1.0000	1.0000	1.0000	1.0000	1.0000	1.0000	0.9999	0.9995	0.9983
	12	1.0000	1.0000	1.0000	1.0000	1.0000	1.0000	1.0000	1.0000	1.0000	0.9999
	13	1.0000	1.0000	1.0000	1.0000	1.0000	1.0000	1.0000	1.0000	1.0000	1.0000

Table 1   Binomial Distribution (continued)

$$P(X \le r) = \sum_{i=0}^{r} \binom{n}{i} p^i (1-p)^{n-i}$$

						p					
n	r	0.05	0.10	0.15	0.20	0.25	0.30	0.35	0.40	0.45	0.50
14	0	0.4877	0.2288	0.1028	0.0440	0.0178	0.0068	0.0024	0.0008	0.0002	0.0001
	1	0.8470	0.5846	0.3567	0.1979	0.1010	0.0475	0.0205	0.0081	0.0029	0.0009
	2	0.9699	0.8416	0.6479	0.4481	0.2811	0.1608	0.0839	0.0398	0.0170	0.0065
	3	0.9958	0.9559	0.8535	0.6982	0.5213	0.3552	0.2205	0.1243	0.0632	0.0287
	4	0.9996	0.9908	0.9533	0.8702	0.7415	0.5842	0.4227	0.2793	0.1672	0.0898
	5	1.0000	0.9985	0.9885	0.9561	0.8883	0.7805	0.6405	0.4859	0.3373	0.2120
	6	1.0000	0.9998	0.9978	0.9884	0.9617	0.9067	0.8164	0.6925	0.5461	0.3953
	7	1.0000	1.0000	0.9997	0.9976	0.9897	0.9685	0.9247	0.8499	0.7414	0.6047
	8	1.0000	1.0000	1.0000	0.9996	0.9978	0.9917	0.9757	0.9417	0.8811	0.7880
	9	1.0000	1.0000	1.0000	1.0000	0.9997	0.9983	0.9940	0.9825	0.9574	0.9102
	10	1.0000	1.0000	1.0000	1.0000	1.0000	0.9998	0.9989	0.9961	0.9886	0.9713
	11	1.0000	1.0000	1.0000	1.0000	1.0000	1.0000	0.9999	0.9994	0.9978	0.9935
	12	1.0000	1.0000	1.0000	1.0000	1.0000	1.0000	1.0000	0.9999	0.9997	0.9991
	13	1.0000	1.0000	1.0000	1.0000	1.0000	1.0000	1.0000	1.0000	1.0000	0.9999
	14	1.0000	1.0000	1.0000	1.0000	1.0000	1.0000	1.0000	1.0000	1.0000	1.0000
15	0	0.4633	0.2059	0.0874	0.0352	0.0134	0.0047	0.0016	0.0005	0.0001	0.0000
	1	0.8290	0.5490	0.3186	0.1671	0.0802	0.0353	0.0142	0.0052	0.0017	0.0005
	2	0.9638	0.8159	0.6042	0.3980	0.2361	0.1268	0.0617	0.0271	0.0107	0.0037
	3	0.9945	0.9444	0.8227	0.6482	0.4613	0.2969	0.1727	0.0905	0.0424	0.0176
	4	0.9994	0.9873	0.9383	0.8358	0.6865	0.5155	0.3519	0.2173	0.1204	0.0592
	5	0.9999	0.9978	0.9832	0.9389	0.8516	0.7216	0.5643	0.4032	0.2608	0.1509
	6	1.0000	0.9997	0.9964	0.9819	0.9434	0.8689	0.7548	0.6098	0.4522	0.3036
	7	1.0000	1.0000	0.9994	0.9958	0.9827	0.9500	0.8868	0.7869	0.6535	0.5000
	8	1.0000	1.0000	0.9999	0.9992	0.9958	0.9848	0.9578	0.9050	0.8182	0.6964
	9	1.0000	1.0000	1.0000	0.9999	0.9992	0.9963	0.9876	0.9662	0.9231	0.8491
	10	1.0000	1.0000	1.0000	1.0000	0.9999	0.9993	0.9972	0.9907	0.9745	0.9408
	11	1.0000	1.0000	1.0000	1.0000	1.0000	0.9999	0.9995	0.9981	0.9937	0.9824
	12	1.0000	1.0000	1.0000	1.0000	1.0000	1.0000	0.9999	0.9997	0.9989	0.9963
	13	1.0000	1.0000	1.0000	1.0000	1.0000	1.0000	1.0000	1.0000	0.9999	0.9995
	14	1.0000	1.0000	1.0000	1.0000	1.0000	1.0000	1.0000	1.0000	1.0000	1.0000
	15	1.0000	1.0000	1.0000	1.0000	1.0000	1.0000	1.0000	1.0000	1.0000	1.0000
16	0	0.4401	0.1853	0.0743	0.0281	0.0100	0.0033	0.0010	0.0003	0.0001	0.0000
	1	0.8108	0.5147	0.2839	0.1407	0.0635	0.0261	0.0098	0.0033	0.0010	0.0003
	2	0.9571	0.7892	0.5614	0.3518	0.1971	0.0994	0.0451	0.0183	0.0066	0.0021
	3	0.9930	0.9316	0.7899	0.5981	0.4050	0.2459	0.1339	0.0651	0.0281	0.0106
	4	0.9991	0.9830	0.9209	0.7982	0.6302	0.4499	0.2892	0.1666	0.0853	0.0384
	5	0.9999	0.9967	0.9765	0.9183	0.8103	0.6598	0.4900	0.3288	0.1976	0.1051
	6	1.0000	0.9995	0.9944	0.9733	0.9204	0.8247	0.6881	0.5272	0.3660	0.2272
	7	1.0000	0.9999	0.9989	0.9930	0.9729	0.9256	0.8406	0.7161	0.5629	0.4018
	8	1.0000	1.0000	0.9998	0.9985	0.9925	0.9743	0.9329	0.8577	0.7441	0.5982
	9	1.0000	1.0000	1.0000	0.9998	0.9984	0.9929	0.9771	0.9417	0.8759	0.7728
	10	1.0000	1.0000	1.0000	1.0000	0.9997	0.9984	0.9938	0.9809	0.9514	0.8949

Table 1   Binomial Distribution (continued)

$$P(X \leq r) = \sum_{i=0}^{r} \binom{n}{i} p^i (1-p)^{n-i}$$

n	r	0.05	0.10	0.15	0.20	0.25	0.30	0.35	0.40	0.45	0.50
	11	1.0000	1.0000	1.0000	1.0000	1.0000	0.9997	0.9987	0.9951	0.9851	0.9616
	12	1.0000	1.0000	1.0000	1.0000	1.0000	1.0000	0.9998	0.9991	0.9965	0.9894
	13	1.0000	1.0000	1.0000	1.0000	1.0000	1.0000	1.0000	0.9999	0.9994	0.9979
	14	1.0000	1.0000	1.0000	1.0000	1.0000	1.0000	1.0000	1.0000	0.9999	0.9997
	15	1.0000	1.0000	1.0000	1.0000	1.0000	1.0000	1.0000	1.0000	1.0000	1.0000
	16	1.0000	1.0000	1.0000	1.0000	1.0000	1.0000	1.0000	1.0000	1.0000	1.0000
17	0	0.4181	0.1668	0.0631	0.0225	0.0075	0.0023	0.0007	0.0002	0.0000	0.0000
	1	0.7922	0.4818	0.2525	0.1182	0.0501	0.0193	0.0067	0.0021	0.0006	0.0001
	2	0.9497	0.7618	0.5198	0.3096	0.1637	0.0774	0.0327	0.0123	0.0041	0.0012
	3	0.9912	0.9174	0.7556	0.5489	0.3530	0.2019	0.1028	0.0464	0.0184	0.0064
	4	0.9988	0.9779	0.9013	0.7582	0.5739	0.3887	0.2348	0.1260	0.0596	0.0245
	5	0.9999	0.9953	0.9681	0.8943	0.7653	0.5968	0.4197	0.2639	0.1471	0.0717
	6	1.0000	0.9992	0.9917	0.9623	0.8929	0.7752	0.6188	0.4478	0.2902	0.1662
	7	1.0000	0.9999	0.9983	0.9891	0.9598	0.8954	0.7872	0.6405	0.4743	0.3145
	8	1.0000	1.0000	0.9997	0.9974	0.9876	0.9597	0.9006	0.8011	0.6626	0.5000
	9	1.0000	1.0000	1.0000	0.9995	0.9969	0.9873	0.9617	0.9081	0.8166	0.6855
	10	1.0000	1.0000	1.0000	0.9999	0.9994	0.9968	0.9880	0.9652	0.9174	0.8338
	11	1.0000	1.0000	1.0000	1.0000	0.9999	0.9993	0.9970	0.9894	0.9699	0.9283
	12	1.0000	1.0000	1.0000	1.0000	1.0000	0.9999	0.9994	0.9975	0.9914	0.9755
	13	1.0000	1.0000	1.0000	1.0000	1.0000	1.0000	0.9999	0.9995	0.9981	0.9936
	14	1.0000	1.0000	1.0000	1.0000	1.0000	1.0000	1.0000	0.9999	0.9997	0.9988
	15	1.0000	1.0000	1.0000	1.0000	1.0000	1.0000	1.0000	1.0000	1.0000	0.9999
	16	1.0000	1.0000	1.0000	1.0000	1.0000	1.0000	1.0000	1.0000	1.0000	1.0000
	17	1.0000	1.0000	1.0000	1.0000	1.0000	1.0000	1.0000	1.0000	1.0000	1.0000
18	0	0.3972	0.1501	0.0536	0.0180	0.0056	0.0016	0.0004	0.0001	0.0000	0.0000
	1	0.7735	0.4503	0.2241	0.0991	0.0395	0.0142	0.0046	0.0013	0.0003	0.0001
	2	0.9419	0.7338	0.4797	0.2713	0.1353	0.0600	0.0236	0.0082	0.0025	0.0007
	3	0.9891	0.9018	0.7202	0.5010	0.3057	0.1646	0.0783	0.0328	0.0120	0.0038
	4	0.9985	0.9718	0.8794	0.7164	0.5187	0.3327	0.1886	0.0942	0.0411	0.0154
	5	0.9998	0.9936	0.9581	0.8671	0.7175	0.5344	0.3550	0.2088	0.1077	0.0481
	6	1.0000	0.9988	0.9882	0.9487	0.8610	0.7217	0.5491	0.3743	0.2258	0.1189
	7	1.0000	0.9998	0.9973	0.9837	0.9431	0.8593	0.7283	0.5634	0.3915	0.2403
	8	1.0000	1.0000	0.9995	0.9957	0.9807	0.9404	0.8609	0.7368	0.5778	0.4073
	9	1.0000	1.0000	0.9999	0.9991	0.9946	0.9790	0.9403	0.8653	0.7473	0.5927
	10	1.0000	1.0000	1.0000	0.9998	0.9988	0.9939	0.9788	0.9424	0.8720	0.7597
	11	1.0000	1.0000	1.0000	1.0000	0.9998	0.9986	0.9938	0.9797	0.9463	0.8811
	12	1.0000	1.0000	1.0000	1.0000	1.0000	0.9997	0.9986	0.9942	0.9817	0.9519
	13	1.0000	1.0000	1.0000	1.0000	1.0000	1.0000	0.9997	0.9987	0.9951	0.9846
	14	1.0000	1.0000	1.0000	1.0000	1.0000	1.0000	1.0000	0.9998	0.9990	0.9962
	15	1.0000	1.0000	1.0000	1.0000	1.0000	1.0000	1.0000	1.0000	0.9999	0.9993
	16	1.0000	1.0000	1.0000	1.0000	1.0000	1.0000	1.0000	1.0000	1.0000	0.9999
	17	1.0000	1.0000	1.0000	1.0000	1.0000	1.0000	1.0000	1.0000	1.0000	1.0000
	18	1.0000	1.0000	1.0000	1.0000	1.0000	1.0000	1.0000	1.0000	1.0000	1.0000

Table 1   Binomial Distribution (continued)

$$P(X \leq r) = \sum_{i=0}^{r} \binom{n}{i} p^i (1-p)^{n-i}$$

						p					
n	r	0.05	0.10	0.15	0.20	0.25	0.30	0.35	0.40	0.45	0.50
19	0	0.3774	0.1351	0.0456	0.0144	0.0042	0.0011	0.0003	0.0001	0.0000	0.0000
	1	0.7547	0.4203	0.1985	0.0829	0.0310	0.0104	0.0031	0.0008	0.0002	0.0000
	2	0.9335	0.7054	0.4413	0.2369	0.1113	0.0462	0.0170	0.0055	0.0015	0.0004
	3	0.9868	0.8850	0.6841	0.4551	0.2631	0.1332	0.0591	0.0230	0.0077	0.0022
	4	0.9980	0.9648	0.8556	0.6733	0.4654	0.2822	0.1500	0.0696	0.0280	0.0096
	5	0.9998	0.9914	0.9463	0.8369	0.6678	0.4739	0.2968	0.1629	0.0777	0.0318
	6	1.0000	0.9983	0.9837	0.9324	0.8251	0.6655	0.4812	0.3081	0.1727	0.0835
	7	1.0000	0.9997	0.9959	0.9767	0.9225	0.8180	0.6656	0.4878	0.3169	0.1796
	8	1.0000	1.0000	0.9992	0.9933	0.9713	0.9161	0.8145	0.6675	0.4940	0.3238
	9	1.0000	1.0000	0.9999	0.9984	0.9911	0.9674	0.9125	0.8139	0.6710	0.5000
	10	1.0000	1.0000	1.0000	0.9997	0.9977	0.9895	0.9653	0.9115	0.8159	0.6762
	11	1.0000	1.0000	1.0000	1.0000	0.9995	0.9972	0.9886	0.9648	0.9129	0.8204
	12	1.0000	1.0000	1.0000	1.0000	0.9999	0.9994	0.9969	0.9884	0.9658	0.9165
	13	1.0000	1.0000	1.0000	1.0000	1.0000	0.9999	0.9993	0.9969	0.9891	0.9682
	14	1.0000	1.0000	1.0000	1.0000	1.0000	1.0000	0.9999	0.9994	0.9972	0.9904
	15	1.0000	1.0000	1.0000	1.0000	1.0000	1.0000	1.0000	0.9999	0.9995	0.9978
	16	1.0000	1.0000	1.0000	1.0000	1.0000	1.0000	1.0000	1.0000	0.9999	0.9996
	17	1.0000	1.0000	1.0000	1.0000	1.0000	1.0000	1.0000	1.0000	1.0000	1.0000
	18	1.0000	1.0000	1.0000	1.0000	1.0000	1.0000	1.0000	1.0000	1.0000	1.0000
	19	1.0000	1.0000	1.0000	1.0000	1.0000	1.0000	1.0000	1.0000	1.0000	1.0000
20	0	0.3585	0.1216	0.0388	0.0115	0.0032	0.0008	0.0002	0.0000	0.0000	0.0000
	1	0.7358	0.3917	0.1756	0.0692	0.0243	0.0076	0.0021	0.0005	0.0001	0.0000
	2	0.9245	0.6769	0.4049	0.2061	0.0913	0.0355	0.0121	0.0036	0.0009	0.0002
	3	0.9841	0.8670	0.6477	0.4114	0.2252	0.1071	0.0444	0.0160	0.0049	0.0013
	4	0.9974	0.9568	0.8298	0.6296	0.4148	0.2375	0.1182	0.0510	0.0189	0.0059
	5	0.9997	0.9887	0.9327	0.8042	0.6172	0.4164	0.2454	0.1256	0.0553	0.0207
	6	1.0000	0.9976	0.9781	0.9133	0.7858	0.6080	0.4166	0.2500	0.1299	0.0577
	7	1.0000	0.9996	0.9941	0.9679	0.8982	0.7723	0.6010	0.4159	0.2520	0.1316
	8	1.0000	0.9999	0.9987	0.9900	0.9591	0.8867	0.7624	0.5956	0.4143	0.2517
	9	1.0000	1.0000	0.9998	0.9974	0.9861	0.9520	0.8782	0.7553	0.5914	0.4119
	10	1.0000	1.0000	1.0000	0.9994	0.9961	0.9829	0.9468	0.8725	0.7507	0.5881
	11	1.0000	1.0000	1.0000	0.9999	0.9991	0.9949	0.9804	0.9435	0.8692	0.7483
	12	1.0000	1.0000	1.0000	1.0000	0.9998	0.9987	0.9940	0.9790	0.9420	0.8684
	13	1.0000	1.0000	1.0000	1.0000	1.0000	0.9997	0.9985	0.9935	0.9786	0.9423
	14	1.0000	1.0000	1.0000	1.0000	1.0000	1.0000	0.9997	0.9984	0.9936	0.9793
	15	1.0000	1.0000	1.0000	1.0000	1.0000	1.0000	1.0000	0.9997	0.9985	0.9941
	16	1.0000	1.0000	1.0000	1.0000	1.0000	1.0000	1.0000	1.0000	0.9997	0.9987
	17	1.0000	1.0000	1.0000	1.0000	1.0000	1.0000	1.0000	1.0000	1.0000	0.9998
	18	1.0000	1.0000	1.0000	1.0000	1.0000	1.0000	1.0000	1.0000	1.0000	1.0000
	19	1.0000	1.0000	1.0000	1.0000	1.0000	1.0000	1.0000	1.0000	1.0000	1.0000
	20	1.0000	1.0000	1.0000	1.0000	1.0000	1.0000	1.0000	1.0000	1.0000	1.0000

Table 2   Poisson Distribution

$$P(X \leq r) = \sum_{i=0}^{r} \frac{\lambda^i e^{-\lambda}}{i!}$$

						λ				
r	0.10	0.20	0.30	0.40	0.50	0.60	0.70	0.80	0.90	1.00
0	0.9048	0.8187	0.7408	0.6703	0.6065	0.5488	0.4966	0.4493	0.4066	0.3679
1	0.9953	0.9825	0.9631	0.9384	0.9098	0.8781	0.8442	0.8088	0.7725	0.7358
2	0.9998	0.9989	0.9964	0.9921	0.9856	0.9769	0.9659	0.9526	0.9371	0.9197
3	1.0000	0.9999	0.9997	0.9992	0.9982	0.9966	0.9942	0.9909	0.9865	0.9810
4	1.0000	1.0000	1.0000	0.9999	0.9998	0.9996	0.9992	0.9986	0.9977	0.9963
5	1.0000	1.0000	1.0000	1.0000	1.0000	1.0000	0.9999	0.9998	0.9997	0.9994
6	1.0000	1.0000	1.0000	1.0000	1.0000	1.0000	1.0000	1.0000	1.0000	0.9999
7	1.0000	1.0000	1.0000	1.0000	1.0000	1.0000	1.0000	1.0000	1.0000	1.0000

r	1.10	1.20	1.30	1.40	1.50	1.60	1.70	1.80	1.90	2.00
0	0.3329	0.3012	0.2725	0.2466	0.2231	0.2019	0.1827	0.1653	0.1496	0.1353
1	0.6990	0.6626	0.6268	0.5918	0.5578	0.5249	0.4932	0.4628	0.4337	0.4060
2	0.9004	0.8795	0.8571	0.8335	0.8088	0.7834	0.7572	0.7306	0.7037	0.6767
3	0.9743	0.9662	0.9569	0.9463	0.9344	0.9212	0.9068	0.8913	0.8747	0.8571
4	0.9946	0.9923	0.9893	0.9857	0.9814	0.9763	0.9704	0.9636	0.9559	0.9473
5	0.9990	0.9985	0.9978	0.9968	0.9955	0.9940	0.9920	0.9896	0.9868	0.9834
6	0.9999	0.9997	0.9996	0.9994	0.9991	0.9987	0.9981	0.9974	0.9966	0.9955
7	1.0000	1.0000	0.9999	0.9999	0.9998	0.9997	0.9996	0.9994	0.9992	0.9989
8	1.0000	1.0000	1.0000	1.0000	1.0000	1.0000	0.9999	0.9999	0.9998	0.9998
9	1.0000	1.0000	1.0000	1.0000	1.0000	1.0000	1.0000	1.0000	1.0000	1.0000

r	2.10	2.20	2.30	2.40	2.50	2.60	2.70	2.80	2.90	3.00
0	0.1225	0.1108	0.1003	0.0907	0.0821	0.0743	0.0672	0.0608	0.0550	0.0498
1	0.3796	0.3546	0.3309	0.3084	0.2873	0.2674	0.2487	0.2311	0.2146	0.1991
2	0.6496	0.6227	0.5960	0.5697	0.5438	0.5184	0.4936	0.4695	0.4460	0.4232
3	0.8386	0.8194	0.7993	0.7787	0.7576	0.7360	0.7141	0.6919	0.6696	0.6472
4	0.9379	0.9275	0.9162	0.9041	0.8912	0.8774	0.8629	0.8477	0.8318	0.8153
5	0.9796	0.9751	0.9700	0.9643	0.9580	0.9510	0.9433	0.9349	0.9258	0.9161
6	0.9941	0.9925	0.9906	0.9884	0.9858	0.9828	0.9794	0.9756	0.9713	0.9665
7	0.9985	0.9980	0.9974	0.9967	0.9958	0.9947	0.9934	0.9919	0.9901	0.9881
8	0.9997	0.9995	0.9994	0.9991	0.9989	0.9985	0.9981	0.9976	0.9969	0.9962
9	0.9999	0.9999	0.9999	0.9998	0.9997	0.9996	0.9995	0.9993	0.9991	0.9989
10	1.0000	1.0000	1.0000	1.0000	0.9999	0.9999	0.9999	0.9998	0.9998	0.9997
11	1.0000	1.0000	1.0000	1.0000	1.0000	1.0000	1.0000	1.0000	0.9999	0.9999
12	1.0000	1.0000	1.0000	1.0000	1.0000	1.0000	1.0000	1.0000	1.0000	1.0000

Table 2   Poisson Distribution (continued)

$$P(X \le r) = \sum_{i=0}^{r} \frac{\lambda^i e^{-\lambda}}{i!}$$

					λ					
r	3.10	3.20	3.30	3.40	3.50	3.60	3.70	3.80	3.90	4.00
0	0.0450	0.0408	0.0369	0.0334	0.0302	0.0273	0.0247	0.0224	0.0202	0.0183
1	0.1847	0.1712	0.1586	0.1468	0.1359	0.1257	0.1162	0.1074	0.0992	0.0916
2	0.4012	0.3799	0.3594	0.3397	0.3208	0.3027	0.2854	0.2689	0.2531	0.2381
3	0.6248	0.6025	0.5803	0.5584	0.5366	0.5152	0.4942	0.4735	0.4532	0.4335
4	0.7982	0.7806	0.7626	0.7442	0.7254	0.7064	0.6872	0.6678	0.6484	0.6288
5	0.9057	0.8946	0.8829	0.8705	0.8576	0.8441	0.8301	0.8156	0.8006	0.7851
6	0.9612	0.9554	0.9490	0.9421	0.9347	0.9267	0.9182	0.9091	0.8995	0.8893
7	0.9858	0.9832	0.9802	0.9769	0.9733	0.9692	0.9648	0.9599	0.9546	0.9489
8	0.9953	0.9943	0.9931	0.9917	0.9901	0.9883	0.9863	0.9840	0.9815	0.9786
9	0.9986	0.9982	0.9978	0.9973	0.9967	0.9960	0.9952	0.9942	0.9931	0.9919
10	0.9996	0.9995	0.9994	0.9992	0.9990	0.9987	0.9984	0.9981	0.9977	0.9972
11	0.9999	0.9999	0.9998	0.9998	0.9997	0.9996	0.9995	0.9994	0.9993	0.9991
12	1.0000	1.0000	1.0000	0.9999	0.9999	0.9999	0.9999	0.9998	0.9998	0.9997
13	1.0000	1.0000	1.0000	1.0000	1.0000	1.0000	1.0000	1.0000	0.9999	0.9999
14	1.0000	1.0000	1.0000	1.0000	1.0000	1.0000	1.0000	1.0000	1.0000	1.0000

r	4.10	4.20	4.30	4.40	4.50	4.60	4.70	4.80	4.90	5.00
0	0.0166	0.0150	0.0136	0.0123	0.0111	0.0101	0.0091	0.0082	0.0074	0.0067
1	0.0845	0.0780	0.0719	0.0663	0.0611	0.0563	0.0518	0.0477	0.0439	0.0404
2	0.2238	0.2102	0.1974	0.1851	0.1736	0.1626	0.1523	0.1425	0.1333	0.1247
3	0.4142	0.3954	0.3772	0.3594	0.3423	0.3257	0.3097	0.2942	0.2793	0.2650
4	0.6093	0.5898	0.5704	0.5512	0.5321	0.5132	0.4946	0.4763	0.4582	0.4405
5	0.7693	0.7531	0.7367	0.7199	0.7029	0.6858	0.6684	0.6510	0.6335	0.6160
6	0.8786	0.8675	0.8558	0.8436	0.8311	0.8180	0.8046	0.7908	0.7767	0.7622
7	0.9427	0.9361	0.9290	0.9214	0.9134	0.9049	0.8960	0.8867	0.8769	0.8666
8	0.9755	0.9721	0.9683	0.9642	0.9597	0.9549	0.9497	0.9442	0.9382	0.9319
9	0.9905	0.9889	0.9871	0.9851	0.9829	0.9805	0.9778	0.9749	0.9717	0.9682
10	0.9966	0.9959	0.9952	0.9943	0.9933	0.9922	0.9910	0.9896	0.9880	0.9863
11	0.9989	0.9986	0.9983	0.9980	0.9976	0.9971	0.9966	0.9960	0.9953	0.9945
12	0.9997	0.9996	0.9995	0.9993	0.9992	0.9990	0.9988	0.9986	0.9983	0.9980
13	0.9999	0.9999	0.9998	0.9998	0.9997	0.9997	0.9996	0.9995	0.9994	0.9993
14	1.0000	1.0000	1.0000	0.9999	0.9999	0.9999	0.9999	0.9999	0.9998	0.9998
15	1.0000	1.0000	1.0000	1.0000	1.0000	1.0000	1.0000	1.0000	0.9999	0.9999
16	1.0000	1.0000	1.0000	1.0000	1.0000	1.0000	1.0000	1.0000	1.0000	1.0000

Table 2   Poisson Distribution (continued)

$$P(X \le r) = \sum_{i=0}^{r} \frac{\lambda^i e^{-\lambda}}{i!}$$

					λ					
r	5.10	5.20	5.30	5.40	5.50	5.60	5.70	5.80	5.90	6.00
0	0.0061	0.0055	0.0050	0.0045	0.0041	0.0037	0.0033	0.0030	0.0027	0.0025
1	0.0372	0.0342	0.0314	0.0289	0.0266	0.0244	0.0224	0.0206	0.0189	0.0174
2	0.1165	0.1088	0.1016	0.0948	0.0884	0.0824	0.0768	0.0715	0.0666	0.0620
3	0.2513	0.2381	0.2254	0.2133	0.2017	0.1906	0.1800	0.1700	0.1604	0.1512
4	0.4231	0.4061	0.3895	0.3733	0.3575	0.3422	0.3272	0.3127	0.2987	0.2851
5	0.5984	0.5809	0.5635	0.5461	0.5289	0.5119	0.4950	0.4783	0.4619	0.4457
6	0.7474	0.7324	0.7171	0.7017	0.6860	0.6703	0.6544	0.6384	0.6224	0.6063
7	0.8560	0.8449	0.8335	0.8217	0.8095	0.7970	0.7841	0.7710	0.7576	0.7440
8	0.9252	0.9181	0.9106	0.9027	0.8944	0.8857	0.8766	0.8672	0.8574	0.8472
9	0.9644	0.9603	0.9559	0.9512	0.9462	0.9409	0.9352	0.9292	0.9228	0.9161
10	0.9844	0.9823	0.9800	0.9775	0.9747	0.9718	0.9686	0.9651	0.9614	0.9574
11	0.9937	0.9927	0.9916	0.9904	0.9890	0.9875	0.9859	0.9841	0.9821	0.9799
12	0.9976	0.9972	0.9967	0.9962	0.9955	0.9949	0.9941	0.9932	0.9922	0.9912
13	0.9992	0.9990	0.9988	0.9986	0.9983	0.9980	0.9977	0.9973	0.9969	0.9964
14	0.9997	0.9997	0.9996	0.9995	0.9994	0.9993	0.9991	0.9990	0.9988	0.9986
15	0.9999	0.9999	0.9999	0.9998	0.9998	0.9998	0.9997	0.9996	0.9996	0.9995
16	1.0000	1.0000	1.0000	0.9999	0.9999	0.9999	0.9999	0.9999	0.9999	0.9998
17	1.0000	1.0000	1.0000	1.0000	1.0000	1.0000	1.0000	1.0000	1.0000	0.9999
18	1.0000	1.0000	1.0000	1.0000	1.0000	1.0000	1.0000	1.0000	1.0000	1.0000

r	6.10	6.20	6.30	6.40	6.50	6.60	6.70	6.80	6.90	7.00
0	0.0022	0.0020	0.0018	0.0017	0.0015	0.0014	0.0012	0.0011	0.0010	0.0009
1	0.0159	0.0146	0.0134	0.0123	0.0113	0.0103	0.0095	0.0087	0.0080	0.0073
2	0.0577	0.0536	0.0498	0.0463	0.0430	0.0400	0.0371	0.0344	0.0320	0.0296
3	0.1425	0.1342	0.1264	0.1189	0.1118	0.1052	0.0988	0.0928	0.0871	0.0818
4	0.2719	0.2592	0.2469	0.2351	0.2237	0.2127	0.2022	0.1920	0.1823	0.1730
5	0.4298	0.4141	0.3988	0.3837	0.3690	0.3547	0.3406	0.3270	0.3137	0.3007
6	0.5902	0.5742	0.5582	0.5423	0.5265	0.5108	0.4953	0.4799	0.4647	0.4497
7	0.7301	0.7160	0.7017	0.6873	0.6728	0.6581	0.6433	0.6285	0.6136	0.5987
8	0.8367	0.8259	0.8148	0.8033	0.7916	0.7796	0.7673	0.7548	0.7420	0.7291
9	0.9090	0.9016	0.8939	0.8858	0.8774	0.8686	0.8596	0.8502	0.8405	0.8305
10	0.9531	0.9486	0.9437	0.9386	0.9332	0.9274	0.9214	0.9151	0.9084	0.9015
11	0.9776	0.9750	0.9723	0.9693	0.9661	0.9627	0.9591	0.9552	0.9510	0.9467
12	0.9900	0.9887	0.9873	0.9857	0.9840	0.9821	0.9801	0.9779	0.9755	0.9730
13	0.9958	0.9952	0.9945	0.9937	0.9929	0.9920	0.9909	0.9898	0.9885	0.9872
14	0.9984	0.9981	0.9978	0.9974	0.9970	0.9966	0.9961	0.9956	0.9950	0.9943
15	0.9994	0.9993	0.9992	0.9990	0.9988	0.9986	0.9984	0.9982	0.9979	0.9976
16	0.9998	0.9997	0.9997	0.9996	0.9996	0.9995	0.9994	0.9993	0.9992	0.9990
17	0.9999	0.9999	0.9999	0.9999	0.9998	0.9998	0.9998	0.9997	0.9997	0.9996
18	1.0000	1.0000	1.0000	1.0000	0.9999	0.9999	0.9999	0.9999	0.9999	0.9999
19	1.0000	1.0000	1.0000	1.0000	1.0000	1.0000	1.0000	1.0000	1.0000	1.0000

Table 2   Poisson Distribution (continued)

$$P(X \leq r) = \sum_{i=0}^{r} \frac{\lambda^i e^{-\lambda}}{i!}$$

						λ					
r	7.10	7.20	7.30	7.40	7.50	7.60	7.70	7.80	7.90	8.00
0	0.0008	0.0007	0.0007	0.0006	0.0006	0.0005	0.0005	0.0004	0.0004	0.0003
1	0.0067	0.0061	0.0056	0.0051	0.0047	0.0043	0.0039	0.0036	0.0033	0.0030
2	0.0275	0.0255	0.0236	0.0219	0.0203	0.0188	0.0174	0.0161	0.0149	0.0138
3	0.0767	0.0719	0.0674	0.0632	0.0591	0.0554	0.0518	0.0485	0.0453	0.0424
4	0.1641	0.1555	0.1473	0.1395	0.1321	0.1249	0.1181	0.1117	0.1055	0.0996
5	0.2881	0.2759	0.2640	0.2526	0.2414	0.2307	0.2203	0.2103	0.2006	0.1912
6	0.4349	0.4204	0.4060	0.3920	0.3782	0.3646	0.3514	0.3384	0.3257	0.3134
7	0.5838	0.5689	0.5541	0.5393	0.5246	0.5100	0.4956	0.4812	0.4670	0.4530
8	0.7160	0.7027	0.6892	0.6757	0.6620	0.6482	0.6343	0.6204	0.6065	0.5925
9	0.8202	0.8096	0.7988	0.7877	0.7764	0.7649	0.7531	0.7411	0.7290	0.7166
10	0.8942	0.8867	0.8788	0.8707	0.8622	0.8535	0.8445	0.8352	0.8257	0.8159
11	0.9420	0.9371	0.9319	0.9265	0.9208	0.9148	0.9085	0.9020	0.8952	0.8881
12	0.9703	0.9673	0.9642	0.9609	0.9573	0.9536	0.9496	0.9454	0.9409	0.9362
13	0.9857	0.9841	0.9824	0.9805	0.9784	0.9762	0.9739	0.9714	0.9687	0.9658
14	0.9935	0.9927	0.9918	0.9908	0.9897	0.9886	0.9873	0.9859	0.9844	0.9827
15	0.9972	0.9969	0.9964	0.9959	0.9954	0.9948	0.9941	0.9934	0.9926	0.9918
16	0.9989	0.9987	0.9985	0.9983	0.9980	0.9978	0.9974	0.9971	0.9967	0.9963
17	0.9996	0.9995	0.9994	0.9993	0.9992	0.9991	0.9989	0.9988	0.9986	0.9984
18	0.9998	0.9998	0.9998	0.9997	0.9997	0.9996	0.9996	0.9995	0.9994	0.9993
19	0.9999	0.9999	0.9999	0.9999	0.9999	0.9999	0.9998	0.9998	0.9998	0.9997
20	1.0000	1.0000	1.0000	1.0000	1.0000	1.0000	0.9999	0.9999	0.9999	0.9999
21	1.0000	1.0000	1.0000	1.0000	1.0000	1.0000	1.0000	1.0000	1.0000	1.0000

Table 2   Poisson Distribution (continued)

$$P(X \le r) = \sum_{i=0}^{r} \frac{\lambda^i e^{-\lambda}}{i!}$$

					λ					
r	8.10	8.20	8.30	8.40	8.50	8.60	8.70	8.80	8.90	9.00
0	0.0003	0.0003	0.0002	0.0002	0.0002	0.0002	0.0002	0.0002	0.0001	0.0001
1	0.0028	0.0025	0.0023	0.0021	0.0019	0.0018	0.0016	0.0015	0.0014	0.0012
2	0.0127	0.0118	0.0109	0.0100	0.0093	0.0086	0.0079	0.0073	0.0068	0.0062
3	0.0396	0.0370	0.0346	0.0323	0.0301	0.0281	0.0262	0.0244	0.0228	0.0212
4	0.0940	0.0887	0.0837	0.0789	0.0744	0.0701	0.0660	0.0621	0.0584	0.0550
5	0.1822	0.1736	0.1653	0.1573	0.1496	0.1422	0.1352	0.1284	0.1219	0.1157
6	0.3013	0.2896	0.2781	0.2670	0.2562	0.2457	0.2355	0.2256	0.2160	0.2068
7	0.4391	0.4254	0.4119	0.3987	0.3856	0.3728	0.3602	0.3478	0.3357	0.3239
8	0.5786	0.5647	0.5507	0.5369	0.5231	0.5094	0.4958	0.4823	0.4689	0.4557
9	0.7041	0.6915	0.6788	0.6659	0.6530	0.6400	0.6269	0.6137	0.6006	0.5874
10	0.8058	0.7955	0.7850	0.7743	0.7634	0.7522	0.7409	0.7294	0.7178	0.7060
11	0.8807	0.8731	0.8652	0.8571	0.8487	0.8400	0.8311	0.8220	0.8126	0.8030
12	0.9313	0.9261	0.9207	0.9150	0.9091	0.9029	0.8965	0.8898	0.8829	0.8758
13	0.9628	0.9595	0.9561	0.9524	0.9486	0.9445	0.9403	0.9358	0.9311	0.9261
14	0.9810	0.9791	0.9771	0.9749	0.9726	0.9701	0.9675	0.9647	0.9617	0.9585
15	0.9908	0.9898	0.9887	0.9875	0.9862	0.9848	0.9832	0.9816	0.9798	0.9780
16	0.9958	0.9953	0.9947	0.9941	0.9934	0.9926	0.9918	0.9909	0.9899	0.9889
17	0.9982	0.9979	0.9977	0.9973	0.9970	0.9966	0.9962	0.9957	0.9952	0.9947
18	0.9992	0.9991	0.9990	0.9989	0.9987	0.9985	0.9983	0.9981	0.9978	0.9976
19	0.9997	0.9997	0.9996	0.9995	0.9995	0.9994	0.9993	0.9992	0.9991	0.9989
20	0.9999	0.9999	0.9998	0.9998	0.9998	0.9998	0.9997	0.9997	0.9996	0.9996
21	1.0000	1.0000	0.9999	0.9999	0.9999	0.9999	0.9999	0.9999	0.9998	0.9998
22	1.0000	1.0000	1.0000	1.0000	1.0000	1.0000	1.0000	1.0000	0.9999	0.9999
23	1.0000	1.0000	1.0000	1.0000	1.0000	1.0000	1.0000	1.0000	1.0000	1.0000

Table 2   Poisson Distribution (continued)

$$P(X \leq r) = \sum_{i=0}^{r} \frac{\lambda^i e^{-\lambda}}{i!}$$

r	λ									
	9.10	9.20	9.30	9.40	9.50	9.60	9.70	9.80	9.90	10.00
0	0.0001	0.0001	0.0001	0.0001	0.0001	0.0001	0.0001	0.0001	0.0001	0.0000
1	0.0011	0.0010	0.0009	0.0009	0.0008	0.0007	0.0007	0.0006	0.0005	0.0005
2	0.0058	0.0053	0.0049	0.0045	0.0042	0.0038	0.0035	0.0033	0.0030	0.0028
3	0.0198	0.0184	0.0172	0.0160	0.0149	0.0138	0.0129	0.0120	0.0111	0.0103
4	0.0517	0.0486	0.0456	0.0429	0.0403	0.0378	0.0355	0.0333	0.0312	0.0293
5	0.1098	0.1041	0.0986	0.0935	0.0885	0.0838	0.0793	0.0750	0.0710	0.0671
6	0.1978	0.1892	0.1808	0.1727	0.1649	0.1574	0.1502	0.1433	0.1366	0.1301
7	0.3123	0.3010	0.2900	0.2792	0.2687	0.2584	0.2485	0.2388	0.2294	0.2202
8	0.4426	0.4296	0.4168	0.4042	0.3918	0.3796	0.3676	0.3558	0.3442	0.3328
9	0.5742	0.5611	0.5479	0.5349	0.5218	0.5089	0.4960	0.4832	0.4705	0.4579
10	0.6941	0.6820	0.6699	0.6576	0.6453	0.6329	0.6205	0.6080	0.5955	0.5830
11	0.7932	0.7832	0.7730	0.7626	0.7520	0.7412	0.7303	0.7193	0.7081	0.6968
12	0.8684	0.8607	0.8529	0.8448	0.8364	0.8279	0.8191	0.8101	0.8009	0.7916
13	0.9210	0.9156	0.9100	0.9042	0.8981	0.8919	0.8853	0.8786	0.8716	0.8645
14	0.9552	0.9517	0.9480	0.9441	0.9400	0.9357	0.9312	0.9265	0.9216	0.9165
15	0.9760	0.9738	0.9715	0.9691	0.9665	0.9638	0.9609	0.9579	0.9546	0.9513
16	0.9878	0.9865	0.9852	0.9838	0.9823	0.9806	0.9789	0.9770	0.9751	0.9730
17	0.9941	0.9934	0.9927	0.9919	0.9911	0.9902	0.9892	0.9881	0.9870	0.9857
18	0.9973	0.9969	0.9966	0.9962	0.9957	0.9952	0.9947	0.9941	0.9935	0.9928
19	0.9988	0.9986	0.9985	0.9983	0.9980	0.9978	0.9975	0.9972	0.9969	0.9965
20	0.9995	0.9994	0.9993	0.9992	0.9991	0.9990	0.9989	0.9987	0.9986	0.9984
21	0.9998	0.9998	0.9997	0.9997	0.9996	0.9996	0.9995	0.9995	0.9994	0.9993
22	0.9999	0.9999	0.9999	0.9999	0.9999	0.9998	0.9998	0.9998	0.9997	0.9997
23	1.0000	1.0000	1.0000	1.0000	0.9999	0.9999	0.9999	0.9999	0.9999	0.9999
24	1.0000	1.0000	1.0000	1.0000	1.0000	1.0000	1.0000	1.0000	1.0000	1.0000

Table 3   Normal Distribution

$$\Phi(z) = P(Z \le z) = \int_{-\infty}^{z} \frac{1}{\sqrt{2\pi}} e^{-t^2/2} \, dt$$

Note: $\Phi(-z) = 1 - \Phi(z)$.

z	0.00	0.01	0.02	0.03	0.04	0.05	0.06	0.07	0.08	0.09
0.0	0.5000	0.5040	0.5080	0.5120	0.5160	0.5199	0.5239	0.5279	0.5319	0.5359
0.1	0.5398	0.5438	0.5478	0.5517	0.5557	0.5596	0.5636	0.5675	0.5714	0.5753
0.2	0.5793	0.5832	0.5871	0.5910	0.5948	0.5987	0.6026	0.6064	0.6103	0.6141
0.3	0.6179	0.6217	0.6255	0.6293	0.6331	0.6368	0.6406	0.6443	0.6480	0.6517
0.4	0.6554	0.6591	0.6628	0.6664	0.6700	0.6736	0.6772	0.6808	0.6844	0.6879
0.5	0.6915	0.6950	0.6985	0.7019	0.7054	0.7088	0.7123	0.7157	0.7190	0.7224
0.6	0.7257	0.7291	0.7324	0.7357	0.7389	0.7422	0.7454	0.7486	0.7517	0.7549
0.7	0.7580	0.7611	0.7642	0.7673	0.7704	0.7734	0.7764	0.7794	0.7823	0.7852
0.8	0.7881	0.7910	0.7939	0.7967	0.7995	0.8023	0.8051	0.8078	0.8106	0.8133
0.9	0.8159	0.8186	0.8212	0.8238	0.8264	0.8289	0.8315	0.8340	0.8365	0.8389
1.0	0.8413	0.8438	0.8461	0.8485	0.8508	0.8531	0.8554	0.8577	0.8599	0.8621
1.1	0.8643	0.8665	0.8686	0.8708	0.8729	0.8749	0.8770	0.8790	0.8810	0.8830
1.2	0.8849	0.8869	0.8888	0.8907	0.8925	0.8944	0.8962	0.8980	0.8997	0.9015
1.3	0.9032	0.9049	0.9066	0.9082	0.9099	0.9115	0.9131	0.9147	0.9162	0.9177
1.4	0.9192	0.9207	0.9222	0.9236	0.9251	0.9265	0.9279	0.9292	0.9306	0.9319
1.5	0.9332	0.9345	0.9357	0.9370	0.9382	0.9394	0.9406	0.9418	0.9429	0.9441
1.6	0.9452	0.9463	0.9474	0.9484	0.9495	0.9505	0.9515	0.9525	0.9535	0.9545
1.7	0.9554	0.9564	0.9573	0.9582	0.9591	0.9599	0.9608	0.9616	0.9625	0.9633
1.8	0.9641	0.9649	0.9656	0.9664	0.9671	0.9678	0.9686	0.9693	0.9699	0.9706
1.9	0.9713	0.9719	0.9726	0.9732	0.9738	0.9744	0.9750	0.9756	0.9761	0.9767
2.0	0.9772	0.9778	0.9783	0.9788	0.9793	0.9798	0.9803	0.9808	0.9812	0.9817
2.1	0.9821	0.9826	0.9830	0.9834	0.9838	0.9842	0.9846	0.9850	0.9854	0.9857
2.2	0.9861	0.9864	0.9868	0.9871	0.9875	0.9878	0.9881	0.9884	0.9887	0.9890
2.3	0.9893	0.9896	0.9898	0.9901	0.9904	0.9906	0.9909	0.9911	0.9913	0.9916
2.4	0.9918	0.9920	0.9922	0.9925	0.9927	0.9929	0.9931	0.9932	0.9934	0.9936
2.5	0.9938	0.9940	0.9941	0.9943	0.9945	0.9946	0.9948	0.9949	0.9951	0.9952
2.6	0.9953	0.9955	0.9956	0.9957	0.9959	0.9960	0.9961	0.9962	0.9963	0.9964
2.7	0.9965	0.9966	0.9967	0.9968	0.9969	0.9970	0.9971	0.9972	0.9973	0.9974
2.8	0.9974	0.9975	0.9976	0.9977	0.9977	0.9978	0.9979	0.9979	0.9980	0.9981
2.9	0.9981	0.9982	0.9982	0.9983	0.9984	0.9984	0.9985	0.9985	0.9986	0.9986
3.0	0.9987	0.9987	0.9987	0.9988	0.9988	0.9989	0.9989	0.9989	0.9990	0.9990
3.1	0.9990	0.9991	0.9991	0.9991	0.9992	0.9992	0.9992	0.9992	0.9993	0.9993
3.2	0.9993	0.9993	0.9994	0.9994	0.9994	0.9994	0.9994	0.9995	0.9995	0.9995
3.3	0.9995	0.9995	0.9995	0.9996	0.9996	0.9996	0.9996	0.9996	0.9996	0.9997
3.4	0.9997	0.9997	0.9997	0.9997	0.9997	0.9997	0.9997	0.9997	0.9997	0.9998
3.5	0.9998	0.9998	0.9998	0.9998	0.9998	0.9998	0.9998	0.9998	0.9998	0.9998
3.6	0.9998	0.9998	0.9999	0.9999	0.9999	0.9999	0.9999	0.9999	0.9999	0.9999
3.7	0.9999	0.9999	0.9999	0.9999	0.9999	0.9999	0.9999	0.9999	0.9999	0.9999
3.8	0.9999	0.9999	0.9999	0.9999	0.9999	0.9999	0.9999	0.9999	0.9999	0.9999
3.9	1.0000	1.0000	1.0000	1.0000	1.0000	1.0000	1.0000	1.0000	1.0000	1.0000
4.0	1.0000	1.0000	1.0000	1.0000	1.0000	1.0000	1.0000	1.0000	1.0000	1.0000

Table 4   Chi-Square Distribution

$$P(X \leq x) = \int_0^x \frac{t^{n/2-1}}{\Gamma\left(\frac{n}{2}\right) 2^{n/2}} e^{-t/2} \, dt$$

Note: **d.f.** $= n$.

					$P(X \leq x)$			
n	0.010	0.025	0.050	0.100	0.900	0.950	0.975	0.990
1	0.000	0.001	0.004	0.016	2.706	3.841	5.024	6.635
2	0.020	0.051	0.103	0.211	4.605	5.991	7.378	9.210
3	0.115	0.216	0.352	0.584	6.251	7.815	9.348	11.345
4	0.297	0.484	0.711	1.064	7.779	9.488	11.143	13.277
5	0.554	0.831	1.145	1.610	9.236	11.070	12.833	15.086
6	0.872	1.237	1.635	2.204	10.645	12.592	14.449	16.812
7	1.239	1.690	2.167	2.833	12.017	14.067	16.013	18.475
8	1.646	2.180	2.733	3.490	13.362	15.507	17.535	20.090
9	2.088	2.700	3.325	4.168	14.684	16.919	19.023	21.666
10	2.558	3.247	3.940	4.865	15.987	18.307	20.483	23.209
11	3.053	3.816	4.575	5.578	17.275	19.675	21.920	24.725
12	3.571	4.404	5.226	6.304	18.549	21.026	23.337	26.217
13	4.107	5.009	5.892	7.042	19.812	22.362	24.736	27.688
14	4.660	5.629	6.571	7.790	21.064	23.685	26.119	29.141
15	5.229	6.262	7.261	8.547	22.307	24.996	27.488	30.578
16	5.812	6.908	7.962	9.312	23.542	26.296	28.845	32.000
17	6.408	7.564	8.672	10.085	24.769	27.587	30.191	33.409
18	7.015	8.231	9.390	10.865	25.989	28.869	31.526	34.805
19	7.633	8.907	10.117	11.651	27.204	30.144	32.852	36.191
20	8.260	9.591	10.851	12.443	28.412	31.410	34.170	37.566
30	14.953	16.791	18.493	20.599	40.256	43.773	46.979	50.892
40	22.164	24.433	26.509	29.051	51.805	55.758	59.342	63.691
50	29.707	32.357	34.764	37.689	63.167	67.505	71.420	76.154
60	37.485	40.482	43.188	46.459	74.397	79.082	83.298	88.379
70	45.442	48.758	51.739	55.329	85.527	90.531	95.023	100.425
80	53.540	57.153	60.391	64.278	96.578	101.879	106.629	112.329
90	61.754	65.647	69.126	73.291	107.565	113.145	118.136	124.116
100	70.065	74.222	77.929	82.358	118.498	124.342	129.561	135.807

Table 5   Student's t Distribution

$$P(X \leq x) = \frac{\Gamma\left(\frac{n+1}{2}\right)}{\sqrt{\pi n}\, \Gamma\left(\frac{n}{2}\right)} \int_{-\infty}^{x} \left(1 + \frac{t^2}{n}\right)^{-(n+1)/2} dt$$

Note: **d.f.= n.**

n	\multicolumn{5}{c}{$P(X \leq x)$}
	0.900	**0.950**	**0.975**	**0.990**	**0.995**
1	3.078	6.314	12.706	31.821	63.657
2	1.886	2.920	4.303	6.965	9.925
3	1.638	2.353	3.182	4.541	5.841
4	1.533	2.132	2.776	3.747	4.604
5	1.476	2.015	2.571	3.365	4.032
6	1.440	1.943	2.447	3.143	3.707
7	1.415	1.895	2.365	2.998	3.499
8	1.397	1.860	2.306	2.896	3.355
9	1.383	1.833	2.262	2.821	3.250
10	1.372	1.812	2.228	2.764	3.169
11	1.363	1.796	2.201	2.718	3.106
12	1.356	1.782	2.179	2.681	3.055
13	1.350	1.771	2.160	2.650	3.012
14	1.345	1.761	2.145	2.624	2.977
15	1.341	1.753	2.131	2.602	2.947
16	1.337	1.746	2.120	2.583	2.921
17	1.333	1.740	2.110	2.567	2.898
18	1.330	1.734	2.101	2.552	2.878
19	1.328	1.729	2.093	2.539	2.861
20	1.325	1.725	2.086	2.528	2.845
21	1.323	1.721	2.080	2.518	2.831
22	1.321	1.717	2.074	2.508	2.819
23	1.319	1.714	2.069	2.500	2.807
24	1.318	1.711	2.064	2.492	2.797
25	1.316	1.708	2.060	2.485	2.787
26	1.315	1.706	2.056	2.479	2.779
27	1.314	1.703	2.052	2.473	2.771
28	1.313	1.701	2.048	2.467	2.763
29	1.311	1.699	2.045	2.462	2.756
30	1.310	1.697	2.042	2.457	2.750
40	1.303	1.684	2.021	2.423	2.704
50	1.299	1.676	2.009	2.403	2.678
60	1.296	1.671	2.000	2.390	2.660
70	1.294	1.667	1.994	2.381	2.648
80	1.292	1.664	1.990	2.374	2.639
90	1.291	1.662	1.987	2.368	2.632
100	1.290	1.660	1.984	2.364	2.626
110	1.289	1.659	1.982	2.361	2.621
120	1.289	1.658	1.980	2.358	2.617
∞	1.282	1.645	1.960	2.326	2.576

Table 6 F Distribution

$$P(X \le x) = \frac{\Gamma\left(\frac{m+n}{2}\right)}{\Gamma\left(\frac{m}{2}\right)\Gamma\left(\frac{n}{2}\right)} \int_0^x \frac{\left(\frac{m}{n}\right) f^{m/2-1}}{\left(1+\frac{mf}{n}\right)^{(m+n)/2}}\, df$$

Note: d.f. = m, denominator d.f. = n.

$P(X \le x) = 0.90$

n	m=1	2	3	4	5	6	7	8	9	10	12	15	20	25	30	40	60	80	100	120
1	39.8	49.5	53.6	55.8	57.2	58.2	58.9	59.4	59.9	60.2	60.7	61.2	61.7	62.1	62.3	62.5	62.8	62.9	63.0	63.1
2	8.53	9.00	9.16	9.24	9.29	9.33	9.35	9.37	9.38	9.39	9.41	9.42	9.44	9.45	9.46	9.47	9.47	9.48	9.48	9.48
3	5.54	5.46	5.39	5.34	5.31	5.28	5.27	5.25	5.24	5.23	5.22	5.20	5.18	5.17	5.17	5.16	5.15	5.15	5.14	5.14
4	4.54	4.32	4.19	4.11	4.05	4.01	3.98	3.95	3.94	3.92	3.90	3.87	3.84	3.83	3.82	3.80	3.79	3.78	3.78	3.78
5	4.06	3.78	3.62	3.52	3.45	3.40	3.37	3.34	3.32	3.30	3.27	3.24	3.21	3.19	3.17	3.16	3.14	3.13	3.13	3.12
6	3.78	3.46	3.29	3.18	3.11	3.05	3.01	2.98	2.96	2.94	2.90	2.87	2.84	2.81	2.80	2.78	2.76	2.75	2.75	2.74
7	3.59	3.26	3.07	2.96	2.88	2.83	2.78	2.75	2.72	2.70	2.67	2.63	2.59	2.57	2.56	2.54	2.51	2.50	2.50	2.49
8	3.46	3.11	2.92	2.81	2.73	2.67	2.62	2.59	2.56	2.54	2.50	2.46	2.42	2.40	2.38	2.36	2.34	2.33	2.32	2.32
9	3.36	3.01	2.81	2.69	2.61	2.55	2.51	2.47	2.44	2.42	2.38	2.34	2.30	2.27	2.25	2.23	2.21	2.20	2.19	2.18
10	3.29	2.92	2.73	2.61	2.52	2.46	2.41	2.38	2.35	2.32	2.28	2.24	2.20	2.17	2.16	2.13	2.11	2.09	2.09	2.08
11	3.23	2.86	2.66	2.54	2.45	2.39	2.34	2.30	2.27	2.25	2.21	2.17	2.12	2.10	2.08	2.05	2.03	2.01	2.01	2.00
12	3.18	2.81	2.61	2.48	2.39	2.33	2.28	2.24	2.21	2.19	2.15	2.10	2.06	2.03	2.01	1.99	1.96	1.95	1.94	1.93
13	3.14	2.76	2.56	2.43	2.35	2.28	2.23	2.20	2.16	2.14	2.10	2.05	2.01	1.98	1.96	1.93	1.90	1.89	1.88	1.88
14	3.10	2.73	2.52	2.39	2.31	2.24	2.19	2.15	2.12	2.10	2.05	2.01	1.96	1.93	1.91	1.89	1.86	1.84	1.83	1.83
15	3.07	2.70	2.49	2.36	2.27	2.21	2.16	2.12	2.09	2.06	2.02	1.97	1.92	1.89	1.87	1.85	1.82	1.80	1.79	1.79
16	3.05	2.67	2.46	2.33	2.24	2.18	2.13	2.09	2.06	2.03	1.99	1.94	1.89	1.86	1.84	1.81	1.78	1.77	1.76	1.75
17	3.03	2.64	2.44	2.31	2.22	2.15	2.10	2.06	2.03	2.00	1.96	1.91	1.86	1.83	1.81	1.78	1.75	1.74	1.73	1.72
18	3.01	2.62	2.42	2.29	2.20	2.13	2.08	2.04	2.00	1.98	1.93	1.89	1.84	1.80	1.78	1.75	1.72	1.71	1.70	1.69
19	2.99	2.61	2.40	2.27	2.18	2.11	2.06	2.02	1.98	1.96	1.91	1.86	1.81	1.78	1.76	1.73	1.70	1.68	1.67	1.67
20	2.97	2.59	2.38	2.25	2.16	2.09	2.04	2.00	1.96	1.94	1.89	1.84	1.79	1.76	1.74	1.71	1.68	1.66	1.65	1.64
21	2.96	2.57	2.36	2.23	2.14	2.08	2.02	1.98	1.95	1.92	1.87	1.83	1.78	1.74	1.72	1.69	1.66	1.64	1.63	1.62
22	2.95	2.56	2.35	2.22	2.13	2.06	2.01	1.97	1.93	1.90	1.86	1.81	1.76	1.73	1.70	1.67	1.64	1.62	1.61	1.60
23	2.94	2.55	2.34	2.21	2.11	2.05	1.99	1.95	1.92	1.89	1.84	1.80	1.74	1.71	1.69	1.66	1.62	1.61	1.59	1.59
24	2.93	2.54	2.33	2.19	2.10	2.04	1.98	1.94	1.91	1.88	1.83	1.78	1.73	1.70	1.67	1.64	1.61	1.59	1.58	1.57
25	2.92	2.53	2.32	2.18	2.09	2.02	1.97	1.93	1.89	1.87	1.82	1.77	1.72	1.68	1.66	1.63	1.59	1.58	1.56	1.56
26	2.91	2.52	2.31	2.17	2.08	2.01	1.96	1.92	1.88	1.86	1.81	1.76	1.71	1.67	1.65	1.61	1.58	1.56	1.55	1.54
27	2.90	2.51	2.30	2.17	2.07	2.00	1.95	1.91	1.87	1.85	1.80	1.75	1.70	1.66	1.64	1.60	1.57	1.55	1.54	1.53
28	2.89	2.50	2.29	2.16	2.06	2.00	1.94	1.90	1.87	1.84	1.79	1.74	1.69	1.65	1.63	1.59	1.56	1.54	1.53	1.52
29	2.89	2.50	2.28	2.15	2.06	1.99	1.93	1.89	1.86	1.83	1.78	1.73	1.68	1.64	1.62	1.58	1.55	1.53	1.52	1.51
30	2.88	2.49	2.28	2.14	2.05	1.98	1.93	1.88	1.85	1.82	1.77	1.72	1.67	1.63	1.61	1.57	1.54	1.52	1.51	1.50
40	2.84	2.44	2.23	2.09	2.00	1.93	1.87	1.83	1.79	1.76	1.71	1.66	1.61	1.57	1.54	1.51	1.47	1.45	1.43	1.42
50	2.81	2.41	2.20	2.06	1.97	1.90	1.84	1.80	1.76	1.73	1.68	1.63	1.57	1.53	1.50	1.46	1.42	1.40	1.39	1.38
60	2.79	2.39	2.18	2.04	1.95	1.87	1.82	1.77	1.74	1.71	1.66	1.60	1.54	1.50	1.48	1.44	1.40	1.37	1.36	1.35
70	2.78	2.38	2.16	2.03	1.93	1.86	1.80	1.76	1.72	1.69	1.64	1.59	1.53	1.49	1.46	1.42	1.37	1.35	1.34	1.32
80	2.77	2.37	2.15	2.02	1.92	1.85	1.79	1.75	1.71	1.68	1.63	1.57	1.51	1.47	1.44	1.40	1.36	1.33	1.32	1.31
90	2.76	2.36	2.15	2.01	1.91	1.84	1.78	1.74	1.70	1.67	1.62	1.56	1.50	1.46	1.43	1.39	1.35	1.32	1.30	1.29
100	2.76	2.36	2.14	2.00	1.91	1.83	1.78	1.73	1.69	1.66	1.61	1.56	1.49	1.45	1.42	1.38	1.34	1.31	1.29	1.28
110	2.75	2.35	2.13	2.00	1.90	1.83	1.78	1.73	1.69	1.66	1.61	1.55	1.49	1.45	1.42	1.37	1.33	1.30	1.28	1.27
120	2.75	2.35	2.13	1.99	1.90	1.82	1.77	1.72	1.68	1.65	1.60	1.55	1.48	1.44	1.41	1.37	1.32	1.29	1.28	1.26

Table 6   F Distribution (continued)

$$P(X \leq x) = \frac{\Gamma\left(\frac{m+n}{2}\right)}{\Gamma\left(\frac{m}{2}\right)\Gamma\left(\frac{n}{2}\right)} \int_0^x \frac{\left(\frac{m}{n}\right) f^{m/2-1}}{\left(1 + \frac{mf}{n}\right)^{(m+n)/2}}\, df$$

Note: d.f. = m, denominator d.f. = n.

$P(X \leq x) = 0.95$

n	m=1	2	3	4	5	6	7	8	9	10	12	15	20	25	30	40	60	80	100	120
1	161.4	199.5	215.7	224.6	230.2	234.0	236.8	238.9	240.5	241.9	243.9	246.0	248.0	249.3	250.1	251.1	252.2	252.7	253.0	253.3
2	18.51	19.00	19.16	19.25	19.30	19.33	19.35	19.37	19.38	19.40	19.40	19.43	19.45	19.46	19.46	19.47	19.48	19.48	19.49	19.49
3	10.13	9.55	9.28	9.12	9.01	8.94	8.89	8.85	8.81	8.79	8.74	8.70	8.66	8.63	8.62	8.59	8.57	8.56	8.55	8.55
4	7.71	6.94	6.59	6.39	6.26	6.16	6.09	6.04	6.00	5.96	5.91	5.86	5.80	5.77	5.75	5.72	5.69	5.67	5.66	5.66
5	6.61	5.79	5.41	5.19	5.05	4.95	4.88	4.82	4.77	4.74	4.68	4.62	4.56	4.52	4.50	4.46	4.43	4.41	4.41	4.40
6	5.99	5.14	4.76	4.53	4.39	4.28	4.21	4.15	4.10	4.06	4.00	3.94	3.87	3.83	3.81	3.77	3.74	3.72	3.71	3.70
7	5.59	4.74	4.35	4.12	3.97	3.87	3.79	3.73	3.68	3.64	3.57	3.51	3.44	3.40	3.38	3.34	3.30	3.29	3.27	3.27
8	5.32	4.46	4.07	3.84	3.69	3.58	3.50	3.44	3.39	3.35	3.28	3.22	3.15	3.11	3.08	3.04	3.01	2.99	2.97	2.97
9	5.12	4.26	3.86	3.63	3.48	3.37	3.29	3.23	3.18	3.14	3.07	3.01	2.94	2.89	2.86	2.83	2.79	2.77	2.76	2.75
10	4.96	4.10	3.71	3.48	3.33	3.22	3.14	3.07	3.02	2.98	2.91	2.85	2.77	2.73	2.70	2.66	2.62	2.60	2.59	2.58
11	4.84	3.98	3.59	3.36	3.20	3.09	3.01	2.95	2.90	2.85	2.79	2.72	2.65	2.60	2.57	2.53	2.49	2.47	2.46	2.45
12	4.75	3.89	3.49	3.26	3.11	3.00	2.91	2.85	2.80	2.75	2.69	2.62	2.54	2.50	2.47	2.43	2.38	2.36	2.35	2.34
13	4.67	3.81	3.41	3.18	3.03	2.92	2.83	2.77	2.71	2.67	2.60	2.53	2.46	2.41	2.38	2.34	2.30	2.27	2.26	2.25
14	4.60	3.74	3.34	3.11	2.96	2.85	2.76	2.70	2.65	2.60	2.53	2.46	2.39	2.34	2.31	2.27	2.22	2.20	2.19	2.18
15	4.54	3.68	3.29	3.06	2.90	2.79	2.71	2.64	2.59	2.54	2.48	2.40	2.33	2.28	2.25	2.20	2.16	2.14	2.12	2.11
16	4.49	3.63	3.24	3.01	2.85	2.74	2.66	2.59	2.54	2.49	2.42	2.35	2.28	2.23	2.19	2.15	2.11	2.08	2.07	2.06
17	4.45	3.59	3.20	2.96	2.81	2.70	2.61	2.55	2.49	2.45	2.38	2.31	2.23	2.18	2.15	2.10	2.06	2.03	2.02	2.01
18	4.41	3.55	3.16	2.93	2.77	2.66	2.58	2.51	2.46	2.41	2.34	2.27	2.19	2.14	2.11	2.06	2.02	1.99	1.98	1.97
19	4.38	3.52	3.13	2.90	2.74	2.63	2.54	2.48	2.42	2.38	2.31	2.23	2.16	2.11	2.07	2.03	1.98	1.96	1.94	1.93
20	4.35	3.49	3.10	2.87	2.71	2.60	2.51	2.45	2.39	2.35	2.28	2.20	2.12	2.07	2.04	1.99	1.95	1.92	1.91	1.90
21	4.32	3.47	3.07	2.84	2.68	2.57	2.49	2.42	2.37	2.32	2.25	2.18	2.10	2.05	2.01	1.96	1.92	1.89	1.88	1.87
22	4.30	3.44	3.05	2.82	2.66	2.55	2.46	2.40	2.34	2.30	2.23	2.15	2.07	2.02	1.98	1.94	1.89	1.86	1.85	1.84
23	4.28	3.42	3.03	2.80	2.64	2.53	2.44	2.37	2.32	2.27	2.20	2.13	2.05	2.00	1.96	1.91	1.86	1.84	1.82	1.81
24	4.26	3.40	3.01	2.78	2.62	2.51	2.42	2.36	2.30	2.25	2.18	2.11	2.03	1.97	1.94	1.89	1.84	1.82	1.80	1.79
25	4.24	3.39	2.99	2.76	2.60	2.49	2.40	2.34	2.28	2.24	2.16	2.09	2.01	1.96	1.92	1.87	1.82	1.80	1.78	1.77
26	4.23	3.37	2.98	2.74	2.59	2.47	2.39	2.32	2.27	2.22	2.15	2.07	1.99	1.94	1.90	1.85	1.80	1.78	1.76	1.75
27	4.21	3.35	2.96	2.73	2.57	2.46	2.37	2.31	2.25	2.20	2.13	2.06	1.97	1.92	1.88	1.84	1.79	1.76	1.74	1.73
28	4.20	3.34	2.95	2.71	2.56	2.45	2.36	2.29	2.24	2.19	2.12	2.04	1.96	1.91	1.87	1.82	1.77	1.74	1.73	1.71
29	4.18	3.33	2.93	2.70	2.55	2.43	2.35	2.28	2.22	2.18	2.10	2.03	1.94	1.89	1.85	1.81	1.75	1.73	1.71	1.70
30	4.17	3.32	2.92	2.69	2.53	2.42	2.33	2.27	2.21	2.16	2.09	2.01	1.93	1.88	1.84	1.79	1.74	1.71	1.70	1.68
40	4.08	3.23	2.84	2.61	2.45	2.34	2.25	2.18	2.12	2.08	2.00	1.92	1.84	1.78	1.74	1.69	1.64	1.61	1.59	1.58
50	4.03	3.18	2.79	2.56	2.40	2.29	2.20	2.13	2.07	2.03	1.95	1.87	1.78	1.73	1.69	1.63	1.58	1.54	1.52	1.51
60	4.00	3.15	2.76	2.53	2.37	2.25	2.17	2.10	2.04	1.99	1.92	1.84	1.75	1.69	1.65	1.59	1.53	1.50	1.48	1.47
70	3.98	3.13	2.74	2.50	2.35	2.23	2.14	2.07	2.02	1.97	1.89	1.81	1.72	1.66	1.62	1.57	1.50	1.47	1.45	1.44
80	3.96	3.11	2.72	2.49	2.33	2.21	2.13	2.06	2.00	1.95	1.88	1.79	1.70	1.64	1.60	1.54	1.48	1.45	1.43	1.41
90	3.95	3.10	2.71	2.47	2.32	2.20	2.11	2.04	1.99	1.94	1.86	1.78	1.69	1.63	1.59	1.53	1.46	1.43	1.41	1.39
100	3.94	3.09	2.70	2.46	2.31	2.19	2.10	2.03	1.97	1.93	1.85	1.77	1.68	1.62	1.57	1.52	1.45	1.41	1.39	1.38
110	3.93	3.08	2.69	2.45	2.30	2.18	2.09	2.02	1.97	1.92	1.84	1.76	1.67	1.61	1.56	1.50	1.44	1.40	1.38	1.36
120	3.92	3.07	2.68	2.45	2.29	2.18	2.09	2.02	1.96	1.91	1.83	1.75	1.66	1.60	1.55	1.50	1.43	1.39	1.37	1.35

Table 6 *F* **Distribution** (continued)

$$P(X \leq x) = \frac{\Gamma\left(\frac{m+n}{2}\right)}{\Gamma\left(\frac{m}{2}\right)\Gamma\left(\frac{n}{2}\right)} \int_0^x \frac{\left(\frac{m}{n}\right) f^{m/2-1}}{\left(1 + \frac{mf}{n}\right)^{(m+n)/2}}\, df$$

Note: d.f. = m, denominator d.f. = n.

$P(X \leq x) = 0.975$

m

n	1	2	3	4	5	6	7	8	9	10	12	15	20	25	30	40	60	80	100	120
1	647.8	799.5	864.2	899.6	921.9	937.1	948.2	956.7	963.3	968.6	976.7	984.9	993.1	998.1	1001	1005	1010	1012	1013	1014
2	38.51	39.00	39.17	39.25	39.30	39.33	39.36	39.37	39.39	39.40	39.41	39.43	39.45	39.46	39.46	39.47	39.48	39.49	39.49	39.49
3	17.44	16.04	15.44	15.10	14.88	14.73	14.62	14.54	14.47	14.42	14.34	14.25	14.17	14.12	14.08	14.04	13.99	13.97	13.96	13.95
4	12.22	10.65	9.98	9.60	9.36	9.20	9.07	8.98	8.90	8.84	8.75	8.66	8.56	8.50	8.46	8.41	8.36	8.33	8.32	8.31
5	10.01	8.43	7.76	7.39	7.15	6.98	6.85	6.76	6.68	6.62	6.52	6.43	6.33	6.27	6.23	6.18	6.12	6.10	6.08	6.07
6	8.81	7.26	6.60	6.23	5.99	5.82	5.70	5.60	5.52	5.46	5.37	5.27	5.17	5.11	5.07	5.01	4.96	4.93	4.92	4.90
7	8.07	6.54	5.89	5.52	5.29	5.12	4.99	4.90	4.82	4.76	4.67	4.57	4.47	4.40	4.36	4.31	4.25	4.23	4.21	4.20
8	7.57	6.06	5.42	5.05	4.82	4.65	4.53	4.43	4.36	4.30	4.20	4.10	4.00	3.94	3.89	3.84	3.78	3.76	3.74	3.73
9	7.21	5.71	5.08	4.72	4.48	4.32	4.20	4.10	4.03	3.96	3.87	3.77	3.67	3.60	3.56	3.51	3.45	3.42	3.40	3.39
10	6.94	5.46	4.83	4.47	4.24	4.07	3.95	3.85	3.78	3.72	3.62	3.52	3.42	3.35	3.31	3.26	3.20	3.17	3.15	3.14
11	6.72	5.26	4.63	4.28	4.04	3.88	3.76	3.66	3.59	3.53	3.43	3.33	3.23	3.16	3.12	3.06	3.00	2.97	2.96	2.94
12	6.55	5.10	4.47	4.12	3.89	3.73	3.61	3.51	3.44	3.37	3.28	3.18	3.07	3.01	2.96	2.91	2.85	2.82	2.80	2.79
13	6.41	4.97	4.35	4.00	3.77	3.60	3.48	3.39	3.31	3.25	3.15	3.05	2.95	2.88	2.84	2.78	2.72	2.69	2.67	2.66
14	6.30	4.86	4.24	3.89	3.66	3.50	3.38	3.29	3.21	3.15	3.05	2.95	2.84	2.78	2.73	2.67	2.61	2.58	2.56	2.55
15	6.20	4.77	4.15	3.80	3.58	3.41	3.29	3.20	3.12	3.06	2.96	2.86	2.76	2.69	2.64	2.59	2.52	2.49	2.47	2.46
16	6.12	4.69	4.08	3.73	3.50	3.34	3.22	3.12	3.05	2.99	2.89	2.79	2.68	2.61	2.57	2.51	2.45	2.42	2.40	2.38
17	6.04	4.62	4.01	3.66	3.44	3.28	3.16	3.06	2.98	2.92	2.82	2.72	2.62	2.55	2.50	2.44	2.38	2.35	2.33	2.32
18	5.98	4.56	3.95	3.61	3.38	3.22	3.10	3.01	2.93	2.87	2.77	2.67	2.56	2.49	2.44	2.38	2.32	2.29	2.27	2.26
19	5.92	4.51	3.90	3.56	3.33	3.17	3.05	2.96	2.88	2.82	2.72	2.62	2.51	2.44	2.39	2.33	2.27	2.24	2.22	2.20
20	5.87	4.46	3.86	3.51	3.29	3.13	3.01	2.91	2.84	2.77	2.68	2.57	2.46	2.40	2.35	2.29	2.22	2.19	2.17	2.16
21	5.83	4.42	3.82	3.48	3.25	3.09	2.97	2.87	2.80	2.73	2.64	2.53	2.42	2.36	2.31	2.25	2.18	2.15	2.13	2.11
22	5.79	4.38	3.78	3.44	3.22	3.05	2.93	2.84	2.76	2.70	2.60	2.50	2.39	2.32	2.27	2.21	2.14	2.11	2.09	2.08
23	5.75	4.35	3.75	3.41	3.18	3.02	2.90	2.81	2.73	2.67	2.57	2.47	2.36	2.29	2.24	2.18	2.11	2.08	2.06	2.04
24	5.72	4.32	3.72	3.38	3.15	2.99	2.87	2.78	2.70	2.64	2.54	2.44	2.33	2.26	2.21	2.15	2.08	2.05	2.02	2.01
25	5.69	4.29	3.69	3.35	3.13	2.97	2.85	2.75	2.68	2.61	2.51	2.41	2.30	2.23	2.18	2.12	2.05	2.02	2.00	1.98
26	5.66	4.27	3.67	3.33	3.10	2.94	2.82	2.73	2.65	2.59	2.49	2.39	2.28	2.21	2.16	2.09	2.03	2.00	1.97	1.95
27	5.63	4.24	3.65	3.31	3.08	2.92	2.80	2.71	2.63	2.57	2.47	2.36	2.25	2.18	2.13	2.07	2.00	1.97	1.94	1.93
28	5.61	4.22	3.63	3.29	3.06	2.90	2.78	2.69	2.61	2.55	2.45	2.34	2.23	2.16	2.11	2.05	1.98	1.94	1.92	1.91
29	5.59	4.20	3.61	3.27	3.04	2.88	2.76	2.67	2.59	2.53	2.43	2.32	2.21	2.14	2.09	2.03	1.96	1.92	1.90	1.89
30	5.57	4.18	3.59	3.25	3.03	2.87	2.75	2.65	2.57	2.51	2.41	2.31	2.20	2.12	2.07	2.01	1.94	1.90	1.88	1.87
40	5.42	4.05	3.46	3.13	2.90	2.74	2.62	2.53	2.45	2.39	2.29	2.18	2.07	1.99	1.94	1.88	1.80	1.76	1.74	1.72
50	5.34	3.97	3.39	3.05	2.83	2.67	2.55	2.46	2.38	2.32	2.22	2.11	1.99	1.92	1.87	1.80	1.72	1.68	1.66	1.64
60	5.29	3.93	3.34	3.01	2.79	2.63	2.51	2.41	2.33	2.27	2.17	2.06	1.94	1.87	1.82	1.74	1.67	1.63	1.60	1.58
70	5.25	3.89	3.31	2.97	2.75	2.59	2.47	2.38	2.30	2.24	2.14	2.03	1.91	1.83	1.78	1.71	1.63	1.59	1.56	1.54
80	5.22	3.86	3.28	2.95	2.73	2.57	2.45	2.35	2.28	2.21	2.11	2.00	1.88	1.81	1.75	1.68	1.60	1.55	1.53	1.51
90	5.20	3.84	3.26	2.93	2.71	2.55	2.43	2.34	2.26	2.19	2.09	1.98	1.86	1.79	1.73	1.66	1.58	1.53	1.50	1.48
100	5.18	3.83	3.25	2.92	2.70	2.54	2.42	2.32	2.24	2.18	2.08	1.97	1.85	1.77	1.71	1.64	1.56	1.51	1.48	1.46
110	5.16	3.82	3.24	2.90	2.68	2.53	2.40	2.31	2.23	2.17	2.07	1.96	1.84	1.76	1.70	1.63	1.54	1.50	1.47	1.45
120	5.15	3.80	3.23	2.89	2.67	2.52	2.39	2.30	2.22	2.16	2.05	1.94	1.82	1.75	1.69	1.61	1.53	1.48	1.45	1.43

Table 6   F Distribution (continued)

$$P(X \leq x) = \frac{\Gamma\left(\frac{m+n}{2}\right)}{\Gamma\left(\frac{m}{2}\right)\Gamma\left(\frac{n}{2}\right)} \int_0^x \frac{\left(\frac{m}{n}\right) f^{m/2-1}}{\left(1 + \frac{mf}{n}\right)^{(m+n)/2}} df$$

Note: **d.f.** $= m$, denominator **d.f.** $= n$.

$P(X \leq x) = 0.99$

n	1	2	3	4	5	6	7	8	9	10	12	15	20	25	30	40	60	80	100	120
1	4052	4999	5403	5625	5764	5859	5928	5981	6022	6056	6106	6157	6209	6240	6261	6287	6313	6326	6334	6339
2	98.50	99.00	99.17	99.25	99.30	99.33	99.36	99.37	99.39	99.40	99.42	99.43	99.45	99.46	99.47	99.47	99.48	99.49	99.49	99.49
3	34.12	30.82	29.46	28.71	28.24	27.91	27.67	27.49	27.35	27.23	27.05	26.87	26.69	26.58	26.50	26.41	26.32	26.27	26.24	26.22
4	21.20	18.00	16.69	15.98	15.52	15.21	14.98	14.80	14.66	14.55	14.37	14.20	14.02	13.91	13.84	13.75	13.65	13.61	13.58	13.56
5	16.26	13.27	12.06	11.39	10.97	10.67	10.46	10.29	10.16	10.05	9.89	9.72	9.55	9.45	9.38	9.29	9.20	9.16	9.13	9.11
6	13.75	10.92	9.78	9.15	8.75	8.47	8.26	8.10	7.98	7.87	7.72	7.56	7.40	7.30	7.23	7.14	7.06	7.01	6.99	6.97
7	12.25	9.55	8.45	7.85	7.46	7.19	6.99	6.84	6.72	6.62	6.47	6.31	6.16	6.06	5.99	5.91	5.82	5.78	5.75	5.74
8	11.26	8.65	7.59	7.01	6.63	6.37	6.18	6.03	5.91	5.81	5.67	5.52	5.36	5.26	5.20	5.12	5.03	4.99	4.96	4.95
9	10.56	8.02	6.99	6.42	6.06	5.80	5.61	5.47	5.35	5.26	5.11	4.96	4.81	4.71	4.65	4.57	4.48	4.44	4.41	4.40
10	10.04	7.56	6.55	5.99	5.64	5.39	5.20	5.06	4.94	4.85	4.71	4.56	4.41	4.31	4.25	4.17	4.08	4.04	4.01	4.00
11	9.65	7.21	6.22	5.67	5.32	5.07	4.89	4.74	4.63	4.54	4.40	4.25	4.10	4.01	3.94	3.86	3.78	3.73	3.71	3.69
12	9.33	6.93	5.95	5.41	5.06	4.82	4.64	4.50	4.39	4.30	4.16	4.01	3.86	3.76	3.70	3.62	3.54	3.49	3.47	3.45
13	9.07	6.70	5.74	5.21	4.86	4.62	4.44	4.30	4.19	4.10	3.96	3.82	3.66	3.57	3.51	3.43	3.34	3.30	3.27	3.25
14	8.86	6.51	5.56	5.04	4.69	4.46	4.28	4.14	4.03	3.94	3.80	3.66	3.51	3.41	3.35	3.27	3.18	3.14	3.11	3.09
15	8.68	6.36	5.42	4.89	4.56	4.32	4.14	4.00	3.89	3.80	3.67	3.52	3.37	3.28	3.21	3.13	3.05	3.00	2.98	2.96
16	8.53	6.23	5.29	4.77	4.44	4.20	4.03	3.89	3.78	3.69	3.55	3.41	3.26	3.16	3.10	3.02	2.93	2.89	2.86	2.84
17	8.40	6.11	5.18	4.67	4.34	4.10	3.93	3.79	3.68	3.59	3.46	3.31	3.16	3.07	3.00	2.92	2.83	2.79	2.76	2.75
18	8.29	6.01	5.09	4.58	4.25	4.01	3.84	3.71	3.60	3.51	3.37	3.23	3.08	2.98	2.92	2.84	2.75	2.70	2.68	2.66
19	8.18	5.93	5.01	4.50	4.17	3.94	3.77	3.63	3.52	3.43	3.30	3.15	3.00	2.91	2.84	2.76	2.67	2.63	2.60	2.58
20	8.10	5.85	4.94	4.43	4.10	3.87	3.70	3.56	3.46	3.37	3.23	3.09	2.94	2.84	2.78	2.69	2.61	2.56	2.54	2.52
21	8.02	5.78	4.87	4.37	4.04	3.81	3.64	3.51	3.40	3.31	3.17	3.03	2.88	2.79	2.72	2.64	2.55	2.50	2.48	2.46
22	7.95	5.72	4.82	4.31	3.99	3.76	3.59	3.45	3.35	3.26	3.12	2.98	2.83	2.73	2.67	2.58	2.50	2.45	2.42	2.40
23	7.88	5.66	4.76	4.26	3.94	3.71	3.54	3.41	3.30	3.21	3.07	2.93	2.78	2.69	2.62	2.54	2.45	2.40	2.37	2.35
24	7.82	5.61	4.72	4.22	3.90	3.67	3.50	3.36	3.26	3.17	3.03	2.89	2.74	2.64	2.58	2.49	2.40	2.36	2.33	2.31
25	7.77	5.57	4.68	4.18	3.85	3.63	3.46	3.32	3.22	3.13	2.99	2.85	2.70	2.60	2.54	2.45	2.36	2.32	2.29	2.27
26	7.72	5.53	4.64	4.14	3.82	3.59	3.42	3.29	3.18	3.09	2.96	2.81	2.66	2.57	2.50	2.42	2.33	2.28	2.25	2.23
27	7.68	5.49	4.60	4.11	3.78	3.56	3.39	3.26	3.15	3.06	2.93	2.78	2.63	2.54	2.47	2.38	2.29	2.25	2.22	2.20
28	7.64	5.45	4.57	4.07	3.75	3.53	3.36	3.23	3.12	3.03	2.90	2.75	2.60	2.51	2.44	2.35	2.26	2.22	2.19	2.17
29	7.60	5.42	4.54	4.04	3.73	3.50	3.33	3.20	3.09	3.00	2.87	2.73	2.57	2.48	2.41	2.33	2.23	2.19	2.16	2.14
30	7.56	5.39	4.51	4.02	3.70	3.47	3.30	3.17	3.07	2.98	2.84	2.70	2.55	2.45	2.39	2.30	2.21	2.16	2.13	2.11
40	7.31	5.18	4.31	3.83	3.51	3.29	3.12	2.99	2.89	2.80	2.66	2.52	2.37	2.27	2.20	2.11	2.02	1.97	1.94	1.92
50	7.17	5.06	4.20	3.72	3.41	3.19	3.02	2.89	2.78	2.70	2.56	2.42	2.27	2.17	2.10	2.01	1.91	1.86	1.82	1.80
60	7.08	4.98	4.13	3.65	3.34	3.12	2.95	2.82	2.72	2.63	2.50	2.35	2.20	2.10	2.03	1.94	1.84	1.78	1.75	1.73
70	7.01	4.92	4.07	3.60	3.29	3.07	2.91	2.78	2.67	2.59	2.45	2.31	2.15	2.05	1.98	1.89	1.78	1.73	1.70	1.67
80	6.96	4.88	4.04	3.56	3.26	3.04	2.87	2.74	2.64	2.55	2.42	2.27	2.12	2.01	1.94	1.85	1.75	1.69	1.65	1.63
90	6.93	4.85	4.01	3.53	3.23	3.01	2.84	2.72	2.61	2.52	2.39	2.24	2.09	1.99	1.92	1.82	1.72	1.66	1.62	1.60
100	6.90	4.82	3.98	3.51	3.21	2.99	2.82	2.69	2.59	2.50	2.37	2.22	2.07	1.97	1.89	1.80	1.69	1.63	1.60	1.58
110	6.87	4.80	3.96	3.49	3.19	2.97	2.81	2.68	2.57	2.49	2.35	2.21	2.05	1.95	1.88	1.78	1.67	1.61	1.58	1.55
120	6.85	4.79	3.95	3.48	3.17	2.96	2.79	2.66	2.56	2.47	2.34	2.19	2.03	1.93	1.86	1.76	1.66	1.60	1.56	1.53

m

Table 7   Wilcoxon Signed-Rank Test

$$P(W^+ \leq a)$$
$$H_0 : \tilde{\mu} = \tilde{\mu}_0$$

n = number of observations $\neq \tilde{\mu}_0$.

					n				
a	3	4	5	6	7	8	9	10	11
0	0.125	0.062	0.031	0.016	0.008	0.004	0.002	0.001	0.000
1	0.250	0.125	0.062	0.031	0.016	0.008	0.004	0.002	0.001
2	0.375	0.188	0.094	0.047	0.023	0.012	0.006	0.003	0.001
3	0.625	0.312	0.156	0.078	0.039	0.020	0.010	0.005	0.002
4	0.750	0.438	0.219	0.109	0.055	0.027	0.014	0.007	0.003
5	0.875	0.562	0.312	0.156	0.078	0.039	0.020	0.010	0.005
6	1.000	0.688	0.406	0.219	0.109	0.055	0.027	0.014	0.007
7		0.812	0.500	0.281	0.148	0.074	0.037	0.019	0.009
8		0.875	0.594	0.344	0.188	0.098	0.049	0.024	0.012
9		0.938	0.688	0.422	0.234	0.125	0.064	0.032	0.016
10		1.000	0.781	0.500	0.289	0.156	0.082	0.042	0.021
11			0.844	0.578	0.344	0.191	0.102	0.053	0.027
12			0.906	0.656	0.406	0.230	0.125	0.065	0.034
13			0.938	0.719	0.469	0.273	0.150	0.080	0.042
14			0.969	0.781	0.531	0.320	0.180	0.097	0.051
15			1.000	0.844	0.594	0.371	0.213	0.116	0.062
16				0.891	0.656	0.422	0.248	0.138	0.074
17				0.922	0.711	0.473	0.285	0.161	0.087
18				0.953	0.766	0.527	0.326	0.188	0.103
19				0.969	0.812	0.578	0.367	0.216	0.120
20				0.984	0.852	0.629	0.410	0.246	0.139
21				1.000	0.891	0.680	0.455	0.278	0.160
22					0.922	0.727	0.500	0.312	0.183
23					0.945	0.770	0.545	0.348	0.207
24					0.961	0.809	0.590	0.385	0.232
25					0.977	0.844	0.633	0.423	0.260
26					0.984	0.875	0.674	0.461	0.289
27					0.992	0.902	0.715	0.500	0.319
28					1.000	0.926	0.752	0.539	0.350
29						0.945	0.787	0.577	0.382
30						0.961	0.820	0.615	0.416
31						0.973	0.850	0.652	0.449
32						0.980	0.875	0.688	0.483
33						0.988	0.898	0.722	0.517
34						0.992	0.918	0.754	0.551
35						0.996	0.936	0.784	0.584
36						1.000	0.951	0.812	0.618
37							0.963	0.839	0.650
38							0.973	0.862	0.681
39							0.980	0.884	0.711

Table 7   Wilcoxon Signed-Rank Test (continued)

$$P(W^+ \leq a)$$
$$H_0 : \tilde{\mu} = \tilde{\mu}_0$$
$$n = \text{number of observations} \neq \tilde{\mu}_0.$$

						n				
a	3	4	5	6	7	8	9	10	11
40							0.986	0.903	0.740
41							0.990	0.920	0.768
42							0.994	0.935	0.793
43							0.996	0.947	0.817
44							0.998	0.958	0.840
45							1.000	0.968	0.861
46								0.976	0.880
47								0.981	0.897
48								0.986	0.913
49								0.990	0.926
50								0.993	0.938
51								0.995	0.949
52								0.997	0.958
53								0.998	0.966
54								0.999	0.973
55								1.000	0.979
56									0.984
57									0.988
58									0.991
59									0.993
60									0.995
61									0.997
62									0.998
63									0.999
64									0.999
65									1.000

					n				
a	12	13	14	15	16	17	18	19	20
0	0.000	0.000	0.000	0.000	0.000	0.000	0.000	0.000	0.000
1	0.000	0.000	0.000	0.000	0.000	0.000	0.000	0.000	0.000
2	0.001	0.000	0.000	0.000	0.000	0.000	0.000	0.000	0.000
3	0.001	0.001	0.000	0.000	0.000	0.000	0.000	0.000	0.000
4	0.002	0.001	0.000	0.000	0.000	0.000	0.000	0.000	0.000
5	0.002	0.001	0.001	0.000	0.000	0.000	0.000	0.000	0.000
6	0.003	0.002	0.001	0.000	0.000	0.000	0.000	0.000	0.000
7	0.005	0.002	0.001	0.001	0.000	0.000	0.000	0.000	0.000
8	0.006	0.003	0.002	0.001	0.000	0.000	0.000	0.000	0.000
9	0.008	0.004	0.002	0.001	0.001	0.000	0.000	0.000	0.000

Table 7   Wilcoxon Signed-Rank Test (continued)

$$P(W^+ \leq a)$$
$$H_0 : \bar{\mu} = \bar{\mu}_0$$

$$n = \text{number of observations} \neq \bar{\mu}_0.$$

a	12	13	14	15	16	17	18	19	20
10	0.010	0.005	0.003	0.001	0.001	0.000	0.000	0.000	0.000
11	0.013	0.007	0.003	0.002	0.001	0.000	0.000	0.000	0.000
12	0.017	0.009	0.004	0.002	0.001	0.001	0.000	0.000	0.000
13	0.021	0.011	0.005	0.003	0.001	0.001	0.000	0.000	0.000
14	0.026	0.013	0.007	0.003	0.002	0.001	0.000	0.000	0.000
15	0.032	0.016	0.008	0.004	0.002	0.001	0.001	0.000	0.000
16	0.039	0.020	0.010	0.005	0.003	0.001	0.001	0.000	0.000
17	0.046	0.024	0.012	0.006	0.003	0.002	0.001	0.000	0.000
18	0.055	0.029	0.015	0.008	0.004	0.002	0.001	0.000	0.000
19	0.065	0.034	0.018	0.009	0.005	0.002	0.001	0.001	0.000
20	0.076	0.040	0.021	0.011	0.005	0.003	0.001	0.001	0.000
21	0.088	0.047	0.025	0.013	0.007	0.003	0.002	0.001	0.000
22	0.102	0.055	0.029	0.015	0.008	0.004	0.002	0.001	0.001
23	0.117	0.064	0.034	0.018	0.009	0.005	0.002	0.001	0.001
24	0.133	0.073	0.039	0.021	0.011	0.005	0.003	0.001	0.001
25	0.151	0.084	0.045	0.024	0.012	0.006	0.003	0.002	0.001
26	0.170	0.095	0.052	0.028	0.014	0.007	0.004	0.002	0.001
27	0.190	0.108	0.059	0.032	0.017	0.009	0.004	0.002	0.001
28	0.212	0.122	0.068	0.036	0.019	0.010	0.005	0.003	0.001
29	0.235	0.137	0.077	0.042	0.022	0.012	0.006	0.003	0.002
30	0.259	0.153	0.086	0.047	0.025	0.013	0.007	0.004	0.002
31	0.285	0.170	0.097	0.053	0.029	0.015	0.008	0.004	0.002
32	0.311	0.188	0.108	0.060	0.033	0.017	0.009	0.005	0.002
33	0.339	0.207	0.121	0.068	0.037	0.020	0.010	0.005	0.003
34	0.367	0.227	0.134	0.076	0.042	0.022	0.012	0.006	0.003
35	0.396	0.249	0.148	0.084	0.047	0.025	0.013	0.007	0.004
36	0.425	0.271	0.163	0.094	0.052	0.028	0.015	0.008	0.004
37	0.455	0.294	0.179	0.104	0.058	0.032	0.017	0.009	0.005
38	0.485	0.318	0.195	0.115	0.065	0.036	0.019	0.010	0.005
39	0.515	0.342	0.213	0.126	0.072	0.040	0.022	0.011	0.006
40	0.545	0.368	0.232	0.138	0.080	0.044	0.024	0.013	0.007
41	0.575	0.393	0.251	0.151	0.088	0.049	0.027	0.014	0.008
42	0.604	0.420	0.271	0.165	0.096	0.054	0.030	0.016	0.009
43	0.633	0.446	0.292	0.180	0.106	0.060	0.033	0.018	0.01
44	0.661	0.473	0.313	0.195	0.116	0.066	0.037	0.020	0.011
45	0.689	0.500	0.335	0.211	0.126	0.073	0.041	0.022	0.012
46	0.715	0.527	0.357	0.227	0.137	0.080	0.045	0.025	0.013
47	0.741	0.554	0.380	0.244	0.149	0.087	0.049	0.027	0.015
48	0.765	0.580	0.404	0.262	0.161	0.095	0.054	0.030	0.016
49	0.788	0.607	0.428	0.281	0.174	0.103	0.059	0.033	0.018

Table 7   Wilcoxon Signed-Rank Test (continued)

$$P(W^+ \leq a)$$
$$H_0 : \tilde{\mu} = \tilde{\mu}_0$$

n = number of observations $\neq \tilde{\mu}_0$.

					n				
a	12	13	14	15	16	17	18	19	20
50	0.810	0.632	0.452	0.300	0.188	0.112	0.065	0.036	0.020
51	0.830	0.658	0.476	0.319	0.202	0.122	0.071	0.040	0.022
52	0.849	0.682	0.500	0.339	0.217	0.132	0.077	0.044	0.024
53	0.867	0.706	0.524	0.360	0.232	0.142	0.084	0.048	0.027
54	0.883	0.729	0.548	0.381	0.248	0.153	0.091	0.052	0.029
55	0.898	0.751	0.572	0.402	0.264	0.164	0.098	0.057	0.032
56	0.912	0.773	0.596	0.423	0.281	0.176	0.106	0.062	0.035
57	0.924	0.793	0.620	0.445	0.298	0.189	0.114	0.067	0.038
58	0.935	0.812	0.643	0.467	0.316	0.202	0.123	0.072	0.041
59	0.945	0.830	0.665	0.489	0.334	0.215	0.132	0.078	0.045
60	0.954	0.847	0.687	0.511	0.353	0.229	0.142	0.084	0.049
61	0.961	0.863	0.708	0.533	0.372	0.244	0.152	0.091	0.053
62	0.968	0.878	0.729	0.555	0.391	0.259	0.162	0.098	0.057
63	0.974	0.892	0.749	0.577	0.410	0.274	0.173	0.105	0.062
64	0.979	0.905	0.768	0.598	0.430	0.290	0.185	0.113	0.066
65	0.983	0.916	0.787	0.619	0.450	0.306	0.196	0.121	0.071
66	0.987	0.927	0.805	0.640	0.470	0.322	0.209	0.129	0.077
67	0.990	0.936	0.821	0.661	0.490	0.339	0.221	0.138	0.082
68	0.992	0.945	0.837	0.681	0.510	0.356	0.234	0.147	0.088
69	0.994	0.953	0.852	0.700	0.530	0.373	0.248	0.156	0.095
70	0.995	0.960	0.866	0.719	0.550	0.391	0.261	0.166	0.101
71	0.997	0.966	0.879	0.738	0.570	0.409	0.275	0.176	0.108
72	0.998	0.971	0.892	0.756	0.590	0.427	0.290	0.187	0.115
73	0.998	0.976	0.903	0.773	0.609	0.445	0.305	0.198	0.123
74	0.999	0.980	0.914	0.789	0.628	0.463	0.320	0.209	0.131
75	0.999	0.984	0.923	0.805	0.647	0.482	0.335	0.221	0.139
76	1.000	0.987	0.932	0.820	0.666	0.500	0.351	0.233	0.147
77		0.989	0.941	0.835	0.684	0.518	0.367	0.245	0.156
78		0.991	0.948	0.849	0.702	0.537	0.383	0.258	0.165
79		0.993	0.955	0.862	0.719	0.555	0.399	0.271	0.174
80		0.995	0.961	0.874	0.736	0.573	0.416	0.284	0.184
81		0.996	0.966	0.885	0.752	0.591	0.433	0.297	0.194
82		0.997	0.971	0.896	0.768	0.609	0.449	0.311	0.205
83		0.998	0.975	0.906	0.783	0.627	0.466	0.325	0.215
84		0.998	0.979	0.916	0.798	0.644	0.483	0.340	0.226
85		0.999	0.982	0.924	0.812	0.661	0.500	0.354	0.237
86		0.999	0.985	0.932	0.826	0.678	0.517	0.369	0.249
87		0.999	0.988	0.940	0.839	0.694	0.534	0.384	0.261
88		1.000	0.990	0.947	0.851	0.710	0.551	0.399	0.273
89			0.992	0.953	0.863	0.726	0.567	0.414	0.285

Table 7   Wilcoxon Signed-Rank Test (continued)

$$P(W^+ \leq a)$$
$$H_0 : \tilde{\mu} = \tilde{\mu}_0$$

n = number of observations $\neq \tilde{\mu}_0$.

a	12	13	14	15	16	17	18	19	20
90			0.993	0.958	0.874	0.741	0.584	0.430	0.298
91			0.995	0.964	0.884	0.756	0.601	0.445	0.311
92			0.996	0.968	0.894	0.771	0.617	0.461	0.324
93			0.997	0.972	0.904	0.785	0.633	0.476	0.337
94			0.997	0.976	0.912	0.798	0.649	0.492	0.351
95			0.998	0.979	0.920	0.811	0.665	0.508	0.364
96			0.998	0.982	0.928	0.824	0.680	0.524	0.378
97			0.999	0.985	0.935	0.836	0.695	0.539	0.392
98			0.999	0.987	0.942	0.847	0.710	0.555	0.406
99			0.999	0.989	0.948	0.858	0.725	0.570	0.420
100			1.000	0.991	0.953	0.868	0.739	0.586	0.435
101				0.992	0.958	0.878	0.752	0.601	0.449
102				0.994	0.963	0.888	0.766	0.616	0.464
103				0.995	0.967	0.897	0.779	0.631	0.478
104				0.996	0.971	0.905	0.791	0.646	0.493
105				0.997	0.975	0.913	0.804	0.660	0.507
106				0.997	0.978	0.920	0.815	0.675	0.522
107				0.998	0.981	0.927	0.827	0.689	0.536
108				0.998	0.983	0.934	0.838	0.703	0.551
109				0.999	0.986	0.940	0.848	0.716	0.565
110				0.999	0.988	0.946	0.858	0.729	0.580
111				0.999	0.989	0.951	0.868	0.742	0.594
112				0.999	0.991	0.956	0.877	0.755	0.608
113				1.000	0.992	0.960	0.886	0.767	0.622
114					0.993	0.964	0.894	0.779	0.636
115					0.995	0.968	0.902	0.791	0.649
116					0.995	0.972	0.909	0.802	0.663
117					0.996	0.975	0.916	0.813	0.676
118					0.997	0.978	0.923	0.824	0.689
119					0.997	0.980	0.929	0.834	0.702
120					0.998	0.983	0.935	0.844	0.715
121					0.998	0.985	0.941	0.853	0.727
122					0.999	0.987	0.946	0.862	0.739
123					0.999	0.988	0.951	0.871	0.751
124					0.999	0.990	0.955	0.879	0.763
125					0.999	0.991	0.959	0.887	0.774
126					0.999	0.993	0.963	0.895	0.785
127					1.000	0.994	0.967	0.902	0.795
128						0.995	0.970	0.909	0.806
129						0.995	0.973	0.916	0.816

Table 7   Wilcoxon Signed-Rank Test (continued)

$$P(W^+ \leq a)$$
$$H_0 : \tilde{\mu} = \tilde{\mu}_0$$

n = **number of observations** $\neq \tilde{\mu}_0$.

a	12	13	14	15	16	17	18	19	20
130						0.996	0.976	0.922	0.826
131						0.997	0.978	0.928	0.835
132						0.997	0.981	0.933	0.844
133						0.998	0.983	0.938	0.853
134						0.998	0.985	0.943	0.861
135						0.998	0.987	0.948	0.869
136						0.999	0.988	0.952	0.877
137						0.999	0.990	0.956	0.885
138						0.999	0.991	0.960	0.892
139						0.999	0.992	0.964	0.899
140						0.999	0.993	0.967	0.905
141						1.000	0.994	0.970	0.912
142							0.995	0.973	0.918
143							0.996	0.975	0.923
144							0.996	0.978	0.929
145							0.997	0.980	0.934
146							0.997	0.982	0.938
147							0.998	0.984	0.943
148							0.998	0.986	0.947
149							0.998	0.987	0.951
150							0.999	0.989	0.955
151							0.999	0.990	0.959
152							0.999	0.991	0.962
153							0.999	0.992	0.965
154							0.999	0.993	0.968
155							0.999	0.994	0.971
156							1.000	0.995	0.973
157								0.995	0.976
158								0.996	0.978
159								0.996	0.980
160								0.997	0.982
161								0.997	0.984
162								0.998	0.985
163								0.998	0.987
164								0.998	0.988
165								0.999	0.989
166								0.999	0.990
167								0.999	0.991
168								0.999	0.992
169								0.999	0.993

Table 7   Wilcoxon Signed-Rank Test (continued)

$$P(W^+ \leq a)$$
$$H_0 : \tilde{\mu} = \tilde{\mu}_0$$

n = **number of observations** $\neq \tilde{\mu}_0$.

						n			
a	12	13	14	15	16	17	18	19	20
170								0.999	0.994
171								1.000	0.995
172									0.995
173									0.996
174									0.996
175									0.997
176									0.997
177									0.998
178									0.998
179									0.998
180									0.998
181									0.999
182									0.999
183									0.999
184									0.999
185									0.999
186									0.999
187									0.999
188									1.000

Table 8   Wilcoxon Rank Sum Test

$$P(W \leq a)$$
$$H_0 : \tilde{\mu}_1 = \tilde{\mu}_2$$
$$n_1 = n, \quad n_2 = m, \quad \text{where } n \leq m$$
$$W = \sum_{i=1}^{m} R_{i2} - \frac{1}{2}m(m+1)$$

n = 3:

a	m							
	3	4	5	6	7	8	9	10
0	0.0500	0.0286	0.0179	0.0119	0.0083	0.0061	0.0045	0.0035
1	0.1000	0.0571	0.0357	0.0238	0.0167	0.0121	0.0091	0.0070
2	0.2000	0.1143	0.0714	0.0476	0.0333	0.0242	0.0182	0.0140
3	0.3500	0.2000	0.1250	0.0833	0.0583	0.0424	0.0318	0.0245
4	0.5000	0.3143	0.1964	0.1310	0.0917	0.0667	0.0500	0.0385
5	0.6500	0.4286	0.2857	0.1905	0.1333	0.0970	0.0727	0.0559
6	0.8000	0.5714	0.3929	0.2738	0.1917	0.1394	0.1045	0.0804
7	0.9000	0.6857	0.5000	0.3571	0.2583	0.1879	0.1409	0.1084
8	0.9500	0.8000	0.6071	0.4524	0.3333	0.2485	0.1864	0.1434
9	1.0000	0.8857	0.7143	0.5476	0.4167	0.3152	0.2409	0.1853
10		0.9429	0.8036	0.6429	0.5000	0.3879	0.3000	0.2343
11		0.9714	0.8750	0.7262	0.5833	0.4606	0.3636	0.2867
12		1.0000	0.9286	0.8095	0.6667	0.5394	0.4318	0.3462
13			0.9643	0.8690	0.7417	0.6121	0.5000	0.4056
14			0.9821	0.9167	0.8083	0.6848	0.5682	0.4685
15			1.0000	0.9524	0.8667	0.7515	0.6364	0.5315
16				0.9762	0.9083	0.8121	0.7000	0.5944
17				0.9881	0.9417	0.8606	0.7591	0.6538
18				1.0000	0.9667	0.9030	0.8136	0.7133
19					0.9833	0.9333	0.8591	0.7657
20					0.9917	0.9576	0.8955	0.8147
21					1.0000	0.9758	0.9273	0.8566
22						0.9879	0.9500	0.8916
23						0.9939	0.9682	0.9196
24						1.0000	0.9818	0.9441
25							0.9909	0.9615
26							0.9955	0.9755
27							1.0000	0.9860
28								0.9930
29								0.9965
30								1.0000

Table 8   Wilcoxon Rank Sum Test (continued)

$$P(W \leq a)$$
$$H_0 : \tilde{\mu}_1 = \tilde{\mu}_2$$
$$n_1 = n, \quad n_2 = m, \quad \text{where } n \leq m$$
$$W = \sum_{i=1}^{m} R_{i2} - \frac{1}{2}m(m + 1)$$

n = 4:

a	m						
	4	5	6	7	8	9	10
0	0.0143	0.0079	0.0048	0.0030	0.0020	0.0014	0.0010
1	0.0286	0.0159	0.0095	0.0061	0.0040	0.0028	0.0020
2	0.0571	0.0317	0.0190	0.0121	0.0081	0.0056	0.0040
3	0.1000	0.0556	0.0333	0.0212	0.0141	0.0098	0.0070
4	0.1714	0.0952	0.0571	0.0364	0.0242	0.0168	0.0120
5	0.2429	0.1429	0.0857	0.0545	0.0364	0.0252	0.0180
6	0.3429	0.2063	0.1286	0.0818	0.0545	0.0378	0.0270
7	0.4429	0.2778	0.1762	0.1152	0.0768	0.0531	0.0380
8	0.5571	0.3651	0.2381	0.1576	0.1071	0.0741	0.0529
9	0.6571	0.4524	0.3048	0.2061	0.1414	0.0993	0.0709
10	0.7571	0.5476	0.3810	0.2636	0.1838	0.1301	0.0939
11	0.8286	0.6349	0.4571	0.3242	0.2303	0.1650	0.1199
12	0.9000	0.7222	0.5429	0.3939	0.2848	0.2070	0.1518
13	0.9429	0.7937	0.6190	0.4636	0.3414	0.2517	0.1868
14	0.9714	0.8571	0.6952	0.5364	0.4040	0.3021	0.2268
15	0.9857	0.9048	0.7619	0.6061	0.4667	0.3552	0.2697
16	1.0000	0.9444	0.8238	0.6758	0.5333	0.4126	0.3177
17		0.9683	0.8714	0.7364	0.5960	0.4699	0.3666
18		0.9841	0.9143	0.7939	0.6586	0.5301	0.4196
19		0.9921	0.9429	0.8424	0.7152	0.5874	0.4725
20		1.0000	0.9667	0.8848	0.7697	0.6448	0.5275
21			0.9810	0.9182	0.8162	0.6979	0.5804
22			0.9905	0.9455	0.8586	0.7483	0.6334
23			0.9952	0.9636	0.8929	0.7930	0.6823
24			1.0000	0.9788	0.9232	0.8350	0.7303
25				0.9879	0.9455	0.8699	0.7732
26				0.9939	0.9636	0.9007	0.8132
27				0.9970	0.9758	0.9259	0.8482
28				1.0000	0.9859	0.9469	0.8801
29					0.9919	0.9622	0.9061
30					0.9960	0.9748	0.9291
31					0.9980	0.9832	0.9471
32					1.0000	0.9902	0.9620
33						0.9944	0.9730
34						0.9972	0.9820

Table 8   Wilcoxon Rank Sum Test (continued)

$$P(W \le a)$$
$$H_0 : \tilde{\mu}_1 = \tilde{\mu}_2$$
$$n_1 = n, \quad n_2 = m, \quad \text{where } n \le m$$
$$W = \sum_{i=1}^{m} R_{i2} - \frac{1}{2}m(m + 1)$$

| | n = 4: | | | | | | |
| | | | | m | | | |
a	4	5	6	7	8	9	10
35						0.9986	0.9880
36						1.0000	0.9930
37							0.9960
38							0.9980
39							0.9990
40							1.0000

| | n = 5: | | | | | | | | n = 5: | | | | | |
| | | | m | | | | | | | | m | | | |
a	5	6	7	8	9	10		a	5	6	7	8	9	10
0	0.0040	0.0022	0.0013	0.0008	0.0005	0.0003		26		0.9848	0.9255	0.8228	0.6968	0.5704
1	0.0079	0.0043	0.0025	0.0016	0.0010	0.0007		27		0.9913	0.9470	0.8578	0.7408	0.6161
2	0.0159	0.0087	0.0051	0.0031	0.0020	0.0013		28		0.9957	0.9634	0.8889	0.7812	0.6607
3	0.0278	0.0152	0.0088	0.0054	0.0035	0.0023		29		0.9978	0.9760	0.9145	0.8182	0.7030
4	0.0476	0.0260	0.0152	0.0093	0.0060	0.0040		30		1.0000	0.9848	0.9363	0.8511	0.7433
5	0.0754	0.0411	0.0240	0.0148	0.0095	0.0063		31			0.9912	0.9534	0.8801	0.7802
6	0.1111	0.0628	0.0366	0.0225	0.0145	0.0097		32			0.9949	0.9674	0.9051	0.8145
7	0.1548	0.0887	0.0530	0.0326	0.0210	0.0140		33			0.9975	0.9775	0.9266	0.8452
8	0.2103	0.1234	0.0745	0.0466	0.0300	0.0200		34			0.9987	0.9852	0.9441	0.8728
9	0.2738	0.1645	0.1010	0.0637	0.0415	0.0276		35			1.0000	0.9907	0.9585	0.8968
10	0.3452	0.2143	0.1338	0.0855	0.0559	0.0376		36				0.9946	0.9700	0.9177
11	0.4206	0.2684	0.1717	0.1111	0.0734	0.0496		37				0.9969	0.9790	0.9354
12	0.5000	0.3312	0.2159	0.1422	0.0949	0.0646		38				0.9984	0.9855	0.9504
13	0.5794	0.3961	0.2652	0.1772	0.1199	0.0823		39				0.9992	0.9905	0.9624
14	0.6548	0.4654	0.3194	0.2176	0.1489	0.1032		40				1.0000	0.9940	0.9724
15	0.7262	0.5346	0.3775	0.2618	0.1818	0.1272		41					0.9965	0.9800
16	0.7897	0.6039	0.4381	0.3108	0.2188	0.1548		42					0.9980	0.9860
17	0.8452	0.6688	0.5000	0.3621	0.2592	0.1855		43					0.9990	0.9903
18	0.8889	0.7316	0.5619	0.4165	0.3032	0.2198		44					0.9995	0.9937
19	0.9246	0.7857	0.6225	0.4716	0.3497	0.2567		45					1.0000	0.9960
20	0.9524	0.8355	0.6806	0.5284	0.3986	0.2970		46						0.9977
21	0.9722	0.8766	0.7348	0.5835	0.4491	0.3393		47						0.9987
22	0.9841	0.9113	0.7841	0.6379	0.5000	0.3839		48						0.9993
23	0.9921	0.9372	0.8283	0.6892	0.5509	0.4296		49						0.9997
24	0.9960	0.9589	0.8662	0.7382	0.6014	0.4765		50						1.0000
25	1.0000	0.9740	0.8990	0.7824	0.6503	0.5235								

Table 8   Wilcoxon Rank Sum Test (continued)

$$P(W \leq a)$$
$$H_0 : \tilde{\mu}_1 = \tilde{\mu}_2$$
$$n_1 = n, \quad n_2 = m, \quad \text{where } n \leq m$$
$$W = \sum_{i=1}^{m} R_{i2} - \frac{1}{2}m(m+1)$$

	n = 6:						n = 6:				
	m						m				
a	6	7	8	9	10	a	6	7	8	9	10
0	0.0011	0.0006	0.0003	0.0002	0.0001	31	0.9870	0.9312	0.8275	0.6965	0.5626
1	0.0022	0.0012	0.0007	0.0004	0.0002	32	0.9924	0.9493	0.8588	0.7357	0.6038
2	0.0043	0.0023	0.0013	0.0008	0.0005	33	0.9957	0.9633	0.8858	0.7720	0.6436
3	0.0076	0.0041	0.0023	0.0014	0.0009	34	0.9978	0.9744	0.9094	0.8058	0.6823
4	0.0130	0.0070	0.0040	0.0024	0.0015	35	0.9989	0.9825	0.9291	0.8362	0.7189
5	0.0206	0.0111	0.0063	0.0038	0.0024	36	1.0000	0.9889	0.9461	0.8639	0.7539
6	0.0325	0.0175	0.0100	0.0060	0.0037	37		0.9930	0.9594	0.8881	0.7861
7	0.0465	0.0256	0.0147	0.0088	0.0055	38		0.9959	0.9704	0.9095	0.8162
8	0.0660	0.0367	0.0213	0.0128	0.0080	39		0.9977	0.9787	0.9277	0.8434
9	0.0898	0.0507	0.0296	0.0180	0.0112	40		0.9988	0.9853	0.9433	0.8683
10	0.1201	0.0688	0.0406	0.0248	0.0156	41		0.9994	0.9900	0.9560	0.8901
11	0.1548	0.0903	0.0539	0.0332	0.0210	42		1.0000	0.9937	0.9668	0.9097
12	0.1970	0.1171	0.0709	0.0440	0.0280	43			0.9960	0.9752	0.9264
13	0.2424	0.1474	0.0906	0.0567	0.0363	44			0.9977	0.9820	0.9411
14	0.2944	0.1830	0.1142	0.0723	0.0467	45			0.9987	0.9872	0.9533
15	0.3496	0.2226	0.1412	0.0905	0.0589	46			0.9993	0.9912	0.9637
16	0.4091	0.2669	0.1725	0.1119	0.0736	47			0.9997	0.9940	0.9720
17	0.4686	0.3141	0.2068	0.1361	0.0903	48			1.0000	0.9962	0.9790
18	0.5314	0.3654	0.2454	0.1638	0.1099	49				0.9976	0.9844
19	0.5909	0.4178	0.2864	0.1942	0.1317	50				0.9986	0.9888
20	0.6504	0.4726	0.3310	0.2280	0.1566	51				0.9992	0.9920
21	0.7056	0.5274	0.3773	0.2643	0.1838	52				0.9996	0.9945
22	0.7576	0.5822	0.4259	0.3035	0.2139	53				0.9998	0.9963
23	0.8030	0.6346	0.4749	0.3445	0.2461	54				1.0000	0.9976
24	0.8452	0.6859	0.5251	0.3878	0.2811	55					0.9985
25	0.8799	0.7331	0.5741	0.4320	0.3177	56					0.9991
26	0.9102	0.7774	0.6227	0.4773	0.3564	57					0.9995
27	0.9340	0.8170	0.6690	0.5227	0.3962	58					0.9998
28	0.9535	0.8526	0.7136	0.5680	0.4374	59					0.9999
29	0.9675	0.8829	0.7546	0.6122	0.4789	60					1.0000
30	0.9794	0.9097	0.7932	0.6555	0.5211						

Table 8   Wilcoxon Rank Sum Test (continued)

$$P(W \leq a)$$
$$H_0 : \tilde{\mu}_1 = \tilde{\mu}_2$$
$$n_1 = n, \quad n_2 = m, \quad \text{where } n \leq m$$
$$W = \sum_{i=1}^{m} R_{i\,2} - \frac{1}{2}m(m + 1)$$

	n = 7:					n = 7:			
	m					m			
a	7	8	9	10	a	7	8	9	10
0	0.0003	0.0002	0.0001	0.0001	36	0.9359	0.8322	0.6968	0.5566
1	0.0006	0.0003	0.0002	0.0001	37	0.9513	0.8595	0.7320	0.5937
2	0.0012	0.0006	0.0003	0.0002	38	0.9636	0.8841	0.7651	0.6302
3	0.0020	0.0011	0.0006	0.0004	39	0.9735	0.9054	0.7961	0.6655
4	0.0035	0.0019	0.0010	0.0006	40	0.9811	0.9240	0.8245	0.6996
5	0.0055	0.0030	0.0017	0.0010	41	0.9869	0.9397	0.8504	0.7319
6	0.0087	0.0047	0.0026	0.0015	42	0.9913	0.9531	0.8739	0.7626
7	0.0131	0.0070	0.0039	0.0023	43	0.9945	0.9639	0.8948	0.7913
8	0.0189	0.0103	0.0058	0.0034	44	0.9965	0.9730	0.9131	0.8181
9	0.0265	0.0145	0.0082	0.0048	45	0.9980	0.9800	0.9292	0.8426
10	0.0364	0.0200	0.0115	0.0068	46	0.9988	0.9855	0.9429	0.8651
11	0.0487	0.0270	0.0156	0.0093	47	0.9994	0.9897	0.9546	0.8852
12	0.0641	0.0361	0.0209	0.0125	48	0.9997	0.9930	0.9644	0.9034
13	0.0825	0.0469	0.0274	0.0165	49	1.0000	0.9953	0.9726	0.9194
14	0.1043	0.0603	0.0356	0.0215	50		0.9970	0.9791	0.9335
15	0.1297	0.0760	0.0454	0.0277	51		0.9981	0.9844	0.9456
16	0.1588	0.0946	0.0571	0.0351	52		0.9989	0.9885	0.9561
17	0.1914	0.1159	0.0708	0.0439	53		0.9994	0.9918	0.9649
18	0.2279	0.1405	0.0869	0.0544	54		0.9997	0.9942	0.9723
19	0.2675	0.1678	0.1052	0.0665	55		0.9998	0.9961	0.9785
20	0.3100	0.1984	0.1261	0.0806	56		1.0000	0.9974	0.9835
21	0.3552	0.2317	0.1496	0.0966	57			0.9983	0.9875
22	0.4024	0.2679	0.1755	0.1148	58			0.9990	0.9907
23	0.4508	0.3063	0.2039	0.1349	59			0.9994	0.9932
24	0.5000	0.3472	0.2349	0.1574	60			0.9997	0.9952
25	0.5492	0.3894	0.2680	0.1819	61			0.9998	0.9966
26	0.5976	0.4333	0.3032	0.2087	62			0.9999	0.9977
27	0.6448	0.4775	0.3403	0.2374	63			1.0000	0.9985
28	0.6900	0.5225	0.3788	0.2681	64				0.9990
29	0.7325	0.5667	0.4185	0.3004	65				0.9994
30	0.7721	0.6106	0.4591	0.3345	66				0.9996
31	0.8086	0.6528	0.5000	0.3698	67				0.9998
32	0.8412	0.6937	0.5409	0.4063	68				0.9999
33	0.8703	0.7321	0.5815	0.4434	69				0.9999
34	0.8957	0.7683	0.6212	0.4811	70				1.0000
35	0.9175	0.8016	0.6597	0.5189					

Table 8   Wilcoxon Rank Sum Test (continued)

$$P(W \le a)$$
$$H_0 : \tilde{\mu}_1 = \tilde{\mu}_2$$
$$n_1 = n, \quad n_2 = m, \quad \text{where } n \le m$$
$$W = \sum_{i=1}^{m} R_{i2} - \frac{1}{2}m(m + 1)$$

	n = 8:				n = 8:				n = 8:		
	m				m				m		
a	8	9	10	a	8	9	10	a	8	9	10
0	0.0001	0.0000	0.0000	28	0.3605	0.2404	0.1577	56	0.9965	0.9768	0.9271
1	0.0002	0.0001	0.0000	29	0.3992	0.2707	0.1800	57	0.9977	0.9820	0.9390
2	0.0003	0.0002	0.0001	30	0.4392	0.3029	0.2041	58	0.9985	0.9863	0.9494
3	0.0005	0.0003	0.0002	31	0.4796	0.3365	0.2299	59	0.9991	0.9897	0.9584
4	0.0009	0.0005	0.0003	32	0.5204	0.3715	0.2574	60	0.9995	0.9924	0.9662
5	0.0015	0.0008	0.0004	33	0.5608	0.4074	0.2863	61	0.9997	0.9944	0.9727
6	0.0023	0.0012	0.0007	34	0.6008	0.4442	0.3167	62	0.9998	0.9961	0.9783
7	0.0035	0.0019	0.0010	35	0.6395	0.4813	0.3482	63	0.9999	0.9972	0.9829
8	0.0052	0.0028	0.0015	36	0.6773	0.5187	0.3809	64	1.0000	0.9981	0.9867
9	0.0074	0.0039	0.0022	37	0.7131	0.5558	0.4143	65		0.9988	0.9897
10	0.0103	0.0056	0.0031	38	0.7473	0.5926	0.4484	66		0.9992	0.9922
11	0.0141	0.0076	0.0043	39	0.7791	0.6285	0.4827	67		0.9995	0.9942
12	0.0190	0.0103	0.0058	40	0.8089	0.6635	0.5173	68		0.9997	0.9957
13	0.0249	0.0137	0.0078	41	0.8359	0.6971	0.5516	69		0.9998	0.9969
14	0.0325	0.0180	0.0103	42	0.8607	0.7293	0.5857	70		0.9999	0.9978
15	0.0415	0.0232	0.0133	43	0.8828	0.7596	0.6191	71		1.0000	0.9985
16	0.0524	0.0296	0.0171	44	0.9026	0.7883	0.6518	72		1.0000	0.9990
17	0.0652	0.0372	0.0217	45	0.9197	0.8148	0.6833	73			0.9993
18	0.0803	0.0464	0.0273	46	0.9348	0.8394	0.7137	74			0.9996
19	0.0974	0.0570	0.0338	47	0.9476	0.8617	0.7426	75			0.9997
20	0.1172	0.0694	0.0416	48	0.9585	0.8821	0.7701	76			0.9998
21	0.1393	0.0836	0.0506	49	0.9675	0.9002	0.7959	77			0.9999
22	0.1641	0.0998	0.0610	50	0.9751	0.9164	0.8200	78			1.0000
23	0.1911	0.1179	0.0729	51	0.9810	0.9306	0.8423	79			1.0000
24	0.2209	0.1383	0.0864	52	0.9859	0.9430	0.8629	80			1.0000
25	0.2527	0.1606	0.1015	53	0.9897	0.9536	0.8815				
26	0.2869	0.1852	0.1185	54	0.9926	0.9628	0.8985				
27	0.3227	0.2117	0.1371	55	0.9948	0.9704	0.9136				

Table 8   Wilcoxon Rank Sum Test (continued)

$$P(W \leq a)$$
$$H_0 : \tilde{\mu}_1 = \tilde{\mu}_2$$
$$n_1 = n, \quad n_2 = m, \quad \text{where } n \leq m$$
$$W = \sum_{i=1}^{m} R_{i2} - \frac{1}{2}m(m+1)$$

| | n = 9: | | | | n = 9: | | | | n = 9: | |
| | m | | | | m | | | | m | |
a	9	10	a	9	10	a	9	10
0	0.0000	0.0000	31	0.2181	0.1388	62	0.9748	0.9218
1	0.0000	0.0000	32	0.2447	0.1577	63	0.9800	0.9333
2	0.0001	0.0000	33	0.2729	0.1781	64	0.9843	0.9436
3	0.0001	0.0001	34	0.3024	0.2001	65	0.9878	0.9526
4	0.0002	0.0001	35	0.3332	0.2235	66	0.9906	0.9606
5	0.0004	0.0002	36	0.3652	0.2483	67	0.9929	0.9674
6	0.0006	0.0003	37	0.3981	0.2745	68	0.9947	0.9733
7	0.0009	0.0005	38	0.4317	0.3019	69	0.9961	0.9783
8	0.0014	0.0007	39	0.4657	0.3304	70	0.9972	0.9825
9	0.0020	0.0011	40	0.5000	0.3598	71	0.9980	0.9860
10	0.0028	0.0015	41	0.5343	0.3901	72	0.9986	0.9890
11	0.0039	0.0021	42	0.5683	0.4211	73	0.9991	0.9914
12	0.0053	0.0028	43	0.6019	0.4524	74	0.9994	0.9934
13	0.0071	0.0038	44	0.6348	0.4841	75	0.9996	0.9949
14	0.0094	0.0051	45	0.6668	0.5159	76	0.9998	0.9962
15	0.0122	0.0066	46	0.6976	0.5476	77	0.9999	0.9972
16	0.0157	0.0086	47	0.7271	0.5789	78	0.9999	0.9979
17	0.0200	0.0110	48	0.7553	0.6099	79	1.0000	0.9985
18	0.0252	0.0140	49	0.7819	0.6402	80	1.0000	0.9989
19	0.0313	0.0175	50	0.8067	0.6696	81	1.0000	0.9993
20	0.0385	0.0217	51	0.8299	0.6981	82		0.9995
21	0.0470	0.0267	52	0.8513	0.7255	83		0.9997
22	0.0567	0.0326	53	0.8710	0.7517	84		0.9998
23	0.0680	0.0394	54	0.8888	0.7765	85		0.9999
24	0.0807	0.0474	55	0.9049	0.7999	86		0.9999
25	0.0951	0.0564	56	0.9193	0.8219	87		1.0000
26	0.1112	0.0667	57	0.9320	0.8423	88		1.0000
27	0.1290	0.0782	58	0.9433	0.8612	89		1.0000
28	0.1487	0.0912	59	0.9530	0.8786	90		1.0000
29	0.1701	0.1055	60	0.9615	0.8945			
30	0.1933	0.1214	61	0.9687	0.9088			

Table 8   Wilcoxon Rank Sum Test (continued)

$$P(W \leq a)$$
$$H_0 : \tilde{\mu}_1 = \tilde{\mu}_2$$
$$n_1 = n, \quad n_2 = m, \quad \text{where } n \leq m$$
$$W = \sum_{i=1}^{m} R_{i2} - \frac{1}{2}m(m + 1)$$

	n = 10:		n = 10:		n = 10:
	m		m		m
a	10	a	10	a	10
0	0.0000	35	0.1399	70	0.9385
1	0.0000	36	0.1575	71	0.9474
2	0.0000	37	0.1763	72	0.9554
3	0.0000	38	0.1965	73	0.9624
4	0.0001	39	0.2179	74	0.9685
5	0.0001	40	0.2406	75	0.9738
6	0.0002	41	0.2644	76	0.9784
7	0.0002	42	0.2894	77	0.9823
8	0.0004	43	0.3153	78	0.9856
9	0.0005	44	0.3421	79	0.9884
10	0.0008	45	0.3697	80	0.9907
11	0.0010	46	0.3980	81	0.9927
12	0.0014	47	0.4267	82	0.9943
13	0.0019	48	0.4559	83	0.9955
14	0.0026	49	0.4853	84	0.9966
15	0.0034	50	0.5147	85	0.9974
16	0.0045	51	0.5441	86	0.9981
17	0.0057	52	0.5733	87	0.9986
18	0.0073	53	0.6020	88	0.9990
19	0.0093	54	0.6303	89	0.9992
20	0.0116	55	0.6579	90	0.9995
21	0.0144	56	0.6847	91	0.9996
22	0.0177	57	0.7106	92	0.9998
23	0.0216	58	0.7356	93	0.9998
24	0.0262	59	0.7594	94	0.9999
25	0.0315	60	0.7821	95	0.9999
26	0.0376	61	0.8035	96	1.0000
27	0.0446	62	0.8237	97	1.0000
28	0.0526	63	0.8425	98	1.0000
29	0.0615	64	0.8601	99	1.0000
30	0.0716	65	0.8763	100	1.0000
31	0.0827	66	0.8912		
32	0.0952	67	0.9048		
33	0.1088	68	0.9173		
34	0.1237	69	0.9284		

Table 9   Spearman's Rho

$$P = P(r_s \le r)$$
$$H_0 : \rho_s = 0$$
$$r_s = 1 - \frac{6\sum_{i=1}^{n} d_i^2}{n(n^2 - 1)}$$

n = 2

r	P
-1.000	0.500
1.000	1.000

n = 3

r	P
-1.000	0.167
-0.500	0.500
0.500	0.833
1.000	1.000

n = 4

r	P
-1.000	0.042
-0.800	0.167
-0.600	0.208
-0.400	0.375
-0.200	0.458
0.000	0.542
0.200	0.625
0.400	0.792
0.600	0.833
0.800	0.958
1.000	1.000

n = 5

r	P
-1.000	0.008
-0.900	0.042
-0.800	0.067
-0.700	0.117
-0.600	0.175
-0.500	0.225
-0.400	0.258
-0.300	0.342
-0.200	0.392
-0.100	0.475
0.000	0.525
0.100	0.608
0.200	0.658
0.300	0.742
0.400	0.775
0.500	0.825
0.600	0.883
0.700	0.933
0.800	0.958
0.900	0.992
1.000	1.000

n = 6

r	P
-1.000	0.001
-0.943	0.008
-0.886	0.017
-0.829	0.029
-0.771	0.051
-0.714	0.068
-0.657	0.087
-0.600	0.121
-0.543	0.149
-0.486	0.178
-0.429	0.210
-0.371	0.249
-0.314	0.282
-0.257	0.329
-0.200	0.357
-0.143	0.401
-0.086	0.460
-0.029	0.500
0.029	0.540
0.086	0.599
0.143	0.643
0.200	0.671
0.257	0.718
0.314	0.751
0.371	0.790
0.429	0.822
0.486	0.851
0.543	0.879
0.600	0.912
0.657	0.932
0.714	0.949
0.771	0.971
0.829	0.983
0.886	0.992
0.943	0.999
1.000	1.000

n = 7

r	P
-1.000	0.000
-0.964	0.001
-0.929	0.003
-0.893	0.006
-0.857	0.012
-0.821	0.017
-0.786	0.024
-0.750	0.033
-0.714	0.044
-0.679	0.055
-0.643	0.069
-0.607	0.083
-0.571	0.100
-0.536	0.118
-0.500	0.133
-0.464	0.151
-0.429	0.177
-0.393	0.198
-0.357	0.222
-0.321	0.249
-0.286	0.278
-0.250	0.297
-0.214	0.331
-0.179	0.357
-0.143	0.391
-0.107	0.420
-0.071	0.453
-0.036	0.482
0.000	0.518
0.036	0.547
0.071	0.580
0.107	0.609
0.143	0.643
0.179	0.669
0.214	0.703
0.250	0.722
0.286	0.751
0.321	0.778
0.357	0.802
0.393	0.823
0.429	0.849
0.464	0.867
0.500	0.882
0.536	0.900
0.571	0.917
0.607	0.931
0.643	0.945
0.679	0.956
0.714	0.967
0.750	0.976
0.786	0.983
0.821	0.988
0.857	0.994
0.893	0.997
0.929	0.999
0.964	1.000
1.000	1.000

n = 8

r	P
-1.000	0.000
-0.976	0.000
-0.952	0.001
-0.929	0.001
-0.905	0.002
-0.881	0.004
-0.857	0.005
-0.833	0.008
-0.810	0.011
-0.786	0.014
-0.762	0.018
-0.738	0.023
-0.714	0.029
-0.690	0.035
-0.667	0.042
-0.643	0.048
-0.619	0.057
-0.595	0.066
-0.571	0.076
-0.548	0.085
-0.524	0.098
-0.500	0.108
-0.476	0.122
-0.452	0.134
-0.429	0.150
-0.405	0.163
-0.381	0.180
-0.357	0.195
-0.333	0.214
-0.310	0.231
-0.286	0.250
-0.262	0.268
-0.238	0.291
-0.214	0.310
-0.190	0.332
-0.167	0.352
-0.143	0.376
-0.119	0.397
-0.095	0.420
-0.071	0.441
-0.048	0.467
-0.024	0.488
0.000	0.512

Table 9 Spearman's Rho (continued)

$$P = P(r_s \le r)$$
$$H_0 : \rho_s = 0$$
$$r_s = 1 - \frac{6\sum_{i=1}^{n}d_i^2}{n(n^2 - 1)}$$

r	P	r	P	r	P	r	P	r	P
0.024	0.533	**n = 9**		−0.333	0.193	0.367	0.844	**n = 10**	
0.048	0.559			−0.317	0.205	0.383	0.854		
0.071	0.580	−1.000	0.000	−0.300	0.218	0.400	0.865	−1.000	0.000
0.095	0.603	−0.983	0.000	−0.283	0.231	0.417	0.875	−0.988	0.000
0.119	0.624	−0.967	0.000	−0.267	0.247	0.433	0.885	−0.976	0.000
0.143	0.648	−0.950	0.000	−0.250	0.260	0.450	0.894	−0.964	0.000
0.167	0.668	−0.933	0.000	−0.233	0.276	0.467	0.903	−0.952	0.000
0.190	0.690	−0.917	0.001	−0.217	0.290	0.483	0.911	−0.939	0.000
0.214	0.709	−0.900	0.001	−0.200	0.307	0.500	0.919	−0.927	0.000
0.238	0.732	−0.883	0.002	−0.183	0.322	0.517	0.926	−0.915	0.000
0.262	0.750	−0.867	0.002	−0.167	0.339	0.533	0.934	−0.903	0.000
0.286	0.769	−0.850	0.003	−0.150	0.354	0.550	0.940	−0.891	0.001
0.310	0.786	−0.833	0.004	−0.133	0.372	0.567	0.946	−0.879	0.001
0.333	0.805	−0.817	0.005	−0.117	0.388	0.583	0.952	−0.867	0.001
0.357	0.820	−0.800	0.007	−0.100	0.405	0.600	0.957	−0.855	0.001
0.381	0.837	−0.783	0.009	−0.083	0.422	0.617	0.962	−0.842	0.002
0.405	0.850	−0.767	0.011	−0.067	0.440	0.633	0.967	−0.830	0.002
0.429	0.866	−0.750	0.013	−0.050	0.456	0.650	0.971	−0.818	0.003
0.452	0.878	−0.733	0.016	−0.033	0.474	0.667	0.975	−0.806	0.004
0.476	0.892	−0.717	0.018	−0.017	0.491	0.683	0.978	−0.794	0.004
0.500	0.902	−0.700	0.022	0.000	0.509	0.700	0.982	−0.782	0.005
0.524	0.915	−0.683	0.025	0.017	0.526	0.717	0.984	−0.770	0.006
0.548	0.924	−0.667	0.029	0.033	0.544	0.733	0.987	−0.758	0.007
0.571	0.934	−0.650	0.033	0.050	0.560	0.750	0.989	−0.745	0.009
0.595	0.943	−0.633	0.038	0.067	0.578	0.767	0.991	−0.733	0.010
0.619	0.952	−0.617	0.043	0.083	0.595	0.783	0.993	−0.721	0.012
0.643	0.958	−0.600	0.048	0.100	0.612	0.800	0.995	−0.709	0.013
0.667	0.965	−0.583	0.054	0.117	0.628	0.817	0.996	−0.697	0.015
0.690	0.971	−0.567	0.060	0.133	0.646	0.833	0.997	−0.685	0.017
0.714	0.977	−0.550	0.066	0.150	0.661	0.850	0.998	−0.673	0.019
0.738	0.982	−0.533	0.074	0.167	0.678	0.867	0.998	−0.661	0.022
0.762	0.986	−0.517	0.081	0.183	0.693	0.883	0.999	−0.648	0.024
0.786	0.989	−0.500	0.089	0.200	0.710	0.900	0.999	−0.636	0.027
0.810	0.992	−0.483	0.097	0.217	0.724	0.917	1.000	−0.624	0.030
0.833	0.995	−0.467	0.106	0.233	0.740	0.933	1.000	−0.612	0.033
0.857	0.996	−0.450	0.115	0.250	0.753	0.950	1.000	−0.600	0.037
0.881	0.998	−0.433	0.125	0.267	0.769	0.967	1.000	−0.588	0.040
0.905	0.999	−0.417	0.135	0.283	0.782	0.983	1.000	−0.576	0.044
0.929	0.999	−0.400	0.146	0.300	0.795	1.000	1.000	−0.564	0.048
0.952	1.000	−0.383	0.156	0.317	0.807			−0.552	0.052
0.976	1.000	−0.367	0.168	0.333	0.821			−0.539	0.057
1.000	1.000	−0.350	0.179	0.350	0.832			−0.527	0.062

Table 9   Spearman's Rho (continued)

$$P = P(r_s \leq r)$$
$$H_0 : \rho_s = 0$$

$$r_s = 1 - \frac{6\sum_{i=1}^{n} d_i^2}{n(n^2 - 1)}$$

r	P	r	P	r	P	r	P
−0.515	0.067	−0.018	0.486	0.479	0.923	0.976	1.000
−0.503	0.072	−0.006	0.500	0.491	0.928	0.988	1.000
−0.491	0.077	0.006	0.514	0.503	0.933	1.000	1.000
−0.479	0.083	0.018	0.527	0.515	0.938		
−0.467	0.089	0.030	0.541	0.527	0.943		
−0.455	0.096	0.042	0.554	0.539	0.948		
−0.442	0.102	0.055	0.567	0.552	0.952		
−0.430	0.109	0.067	0.581	0.564	0.956		
−0.418	0.116	0.079	0.594	0.576	0.960		
−0.406	0.124	0.091	0.607	0.588	0.963		
−0.394	0.132	0.103	0.621	0.600	0.967		
−0.382	0.139	0.115	0.633	0.612	0.970		
−0.370	0.148	0.127	0.646	0.624	0.973		
−0.358	0.156	0.139	0.659	0.636	0.976		
−0.345	0.165	0.152	0.672	0.648	0.978		
−0.333	0.174	0.164	0.684	0.661	0.981		
−0.321	0.184	0.176	0.696	0.673	0.983		
−0.309	0.193	0.188	0.708	0.685	0.985		
−0.297	0.203	0.200	0.720	0.697	0.987		
−0.285	0.214	0.212	0.732	0.709	0.988		
−0.273	0.224	0.224	0.743	0.721	0.990		
−0.261	0.235	0.236	0.754	0.733	0.991		
−0.248	0.246	0.248	0.765	0.745	0.993		
−0.236	0.257	0.261	0.776	0.758	0.994		
−0.224	0.268	0.273	0.786	0.770	0.995		
−0.212	0.280	0.285	0.797	0.782	0.996		
−0.200	0.292	0.297	0.807	0.794	0.996		
−0.188	0.304	0.309	0.816	0.806	0.997		
−0.176	0.316	0.321	0.826	0.818	0.998		
−0.164	0.328	0.333	0.835	0.830	0.998		
−0.152	0.341	0.345	0.844	0.842	0.999		
−0.139	0.354	0.358	0.852	0.855	0.999		
−0.127	0.367	0.370	0.861	0.867	0.999		
−0.115	0.379	0.382	0.868	0.879	0.999		
−0.103	0.393	0.394	0.876	0.891	1.000		
−0.091	0.406	0.406	0.884	0.903	1.000		
−0.079	0.419	0.418	0.891	0.915	1.000		
−0.067	0.433	0.430	0.898	0.927	1.000		
−0.055	0.446	0.442	0.904	0.939	1.000		
−0.042	0.459	0.455	0.911	0.952	1.000		
−0.030	0.473	0.467	0.917	0.964	1.000		

Table 10   Kruskal-Wallis Test

$$H_0 : \tilde{\mu}_1 = \tilde{\mu}_2 = \cdots = \tilde{\mu}_k$$

$$P = P(H \le h)$$

$$H = \frac{12}{n(n+1)} \sum_{i=1}^{k} n_i \left(R_{i\cdot} - \frac{n+1}{2} \right)^2$$

h	P	h	P	h	P	h	P	h	P
(1, 1, 2)		**(1, 2, 2)**		0.643	0.200	1.533	0.440	**(1, 3, 4)**	
				0.696	0.219	1.783	0.464		
0.300	0.167	0.000	0.065	1.018	0.257	1.800	0.512	0.056	0.014
1.800	0.500	0.400	0.129	1.071	0.295	1.917	0.536	0.056	0.029
2.700	1.000	0.600	0.258	1.125	0.333	2.050	0.571	0.097	0.050
		1.400	0.387	1.286	0.371	2.333	0.607	0.208	0.079
		2.000	0.516	1.393	0.410	2.450	0.702	0.333	0.100
h	**P**	2.400	0.645	1.446	0.467	2.717	0.714	0.431	0.129
(1, 1, 3)		3.000	0.774	1.875	0.505	2.800	0.786	0.500	0.157
		3.600	0.968	2.036	0.524	2.867	0.798	0.556	0.214
0.533	0.200	8.400	1.000	2.143	0.543	3.133	0.810	0.764	0.257
0.800	0.300			2.250	0.600	3.333	0.821	0.875	0.279
2.133	0.700			2.411	0.695	3.383	0.869	1.097	0.293
3.200	1.000	**h**	**P**	2.571	0.714	3.783	0.881	1.208	0.321
		(1, 2, 3)		2.786	0.733	4.050	0.905	1.222	0.371
				2.893	0.810	4.200	0.929	1.389	0.443
h	**P**	0.095	0.067	3.161	0.829	4.450	0.952	1.431	0.464
(1, 1, 4)		0.238	0.100	3.696	0.867	5.000	0.964	1.764	0.486
		0.429	0.167	3.750	0.886	5.250	1.000	1.833	0.529
0.143	0.067	0.810	0.200	4.018	0.924			1.875	0.543
0.786	0.200	0.857	0.300	4.500	0.943			2.097	0.557
1.000	0.333	1.238	0.400	4.821	1.000	**h**	**P**	2.208	0.571
1.286	0.400	1.381	0.433			**(1, 3, 3)**		2.333	0.629
2.143	0.533	1.952	0.467					2.431	0.700
2.500	0.800	2.143	0.567	**h**	**P**	0.000	0.014	2.722	0.771
3.571	1.000	2.381	0.733	**(1, 2, 5)**		0.143	0.043	2.764	0.779
		3.095	0.800			0.286	0.129	3.000	0.786
		3.524	0.867	0.050	0.036	0.571	0.229	3.097	0.800
h	**P**	3.857	0.900	0.133	0.060	1.000	0.257	3.208	0.843
(1, 1, 5)		4.286	1.000	0.200	0.095	1.143	0.400	3.222	0.864
				0.450	0.155	1.286	0.429	3.764	0.871
0.257	0.095			0.467	0.179	1.571	0.486	3.889	0.907
0.429	0.143	**h**	**P**	0.583	0.202	2.000	0.514	4.056	0.914
1.029	0.238	**(1, 2, 4)**		0.667	0.226	2.286	0.671	4.097	0.921
1.114	0.333			0.717	0.250	2.571	0.757	4.208	0.929
1.457	0.429	0.000	0.029	1.000	0.262	3.143	0.843	4.764	0.943
1.714	0.476	0.161	0.067	1.117	0.286	3.286	0.871	5.000	0.950
2.314	0.667	0.268	0.105	1.200	0.345	4.000	0.900	5.208	0.964
2.829	0.857	0.321	0.143	1.250	0.381	4.571	0.957	5.389	0.979
3.857	1.000	0.536	0.181	1.383	0.417	5.143	1.000	5.833	1.000

Table 10 Kruskal-Wallis Test (continued)

$$H_0 : \tilde{\mu}_1 = \tilde{\mu}_2 = \cdots = \tilde{\mu}_k$$

$$P = P(H \le h)$$

$$H = \frac{12}{n(n+1)} \sum_{i=1}^{k} n_i \left(R_{i\cdot} - \frac{n+1}{2} \right)^2$$

(1, 3, 5)

h	P
0.000	0.008
0.071	0.028
0.160	0.048
0.178	0.071
0.284	0.111
0.338	0.131
0.551	0.147
0.604	0.167
0.640	0.230
0.711	0.250
0.818	0.270
0.960	0.306
1.084	0.317
1.138	0.349
1.351	0.389
1.404	0.409
1.440	0.429
1.511	0.440
1.600	0.480
1.671	0.512
1.778	0.520
1.884	0.532
1.938	0.548
2.044	0.563
2.204	0.587
2.400	0.595
2.418	0.659
2.560	0.742
2.844	0.782
2.951	0.790
3.040	0.810
3.218	0.817
3.271	0.857
3.378	0.865
3.484	0.869
3.804	0.877
3.840	0.905
4.018	0.917
4.284	0.921
4.338	0.925
4.551	0.944
4.711	0.948
4.871	0.952
4.960	0.956
5.404	0.964
5.440	0.972
5.760	0.980
6.044	0.988
6.400	1.000

(1, 4, 4)

h	P
0.000	0.013
0.067	0.032
0.167	0.070
0.267	0.089
0.300	0.127
0.567	0.165
0.600	0.197
0.667	0.241
0.867	0.279
0.967	0.311
1.067	0.324
1.200	0.356
1.367	0.400
1.500	0.463
1.667	0.502
1.767	0.540
2.167	0.590
2.267	0.616
2.400	0.651
2.467	0.695
2.667	0.740
2.700	0.765
2.967	0.778
3.000	0.822
3.267	0.829
3.367	0.848
3.467	0.879
3.867	0.892
3.900	0.898
4.067	0.917
4.167	0.930
4.267	0.933
4.800	0.946
4.867	0.952
4.967	0.959
5.100	0.965
5.667	0.971
6.000	0.978
6.167	0.990
6.667	1.000

(1, 4, 5)

h	P
0.033	0.017
0.060	0.032
0.104	0.048
0.185	0.062
0.273	0.078
0.278	0.094
0.295	0.110
0.360	0.125
0.409	0.152
0.540	0.179
0.622	0.194
0.731	0.206
0.758	0.222
0.796	0.238
0.818	0.270
0.905	0.281
0.933	0.310
0.976	0.324
1.151	0.335
1.167	0.349
1.195	0.360
1.233	0.375
1.342	0.386
1.369	0.394
1.495	0.411
1.500	0.438
1.587	0.465
1.604	0.483
1.669	0.502
1.778	0.517
1.805	0.532
1.849	0.540
1.931	0.559
2.040	0.568
2.067	0.581
2.105	0.594
2.242	0.600
2.285	0.606
2.455	0.646
2.460	0.654
2.504	0.700
2.591	0.714
2.651	0.749
2.896	0.778
2.913	0.784
2.940	0.792
3.000	0.806
3.087	0.813
3.158	0.817
3.240	0.849
3.349	0.854
3.524	0.862
3.595	0.868
3.682	0.890
3.813	0.898
3.960	0.902
3.987	0.905
4.205	0.913
4.222	0.929
4.287	0.933
4.549	0.937
4.636	0.940
4.724	0.941
4.833	0.944
4.860	0.956
4.985	0.959
5.078	0.962
5.160	0.963
5.515	0.965
5.558	0.967
5.596	0.973
5.733	0.975
5.776	0.976
5.858	0.978
5.864	0.979
5.967	0.981
6.431	0.984
6.578	0.987
6.818	0.989
6.840	0.992
6.955	0.995
7.364	1.000

(1, 5, 5)

h	P
0.000	0.006
0.036	0.018
0.109	0.044
0.145	0.056
0.182	0.080
0.327	0.115
0.400	0.128
0.436	0.153
0.545	0.198
0.582	0.208
0.727	0.229
0.836	0.248

Table 10   Kruskal-Wallis Test (continued)

$$H_0 : \tilde{\mu}_1 = \tilde{\mu}_2 = \cdots = \tilde{\mu}_k$$

$$P = P(H \le h)$$

$$H = \frac{12}{n(n+1)} \sum_{i=1}^{k} n_i \left(R_{i\cdot} - \frac{n+1}{2} \right)^2$$

h	P	h	P	h	P	h	P	h	P
0.909	0.284	5.782	0.978	1.357	0.436	2.125	0.637	1.440	0.457
0.982	0.331	6.000	0.981	1.464	0.474	2.458	0.665	1.493	0.483
1.127	0.354	6.145	0.982	1.607	0.531	2.667	0.684	1.533	0.505
1.200	0.370	6.509	0.985	1.929	0.559	2.792	0.703	1.693	0.526
1.309	0.395	6.545	0.986	2.000	0.578	2.833	0.722	1.800	0.547
1.345	0.416	6.582	0.988	2.214	0.616	3.000	0.751	2.133	0.555
1.600	0.429	6.727	0.989	2.429	0.635	3.125	0.770	2.160	0.597
1.636	0.491	6.836	0.991	2.464	0.673	3.167	0.789	2.173	0.618
1.709	0.507	7.309	0.992	2.750	0.711	3.458	0.808	2.293	0.634
1.745	0.532	7.527	0.995	2.857	0.730	3.667	0.817	2.333	0.655
1.782	0.538	7.745	0.998	3.179	0.749	4.000	0.846	2.373	0.682
1.927	0.562	8.182	1.000	3.429	0.758	4.125	0.893	2.693	0.703
2.000	0.578			3.607	0.777	4.167	0.898	2.760	0.724
2.145	0.589			3.750	0.815	4.458	0.907	2.973	0.734
2.182	0.621	h	P	3.929	0.891	4.500	0.945	3.093	0.745
2.327	0.626	**(2, 2, 2)**		4.464	0.929	5.125	0.964	3.133	0.761
2.436	0.639			4.500	0.948	5.333	0.974	3.240	0.793
2.509	0.686	0.000	0.066	4.714	0.967	5.500	0.983	3.333	0.803
2.582	0.714	0.286	0.198	5.357	0.995	6.000	0.998	3.360	0.814
2.727	0.758	0.857	0.330	6.429	1.000	6.125	1.000	3.573	0.824
2.909	0.773	1.143	0.462					3.773	0.835
2.945	0.812	2.000	0.593					3.840	0.840
3.236	0.832	2.571	0.659	h	P	h	P	3.973	0.851
3.345	0.839	3.429	0.791	**(2, 2, 4)**		**(2, 2, 5)**		4.093	0.861
3.382	0.859	3.714	0.923					4.200	0.877
3.527	0.868	4.571	0.989	0.000	0.029	0.000	0.016	4.293	0.909
3.600	0.884	7.143	1.000	0.125	0.086	0.093	0.063	4.373	0.914
3.636	0.887			0.167	0.109	0.133	0.087	4.573	0.935
3.927	0.895	h	P	0.333	0.138	0.240	0.119	4.800	0.938
4.036	0.914	**(2, 2, 3)**		0.458	0.185	0.360	0.156	4.893	0.943
4.109	0.918			0.500	0.242	0.373	0.193	5.040	0.964
4.182	0.924	0.000	0.028	0.667	0.266	0.533	0.209	5.160	0.970
4.400	0.926	0.179	0.104	0.792	0.304	0.573	0.240	5.693	0.980
4.545	0.944	0.214	0.142	1.000	0.342	0.773	0.277	5.973	0.982
4.800	0.947	0.500	0.199	1.125	0.418	0.840	0.314	6.000	0.987
4.909	0.954	0.607	0.256	1.333	0.447	0.893	0.346	6.133	0.992
5.127	0.961	0.714	0.313	1.500	0.485	0.960	0.362	6.533	1.000
5.236	0.967	0.857	0.341	1.792	0.513	1.093	0.394		
5.636	0.970	1.179	0.379	1.833	0.551	1.200	0.410		
5.709	0.973			2.000	0.589	1.373	0.436		

Table 10   Kruskal-Wallis Test (continued)

$$H_0 : \tilde{\mu}_1 = \tilde{\mu}_2 = \cdots = \tilde{\mu}_k$$

$$P = P(H \le h)$$

$$H = \frac{12}{n(n+1)} \sum_{i=1}^{k} n_i \left(\bar{R}_{i\cdot} - \frac{n+1}{2} \right)^2$$

h	P	h	P	h	P	h	P	h	P
(2,3,3)		(2,3,4)		2.200	0.624	4.811	0.927	0.713	0.286
				2.211	0.632	4.878	0.929	0.724	0.297
0.028	0.032	0.000	0.013	2.244	0.643	4.900	0.941	0.767	0.308
0.111	0.054	0.078	0.035	2.378	0.654	4.978	0.943	0.887	0.320
0.222	0.104	0.100	0.056	2.400	0.662	5.078	0.946	0.942	0.341
0.250	0.136	0.111	0.078	2.411	0.671	5.144	0.948	1.015	0.352
0.472	0.193	0.244	0.098	2.444	0.679	5.378	0.949	1.058	0.362
0.556	0.243	0.278	0.119	2.500	0.706	5.400	0.954	1.091	0.384
0.694	0.314	0.311	0.138	2.778	0.716	5.444	0.960	1.149	0.407
1.000	0.329	0.344	0.156	2.800	0.729	5.500	0.968	1.178	0.421
1.111	0.400	0.400	0.171	2.911	0.738	5.611	0.970	1.276	0.431
1.139	0.436	0.444	0.189	2.944	0.744	5.800	0.976	1.324	0.463
1.361	0.461	0.544	0.206	3.011	0.749	6.000	0.979	1.378	0.471
1.444	0.489	0.600	0.230	3.100	0.762	6.111	0.986	1.451	0.481
1.806	0.554	0.611	0.244	3.111	0.768	6.144	0.989	1.585	0.490
1.889	0.575	0.700	0.278	3.244	0.775	6.300	0.992	1.596	0.498
2.000	0.604	0.778	0.297	3.278	0.784	6.444	0.995	1.615	0.517
2.028	0.632	0.811	0.311	3.300	0.797	7.000	1.000	1.713	0.526
2.250	0.643	0.900	0.327	3.311	0.803			1.727	0.541
2.472	0.671	0.978	0.340	3.444	0.810	h	P	1.760	0.549
2.694	0.693	1.000	0.373	3.478	0.816	(2,3,5)		1.815	0.556
2.778	0.714	1.078	0.386	3.544	0.825			1.858	0.571
2.889	0.757	1.111	0.398	3.600	0.832	0.015	0.019	1.876	0.580
3.139	0.779	1.178	0.414	3.811	0.837	0.069	0.034	2.022	0.597
3.222	0.793	1.244	0.429	3.844	0.841	0.113	0.049	2.033	0.604
3.361	0.800	1.344	0.441	3.911	0.844	0.131	0.068	2.076	0.611
3.778	0.821	1.378	0.452	3.978	0.851	0.142	0.083	2.105	0.618
3.806	0.836	1.411	0.463	4.000	0.860	0.273	0.099	2.196	0.625
4.028	0.871	1.500	0.489	4.078	0.863	0.276	0.114	2.251	0.632
4.111	0.879	1.600	0.498	4.200	0.876	0.305	0.131	2.295	0.638
4.250	0.900	1.611	0.522	4.278	0.892	0.331	0.145	2.367	0.644
4.556	0.907	1.678	0.532	4.311	0.895	0.364	0.177	2.455	0.650
4.694	0.925	1.711	0.543	4.378	0.898	0.451	0.193	2.458	0.664
5.000	0.939	1.778	0.552	4.444	0.902	0.549	0.206	2.469	0.670
5.139	0.968	1.844	0.563	4.511	0.914	0.567	0.219	2.487	0.679
5.361	0.975	1.944	0.583	4.544	0.917	0.622	0.231	2.545	0.706
5.556	0.989	2.144	0.594	4.611	0.921	0.636	0.257	2.633	0.713
6.250	1.000	2.178	0.602	4.711	0.924			2.749	0.721

Table 10   Kruskal-Wallis Test (continued)

$$H_0 : \tilde{\mu}_1 = \tilde{\mu}_2 = \cdots = \tilde{\mu}_k$$

$$P = P(H \le h)$$

$$H = \frac{12}{n(n+1)} \sum_{i=1}^{k} n_i \left(R_{i\cdot} - \frac{n+1}{2} \right)^2$$

h	P	h	P	h	P	h	P	h	P
2.818	0.731	4.942	0.940	0.409	0.180	4.418	0.897	0.255	0.109
2.895	0.737	5.076	0.947	0.491	0.221	4.445	0.902	0.273	0.120
2.924	0.743	5.087	0.948	0.627	0.243	4.555	0.906	0.300	0.134
2.949	0.748	5.105	0.951	0.736	0.288	4.582	0.920	0.323	0.145
2.978	0.752	5.251	0.954	0.764	0.315	4.691	0.925	0.368	0.168
3.022	0.757	5.349	0.956	0.873	0.329	4.773	0.929	0.405	0.177
3.069	0.763	5.513	0.957	0.955	0.349	4.855	0.935	0.505	0.188
3.167	0.767	5.524	0.959	1.091	0.362	4.991	0.943	0.518	0.199
3.185	0.778	5.542	0.963	1.145	0.404	5.127	0.948	0.541	0.209
3.273	0.789	5.727	0.966	1.173	0.423	5.236	0.954	0.564	0.219
3.331	0.794	5.742	0.967	1.282	0.441	5.455	0.956	0.573	0.241
3.342	0.799	5.785	0.967	1.309	0.463	5.509	0.958	0.614	0.251
3.385	0.807	5.804	0.974	1.364	0.474	5.536	0.961	0.618	0.260
3.415	0.811	5.949	0.975	1.582	0.490	5.645	0.966	0.655	0.270
3.505	0.817	6.004	0.976	1.636	0.512	5.727	0.972	0.723	0.280
3.545	0.825	6.033	0.979	1.718	0.559	5.945	0.975	0.791	0.290
3.604	0.829	6.091	0.980	1.827	0.574	6.082	0.976	0.841	0.299
3.676	0.833	6.124	0.983	1.964	0.600	6.327	0.978	0.864	0.309
3.767	0.841	6.295	0.984	2.045	0.614	6.409	0.980	0.891	0.317
3.778	0.844	6.385	0.985	2.236	0.625	6.545	0.983	0.905	0.326
3.822	0.848	6.415	0.988	2.264	0.637	6.600	0.984	0.914	0.343
3.909	0.854	6.818	0.990	2.373	0.662	6.627	0.989	0.950	0.351
3.942	0.861	6.822	0.991	2.455	0.683	6.873	0.994	0.955	0.360
3.996	0.863	6.909	0.994	2.509	0.699	7.036	0.996	1.018	0.368
4.058	0.868	6.949	0.996	2.673	0.719	7.282	0.998	1.023	0.377
4.069	0.871	7.182	0.998	2.809	0.728	7.855	1.000	1.050	0.386
4.204	0.875	7.636	1.000	2.918	0.737			1.091	0.393
4.215	0.878			2.945	0.761			1.200	0.401
4.233	0.880			3.055	0.772	**h**	**P**	1.205	0.408
4.258	0.883	**h**	**P**	3.136	0.780	**(2, 4, 5)**		1.268	0.424
4.331	0.887	**(2, 4, 4)**		3.327	0.790			1.291	0.431
4.378	0.899			3.355	0.808	0.000	0.008	1.314	0.438
4.495	0.909	0.000	0.012	3.464	0.815	0.041	0.021	1.318	0.446
4.651	0.911	0.055	0.030	3.491	0.820	0.064	0.035	1.391	0.463
4.695	0.913	0.082	0.060	3.682	0.834	0.068	0.048	1.414	0.471
4.724	0.915	0.191	0.090	3.764	0.848	0.141	0.061	1.450	0.479
4.727	0.929	0.218	0.107	3.818	0.858	0.155	0.074	1.473	0.493
4.815	0.933	0.273	0.121	4.009	0.875	0.164	0.087	1.518	0.501
4.869	0.937	0.327	0.152	4.364	0.880	0.223	0.098	1.591	0.509
4.913	0.938								

Table 10   Kruskal-Wallis Test (continued)

$$H_0 : \bar{\mu}_1 = \bar{\mu}_2 = \cdots = \bar{\mu}_k$$

$$P = P(H \leq h)$$

$$H = \frac{12}{n(n+1)} \sum_{i=1}^{k} n_i \left(R_{i\cdot} - \frac{n+1}{2} \right)^2$$

h	P	h	P	h	P	h	P	h	P
(2, 4, 5)		3.064	0.774	4.768	0.922	6.905	0.987	0.908	0.339
		3.118	0.779	4.791	0.924	6.914	0.987	0.931	0.362
1.618	0.515	3.164	0.783	4.800	0.926	7.000	0.988	1.115	0.389
1.641	0.521	3.268	0.786	4.818	0.928	7.018	0.988	1.154	0.407
1.664	0.528	3.314	0.792	4.841	0.929	7.064	0.990	1.185	0.431
1.705	0.535	3.341	0.800	4.868	0.937	7.118	0.991	1.277	0.442
1.750	0.541	3.364	0.803	4.950	0.939	7.205	0.991	1.300	0.448
1.755	0.548	3.414	0.807	5.073	0.941	7.255	0.992	1.362	0.461
1.814	0.568	3.455	0.810	5.155	0.947	7.291	0.993	1.431	0.472
1.823	0.573	3.523	0.813	5.164	0.948	7.450	0.993	1.485	0.484
1.973	0.580	3.564	0.816	5.255	0.949	7.500	0.994	1.523	0.494
2.005	0.597	3.568	0.819	5.268	0.951	7.568	0.995	1.554	0.504
2.018	0.602	3.573	0.822	5.273	0.952	7.573	0.996	1.646	0.514
2.073	0.608	3.618	0.825	5.300	0.954	7.773	0.997	1.669	0.537
2.114	0.613	3.641	0.830	5.314	0.955	7.814	0.998	1.731	0.555
2.118	0.619	3.655	0.836	5.414	0.957	8.018	0.999	1.854	0.566
2.141	0.625	3.700	0.840	5.518	0.958	8.114	0.999	1.915	0.576
2.164	0.629	3.705	0.843	5.523	0.962	8.591	1.000	1.923	0.593
2.223	0.634	3.791	0.849	5.564	0.963			2.015	0.602
2.255	0.639	3.800	0.852	5.641	0.964			2.038	0.621
2.291	0.649	3.818	0.855	5.664	0.965	**h**	**P**	2.223	0.626
2.318	0.654	3.823	0.857	5.755	0.966			2.262	0.637
2.323	0.665	3.864	0.861	5.823	0.968	**(2, 5, 5)**		2.285	0.647
2.391	0.671	4.041	0.865	5.891	0.970			2.292	0.655
2.455	0.676	4.064	0.867	5.955	0.971	0.008	0.012	2.385	0.670
2.473	0.680	4.073	0.870	5.973	0.974	0.046	0.022	2.408	0.677
2.505	0.685	4.091	0.872	6.005	0.975	0.069	0.034	2.469	0.685
2.550	0.689	4.141	0.874	6.041	0.975	0.077	0.053	2.538	0.700
2.618	0.694	4.155	0.877	6.068	0.976	0.169	0.072	2.592	0.708
2.700	0.699	4.200	0.879	6.118	0.977	0.192	0.104	2.662	0.714
2.723	0.704	4.223	0.881	6.141	0.978	0.254	0.123	2.754	0.721
2.755	0.715	4.250	0.884	6.223	0.979	0.323	0.141	2.777	0.724
2.768	0.727	4.323	0.886	6.368	0.979	0.377	0.170	2.908	0.730
2.773	0.733	4.364	0.888	6.391	0.980	0.415	0.178	2.962	0.757
2.868	0.738	4.368	0.890	6.473	0.980	0.446	0.193	3.023	0.766
2.891	0.742	4.405	0.896	6.505	0.983	0.538	0.225	3.031	0.772
2.905	0.751	4.500	0.899	6.541	0.983	0.562	0.241	3.123	0.782
2.914	0.754	4.518	0.902	6.550	0.984	0.623	0.251	3.146	0.790
2.973	0.763	4.541	0.910	6.564	0.984	0.692	0.265	3.331	0.797
3.023	0.766	4.614	0.912	6.655	0.985	0.746	0.281	3.369	0.802
3.050	0.769	4.664	0.921	6.723	0.986	0.808	0.312	3.392	0.810
						0.815	0.326		

Table 10   Kruskal-Wallis Test (continued)

$$H_0 : \tilde{\mu}_1 = \tilde{\mu}_2 = \cdots = \tilde{\mu}_k$$

$$P = P(H \le h)$$

$$H = \frac{12}{n(n+1)} \sum_{i=1}^{k} n_i \left(R_{i\cdot} - \frac{n+1}{2} \right)^2$$

Continuation (main table):

h	P	h	P
3.492	0.814	6.354	0.980
3.515	0.819	6.446	0.981
3.577	0.831	6.469	0.983
3.646	0.835	6.654	0.984
3.738	0.837	6.692	0.985
3.769	0.850	6.815	0.986
3.862	0.854	6.838	0.987
3.885	0.864	6.969	0.987
4.015	0.868	7.023	0.988
4.069	0.870	7.185	0.989
4.131	0.873	7.208	0.990
4.138	0.876	7.269	0.990
4.231	0.886	7.338	0.991
4.254	0.894	7.392	0.992
4.438	0.897	7.462	0.993
4.477	0.900	7.577	0.993
4.508	0.903	7.762	0.994
4.623	0.908	7.923	0.994
4.685	0.916	8.008	0.994
4.754	0.919	8.077	0.995
4.808	0.927	8.131	0.997
4.846	0.932	8.169	0.997
4.877	0.934	8.292	0.998
4.992	0.940	8.377	0.998
5.054	0.943	8.562	0.999
5.177	0.946	8.685	0.999
5.238	0.949	8.938	1.000
5.246	0.953	9.423	1.000
5.338	0.955		
5.546	0.959		
5.585	0.960		
5.608	0.961		
5.615	0.963		
5.708	0.964		
5.731	0.968		
5.792	0.970		
5.915	0.972		
5.985	0.973		
6.077	0.974		
6.231	0.975		
6.346	0.979		

(3, 3, 3)

h	P
0.000	0.007
0.089	0.071
0.267	0.121
0.356	0.171
0.622	0.278
0.800	0.336
1.067	0.371
1.156	0.457
1.422	0.489
1.689	0.560
1.867	0.617
2.222	0.639
2.400	0.660
2.489	0.703
2.756	0.746
3.200	0.767
3.289	0.803
3.467	0.832
3.822	0.860
4.267	0.867
4.356	0.899
4.622	0.914
5.067	0.928
5.422	0.949
5.600	0.971
5.689	0.974
5.956	0.989
6.489	0.996
7.200	0.999
8.622	1.000

(3, 3, 4)

h	P
0.018	0.016
0.045	0.030
0.118	0.059
0.164	0.075
0.200	0.105
0.336	0.131
0.345	0.158
0.409	0.170
0.455	0.183
0.482	0.209
0.636	0.236
0.700	0.283
0.745	0.310
0.891	0.344
1.064	0.367
1.073	0.389
1.136	0.398
1.182	0.418
1.209	0.459
1.427	0.477
1.473	0.487
1.573	0.503
1.618	0.519
1.655	0.553
1.791	0.567
1.800	0.585
1.864	0.598
2.091	0.611
2.200	0.632
2.227	0.649
2.300	0.674
2.382	0.686
2.518	0.697
2.527	0.709
2.664	0.719
2.882	0.727
2.927	0.747
2.955	0.756
3.027	0.766
3.073	0.780
3.109	0.788
3.255	0.797
3.364	0.804
3.391	0.812
3.609	0.820
3.682	0.822
3.755	0.835
3.800	0.850
3.836	0.857
3.973	0.868
4.045	0.874
4.091	0.877
4.273	0.883
4.336	0.889
4.382	0.894
4.564	0.899
4.700	0.908
4.709	0.915
4.818	0.919
4.845	0.926
5.000	0.930
5.064	0.932
5.109	0.936
5.255	0.938
5.436	0.944
5.500	0.947
5.573	0.950
5.727	0.954
5.791	0.964
5.936	0.966
5.982	0.973
6.018	0.975
6.155	0.977
6.300	0.983
6.564	0.986
6.664	0.987
6.709	0.990
6.745	0.994
7.000	0.996
7.318	0.998
7.436	0.999
8.018	1.000

(3, 3, 5)

h	P
0.000	0.006
0.048	0.030
0.061	0.042
0.133	0.052
0.170	0.063
0.170	0.074
0.194	0.098
0.242	0.110
0.315	0.132

Table 10 Kruskal-Wallis Test (continued)

$$H_0 : \bar{\mu}_1 = \bar{\mu}_2 = \cdots = \bar{\mu}_k$$

$$P = P(H \le h)$$

$$H = \frac{12}{n(n+1)} \sum_{i=1}^{k} n_i \left(R_i. - \frac{n+1}{2} \right)^2$$

h	P	h	P	h	P	h	P	h	P
0.376	0.153	3.103	0.782	6.194	0.974	0.727	0.296	3.326	0.793
0.412	0.174	3.333	0.785	6.303	0.979	0.848	0.315	3.386	0.799
0.436	0.196	3.382	0.791	6.315	0.980	0.894	0.332	3.394	0.805
0.533	0.206	3.394	0.804	6.376	0.981	0.932	0.349	3.417	0.810
0.545	0.217	3.442	0.816	6.533	0.981	0.962	0.365	3.477	0.816
0.594	0.235	3.467	0.821	6.594	0.986	1.053	0.380	3.576	0.822
0.679	0.275	3.503	0.827	6.715	0.987	1.076	0.396	3.598	0.827
0.776	0.314	3.576	0.833	6.776	0.988	1.136	0.403	3.659	0.838
0.848	0.332	3.648	0.838	6.861	0.989	1.144	0.418	3.682	0.840
0.970	0.359	3.709	0.844	6.982	0.991	1.295	0.432	3.727	0.846
1.042	0.376	3.879	0.851	7.079	0.992	1.303	0.447	3.803	0.850
1.079	0.391	3.927	0.856	7.333	0.992	1.326	0.461	3.848	0.855
1.103	0.406	4.012	0.861	7.467	0.994	1.394	0.476	3.932	0.860
1.200	0.413	4.048	0.865	7.503	0.995	1.417	0.490	3.962	0.865
1.212	0.429	4.170	0.874	7.515	0.996	1.500	0.497	4.144	0.869
1.261	0.461	4.194	0.878	7.636	0.997	1.545	0.510	4.167	0.871
1.442	0.474	4.242	0.883	7.879	0.998	1.598	0.523	4.212	0.875
1.503	0.488	4.303	0.887	8.048	0.999	1.636	0.530	4.295	0.879
1.515	0.495	4.315	0.891	8.242	0.999	1.682	0.543	4.303	0.884
1.527	0.509	4.412	0.903	8.727	1.000	1.750	0.556	4.326	0.887
1.576	0.522	4.533	0.906			1.803	0.579	4.348	0.894
1.648	0.550	4.679	0.910			1.909	0.591	4.409	0.898
1.745	0.563	4.776	0.913	h	P	1.962	0.612	4.477	0.901
1.770	0.575	4.800	0.915	(3, 4, 4)		2.053	0.622	4.545	0.903
1.867	0.586	4.848	0.918			2.144	0.632	4.576	0.907
2.012	0.597	4.861	0.921	0.000	0.007	2.227	0.646	4.598	0.910
2.048	0.607	4.909	0.923	0.045	0.019	2.295	0.656	4.712	0.913
2.061	0.618	5.042	0.931	0.053	0.041	2.303	0.666	4.750	0.916
2.133	0.633	5.079	0.933	0.144	0.063	2.326	0.675	4.894	0.922
2.170	0.642	5.103	0.935	0.167	0.075	2.394	0.685	5.053	0.927
2.182	0.648	5.212	0.938	0.182	0.087	2.417	0.694	5.144	0.932
2.194	0.658	5.261	0.942	0.212	0.110	2.598	0.710	5.182	0.934
2.315	0.666	5.345	0.945	0.326	0.130	2.636	0.719	5.212	0.937
2.376	0.685	5.442	0.947	0.348	0.150	2.667	0.724	5.295	0.939
2.594	0.694	5.503	0.949	0.386	0.171	2.712	0.731	5.303	0.942
2.667	0.702	5.515	0.951	0.409	0.181	2.848	0.739	5.326	0.946
2.679	0.709	5.648	0.953	0.477	0.201	2.894	0.746	5.386	0.948
2.715	0.733	5.770	0.958	0.576	0.221	2.909	0.750	5.500	0.949
2.836	0.742	5.867	0.960	0.598	0.239	2.932	0.757	5.576	0.951
2.861	0.758	6.012	0.967	0.659	0.258	2.962	0.770	5.598	0.953
2.970	0.761	6.061	0.968	0.667	0.269	3.076	0.782	5.667	0.955
3.079	0.768	6.109	0.973	0.712	0.287	3.136	0.788	5.803	0.957

Table 10   Kruskal-Wallis Test (continued)

$$H_0 : \tilde{\mu}_1 = \tilde{\mu}_2 = \cdots = \tilde{\mu}_k$$

$$P = P(H \le h)$$

$$H = \frac{12}{n(n+1)} \sum_{i=1}^{k} n_i \left(R_i \cdot - \frac{n+1}{2} \right)^2$$

h	P	h	P	h	P	h	P	h	P
5.932	0.959	0.081	0.037	1.241	0.438	2.491	0.706	3.810	0.858
5.962	0.960	0.092	0.047	1.246	0.451	2.522	0.712	3.831	0.860
6.000	0.961	0.118	0.056	1.260	0.456	2.573	0.716	3.865	0.862
6.045	0.965	0.138	0.066	1.349	0.462	2.579	0.719	3.876	0.864
6.053	0.968	0.173	0.075	1.414	0.467	2.641	0.722	3.958	0.868
6.144	0.969	0.179	0.084	1.445	0.483	2.645	0.725	4.015	0.871
6.167	0.970	0.214	0.092	1.465	0.488	2.676	0.728	4.029	0.874
6.182	0.973	0.241	0.101	1.472	0.499	2.677	0.735	4.060	0.876
6.348	0.974	0.256	0.110	1.487	0.509	2.737	0.741	4.122	0.878
6.386	0.975	0.265	0.119	1.506	0.515	2.829	0.744	4.154	0.883
6.394	0.977	0.276	0.127	1.558	0.525	2.887	0.747	4.179	0.884
6.409	0.978	0.323	0.136	1.568	0.530	2.908	0.756	4.195	0.887
6.417	0.979	0.337	0.144	1.599	0.540	2.949	0.759	4.235	0.889
6.545	0.980	0.368	0.161	1.615	0.545	2.953	0.767	4.241	0.891
6.659	0.981	0.426	0.169	1.718	0.548	2.964	0.769	4.276	0.893
6.712	0.982	0.429	0.177	1.718	0.550	3.010	0.772	4.318	0.897
6.727	0.983	0.462	0.185	1.733	0.555	3.035	0.775	4.327	0.898
6.962	0.984	0.491	0.189	1.753	0.559	3.087	0.786	4.368	0.899
7.000	0.986	0.491	0.193	1.779	0.564	3.092	0.788	4.419	0.901
7.053	0.989	0.503	0.218	1.814	0.569	3.106	0.791	4.426	0.902
7.076	0.989	0.542	0.225	1.856	0.578	3.137	0.794	4.487	0.904
7.136	0.990	0.549	0.233	1.906	0.583	3.195	0.799	4.522	0.905
7.144	0.991	0.626	0.249	1.927	0.587	3.256	0.801	4.523	0.909
7.212	0.994	0.645	0.256	1.938	0.606	3.260	0.804	4.549	0.911
7.477	0.996	0.692	0.264	1.964	0.610	3.312	0.808	4.564	0.914
7.598	0.996	0.727	0.287	1.968	0.614	3.318	0.811	4.645	0.915
7.636	0.997	0.737	0.294	1.985	0.619	3.353	0.813	4.676	0.918
7.682	0.997	0.799	0.307	2.019	0.623	3.414	0.815	4.754	0.919
7.848	0.998	0.829	0.314	2.029	0.627	3.445	0.817	4.788	0.920
8.227	0.999	0.831	0.336	2.060	0.631	3.462	0.820	4.810	0.925
8.326	0.999	0.856	0.343	2.103	0.640	3.496	0.824	4.829	0.926
8.909	1.000	0.953	0.349	2.112	0.648	3.503	0.826	4.856	0.927
		1.004	0.363	2.169	0.652	3.506	0.828	4.881	0.931
		1.041	0.376	2.272	0.656	3.568	0.830	4.891	0.933

h	P
(3, 4, 5)	

h	P	h	P	h	P	h	P	h	P
		1.045	0.382	2.308	0.660	3.579	0.835	4.938	0.934
		1.062	0.389	2.337	0.663	3.599	0.839	4.953	0.935
		1.103	0.395	2.349	0.671	3.626	0.843	4.983	0.936
0.010	0.005	1.106	0.401	2.368	0.674	3.703	0.845	5.041	0.937
0.010	0.010	1.118	0.414	2.388	0.685	3.722	0.847	5.045	0.938
0.029	0.014	1.137	0.420	2.395	0.689	3.753	0.848	5.106	0.940
0.029	0.019	1.164	0.426	2.472	0.696	3.773	0.852	5.137	0.941
0.060	0.028	1.188	0.432	2.481	0.699	3.785	0.856	5.158	0.943

Table 10   Kruskal-Wallis Test (continued)

$$H_0 : \tilde{\mu}_1 = \tilde{\mu}_2 = \cdots = \tilde{\mu}_k$$

$$P = P(H \le h)$$

$$H = \frac{12}{n(n+1)} \sum_{i=1}^{k} n_i \left(R_{i\cdot} - \frac{n+1}{2} \right)^2$$

h	P	h	P	h	P	h	P	h	P
5.179	0.945	6.635	0.983	8.481	0.999	0.879	0.347	2.637	0.724
5.291	0.946	6.676	0.984	8.503	0.999	0.949	0.352	2.716	0.729
5.308	0.947	6.703	0.984	8.573	0.999	1.029	0.357	2.752	0.733
5.342	0.948	6.779	0.985	8.626	0.999	1.037	0.368	2.778	0.738
5.349	0.950	6.785	0.986	8.795	1.000	1.055	0.389	2.848	0.743
5.353	0.950	6.799	0.986	9.035	1.000	1.064	0.398	2.857	0.745
5.414	0.951	6.829	0.986	9.118	1.000	1.116	0.408	2.884	0.754
5.426	0.953	6.891	0.987	9.199	1.000	1.134	0.417	2.936	0.759
5.549	0.954	7.004	0.987	9.692	1.000	1.143	0.427	2.963	0.763
5.568	0.955	7.010	0.987			1.248	0.437	3.095	0.776
5.619	0.956	7.096	0.988			1.266	0.446	3.112	0.780
5.631	0.958	7.106	0.988	h	P	1.292	0.450	3.121	0.784
5.656	0.959	7.188	0.989	(3, 5, 5)		1.371	0.459	3.165	0.792
5.660	0.959	7.195	0.989			1.407	0.475	3.191	0.794
5.677	0.961	7.256	0.990	0.000	0.004	1.407	0.486	3.279	0.798
5.718	0.962	7.260	0.990	0.026	0.011	1.451	0.494	3.305	0.805
5.722	0.962	7.272	0.990	0.035	0.026	1.459	0.503	3.429	0.809
5.753	0.963	7.291	0.990	0.088	0.037	1.512	0.520	3.464	0.813
5.779	0.965	7.318	0.990	0.088	0.041	1.565	0.528	3.516	0.827
5.804	0.966	7.395	0.991	0.105	0.049	1.688	0.540	3.622	0.833
5.814	0.966	7.445	0.992	0.114	0.052	1.723	0.547	3.648	0.836
5.862	0.967	7.465	0.992	0.114	0.056	1.741	0.555	3.666	0.839
5.876	0.968	7.477	0.993	0.141	0.070	1.749	0.562	3.745	0.842
5.964	0.968	7.523	0.994	0.193	0.084	1.802	0.569	3.780	0.848
6.026	0.969	7.568	0.994	0.220	0.098	1.829	0.580	3.798	0.853
6.029	0.969	7.641	0.994	0.237	0.105	1.855	0.587	3.807	0.856
6.060	0.970	7.708	0.995	0.264	0.120	1.934	0.607	3.912	0.858
6.087	0.971	7.753	0.995	0.316	0.134	1.978	0.614	3.965	0.861
6.164	0.972	7.810	0.995	0.352	0.150	2.066	0.620	3.991	0.864
6.173	0.972	7.876	0.995	0.352	0.160	2.136	0.623	4.114	0.865
6.231	0.973	7.887	0.996	0.422	0.181	2.145	0.630	4.141	0.868
6.265	0.974	7.906	0.996	0.457	0.187	2.163	0.636	4.149	0.873
6.272	0.975	7.927	0.996	0.484	0.200	2.198	0.649	4.202	0.875
6.337	0.975	8.029	0.996	0.536	0.212	2.251	0.661	4.220	0.883
6.368	0.976	8.060	0.996	0.563	0.237	2.321	0.673	4.255	0.888
6.369	0.977	8.077	0.997	0.580	0.249	2.374	0.679	4.308	0.890
6.395	0.978	8.118	0.997	0.659	0.255	2.409	0.685	4.352	0.893
6.410	0.979	8.122	0.997	0.695	0.267	2.479	0.690	4.378	0.895
6.491	0.980	8.215	0.998	0.721	0.279	2.488	0.695	4.457	0.896
6.522	0.980	8.256	0.998	0.774	0.302	2.514	0.701	4.466	0.898
6.542	0.982	8.429	0.998	0.791	0.314	2.593	0.706	4.536	0.900
6.579	0.982	8.446	0.998	0.826	0.325	2.620	0.711	4.545	0.902

Table 10   Kruskal-Wallis Test (continued)

$$H_0 : \tilde{\mu}_1 = \tilde{\mu}_2 = \cdots = \tilde{\mu}_k$$

$$P = P(H \le h)$$

$$H = \frac{12}{n(n+1)} \sum_{i=1}^{k} n_i \left(R_{i\cdot} - \frac{n+1}{2} \right)^2$$

h	P	h	P	h	P	h	P	h	P
4.571	0.906	6.655	0.978	9.284	0.999	2.923	0.766	8.654	0.999
4.695	0.908	6.734	0.979	9.336	0.999	3.038	0.781	8.769	0.999
4.774	0.911	6.752	0.981	9.398	1.000	3.115	0.788	9.269	1.000
4.826	0.912	6.866	0.982	9.521	1.000	3.231	0.803	9.846	1.000
4.835	0.918	6.892	0.982	9.635	1.000	3.500	0.827		
4.888	0.921	6.945	0.983	9.916	1.000	3.577	0.838		
4.914	0.923	6.963	0.985	10.057	1.000	3.731	0.849	h	P
4.941	0.925	6.998	0.985	10.549	1.000	3.846	0.855		
4.993	0.928	7.051	0.986			3.962	0.864	(4, 4, 5)	
5.020	0.930	7.121	0.986			4.154	0.869	0.000	0.004
5.064	0.933	7.209	0.988	h	P	4.192	0.878	0.030	0.019
5.152	0.935	7.226	0.988	(4, 4, 4)		4.269	0.886	0.033	0.026
5.169	0.935	7.288	0.988			4.308	0.896	0.086	0.033
5.222	0.937	7.305	0.989	0.000	0.006	4.500	0.903	0.096	0.048
5.284	0.938	7.314	0.989	0.038	0.032	4.654	0.906	0.119	0.063
5.363	0.941	7.437	0.990	0.115	0.059	4.769	0.914	0.132	0.070
5.407	0.943	7.543	0.990	0.154	0.087	4.885	0.920	0.201	0.084
5.486	0.944	7.578	0.991	0.269	0.136	4.962	0.926	0.218	0.097
5.495	0.945	7.622	0.991	0.346	0.160	5.115	0.937	0.227	0.111
5.521	0.947	7.736	0.992	0.462	0.185	5.346	0.943	0.267	0.125
5.574	0.949	7.763	0.992	0.500	0.230	5.538	0.945	0.297	0.131
5.600	0.949	7.780	0.993	0.615	0.254	5.654	0.951	0.343	0.138
5.626	0.954	7.859	0.993	0.731	0.294	5.692	0.956	0.376	0.151
5.705	0.955	7.895	0.993	0.808	0.333	5.808	0.960	0.382	0.164
5.802	0.958	7.912	0.994	0.962	0.352	6.000	0.963	0.399	0.177
5.837	0.960	8.026	0.994	1.038	0.370	6.038	0.967	0.425	0.189
5.934	0.961	8.079	0.994	1.077	0.408	6.269	0.970	0.475	0.202
5.943	0.962	8.105	0.995	1.192	0.443	6.500	0.974	0.527	0.208
6.022	0.963	8.237	0.995	1.385	0.460	6.577	0.976	0.544	0.220
6.048	0.965	8.264	0.995	1.423	0.490	6.615	0.979	0.597	0.232
6.198	0.966	8.316	0.995	1.500	0.520	6.731	0.981	0.610	0.243
6.207	0.966	8.334	0.996	1.654	0.548	6.962	0.982	0.613	0.255
6.251	0.967	8.545	0.996	1.846	0.564	7.038	0.984	0.640	0.266
6.259	0.969	8.571	0.996	1.885	0.603	7.269	0.985	0.689	0.277
6.286	0.970	8.580	0.997	2.000	0.630	7.385	0.987	0.742	0.289
6.312	0.970	8.651	0.997	2.192	0.652	7.423	0.989	0.771	0.294
6.365	0.972	8.659	0.998	2.346	0.673	7.538	0.992	0.804	0.305
6.391	0.973	8.791	0.998	2.423	0.693	7.654	0.993	0.824	0.310
6.435	0.975	8.809	0.998	2.462	0.704	7.731	0.995	0.860	0.321
6.488	0.976	8.949	0.998	2.577	0.723	8.000	0.997	0.870	0.332
6.549	0.976	9.002	0.999	2.808	0.740	8.115	0.998	0.903	0.342
6.593	0.978	9.055	0.999	2.885	0.748	8.346	0.999	0.910	0.353

Table 10   Kruskal-Wallis Test (continued)

$$H_0 : \tilde{\mu}_1 = \tilde{\mu}_2 = \cdots = \tilde{\mu}_k$$

$$P = P(H \le h)$$

$$H = \frac{12}{n(n+1)} \sum_{i=1}^{k} n_i \left(R_{i\cdot} - \frac{n+1}{2} \right)^2$$

h	P	h	P	h	P	h	P	h	P
0.940	0.363	2.440	0.690	4.025	0.866	5.476	0.944	7.058	0.983
1.019	0.373	2.443	0.695	4.042	0.868	5.486	0.944	7.075	0.983
1.058	0.383	2.453	0.701	4.068	0.870	5.489	0.946	7.101	0.984
1.068	0.393	2.558	0.707	4.075	0.873	5.519	0.948	7.124	0.984
1.124	0.402	2.575	0.712	4.118	0.875	5.568	0.949	7.190	0.985
1.167	0.411	2.601	0.717	4.170	0.878	5.571	0.950	7.203	0.985
1.187	0.416	2.667	0.721	4.200	0.879	5.618	0.951	7.233	0.986
1.190	0.426	2.670	0.729	4.233	0.881	5.657	0.952	7.240	0.986
1.203	0.435	2.733	0.733	4.253	0.883	5.687	0.953	7.256	0.986
1.256	0.444	2.756	0.738	4.273	0.886	5.756	0.954	7.418	0.987
1.273	0.452	2.799	0.743	4.289	0.888	5.782	0.955	7.467	0.987
1.299	0.461	2.881	0.747	4.332	0.892	5.815	0.957	7.470	0.987
1.371	0.466	2.904	0.751	4.381	0.894	5.819	0.958	7.497	0.988
1.404	0.474	2.918	0.755	4.447	0.896	5.914	0.958	7.503	0.988
1.414	0.482	2.967	0.760	4.497	0.898	6.003	0.959	7.586	0.988
1.454	0.491	2.987	0.764	4.553	0.900	6.013	0.960	7.596	0.989
1.503	0.499	2.997	0.772	4.619	0.902	6.030	0.961	7.714	0.989
1.530	0.507	3.013	0.776	4.668	0.904	6.096	0.962	7.744	0.991
1.533	0.515	3.086	0.779	4.685	0.906	6.119	0.963	7.760	0.991
1.586	0.523	3.119	0.783	4.701	0.908	6.132	0.964	7.767	0.991
1.596	0.531	3.129	0.786	4.711	0.909	6.201	0.966	7.797	0.991
1.615	0.535	3.168	0.790	4.727	0.911	6.214	0.967	7.810	0.992
1.668	0.542	3.218	0.794	4.747	0.912	6.227	0.968	7.833	0.993
1.701	0.550	3.260	0.798	4.760	0.914	6.267	0.969	7.942	0.993
1.718	0.557	3.297	0.800	4.813	0.916	6.310	0.970	7.981	0.994
1.744	0.564	3.330	0.803	4.830	0.918	6.343	0.971	8.047	0.994
1.810	0.571	3.382	0.810	4.833	0.919	6.382	0.972	8.113	0.994
1.876	0.578	3.432	0.813	4.896	0.923	6.399	0.973	8.130	0.995
1.899	0.586	3.442	0.817	4.975	0.924	6.462	0.973	8.140	0.995
1.929	0.592	3.481	0.820	5.014	0.926	6.544	0.974	8.156	0.995
1.942	0.599	3.511	0.824	5.024	0.927	6.547	0.974	8.189	0.996
1.958	0.612	3.590	0.830	5.027	0.929	6.597	0.976	8.403	0.996
2.047	0.625	3.613	0.833	5.090	0.931	6.673	0.976	8.440	0.996
2.110	0.629	3.630	0.836	5.173	0.932	6.676	0.977	8.456	0.997
2.140	0.635	3.640	0.840	5.196	0.934	6.804	0.978	8.525	0.997
2.143	0.638	3.656	0.843	5.225	0.935	6.860	0.978	8.558	0.997
2.176	0.644	3.696	0.846	5.344	0.937	6.870	0.979	8.571	0.997
2.196	0.656	3.758	0.849	5.360	0.938	6.887	0.979	8.575	0.997
2.275	0.662	3.827	0.854	5.370	0.939	6.890	0.980	8.604	0.997
2.387	0.668	3.910	0.857	5.387	0.940	6.943	0.980	8.703	0.998
2.390	0.673	3.986	0.859	5.410	0.941	6.953	0.981	8.733	0.998
2.403	0.684	3.989	0.861	5.440	0.943	6.976	0.982	8.782	0.998

Table 10   Kruskal-Wallis Test (continued)

$$H_0 : \tilde{\mu}_1 = \tilde{\mu}_2 = \cdots = \tilde{\mu}_k$$

$$P = P(H \le h)$$

$$H = \frac{12}{n(n+1)} \sum_{i=1}^{k} n_i \left(R_i. - \frac{n+1}{2} \right)^2$$

h	P	h	P	h	P	h	P	h	P
8.868	0.999	0.591	0.248	1.871	0.586	3.243	0.791	4.660	0.908
8.997	0.999	0.600	0.258	1.963	0.591	3.266	0.797	4.706	0.911
9.053	0.999	0.691	0.262	1.971	0.602	3.286	0.800	4.806	0.912
9.099	0.999	0.706	0.271	1.986	0.607	3.311	0.803	4.843	0.914
9.129	0.999	0.726	0.280	2.006	0.618	3.343	0.812	4.851	0.916
9.168	0.999	0.751	0.289	2.031	0.623	3.380	0.813	4.866	0.917
9.396	0.999	0.771	0.307	2.051	0.628	3.403	0.816	4.886	0.921
9.527	0.999	0.783	0.316	2.063	0.631	3.471	0.824	4.911	0.922
9.590	1.000	0.843	0.325	2.100	0.636	3.540	0.826	4.943	0.924
9.613	1.000	0.863	0.333	2.143	0.646	3.571	0.830	4.980	0.925
9.758	1.000	0.866	0.342	2.191	0.651	3.586	0.833	5.023	0.926
10.118	1.000	0.966	0.346	2.246	0.656	3.651	0.838	5.071	0.927
10.187	1.000	0.980	0.350	2.280	0.661	3.743	0.840	5.126	0.930
10.681	1.000	1.000	0.358	2.306	0.665	3.746	0.845	5.163	0.931
		1.003	0.374	2.351	0.670	3.791	0.847	5.171	0.932
		1.011	0.383	2.371	0.674	3.800	0.849	5.186	0.933
h	P	1.046	0.390	2.383	0.678	3.846	0.852	5.206	0.934
(4, 5, 5)		1.071	0.406	2.420	0.681	3.883	0.856	5.231	0.936
		1.140	0.413	2.443	0.693	3.891	0.858	5.263	0.937
0.006	0.006	1.183	0.421	2.463	0.698	3.906	0.860	5.323	0.939
0.020	0.012	1.186	0.428	2.466	0.702	3.926	0.863	5.400	0.941
0.043	0.024	1.286	0.447	2.511	0.706	3.951	0.865	5.446	0.942
0.051	0.030	1.300	0.453	2.520	0.708	3.971	0.867	5.460	0.943
0.086	0.042	1.323	0.460	2.566	0.712	4.043	0.869	5.483	0.944
0.111	0.054	1.331	0.468	2.600	0.716	4.063	0.873	5.491	0.944
0.131	0.065	1.346	0.475	2.626	0.720	4.166	0.876	5.526	0.945
0.143	0.071	1.366	0.482	2.691	0.724	4.200	0.878	5.571	0.948
0.180	0.077	1.411	0.488	2.740	0.728	4.203	0.880	5.583	0.949
0.203	0.088	1.423	0.495	2.783	0.732	4.246	0.882	5.620	0.950
0.223	0.099	1.483	0.502	2.786	0.743	4.271	0.885	5.643	0.951
0.226	0.110	1.551	0.508	2.831	0.746	4.291	0.887	5.666	0.952
0.271	0.121	1.560	0.515	2.840	0.750	4.303	0.889	5.711	0.952
0.280	0.126	1.606	0.521	2.886	0.754	4.363	0.890	5.780	0.953
0.326	0.137	1.620	0.530	2.931	0.761	4.383	0.892	5.803	0.954
0.360	0.148	1.643	0.542	2.946	0.764	4.386	0.894	5.811	0.955
0.371	0.159	1.651	0.545	2.966	0.768	4.486	0.895	5.871	0.957
0.386	0.179	1.686	0.551	2.991	0.771	4.500	0.899	5.903	0.958
0.463	0.195	1.711	0.557	3.023	0.776	4.520	0.901	5.963	0.958
0.500	0.200	1.731	0.563	3.083	0.779	4.523	0.902	5.983	0.959
0.523	0.210	1.743	0.569	3.103	0.782	4.531	0.904	5.986	0.960
0.543	0.219	1.803	0.575	3.160	0.785	4.591	0.905	6.031	0.960
0.546	0.229	1.826	0.580	3.240	0.789	4.611	0.907	6.086	0.962

Table 10   Kruskal-Wallis Test (continued)

$$H_0 : \tilde{\mu}_1 = \tilde{\mu}_2 = \cdots = \tilde{\mu}_k$$

$$P = P(H \le h)$$

$$H = \frac{12}{n(n+1)} \sum_{i=1}^{k} n_i \left(R_{i\cdot} - \frac{n+1}{2} \right)^2$$

h	P	h	P	h	P	h	P	h	P
6.100	0.963	7.503	0.987	9.026	0.998	0.420	0.193	3.020	0.777
6.123	0.963	7.563	0.988	9.071	0.998	0.500	0.206	3.120	0.784
6.146	0.965	7.586	0.988	9.103	0.998	0.540	0.217	3.140	0.792
6.166	0.965	7.631	0.989	9.163	0.998	0.560	0.241	3.260	0.799
6.211	0.966	7.640	0.989	9.231	0.998	0.620	0.264	3.380	0.810
6.223	0.966	7.686	0.989	9.286	0.999	0.720	0.275	3.420	0.816
6.283	0.967	7.720	0.990	9.323	0.999	0.740	0.297	3.440	0.823
6.303	0.968	7.766	0.990	9.411	0.999	0.780	0.319	3.500	0.829
6.351	0.969	7.791	0.990	9.503	0.999	0.860	0.340	3.620	0.835
6.406	0.970	7.823	0.990	9.506	0.999	0.960	0.350	3.660	0.841
6.440	0.971	7.860	0.991	9.606	0.999	0.980	0.380	3.780	0.847
6.451	0.971	7.903	0.991	9.643	0.999	1.040	0.399	3.840	0.850
6.486	0.972	7.906	0.991	9.651	0.999	1.140	0.418	3.860	0.855
6.531	0.972	8.006	0.991	9.686	0.999	1.220	0.436	3.920	0.863
6.543	0.973	8.043	0.992	9.926	1.000	1.260	0.453	3.980	0.868
6.603	0.974	8.051	0.992	9.986	1.000	1.280	0.462	4.020	0.873
6.623	0.974	8.066	0.992	10.051	1.000	1.340	0.479	4.160	0.877
6.626	0.975	8.086	0.992	10.063	1.000	1.460	0.495	4.220	0.882
6.671	0.975	8.131	0.992	10.100	1.000	1.500	0.503	4.340	0.890
6.760	0.976	8.143	0.993	10.260	1.000	1.520	0.519	4.380	0.895
6.763	0.976	8.223	0.993	10.511	1.000	1.580	0.534	4.460	0.898
6.771	0.977	8.226	0.993	10.520	1.000	1.620	0.541	4.500	0.900
6.786	0.978	8.271	0.994	10.566	1.000	1.680	0.556	4.560	0.904
6.806	0.978	8.280	0.994	10.646	1.000	1.820	0.584	4.580	0.908
6.831	0.979	8.340	0.994	11.023	1.000	1.860	0.597	4.740	0.911
6.900	0.980	8.363	0.995	11.083	1.000	1.940	0.610	4.820	0.915
6.943	0.981	8.371	0.995	11.571	1.000	2.000	0.617	4.860	0.916
7.000	0.981	8.386	0.995			2.060	0.629	4.880	0.919
7.046	0.982	8.431	0.995			2.160	0.635	4.940	0.925
7.080	0.982	8.463	0.995	h	P	2.180	0.647	5.040	0.928
7.106	0.982	8.523	0.995			2.220	0.658	5.120	0.930
7.171	0.983	8.543	0.996	(5, 5, 5)		2.240	0.670	5.180	0.935
7.183	0.983	8.546	0.996			2.340	0.681	5.360	0.937
7.220	0.983	8.683	0.996	0.000	0.002	2.420	0.686	5.420	0.940
7.243	0.984	8.691	0.996	0.020	0.017	2.480	0.696	5.460	0.945
7.266	0.985	8.726	0.996	0.060	0.032	2.540	0.706	5.540	0.947
7.311	0.985	8.751	0.996	0.080	0.046	2.580	0.716	5.580	0.949
7.320	0.985	8.771	0.997	0.140	0.075	2.660	0.735	5.660	0.951
7.426	0.986	8.966	0.997	0.180	0.089	2.780	0.744	5.780	0.952
7.446	0.986	8.980	0.997	0.240	0.102	2.880	0.748	5.820	0.954
7.471	0.986	9.000	0.997	0.260	0.129	2.940	0.761	5.840	0.956
7.491	0.987	9.011	0.997	0.320	0.142	2.960	0.769	6.000	0.957
				0.380	0.168				

Table 10   Kruskal-Wallis Test (continued)

$$H_0 : \tilde{\mu}_1 = \tilde{\mu}_2 = \cdots = \tilde{\mu}_k$$
$$P = P(H \leq h)$$
$$H = \frac{12}{n(n+1)} \sum_{i=1}^{k} n_i \left(R_{i\cdot} - \frac{n+1}{2} \right)^2$$

h	P	h	P
6.020	0.960	9.060	0.997
6.080	0.962	9.140	0.997
6.140	0.964	9.260	0.997
6.180	0.965	9.360	0.997
6.260	0.967	9.380	0.998
6.320	0.968	9.420	0.998
6.480	0.969	9.500	0.998
6.500	0.970	9.620	0.999
6.540	0.972	9.680	0.999
6.620	0.973	9.740	0.999
6.660	0.974	9.780	0.999
6.720	0.975	9.920	0.999
6.740	0.976	9.980	0.999
6.860	0.979	10.140	0.999
6.980	0.980	10.220	1.000
7.020	0.981	10.260	1.000
7.220	0.982	10.500	1.000
7.260	0.982	10.580	1.000
7.280	0.984	10.640	1.000
7.340	0.985	10.820	1.000
7.440	0.985	11.060	1.000
7.460	0.986	11.180	1.000
7.580	0.987	11.520	1.000
7.620	0.988	11.580	1.000
7.740	0.988	12.020	1.000
7.760	0.989	12.500	1.000
7.940	0.989		
7.980	0.991		
8.000	0.991		
8.060	0.992		
8.180	0.992		
8.240	0.993		
8.340	0.993		
8.420	0.994		
8.540	0.994		
8.640	0.994		
8.660	0.995		
8.720	0.995		
8.780	0.995		
8.820	0.996		
8.880	0.996		
8.960	0.996		

Table 11   Friedman Test

$$H_0 : \bar{\mu}_1 = \bar{\mu}_2 = \cdots = \bar{\mu}_k$$

$$P = P(S \le s)$$

$$S = \frac{12km}{k(k+1)} \sum_{i=1}^{k} \left[R_{i\cdot} - \frac{1}{2}(k+1) \right]^2$$

Note: k = **number of treatment levels,** m = **number of blocks.**

s	P	s	P	s	P	s	P	s	P
k = 3, m = 2		2.800	0.818	3.429	0.808	13.000	1.000	**k = 3, m = 10**	
		3.600	0.876	3.714	0.888	14.250	1.000		
0.000	0.167	4.800	0.907	4.571	0.915	16.000	1.000	0.000	0.026
1.000	0.500	5.200	0.961	5.429	0.949			0.200	0.170
3.000	0.833	6.400	0.976	6.000	0.973			0.600	0.290
4.000	1.000	7.600	0.992	7.143	0.979	s	P	0.800	0.399
		8.400	0.999	7.714	0.984	**k = 3, m = 9**		1.400	0.564
		10.000	1.000	8.000	0.992			1.800	0.632
s	P			8.857	0.996	0.000	0.029	2.400	0.684
k = 3, m = 3				10.286	0.997	0.222	0.186	2.600	0.778
		s	P	10.571	0.999	0.667	0.315	3.200	0.813
0.000	0.056	**k = 3, m = 6**		11.143	1.000	0.889	0.431	3.800	0.865
0.667	0.472			12.286	1.000	1.556	0.602	4.200	0.908
2.000	0.639	0.000	0.044	14.000	1.000	2.000	0.672	5.000	0.922
2.667	0.806	0.333	0.260			2.667	0.722	5.400	0.934
4.667	0.972	1.000	0.430			2.889	0.813	5.600	0.954
6.000	1.000	1.333	0.570	s	P	3.556	0.846	6.200	0.970
		2.333	0.748	**k = 3, m = 8**		4.222	0.893	7.200	0.974
		3.000	0.816			4.667	0.931	7.400	0.982
s	P	4.000	0.858	0.000	0.033	5.556	0.943	7.800	0.988
k = 3, m = 4		4.333	0.928	0.250	0.206	6.000	0.952	8.600	0.993
		5.333	0.948	0.750	0.346	6.222	0.969	9.600	0.994
0.000	0.069	6.333	0.971	1.000	0.469	6.889	0.981	9.800	0.997
0.500	0.347	7.000	0.988	1.750	0.645	8.000	0.984	10.400	0.998
1.500	0.569	8.333	0.992	2.250	0.715	8.222	0.990	11.400	0.999
2.000	0.727	9.000	0.994	3.000	0.764	8.667	0.994	12.200	0.999
3.500	0.875	9.333	0.998	3.250	0.851	9.556	0.996	12.600	0.999
4.500	0.931	10.333	1.000	4.000	0.880	10.667	0.997	12.800	1.000
6.000	0.958	12.000	1.000	4.750	0.921	10.889	0.999	13.400	1.000
6.500	0.995			5.250	0.953	11.556	0.999	14.600	1.000
8.000	1.000			6.250	0.962	12.667	1.000	15.000	1.000
		s	P	6.750	0.970	13.556	1.000	15.200	1.000
		k = 3, m = 7		7.000	0.982	14.000	1.000	15.800	1.000
s	P			7.750	0.990	14.222	1.000	16.200	1.000
k = 3, m = 5		0.000	0.036	9.000	0.992	14.889	1.000	16.800	1.000
		0.286	0.232	9.250	0.995	16.222	1.000	18.200	1.000
0.000	0.046	0.857	0.380	9.750	0.998	18.000	1.000	20.000	1.000
0.400	0.309	1.143	0.514	10.750	0.999				
1.200	0.478	2.000	0.695	12.000	0.999				
1.600	0.633	2.571	0.763	12.250	1.000				

Table 11   Friedman Test (continued)

$$H_0 : \tilde{\mu}_1 = \tilde{\mu}_2 = \cdots = \tilde{\mu}_k$$

$$P = P(S \leq s)$$

$$S = \frac{12km}{k(k+1)} \sum_{i=1}^{k} \left[R_i. - \frac{1}{2}(k+1) \right]^2$$

Note: k = number of treatment levels, m = number of blocks.

s	P	s	P	s	P	s	P	s	P
k = 3, m = 11		20.182	1.000	15.167	1.000	8.000	0.988	**k = 3, m = 10**	
		22.000	1.000	15.500	1.000	8.769	0.991		
0.000	0.024			16.167	1.000	9.385	0.993	0.571	0.306
0.182	0.156			16.667	1.000	9.692	0.995	1.000	0.449
0.545	0.268	s	P	17.167	1.000	9.846	0.996	1.286	0.511
0.727	0.371	**k = 3, m = 12**		18.000	1.000	10.308	0.997	1.714	0.562
1.273	0.530			18.167	1.000	11.231	0.998	1.857	0.656
1.636	0.597	0.000	0.022	18.500	1.000	11.538	0.998	2.286	0.695
2.182	0.649	0.167	0.144	18.667	1.000	11.692	0.999	2.714	0.758
2.364	0.744	0.500	0.249	19.500	1.000	12.154	0.999	3.000	0.812
2.909	0.781	0.667	0.346	20.167	1.000	12.462	0.999	3.571	0.833
3.455	0.837	1.167	0.500	20.667	1.000	12.923	0.999	3.857	0.850
3.818	0.884	1.500	0.566	22.167	1.000	14.000	1.000	4.000	0.884
4.545	0.900	2.000	0.617	24.000	1.000	14.308	1.000	4.429	0.910
4.909	0.913	2.167	0.713			14.923	1.000	5.143	0.920
5.091	0.938	2.667	0.751			15.385	1.000	5.286	0.937
5.636	0.957	3.167	0.809	s	P	15.846	1.000	5.571	0.952
6.545	0.962	3.500	0.859	**k = 3, m = 13**		16.615	1.000	6.143	0.963
6.727	0.973	4.167	0.877			16.769	1.000	6.857	0.967
7.091	0.981	4.500	0.892	0.000	0.020	17.077	1.000	7.000	0.978
7.818	0.987	4.667	0.920	0.154	0.134	17.231	1.000	7.429	0.983
8.727	0.989	5.167	0.942	0.462	0.233	18.000	1.000	8.143	0.987
8.909	0.994	6.000	0.949	0.615	0.325	18.615	1.000	8.714	0.990
9.455	0.996	6.167	0.962	1.077	0.473	19.077	1.000	9.000	0.992
10.364	0.997	6.500	0.973	1.385	0.537	19.538	1.000	9.143	0.993
11.091	0.998	7.167	0.980	1.846	0.588	19.846	1.000	9.571	0.995
11.455	0.999	8.000	0.983	2.000	0.684	20.462	1.000	10.429	0.996
11.636	0.999	8.167	0.989	2.462	0.722	21.385	1.000	10.714	0.997
12.182	0.999	8.667	0.993	2.923	0.783	22.154	1.000	10.857	0.998
13.273	1.000	9.500	0.995	3.231	0.835	22.615	1.000	11.286	0.998
13.636	1.000	10.167	0.996	3.846	0.855	24.154	1.000	11.571	0.998
13.818	1.000	10.500	0.997	4.154	0.871	26.000	1.000	12.000	0.999
14.364	1.000	10.667	0.998	4.308	0.902			13.000	0.999
14.727	1.000	11.167	0.998	4.769	0.927			13.286	1.000
15.273	1.000	12.167	0.999	5.538	0.935	s	P	13.857	1.000
16.545	1.000	12.500	0.999	5.692	0.950	**k = 3, m = 14**		14.286	1.000
16.909	1.000	12.667	0.999	6.000	0.963			14.714	1.000
17.636	1.000	13.167	1.000	6.615	0.972	0.000	0.019	15.429	1.000
18.182	1.000	13.500	1.000	7.385	0.975	0.143	0.126	15.571	1.000
18.727	1.000	14.000	1.000	7.538	0.984	0.429	0.219	15.857	1.000

Table 11   Friedman Test (continued)

$$H_0 : \tilde{\mu}_1 = \tilde{\mu}_2 = \cdots = \tilde{\mu}_k$$

$$P = P(S \le s)$$

$$S = \frac{12km}{k(k+1)} \sum_{i=1}^{k} \left[R_{i\cdot} - \frac{1}{2}(k+1) \right]^2$$

Note: k = number of treatment levels, m = number of blocks.

s	P
16.000	1.000
16.714	1.000
17.286	1.000
17.714	1.000
18.143	1.000
18.429	1.000
19.000	1.000
19.857	1.000
20.571	1.000
21.000	1.000
21.143	1.000
21.571	1.000
22.286	1.000
22.429	1.000
23.286	1.000
24.143	1.000
24.571	1.000
26.143	1.000
28.000	1.000

k = 3, m = 15

s	P
0.000	0.018
0.133	0.118
0.400	0.206
0.533	0.289
0.933	0.427
1.200	0.487
1.600	0.537
1.733	0.631
2.133	0.670
2.533	0.733
2.800	0.789
3.333	0.811
3.600	0.830
3.733	0.865
4.133	0.894
4.800	0.904

s	P
4.933	0.923
5.200	0.940
5.733	0.953
6.400	0.958
6.533	0.970
6.933	0.977
7.600	0.982
8.133	0.986
8.400	0.989
8.533	0.990
8.933	0.993
9.733	0.994
10.000	0.995
10.133	0.996
10.533	0.997
10.800	0.997
11.200	0.998
12.133	0.999
12.400	0.999
12.933	0.999
13.333	0.999
13.733	1.000
14.400	1.000
14.533	1.000
14.800	1.000
14.933	1.000
15.600	1.000
16.133	1.000
16.533	1.000
16.933	1.000
17.200	1.000
17.733	1.000
18.533	1.000
19.200	1.000
19.600	1.000
19.733	1.000
20.133	1.000
20.800	1.000
20.933	1.000
21.733	1.000

s	P
22.533	1.000
22.800	1.000
22.933	1.000
23.333	1.000
24.133	1.000
24.400	1.000
25.200	1.000
26.133	1.000
26.533	1.000
28.133	1.000
30.000	1.000

k = 4, m = 2

s	P
0.000	0.042
0.600	0.167
1.200	0.208
1.800	0.375
2.400	0.458
3.000	0.542
3.600	0.625
4.200	0.792
4.800	0.833
5.400	0.958
6.000	1.000

k = 4, m = 3

s	P
0.200	0.042
0.600	0.090
1.000	0.273
1.800	0.392
2.200	0.476
2.600	0.554
3.400	0.658
3.800	0.700

s	P
4.200	0.793
5.000	0.825
5.400	0.852
5.800	0.925
6.600	0.946
7.000	0.967
7.400	0.983
8.200	0.998
9.000	1.000

k = 4, m = 4

s	P
0.000	0.008
0.300	0.072
0.600	0.100
0.900	0.200
1.200	0.246
1.500	0.323
1.800	0.351
2.100	0.476
2.400	0.492
2.700	0.568
3.000	0.611
3.300	0.645
3.600	0.676
3.900	0.758
4.500	0.800
4.800	0.810
5.100	0.842
5.400	0.859
5.700	0.895
6.000	0.906
6.300	0.923
6.600	0.932
6.900	0.946
7.200	0.948
7.500	0.964
7.800	0.967

k = 3, m = 10

s	P
8.100	0.981
8.400	0.986
8.700	0.988
9.300	0.993
9.600	0.994
9.900	0.997
10.200	0.998
10.800	0.999
11.100	1.000
12.000	1.000

k = 4, m = 5

s	P
0.120	0.025
0.360	0.056
0.600	0.143
1.080	0.229
1.320	0.291
1.560	0.348
2.040	0.439
2.280	0.479
2.520	0.555
3.000	0.592
3.240	0.628
3.480	0.702
3.960	0.740
4.200	0.774
4.440	0.790
4.920	0.838
5.160	0.849
5.400	0.877
5.880	0.893
6.120	0.907
6.360	0.925
6.840	0.933
7.080	0.945

Table 11   Friedman Test (continued)

$$H_0 : \bar{\mu}_1 = \bar{\mu}_2 = \cdots = \bar{\mu}_k$$

$$P = P(S \le s)$$

$$S = \frac{12km}{k(k+1)} \sum_{i=1}^{k} \left[R_{i\cdot} - \frac{1}{2}(k+1) \right]^2$$

Note: k = number of treatment levels, m = number of blocks.

s	P	s	P	s	P	s	P	s	P
7.320	0.956	2.400	0.488	10.800	0.994	1.629	0.348	10.543	0.992
7.800	0.966	2.600	0.569	11.000	0.996	1.800	0.410	10.714	0.993
8.040	0.969	3.000	0.614	11.400	0.997	2.143	0.443	11.057	0.995
8.280	0.977	3.200	0.625	11.600	0.997	2.314	0.476	11.229	0.996
8.760	0.980	3.400	0.662	11.800	0.998	2.486	0.544	11.400	0.996
9.000	0.983	3.600	0.683	12.000	0.998	2.829	0.582	11.743	0.997
9.240	0.988	3.800	0.730	12.200	0.999	3.000	0.618	11.914	0.997
9.720	0.991	4.000	0.744	12.600	0.999	3.171	0.634	12.086	0.998
9.960	0.993	4.200	0.770	12.800	0.999	3.514	0.690	12.429	0.998
10.200	0.995	4.400	0.782	13.000	0.999	3.686	0.703	12.600	0.998
10.680	0.997	4.600	0.803	13.200	0.999	3.857	0.738	12.771	0.999
10.920	0.998	4.800	0.806	13.400	1.000	4.200	0.761	13.114	0.999
11.160	0.998	5.000	0.837	13.600	1.000	4.371	0.780	13.286	0.999
11.640	0.998	5.200	0.845	13.800	1.000	4.543	0.805	13.457	0.999
11.880	0.999	5.400	0.873	14.000	1.000	4.886	0.820	13.800	0.999
12.120	0.999	5.600	0.886	14.400	1.000	5.057	0.839	13.971	0.999
12.600	1.000	5.800	0.892	14.600	1.000	5.229	0.857	14.143	1.000
12.840	1.000	6.200	0.911	14.800	1.000	5.571	0.878	14.486	1.000
13.080	1.000	6.400	0.912	15.000	1.000	5.743	0.882	14.657	1.000
13.560	1.000	6.600	0.927	15.200	1.000	5.914	0.900	14.829	1.000
14.040	1.000	6.800	0.934	15.400	1.000	6.257	0.907	15.171	1.000
15.000	1.000	7.000	0.940	15.800	1.000	6.429	0.915	15.343	1.000
		7.200	0.944	16.000	1.000	6.600	0.927	15.514	1.000
		7.400	0.957	16.200	1.000	6.943	0.937	15.857	1.000
s	**P**	7.600	0.959	16.400	1.000	7.114	0.944	16.029	1.000
k = 4, m = 6		7.800	0.963	17.000	1.000	7.286	0.948	16.200	1.000
		8.000	0.965	18.000	1.000	7.629	0.959	16.543	1.000
0.000	0.004	8.200	0.968			7.800	0.962	16.714	1.000
0.200	0.043	8.400	0.971			7.971	0.965	16.886	1.000
0.400	0.060	8.600	0.977	**s**	**P**	8.314	0.967	17.229	1.000
0.600	0.126	8.800	0.978	**k = 4, m = 7**		8.486	0.970	17.400	1.000
0.800	0.156	9.000	0.983			8.657	0.977	17.571	1.000
1.000	0.211	9.400	0.986	0.086	0.016	9.000	0.980	17.914	1.000
1.200	0.228	9.600	0.987	0.257	0.037	9.171	0.983	18.257	1.000
1.400	0.321	9.800	0.990	0.429	0.094	9.343	0.985	18.771	1.000
1.600	0.332	10.000	0.990	0.771	0.155	9.686	0.987	18.943	1.000
1.800	0.391	10.200	0.991	0.943	0.200	9.857	0.988	19.286	1.000
2.000	0.426	10.400	0.993	1.114	0.243	10.029	0.990	19.971	1.000
2.200	0.459	10.600	0.994	1.457	0.315	10.371	0.991	21.000	1.000

Table 11 Friedman Test (continued)

$$H_0 : \bar{\mu}_1 = \bar{\mu}_2 = \cdots = \bar{\mu}_k$$

$$P = P(S \le s)$$

$$S = \frac{12km}{k(k+1)} \sum_{i=1}^{k} \left[R_{i\cdot} - \frac{1}{2}(k+1) \right]^2$$

Note: k = number of treatment levels, m = number of blocks.

s	P	s	P	s	P	s	P
k = 4, m = 8		5.850	0.890	12.300	0.997	18.600	1.000
		6.000	0.894	12.450	0.998	18.750	1.000
0.000	0.002	6.150	0.900	12.600	0.998	19.050	1.000
0.150	0.029	6.300	0.906	12.750	0.998	19.200	1.000
0.300	0.041	6.450	0.919	12.900	0.998	19.350	1.000
0.450	0.088	6.600	0.921	13.050	0.998	19.500	1.000
0.600	0.110	6.750	0.932	13.200	0.999	19.650	1.000
0.750	0.151	7.050	0.940	13.350	0.999	19.800	1.000
0.900	0.163	7.200	0.942	13.500	0.999	19.950	1.000
1.050	0.235	7.350	0.949	13.650	0.999	20.250	1.000
1.200	0.243	7.500	0.951	13.800	0.999	20.400	1.000
1.350	0.290	7.650	0.954	13.950	0.999	20.550	1.000
1.500	0.319	7.800	0.958	14.250	0.999	20.700	1.000
1.650	0.346	7.950	0.962	14.400	0.999	20.850	1.000
1.800	0.371	8.100	0.963	14.550	0.999	21.150	1.000
1.950	0.442	8.250	0.969	14.700	1.000	21.600	1.000
2.250	0.483	8.550	0.972	14.850	1.000	21.750	1.000
2.400	0.493	8.700	0.975	15.000	1.000	21.900	1.000
2.550	0.529	8.850	0.977	15.150	1.000	22.200	1.000
2.700	0.550	9.000	0.978	15.300	1.000	22.950	1.000
2.850	0.596	9.150	0.981	15.450	1.000	24.000	1.000
3.000	0.611	9.450	0.984	15.600	1.000		
3.150	0.638	9.600	0.985	15.750	1.000		
3.300	0.650	9.750	0.986	15.900	1.000	**s**	**P**
3.450	0.674	9.900	0.986	16.050	1.000	**k = 5, m = 2**	
3.600	0.677	10.050	0.989	16.200	1.000		
3.750	0.713	10.200	0.989	16.350	1.000	0.000	0.008
3.900	0.722	10.350	0.991	16.650	1.000	0.400	0.042
4.050	0.758	10.500	0.991	16.800	1.000	0.800	0.067
4.200	0.774	10.650	0.992	16.950	1.000	1.200	0.117
4.350	0.781	10.800	0.992	17.100	1.000	1.600	0.175
4.650	0.807	10.950	0.994	17.250	1.000	2.000	0.225
4.800	0.809	11.100	0.994	17.400	1.000	2.400	0.258
4.950	0.832	11.250	0.995	17.550	1.000	2.800	0.342
5.100	0.842	11.400	0.995	17.700	1.000	3.200	0.392
5.250	0.852	11.550	0.996	17.850	1.000	3.600	0.475
5.400	0.859	11.850	0.996	18.150	1.000	4.000	0.525
5.550	0.879	12.000	0.996	18.300	1.000	4.400	0.608
5.700	0.883	12.150	0.997	18.450	1.000	4.800	0.658

s	P
5.200	0.742
5.600	0.775
6.000	0.825
6.400	0.883
6.800	0.933
7.200	0.958
7.600	0.992
8.000	1.000

s	P
k = 5, m = 3	
0.000	0.000
0.267	0.012
0.533	0.028
0.800	0.059
1.067	0.086
1.333	0.155
1.600	0.169
1.867	0.232
2.133	0.280
2.400	0.318
2.667	0.351
2.933	0.405
3.200	0.441
3.467	0.507
3.733	0.525
4.000	0.568
4.267	0.594
4.533	0.653
4.800	0.674
5.067	0.709
5.333	0.747
5.600	0.764
5.867	0.787
6.133	0.828
6.400	0.837
6.667	0.873

Table 11   Friedman Test (continued)

$$H_0 : \tilde{\mu}_1 = \tilde{\mu}_2 = \cdots = \tilde{\mu}_k$$

$$P = P(S \le s)$$

$$S = \frac{12km}{k(k+1)} \sum_{i=1}^{k} \left[R_{i\cdot} - \frac{1}{2}(k+1) \right]^2$$

Note: k = number of treatment levels, *m* = number of blocks.

s	P	s	P	s	P
6.933	0.883	3.200	0.448	11.000	0.992
7.200	0.904	3.400	0.500	11.200	0.993
7.467	0.920	3.600	0.521	11.400	0.994
7.733	0.937	3.800	0.558	11.600	0.995
8.000	0.944	4.000	0.587	11.800	0.996
8.267	0.955	4.200	0.605	12.000	0.996
8.533	0.962	4.400	0.630	12.200	0.997
8.800	0.972	4.600	0.671	12.400	0.998
9.067	0.974	4.800	0.683	12.600	0.998
9.333	0.983	5.000	0.714	12.800	0.999
9.600	0.985	5.200	0.725	13.000	0.999
9.867	0.992	5.400	0.751	13.200	0.999
10.133	0.995	5.600	0.773	13.400	0.999
10.400	0.996	5.800	0.795	13.600	1.000
10.667	0.997	6.000	0.803	13.800	1.000
10.933	0.999	6.200	0.822	14.000	1.000
11.467	1.000	6.400	0.839	14.200	1.000
12.000	1.000	6.600	0.857	14.400	1.000
		6.800	0.864	14.600	1.000
		7.000	0.879	14.800	1.000
s	P	7.200	0.887	15.200	1.000
k = 5, m = 4		7.400	0.905	15.400	1.000
		7.600	0.914	16.000	1.000
0.000	0.001	7.800	0.920		
0.200	0.009	8.000	0.928		
0.400	0.020	8.200	0.937		
0.600	0.041	8.400	0.940		
0.800	0.060	8.600	0.951		
1.000	0.094	8.800	0.957		
1.200	0.105	9.000	0.962		
1.400	0.150	9.200	0.965		
1.600	0.185	9.400	0.972		
1.800	0.215	9.600	0.975		
2.000	0.241	9.800	0.979		
2.200	0.285	10.000	0.981		
2.400	0.315	10.200	0.983		
2.600	0.370	10.400	0.986		
2.800	0.388	10.600	0.989		
3.000	0.421	10.800	0.990		

B

Probability Distributions

Table 1   Discrete Random Variables

Distribution	Probability Function	Mean $\mu = E[X]$	Variance $\sigma^2 = E[(X - \mu)^2]$	Moment Generating Function $M_X(t) = E[e^{tX}]$
Hypergeometric	$\dfrac{\binom{M}{a}\binom{N-M}{n-a}}{\binom{N}{n}}$ $a = 0, 1, \ldots, \min(n, M)$ $M = 0, 1, \ldots, N$	$\dfrac{nM}{N}$	$\dfrac{N-n}{N-1}\dfrac{nM}{N}\left(1 - \dfrac{M}{N}\right)$	Not Helpful
Binomial	$\binom{n}{a} p^a (1-p)^{n-a}$ $a = 0, 1, \ldots, n$ $0 \le p \le 1$	np	$np(1-p)$	$\left(1 - p + pe^t\right)^n$ $-\infty < t < \infty$
Geometric	$p(1-p)^{a-1}$ $a = 1, 2, 3, \ldots$ $0 < p < 1$	$\dfrac{1}{p}$	$\dfrac{1-p}{p^2}$	$\dfrac{pe^t}{1 - (1-p)e^t}$ $(1-p)e^t < 1$

Table 1   Discrete Random Variables (continued)

Distribution	Probability Function	Mean $\mu = E[X]$	Variance $\sigma^2 = E[(X - \mu)^2]$	Moment Generating Function $M_X(t) = E[e^{tX}]$
Negative Binomial	$\binom{a-1}{r-1} p^r (1 - p)^{a-r}$ $a = r, r+1, r+2, \ldots$ $r = 1, 2, \ldots, \quad 0 < p < 1$	$\dfrac{r}{p}$	$\dfrac{r(1-p)}{p^2}$	$\left(\dfrac{pe^t}{1 - (1-p)e^t} \right)^r$ $(1-p)e^t < 1$
Multinomial	$\binom{n}{a_1 \cdots a_k} \left(p_1^{a_1} p_2^{a_2} \cdots p_k^{a_k} \right)$ $a_i = 0, 1, \ldots, n, \ \sum_{i=1}^{k} a_i = n$ $0 \le p_i \le 1, \ \sum_{i=1}^{k} p_i = 1$	$E[X_i] = np_i$	$\text{Var}(X_i) = np_i(1 - p_i)$	$\left(p_1 e^{t_1} + \ldots + p_k e^{t_k} \right)^n$ $-\infty < t_i < \infty$
Poisson	$\dfrac{\lambda^a e^{-\lambda}}{a!}$ $a = 0, 1, 2, \ldots$ $\lambda > 0$	λ	λ	$e^{\lambda(e^t - 1)}$ $-\infty < t < \infty$

Table 2   Continuous Random Variables

Distribution	Probability Density Function	Mean $\mu = E[X]$	Variance $\sigma^2 = E[(X - \mu)^2]$	Moment Generating Function $M_X(t) = E[e^{tX}]$
Uniform	$\dfrac{1}{b-a}$ $a \leq x \leq b$ $-\infty < a < b < \infty$	$\dfrac{b+a}{2}$	$\dfrac{(b-a)^2}{12}$	$\dfrac{e^{bt} - e^{at}}{t(b-a)}$ $-\infty < t < \infty$
Normal	$\dfrac{1}{\sqrt{2\pi\sigma^2}} e^{-(x-\mu)^2/(2\sigma^2)}$ $-\infty < x < \infty$ $-\infty < \mu < \infty, \, 0 < \sigma^2$	μ	σ^2	$e^{\mu t + (1/2)\sigma^2 t^2}$ $-\infty < t < \infty$
Gamma	$\dfrac{x^{\alpha-1}}{\Gamma(\alpha)\beta^\alpha} e^{-x/\beta}$ $0 < x$ $0 < \alpha, \, 0 < \beta$	$\alpha\beta$	$\alpha\beta^2$	$\left(\dfrac{1}{1-\beta t}\right)^\alpha$ $t < \dfrac{1}{\beta}$
Weibull	$\left(\dfrac{\alpha}{\beta}\right)\left(\dfrac{x-\theta}{\beta}\right)^{\alpha-1} e^{[(x-\theta)/\beta]^\alpha}$ $\theta < x$ $-\infty < \theta < \infty, \, 0 < \alpha, \, 0 < \beta$	$\Gamma\left(\dfrac{\alpha+1}{\alpha}\right)$	$\Gamma\left(\dfrac{2+\alpha}{\alpha}\right) - \left[\Gamma\left(\dfrac{\alpha+1}{\alpha}\right)\right]^2$	Not Helpful
Beta	$\dfrac{\Gamma(\alpha+\beta)}{\Gamma(\alpha)\Gamma(\beta)} x^{\alpha-1}(1-x)^{\beta-1}$ $0 \leq x \leq 1$ $0 < \alpha, \, 0 < \beta$	$\dfrac{\alpha}{\alpha+\beta}$	$\dfrac{\alpha\beta}{(\alpha+\beta)^2(\alpha+\beta+1)}$	Not Helpful
Cauchy	$\dfrac{1}{\pi[1+(x-\theta)^2]}$ $-\infty < x < \infty$ $-\infty < \theta < \infty$	Doesn't Exist	Doesn't Exist	Doesn't Exist
Double Exponential	$\dfrac{1}{2\sigma} e^{-|x-\theta|/\sigma}$ $-\infty < x < \infty$ $-\infty < \theta < \infty, \, 0 < \sigma$	θ	$2\sigma^2$	$\dfrac{e^{t\theta}}{1-\sigma^2 t^2}$ $t < \dfrac{1}{\sigma}$

Table 2   Continuous Random Variables (continued)

Distribution	Probability Density Function	Mean $\mu = E[X]$	Variance $\sigma^2 = E[(X-\mu)^2]$	Moment Generating Function $M_X(t) = E[e^{tX}]$
Logistic	$\dfrac{e^{-(x-\theta)/\sigma}}{\sigma\left[1+e^{-(x-\theta)/\sigma}\right]^2}$ $-\infty < x < \infty$ $-\infty < \theta < \infty,\ 0 < \sigma$	θ	$\dfrac{\sigma^2\pi^2}{3}$	$e^{t\theta}\,\pi\sigma t\,\csc(\pi\sigma t)$ $\|t\| < \dfrac{1}{2\sigma}$
Bivariate Normal	$\dfrac{1}{2\pi\sqrt{\sigma_X^2\sigma_Y^2(1-\rho^2)}}e^{-Q^2/2}$ $Q = \dfrac{1}{1-\rho^2}\left[\left(\dfrac{x-\mu_X}{\sigma_X}\right)^2\right.$ $-2\rho\left(\dfrac{x-\mu_X}{\sigma_X}\right)\left(\dfrac{y-\mu_Y}{\sigma_Y}\right)+\left.\left(\dfrac{y-\mu_Y}{\sigma_Y}\right)^2\right]$ $-\infty < x < \infty,\ -\infty < y < \infty$	$E[X] = \mu_x$ $E[Y] = \mu_Y$	$\text{Var}(X) = \sigma_X^2$ $\text{Var}(Y) = \sigma_Y^2$	$\exp\left[t_1\mu_X + t_2\mu_Y\right.$ $+\tfrac{1}{2}\left[(t_1\sigma_X^2 + 2t_1t_2\rho\sigma_X\sigma_y\right.$ $\left.\left.+(t_2\sigma_Y)^2\right]\right]$ $-\infty < t_1 < \infty,\ -\infty < t_2 < \infty$

C Bibliography

Andrews, D. F. et al. *Robust Estimates of Location.* Princeton, N. J.: Princeton University Press, 1972.

Berman G., and K. D Fryer. *Introduction to Combinatorics.* New York: Academic Press.

Blalock, H. M. *Social Statistics,* New York: McGraw-Hill, 1972.

Cochran, W. G. *Sampling Techniques.* New York: John Wiley and Sons, 1963.

Cramér, H. *Mathematical Methods of Statistics.* Princeton, N. J.: Princeton University Press, 1946.

David, H. A. *Order Statistics.* New York: John Wiley and Sons, 1970.

Davies, O. L., ed. *Design and Analysis of Industrial Experiments.* London: Oliver and Boyd, Ltd., 1954.

Draper, N. R., and H. Smith. *Applied Regression Analysis.* New York: John Wiley and Sons, 1966.

Feller, W. *An Introduction to Probability Theory and Its Applications.* Vol. 1. New York: John Wiley and Sons, 1968.

Golub, G. H., and C. F. Van Loan. *Matrix Computations.* Baltimore, Md.: The John Hopkins University Press, 1983.

Graybill, F. A. *An Introduction to Linear Statistical Models.* Vol. 1. New York: McGraw-Hill, 1961.

Halmos, P. R. *Naive Set Theory.* New York: Springer-Verlag, 1974.

Hollander, M., and D. A. Wolfe. *Nonparametric Statistical Methods.* New York: John Wiley and Sons, 1973.

Johnson, N. L., and S. Kotz. *Distributions in Statistics, Discrete Distributions*. New York: John Wiley and Sons, 1969.

Johnson, N. L., and S. Kotz. *Distributions in Statistics, Continuous Univariate Distributions—1*. New York: John Wiley and Sons, 1970.

Johnson, N. L., and S. Kotz. *Distributions in Statistics, Continuous Univariate Distributions—2*. New York: John Wiley and Sons, 1970.

Johnson N. L., and S. Kotz. *Distributions in Statistics, Continuous Multivariate Distributions*. New York: John Wiley and Sons, 1972.

Kempthorne, O. *The Design and Analysis of Experiments*. New York: John Wiley and Sons, 1952.

Kendall, M. G., and A. Stuart. *The Advanced Theory of Statistics*. Vol. 2. London: Charles Griffin and Company, Ltd., 1973.

Lehmann, E. L. *Testing Statistical Hypotheses*. New York: John Wiley and Sons, 1959.

Lehmann, E. L. *Nonparametrics: Statistical Methods Based on Ranks*. San Francisco: Holden-Day, 1975.

Lehmann, E. L. *Theory of Point Estimation*. New York: John Wiley and Sons, 1983.

Ostle, B. and R. W. Mensing. *Statistics in Research*. Ames, Iowa: Iowa State University Press, 1975

Snedecor G., and W. Cochran. *Statistical Methods*. Iowa State University Press, 1989.

Stigler, S. M. "Kruskal's Proof of the Joint Distribution of $\bar{X}$ and s^2. *The American Statistician* 38, (1984): pp 134–5.

Chapter 1

Section 1.2

1.
$$A = \{a, e, i, o, u\} \qquad B = \{1, a, t, e, r\}$$
$$C = \{s, t, u, v, w, x, y, z\} \qquad D = \{a, e, l, r, t\}$$
$$B = D$$

3. (a) $B \subset A$ (b) $A \subset C$ (c) $A \subset U$
   (d) $\varnothing \subset B$ (e) $(0, 1] \subset B$

Section 1.3

1. (a) $A \cap B \cap C$ (b) $A \cup B \cup C$
   (c) $(A \cup B \cup C)'$
   (d) $(A \cap B) \cup (A \cap C) \cup (B \cap C)$
   (e) $A \cap (B \cup C)$
   (f) $(A \cup B) \cap (A \cap B)'$
3. (a) $\{(1,3), (3,3), (5,3)\}$ (b) $\varnothing$
   (c) $\{(1,3), (1,4), (1,6), (2,3), (2,5), (3,2), (3,3),$
   $(3,4), (3.6), (4,1), (4,3), (4,5), (5,2), (5,3),$
   $(5,4), (5,6), (6,1), (6,3), (6,5)\}$
   (d) $\{(1,2), (2,1)\}$
5. (a) $S = \{(p_1, n_1), (p_1, n_2), (p_1, d_1), (p_1, d_2),$
   $(p_2, n_1), (p_2, n_2), (p_2, d_1), (p_2, d_2), (n, n_1),$
   $(n, n_2), (n, d_1), (n, d_2), (d, n_1), (d, n_2), (d, d_1),$
   $(d, d_2)\}$
   (b) $\{(p_1, n_1), (p_1, n_2), (p_2, n_1), (p_2, n_2), (n, n_1),$
   $(n, n_2), (n, d_1), (n, d_2), (d, n_1), (d, n_2)\}$
   (c) $\{(p_1, n_1), (p_1, n_2), (p_2, n_1), (p_2, n_2)\}$
7. (a) $S = [0, L]$ (b) $\left[\frac{1}{3}, \frac{2}{3}\right]$

Section 1.4

13. (a) 0.5 (b) 0.5 (c) 0.3
15. (a) 0.7 (b) 0.9
    (c) 0.3 (d) 0.1

Section 1.5

1. $\sum_{j=1}^{i} n_j$
3. (a) $\frac{7!}{4!3!}$ (b) $4(3!)^2$
7. (a) $\frac{9!}{2 \cdot 2 \cdot 2}$ (b) $\frac{10!}{3!3!2!}$
11. 40
19. 3^8
21. $\binom{12}{3}$
23. $5ab^4 + 10a^3c^2 + 30a^2b^2c$
25. (a) 8! (b) 7!
    (c) 144 (d) $3!2^4$

Section 1.6

1. (a) $\frac{38}{100}, \frac{1}{2}$ (b) $\frac{28}{90}, \frac{1}{2}$
3. $\frac{1}{25}$
5. $1 - \left(\frac{5}{6}\right)^{10}$
7. (a) $\dfrac{\binom{13}{1}\binom{48}{1}}{\binom{52}{5}}$ (b) $\dfrac{\binom{13}{2}\binom{4}{3}\binom{4}{2}}{\binom{52}{5}}$
   (c) $\dfrac{\binom{4}{2}\binom{13}{1}\binom{12}{3}\binom{4}{1}^3}{\binom{52}{5}}$

11. (a) 0.969   (b) 0.021

13. $\sqrt{\frac{1}{8}} \left(\frac{1}{16}\right)^{20} \left(\frac{5}{4}\right)^{60} 10^{20}$

15. $1 - \left(\frac{9}{10}\right)^n, 7$

17. $m = n$ is one.

19. (a) $1 - \dfrac{\binom{m-1}{m-n}}{\binom{n+m-1}{m}}$ (b) $\dfrac{n\binom{m+n-2}{m}}{\binom{n+m-1}{m}}$

   (c) $\dfrac{\binom{n}{i}\binom{m+n-i-1}{m}}{\binom{n+m-1}{m}}$

Review Exercises—Chapter 1

3. (a) False   (b) False
   (c) False   (d) True

5. $\frac{32}{663}$

7. (a) $\dfrac{\binom{5}{3}\binom{4}{2}\binom{2}{1}}{\binom{11}{5}}$

   (b) $1 - \left[\dfrac{\binom{6}{5}}{\binom{11}{5}} + \dfrac{\binom{5}{1}\binom{6}{4}}{\binom{11}{5}}\right]$

9. $\dfrac{\binom{2}{1}\binom{2}{1}}{\binom{4}{2}}$

11. $1 - \dfrac{\binom{4}{1}\binom{4}{1}\binom{3}{1}\binom{2}{1}}{\binom{13}{4}}$

13. (a) $\frac{1}{2^{n-1}}$

   (b)
   $$P(r = 2k) = \frac{2\binom{m-1}{k-1}\binom{n-m-1}{k-1}}{\binom{n}{m}}$$

   $$P(r = 2k+1) = \frac{\binom{m-1}{n-1}\binom{n-m-1}{k} + \binom{m-1}{k}\binom{n-m-1}{k-1}}{\binom{n}{m}}$$

Chapter 2
Section 2.2

1. (a) $B \cap C'$ (b) $C \mid A$
   (c) $B \mid (A \cap C)$ (d) $(B \cap C') \mid A'$

3. (a) True   (b) True
   (c) False   (d) False

5. (a) $\dfrac{\binom{5}{2} + \binom{3}{2}}{\binom{8}{2}}$ (b) $\dfrac{5 \cdot 5 + 3 \cdot 3}{8 \cdot 8}$

7. (a) $\frac{3}{7}$ (b) $\dfrac{5\binom{5}{2}}{\binom{9}{5}}$

9. $\dfrac{(4!)^2}{24 \cdot 23 \cdot 22 \cdot 21}$

11. $\dfrac{\binom{5}{2} + \binom{4}{2}}{2\binom{7}{2}}$

13. $\frac{1}{25}$

15. (a) $\dfrac{\binom{3}{1}\binom{48}{3} + \binom{3}{2}\binom{48}{2} + \binom{48}{1}}{\binom{51}{4}}$

   (b) $\dfrac{\binom{12}{2}\binom{39}{2} + \binom{12}{3}\binom{39}{1} + \binom{12}{1}}{\binom{51}{4}}$

Section 2.3

1. 0.6
3. 0.206
5. 0.49
7. $\frac{28}{127}$
9. 0.305

Section 2.4

3. 3
5. (a) $(0.25)^{10}$ (b) $1 - (0.75)^{10}$
7. (a) $\left(\frac{2}{3}\right)^3$ (b) $1 - \left(\frac{2}{3}\right)^3$
   (c) $3\left(\frac{1}{3}\right)\left(\frac{2}{3}\right)^2$
9. (a) False   (b) False
   (c) True   (d) False
11. (a) $[1 - P(A)][1 - P(B)][1 - P(C)]$
   (b) $P(A)P(B)[1 - P(C)] + P(A)P(C)[1 - P(B)] + P(B)P(C)[1 - P(A)]$
   (c) $P(A)P(B)[1 - P(C)] + P(A)P(C)[1 - P(B)] + P(B)P(C)[1 - P(A)] + P(A)P(B)P(C)$
   (d) $1 - P(A)P(B)P(C)$
13. (a) $\frac{3}{10}$ (b) $\frac{2}{5}$ (c) $\frac{19}{40}$
15. $P_1[1 - (1 - P_2)(1 - P_3)(1 - P_4)(1 - P_5)]$
17. (a) $\frac{1}{2}$ (b) 8   (c) $\left(\frac{1}{2}\right)^k$
19. (a) $\frac{5}{12}$ (b) $\frac{31}{60}$ (c) $\frac{29}{60}$

Review Exercises—Chapter 2

1. (a) $\frac{1}{4}$ (b) $\frac{3}{4}$
3. (a) $\dfrac{w}{w+b+c}$ (b) $\dfrac{w+c}{w+b+c}$ (c) $\dfrac{w}{w+b}$
5. $\frac{5}{54}$
9. (a) 0.895   (b) 0.395
11. (a) $\dfrac{1}{\binom{w+b}{w}}$ (b) $\displaystyle\sum_{i=0}^{b-1} \dfrac{\binom{b}{i}}{\binom{w+b}{w+i}}$
13. (a) $\frac{7-i}{21}$ (b) $\frac{13}{36}$
15. $\frac{131}{243}$

Chapter 3

Section 3.1

1. (a) $-1, 0, 2$
   (b) $P(X = -1) = \frac{1}{6}, P(X = 0) = \frac{2}{6}, P(X = 2) = \frac{3}{6}$
   (c) $\frac{1}{6}$

3. (a) $0, 1, 2, 3, 4, 5$
   (b) $P(X = i) = \frac{6-i}{21}, i = 0, 1, 2, 3, 4, 5$

5. (a) $(0, \infty)$ (b) $(0, \infty)$
   (c) $(-\infty, \infty)$ (d) $(0, 1)$

Section 3.2

1.
$$F_X(a) = \begin{cases} 0, & a < -1 \\ \frac{1}{6}, & -1 \le a < 0 \\ \frac{1}{2}, & 0 \le a < 2 \\ 1, & 2 \le a \end{cases}$$

3.
$$F_X(a) = \begin{cases} 0, & a < 0 \\ \frac{6}{21}, & 0 \le a < 1 \\ \frac{11}{21}, & 1 \le a < 2 \\ \frac{15}{21}, & 2 \le a < 3 \\ \frac{18}{21}, & 3 \le a < 4 \\ \frac{20}{21}, & 4 \le a < 5 \\ 1, & 5 \le a \end{cases}$$

7. (a) $\frac{3}{4}$ (b) $\frac{1}{4}$ (c) $\frac{1}{2}$
   (d) 0 (e) $\frac{1}{8}$ (f) 0

9. $\gamma_{25} = \gamma_{50} = 0, \gamma_{75} = 2$
11. $\gamma_{25} = \gamma_{50} = 1, \gamma_{75} = 3$
13. (a) 1 (b) $\frac{1}{2}$
    (c) 3 (d) 4

Section 3.3

1. (a)
$$p_X(a) = \begin{cases} \frac{1}{35}, & a = 0 \\ \frac{12}{35}, & a = 1 \\ \frac{18}{35}, & a = 2 \\ \frac{4}{35}, & a = 3 \\ 0, & \text{otherwise} \end{cases}$$

   (b) $\frac{1}{35}$ (c) $\frac{22}{35}$ (d) 2

3. (a)
$$p_X(a) = \begin{cases} \dfrac{\binom{a-1}{2}}{\binom{10}{3}}, & a = 1, \ldots, 8 \\ 0, & \text{otherwise} \end{cases}$$

   (b)
$$p_X(a) = \begin{cases} \dfrac{\binom{a+2-1}{2}}{\binom{11}{3}}, & a = 1, \ldots, 10 \\ 0, & \text{otherwise} \end{cases}$$

5. (a) $\frac{1}{45}$ (b) $\frac{3}{45}$ (c) $\frac{20}{45}$
   (d)
$$F_X(a) = \begin{cases} 0, & a < 1 \\ \frac{1}{45}, & 1 \le a < 2 \\ \frac{3}{45}, & 2 \le a < 3 \\ \frac{6}{45}, & 3 \le a < 4 \\ \frac{10}{45}, & 4 \le a < 5 \\ \frac{15}{45}, & 5 \le a < 6 \\ \frac{21}{45}, & 6 \le a < 7 \\ \frac{28}{45}, & 7 \le a < 8 \\ \frac{36}{45}, & 8 \le a < 9 \\ 1, & 9 \le a \end{cases}$$

   (e) 7

7. Assume X = no. vowels $-$ no. consonants.
   (a)
$$p_X(a) = \begin{cases} \dfrac{\binom{55}{5}}{\binom{100}{5}}, & a = -5 \\[2mm] \dfrac{\binom{45}{1}\binom{55}{4}}{\binom{100}{5}}, & a = -3 \\[2mm] \dfrac{\binom{45}{2}\binom{55}{3}}{\binom{100}{5}}, & a = -1 \\[2mm] \dfrac{\binom{45}{3}\binom{55}{2}}{\binom{100}{5}}, & a = 1 \\[2mm] \dfrac{\binom{45}{4}\binom{55}{1}}{\binom{100}{5}}, & a = 3 \\[2mm] \dfrac{\binom{45}{5}}{\binom{100}{5}}, & a = 5 \\[2mm] 0, & \text{otherwise} \end{cases}$$

   (b) $p_X(1) + p_X(3) + p_X(5)$

9. (a) 0.3 (b) 0.75 (c) $\frac{9}{13}$

11.
$$0.75 \sum_{a=9}^{12} \binom{12}{a} (0.8)^a (0.2)^{12-a} +$$
$$0.25 \sum_{a=0}^{8} \binom{12}{a} (0.3)^a (0.7)^{12-a}$$

Section 3.4

1. (a) Yes   (b) No
   (c) Yes   (d) Yes
3. (b) $F_X(a) = \frac{1}{2} + \frac{1}{\pi} \tan^{-1}(a), -\infty < a < \infty$
   (c) $\frac{1}{2}$ (d) 0.83

5. (a) $\frac{1}{2}$ (c) $\frac{1}{2}$ (d) $\frac{1}{\sqrt{5}}$

7. (a) $\frac{1}{4}$ (b) $\frac{175}{256}$

9. $\frac{1}{3}$

11. (b)
$$f_X(x) = \begin{cases} xe^{-x}, & 0 \le x \\ 0, & \text{otherwise} \end{cases}$$

(c) $2e^{-1} - 3e^{-2}$

Section 3.5

1. (a)
$$p_{X,Y}(a,b) = \begin{cases} \frac{2}{15}, & (a,b) = (0,0) \\ \frac{4}{15}, & (a,b) = (1,0),(0,1) \\ \frac{5}{15}, & (a,b) = (1,1) \\ 0, & \text{otherwise} \end{cases}$$

(b)
$$p_X(a) = \begin{cases} \frac{6}{15}, & a = 0 \\ \frac{9}{15}, & a = 1 \\ 0, & \text{otherwise} \end{cases}$$

(c)
$$p_Y(a) = \begin{cases} \frac{9}{15}, & a = 0 \\ \frac{6}{15}, & a = 1 \\ 0, & \text{otherwise} \end{cases}$$

(d) $F_{X,Y}(0,0) = 0, F_{X,Y}(0,1) = F_{X,Y}(1,0) = \frac{2}{15}, F_{X,Y}(1,1) = 1$

3. (a)
$$p_{X,Y}(a,b) = \begin{cases} \frac{1}{28}, & (a,b) = \begin{array}{l} (1,1),(1,2),(1,3),(1,4), \\ (1,5),(1,6),(1,7),(2,1), \\ (2,2),(2,3),(2,4),(2,5), \\ (2,6),(3,1),(3,2),(3,3), \\ (3,4),(3,5),(4,1),(4,2), \\ (4,3),(4,4),(5,1),(5,2), \\ (5,3),(6,1),(6,2),(7,1) \end{array} \\ 0, & \text{otherwise} \end{cases}$$

(b) $\frac{3}{4}$

5. (a) $\frac{48}{17}$

(b)
$$f_X(x) = \begin{cases} \frac{48}{17}\left[x^2 - \frac{x}{2}\right], & \frac{1}{2} < x < 1 \\ \frac{48}{17}\left[1 - \frac{x}{2}\right], & 1 \le x < 2 \\ 0, & \text{otherwise} \end{cases}$$

(c)
$$f_Y(y) = \begin{cases} \frac{24}{17}\left[1 - \frac{1}{y^4}\right], & 1 < y < 2 \\ 0, & \text{otherwise} \end{cases}$$

9. (a)
$$f_X(x) = \begin{cases} e^{-x}, & 0 \le x \\ 0, & \text{otherwise} \end{cases}$$

(b)
$$f_Y(y) = \begin{cases} \frac{1}{(1+y)^2}, & 0 \le y \\ 0, & \text{otherwise} \end{cases}$$

(c) $1 + e^{-2} - \frac{1}{3}e^{-6}$

13. (a) 48

(b)
$$f_X(x) = \begin{cases} 6x(x^4 - 2x^2 + 1), & 0 < x < 1 \\ 0, & \text{otherwise} \end{cases}$$

(c)
$$f_Y(y) = \begin{cases} 12y^3(1 - y^2), & 0 < z < 1 \\ 0, & \text{otherwise} \end{cases}$$

(d)
$$f_Z(z) = \begin{cases} 6z^5, & 0 < z < 1 \\ 0, & \text{otherwise} \end{cases}$$

(e) $\frac{3}{4}$

15. (a) $\frac{5}{6}$ (b) $\frac{1}{4}$ (c) $\frac{1}{6}$

Section 3.6

1. Independent
3. Not independent
5. Independent
7. Independent
9. (a) Not independent
   (b) Not independent
11. Independent
13. (a) 2   (b) 3
15. $3e^{-2}$

Section 3.7

1. (a)
$$p_{X|Y}(x \mid 1) = \begin{cases} \frac{2x-1}{7}, & x = 1,2,3 \\ 0, & \text{otherwise} \end{cases}$$
$$p_{X|Y}(x \mid 2) = \begin{cases} \frac{2x-2}{4}, & x = 1,2,3 \\ 0, & \text{otherwise} \end{cases}$$

(b)
$$p_{Y|X}(y \mid 1) = \begin{cases} 2 - y, & y = 1,2 \\ 0, & \text{otherwise} \end{cases}$$
$$p_{Y|X}(y \mid 2) = \begin{cases} \frac{4-y}{5}, & y = 1,2 \\ 0, & \text{otherwise} \end{cases}$$
$$p_{Y|X}(y \mid 3) = \begin{cases} \frac{6-y}{9}, & y = 1,2 \\ 0, & \text{otherwise} \end{cases}$$

3. (a)
$$f_{X|Y}(x \mid y) = \begin{cases} \frac{3(y^2 - x^2)}{2y^3}, & 0 < x < y \\ 0, & \text{otherwise} \end{cases}$$

(b)
$$f_{Y|X}(y \mid x) = \begin{cases} \frac{3(y^2 - x^2)}{2x^3 - 3x^2 + 1}, & x < y < 1 \\ 0, & \text{otherwise} \end{cases}$$

7. (a) $\frac{1}{400}$

(b)
$$f_{X|Y}(x \mid y) = \begin{cases} \frac{1}{20}, & 0 \le x \le 20 \\ 0, & \text{otherwise} \end{cases}$$

Review Exercises—Chapter 3

1. (a)
$$p_Y(a) = \begin{cases} \frac{21}{45}, & a = 0, 1 \\ \frac{3}{45}, & a = 2 \\ 0, & \text{otherwise} \end{cases}$$

(b)
$$p_Y(a) = \begin{cases} \frac{49}{100}, & a = 0 \\ \frac{42}{100}, & a = 1 \\ \frac{9}{100}, & a = 2 \\ 0, & \text{otherwise} \end{cases}$$

3. (a) 21   (b) $\frac{1}{3025}$ (c) 3
5. (b) Not independent
7. $\frac{31}{32}$
9. (b)
$$f_X(x) = \begin{cases} \frac{2}{\pi}\sqrt{1 - x^2}, & -1 \le x \le 1 \\ 0, & \text{otherwise} \end{cases}$$

(c) $\frac{1}{2}$

Chapter 4
Section 4.2

1. (a) 3   (b) Does not exist.
(c) $\frac{2n+1}{3}$
5. $\frac{23}{15}$
7. 4
9. 1

Section 4.3

3. 11.02
5. 4
7. $\$\frac{35}{16}D$
9. $\frac{2}{\pi}\left(1 - \ln\frac{1}{\sqrt{2}}\right)$
11. $n[1 - (1 - p)^8 - 8p(1 - p)^7]$
17. -25

Section 4.4

1. (a) 2   (b) Does not exist.
(c) $\frac{(n+2)(n-1)}{18}$
3. $\frac{1-p}{p}$
5. $E[R] = \frac{2}{3}a$, $\text{Var}(R) = \frac{1}{18}a^2$
7. Mean $= \mu_N \mu_A$, Variance $= \sigma_N^2 \mu_A^2 + \sigma_A^2 \mu_N^2 + \sigma_N^2 \sigma_A^2$
9. $\rho = 0$, not independent.

13. $\rho_{T,U} = \rho_{X,Y}$
15. $\sqrt{\frac{45}{48}}$

Section 4.5

1. (b) np (c) $np(1 - p)$
3. $E[X] = 0$, $\text{Var}(X) = 1$, $\gamma_1 = 0$, $\gamma_2 = 3$
5. (a) $(r + pe^{t_1} + qe^{t_2})^n$ (b) $(r + q + pe^{t_1})^n$
7.
$$E[X] = \frac{\partial}{\partial t_1} M_{X,Y}(0,0)$$

$$E[Y] = \frac{\partial}{\partial t_2} M_{X,Y}(0,0)$$

$$E[X^2] = \frac{\partial^2}{\partial t_1^2} M_{X,Y}(0,0)$$

$$E[XY] = \frac{\partial^2}{\partial t_1 \partial t_2} M_{X,Y}(0,0)$$

Section 4.6

1. (a) $\frac{13}{4}$ (b) $\frac{13}{4}$
3. $E[Y \mid X = i] = \frac{8+4i}{10}$, $i = 0, 1, 2$
5. (a) $\frac{4-6x}{9-6x}$ (b) $\frac{5-4x}{18-12x}$
(c) $\frac{5-4x}{18-12x} - \left[\frac{4-6x}{9-6x}\right]^2$

Review Exercises—Chapter 4

1. (a) True   (b) True
(c) True
3. (a) $\frac{\partial^3}{\partial t_1 \partial t_2^2} M_{X,Y}(0,0)$ (b) $\frac{\partial^3}{\partial t_2^3} M_{X,Y}(0,0)$
(c) $\frac{\partial^4}{\partial t_1^2 \partial t_2^2} M_{X,Y}(0,0)$
5. 0
7. $\gamma_{1X} = \gamma_{1Y}$ $\gamma_{2X} = \gamma_{2Y}$
9. $E[X] = 1$, $\text{Var}(X) = \frac{1}{6}$, $M_X(t) = \frac{1}{t^2}(e^{2t} - 2e^t + 1)$

Chapter 5
Section 5.2

5. $\dfrac{\binom{4}{3}\binom{6}{0}}{\binom{10}{3}}$

7. $\frac{11}{15}$

Section 5.3

5. (a) $\binom{4}{2}\left(\frac{1}{2}\right)^2\left(\frac{3}{4}\right)^2$ (b) $\binom{3}{1}\left(\frac{1}{4}\right)^2\left(\frac{3}{4}\right)^2$
7. 0.6242

9. (a) $4(0.6)^4(0.4)$
   (b) $(0.6)^4[1 + 4(0.4) + 10(0.4)^2 + 20(0.4)^3]$

11. $1 - \sum_{a=1}^{5} (0.623)^{a-1}(0.377)$

13. (a) $1 - \sum_{a=1}^{4} \left(\frac{1}{6}\right)\left(\frac{5}{6}\right)^{a-1}$

   (b) $\sum_{a=1}^{3} \left(\frac{1}{6}\right)\left(\frac{5}{6}\right)^{a-1}$

15. $\sum_{a=3}^{5} \binom{5}{a}\left(\frac{1}{3}\right)^a\left(\frac{2}{3}\right)^{5-a}$

17. 0.6238

Section 5.4

3. (a)

$$p_{X,Y}(a,b) = \frac{5!}{a!b!(5-a-b)!}\left(\frac{12}{52}\right)^a\left(\frac{20}{52}\right)^b\left(\frac{20}{52}\right)^{5-a-b}$$

   (b) $20\left(\frac{3}{13}\right)^3\left(\frac{5}{13}\right)^2$ (c) $\binom{4}{2}\left(\frac{1}{2}\right)^4$

5. (a) $10(0.4)^2(0.5)^3$

   (b) $\sum_{a=0}^{2}\sum_{b=a+1}^{5-a} \frac{5!}{a!b!(5-a-b)!}(0.4)^a(0.5)^b(0.1)^{5-a-b}$

7. (a) $\frac{10!}{3!3!3!}(0.25)^6(0.15)^4$

   (b) $\frac{10!}{6!}(0.25)^2(0.15)^2(0.2)^6$ (c) 0.0328

Section 5.5

1. 0.2642
3. (a) 0.8153   (b) 0.8009
   (c) 0.7674   (d) 3
7. 230.26
9. 0.1181
11. (a) 0.1032   (b) 0.0183   (c) 4
13. (a) 0   (b) 0.7245
15. 0.54 mi

Section 5.6

1. (a) $\frac{1}{2}$ (b) $\frac{1}{4}$ (c) $\frac{5}{8}$
3. $\gamma_1 = 0, \gamma_2 = 1.8$
5. $\frac{2L}{\pi D}\left[1 - \sin\left(\cos^{-1}\frac{D}{L}\right)\right] + \frac{2}{\pi}\cos^{-1}\frac{D}{L}$
9. $\frac{4}{9}$
11. $\frac{1}{\pi}$

Section 5.7

5. (a) 0.6915   (b) 0.5987   (c) 0.0679
   (d) 0.0228   (e) 25.12
7. 0.905   9. 0.0019
11. (a) 0.9671   (b) 0.8461   (c) 0.0329

13. 0.9544
15. Yes

Section 5.8

5. $\beta(\alpha - 1)$
9. 0.1847
11. (a) 15 sec   (b) 0.453

Section 5.9

1. $$F_{X(x)} = \begin{cases} 0 & , x \le \theta \\ 1 - e^{-[(x-\theta)/\beta]^\alpha} & , \theta < x \end{cases}$$

3. 0
7. 0.9452
9. $F_X(x) = \frac{1}{2} + \frac{1}{\pi}\tan^{-1}(x - \theta), -\infty < x < \infty$
11. $$\gamma_{100p} = \begin{cases} \ln 2p & , 0 < p < \frac{1}{2} \\ -\ln 2(1 - p), & \frac{1}{2} \le p < 1 \end{cases}$$

13. 6
15. 0.7311
17. 4.2
19. (a) 0.018   (b) 0.612   (c) 0.762

Section 5.10

3. 0.7454
5. $\mu_X = 1, \mu_Y = 0, \sigma_X^2 = 2, \sigma_Y^2 = 1, \rho = \frac{1}{\sqrt{2}}$

Review Exercises—Chapter 5

1. More with the normal distribution.
5. (a) $\frac{37}{64}$ (b) $\frac{67}{256}$ (c) $\frac{29}{128}$ (d) $\frac{137}{228}$
7. 0.4013
9. $\frac{(1-p)^2}{2-p}$
11. 0.2873
13. 0.4679

Chapter 6
Section 6.2

1.
$$p(a) = \begin{cases} \frac{5}{15} & , a = \frac{4}{5} \\ \frac{4}{15} & , a = 1 \\ \frac{3}{15} & , a = \frac{4}{3} \\ \frac{2}{15} & , a = 2 \\ \frac{1}{15} & , a = 4 \\ 0 & , \text{otherwise} \end{cases}$$

3.
$$p_Y(a) = \begin{cases} \frac{2}{3n+1} & , a = 1, 4, \ldots, n^2 \\ \frac{1}{3n+1} & , a = 0(n+1)^2, \ldots, (2n)^2 \\ 0 & , \text{otherwise} \end{cases}$$

5.
$$p_T(a) = \begin{cases} (a-1)(0.25)^2(0.75)^{a-2} \, , a = 2,3,\ldots \\ 0 \qquad\qquad\qquad\text{, otherwise} \end{cases}$$

9.
$$P_Y(a) = \begin{cases} \frac{1}{5} \, , a = 0, \ln 3, \ln 5, \ln 7, \ln 9 \\ 0 \, , \text{otherwise} \end{cases}$$

Section 6.3

1.
$$f_Y(y) = \begin{cases} \frac{y^{n/2-1}n^{n/2}}{\Gamma\left(\frac{n}{2}\right)2^{n/2}}e^{-ny/2} \, , 0 < y \\ 0 \qquad\qquad\text{, otherwise} \end{cases}$$

3. $Y \sim G(1,1)$
5. $Y \sim U(0,1)$
7. $Y \sim G(1,1)$

Section 6.4

5.
$$f_Y(y) = \begin{cases} \sqrt{\frac{2}{\pi}}e^{-y^2/2} \, , 0 \le y \\ 0 \qquad\quad\text{, otherwise} \end{cases}$$

7. $A \sim U(0, \pi)$

Section 6.5

1. (a)
$$p_Y(a) = \begin{cases} \frac{1}{5} \, , a = 0, 7, 8, 9, 16 \\ 0 \, , \text{otherwise} \end{cases}$$

(b)
$$p_Y(a) = \begin{cases} \frac{1}{5} \, , a = 0 \\ \frac{2}{5} \, , a = 1, 2 \\ 0 \, , \text{otherwise} \end{cases}$$

(c)
$$p_Y(a) = \begin{cases} \frac{1}{5} \, , a = 1 \\ \frac{2}{5} \, , a = -1, 0 \\ 0 \, , \text{otherwise} \end{cases}$$

5. $T \sim N(a\mu + b\eta, a^2\sigma^2 + b^2\tau^2 + 2ab\rho\sigma\tau)$
7. $Y \sim G(1,1)$

Section 6.6

1.
$$f_T(t) = \begin{cases} \frac{t}{2} \, , 0 < t < 1 \\ \frac{1}{2} \, , 1 \le t < 2 \\ \frac{3}{2} - \frac{t}{2} \, , 2 \le t < 3 \\ 0 \, , \text{otherwise} \end{cases}$$

3.
$$f_T(t) = \begin{cases} \frac{1}{2}t^2 \, , 0 < 1 < 1 \\ 3t - t^2 - \frac{3}{2} \, , 1 \le t < 2 \\ \frac{1}{2}t^2 - 3t + \frac{9}{2} \, , 2 \le t < 3 \\ 0 \, , \text{otherwise} \end{cases}$$

5.
$$f_T(t) = \begin{cases} t^3\left[(1-t)^2 + \frac{9}{4}t(1-t) + \frac{9}{5}t^2\right] \, , 0 < t < 1 \\ 3(1-t)^2 + \frac{9}{4}(1-t) + \frac{9}{5} + \\ \qquad \frac{51}{20}(1-t)^5 \, , 1 \le t < 2 \\ 0 \qquad\qquad\qquad\qquad\text{, otherwise} \end{cases}$$

Section 6.7

1. (a)
$$f_{X-Y}(t) = \begin{cases} 1+t \, , -1 < t < 0 \\ 1-t \, , 1 \le t < 1 \\ 0 \, , \text{otherwise} \end{cases}$$

(b)
$$f_{XY}(t) = \begin{cases} -\ln t \, , 0 < t < 1 \\ 0 \qquad\text{, otherwise} \end{cases}$$

(c)
$$f_{X|Y}(t) = \begin{cases} \frac{1}{2} \, , 0 < t < 1 \\ \frac{1}{2t^2} \, , 1 \le t \\ 0 \, , \text{otherwise} \end{cases}$$

3. $f_{X,Y}(x,y) = \frac{1}{2\pi}e^{-(x^2+y^2)/2}, -\infty < x < \infty,$
   $-\infty < y < \infty.$ Independent.

5.
$$f_{R,S,T}(r,s,t) = \begin{cases} r^2t^2e^{-r} \, , 0 < r, 0 < s < 1, 0 < t < 1 \\ 0 \qquad\text{, otherwise} \end{cases}$$

   Independent.

7.
$$f_{U,V}(u,v) = \begin{cases} \frac{v}{\pi}e^{-v^2(1+u^2)/2} \, , -\infty < u < \infty, 0 < v \\ 0 \qquad\qquad\text{, otherwise} \end{cases}$$

9. (a)
$$f_{T,U,V}(t,u,v) = \begin{cases} 6e^{-v} \, , 0 < t < \frac{u}{2}, 0 < u, 2u - t < v \\ 0 \quad\text{, otherwise} \end{cases}$$

(b)
$$f_U(u) = \begin{cases} 6e^{-2u}(e^{u/2}-1) \, , 0 < u \\ 0 \qquad\qquad\text{, otherwise} \end{cases}$$

Review Exercises—Chapter 6

1. 215
3. 0.4833
5. No
7. (a)
$$p_{5X+1}(a) = \begin{cases} p(1-p)^{(a-1)/5} \, , a = 6, 11, 16, \ldots \\ 0 \qquad\qquad\text{, otherwise} \end{cases}$$

(b)
$$p_{(X-1)^2}(a) = \begin{cases} p(1-p)^{\sqrt{a}} \, , a = 01, 4, \ldots \\ 0 \qquad\quad\text{, otherwise} \end{cases}$$

9. $f_Y(y) = \frac{1}{y^2} f_X\left(\frac{1}{y}\right)$

11. (a)
$$f_{T,U}(t,u) = \begin{cases} 8(1-t)(1-u) , & 0 < u < t < 1 \\ 0 & , \text{ otherwise} \end{cases}$$

(b)
$$f_R(r) = \begin{cases} \frac{4}{3}r^3 - 4r + \frac{8}{3} , & 0 < r < 1 \\ 0 & , \text{ otherwise} \end{cases}$$

13.
$$p_Y(a) = \begin{cases} \frac{1}{3} , & a = 0, 2, 6 \\ 0 , & \text{otherwise} \end{cases}$$

15.
$$f_{T,U}(t,u) = \begin{cases} \frac{1}{2}tu^3 , & 0 < t < 1, 0 < u < 2 \\ 0 , & \text{otherwise} \end{cases}$$

Chapter 7
Section 7.2
1. $P(X \geq 475) \leq \frac{450}{475}$

5.
$$p_{X_n}(a) = \begin{cases} 1 - \frac{2}{n^2} , & a = 0 \\ \frac{1}{n^2} , & a = \pm n \\ 0 , & \text{otherwise} \end{cases}$$

15. $\frac{3}{4}$

Section 7.3
1. 0.8414
3. 0.9868
5. 0.1788
11. 0
15. 0.3085

Review Exercises—Chapter 7
1. 5294
5. (a) ≤ 0.04 (b) 0
7. 0.2776
9. 0.0668
11. (c) 1

Chapter 8
Section 8.2
5. Answers vary.

Section 8.3
1. (a) $\sum_{i=1}^{n} x_i^3 - 3\bar{x}\sum_{i=1}^{n} x_i^2 + 2n\bar{x}^3$

(b) $\sum_{i=1}^{n} x_i^4 - 4\bar{x}\sum_{i=1}^{n} x_i^3 + 6\bar{x}^2 \sum_{i=1}^{n} x_i^2 - 3n\bar{x}^4$

3. $\bar{X} = .1665, \tilde{X} = .1665, s^2 = 2.3 \times 10^{-6}$,
MAD = 0.0015, SIR = 0.001375

5. $\bar{X} = 46.6, \tilde{X} = 39.5, s^2 = 992.144$,
MAD = 28, SIR = 36.25

7. $\bar{X} = 40.05, \tilde{X} = 40, s^2 = 9.382, \text{MAD} = 2.5$,
SIR = 2.5, $b_3 = 0.184, b_4 = 2.10$

Review Exercises—Chapter 8
7. $\bar{X} = 49.975, \tilde{X} = 50, s^2 = 0.0713, \text{MAD} = 7.924$,
SIR = 2, $b_3 = -0.0828, b_4 = 2.3375$

Chapter 9
Section 9.2
7. $\sum_{i=1}^{n} a_i b_i = 0$

9. (a) 0.95   (b) $a = 9.39, b = 28.869$
(c) 31.526

Section 9.3
5. 0.94

Section 9.4
5.
$$f_{Y_1}(y) = \begin{cases} 2e^{-2(y-5)} , & 5 \leq y \\ 0 & , \text{ otherwise} \end{cases}$$

$$E[Y_1] = \frac{11}{2}$$

7. (a)
$$f_{Y_1}(y) = \begin{cases} 15y^2(1-y^3)^4 , & 0 \leq y \leq 1 \\ 0 & , \text{ otherwise} \end{cases}$$

(b)
$$f_{Y_3}(y) = \begin{cases} 810y^6(1-y^3)^2 , & 0 \leq y \leq 1 \\ 0 & , \text{ otherwise} \end{cases}$$

9. (a)
$$f_{Y_6}(y) = \begin{cases} 6\left(1-\frac{1}{y}\right)^5 \frac{1}{y^2} , & 1 \leq y \\ 0 & , \text{ otherwise} \end{cases}$$

(b)
$$f_{Y_3}(y) = \begin{cases} 60\left(1-\frac{1}{y}\right)^2 \frac{1}{y^5} , & 1 \leq y \\ 0 & , \text{ otherwise} \end{cases}$$

13. $N(0, \frac{1}{n})$

15. $\frac{19}{4}$

Review Exercises—Chapter 9
1. d.f. = $(1, \nu)$
3. $E[Y_i] = a + b\mu$, $\text{Var}(Y_i) = b^2\sigma^2$, $\bar{Y} = a + b\bar{X}$,
$s^2 = \frac{b^2}{n-1}\sum_{i=1}^{n}(X_i - \bar{X})^2$

5. (b) 38.77 qt
9. (a) 0.1   (b) 0.05
(c) $a = 8.26, b = 31.41$ (d) 37.566
11. (a) 0.975   (b) $\frac{1}{2.06}$

13. (a) $N\left(\frac{1}{\sqrt{2}}, \frac{1}{8n}\right)$

(b) Bivariate normal, $\mu_{Q_1} = \frac{1}{2}$, $\mu_{Q_3} = \frac{\sqrt{3}}{2}$,

$\sigma_{Q_1}^2 = \frac{3}{16n}$, $\sigma_{Q_3}^2 = \frac{1}{16n}$, $\rho = \frac{1}{3}$

(c) $N\left(\frac{\sqrt{3}-1}{4}, \frac{12-\sqrt{3}}{192n}\right)$

Chapter 10
Section 10.2

3. $\frac{-n}{\sum_{i=1}^{n} \ln X_i}$

5. Numerical solution to $\sum_{i=1}^{n} \frac{2(x_i - \theta)}{1 + (x_i - \theta)^2} = 0$.

7. Numerical solution to $\sum_{i=1}^{n} \ln x_i - n\frac{\Gamma'(\alpha)}{\Gamma(\alpha)} = 0$.

9. $e^{-\bar{X}}$

11. $\hat{a} = Y_1, \hat{b} = Y_n, \widehat{b-a} = Y_n - Y_1$

13. $\hat{p} = \frac{r}{x}$

Section 10.3

1. $\bar{X}$

3. $\frac{1}{\bar{X}}$

5. $\bar{X}$

7. $\hat{p}_1 = \frac{x}{n}, \hat{p}_2 = \frac{y}{n}, \hat{p}_3 = \frac{z}{n}$

Section 10.4

1. $\text{Var}(\bar{X}) = \frac{\lambda}{n}$

5. $\bar{X}^2 + \frac{1}{n}s^2$, unbiased.

7. $2Y_3, 5$

9. No

15. (a) $\frac{\theta^2}{n}$ (b) $\prod_{i=1}^{n} X_i$ (c) $\frac{\bar{X}}{1-\bar{X}}$

21. $\sum_{i=1}^{n} X_i$

23. (Y_1, Y_n)

29. $\bar{X}$

31. $Y_1 - \frac{1}{n}$

Section 10.5

3. Yes

Section 10.6

1. $h(\lambda \mid x) \sim G\left(x + 1, \frac{1}{\theta + 1}\right), \frac{x+1}{\theta + 1}$

3. (a) $h(\mu \mid x) \sim N\left(\frac{x}{\sigma^2 + 1}, \frac{\sigma^2}{\sigma^2 + 1}\right)$

(b) $\frac{x}{\sigma^2 + 1}$

5. $\hat{\theta} = \sqrt{2}$, if $x < 1$ and $\hat{\theta} = \sqrt{2}x$, if $x \geq 1$

Section 10.7

1. 10.2

Review Exercises—Chapter 10

1. (a) $\hat{\beta} = \frac{1}{\bar{X} - Y_1}, \hat{\theta} = Y_1$

(b) $\hat{\beta} = \frac{1}{\sqrt{m_2 - 2\bar{x}^2}}, \hat{\theta} = \bar{X} - \sqrt{m_2 - 2\bar{x}^2}$,

where $m_2 = \frac{1}{n}\sum_{i=1}^{n} X_i^2$

3. $\hat{\mu} = \bar{X}, \hat{\tau} = \bar{Y}$,

$\hat{\sigma}^2 = \frac{1}{m+n}\left[\sum_{i=1}^{n}(X_i - \bar{X})^2 + \sum_{j=1}^{m}(Y_j - \bar{Y})^2\right]$

5. (b) $\frac{n-1}{n}Y_1$

7. (b) $\bar{X}$

9. $\frac{x}{2}$

Chapter 11
Section 11.2

3. $\frac{(n+m-2)\widehat{\sigma^2}}{\chi^2(n+m-2)_{\alpha/2}} < \sigma^2 < \frac{(n+m-2)\widehat{\sigma^2}}{\chi^2(n+m-2)_{1-\alpha/2}}$,

where $\widehat{\sigma^2} = \frac{(n-1)s_X^2 + (m-1)s_Y^2}{n+m-2}$

5. $0.556 < \mu < 7.444$

7. $8.82 < \mu < 10.58$

9. $4.29 < \mu < 7.33$

11. $-10.474 < \mu_A - \mu_B < 6.426$

15. $0.372 < \frac{\sigma_1^2}{\sigma_2^2} < 3.114$

Section 11.3

3. $0.355 < p < 0.56$

5. (a) $\frac{\ln\left(1 - \frac{\alpha}{2}\right)}{\ln x} < \theta < \frac{\ln\left(\frac{\alpha}{2}\right)}{\ln x}$

(b) $2.10 < \theta < 3.06$

7. $x\sqrt{\frac{\alpha}{2}} < \theta < x\sqrt{\frac{1-\alpha}{2}}$

Section 11.4

1. $\bar{X} - z_{\alpha/2}\sqrt{\frac{\bar{X}}{n}} < \lambda < \bar{X} + z_{\alpha/2}\sqrt{\frac{\bar{X}}{n}}$

3. $\frac{-n}{\sum_{i=1}^{n} \ln X_i} + z_{\alpha/2}\frac{\sqrt{n}}{\sum_{i=1}^{n} \ln X_i} < \theta < \frac{-n}{\sum_{i=1}^{n} \ln X_i} - z_{\alpha/2}\frac{\sqrt{n}}{\sum_{i=1}^{n} \ln X_i}$

5. $s^2 - z_{\alpha/2}s^2\sqrt{\frac{2}{n}} < \sigma^2 < s^2 + z_{\alpha/2}s^2\sqrt{\frac{2}{n}}$

7. $227.001 < \beta < 272.999$

Section 11.5

1. $41.1 < \tilde{\mu} < 45.5$
3. $36 < \tilde{\mu} < 43, 99.06\%$

Review Exercises—Chapter 11

1. (a) Decrease   (b) Increase
   (c) Increase
3. 107
5. 1068
7. (a) $x\sqrt{\frac{2}{2-\alpha}} < \theta < x\sqrt{\frac{2}{\alpha}}$
   (b) $0.769 < \theta < 3.354$
9. $7 < \tilde{\mu} < 14$
11. $22 < \tilde{\mu} < 25$
13. $0.069 < \frac{\sigma_A^2}{\sigma_B^2} < 2.02$
15. (c) $\frac{\frac{\bar{X}}{\bar{Y}}}{F(2n,2m)_{\alpha/2}} < \frac{\beta_X}{\beta_Y} < \frac{\frac{\bar{X}}{\bar{Y}}}{F(2n,2m)_{1-\alpha/2}}$

Chapter 12
Section 12.2

1. $\alpha = \frac{4}{19}, \beta = \frac{137}{228}$
3. (a) 0.0037
   (b)

λ	0.1	0.2	0.3	0.4	0.5
$\eta(\lambda)$	0.0037	0.0527	0.1847	0.3712	0.5595

5. Reject H_0 if $\sum_{i=1}^{n} x_i \geq 5$ or if $\sum_{i=1}^{n} x_i = 4$ with probability 0.412.
7. (a) $a = 3$
   (b) $a > 3$ or $a = 3$ with probability 0.475.

Section 12.3

1. Let X = no. red balls.
   (a) $x = 1, x = 2$ (b) $x = 2$
3. (b) 15.987
5. $\prod_{i=1}^{n} x_i > a$

Section 12.4

1. (c) $a = 0.1645$
3. (c) $a = 28.412$
7. $\bar{X} \leq c$

Section 12.5

7. Reject H_0 if $\sum_{i=1}^{n} X_i^2 \leq a$ or $\sum_{i=1}^{n} x_i^2 \geq b$.
11. Reject H_0 if $\frac{s_X^2}{s_Y^2} \leq 0.224$ or $\frac{s_X^2}{s_Y^2} \geq 8.84$.
13. $t = 2.5606$, reject H_0.

Section 12.6

1. Results depend on sample.
3. $Q = 2.859$, do not reject H_0.
5. $Q = 2.6$, yes.
7. $Q = 2.813$, do not reject H_0.

Section 12.7

3. $x = -1.00$, no.
5.
$$Z = \frac{\frac{-n}{\sum_{i=1}^{n} \ln X_i} - \theta_0}{\frac{\theta_0}{\sqrt{n}}}$$

Reject H_0 if $|Z| \geq z_{\alpha/2}$.

Section 12.8

3. $a_n = \frac{n \ln 10 - \ln 8}{9}, b_n = \frac{n \ln 10 + \ln 9^2}{9}$

Section 12.9

1. (a) $R^+ = 5$, do not reject.
   (b) $W^+ = 9.5$, do not reject.
3. $W^+ = 1.5$, reject H_0.
7. (a) $Z = -0.655$, do not reject.
   (b) $Z = 1.15$, do not reject.

Review Exercises—Chapter 12

1. (a) $H_0 : b = 5, H_1 : b = 8$
   (b) X = no. black balls.
   (c) $X = 2$ (d) $\alpha = \frac{1}{9}, \beta = \frac{17}{45}$
3. $Y_n \geq a$
7. $Q = 28.6$, no.
11. $Z = -19.61$, no.
13. $Z = 5.21$, yes.
15. $Z = 1.25$, no.

Chapter 13
Section 13.2

7. $\bar{X} = 0$
13. $\hat{Y} = 101.549 - 0.99X, \hat{Y} = 44.129$
15. $t = 0.569$, do not reject H_0.
17. $-1.094 < b < -.887$
19. $25.22 < \mu_{Y|72} < 35.29$
21. $\hat{Y} = -4.7087 + 0.15895X, \hat{Y} = 0.378$
23. $t = -4.364$, reject H_0.
25. $0.1245 < b < 0.1934$
27. $0.977 < \mu_{Y|37.5} < 1.527$

Section 13.3

1. (a)
$$A = \begin{pmatrix} 1 & 0 & 0 & 0 & 0 \\ 0 & 1 & 0 & 0 & 0 \\ 0 & 0 & 1 & 0 & 0 \\ 0 & 0 & 0 & 1 & 0 \\ 0 & 0 & 0 & 0 & 1 \end{pmatrix}$$

(b)
$$A = \frac{1}{5} \begin{pmatrix} 1 & 1 & 1 & 1 & 1 \\ 1 & 1 & 1 & 1 & 1 \\ 1 & 1 & 1 & 1 & 1 \\ 1 & 1 & 1 & 1 & 1 \\ 1 & 1 & 1 & 1 & 1 \end{pmatrix}$$

(c)
$$A = \begin{pmatrix} \frac{4}{5} & \frac{-1}{5} & \frac{-1}{5} & \frac{-1}{5} & \frac{-1}{5} \\ \frac{-1}{5} & \frac{4}{5} & \frac{-1}{5} & \frac{-1}{5} & \frac{-1}{5} \\ \frac{-1}{5} & \frac{-1}{5} & \frac{4}{5} & \frac{-1}{5} & \frac{-1}{5} \\ \frac{-1}{5} & \frac{-1}{5} & \frac{-1}{5} & \frac{4}{5} & \frac{-1}{5} \\ \frac{-1}{5} & \frac{-1}{5} & \frac{-1}{5} & \frac{-1}{5} & \frac{4}{5} \end{pmatrix}$$

3. $$\Sigma = \frac{1}{5}\begin{pmatrix} 3 & 1 \\ 1 & 2 \end{pmatrix} \quad \rho = \frac{1}{\sqrt{6}}$$

7. $$(X'X)^{-1} = \frac{1}{19024}\begin{pmatrix} 19208 & -416 \\ -416 & 10 \end{pmatrix}$$

Section 13.4

1. $E[Y] = 15.59 + 1.93X_1 - 1.57Y_2, \hat{Y} = 15.85$
3. $-1.837 < \beta_2 < -1.298$
5. $t = 18.51$, reject H_0.
7. 0.988
11. 395.25
13. $F = 2.654$, do not reject H_0.
15. 0.57
21. $\hat{Y} = 9.6371 - 7.9846X + 3.3916X^2$

Section 13.5

1. $r = -0.242$
3. $r = 0.921$
5. $-0.935 < \rho < -0.781$

Section 13.6

1. 1.333
3. $r_S = 0.973$
5. $Z = 2.919$, reject H_0.

Review Exercises—Chapter 13

1. $\hat{Y} = 0.480 + 0.3335X_1 + 0.000991X_2, \hat{Y} = 2.65$
3. $F = 0.358$, do not reject H_0.

5. -0.0865
7. $r_S = 0.119$, do not reject H_0.

Chapter 14
Section 14.2

5. $$\begin{bmatrix} A & B & C \\ B & C & A \\ C & A & B \end{bmatrix} \begin{bmatrix} A & B & C \\ C & A & B \\ B & C & A \end{bmatrix}, 12$$

7. 4 distinct, may permute these in 576 ways.

Section 14.3

3. $$A = \begin{pmatrix} A_1 & 0 & 0 & \cdots & 0 \\ 0 & A_2 & 0 & \cdots & 0 \\ 0 & 0 & 0 & \cdots & 0 \\ \vdots & \vdots & \vdots & \ddots & \vdots \\ 0 & 0 & 0 & \cdots & A_k \end{pmatrix},$$

where $A_{i_{jj}} = 1 - \frac{1}{n_i}$, and $A_{i_{jl}} = \frac{1}{n_i}$.

7. $F = 2.84$, do not reject H_0.
9. $F = 4.80$, do not reject H_0.

Section 14.4

5. Paint: $F = 1.72$, Block: $F = 0.129$.
9. IQ: $F = 13.81$, Distraction: $F = 3.03$, Interaction: $F = 3.65$.

Section 14.5

1. $\widehat{\tau_1 - \tau_3} = -1.277$
3. 1.605
5. 27
7. 55.8
9. $3\tau_2 - \tau_1 - \tau_3 - \tau_4$

$$t = \frac{3\bar{Y}_2 - \bar{Y}_1 - \bar{Y}_3 - \bar{Y}_4}{\sqrt{\widehat{\sigma^2}\left[\frac{1}{n_1} + \frac{9}{n_2} + \frac{1}{n_3} + \frac{1}{n_4}\right]}}$$

Section 14.6

3. $H = 1.423$, do not reject.
7. $S = 5.7$, do not reject.
9. $H = 3.785$, do not reject.

Review Exercises—Chapter 14

1. (a) $F = 17.73$ (b) $H = 11.18$
   (c) $-1952.79 < \tau_3 - \tau_1 < 889.61$
3. Type: $F = 10.06$, Region: $F = 25.46$, Interaction: $F = 11.85$.

5. $F = 822.93$

7. Form: $F = 2.25$, Block: $F = 0.37$.

9. Medium: $F = 0.02$, Fertilizer: $F = 10.52$,
   Interaction: $F = 1.14$.

11. $H = 0.9615$

13. $S = 4.3$

15.
$$Y_{i,j} = \mu + \tau_i + \beta_j + e_{ij}, \begin{cases} i = 1,2,3,4 \\ j = 1,2,3,4,5 \end{cases}$$

SV	d.f.	E[MS]
T	3	$\sigma^2 + 4\sum_{i=1}^{4} \tau_i^2$
B	4	$\sigma^2 + \frac{4}{5}\sum_{j=1}^{5} \beta_j^2$
E	7	σ^2
Total	19	–

Index